TELECOMMUNICATION
TRANSMISSION
HANDBOOK

WILEY SERIES IN TELECOMMUNICATIONS

Donald L. Schilling, Editor
City College of New York

Digital Telephony, 2nd Edition
John Bellamy

Elements of Information Theory
Thomas M. Cover and Joy A. Thomas

Telecommunication System Engineering, 2nd Edition
Roger L. Freeman

Telecommunication Transmission Handbook, 3rd Edition
Roger L. Freeman

Introduction to Communications Engineering, 2nd Edition
Robert M. Gagliardi

Expert System Applications to Telecommunications
Jay Liebowitz

Synchronization in Digital Communications, Volume 1
Heinrich Meyr and Gerd Ascheid

Synchronization in Digital Communications, Volume 2
Heinrich Meyr and Gerd Ascheid (in preparation)

Computational Methods of Signal Recovery and Recognition
Richard J. Mammone (in preparation)

Business Earth Stations for Telecommunications
Walter L. Morgan and Denis Rouffet

Satellite Communications: The First Quarter Century of Service
David W. E. Rees

Worldwide Telecommunications Guide for the Business Manager
Walter L. Vignault

TELECOMMUNICATION TRANSMISSION HANDBOOK

THIRD EDITION

Roger L. Freeman

A WILEY-INTERSCIENCE PUBLICATION

JOHN WILEY & SONS, INC.

New York / Chichester / Brisbane / Toronto / Singapore

Copyright © 1991 by Roger L. Freeman

Published by John Wiley & Sons, Inc.

Library of Congress Cataloging in Publication Data:
Freeman, Roger L.
    Telecommunication transmission handbook / Roger L. Freeman.
        p.   cm.—(Wiley series in telecommunications)
    "A Wiley-Interscience publication".
    Includes bibliographical references.
    ISBN 0-471-51816-6
    1. Telecommunication.   I. Title.   II. Series.
TK5101.F66   1991
621.382—dc20   91-14668
                                                          CIP

Printed in the United States of America

10 9 8 7 6 5 4 3 2 1

**To Paquita,
Cristi, Bob, and Rossi**

CONTENTS

PREFACE

Engineering today encompasses a large array of disciplines, each of which may exhibit further specialization. This is particularly true in the broad field of telecommunications. There are traffic engineers and radio specialists, for example. Radio itself includes satellite communications, line-of-sight microwave, high frequency, tropospheric scatter, and so forth. There are switching engineers, experts in signaling, and telephone transmission engineers who may know little if anything about data communications. There are the disciplines of outside plant, plant extension, and plant operations. Each of these disciplines should be treated as part of a larger system rather than in isolation. It is the intention of this book to provide the necessary guidance to incorporate the multidisciplines involved in transmission and coordinate them into an optimal operational system.

The term *telecommunication* has grown in acceptance and usage over the past 30 years. Yet, it is still ambiguous to some people. For this book, telecommunication is defined as a service that permits people or machines to communicate at a distance.

My primary concern when preparing this third edition, as it has been for the previous editions, was to emphasize the system approach. *System*, as used in this context, means the interaction of one discipline with another to reach a definite, practical end-product that serves the needs of a specific user group. This text emphasizes point-to-point transmission systems in order to keep the subject under discussion within manageable limits. Television signals are treated only in the context of those special systems that are necessary for the point-to-point transport of a video signal and its related aural channel, or where television has direct application in the telecommunication user community, such as video conferencing.

This book has been prepared with three objectives in mind—first, as an applications handbook, second, as a tutorial text, and finally, as a reference. The book is addressed to the electrical engineer who wishes to learn about transmission and to the graduate engineering student who wishes to focus on the practical implementation of a transmission system. In addition, the book is also addressed to the nonengineer who would like an introduction to transmission engineering. Technicians working in the field, for example, along with telecommunication managers and other corporate and military staff responsible for electrical communications would be included in this latter group. To carry out these objectives, I have made every effort to provide clear and concise explanations. My basic aim is to explain how to effectively design a transmission subsystem as part of an overall telecommunication system. I have provided all necessary information, including background material, for a successful first cut at the design of an operational transmission system using the book as a background reference.

To best profit from the material in the book, the typical reader should have a solid background in algebra, trigonometry, logarithms and electricity. He or she should also be familiar with electrical communications, particularly modulation and the electromagnetic spectrum. In this regard, Chapter 1 was organized to introduce the reader to a number of specialized units of measure used in transmission, such as the decibel and its derived units, as well as basic transmission impairments. For many readers, Chapter 1 may serve as a review, and I have included it in order to refer to it in later chapters of the book.

Throughout I have endeavored to keep the reader in mind. For example, to analyze how the change in several decibels of signal-to-noise ratio at a line-of-sight microwave repeater will change the picture reception quality for a TV viewer, I have purposefully avoided involving broadcast, CATV, air-ground, telemetry and marine and other mobile transmission systems in order to focus on the basic principles addressed in the book.

The order of the chapters signifies a movement from the voice channel to wire systems and through frequency and time-division multiplex. With the preceding material serving as a background, broadband radio systems are presented. A thorough grounding is provided in line-of-sight microwave, both analog and digital, before proceeding into tropospheric scatter and satellite communications. It is my firm belief that once we can handle line-of-sight microwave transmission, tropospheric scatter and satellite communications will be easily understood. I then move into narrow-band radio systems, including high-frequency and meteor burst. The chapters that follow discuss broadband cable, including coaxial and fiber optic. The final three chapters cover the transmission of information other than voice over the telecommunication network. This last group includes data, video and facsimile transmission. Appendix A presents a brief background on the Asynchronous Transfer Mode (ATM), which I expect to become a cornerstone of bulk information transfer systems and which will provide bandwidth on demand. A glossary of abbreviations and acronyms used throughout the book is also included.

A book of this size can only treat a discipline to a limited depth. Therefore, references and bibliography are given at the end of each chapter to assist readers who want more information as well as to refer them to sources used in the preparation of the text.

Sudbury, Massachusetts   ROGER L. FREEMAN
June 1991

ACKNOWLEDGMENTS

I am deeply indebted to a large group of friends and colleagues who have reviewed and suggested changes and improvements to all three editions of the book. I am particularly grateful to Marshall Cross, president of Radio General Company, Stow, MA for his extensive review of the revised HF chapter. David Kocyba, vice president of Meteor Communications Corporation (MCC), Kent, WA, Dale Smith and Tom Donich, MCC technical staff, were extremely helpful, providing material and review of the meteor burst chapter. I am grateful to Dr. Len Wagner of the Naval Research Laboratories for material he provided on HF propagation and to Dan Odom of Raytheon Equipment Division who tutored me on some of the finer points of propagation. Don Hastings and Mark Kiryelejza of Raytheon Communication Systems Directorate (CSD) were there when I needed them. Zaheer Ali, also of CSD, put me on the right track regarding VSAT networks. One's colleagues can provide a wealth of information and I have taken full advantage of their good will.

The video transmission chapter was greatly enhanced by the contributions of PictureTel, Peabody, MA, Colorado Video, Boulder, CO, and ABL Engineering, Mentor, OH. Jack Dicks, director of System Engineering for INTELSAT, provided me with excellent updating on satellite communications and INTELSAT developments.

I greatly appreciate the help provided by Dr. Ron Brown and John Lawlor, independent consultants in telecommunications, in the area of background data and suggestions dealing with several chapters. Dr. Don Schilling, professor emeritus in electrical engineering of City College of New York and president of SCS Telecom for his encouragement; and Don Marsh, vice president of CONTEL and an old friend, for his many suggestions which enabled me to enhance the structure and organization of the book. My wife, Paquita, endured my many hours on the PC during the laborious preparation of the book. She was always there when I was down with a smile and words of encouragement. I thank you all.

R. L. F.

1

INTRODUCTORY CONCEPTS

1.1 THE TRANSMISSION PROBLEM

The word *transmission* is often misunderstood as we try to compartmentalize
telecommunications into neatly separated disciplines. A transmission engineer in telecommunications must develop an electrical signal from a source
and deliver it to the destination to the satisfaction of a customer. In a broad
sense we may substitute the words *transmitter* for source and *receiver* for
destination. Of major concern are those phenomena, conditions, and factors
that distort or otherwise make the signal at the destination such that the
customer is dissatisfied. To understand the problems of transmission, we
must do away with some of the compartmentalization. Besides switching and
signaling, transmission engineers must have some knowledge of maps, civil
engineering, and power. A familiarity with basic traffic engineering concepts,
such as busy hour, activity factor, and so forth, is also helpful.

Transmission system engineering deals with the production, transport, and
delivery of a quality signal from source to destination. The following chapters
describe methods of carrying out this objective.

1.2 A SIMPLIFIED TRANSMISSION SYSTEM

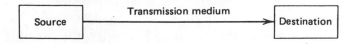

The simple drawing illustrates a transmission system. The source may be a
telephone mouthpiece (transmitter) and the destination may be the tele-

phone earpiece (receiver). The source converts the human intelligence, such as voice, data information, or video, into an electrical equivalent or electrical signal. The destination accepts the electrical signal and reconverts it to an approximation of the original human intelligence. The source and destination are electrical transducers. In the case of printer telegraphy, the source may be a keyboard, where each key, when depressed, transmits to the destination distinct electrical impulses. The destination in this case may be a teleprinter, which converts each impulse grouping back to the intended character keyed or depressed at the source.

The transmission media can be represented as a network:

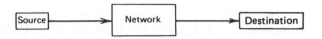

or as a series of electrical networks

Some networks show a gain in level, others a loss. We must be prepared to discuss these gains and losses as well as electrical signal levels and the level of disturbing effects such as noise and/or distortion. To do this we must have a firm and solid knowledge of the decibel and related measurement units.

1.3   THE DECIBEL

The decibel is a unit that describes a ratio. It is a logarithm with a base of 10. Consider first a power ratio. The number of decibels (dB) = $10 \log_{10}$ (the power ratio).

Let us look at the following network:

$$\xrightarrow{1\ W}\ \boxed{\text{Network}}\ \xrightarrow{2\ W}$$

The input is 1 W and its output 2 W, in the power domain. Therefore we can say the network has approximately a 3-dB gain. In this case

$$\text{Gain (dB)} = 10 \log \frac{\text{output}}{\text{input}} = 10 \log \tfrac{2}{1} = 10(0.3013) = 3.0103 \text{ dB}$$

or approximately, a 3-dB gain.

Now let us look at another network:

In this case there is a loss of 30 dB:

$$\text{Loss (dB)} = 10 \log \frac{\text{input}}{\text{output}} = 10 \log \frac{1}{1000} = -30 \text{ dB}$$

(Note that the negative sign shows that it is a loss.) Or in general we can state

$$\text{Power (dB)} = 10 \log \frac{P_2}{P_1} \tag{1.1}$$

where P_1 = input level and P_2 = output level.

A network with an input of 5 W and an output of 10 W is said to have a 3-dB gain:

$$\text{Gain (dB)} = 10 \log \tfrac{10}{5} = 10 \log \tfrac{2}{1} = 10(0.30103)$$
$$= 3.0103 \text{ dB} \simeq 3 \text{ dB}$$

This is a good figure to remember. Doubling the power means a 3-dB gain; likewise, halving the power means a 3-dB loss.

Consider another example, a network with a 13-dB gain:

$$\text{Gain (dB)} = 10 \log \frac{P_2}{P_1} = 10 \log \frac{P_2}{0.1} = 13 \text{ dB}$$

Then

$$P_2 \simeq 2 \text{ W}$$

Table 1.1 may be helpful. All values in the power ratio column are $X/1$, or compared to 1.

Table 1.1

Power Ratio		dB	Power Ratio	dB
10^1	(10)	+10	10^{-1} (1/10)	−10
10^2	(100)	+20	10^{-2} (1/100)	−20
10^3	(1,000)	+30	10^{-3} (1/1,000)	−30
10^4	(10,000)	+40	10^{-4} (1/10,000)	−40
10^5	(100,000)	+50	10^{-5} (1/100,000)	−50
10^6	(1,000,000)	+60	10^{-6} (1/1,000,000)	−60

It is useful to be able to work with decibels without pencil and paper or a calculator. Relationships of 10 and 3 have been reviewed. Now consider the following:

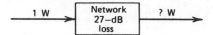

What is the power output of this network? To do this without a calculator, we would proceed as follows. Suppose that the network attenuated the signal 30 dB. Then the output would be 1/1000 of the input, or 1 mW. 27 dB is 3 dB less than 30 dB. Thus the output would be twice 1 mW, or 2 mW. It really is quite simple. If we have multiples of 10, as in Table 1.1, or 3 up or 3 down from these multiples, we can work it out in our heads, without a calculator.

Look at this next example:

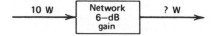

Working it out with a calculator, we see that the output is approximately 40 W. Here we have a multiple of 4. A 6-dB gain represents approximately a fourfold power gain. Likewise, a 6-dB loss would represent approximately one-fourth the power output. Now we should be able to work out many combinations without resorting to a calculator.

Consider a network with a 33-dB gain with an input level of 0.15 W. What would the output be? 30 dB represents multiplying the input power by 1000, and 3 additional decibels doubles it. In this case the input power is multiplied by 2000. Thus the answer is 0.15 × 2000 = 300 W.

The following table may further assist the reader regarding the use of decibels as power ratios.

	Approximate Power Ratio	
Decibels	Losses	Gains
1	0.8	1.25
2	0.63	1.6
3	0.5	2.0
4	0.4	2.5
5	0.32	3.2
6	0.25	4.0
7	0.2	5.0
8	0.16	6.3
9	0.125	8.0
10	0.1	10.0

The decibel is a useful tool for calculating gains, losses, and power levels of networks in series. Consider this example. There are three networks in series. The first is an attenuator with a 12-dB loss (−12 dB), the second network is an amplifier with a 35-dB gain, and the third has an insertion loss of 10 dB. The input to the first network is 4 mW; what is the output of the third network in watts?

There are several ways to calculate the output. One of the simplest is to algebraically add the decibel values (here, −12, +35, and −10), which in this case equals +13 dB. Note that the plus sign indicates gain and the minus sign, loss. Thus we have

+13 dB can be represented by a multiplier of 10 and a multiplier of 2 if we learn the 10- and 3-dB rules. We now have 10×4 mW = 40 mW and $40 \times 2 = 80$ mW or 0.08 W.

When applying decibels in the current and voltage regimes, we handle the problem somewhat differently. Turn to the power law formula in electricity where

$$\text{Power (watts)} = I^2 R = \frac{E^2}{R} \tag{1.2}$$

where I is the current in amperes, R is the resistance in ohms, and E is the voltage in volts.

Thus

$$\text{dB (voltage)} = 20 \log \frac{E_2}{E_1} \tag{1.3}$$

$$\text{dB (current)} = 20 \log \frac{I_2}{I_1} \tag{1.4}$$

The relationships between E_2 and E_1 and between I_2 and I_2 are as follows:

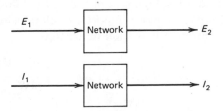

The network is any device that has a gain or a loss such as an amplifier or an attenuator.

When using the current and voltage relationships shown above, keep in mind that they must be compared against like impedances. For instance, E_2 may not be taken at a point of 600-ohm (Ω) impedance and E_1 at a point of 900 Ω.

Example 1. How many decibels correspond to a voltage ratio of 100?

$$dB = 20 \log \frac{E_2}{E_1}$$

When $E_2/E_1 = 100$,

$$dB = 20 \log 100 = 40 \text{ dB}$$

(Same impedances assumed.)

Example 2. If an amplifier has a 30-dB gain, what voltage ratio does the gain represent? Assume equal impedances at input and output of the amplifier.

$$30 = 20 \log \frac{E_2}{E_1}$$

$$\frac{E_2}{E_1} = 31.6$$

Thus the ratio is 31.6 : 1. (Again same impedances assumed.)

1.4   BASIC DERIVED DECIBEL UNITS

1.4.1   The dBm

Up to now all reference to decibels has been made in terms of ratios or relative units. We *cannot* say the output of an amplifier is 33 dB. We *can* say that an amplifier has a gain of 33 dB or that a certain attenuator has a 6-dB - loss. These figures or units give no idea whatsoever of the absolute level. Several derived decibel units do.

Perhaps dBm is the most common of these. By definition dBm is a power level related to 1 mW. A most important relationship to remember is 0 dBm = 1 mW. The formula may then be written:

$$\text{Power (dBm)} = 10 \log \frac{\text{power (mW)}}{1 \text{ mW}} \tag{1.5}$$

Example 1. An amplifier has an output of 20 W. What is its output in dBm?

$$\text{Power (dBm)} = 10\log\frac{20\text{ W}}{1\text{ mW}}$$

$$= 10\log\frac{20\times10^3\text{ mW}}{1\text{ mW}} \simeq +43\text{ dBm}$$

(The plus sign indicates that the quantity is above the level of reference, 0 dBm.)

Example 2. The input to a network is 0.0004 W. What is the input in dBm?

$$\text{Power (dBm)} = 10\log\frac{0.0004\text{ W}}{1\text{ mW}}$$

$$= 10\log 4\times10^{-1}\text{ mW} \simeq -4\text{ dBm}$$

(The minus sign in this case tells us that the level is below reference, 0 dBm or 1 mW.)

1.4.2 The dBW

The dBW is used extensively in microwave applications. It is an absolute decibel unit and may be defined as decibels referred to 1 W:

$$\text{Power level (dBW)} = 10\log\frac{\text{power (W)}}{1\text{ W}} \qquad (1.6)$$

Remember the following relationships:

$$+30\text{ dBm} = 0\text{ dBW} \qquad (1.7)$$

$$-30\text{ dBW} = 0\text{ dBm} \qquad (1.8)$$

Consider this network:

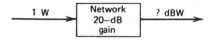

Its output level in dBW is +20 dBW. Remember that the gain of the network is 20 dB or 100. This output is 100 W or +20 dBW.

Another way to look at the network is to convert the input to dBW and add dB values to get the output. One watt equals 0 dBW. Then 0 dBW + 20 dB = +20 dBW. Table 1.2, a table of equivalents, may be helpful.

Table 1.2

dBm	dBW	Watts	dBm	dBW	Milliwatts
+ 66	+ 36	4000	+ 30	0	1000
+ 63	+ 33	2000	+ 27	− 3	500
+ 60	+ 30	1000	+ 23	− 7	200
+ 57	+ 27	500	+ 20	− 10	100
+ 50	+ 20	100	+ 17	− 13	50
+ 47	+ 17	50	+ 13	− 17	20
+ 43	+ 13	20	+ 10	− 20	10
+ 40	+ 10	10	+ 7	− 23	5
+ 37	+ 7	5	+ 6	− 24	4
+ 33	+ 3	2	+ 3	− 27	2
+ 30	0	1	0	− 30	1
			− 3	− 33	0.5
			− 6	− 36	0.25
			− 7	− 37	0.20
			− 10	− 40	0.1

1.4.3 The dBmV

The absolute decibel unit dBmV is used widely in video transmission. A voltage level may be expressed in decibels above or below 1 mV across 75 Ω, which is said to be the level in decibel-millivolts or dBmV. In other words,

$$\text{Voltage level (dBmV)} = 20 \log_{10} \frac{\text{voltage (mV)}}{1 \text{ mV}} \qquad (1.9)$$

Table 1.3

Root-Mean-Square Voltage across 75 Ω	dBmV
10 V	+ 80
2 V	+ 66
1 V	+ 60
10 mV	+ 20
2 mV	+ 6
1 mV	0
500 μV	− 6
316 μV	− 10
200 μV	− 14
100 μV	− 20
10 μV	− 40
1 μV	− 60

* when the voltage is measured at the 75-Ω *impedance level*. Simplified,

$$\text{dBmV} = 20 \log_{10} \text{ (voltage in milivolts at 75-}\Omega \text{ impedance)} \quad (1.10)$$

Table 1.3 may prove helpful.

1.5 THE NEPER

A transmission unit used in a number of northern European countries as an alternative to the decibel is the neper (Np). To convert decibels to nepers, multiply the number of decibels by 0.1151. To convert nepers to decibels, multiply the number of nepers by 8.686. Mathematically,

$$\text{Np} = \tfrac{1}{2} \log_e \frac{P_2}{P_1} \quad (1.11)$$

where P_2, P_1 = higher and lower powers, respectively, and e = 2.718, the base of the natural or Naperian logarithm. A common derived unit is the decineper (dNp). A decineper is one-tenth of a neper.

1.6 ADDITION OF POWER LEVELS IN dB
(dBm/dBW) OR SIMILAR ABSOLUTE LOGARITHMIC UNITS

Adding decibels corresponds to the multiplying of power ratios. Care must be taken when adding or subtracting absolute decibel units such as dBm, dBW, and some noise units. Consider the combining network below, which is theoretically lossless:

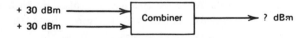

What is the resultant output? Answer: +33 dBm.

Figure 1.1 is a curve for directly determining the level in absolute decibel units corresponding to the sum or difference of two levels, the values of which are known in terms of decibels with respect to some reference.

As an example, let us add two power levels, 10 dBm and 6.0 dBm. Take the difference between them, 4 dB. Spot this value on the horizontal scale (the abscissa) on the curve. Project the point upward to where it intersects the "addition" curve (the upper curve). Take the corresponding number to the right and add it to the larger level. Thus

$$10 \text{ dBm} + 1.45 \text{ dB(m)} = 11.45 \text{ dBm}$$

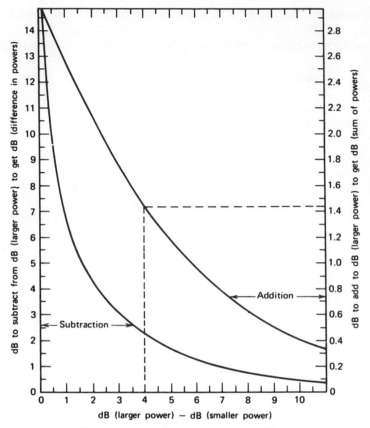

Figure 1.1 Decibels corresponding to the sum or difference of two levels.

Suppose we subtract the 6.0-dBm signal from the 10-dBm signal. Again the difference is 4 dB. Spot this value on the horizontal scale as before. Project the point upward to where it meets the "subtraction" curve (the lower curve). Take the corresponding number and subtract it from the larger level. Thus

$$10 \text{ dBm} - 2.3 \text{ dB(m)} = 7.7 \text{ dBm}$$

When it is necessary to add equal absolute levels expressed in decibels, add 10 log (the number of equal powers) to the level value. For example, add four signals of +10 dBm each. Thus

$$10 \text{ dBm} + 10 \log 4 = 10 \text{ dB(m)} + 6 \text{ dB} = +16 \text{ dBm}$$

When there are more than two levels to be added and they are not of equal value, proceed as follows. Pair them and sum the pairs, using Figure 1.1. Sum the resultants of the pairs in the same manner until one single resultant is obtained.

Another, more exact method of summing power levels expressed in dB is to convert the dB value to the equivalent numeric value, add, and then convert back to the equivalent decibel value. Consider the example below:

Convert the input dBm values to their numeric equivalent.

$$\begin{array}{rl} \log^{-1}(7/10) & = 5.01 \text{ mW} \\ \log^{-1}(11/10) & = 12.59 \text{ mW} \\ \hline \text{Sum} & = 17.60 \text{ mW} \end{array}$$

Convert the sum to the equivalent dBm value or $10 \log 17.60$ mW $= +12.45$ dBm. If the network in question had an insertion loss of 3 dB, the output would be 3 dB lower or $+9.45$ dBm.

If we wish to convert this value back to milliwatts, we would use our calculator. Divide the 9.45 mW by 10 and then get its antilog. This is represented by the equation

$$\text{Power (mW)} = \log^{-1}\left(\frac{9.45}{10}\right) = 8.81 \text{ mW}$$

1.7 NORMAL DISTRIBUTION — STANDARD DEVIATION

A normal or Gaussian distribution is a binomial distribution where n, the number of points plotted (number of events), approaches infinity (Ref. 11, pp. 121, 122, and 143). A distribution is an arrangement of data. A frequency distribution is an arrangement of numerical data according to size and magnitude. The normal distribution curve (Figure 1.2) is a symmetrical distribution. A nonsymmetrical frequency distribution curve is one in which the distributions extend further in one direction than in the other. This type of distortion is called skew. The peak of the normal distribution curve is called the point of central tendency, and its measure is its average. This is the point where the group of values tends to cluster.

The dispersion is the variation, scatteration of data, or the lack of tendency to congregate (Ref. 17). The range is the simplest measure of dispersion and is the difference between maximum and minimum values of a series. The mean deviation is another measure of dispersion. In a frequency distribution, ignoring signs, it is the average distance of items from a measure of the central tendency.

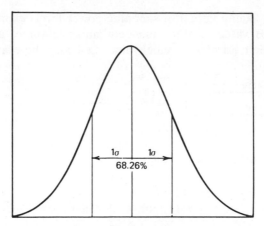

Figure 1.2 A normal distribution curve showing one standard deviation measured off either side of the arithmetic mean.

The standard deviation is the root mean square (rms) of the deviations from the arithmetic mean and is expressed by the small Greek letter sigma:

$$\sigma = \sqrt{\frac{\Sigma(X'^2)}{N}} \qquad (1.12)$$

where X' = deviations from the arithmetic mean $(X - \bar{X})$, and N = total number of items.

The following expressions are useful when working with standard deviations: they refer to a "normal" distribution:

- The mean deviation = 0.7979σ.
- Measure off on both sides of the arithmetic mean one standard deviation; then 68.26% of all samples will be included within the limits.
- For two standard deviations measured off, 95.46% of all values will be included.
- For three standard deviations, 99.73% will be included.

These last three items relate to exact normal distributions. In cases where the distribution has moderate skew, approximate values are used, such as 68% for 1σ, 95% for 2σ, and so on.

1.8   THE SIMPLE TELEPHONE CONNECTION

Two people may speak to one another over a distance by connecting two telephone subsets together with a pair of wires and a common microphone battery supply. As the wires are extended (i.e., the distance between the talkers is increased), the speech power level decreases until at some point, depending on the distance, the diameter of the wire, and the mutual capacitance between each wire in the pair, communication becomes unacceptable. For example, in the early days of telephony in the United States it was noted that a telephone connection including as much as 30 mi (48 km) of 19-gauge nonloaded cable was at about the limit of useful transmission.

Suppose that several people want to join the network. We could add them in parallel (bridge them together). As each is added, however, the efficiency decreases because we have added subsets in parallel and, as a result, the impedance match between subset and line deteriorates. Besides, each party can overhear what is being said between any two others. Lack of privacy may be a distinct disadvantage at times.

This can be solved by using a switch so that the distant telephone may be selected. Now a signaling system must be developed so that the switch can connect the caller to the distant telephone. A system of monitoring or supervision will also be required so that on-hook (idle) and off-hook (busy) conditions may be known by the switch as well as to permit line seizure by a subscriber.

Now extend the system again, allowing several switches to be used interconnected by trunks (junctions). Because of the extension of the two-wire system without amplifiers, a reduced signal level at many of the subscribers' telephone subsets may be experienced. Now we start to reach into the transmission problem. A satisfactory signal is not being delivered to some subscribers owing to line losses because of excessive wire line lengths. Remember that line loss increases with length.

Before delving into methods of improving subscriber signal level and satisfactory signal-to-noise ratio, we must deal with basic voice channel criteria. In other words, just what are we up against? Consider also that we may want to use these telephone facilities for other types of communication such as telegraph, data, facsimile, and video transmission. The voice channel (telephone channel) criteria covered below are aimed essentially at speech transmission. However, many parameters affecting speech most certainly have bearing on the transmission of other types of signals, and other specialized criteria are peculiar to these other types of transmission. These are treated in depth in later chapters, where they become more meaningful. Where possible, cross reference is made.

Before going on, refer to the simplified sketch (Figure 1.3) of a basic telephone connection. The sketch contains all the basic elements that will deteriorate the signal from source to destination. The medium may be wire,

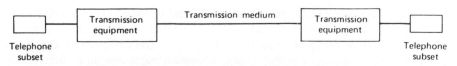

Figure 1.3   A simplified telephone transmission system.

coaxial cable, optical fiber, radio, or combinations of the four. Other transmission equipment may be used to enhance the medium by extending or expanding it. This equipment might consist of amplifiers, multiplex devices, and other signal processors such as compandors, voice terminals, and so forth.

1.9 THE PRACTICAL TRANSMISSION OF SPEECH

The telephone channel, hereafter called the voice channel, may be described technically using the following parameters:

- Nominal bandwidth
- Attenuation distortion (frequency response)
- Phase distortion
- Noise and signal-to-noise ratio
- Level

Return loss, singing, stability, echo, reference equivalent, and some other parameters deal more with the voice channel in a network and are discussed at length when we look at a transmission network later.

1.9.1 Bandwidth

The range between the lowest and highest frequencies used for a particular purpose may be defined as bandwidth. For our purposes we should consider bandwidth as those frequencies within which a performance characteristic of a device is above certain specified limits. For filters, attenuators, and amplifiers, these limits are generally taken where a signal will fall 3 dB below the average level in the passband or below the level at a reference frequency. The voice channel is a notable exception. In North America it is specifically defined at the 10-dB points about the reference frequency of 1000 Hz, approximately the band 200–3300 Hz. The International Consultive Committee for Telephone and Telegraph (CCITT) traditionally uses the reference frequency of 800 Hz and the CCITT voice channel occupies the band 300–3400 Hz at implied 10-dB points. In either case, we often refer to the nominal 4-kHz voice channel.

1.9.2 Speech Transmission — The Human Factor

Frequency components of speech may be found between 20 Hz and 20 kHz. The frequency response of the ear (i.e., how it reacts to different frequencies) is a nonlinear function between 30 Hz and 30 kHz; however, the major intelligence and energy content exists in a much narrower band. For energy

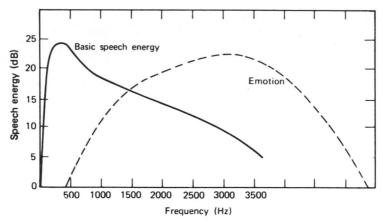

Figure 1.4   Energy and emotion distribution in speech. (From *Bell Syst. Tech. J.*, July 1931.)

distribution see Figure 1.4. The emotional content, which transfers intelligence, is carried in a band that lies above the main energy portion. Tests have shown that low frequencies up to 600–700 Hz add very little to the intelligibility of a signal to the human ear, but in this very band much of the voice energy is transferred (solid line in Figure 1.4). The dashed line in Figure 1.4 shows the portion of the frequency band that carries emotion. From this it can be seen that for economical transfer of speech intelligence, a band much narrower than 20 Hz to 20 kHz is necessary. In fact the standard bandwidth of a voice channel is 300–3400 Hz (CCITT Recs. G.132 and G.151). However, the generally accepted voice channel frequency band in North America is 200–3200 Hz (Ref. 16 p. 32). As is shown later, this bandwidth is a compromise between what telephone subscribers demand (Figure 1.4) and what can be provided to them economically. However, many telephone subsets have a response range no greater than approximately 500–3000 Hz. This is shown in Figure 1.5, where the response of the more modern Bell System 500 telephone set is compared to the older 302 set.

1.9.3 Attenuation Distortion

A signal transmitted over a voice channel suffers various forms of distortion. That is, the output signal from the channel is distorted in some manner such that it is not an exact replica of the input. One form of distortion is called attenuation distortion and this is the result of less than perfect amplitude–frequency response. If attenuation distortion is to be avoided, all frequencies within the passband should be subjected to the same loss (or gain). On typical wire systems higher frequencies in the passband are attenuated more than lower ones. In carrier equipment the filters used tend to

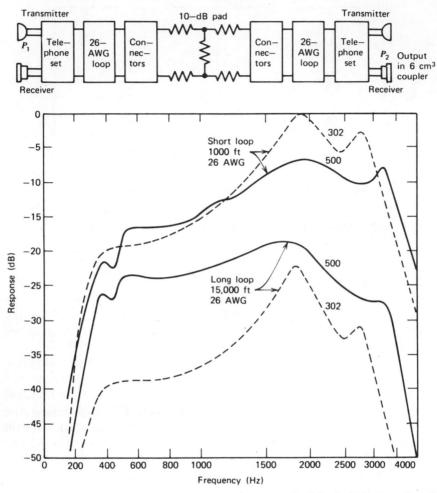

Figure 1.5 Comparison of overall response. (From: W. F. Tuffnell, "500-Type Telephone Set," *Bell Lab. Rec.*, vol. 29, 414–418, Sept. 1951; copyright ©1951 by Bell Telephone Laboratories.)

attenuate frequencies around band center the least, and attenuation increases as the band edges are approached. Figure 1.6 is a good example of this. The cross-hatched areas in the figure express the specified limits of attenuation distortion, and the solid line shows measured distortion on typical carrier (multiplex) equipment for the channel band (see Chapter 3). It should be remembered that any practical communication channel will suffer some form of attenuation distortion.

Attenuation distortion across the voice channel is measured compared to a reference frequency. CCITT specifies the reference frequency at 800 Hz. However, 1000 Hz is used more commonly in North America (see Figure 1.6).

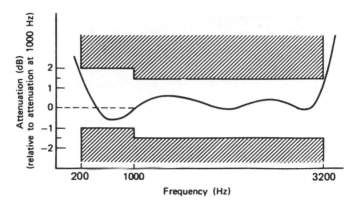

Figure 1.6 Typical attenuation distortion curve for a voice channel. Note that the reference frequency in this case is 1000 Hz.

For example, one requirement may state that between 600 and 2800 Hz the level will vary by not more than -1, $+2$ dB, where the plus sign means more loss and the minus sign means less loss. Thus if a signal at -10 dBm is placed at the input of the channel, we would expect -10 dBm at the output at 800 Hz (if there was no overall loss or gain), but at other frequencies we could expect a variation between -1 and $+2$ dB. For instance, we might measure the level at the output at 2500 Hz at -11.9 dBm and at 1000 Hz at -9 dBm.

CCITT recommendations for attenuation distortion may be found in volume III, Recs. G.132, G.151, and G.232. Figure 2.19 is taken from Rec. G.132 and shows permissible variation of attenuation between 300 and 3400 Hz. Often the requirement is stated as a slope in decibels. The slope is the maximum excursion that levels may vary in a band of interest about a certain frequency. A slope of 5 dB may be a curve with an excursion from -0.5 to $+4.5$ dB, -3 to $+2$ dB, and so forth. As links in a system are added in tandem, to maintain a fixed attenuation distortion across the system, the slope requirement for each link becomes more severe.

1.9.4   Phase Distortion and Envelope Delay Distortion

One may look at a voice channel or any bandpass as a bandpass filter. A signal takes a finite time to pass through the filter. This time is a function of the velocity of propagation. The velocity of propagation tends to vary with frequency, increasing toward band center and decreasing toward band edge, usually in the form of a parabola (see Figure 1.7).

The finite time it takes a signal to pass through the total extension of a voice channel or any other network is called delay. Absolute delay is the delay a signal experiences passing through the channel at a reference frequency. But we see that the propagation time is different for different

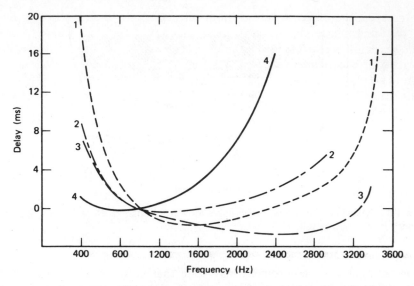

Figure 1.7 Comparison of envelope delay in some typical voice channels. Curves 1 and 3 represent the delay in several thousand miles of a toll-quality carrier system. Curve 2 shows the delay produced by 100 mi of loaded cable. Curve 4 shows the delay in 200 mi of heavily loaded cable. (Courtesy of GTE Lenkurt Inc., San Carlos, CA.) (Ref. 8)

frequencies. This is equivalent to phase shift. If the phase shift changes uniformly with frequency, the output signal will be a perfect replica of the input and there will be no distortion, whereas if the phase shift is nonlinear with respect to frequency, the output signal is distorted (i.e., it is not a perfect replica of the input). Delay distortion, or phase distortion as it is often called, is usually expressed in milliséconds or microseconds about a reference frequency.

We can relate phase distortion to phase delay. If the phase shift characteristic is known, the phase delay T_p at any frequency β_1 can be calculated as follows:

$$T_p = \frac{\beta_1 \, (\text{rad})}{\omega_1 \, (\text{rad/s})} \tag{1.13}$$

The difference between phase delays at two frequencies in a band of interest is called delay distortion T_d and can be expressed as

$$T_d = \frac{\beta_2}{\omega_2} - \frac{\beta_1}{\omega_1} \tag{1.14}$$

where β_2/ω_2 and β_1/ω_1 are the phase delays at ω_2 and ω_1, frequency being expressed as angular frequency ($2\pi f$ rad).

In essence, therefore, we are dealing with the phase linearity of a circuit. The resulting phase distortion is best measured by a parameter called envelope delay distortion (EDD). Mathematically envelope delay is the derivative of the phase shift with respect to frequency and expresses the instantaneous slope of the phase shift characteristic. Envelope delay T_{ed} can be stated as follows:

$$T_{ed} = \frac{d\beta}{d\omega} \tag{1.15}$$

and the expression is valid for very small bandwidths, often referred to as apertures. For instance, the U.S. Bell System uses 166 2/3 Hz and CCITT, 83 1/3 Hz as standard apertures in measurement equipment.

The measurement of envelope delay is useful in television and facsimile transmission systems and is used in data transmission as some measure of intersymbol interference. It is a major limitation to maximum bit rate over a transmission channel. EDD is discussed in greater detail in Chapter 12.

In commercial telephony the high-frequency (HF) harmonic components, produced by the discontinuous nature of speech sounds, arrive later than the fundamental components and produce sounds that may be annoying, but do not appreciably reduce intelligibility. With present handset characteristics the evidence is that the human ear is not very sensitive to phase distortions that develop in the circuit. Although a phase delay of 12 ms between the band limits is noticeable, the transmission in commercial telephone systems often contains distortions greatly in excess of this minimum.

Owing to the large amount of delay distortion in a telephone channel, as measured in its band and relative to a point of minimum delay, the usefulness of the entire telephone channel between its 3-dB cutoff points is severely restricted for the transmission of other than voice signals (e.g., data—see Chapter 12).

For the transmission of information that is sensitive to delay distortion, such as medium-speed digital signals, it is necessary to restrict occupancy to that part of the telephone channel in which the delay distortion can be tolerated or equalized at reasonable cost.

The applicable CCITT recommendation is Rec. G.133.

1.9.5 Level

General In most systems when we refer to level, we refer to a power level that may well be in dBm, dBW, or other power units. One notable exception is video, which uses voltage, usually measured in dBmV.

Level is an important system parameter. If levels are maintained at too high a point, amplifiers become overloaded, with resulting increases in intermodulation products or crosstalk. If levels are too low, customer satisfaction may suffer.

Reference Level Points   System levels usually are taken from a level chart or reference system drawing made by a planning group or as part of an engineered job. On the chart a 0 TLP (test level point) is established. A TLP is the location in a circuit or system at which a specified test tone level is expected during alignment. A 0 TLP is the point at which the test tone level should be 0 dBm.

From the 0 TLP other points may be shown using the unit dBr (dB reference). A minus sign shows that the level is so many decibels below reference, and a positive sign that the level is so many decibels above reference. The unit dBm0 is an absolute unit of power in dBm referred to the 0 TLP. dBm can be related to dBr and dBm0 by the following formula:

$$dBm = dBm0 + dBr \qquad (1.16)$$

For instance, a value of -32 dBm at a -22-dBr point corresponds to a referenced level of -10 dBm0. A -10-dBm0 signal introduced at the 0-dBr point (0 TLP) has an absolute value of signal level of -10 dBm.

In North American practice the 0 TLP was originally defined at the transmission jack of a toll (long-distance) switchboard. Many technical changes, of course, have occurred since the days of manual switchboards. Nevertheless it was deemed desirable to maintain the 0 TLP concept. As a result, the outgoing side of a switch to which an intertoll trunk is connected (see Figure 2.11) is designated a -2-dB TLP, and the outgoing side of the switch at which a local area trunk is terminated is defined as 0 TLP (Ref. 18).

To quote from Ref. 18 in part:

> In the layout of four-wire trunks, a patch bay, called the four-wire patch bay, is usually provided to facilitate test, maintenance, and circuit rearrangements between trunks and the switching machine terminations. TLPs at these four-wire patch bays have been standardized for all four-wire trunks. On the transmitting side the TLP is -16 dB, and on the receiving side the TLP is $+7$ dB. Thus a four-wire trunk, whether derived from voice frequency or from carrier facilities, must be designed to have 23-dB gain between four-wire patch bays. These standard TLPs are necessary to permit flexible telephone plant administration.

> In four-wire circuits, the TLP concept is easily understood and applied because each transmission path has only one direction of transmission. In two-wire circuits, however, confusion or ambiguity may be introduced by the fact that a single point may be properly designated as two different TLPs, each depending on the assumed direction of transmission.

Refer to Section 2.6.3 for a discussion of two-wire and four-wire transmission and to CCITT Rec. G.141 for transmission reference point, and to Figure 2.9 for a definition of virtual switching points a–t–b.

The Volume Unit (VU) .One measure of level is the volume unit (VU). Such a unit is used to measure the power level (volume) of program channels (broadcast) and certain other types of speech or music. VU meters are usually kept on line to measure volume levels of program or speech material being transmitted. If a simple dB meter or voltmeter is bridged across the circuit to monitor the program volume level, the indicating needle tries to follow every fluctuation of speech or program power and is difficult to read; besides, the reading will have no real meaning. To further complicate matters, different meters made by different manufacturers will probably read differently because of differences in their damping and ballistic characteristics.

The indicating instrument used in VU meters is a dc millimeter having a slow response time and damping slightly less than critical. If a steady sine wave is suddenly impressed on the VU meter, the pointer or needle will move to within 90% of the steady-state value in 0.3 s and overswing the steady-state value by no more than 1.5%.

The standard volume indicator (U.S.), which includes the meter and an associated attenuator, is calibrated to read 0 VU when connected across a 600-Ω circuit (voice pair) carrying a 1-mW sine wave power at any frequency between 35 and 10,000 Hz. For complex waves such as music and speech, a VU meter will read some value between average and peak of the complex wave. The reader must remember that there is no simple relationship between the volume measured in VUs and the power of a complex wave. It can be said, however, that for a continuous sine wave signal across 600 Ω, 0 dBm = 0 VU by definition, or that the readings in dBm and VU are the same for continuous simple sine waves in the voice-frequency (VF) range. For a complex signal subtract 1.4 from the VU reading, and the result will be approximate talker power in dBm.

Talker volumes, or levels of a talker at the telephone subset, vary over wide limits for both long-term average power and peak power. Based on comprehensive tests by Holbrook and Dixon (Ref. 1), "mean talker" average power varies between -10 and -15 VU, with a mean of -13 VU.

1.9.6   Noise

General   "Noise, in its broadest definition, consists of any undesired signal in a communication circuit" (Ref. 2). The subject of noise and its reduction is probably the most important that a transmission engineer must face. It is noise that is the major limiting factor in telecommunication system performance. Noise may be divided into four categories:

1. Thermal noise
2. Intermodulation noise
3. Crosstalk
4. Impulse noise

nonlinear device. The coefficients indicate first, second, or third harmonics:

- Second-order products $2F_1, 2F_2, F_1 \pm F_2$
- Third-order products $2F_1 \pm F_2; 2F_2 \pm F_1$
- Fourth-order products $2F_1 \pm 2F_2; 3F_1 \pm F_2$

Devices passing multiple signals, such as multichannel radio equipment, develop intermodulation products that are so varied that they resemble white noise.

Intermodulation noise may result from a number of causes:

- Improper level setting; too high a level input to a device drives the device into its nonlinear operating region (overdrive)
- Improper alignment causing a device to function nonlinearly
- Nonlinear envelope delay

To sum up, intermodulation noise results from either a nonlinearity or a malfunction having the effect of nonlinearity. The cause of intermodulation noise is different from that of thermal noise; however, its detrimental effects and physical nature are identical to those of thermal noise, particularly in multichannel systems carrying complex signals.

Crosstalk   Crosstalk refers to unwanted coupling between signal paths. Essentially there are three causes of crosstalk. The first is the electrical coupling between transmission media, for example, between wire pairs on a VF cable system. The second is poor control of frequency response (i.e., defective filters or poor filter design), and the third is the nonlinearity performance in analog (FDM) multiplex systems. Crosstalk has been categorized into two types:

1. *Intelligible Crosstalk*.   At lest four words are intelligible to the listener from extraneous conversation(s) in a 7-s period (Ref. 16, 3rd ed., p. 46).
2. *Unintelligible Crosstalk*.   Any other form of disturbing effects of one channel upon another. Babble is one form of unintelligible crosstalk.

Intelligible crosstalk presents the greatest impairment because of its distraction to the listener. One point of view is that the distraction is caused by fear of loss of privacy. Another is that the annoyance is caused primarily by the user of the primary line consciously or unconsciously trying to understand what is being said on the secondary or interfering circuits; this would be true for any interference that is syllabic in nature.

Received crosstalk varies with the volume of the disturbing talker, the loss from the disturbing talker to the point of crosstalk, the coupling loss between the two circuits under consideration, and the loss from the point of crosstalk to the listener.

As far as this discussion is concerned, the controlling element is the coupling loss between the two circuits under consideration. Talker volume or level is covered in Section 1.9.5. The effects of crosstalk are subjective, and other factors also have to be considered when the crosstalk impairment is to be measured. Among these factors are the type of people who use the channel, the acuity of listeners, traffic patterns, and operating practices.

Crosstalk coupling loss can be measured quantitatively with precision between a given sending point on a disturbing circuit and a given receiving point on a disturbed circuit. Essentially, then, it is the simple measurement of transmission loss in decibels between the two points. Between carrier circuits, crosstalk coupling is most usually flat. In other words, the amount of coupling experienced at one frequency will be nearly the same for every other frequency in the voice channel. For speech pairs, the coupling is predominantly capacitive, and coupling loss usually has an average slope of 6 dB/octave. See CCITT Rec. G.227 and the annex to CCITT Rec. G.134, which treat crosstalk measurements.

Two other units are also commonly used to measure crosstalk. One expresses the coupling in decibels above "reference coupling" and uses dBx for the unit of measure. The dBx was invented to allow crosstalk coupling to be expressed in positive units. The reference coupling is taken as a coupling loss of 90 dB between disturbing and disturbed circuits. Thus crosstalk coupling in dBx is equal to 90 minus the coupling loss in dB. If crosstalk coupling loss is 60 dB, we have 30 dBx crosstalk coupling. Thus, by definition, 0 dBx = -90 dBm at 1000 Hz.

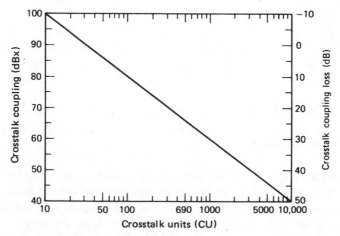

Figure 1.8 Relations between crosstalk measuring units. (From Ref. 10; copyright ©1961 by American Telephone and Telegraph Company.)

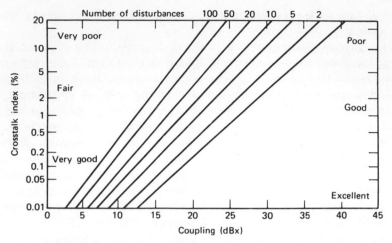

Figure 1.9 Crosstalk judgment curves. (From Ref. 16; 3rd ed., copyright ©1964 by Bell Telephone Laboratories.)

The second unit is the crosstalk unit (CU). When the impedances of the disturbed and disturbing circuits are the same, the number of CUs is one million times the ratio of the induced crosstalk voltage or current to the disturbing voltage or current. When the impedances are not equal,

$$CU = 10^6 \sqrt{\frac{\text{crosstalk signal power}}{\text{disturbing signal power}}} \qquad (1.21)$$

Figure 1.8 relates all three units used in measuring crosstalk. CCITT recommends crosstalk criteria in Rec. G.151 (p. 4).

The percentage change of intelligible crosstalk on a circuit is defined by the crosstalk index. North American practice is to allow arbitrarily that on no more than 1% of calls will a customer hear a foreign conversation which we have defined as intelligible crosstalk. The design objective is 0.5%. The graph in Figure 1.9 may be used for guidance. It relates customer reaction to crosstalk, and crosstalk index to crosstalk coupling.

All forms of unintelligible crosstalk are covered above; its nature is very similar to intermodulation noise.

Impulse Noise This type of noise is noncontinuous, consisting of irregular pulses or noise spikes of short duration and of relatively high amplitude. Often these spikes are called "hits." Impulse noise degrades voice telephony only marginally, if at all. However, it may seriously degrade the error rate on a data transmission circuit, and the subject is covered in more depth in Chapter 12.

Noise Measurement Units The interfering effect of noise on speech telephony is a function of the response of the human ear to specific frequencies in the voice channel as well as of the type of subset used.

When noise measurement units were first defined, it was decided that it would be convenient to measure the relative interfering effect of noise on the listener as a positive number. The level of a 1000-Hz tone at −90 dBm or 10^{-12} W (1 pW) was chosen by the U.S. Bell System because a tone whose level is less than −90 dBm is not ordinarily audible. Such a negative threshold meant that all noise measurements used in telephony would be greater than this number, or positive. The telephone subset then in early universal use in North America was the Western Electric 144 handset. The noise measurement unit was the dB-rn or dBrn, the rn standing for reference noise. 0 dBrn = −90 dBm at 1000 Hz.

With the 144-type handset as a test receiver and with a wide distribution of "average" listeners, it was found that a 500-Hz sinusoidal signal had to have its level increased by 15 dB to have the same interfering effect on the "average" listener over the 1000-Hz reference. A 3000-Hz signal required an 18-dB increase to have the same interfering effect, 6 dB at 800 Hz, and so on. A curve showing the relative interfering effects of sinusoidal tones compared to a reference frequency is called a weighting curve. Artificial filters are made with a response resembling the weighting curve. These filters, normally used on noise measurement sets, are called weighting networks.

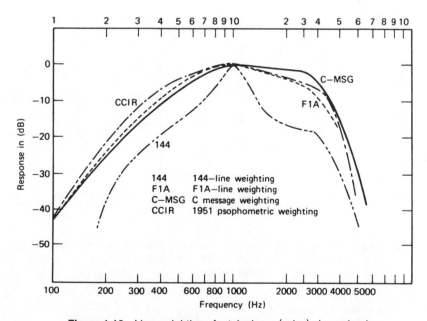

Figure 1.10 Line weightings for telephone (voice) channel noise.

Table 1.4   Conversion chart, psophometric, F1A, and C-message noise units

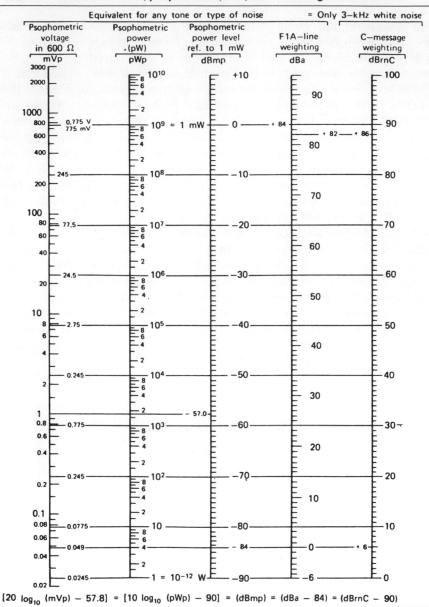

Equivalent for any tone or type of noise = Only 3–kHz white noise

Psophometric voltage in 600 Ω (mVp)	Psophometric power ·(pW) (pWp)	Psophometric power level ref. to 1 mW (dBmp)	F1A–line weighting (dBa)	C–message weighting (dBrnC)

$$[20\ \log_{10}\ (\text{mVp}) - 57.8] = [10\ \log_{10}\ (\text{pWp}) - 90] = (\text{dBmp}) = (\text{dBa} - 84) = (\text{dBrnC} - 90)$$

Chart basis:
   dBmp = dBa − 84
   1 mW unweighted 3-kHz white noise reads 82 dBa = 88.5 dBrnC (C-message) rounded off to 88.0 dBrnC. 1 mW into 600 Ω = 775 mV = 0 dBm = 10^9 pW.
Readings of noise measuring sets when calibrated on 1-mW test tone:
   F1A at 1000 Hz reads 85 dBa
   C-message at 1000 Hz reads 90 dBrn
   Psophometer at 800 Hz reads 0 dBm.

Subsequent to the 144 handset, the Western Electric Company developed the F1A handset, which had a considerably broader response than the older handset, but was 5 dB less sensitive at 1000 Hz. The reference level for this type of handset was −85 dBm. The new weighting curve and its noise measurement weighting network were denoted by F1A (i.e., an F1A line weighting curve, and F1A weighting network). The noise measurement unit was the dB adjusted (dBa).

A third, more sensitive handset (500 type) is now in use in North America, giving rise to the C-message line weighting curve and its companion noise measurement unit, the dBrnC. It is 3.5 dB more sensitive at the 1000-Hz reference frequency than the F1A, and 1.5 dB less sensitive than the 144-type weighting. Rather than choosing a new reference power level (−88.5 dBm), the reference power level of −90 dBm was maintained.

Figure 1.10 compares the various noise weighting curves now in use. Table 1.4 compares weighted noise units.

One important weighting curve and noise measurement unit has yet to be mentioned. The curve is the CCIR (CCITT) psophometric weighting curve. The noise measurement units associated with this curve are dBmp and pWp (dBm psophometrically weighted and picowatts psophometrically weighted, respectively). The reference frequency in this case is 800 Hz rather than 1000 Hz.

Consider now a 3-kHz band of white noise (flat, i.e., not weighted). Such a band is attenuated 8 dB when measured by a noise measurement set using a 144 weighting network, 3 dB using F1A weighting, 2.5 dB for CCIR/CCITT weighting, and 1.5 dB rounded off to 2.0 dB for C-message weighting. Table 1.4 may be used to convert from one noise measurement unit to another.

CCITT states in Rec. G.223 that

> If uniform-spectrum random noise is measured in a 3.1-kHz band with a flat attenuation frequency characteristic, the noise level must be reduced 2.5 dB to obtain a psophometric power level. For another bandwidth B, the weighting factor will be equal to

$$2.5 + 10 \log B/3.1 \text{ dB} \tag{1.22}$$

When $B = 4$ kHz, for example, this formula gives a weighting factor of 3.6 dB.

1.10 SIGNAL-TO-NOISE RATIO

The transmission system engineer deals with signal-to-noise ratio probably more frequently than with any other criterion when engineering a telecommunication system.

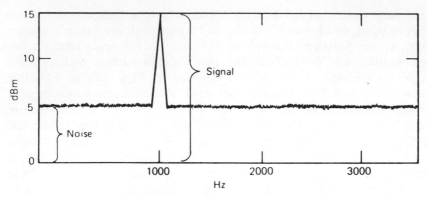

Figure 1.11   Signal-to-noise ratio S/N.

The signal-to-noise ratio expresses in decibels the amount by which a signal level exceeds its corresponding noise.

As we review the several types of material to be transmitted, each will require a minimum signal-to-noise ratio to satisfy the customer or to make the receiving-end instrument function within certain specified criteria. We might require the following signal-to-noise ratios with corresponding end instruments:

Voice   30 dB ⎫ based on customer satisfaction
Video   45 dB ⎭

Data   15 dB   based on a specified error rate and type of modulation

In Figure 1.11 the 1000-Hz signal has a signal-to-noise ratio S/N of 10 dB. The level of the noise is 5 dBm and the signal, 15 dBm. Thus

$$(S/N)_{\text{dB}} = \text{level}_{\text{signal (dBm)}} - \text{level}_{\text{noise (dBm)}} \tag{1.23}$$

1.11   THE EXPRESSION E_b/N_0

For digital transmission systems the expression E_b/N_0 is a more convenient term to qualify a received digital signal than the signal-to-noise ratio under many circumstances. E_b/N_0 expresses the received signal energy per bit per hertz of thermal noise. Thus

$$\frac{E_b}{N_0} = \frac{C}{kT\ (\text{bit rate})} \tag{1.24}$$

where C = the receive signal level (RSL). Expressed in decibel notation,

$$\frac{E_b}{N_0} = C_{dBW} - 10 \log \text{(bit rate)} - (-228.6 \text{ dBW}) - 10 \log T_e \quad (1.25)$$

where T_e = effective noise temperature of the receiving system (see Section 1.9.6).

Example. If the RSL for a particular digital system is -151 dBW and the receiver system effective noise temperature is 1500 K, what is E_b/N_0 for a link transmitting 2400 bps?

$$\frac{E_b}{N_0} = -151 \text{ dBW} - 10 \log 2400 - 10 \log 1500 + 228.6 \text{ dBW}$$

$$= 12 \text{ dB}$$

Depending on the modulation scheme, the type of detector used, and the coding of the transmitted signal, E_b/N_0 required for a given bit error rate (BER) may vary from 4 to 25 dB.

1.12 NOISE FIGURE

It has been established that all networks, whether passive or active, and all other forms of transmission media contribute noise to a transmission system. The noise figure is a measure of the noise produced by a practical network compared to an ideal network (i.e., one that is noiseless). For a linear system, the noise figure (NF) is expressed by

$$NF = \frac{(S/N)_{in}}{(S/N)_{out}} \quad (1.26)$$

It simply relates the signal-to-noise ratio of the output signal from the network to the signal-to-noise ratio of the input signal. From equation 1.18 the thermal noise may be expressed by the basic formula kTB, where $T = 290$ K (room temperature). As we can see, NF can be interpreted as the degradation of the signal-to-noise ratio by the network.

By letting the gain of the network G equal S_{out}/S_{in},

$$NF = \frac{N_{out}}{kTBG} \quad (1.27)$$

It should be noted that we defined the network as fully linear, so NF has not been degraded by intermodulation noise. NF more commonly is expressed in decibels, where

$$NF_{dB} = 10 \log_{10} NF \qquad (1.28)$$

Example. Consider a receiver with an NF of 10 dB. Its output signal-to-noise ratio is 50 dB. What is its input equivalent signal-to-noise ratio?

$$NF_{dB} = (S/N)_{dB\ input} - (S/N)_{dB\ output}$$

$$10\ dB = (S/N)_{input} - 50\ dB$$

$$(S/N)_{input} = 60\ dB$$

1.13 RELATING NOISE FIGURE TO NOISE TEMPERATURE

The noise temperature of a two-port device, a receiver, for instance, is the thermal noise that that device adds to a system. If the device is connected to a noise-free source, its equivalent noise temperature

$$T_e = \frac{P_{ne}}{Gk\,df} \qquad (1.29)$$

where G = gain and df = specified small band of frequencies. T_e is referred to as the effective input noise temperature of the network and is a measure of the internal noise sources of the network, and P_{ne} is the available noise power of the device (Ref. 16).

The noise temperature of a device and its NF are analytically related. Thus

$$NF = 1 + \frac{T_e}{T_0} \qquad (1.30)$$

where T_0 = equivalent room temperature or 290 K.

$$T_e = T_0\,(NF - 1) \qquad (1.31)$$

To convert NF in decibels to equivalent noise temperature T_e in kelvins, use the following formula:

$$NF_{dB} = 10 \log_{10}\left(1 + \frac{T_e}{290}\right) \qquad (1.32)$$

Table 1.5   Noise figure – noise temperature conversion

NF_{dB}	$T(K)$ (approx.)	NF_{dB}	$T(K)$ (approx.)
15	8950	6	865
14	7000	5	627
13	5500	4	439
12	4300	3	289
11	3350	2.5	226
10	2610	2.0	170
9	2015	1.5	120
8	1540	1.0	75
7	1165	0.5	35.4

Example 1. Consider a receiver with an equivalent noise temperature of 290 K. What is its NF?

$$NF_{dB} = 10 \log\left(1 + \frac{290}{290}\right)$$

$$NF = 10 \log 2 = 3 \text{ dB}$$

Example 2. A receiver has an NF of 10 dB. What is its equivalent noise temperature in kelvins?

$$10 \text{ dB} = 10 \log\left(1 + \frac{T_e}{290}\right)$$

$$10 = 10 \log X$$

where $X = 1 + (T_e/290)$. Thus $\log X = 1$, $X = 10$, and

$$T_e = 2900 - 290 = 2610 \text{ K}$$

Several NFs are given with their corresponding equivalent noise temperatures in Table 1.5.

1.14   EFFECTIVE ISOTROPICALLY RADIATED POWER (EIRP)

EIRP is a tool we use to describe the performance of a radio transmitting system. There are three basic elements in the system: a transmitter with a certain output power, an antenna with a gain (or perhaps a loss), and a transmission line that connects the transmitter to the antenna. The transmission line has a loss. It is convenient to express all values using a decibel notation. Thus

$$\text{EIRP}_{\text{dBW}} = P_t + G_{\text{ant}} - L_L \qquad (1.33)$$

where P_t = the RF output power of the transmitter in dB units
        G_{ant} = the gain of the antenna (or loss) in dB
        L_L = the transmission line loss in dB

Equation 1.33 is written using the dB power notation in dBW. It also can be written in dBm. Consistency is urged. If dBW is used, then dBW must be used throughout any related calculation procedure; if dBm is used, we have to be equally consistent.

Consider the following worked examples.

Example 1. A high-frequency (HF) transmitter has an output of 20 kW; its associated rhombic antenna has a gain of 12 dB and the balanced transmission line has a loss of 1.1 dB. What is the EIRP of this transmitting installation? 20 kW = +43 dBW.

$$\text{EIRP}_{\text{dBW}} = +43 \text{ dBW} + 12 \text{ dB} - 1.1 \text{ dB}$$

$$= +53.9 \text{ dBW}$$

Example 2. A microwave transmitter has an output of 1 W; the waveguide connecting this transmitter to its antenna has a loss of 4.6 dB and the antenna has a 36-dB gain. What is the EIRP in dBm? 1 W = +30 dBm.

$$\text{EIRP}_{\text{dBm}} = +30 \text{ dBm} + 36 \text{ dB} - 4.6 \text{ dB}$$

$$= +61.4 \text{ dBm}$$

1.14.1 The Concept of the Isotropic Antenna

An isotropic antenna is an antenna with a gain of 1 (0 dB) that radiates uniformly in all three dimensions. It is a fictitious antenna because we cannot build an antenna with such characteristics. It is, however, a very useful tool to describe the performance of a real antenna when compared to an isotropic. Gain of an antenna is most often given in dBi. This means the number of dB above or below the isotropic reference. The plus or minus sign in front of the dB value indicates whether the gain is greater or less than an isotropic.

Some texts and some of our peers will specify or give antenna gain in dB. We must remember that decibels express a ratio. We must then ask, dB relative to what? The isotropic is most convenient, but in some cases a dipole is inferred. Sometimes we will find ERP (effective radiated power) used rather than EIRP. That power may be relative to an isotropic or a dipole, or even some other antenna. To relate a half-wave dipole (in free space) to an isotropic in free space, we should keep in mind that the dipole has a gain of 2.15 dBi (decibels over an isotropic).

1.15 WAVELENGTH–FREQUENCY RELATIONSHIP

If we are given the frequency of an RF wave, we can calculate its wavelength; and conversely, given the wavelength, we can calculate the frequency by the following equation:

$$F\lambda = 3 \times 10^8 \text{ m/s} \tag{1.34}$$

where F = the frequency in hertz and λ = the wavelength in meters.

Examples: If the wavelength of an RF emission is 40 m, what is the equivalent frequency?

$$40F = 3 \times 10^8 \text{ m/s}$$

$$F = 3 \times 10^8/40$$

$$= 7{,}500{,}000 \text{ Hz or } 7.5 \text{ MHz}$$

If the wavelength of the international calling and distress frequency is 600 m, what is its equivalent frequency?

$$600F = 3 \times 10^8 \text{ m/s}$$

$$F = 500 \text{ kHz}$$

1.16 LONGITUDINAL BALANCE

Longitudinal balance is an important transmission parameter when dealing with transmission lines carrying baseband signals such as subscriber loops, metallic trunks, and metallic switching circuits. Longitudinal balance is a measure of each leg of a balanced circuit's symmetry to ground. With good balance, induced noise can be reduced or nearly eliminated entirely.

We define a balanced circuit as one in which two branches are electrically alike and symmetrical with respect to a common reference point, usually ground. A metallic circuit is defined as one where ground forms no part of that circuit. Longitudinal balance is measured in dB and is expressed by the following formula:

$$\text{Longitudinal balance (dB)} = 20 \log\left(\frac{\text{open circuit longitudinal voltage}}{\text{metallic voltage}}\right)$$

$$(1.35)$$

The IEEE (Ref. 22) defines the degree of longitudinal balance measured in dB as the ratio of the disturbing longitudinal voltage and the resulting metallic voltage of the network under test.

These concepts are shown graphically in Figure 1.12, where we distinguish between longitudinal and metallic current. The figure shows a telephone switch which is the terminating circuit (network) connected to a signal generator (E) by means of a balanced transmission line. Examples of such transmission lines are subscriber loops, metallic pair trunks, and talk paths terminating in switch signal ports in the case of a digital switch and through-switch ports in the case of analog space division switches.

The figure shows that currents that flow in the same direction on the two conductors of the transmission line are called longitudinal currents (Figure 1.12*a*), while currents flowing in opposite directions in the two conductors are called metallic currents (Figure 1.12*b*).

In Figure 1.12, the signal generator (voltage source) is coupled to the balanced transmission line by a coupling mechanism Z_c. The resulting currents are transmitted to the switch through a load impedance Z_L through common battery feed circuits through impedance Z_s. Z_s may represent switch components such as relays, transformers, and common battery leads.

Figure 1.12*a* shows a perfectly balanced condition, meaning that the balanced leads are electrically alike and symmetrical with respect to ground. In this case interference voltages cause no interference current through Z_L. However, if the Z_L and Z_s networks are not balanced, unequal currents flow on the two sides of the circuit. The difference between them is the metallic current that appears in load Z_L as interference. This metallic component of the current can be represented as originating in the equivalent circuit, as shown in Figure 1.12*b*, where E may represent the source of the unbalanced current, some system-generated interference, or a wanted signal source.

Some examples of good longitudinal balance follow. Subscriber loop: over 50 dB (Ref. 18, vol. 3) and four-wire switches (space division) should exceed 53 dB (Ref. 21).

When one reviews vendor longitudinal balance values for a certain equipment, care must be taken to assure the test method used. Our experience shows that the most widely accepted longitudinal balance test method is presented in ANSI/IEEE Std. 455-1985 (Ref. 22).

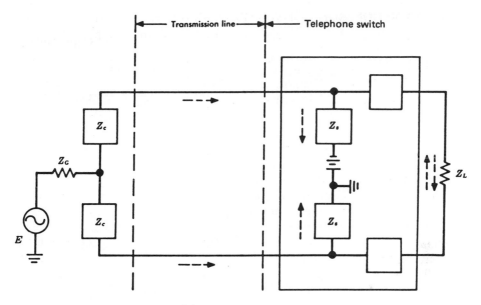

(*a*) Longitudinal current

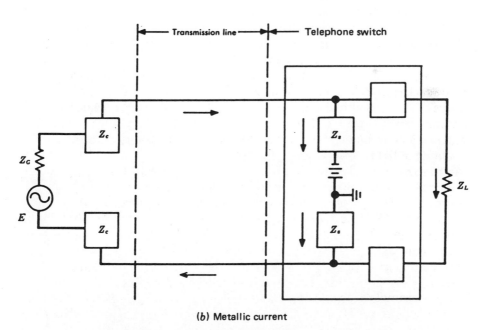

(*b*) Metallic current

Figure 1.12 Longitudinal and metallic current distinguished.

1.17 SOME COMMON CONVERSION FACTORS

To Convert	Into	Multiply By	Conversely, Multiply By
acres	hectares	0.4047	2.471
Btu	kilogram-calories	0.2520	3.969
°Celsius	°Fahrenheit	$9°C/5 = °F - 32$ $9(°C + 40)/5$ $= (°F + 40)$	
circular mils	square centimeters	5.067×10^{-6}	1.973×10^5
circular mils	square mils	0.7854	1.273
degrees (angle)	radians	1.745×10^{-2}	57.30
kilometers	feet	3281	3.048×10^{-4}
kilowatt-hours	Btu	3413	2.930×10^{-4}
liters	gallons (liq. U.S.)	0.2642	3.785
$\log_e$ or ln	$\log_{10}$	0.4343	2.303
meters	feet	3.281	0.3048
miles (nautical)	meters	1852	5.400×10^{-4}
miles (nautical)	miles (statute)	1.1508	0.8690
miles (statute)	feet	5280	1.890×10^{-4}
miles (statute)	kilometers	1.609	0.6214
nepers	decibels	8.686	0.1151
square inches	circular mils	1.273×10^6	7.854×10^{-7}
square millimeters	circular mils	1973	5.067×10^{-4}

Boltzmann's constant $(1.38044 \pm 0.00007) \times 10^{-16}$ erg/deg

velocity of light in free space 2.998×10^8 m/s

186,280 mi/s
984×10^6 ft/s

1 degree of longitude at the equator 68.703 statute mi or 59.661 nautical mi

1 rad $180°/\pi = 57.2958°$

1 m 39.3701 in. = 3.28084 ft

1° 17.4533 mrad

e 2.71828

REVIEW EXERCISES

1. Distinguish *transmission* from *switching*. List four subdisciplines of each.

2. A network has an input of 12 mW and a gain of 26 dB. What is the output of the network in dBm?

3. There are four networks in series. The first network has a gain of 15 dB, the second a loss of 4 dB, the third a gain of 35 dB, and the fourth a loss of 5 dB. The input of the first network is +3 dBm. What is the output to the last network in mW?

4. There are three networks in series. The first network has a gain of 19 dB, the second a loss of 23 dB, and the third a gain of 11 dB. The output of the third network is +23 dBm. What is the input to the first network in mW?

5. A combining network has two inputs: +20 dBm and +6 dBm. It has an insertion loss of 3 dB. What is the combined output in dBm?

6. What is the equivalent mW value of +23.65 dBm?

7. In a population of 3500 subscribers, 2σ of them are satisfied with the transmission level. How many subscribers are unsatisfied with transmission level? The 3σ of the subscriber lines have a return loss of 11 dB or better. How many subscriber lines in this population have a return loss of less than 11 dB?

8. As a subscriber loop is extended in length, what are the two electrical constraints on loop length and how do these constraints affect the subscriber?

9. What are the three basic *impairments* (not echo or singing) we have to deal with regarding the voice channel?

10. Describe the basic CCITT voice channel with regard to bandwidth at the 10-dB points. It occupies the band _____ Hz to _____ Hz.

11. Explain the cause of phase distortion.

12. Phase distortion, in general, does not impair speech transmission. It does impair one important type of information transmitted across the voice channel. What type of information is this?

13. Define 0 TLP.

14. What is the more common unit of measure of speech level? Relate the unit to dBm. Include certain restrictions placed on the relationship.

15. Give the four basic types of noise that we must deal with in transmission as provided in the text.

16. What is the thermal noise threshold of a *perfect* receiver operating at absolute zero? At room temperature? Use the unit dBW in the answers.

17. What is the thermal noise threshold of a receiver with a 2-MHz bandwidth and an effective noise temperature of 2000 K?

18. Two signals, A and B, mix in a nonlinear device. Give the two most probable values of third-order products. Use the notation given in the text.

19. Of the four types of noise discussed, one type is not generally an impairment to speech transmission, but can seriously affect data bit error rate. What type of noise is this?

20. Why are dB-related noise units referenced to -90 dBm and not some other value?

21. Give the dB difference between psophometric noise weighting and flat weighting when dealing with the standard voice channel.

22. It is often useful to give S/N using a formula based on dB-derived units. Write the formula for S/N using dBm.

23. $+13$ dBW $= ?$ dBm. -3 dBm $= ?$ dBW.

24. Define noise figure in an equation with decimal values (in some texts this is called noise factor). Now define noise figure using dB units.

25. (*a*) A certain receiver has a 3-dB noise figure. Give its equivalent noise temperature in K. (*b*) The effective noise temperature of a receiver is 2000 K. What is its equivalent noise figure in dB?

26. An antenna has a 15-dB gain and is fed by a transmitter with 2-kW output. Transmission line losses are 0.6 dB. What is the EIRP (dBW) of the peak main beam of the antenna?

27. A transmitter has an output of 500 mW. Transmission line losses are 3.7 dB and the gain of the antenna is 36 dB. What is the EIRP of the peak main beam?

28. Discuss the importance of longitudinal balance and its measurement. Name at least three places in the telecommunications plant where longitudinal balance would be important to the transmission engineer.

29. (*a*) A transmitter has an output at 3 MHz. What is the equivalent wavelength of the signal? (*b*) A certain signal has a wavelength of 6 cm. What is its equivalent frequency?

REFERENCES AND BIBLIOGRAPHY

1. B. D. Holbrook and J. T. Dixon, "Load Rating Theory for Multichannel Amplifiers," *Bell Syst. Tech. J.*, vol. 18, 624–644, Oct. 1939.

2. *Reference Data for Radio Engineers*, 6th ed., Howard W. Sams, Indianapolis, IN, 1977.

3. CCITT, Blue Books, XIXth Plenary Assembly, Melbourne, 1988, vol. III, G recommendations.

4. *DCS Engineering Installation Manual*, DCAC 330-175-1, through Change 9, U.S. Department of Defense, Washington, DC.

5. MIL-STD-188C, U.S. Department of Defense, Washington, DC.

6. W. Oliver, *White Noise Loading of Multichannel Communication Systems*, Marconi Instruments Ltd., St. Albans, UK, Sept. 1964.

7. N. Kramer, "Communication Needs versus Existing Facilities," lecture given at the 1964 Planning Seminar of the North Jersey Section, IEEE.

8. *Lenkurt Demodulator*, Lenkurt Electric Corp., San Carlos, CA, Dec. 1964, June 1965, Sept. 1965.

9. F. R. Connor, *Introductory Topics in Electronics and Telecommunication—Modulation*, Edward Arnold, London, 1973.

10. *Principles of Electricity Applied to Telephone and Telegraph Work*, American Telephone and Telegraph Company, New York, 1961.

11. H. Arkin and R. R. Colton, *Statistical Methods*, 5th ed., Barnes and Noble College Outline Series, New York.

12. C. E. Smith, *Applied Mathematics for Radio and Communication Engineers*, Dover, New York, 1945.

13. S. Goldman, *Information Theory*, Dover, New York, 1953.

14. H. H. Smith, "Noise Transmission Level Terms in American and International Practice," paper, ITT Communication Systems, Paramus, NJ, 1964.

15. M. M. Rosenfeld, "Noise in Aerospace Communication," *Electro-Technology*, May 1965.

16. *Transmission Systems for Communications*, 5th ed., Bell Telephone Laboratories, Holmdel, NJ, 1982.

17. R. C. James (James and James), *Mathematics Dictionary*, 3rd ed., Van Nostrand, Princeton, NJ.

18. *Telecommunication Transmission Engineering*, vols. 1–3, 2nd ed., American Telephone and Telegraph Co., New York, 1977.

19. IEEE Std. 100-1988, *Dictionary of Electrical and Electronics Terms*, 4th ed., IEEE, New York, 1988.

20. Roger L. Freeman, *Reference Manual for Telecommunications Engineering*, Wiley, New York, 1985.

21. USITA Symposium, Apr. 1970, Open Questions 18–37.

22. "IEEE Standard Test Procedure Measuring Longitudinal Balance of Telephone Equipment Operating in the Voice Band," ANSI/IEEE Std. 455-1985.

2

TELEPHONE TRANSMISSION

2.1 GENERAL

Section 1.8 introduced the simple telephone connection. This chapter delves into telephony and problems of telephone transmission more deeply. It exclusively treats speech transmission over wire systems. Other transmission media are treated only in the abstract so that we can consider problems in telephone networks. The subscriber loop, an important segment of the telephone network, is also covered.

2.2 THE TELEPHONE INSTRUMENT

The input–output (I/O) device that provides the human interface with the telephone network is the telephone instrument or subset. It converts sound energy into electrical energy, and vice versa. The degree of efficiency and fidelity with which it performs these functions has a vital effect upon the quality of telephone service provided. The modern telephone subset consists of a transmitter (mouthpiece), a receiver (earpiece), and an electrical network for equalization, sidetone circuitry, and devices for signaling and supervision. All these items are contained in a device that, when mass-produced, sells for about $40.

Let us discuss transmitters and receivers for a moment.

2.2.1 Transmitters

The transmitter converts acoustic energy into electric energy by means of a carbon granule transmitter. The transmitter requires a dc potential, usually on the order of 3–5 V, across its electrodes. We call this the talk battery, and in modern systems it is supplied over the line (central battery) from the switch (see Section 1.8). Current from the battery flows through the carbon granules or grains when the telephone is lifted off its cradle (off-hook). When sound impinges on the diaphragm of the transmitter, variations of air pressure are transferred to the carbon, and the resistance of the electrical path through the carbon changes in proportion to the pressure. A pulsating direct current results. The frequency response of carbon transmitter peaks between 800 and 1000 Hz.

2.2.2 Receivers

A typical receiver consists of a diaphragm of magnetic material, often soft iron alloy, placed in a steady magnetic field supplied by a permanent magnet, and a varying magnetic field, caused by the voice currents flowing through the voice coils. Such voice currents are alternating (ac) in nature and originate at the far-end telephone transmitter. These currents cause the magnetic field of the receiver to alternately increase and decrease, making the diaphragm move and respond to the variations. As a result an acoustic pressure wave is set up, reproducing, more or less exactly, the original sound wave from the distant telephone transmitter. The telephone receiver, as a converter of electrical energy to acoustic energy, has a comparatively low efficiency, on the order of 2–3%.

Sidetone is the sound of the talker's voice heard in his own receiver. The sidetone level must be controlled. When the level is high, the natural human reaction is for the talker to lower his voice. Thus by regulating the sidetone, talker levels can be regulated. If too much sidetone is fed back to the receiver, the output level of the transmitter is reduced owing to the talker lowering his or her voice, thereby reducing the level (voice volume) at the distant receiver, deteriorating performance.

2.3 THE TELEPHONE LOOP

We speak of the telephone subscriber as the user of the subset. As mentioned in Section 1.8, subscribers' telephone sets are interconnected via a switch or network of switches. Present commercial telephone service provides for transmission and reception on the same pair of wires that connect the subscriber to the local switch. Let us now define some terms.

The pair of wires connecting the subscriber to the local switch that serves him is the *subscriber loop*. It is a dc loop in that it is a wire pair typically supplying a metallic path for the following:

1. Talk battery for the telephone transmitter.
2. An ac ringing voltage for the bell on the telephone instrument supplied from a special ringing source voltage.

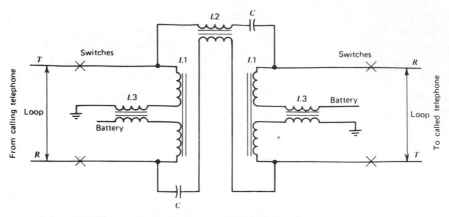

Figure 2.1 Battery feed circuit. *Note*: Battery and ground are fed through inductors *L*3 and *L*1 through switch to loops. (From Ref. 7; copyright © 1961 by Bell Telephone Laboratories.)

3. Current to flow through the loop when the telephone instrument is taken out of its cradle, telling the switch that it requires "access" and causing line seizure at the switching center.

4. The telephone dial that, when operated, makes and breaks the dc current on the closed loop, which indicates to the switching equipment the number of the distant telephone with which communication is desired.

The typical subscriber loop is supplied battery by means of a battery feed circuit at the switch. Such a circuit is shown in Figure 2.1. One important aspect of battery feed is the line balance. The telephone battery voltage has been fairly well-standardized at -48 V. It is a negative voltage to minimize cathodic reaction.

Figure 2.2 shows the functional elements of a subscriber loop, local switch termination, and subscriber subset.

2.4 TELEPHONE LOOP LENGTH LIMITS

It is desirable from an economic viewpoint to permit subscriber loop lengths to be as long as possible. Thus the subscriber area served by a single switching center may be much larger. As a consequence, the total number of switches or telephone central offices may be reduced to a minimum. For instance, if loops were limited to 4 km in length, a switching center could

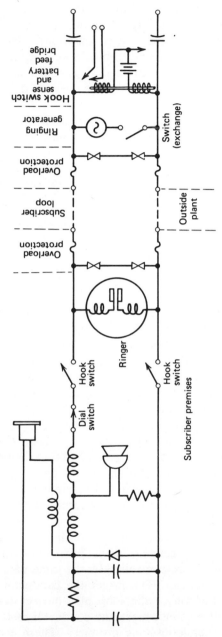

Figure 2.2 Signaling with a conventional telephone subset. Note functions of hook switch, dial, and ringer.

serve all subscribers within a radius of something less than 4 km. If 10 km were the maximum loop length, the radius of an equivalent area that one office could cover would be extended an additional 6 km, out to a total of nearly 10 km. It is evident that to serve a large area, fewer switches (switching centers) are required for the 10-km situation than for the 4-km. The result is fewer buildings, less land to buy, fewer locations where maintenance is required, and all the benefits accruing from greater centralization, which become even more evident as subscriber density decreases, such as in rural areas.

The two basic criteria that must be considered when designing subscriber loops, and that limit their length, are the following:

- Attenuation limits
- Signaling limits

Attenuation in this case refers to loop loss in decibels (or nepers) at

- 1000 Hz in North America
- 800 Hz in Europe and many other parts of the world

As a loop is extended in length, its loss at reference frequency increases. It follows that at some point as the loop is extended, the level will be attenuated such that the subscriber cannot hear sufficiently well.

Likewise, as a loop is extended in length, some point is reached where signaling (supervision) is no longer effective. This limit is a function of the IR drop of the line. We know that R increases as length increases. With today's modern telephone sets, the first to suffer is usually the "supervision." This is a signal sent to the switching equipment requesting "seizure" of a switch circuit and, at the same time, indicating the line is busy. *Off-hook* is a term more commonly used to describe this signal condition. When a telephone is taken "off-hook" (i.e., out of its cradle), the telephone loop is closed and current flows, closing a relay at the switch. If current flow is insufficient, the relay will not close or it will close and open intermittently (chatter) such that line seizure cannot be effected.

Signaling (supervision) limits are a function of the conductivity of the cable conductor and its diameter or gauge. For this introductory discussion we can consider that the loss limits are controlled by the same parameters.

Consider a copper conductor. The larger the conductor, the higher the conductivity, and thus the longer the loop may be for signaling purposes. Copper is expensive, so we cannot make the conductor as large as we would wish and extend subscriber loops long distances. These economic limits of loop length are discussed in detail below. First we must describe what a subscriber considers as hearing sufficiently well, which is embodied in "transmission loss design" (regarding subscriber loop).

2.5 TRANSMISSION FACTORS IN SPEECH TELEPHONY

2.5.1 Loudness

Hearing "sufficiently well" on a telephone connection is a subjective matter under the blanket heading of customer satisfaction. Various methods have been devised over the years to rate telephone connections regarding customer (subscriber) satisfaction. Subscriber satisfaction will be affected by the following regarding the received telephone signal:

- Level (see Section 1.9.5)
- Signal-to-noise ratio (see Section 1.10)
- Response or attenuation distortion (see Section 1.9.3)

2.5.2 Reference Equivalent

2.5.2.1 Definition Of course an essential purpose of a telephone connection is to provide a transmission path for speech between a talker's mouth and the ear of a listener. The level or loudness of the received speech signal depends on acoustic pressure provided by the talker and the loudness loss of the acoustic-to-acoustic path from input to a telephone microphone at one end of the connection to the output of telephone receiver at the other end of a connection. The effectiveness of speech communication over telephone connections and customer satisfaction depend, to a large extent, on the loudness loss of a connection. As the loudness loss is increased from a preferred range, the listening effort is increased and customer satisfaction decreases. At still higher levels of loudness loss, the intelligibility decreases and it takes longer to convey a given quantity of information. On the other hand, if too little loudness loss is provided, customer satisfaction decreases because the received speech is too loud.

Over the years various methods have been used by transmission engineers to measure and express loudness loss of telephone connections. The *reference equivalent* system, a common rating system used to grade customer satisfaction, is one method that has been used for years internationally. It is covered in CCITT Recs. P.42 and 72 and is described briefly below.

When *reference equivalents* have been used as one measure of loudness loss, they typically have been stated in terms of the planning values of the overall reference equivalent (ORE) of a complete connection, which was defined for one direction of transmission as the sum of the following quantities (measured in dB):

- The nominal values of the reference equivalents of the sending and receiving local systems
- The nominal value of the losses at 800 or 1000 Hz of the chain of links and exchanges (switches) interconnecting the two local systems.

Because difficulties have been encountered in the use of reference equivalents, the planning value of the ORE has been replaced by the corrected reference equivalent (CRE). This has required some adjustments in the recommended values of loudness loss for complete connections.

The reference equivalent system considers only the first criterion listed above under "Loudness" (Section 2.5.1), namely, level. It must be emphasized that subscriber satisfaction is subjective. To measure satisfaction, the world regulative body for telecommunications, the International Telecommunication Union (ITU), devised a system of rating sufficient level to "satisfy," using the familiar decibel as the unit of measurement. It is particularly convenient in that, first disregarding the subscriber telephone subset, essentially we can add losses and gains (measured at 800 Hz) in the intervening network end to end, and determine the reference equivalent of a circuit by then adding this sum to a decibel value assigned to the subset, or to a subset plus a fixed subscriber loop length with wire gauge stated.

Let us look at how the reference equivalent system was developed, keeping in mind again that it is a subjective measurement dealing with the likes and dislikes of the "average" human being. Development took place in Europe. A standard for reference equivalent was determined using a team of qualified personnel in a laboratory. A telephone connection was established in the laboratory which was intended to be the most efficient telephone system known. The original reference system or unique master reference consisted of the following:

- A solid-back telephone transmitter
- A Bell telephone receiver
- Interconnecting these, a "zero decibel loss" subscriber loop
- Connecting the loop, a manual, central battery, 22-V dc telephone exchange (switch)

The test team, to avoid ambiguity of language, used a test language that consisted of logatoms. A logatom is a one-syllable word consisting of a consonant, a vowel, and another consonant.

More accurate measurement methods have evolved since then. A more modern reference system is now available in the ITU laboratory in Geneva, Switzerland, called the NOSFER. From this master reference, field test standards are available to telephone companies, administrations, and industry to establish the reference equivalent of telephone subsets in use. These field test standards are equivalent to the NOSFER.

The NOSFER is made up of a standard telephone transmitter, receiver, and network. The reference equivalent of a subscriber's subset, together with the associated subscriber line and feeding bridge, is a quantity obtained by balancing the loudness of received speech signals and is expressed relative to the whole or a corresponding part of the NOSFER (or field) reference system.

2.5.2.2 Application Essentially, as mentioned earlier, type tests are run on subscriber subsets or on the subsets plus a fixed length of subscriber loop of known characteristics. These are subjective tests carried out in a laboratory to establish the reference equivalent of a specific subset as compared to a reference standard. The microphone or transmitter and the earpiece or receiver are each related separately and are called, respectively,

- Transmit reference equivalent (TRE)
- Receive reference equivalent (RRE)

Note: Negative values indicate that the reference equivalent is better than the laboratory standard.

Consider the following simplified telephone network:

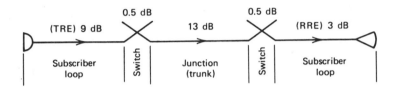

The reference equivalent for this circuit is 26 dB including a 0.5-dB loss for each switch. Junction here takes on the meaning of a circuit connecting two adjacent local or metropolitan switches.

The above circuit may be called a small transmission plan. For this discussion we can define a transmission plan as a method of assigning losses end to end on a telephone circuit. Later in this chapter we discuss why all telephone circuits that have two-wire telephone subscribers must be lossy. The reference equivalent is a handy device to rate such a plan regarding subscriber satisfaction.

When studying transmission plans or developing them, we usually consider that all sections of a circuit in a plan are symmetrical. Let us examine this. On each end of a circuit we have a subscriber loop. Thus in the plan the same loss is assigned to each loop, which may not be the case at all in real life. From the local exchange to the first long-distance exchange, called variously junctions or toll-connecting trunks, a loss is assigned that is identical at each end, and so forth.

To maintain this symmetry regarding reference equivalent of telephone subsets, we use the term $(T + R)/2$. As we see from the above drawing, the TRE and the RRE of the subset have different values. We get the $(T + R)/2$ by summing the TRE and the RRE and dividing by 2. This is done to arrive at the desired symmetry. Table 2.1 gives reference equivalent data on a number of standard subscriber sets used in various parts of the world.

The ORE is the overall reference equivalent and equals the sum of the TRE, the RRE, and all the intervening losses of a telephone connection end

Table 2.1 Reference equivalents for subscriber sets in various countries

Country	Sending (dB)	Receiving (dB)
With limiting subscriber lines and exchange feeding bridges		
Australia	14[a]	6[a]
Austria	11	2.6
France	11	7
Norway	12	7
Germany	11	2
Hungary	12	3
Netherlands	17	4
United Kingdom	12	1
South Africa	9	1
Sweden	13	5
Japan	7	1
New Zealand	11	0
Spain	12	2
Finland	9.5	0.9
With no subscriber lines		
Italy	2	−5
Norway	3	−3
Sweden	3	−3
Japan	2	−1
United States (loop length 1000 ft, 83 Ω)	5	−1[b]

Source: CCITT, *Local Telephone Networks*, ITU, Geneva, July 1968, and *National Telephone Networks for the Automatic Service*. Courtesy of ITU–CCITT.

[a]Minimum acceptable performance.

[b]Ref. 8.

to end with reference to 800 Hz. We would arrive at the same figure if we added twice the $(T + R)/2$ and the intervening losses at 800 Hz. On all these calculations we assume that the same telephone set is used on either end.

CCITT Rec. G.121 states that the reference equivalent from the subscriber set to an international connection should not exceed 20.8 dB (TRE), and to the subscriber set at the other end from the same point of reference RRE should not exceed 12.2 dB. (*Note*: The intervening losses already are included in these figures.) By adding 12.2 and 20.8 dB, we find 33 dB to be the ORE recommended as a maximum* for an international connection. In this regard Table 2.2 should be of interest. It should also be noted that as the reference equivalent (overall, end to end) drops to about 6 dB, the subscriber begins to complain that the call is too loud.

It is noted in Table 2.2 that the 33-dB ORE discussed above is unsatisfactory for more than 10% of calls. Therefore the tendency in many telephone

*For 97% of the connections made in a country of average size.

Table 2.2 British post office survey of subscribers for percentage of unsatisfactory calls

Overall Reference Equivalent (dB)	Percentage of Unsatisfactory Calls
40	33.6
36	18.9
32	9.7
28	4.2
24	1.7
20	0.67
16	0.228

administrations is to reduce this figure as much as possible. In fact the long-term design objective of CCITT for traffic-weighted mean values of the reference equivalent should lie in the range of 13–18 dB for international connections (CCITT Rec. G.111b). As we shall see later, this process is difficult and can prove costly.

2.5.3 Corrected Reference Equivalent

It has been found by the CCITT that if a local system were to be connected with a circuit having a loss x and without distortion, the reference equivalent of the system increases by a value smaller than x. As a consequence, the overall reference equivalent (Section 2.5.2) obtained by summing does not correspond to any physically well-defined quantity that can be determined directly by subjective tests or by calculation from objective measurements.

As a result, let's turn to Figure 2.3 to introduce a method of reference equivalent measurement so a distortionless circuit will have a rating equal to x. The circuits shown in the figure use the "R25" equivalents in paths 2–6. In path 2, X_2 (the NOSFER) is set to 25 dB, its R25 equivalent.

CCITT, in Rec. G.111, states that the introduction of a new subjective test method is not justified, since reference equivalent values (q) are available for many local systems and the corresponding R25 equivalents may be calculated by the formula below.

The corrected reference equivalent (CRE) of a local system or complete system is termed y. Now:

$$y = 0.0082q^2 + 1.148q - 0.48 \text{ dB} \qquad (2.1)$$

where, again, q is the conventional value in dB for reference equivalent discussed in Section 2.5.2.

Table 2.3 gives some values of y of CRE for integer values of q within the range useful for network planning.

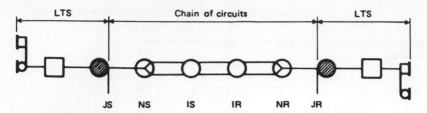

JS   NS   IS   IR   NR   JR

Path 1 — Complete connection

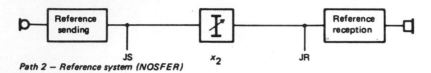

JS x_2 JR

Path 2 — Reference system (NOSFER)

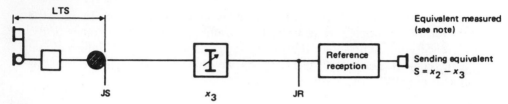

LTS

Equivalent measured (see note)

Sending equivalent
$S = x_2 - x_3$

JS x_3 JR

Path 3 — For the determination of a sending equivalent

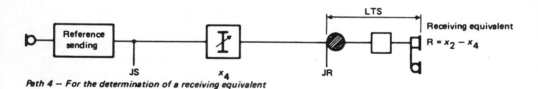

LTS

Receiving equivalent
$R = x_2 - x_4$

JS x_4 JR

Path 4 — For the determination of a receiving equivalent

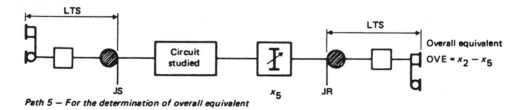

LTS

Overall equivalent
$OVE = x_2 - x_5$

JS x_5 JR

Path 5 — For the determination of overall equivalent

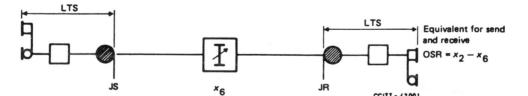

LTS

Equivalent for send and receive
$OSR = x_2 - x_6$

JS x_6 JR

CCITT - 42001

Path 6 — Overall system, send and receive

Figure 2.3 Connections and systems considered for the definition of reference equivalents and R25 equivalents. (From CCITT Rec. G.111; Red Books, Fascicle III.1, Page 105, Figure A-1 / 6.111, VIIIth Plenary Assy., Malaga-Torremolinos 1984. See notes for this figure on page 53.

Table 2.3   Values of y (CRE) as a function of q (RE)

q (dB)		-10	-9	-8	-7	-6	-5	-4
y (dB)		-10.18	-9.18	-8.18	-7.15	-6.11	-5.06	-3.98
q (dB)	-3	-2	-1	0	1	2	3	4
y (dB)	-2.89	-1.78	-0.66	0.48	1.64	2.81	4.00·	5.20
q (dB)	5	6	7	8	9	10	11	12
y (dB)	6.43	7.66	8.92	10.19	11.48	12.78	14.10	15.44
q (dB)	13	14	15	16	17		17.50	18
y (dB)	16.79	18.16	19.55	20.95	22.37		23.08	23.80

Source: CCITT Rec. G.111.
Red Books, Fascicle III.1, Page 106, Table A-1/6.111, VIIIth Plenary Assy, Malaga-Torre-molinos 1984.

Note 1: This formula is applicable to a receiving system and to a (sending or complete) system comprising a linear microphone; in the case of a carbon microphone, experience shows that y should be reduced by 1 dB.

Note 2: Values were computed with two decimal figures to make interpolation easier; the result may be rounded to the nearest 0.5 dB.

Table 2.4 provides guidance for nominal maximum CREs for national systems. Table 2.5 gives values of reference equivalent (q) and CRE (y) for various connections using information provided in CCITT Recs. G.111 and 121. Table 2.6 gives customer opinion results for overall corrected reference equivalents.

2.5.4 Loudness Rating

CCITT in its IXth Plenary Assembly (1988) revised its standard for telephone speech quality and now recommends the use of "loudness rating" (LR). It is

Notes for Figure 2.3

JS	= junction sending side
JR	= junction receiving side
NS	= national sending side
IS	= international sending side
IR	= international receiving side
NR	= national receiving side
LTS	= local telephone system studied (telephone set + line + feeding bridge)

Junction = local trunk in North America

Note: x is obtained in each case by balancing the path in question with NOSFER. The equivalents measured are:

(a) reference equivalents if x_2 is varied so as to obtain the balance
(b) R25 equivalents if x_2 is set at 25 dB

Table 2.4 Nominal maximum CREs in dB recommended for national systems

		CRE
For an Average-Sized Country (see Figure 1/G.121)	Sending	25
	Receiving	14
For large countries: —If a fourth national circuit is part of the 4-wire chain —If five national circuits form part of the 4-wire chain	Sending	25.5
	Receiving	14.5
	Sending	26
	Receiving	15

Source: CCITT Rec. G.121. Red Books, Fascicle III.1, page 152, Table 1/G.111, VIIIth Plenary Assy., Malaga-Torremolinos 1984.

Table 2.5 Values (dB) of reference equivalent (q) and CRE (y) for various connections cited in CCITT RECs. G.111 and G.121

		Previously Recommended RE (q)	Presently Recommended CRE (y)
Optimum range for a connection (Rec. G.111, § 3.2)	Minimum	6	5[a]
	Optimum	9	7[a] to 11
	Maximum	18	16
Traffic weighted mean values Long-term objectives			
—Connection (Rec. G.111, § 3.2)	Minimum	13	13
	Maximum	18	16
—National system send (Rec. G.121, § 1)	Minimum	10	11.5
	Maximum	13	13
—National system receive (Rec. G.121, § 1)	Minimum	2.5	2.5
	Maximum	4.5	4
Short-term objectives —Connection (Rec. G.111, § 3.2)	Maximum	23	25.5
—National system send (Rec. G.121, § 1)	Maximum	16	19
—National system receive (Rec. G.121, § 1)	Maximum	6.5	7.5
Maximum values for national system (Rec. G.121, § 2.1) of an average-sized country	Sending	21	25
	Receiving	12	14
Minimum for the national sending system (Rec. G.121, § 3)		6	7

Source: CCITT Rec. P.11, Blue Books, Vol. V, Page 17, Table 2a/P.11, IXth Plenary Assy., Melbourne 1988.

[a]These values apply for conditions free from echo; customers may prefer slightly larger values if some echo is present.

Table 2.6 Customer opinion ratings for various values of CREs

Planning Value of the Overall Corrected Reference Equivalent (dB)	Representative Opinion Results	
	Percent "Good Plus Excellent"	Percent "Poor Plus Bad"
5–15	> 90	< 1
20	80	2
25	65	5
30	40	15

Source: CCITT Rec. P.11, Red Books, Vol. V, Page 14, VIIIth Plenary Assy., Malaga-Torremolinos 1984.

[a]Based on a composite opinion model.

conceptually similar to ORE (overall reference equivalent). OLR (overall loudness rating) has become the international standard for measuring customer satisfaction of a speech telephone connection. Table 2.7 gives representative opinion results of OLR. OLR is defined in Figure 2.4, and its unit of measure is again the dB. Methods of measuring OLR are given in CCITT Rec. P.79 (Blue Books, Ref. 4).

2.5.5 Circuit Noise

Circuit noise is another impairment to speech transmission. Such noise can be made up of components of thermal noise, intermodulation noise, impulse noise, and single-frequency tones.

The subjective effect of circuit noise measured at a particular point in a telephone connection depends on the electrical-acoustical loss or gain from the point of measurement to the output of the telephone receiver. Circuit noise is frequently referred to the input of receiving system with a specified CRE. A common reference point is the input of a receiving system having a

Table 2.7 Representative opinion results for overall loudness rating (OLR)

Overall Loudness Rating (dB)	Representative Opinion Results	
	Percent "Good Plus Excellent"	Percent "Poor Plus Bad"
5–15	> 90	< 1
20	80	4
25	65	10
30	45	20

Source: CCITT Rec. P.11, Blue Books, Vol. V, Table 1/P.11, Page 16, IXth Plenary Assy., Melbourne 1988.

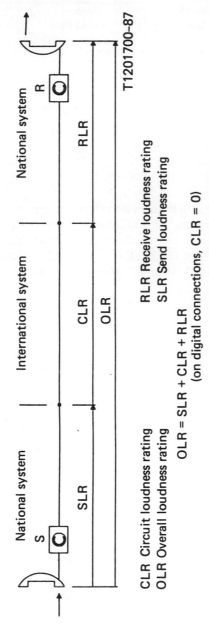

Figure 2.4 Designation of LRs on an international connection. From CCITT Rec. P.11, Blue Books, Vol. V, Figure I / P.11, Page 18, IXth Plenary Assy., Melbourne 1988.

CLR Circuit loudness rating
OLR Overall loudness rating

RLR Receive loudness rating
SLR Send loudness rating

$$OLR = SLR + CLR + RLR$$
$$\text{(on digital connections, CLR = 0)}$$

T1201700-87

Table 2.8   Customer opinion ratings of circuit noise as a transmission impairment

Circuit Noise at Point of 0 dB Receiving CRE (dBmp)[b]	Representative Opinion Results[a]	
	Percent "Good Plus Excellent"	Percent "Poor Plus Bad"
− 65	90	< 1
− 60	85	1
− 55	70	3
− 50	50	10
− 45	30	20

Source: CCITT Rec. P.11, Blue Books, Vol. V, Table 3/P.11, Page 19, IXth Plenary Assy., Melbourne 1988.

[a]Based on a composite opinion model (see Annex A).

[b]The noise values apply for a receiving CRE = 0.46 dB which has been rounded off to 0 dB for simplicity and to take account of the precision of the calculations.

receiving CRE of 0 dB. When circuit noise is referred to this point, circuit noise values of less than − 65 dBmp have little effect on transmission quality in typical room noise environments. Transmission quality decreases with circuit noise values higher than − 65 dBmp.

Table 2.8 presents opinion results representative of laboratory conversation tests and illustrates the effect of circuit noise when other characteristics such as loudness (level) introduce little additional impairment. When loudness loss is greater than the preferred range, the effect of a given level of circuit noise becomes more severe.

2.6   TELEPHONE NETWORKS

2.6.1   General

The next logical step in our discussion of telephone transmission is to consider the large-scale interconnection of telephones. As we have seen, subscribers within a reasonable distance of one another can be interconnected by wire lines and we can still expect satisfactory communication. A switch is used so that a subscriber can speak with some other discrete subscriber as he or she chooses. As we extend the network to include more subscribers and circumscribe a wider area, two technical/economic factors must be taken into account:

1. More than one switch must be used.
2. Wire pair transmission losses on longer circuits must be offset by amplifiers, or the pairs must be replaced by other, more efficient means.

Let us accept item 1. The remainder of this section concentrates on item 2. The reason for the second statement becomes obvious when the salient point of Section 2.5 regarding reference equivalents and loudness loss are reviewed.

Example. How far can a two-wire line be run without amplifiers following the rules of reference equivalent? Allow, in this case, no more than an ORE of 33 dB, a high value for a design goal. Referring to Table 2.1 and using Spain as an example, the TRE + RRE for telephone subsets sums to 14 dB. This leaves us with a limiting loss of 19 dB (33 − 14) for the remaining network. Use 19-gauge (0.91-mm) telephone cable, typical for "long-distance" communication (Table 2.13). This cable has a loss of 0.71 dB/km. Therefore the total extension of the network will be about 26 km, allowing no loss for a switch.

2.6.2 Basic Considerations

What are some of the more common approaches that may be used to extend the network? We may use

1. Coarser gauge cable (larger-diameter conductors)
2. Amplifiers in the present wire pair system; use of inductive loading
3. Carrier transmission techniques (Chapter 3)
4. Radio transmission techniques (Chapters 4–6)
5. Fiber optics transmission (Chapter 11).

Items 1 and 2 are used quite widely but often become unattractive from an economic point of view as the length increases. Items 3 and 4 become attractive for multichannel transmission over longer distances. Discussion of the formation of a multichannel signal is left for Chapter 3. Such a multichannel transmission technique is referred to as *carrier* transmission.

For our purposes we can consider carrier as a method of high-velocity transmission of bands of frequency above the VF region (i.e., above 4000 Hz) over wire, optical fiber, or radio.

2.6.3 Two-Wire / Four-Wire Transmission

Two-Wire Transmission   By its basic nature a telephone conversation requires transmission in both directions. When both directions are carried on the same wire pair, we call it two-wire transmission. The telephones in our home and office are connected to a local switching center by means of two-wire circuits. A more proper definition for transmitting and switching purposes is that when oppositely directed portions of a single telephone

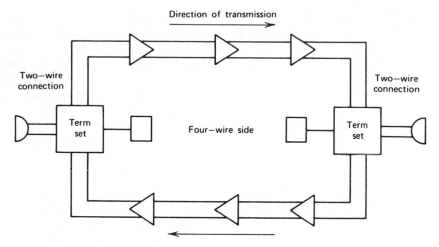

Figure 2.5 Typical long-distance telephone connection.

conversation occur over the same electrical transmission channel or path, we call this two-wire operation.

Four-Wire Transmission   Carrier and radio systems require that oppositely directed portions of a single conversation occur over separate transmission channels or paths (or using mutually exclusive time periods). Thus we have two wires for the transmit path and two wires for the receive path, or a total of four wires for a full-duplex (two-way) telephone conversation. For almost all operational telephone systems, the end instrument (i.e., the telephone subset) is connected to its intervening network on a two-wire basis.*

Nearly all long-distance telephone connections traverse four-wire links. From the near-end user the connection to the long-distance network is two wire. Likewise, the far-end user is also connected to the long-distance network via a two-wire link. Such a long-distance connection is shown in Figure 2.5. Schematically the four-wire interconnection is shown as if it were wire line, single channel with amplifiers. More likely, it would be multichannel carrier on cable and/or multiplex on radio. However, the amplifiers in the figure serve to convey the ideas that this chapter considers.

As shown in Figure 2.5, conversion from two-wire to four-wire operation is carried out by a terminating set, more commonly referred to in the industry as a *term set*. This set contains a four-winding balanced transformer (a hybrid) or a resistive network, the latter being less common.

---

*A notable exception is the U.S. military telephone network Autovon, where end users are connected on a four-wire basis.

Operation of a Hybrid   A hybrid, for telephone work (at VF), is a transformer. For a simplified description, a hybrid may be viewed as a power splitter with four sets of wire pair connections. A functional block diagram of a hybrid device is shown in Figure 2.6. Two of these wire pair connections belong to the four-wire path, which consists of a transmit pair and a receive pair. The third pair is the connection to the two-wire link (L) to the subscriber subset. The last wire pair connects the hybrid to a resistance–capacitance balancing network (N) which electrically balances the hybrid with the two-wire connection to the subscriber's subset over the frequency range of the balancing network. An artificial line may also be used for this purpose.

The hybrid function permits signals to pass from any pair through the transformer to both adjacent pairs but blocks signals to the opposite pairs (as shown in Figure 2.6). Signal energy entering from the four-wire side divides equally, half dissipating into the balancing network and half going to the desired two-wire connection. Ideally no signal energy in this path crosses over the four-wire transmit side. This is an important point, which we take up later.

Signal energy entering from the two-wire subset connection divides equally, half of it dissipating in the impedance of the four-wire side receive path, and half going to the four-wire side transmit path. Here the *ideal* situation is that no energy is to be dissipated by the balancing network (i.e., there is a perfect balance). The balancing network is supposed to display the characteristic impedance of the two-wire line (subscriber connection) to the hybrid.

The reader notes that in the description of the hybrid, in every case, ideally half of the signal energy entering the hybrid is used to advantage and half is dissipated, wasted. Also keep in mind that any passive device inserted in a circuit such as a hybrid has an insertion loss. As a rule of thumb we say that the insertion loss of a hybrid is 0.5 dB. Hence there are two losses here of which the reader must not lose sight:

Hybrid insertion loss   0.5 dB
Hybrid dissipation loss   3.0 dB (half power)
                              3.5 dB total

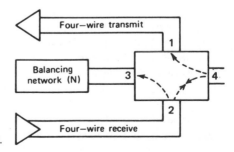

Figure 2.6   Operation of a hybrid transformer.

As far as this chapter is concerned, any signal passing through a hybrid suffers a 3.5-dB loss. Some hybrids used on short subscriber loops purposely have higher losses, as do special resistance type hybrids.

An important parameter is *transhybrid loss* or transhybrid balance, which is the isolation between ports 1 and 2 in Figure 2.6. When ports 3 and 4 in the figure are perfectly balanced, transhybrid loss may reach 50 dB (Ref. 7).

2.6.4   Echo, Singing, and Design Loss

General   The operation of the hybrid with its two-wire connection on one end and four-wire connection on the other leads us to the discussion of two phenomena that, if not properly designed for, may lead to major impairments in communication. These impairments are echo and singing.

Echo. As the name implies, echo in telephone systems is the return of a talker's voice. The returned voice, to be an impairment, must suffer some noticeable delay.

Thus we can say that echo is a reflection of the voice. Analogously, it may be considered as that part of the voice energy that bounces off obstacles in a telephone connection. These obstacles are impedance irregularities, more properly called impedance mismatches.

Echo is a major annoyance to the telephone user. It affects the talker more than the listener. Two factors determine the degree of annoyance of echo: its loudness and how long it is delayed.

Singing. Singing is the result of sustained oscillations due to positive feedback in telephone amplifiers or amplifying circuits. Circuits that sing are unusable and promptly overload multichannel carrier equipment (FDM, see Chapter 3).

Singing may be thought of as echo that is completely out of control. This can occur at the frequency at which the circuit is resonant. Under such conditions the circuit losses at the singing frequency are so low that oscillation will continue even after the impulse that started it ceases to exist.

The primary cause of echo and singing generally can be attributed to the mismatch between the balancing network and its two-wire connection associated with the subscriber loop. It is at this point that the major impedance mismatch usually occurs and an echo path exists. To understand the cause of the mismatch, remember that we always have at least one two-wire switch between the hybrid and the subscriber. Ideally the hybrid balancing network must match each and every subscriber line to which it may be switched. Obviously the impedances of the four-wire trunks (lines) may be kept fairly uniform. However, the two-wire subscriber lines may vary over a wide range. The subscriber loop may be long or short, may or may not have inductive loading (see Section 2.8.4), and may or may not be carrier derived (see Chapter 3). The hybrid imbalance causes signal reflection or signal "return."

The better the match, the more the return signal is attenuated. The amount that the return signal (or reflected signal) is attenuated is called the *return loss* and is expressed in decibels. The reader should remember that any four-wire circuit may be switched to hundreds or even thousands of different subscribers. If not, it would be a simple matter to match the four-wire circuit to its single subscriber through the hybrid. This is why the hybrid to which we refer has a compromise balancing network rather than a precision network. A compromise network is usually adjusted for a compromise in the range of impedance that is expected to be encountered on the two-wire side.

Let us consider now the problem of match. For our case the impedance match is between the balancing network N and the two-wire line L (see Figure 2.6). With this in mind,

$$\text{Return loss}_{dB} = 20 \log_{10} \frac{Z_N + Z_L}{Z_N - Z_L} \qquad (2.2)$$

If the network perfectly balances the line, $Z_N = Z_L$, and the return loss would be infinite.

The return loss may also be expressed in terms of the reflection coefficient:

$$\text{Return loss}_{dB} = 20 \log_{10} \frac{1}{\text{reflection coefficient}} \qquad (2.3)$$

where the reflection coefficient is the ratio of reflected signal to incident signal.

The CCITT uses the term *balance return loss* (see CCITT Rec. G.122) and classifies it as two types:

1. Balance return loss from the point of view of echo.* This is the return loss across the band of frequencies from 300 to 2500 Hz.
2. Balance return loss from the point of view of stability. This is the return loss between 0 and 4000 Hz.

The band of frequencies that is most important from the standpoint of echo for the voice channel is that between 300 and 2500-Hz. A good value for the echo return loss (ERL) for a toll telephone plant is 11 dB, with values on some connections dropping to as low as 6 dB. For the local telephone network, CCITT recommends better than 6 dB, with a standard deviation of 2.5 dB (Ref. 18).

For frequencies outside the 300–2500 Hz band, return loss values often are below the desired 11 dB. For these frequencies we are dealing with return loss from the point of view of stability. CCITT recommends that balance return loss from the point of view of stability (singing) should have a

---

*Called echo return loss (ERL) in via net loss (North American practice, Section 2.6.5), both use a weighted distribution of level.

value of not less than 2 dB for all terminal conditions encountered during normal operation (CCITT Rec. G.122, p.2.2). For further information the reader should consult Appendix A of CCITT Recs. G.122 and G.131.

Echo and singing may be controlled by

- Improved return loss at the term set (hybrid)
- Adding loss on the four-wire side (or on the two-wire side)
- Reducing the gain of the individual four-wire amplifiers

The annoyance of echo to a subscriber is also a function of its delay. Delay is a function of the velocity of propagation of the intervening transmission facility. A telephone signal requires considerably more time to traverse 100 km of a voice pair cable facility, particularly if it has inductive loading, than 100 km of radio facility.

Delay is expressed in one-way or round-trip propagation time measured in milliseconds. CCITT recommends that if the mean round-trip propagation time exceeds 50 ms for a particular circuit, an echo suppressor should be used. Bell System practices in North America use 45 ms as a dividing line. In other words, where the echo delay is less than that stated above, the echo will be controlled by adding loss.

An echo suppressor is an electronic device inserted in a four-wire circuit which effectively blocks passage of reflected signal energy. The device is voice operated, with a sufficiently fast reaction time to "reverse" the direction of transmission, depending on which subscriber is talking at the moment. The blocking of reflected energy is carried out by simply inserting a high loss in the return four-wire path.

Figure 2.7 shows the echo path on a four-wire circuit.

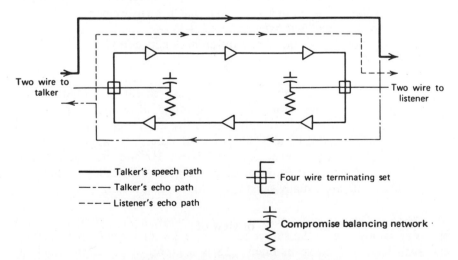

Figure 2.7 Echo paths in a four-wire circuit.

Echo Return Loss, North American Definition. Return loss, as we discussed above, is a measure of impedance match or mismatch. Return losses on the order of 20–33 dB are indicative of a good match, whereas match deteriorates as the dB value gets smaller.

Our main area of interest here is the mismatch between a hybrid (term set) and a·telephone channel. Return loss is defined rigidly in terms of the ratio of sum and difference of the complex impedances where we would expect an impedance discontinuity (equation 2.2). Because of the complexity of phase relationships in the incident and reflected voltage waves, it makes it impractical to express return loss over a band of frequencies except by averaging the performance over the band of interest. Echo return loss (ERL), then, is weighted *power* averaged over the band 500–2500 Hz.

One method of determining return loss in this situation is the measurement of single-frequency return loss in 500-Hz increments from 500 to 2500 Hz. Other methods sweep each 500-Hz subband or use random noise measurements. One weighting method halves the value of each frequency extreme (e.g., 500 and 2500 Hz). If we look at the four bands of interest, calling them $B(1)$, $B(2)$, $B(3)$, and $B(4)$, and the five frequencies $f(1) \cdots f(5)$, we can then define ERL as

$$\text{ERL} = -10 \log \frac{B}{4B} \left(\frac{\text{RL}_{f1} + \text{RL}_{f5}}{2} \text{RL}_{f2} + \text{RL}_{f3} + \text{RL}_{f4} \right) \text{dB} \quad (2.4)$$

where $B = 500$ Hz and $\text{RL}(fn)$ is the return loss for each of the five frequencies expressed as a *power* ratio (Ref. 21).

In the North American switched network, ERL measurements are made at various switching centers (exchanges) and impedances are corrected to improve performance where required. The measurements are made against standard impedances (e.g., $600/900 \ \Omega$). Such measurements are called *through balance* and *terminal balance*.

The reader should note the subtle differences between CCITT and North American practice.

Singing Return Loss. Singing return loss measurements are made to give assurance of the necessary stability and are made at all sample frequencies at which a circuit may display instability. Instability usually occurs in the bands from 200 to 500 Hz and 2500 to 3200 Hz (North American practice)—for the CCITT voice channel, the lower frequency is 300 Hz and the higher is 3400 Hz. It has been found that frequencies in the band 500–2500 Hz that meet ERL requirements are usually satisfactory for singing return loss.

Transmission Design to Control Echo and Singing. As stated previously, echo is an annoyance to the subscriber. Figure 2.8 relates echo path delay to echo path loss. The curve in Figure 2.8 is a group of points at which the average subscriber will tolerate echo as a function of its delay. Remember that the greater the return signal delay, the more annoying it is to the

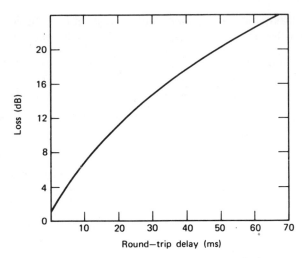

Figure 2.8   Talker echo tolerance for average telephone users.

telephone talker (i.e., the more the echo signal must be attenuated). For instance, if the echo path delay on a particular circuit is 20 ms, an 11-dB loss must be inserted to make echo tolerable to the talker. The careful reader will note that the 11 dB designed into the circuit will increase the end-to-end reference equivalent by that amount, something quite undesirable. The effect of loss design on reference equivalents and the trade-offs available are discussed below.

To control singing, all four-wire paths must have some loss. Once they go into a gain condition, and we refer here to overall circuit gain, we will have positive feedback and the amplifiers will begin to oscillate or "sing." North American practice calls for a 4-dB loss on all four-wire circuits to guard against singing.

Almost all four-wire circuits have some form of amplifier and level control. Often such amplifiers are embodied in the channel banks of the carrier equipment. For a discussion of carrier equipment, see Chapter 3.

An Introduction to Transmission Loss Planning.   One major aspect of transmission system design for a telephone network is to establish a transmission loss plan on a national basis. Such a plan, when implemented, is formulated to accomplish three goals:

- Control singing
- Keep echo levels within limits tolerable to the subscriber
- Provide an acceptable overall reference equivalent to the subscriber

For North America the via net loss (VNL) concept embodies the transmission loss plan idea. VNL is covered in Section 2.6.5.

From our discussions above, we have much of the basic background necessary to develop a transmission loss plan. We know the following:

1. A certain minimum loss must be maintained in four-wire circuits to prevent singing.
2. Up to a certain limit of round-trip delay, echo is controlled by loss.
3. It is desirable to limit these losses as much as possible to improve the reference equivalent.

National transmission plans vary considerably. Obviously the length of circuit, as well as the velocity of propagation of the transmission media, is important. Two approaches are available in the preparation of a loss plan:

- Variable loss plan (i.e., VNL)
- Fixed loss plan (i.e., as used in Europe and recommended in CCITT Rec. G.122)

A national transmission loss plan for a small country (i.e., small in extension) such as Belgium could be quite simple. Assume that a 4-dB loss is inserted in all four-wire circuits to prevent singing (see Figure 2.8). Here 4 dB allows for 5 ms of round-trip delay. If we assume carrier transmission for the entire length of the connection and use 105,000 mi/s for the velocity of propagation, we can then satisfy Belgium's echo problem. The velocity of propagation used comes out to 105 mi (168 km)/ms. By simple arithmetic we see that a 4-dB loss on all four-wire circuits will make echo tolerable for all circuits extending 262 mi (420 km). This is an application of the fixed-loss type of transmission plan. In the case of small countries or telephone companies operating over a small geographical extension, the minimum loss inserted to control singing controls echo as well for the entire country.

Let us try another example. Assume that all four-wire connections have a 7-dB loss. Figure 2.8 indicates that 7 dB permits an 11-ms round-trip delay. Assume that the velocity of propagation is 105,000 mi/s. Remember that we deal with round-trip delay. The talker's voice goes out to the far-end hybrid and is then reflected back. This means that the signal traverses the system twice, as shown:

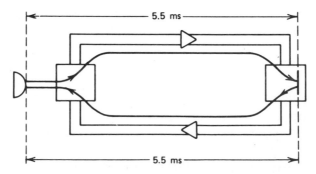

In this example the round-trip delay is 5.5 + 5.5 = 11 ms.

Thus 7 dB of loss for the velocity of propagation given allows about 578 mi of extension (5.5 × 105) or, for all intents and purposes, the distance between subscribers.

It has become evident by now that we cannot continue increasing losses indefinitely to compensate for echo on longer circuits. Most telephone companies and administrations have set the 45- or 50-ms round-trip delay criterion. This sets a top figure above which echo suppressors shall be used.

One major goal of the transmission loss plan is to improve overall reference equivalents to apportion more loss to the subscriber plant so that subscriber loops can be longer, or to allow the use of less copper (i.e., smaller diameter conductors). The question is, what measures can be taken to reduce losses and still keep echo within tolerable limits? One obvious target is to improve return losses at the hybrids. If all hybrid return losses are improved, then the echo tolerance curve gets shifted. This is so because improved return losses reduce the intensity of the echo returned to the talker. Thus the subscriber is less annoyed by the echo effect.

One way of improving the return loss is to make all two-wire lines out of hybrid look alike, that is, have the same impedance. The switch at the other end of the hybrid (i.e., on the two-wire side) connects two-wire loops of varying lengths, causing the resulting impedances to vary greatly. One approach is to extend four-wire transmission to the local office such that each hybrid can be better balanced. This is being carried out with success in Japan. The U.S. Department of Defense has its Autovon (automatic voice network) in which every subscriber line is operated on a four-wire basis. Two-wire subscribers connect through the system on a private automatic branch exchange (PABX).

Let us return to standard telephone networks using two-wire switches in the subscriber area. Suppose that the balance return loss could be improved to 27 dB. Therefore the minimum loss to prevent singing could be reduced to 0.4 dB. Suppose that we distributed this loss across four four-wire circuits in tandem. Hence each four-wire circuit would be assigned a 0.1-dB loss. If we have gain in the network, singing will result. The safety factor between loss and gain is 0.4 dB. The loss in each circuit or link is maintained by amplifiers. It is difficult to adjust the gain of an amplifier to 0.1 dB, much less keep it there over long periods, even with good automatic regulation. *Stability* (or *gain stability*) is the term used to describe how well a circuit can maintain a desired level. Of course, in this case we refer to a test-tone level. In the example above it would take only one amplifier to shift 0.4 dB, two to shift in the positive direction 0.2 dB, and so forth. The importance of stability, then, becomes evident.

The stability of a telephone connection depends on

- The variation of transmission level with time
- The attenuation–frequency characteristics of the links in tandem
- The distribution of balance return loss

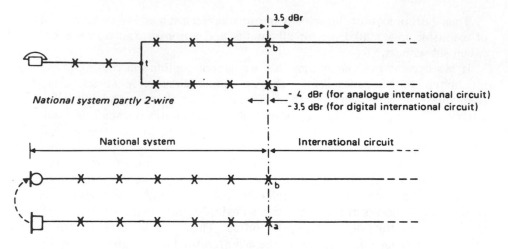

National system wholly 4-wire

Note — a, b are the virtual analogue switching points of the international circuit.

Figure 2.9 Definition of point *a* and *b* and *a–t–b*, virtual analog switching point, and national system representation. (From CCITT Rec. G.122; Blue Books, Figure 1 / G.122, Page 126, Fascicle III.1, Page 12, IXth Plenary Assy., Melbourne 1988.

Each of these criteria becomes magnified when circuits are switched in tandem. To handle the problem properly, we must talk about statistical methods and standard distributions.

Returning to the criteria above, in the case of the first two items we refer to the tandeming of four-wire circuits. The last item refers to switching subscriber loops/hybrid combinations that will give a poorer return loss than the 11 dB stated above. Return losses on some connections can drop to 3 dB or less.

CCITT Recs. G.122, G.131, and G.151 (p. 3) treat stability. In essence the loss through points *a–t–b* in Figure 2.9 shall have a value not less than $(6 + N)$ dB, where N is the number of four-wire circuits in the national chain. Thus the minimum loss is stated (CCITT Rec. G.122). Rec. G.131 is quoted in part below:

> The standard deviation of transmission loss among [analog] international circuits routed in groups equipped with automatic regulation is 1 dB.... This accords with...that the tests...indicate that this target is being approached in that 1.1 dB was the standard deviation of the recorded data.... .

CCITT Rec. G.131 continues:

> It is also evident that those national networks which can exhibit no better stability balance return loss than 3 dB, 1.5 dB standard deviation, are unlikely

to seriously jeopardize the stability of international connections as far as oscillation is concerned. However, the near-singing (rain barrel effect) distortion and echo effects that may result give no grounds for complacency in this matter.

Stability requirements in regard to North American practice are embodied in the VNL concept discussed in the next section.

2.6.5 Via Net Loss (VNL)

VNL is a concept or method of transmission planning that permits a relatively close approach to an overall zero transmission loss in the telephone network (lowest practicably attainable) and maintains singing and echo within specified limits. The two criteria that follows are basic to VNL design:

1. The customer-to-customer talker echo shall be satisfactorily low on more than 99% of all telephone connections that encounter the maximum delay likely to be experienced.
2. The total amount of overall loss is distributed throughout the trunk segments of the connection by the allocation of loss to the echo characteristics of each segment.

One important concept in the development of the discussion of VNL is ERL (see Section 2.6.4). By using ERL measurements it is possible to arrive at a basic factor for the development of the VNL formula. This design factor states that the average return loss at Class 5 offices (local offices) is 11 dB, with a standard deviation of 3 dB. Considering then a standard distribution curve and the 1, 2, or 3σ points on the curve, we could therefore expect practically all measurements of ERL to fall between 2 and 20 dB at Class 5 offices (local offices). VNL also considers that reflection occurs at the far end in relation to the talker where the toll-connecting trunks are switched to the intertoll trunks. (See Ref. 22, Chapters 2 and 3.)

The next concept in the development of the VNL discussion is that of overall connection loss (OCL), which is the value of one-way trunk loss between two end (local) offices (not) subscribers).

$$\text{Echo path loss} = 2 \times \text{trunk loss (one way)} + \text{return loss (hybrid)} \quad (2.5)$$

Now let us consider the average tolerance for a particular echo path loss. The average echo tolerance is taken from the curve in Figure 2.8. Therefore

$$\text{OCL} = \frac{\text{average echo tolerance (loss)} - \text{return loss}}{2} \quad (2.6)$$

Return loss in this case is the average echo return loss that must be maintained at the distant Class 5 office—the 11 dB given above.

An important variability factor has not been considered in the formula, namely, trunk stability. This factor defines how close assigned levels are maintained on a trunk. VNL practice dictates trunk stability to be maintained with a normal distribution of levels and a standard deviation of 1 dB in each direction. For a round-trip echo path the deviation is taken as 2 dB. This variability applies to each trunk in a tandem connection. If there are three trunks in tandem, this deviation must be applied to each of them.

The reader will recall that the service requirement in VNL practice is satisfactory echo performance for 99% of all connections. This may be considered as a cumulative distribution or +2.33 standard deviations summing from negative infinity towards the positive direction.

The OCL formula may now be rewritten:

$$\text{OCL} = \frac{\text{average echo tolerance} - \text{average return loss} + 2.33\sigma'}{2} \quad (2.7)$$

where σ' = composite standard deviation of all functions, namely,

$$\sigma' = \sqrt{\sigma_t^2 + \sigma_{r1}^2 + N\sigma_1^2}$$

with σ_t = standard deviation of the distribution of echo tolerance among a large group of observers, 2.5 dB

σ_{r1} = standard deviation of the distribution of return loss, 3 dB

σ_1 = standard deviation of the distribution of the variability of trunk loss for a round-trip echo path, given as 2 dB

N = number of trunks switched in tandem to form a connection: Class 5 office to Class 5 office

Consider now several trunks in tandem. Then it can be calculated that at just about any given echo path delay, the OCL increases approximately 0.4 dB for each trunk added. With this simplification, once we have the OCL for one trunk, all that is needed to compute the OCL for additional trunks is to add 0.4 dB times the number of trunks added in tandem. This loss may be regarded as an additional constant needed to compensate for variations in trunk loss in the VNL formula.

Figure 2.10 relates echo path delay (round trip) to OCL (for one trunk, then for a second trunk in tandem, and for four and six trunks in tandem). The straight-line curve has been simplified, yet the approximation is sufficient for engineering VNL circuits. On examining the straight-line curve in Figure 2.10 it will be noted that that curve cuts the Y axis at 4.4 dB where round-trip delay is 0. This 4.4 dB is made up of two elements, namely, that all trunks have a minimum of 4 dB to control singing and 0.4-dB protection against negative variation of trunk loss.

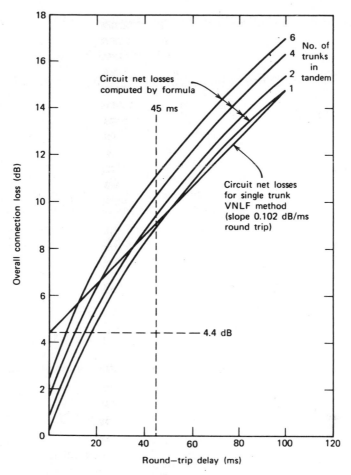

Figure 2.10 Approximate relationship between round-trip echo delay and overall connection loss (OCL).

Another important point to be defined on the linear curve in Figure 2.10 is that of a round-trip delay of 45 ms, which corresponds to an OCL of 9.3 dB. Empirically it has been determined that for delays greater than 45 ms, echo suppressors or cancellers must be used.

From the linear curve in the figure the following formula for OCL may be derived:

$$\text{OCL} = (0.102)(\text{path delay in ms})$$
$$+ (0.4 \text{ dB})(\text{number of trunks in tandem}) + 4 \text{ dB} \quad (2.8)$$

The 4 dB derives from Figure 2.10 but did not include local switching loss, which is a nominal 0.5 dB. The new OCL equation has changed the 4-dB value to 5 dB or 2.5 dB at each connection extremity. Thus

$$OCL = 0.102D + 0.4N + 5 \text{ dB}$$

where N = number of trunks in tandem and D = the path delay in milliseconds.

A word now about the last term in the OCL equation, a constant (4 dB): usually the 4 dB is applied to the extremity of each trunk network, namely, to the toll-connecting trunks, 2 dB to each.

OCL deals with the losses of an entire network consisting of trunks in tandem. VNL deals with the losses assigned to one trunk. The VNL formula follows from the OCL formula. The key here is the round-trip delay on the trunk in question. The delay time for a transmission facility employing only one particular medium is equal to the reciprocal of the velocity of propagation of the medium multiplied by the length of the trunk. To obtain round-trip time, this figure must be multiplied by 2. Therefore

$$VNL = 0.102 \times 2 \times \left(\frac{1}{\text{velocity of propagation}} \right)$$

$$\times (\text{one-way length of the trunk}) + 0.4 \text{ dB} \qquad (2.9)$$

Often another term is introduced to simplify the equation, the via net loss factor (VNLF):

$$VNL = VNLF \times (\text{one-way length of trunk in miles}) + 0.4 \text{ dB} \quad (2.10)$$

$$VNLF = \frac{2 \times 0.102}{\text{velocity of propagation of medium}} \ (\text{dB/mi}) \qquad (2.11)$$

It should be noted that the velocity of propagation of the medium used here must be modified by such things as delays caused by repeaters, intermediate modulation points, and facility terminals.

VNLFs for loaded two-wire facilities are 0.03 dB/mi with H-88 loading on 19-gauge wire and increase to 0.04 dB/mi on B-88 loading. On H-44 facilities they are 0.02 dB/mi. On four-wire carrier and radio facilities the factor improves to 0.0015 dB/mi.

For connections with round-trip delay times in excess of 45 ms, the standard VNL approach must be modified. As mentioned previously, these circuits use echo suppressors which automatically switch about 50 dB into the echo return path. The switch actuates when speech is received in the "return" path, switching the pad into the "go" path. Hence whenever delay exceeds 45 ms, echo suppressors or cancellers are used. In North America they are used on interregional high-usage intertoll trunks (Figure 2.11) and on interregional toll-connecting and end-office (local exchange) toll trunks more than 1850 mi long.

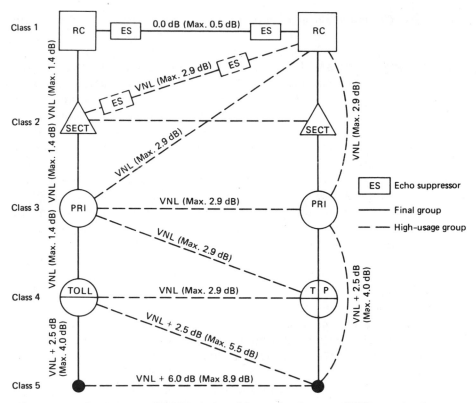

Figure 2.11 Trunk losses with VNL design. RC = regional center; SECT = sectional center; PRI = primary center; TOLL = toll center; TP = toll point. (Copyright © 1977 by American Telephone and Telegraph Company.) Note network's hierarchical structure.

When dealing with VNL in the toll network (long-distance network), trunk losses are stated in terms of inserted connection loss (ICL), which is defined as the 1000-Hz loss inserted by switching the trunk into actual operating connection. The losses shown in Figure 2.11 are all ICL.

In the Bell System of North America* nearly all local exchanges (Class 5) use an impedance of 900 Ω, whereas toll (long-distance) exchanges use 600 Ω. However, tandem exchanges using No. 5 crossbar switches use the 900-Ω impedance value. These values are important for terminal and through balance to meet return loss objectives.

In summary, in VNL design we have three types of losses that may be assigned to a trunk:

Type	Loss
Toll-connecting trunk	VNL + 2.5 dB
Intertoll trunk (no echo suppressor)	VNL
Intertoll trunk (with echo suppressor)	0 dB

*Now Regional Bell Operating Companies or RBOCs.

Note: See discussion on echo suppressors for an explanation of the 0-dB figure.

VNL Penalty Factors Most of the toll network using the VNL concept utilizes two-wire switches even though the network is considered four wire. At each point where the network has a two-wire to four-wire transformation, another source of echo occurs. Again the amount of echo is a function of the ERL at each point of transformation. If the return loss at each of these intermediate points, often called *through return loss*, is high enough, the point is transparent regarding echo. Echo, then, can be considered only as a function of the return loss at the terminating points of transformation, often called *terminating return loss*. If a two-wire toll switch has an ERL of 27 dB on at least 50% of the through connections, it meets through return loss objectives, and as far as VNL design is concerned, it may be considered transparent regarding echo.

However, a number of two-wire toll exchanges do not meet this minimum criterion. Therefore a penalty is assigned to each in the VNL design. This penalty is called the *B* factor, which is the amount of loss that must be added to the VNL value of each trunk (incoming as well as outgoing) to compensate for the excessive amount of echo and singing current reflection that this office will create in an intertoll connection. The following table provides median office ERL and the corresponding *B* factors:

Median Office ERL (dB)	*B* Factor (dB)
27	0
21	0.3
18	0.6
16	0.9
15	1.2
14	1.5

2.6.6 Fixed-Loss Plan

VNL is a variable-loss plan for the North American network. It is one valid approach for a minimum-loss network to reasonably meet subscriber received-volume objectives on the one hand and echo performances objectives on the other. The North American network is now in a stage of evolvement from analog to digital, and a fixed-loss plan is more appropriate once the network is all digital.

The fixed-loss plan being introduced in North America by AT&T and BELLCORE is based on a 0-dB loss on all-digital intertoll trunks, that is, on digital transmission facilities interconnecting digital toll exchanges. The plan specifies a 6-dB trunk loss between local exchanges regardless of the connection mileage. This 6-dB loss provides an acceptable compromise between loss and echo performance over a wide range of connection lengths and loop losses found in this environment. Toll-connecting trunks are assigned a 3-dB loss under the plan.

It will take some years for the present predominantly analog network to be converted to a fully digital network. In the meantime the network will be a mix of analog and digital facilities which will increasingly trend toward the digital. For the combined analog–digital network of the interim period, the following principal characteristics and constraints will be adhered to:

1. The expected measured loss and inserted connection loss of each trunk must be the same for both directions of transmission.
2. The -2-dB TLP at the outgoing side of analog toll switches and the 0-dB TLP at the Class 5 (local) exchanges are retained, and a -3-dB TLP is established for digital toll exchanges.
3. The -16-dB and $+7$-dB TLPs at carrier system input and output are retained.
4. Existing test and lineup procedures for digital channel banks are retained.
5. Combination intertoll trunks, namely, those terminating in digital terminals at a digital (i.e., No. 4 ESS) switching machine at one end and in D-type channel banks at an analog switch at the other end, are designed to have a 1-dB ICL.
6. Analog intertoll trunks are designed according to the VNL plan.

2.7 NOTES ON NETWORK HIERARCHY AND ROUTING

Telephone networks require some form of organization regarding switches and architecture to route traffic effectively and economically. Prior to about 1987 the North American network and the CCITT-recommended international network were based on a hierarchical structure. Such a structure is shown in Figure 2.11.

Connectivities among international switching centers (ISCs) are direct/tandem connections, and the hierarchical structure has been done away with. North America (AT&T) is moving toward *dynamic nonhierarchical routing* (DNHR). Such a network based on this concept is shown in Figure 2.12. Ref. 27 states that switching nodes in a DNHR network should be stored program control (SPC) and utilize common channel signaling (e.g., CCITT Signaling System No. 7). A DNHR network has only one level of tandem switching, and local exchange carriers (LEC) end offices (local switches) and customer equipment are directly connected to or "home" to the higher-level tandem switches in the DNHR environment in a manner similar to a hierarchical network at the lowest levels. The LEC end offices

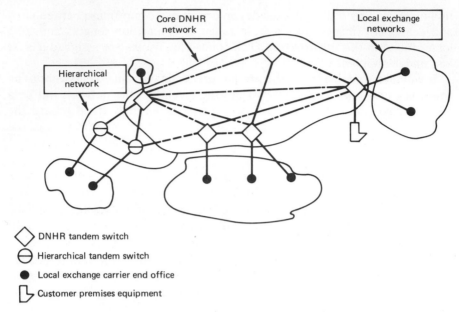

Figure 2.12 DNHR network topology. (From Ref. 27; reprinted with permission of *IEEE Communications Magazine*, Oct. 1990.)

and customer premises equipment switch subscriber loops to each other or to the interexchange carrier network. The DNHR portion of the ATT intercity network consists of 4ESS switches (digital SPC switches) interconnected by the common channel signaling network. Dynamic routing rules are used only among DNHR switches, and conventional hierarchical routing rules are used among all other switch pairs. A number of small switching subsystems use only hierarchical rules. As shown in Figure 2.12, many of these smaller systems are designated "hierarchical tandem switches" and home directly on DNHR switches. If we look at this network as a hierarchy, the DNHR switches appear as a large network of the highest level hierarchical centers. From this we can interpret the evolving North American network as a two-level hierarchy. The lowest level consists of the LEC switches with direct or tandem connectivity to the higher-level DNHR network. DNHR switches are interconnected by direct/high-usage routes or tandem routes.

The dynamic routing is based on selection of minimum-cost paths between originating and terminating switches and design of optimal, time-varying routing patterns to achieve minimum-cost trunking by capitalizing on noncoincident network busy periods. Both engineered and real-time routing are utilized.

The DNHR algorithm combines into one unified approach several techniques for achieving network design savings. As stated in Ref. 27, the approach

- Uses time-varying routing so that trunks are shared to the greatest extent possible with the variation of loads in the network.
- Equalizes the loads on links throughout the busy periods on the network.
- Routes traffic on least costly paths.
- Selects an optimum number of trunks (or equivalently, an optimum level of blocking) on each link to optimally divide the traffic between the direct link and alternate paths used by the traffic. (See Ref. 22 for a discussion of hierarchical networks and routing.)

With DNHR routing, most call connectivities involve only one DNHR link, with a maximum of two DNHR links if tandem routing is used. This is in keeping with CCITT Rec. E.171, which limits links in tandem on a national connection to four. On an international call, 12 links in tandem are permitted: 4 on each national network at the originating and terminating ends and 4 for the international network. The reason for this limitation of links in tandem is to limit accumulation of transmission impairments and limit postdial delay. We again refer to Ref. 22 for further explanation of these system issues.

2.8   DESIGN OF SUBSCRIBER LOOP

2.8.1   Introduction

The subscriber loop connects a subscriber telephone subset with a local switching center (Class 5 office in Figure 2.11). A subscriber loop in nearly all cases is two wire with simultaneous transmission in both directions. The simplified drawing below will help to illustrate the problem:

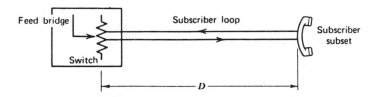

Distance D, the loop length, is most important. By Section 2.4, D must be limited in length owing to (1) attenuation of the voice signal and (2) dc resistance for signaling.

The attenuation is taken from the national transmission plan covered in Section 2.6.4. For our discussion we shall assign 6 dB as the loop attenuation limit (referred to 800 Hz). For the loop resistance limit, many crossbar switches will accept up to 1300 Ω.* From this figure we subtract 300 Ω, the nominal resistance for the telephone subset in series with the loop, leaving us with a 1000-Ω limit for the wire pair (bridge resistance disregarded). Therefore in the paragraphs that follow we shall use the following figures:

6 dB (attenuation limit for a loop)[†]

1000 Ω (resistance limit)

2.8.2   Calculation of Resistance Limit

To calculate the dc loop resistance for copper conductors, the following formula is applicable:

$$R_{dc} = \frac{0.1095}{d^2} \qquad (2.12)$$

where R_{dc} = loop resistance (Ω/mi) and d = diameter of the conductor (in.).

If we want a 10-mi loop and allow 100 Ω/mi of loop (for the 1000-Ω limit), what diameter of copper wire would we need?

$$100 = \frac{0.1095}{d^2}$$

$$d^2 = \frac{0.1095}{100} = 0.001095$$

$$d = 0.033 \text{ in. or } 0.84 \text{ mm (rounded off to } 0.80 \text{ mm)}$$

Using Table 2.9 we can compute maximum loop lengths for 1000-Ω signaling resistance. As an example, for a 26-gauge loop:

$$\frac{1000}{83.5} = 11.97 \text{ kft} \quad \text{or} \quad 11,970 \text{ ft}$$

This, then, is the signaling limit and not the loss (attenuation) limit, or what some call the "transmission limit."

To assist relating American wire gauge (AWG) to cable diameter in millimeters, Table 2.10 is presented.

Another guideline in the design of subscriber loops is the minimum loop current off-hook for effective subset operation. For instance, the Bell System 500-type subset requires at least 23 mA for efficient operation.

---

*Many semielectronic switches will accept 1800-Ω loops, and with special line equipment, 2400-Ω loops.

[†]In the United States this value may be as high as 9 dB.

Table 2.9

Gauge of Conductor	$\Omega/1000$ ft of loop	Ω/mi of loop	Ω/km of loop
26	83.5	440	268
24	51.9	274	168.5
22	32.4	171	106
19	16.1	85	53

2.8.3 Calculation of Loss (Attenuation) Limit

The second design consideration mentioned above was attenuation or loss. The attenuation of a wire pair used on a subscriber loop varies with frequency, resistance, inductance, capacitance, and leakage conductance. Resistance of the line will depend on the temperature. For open-wire lines attenuation may vary $\pm 12\%$ between winter and summer conditions. For

Table 2.10 American wire gauge (B & S) versus wire diameter and resistance

American Wire Gauge	Diameter (mm)	Resistance $(\Omega/km)^a$ at 20°C
11	2.305	4.134
12	2.053	5.210
13	1.828	6.571
14	1.628	8.284
15	1.450	10.45
16	1.291	13.18
17	1.150	16.61
18	1.024	20.95
19	0.9116	26.39
20	0.8118	33.30
21	0.7229	41.99
22	0.6439	52.95
23	0.5733	66.80
24	0.5105	84.22
25	0.4547	106.20
26	0.4049	133.9
27	0.3607	168.9
28	0.3211	212.9
29	0.2859	268.6
30	0.2547	338.6
31	0.2268	426.8
32	0.2019	538.4

[a]These figures must be doubled for loop/km. Remember it has a "go" and "return" path.

Table 2.11   Loss per unit length of subscriber cable[a]

Cable Gauge	Loss/1000 ft (dB)	dB/km	dB/mi
26	0.51	1.61	2.69
24	0.41	1.27	2.16
22	0.32	1.01	1.69
19	0.21	0.71	1.11
16	0.14	0.46	0.74

[a]Cable is low capacitance type (i.e., under 0.075 nF/mi).

buried cable, with which we are more concerned, loss variations due to temperature are much less.

Table 2.11 gives losses of some common subscriber cable per 1000 ft. If we are limited to a 6-dB (loss) subscriber loop, then by simple division we can derive the maximum loop length permissible for transmission design considerations for the wire gauges shown:

$$26 \quad \frac{6}{0.51} = 11.7 \text{ kft}$$

$$24 \quad \frac{6}{0.41} = 14.6 \text{ kft}$$

$$22 \quad \frac{6}{0.32} = 19.0 \text{ kft}$$

$$19 \quad \frac{6}{0.21} = 28.5 \text{ kft}$$

$$16 \quad \frac{6}{0.14} = 42.8 \text{ kft}$$

2.8.4   Loading

In many situations it is desirable to extend subscriber loop lengths beyond the limits described in Section 2.8.3. Common methods to attain longer loops without exceeding loss limits include

1. Increasing conductor diameter
2. Using amplifiers and/or loop extenders*
3. Using inductive loading
4. Using carrier equipment (Chapter 3)

Loading tends to reduce the transmission loss on subscriber loops and other types of voice pairs at the expense of good attenuation–frequency response

*A loop extender is a device that increases the battery voltage on a loop, extending its signaling range. It may also contain an amplifier, thereby extending the transmission loss limits.

Table 2.12   Code for load coil spacing

Code Letter	Spacing (ft)	Spacing (m)
A	700	213.5
B	3000	915
C	929	283.3
D	4500	1372.5
E	5575	1700.4
F	2787	850
H	6000	1830
X	680	207.4
Y	2130	649.6

beyond 3000–3400 Hz. Loading a particular voice pair loop consists of inserting series inductances (loading coils) into the loop at fixed intervals. Adding load coils tends to

- Decrease the velocity of propagation
- Increase impedance

Loaded cables are coded according to the spacing of the load coils. The standard code for load coils regarding spacing is shown in Table 2.12. Loaded cables typically are designated 19H44, 24B88, and so forth. The first number indicates the wire gauge, the letter is taken from Table 2.12 and is indicative of the spacing, and the third item is the inductance of the coil in millihenrys (mH). 19H66 is a cable commonly used for long-distance operation in Europe. Thus the cable has 19-gauge voice pairs loaded at 1830-m intervals with coils of 66-mH inductance. The most commonly used spacings are B, D, and H.

Table 2.13 will be useful in calculating the attenuation of loaded loops for a given length. For example, for 19H88 (last entry in table) cable, the attenuation per kilometer is 0.26 dB (0.42 dB/statute mi). Thus for our 6-dB loop loss limit, we have 6/0.26, limiting the loop to 23 km in length (14.3 statute mi).

When determining signaling limits in loop design, add about 15 Ω per load coil as series resistors.

The tendency in many administrations is to use a new loading technique. This has been taken from "unigauge design" discussed in the next section. With this technique no loading is required on any loop less than 5000 m long (15,000 ft). For loops longer than 5000 m, loading starts at the 4200-m point, and load coils are installed at 1830-m intervals from there on. The loading intervals should not vary by more than 2%.

Table 2.13   Some properties of cable conductors

Diameter (mm)	AWG No.	Mutual Capacitance (nF/km)	Type of Loading	Loop Resistance (Ω/km)	Attenuation at 1000 Hz (dB/km)
0.32	28	40	None	433	2.03
		50	None		2.27
0.40		40	None	277	1.62
		50	H66		1.42
		50	H88		1.24
0.405	26	40	None	270	1.61
		50	None		1.79
		40	H66	273	1.25
		50	H66		1.39
		40	H88	274	1.09
		50	H88		1.21
0.50		40	None	177	1.30
		50	H66	180	0.92
		50	H88	181	0.80
0.511	24	40	None	170	1.27
		50	None		1.42
		40	H66	173	0.79
		50	H66		0.88
		40	H88	174	0.69
		50	H88		0.77
0.60		40	None	123	1.08
		50	None		1.21
		40	H66	126	0.58
		50	H88	127	0.56
0.644	22	40	None	107	1.01
		50	None		1.12
		40	H66	110	0.50
		50	H66		0.56
		40	H88	111	0.44
0.70		40	None	90	0.92
		50	H66		0.48
		40	H88	94	0.37
0.80		40	None	69	0.81
		50	H66	72	0.38
		40	H88	73	0.29
0.90		40	None	55	0.72
0.91	19	40	None	53	0.71
		50	None		0.79
		40	H44	55	0.31
		50	H66	56	0.29
		50	H88	57	0.26

Source: ITT, *Outside Plant*, Telecommunication Planning Documents. Courtesy of ITT.

2.8.5   Some Standard Approaches to Subscriber Loop Design, North American Practice

2.8.5.1 Introduction   Between 30 and 50% of a telephone company's investment is tied up in what is generally referred to as "outside plant." Outside plant, for this discussion, can be defined as that part of the telephone plant that takes the signal from the local switch and delivers it to the subscriber. Much of this expense is attributable to copper in the subscriber cable. Another important expense is cable installation, such as that incurred in tearing up city streets to augment present installation or install new plant. Much work today is being done to devise methods to reduce these expenses. Among these methods are unigauge design, dedicated plant, and fine-gauge design. All have a direct impact on the outside plant transmission engineer.

2.8.5.2 Unigauge Design   Unigauge is a concept developed by BELL-CORE of North America to save on the expense of copper in the subscriber loop plant. This is done by reducing the gauge (diameter) of wire pairs as much as possible while retaining specific resistance and transmission limits. The description that follows is an attempt at applying unigauge to the more general case.

To start with, let us review the basic rules set down up to this point for subscriber loop design:

- Maximum loop resistance 1300 Ω
- Maximum loop loss 6 dB or a figure taken from the national transmission plan (the RBOC objective is 8 dB)

To these, let us add two more:

- The use of modern equalized telephone subsets (or at least on all loops longer than 10,000 ft [3000 m])
- Cable gauges limited to AWG 19, 22, 24, and 26, while maintaining the 1300-Ω limit or whichever limit needs to be established inside which signaling can be effected

All further argument will be based on minimizing the amount of copper used (e.g., using the smallest diameter wire possible within the limits prescribed above).

The reader should keep in mind that many telephone companies must install subscriber loops in excess of 15,000 ft (5000 m) and thus must use a conductor size (or sizes) larger than 26 gauge. BELLCORE, for example, found that 20% of its subscriber loops exceeded 15,000 ft in 1964. To meet transmission and signaling objectives, the cost of such loops tends to mount excessively for those loops over 15,000 ft in length.

The concept of unigauge design basically is one that takes advantage of relatively inexpensive voice-frequency-gain devices and range (loop) extenders to permit the use of 26-gauge conductors on the greater percentage of its longer loops. Also, unigauge design makes it mandatory that a unified design approach be used. In other words, the design of the subscriber loop plant is an integrated process and not one of piece-by-piece engineering. Unigauge allows the use of 26-gauge cable on subscriber loops up to 30,000 ft in length. This results in very significant savings in copper and a general overall economy in present-day outside plant installation.

Application of Unigauge Design. A typical layout of a subscriber plant based on unigauge design, is shown in Figure 2.13 (Ref. 10). In the figure it will be seen that subscribers within 15,000 ft of the switch are connected over loops made up of 26-gauge nonloaded cable with standard 48-V battery. Their connection at the switch is conventional. It is also seen that 80% of Bell System subscribers are within this radius. Loops 15,000–30,000 ft long are called unigauge loops. Subscribers in the range of 15,000–24,000 ft from the switch are connected by 26-gauge nonloaded cable as well, but require a range extender to provide sufficient voltage for signaling and supervision. In the drawing a 72-V range extender is shown equipped with an amplifier that gives a midband gain of 5 dB. The output of the amplifier "emphasizes" the higher frequencies. This offsets that additional loss suffered at the higher frequencies of the voice channel on the long nonloaded loops. To extend the loops to the full 30,000 ft, the Bell System adds 88-mH loading coils at the 15,000- and 21,000-ft points.

For long loops (more than 15,000 ft long) the range extender–amplifier combinations are not connected on a line-for-line basis. It is standard practice to equip four or five subscriber loops with only one range extender–amplifier used on a shared basis. When the subscriber goes off-hook on a long unigauge loop, a line is seized, and the range extender is switched in. This concentration is another point in favor of unigauge because of the economics involved. It should be noted that the long 15,000-ft nonloaded sections, into which the switch faces, provide a fairly uniform impedance for all conditions when an active amplifier is switched in. This is a positive factor with regard to stability.

Loops more than 30,000 ft long may also use the unigauge principle and often are referred to as *extended unigauge*. Such a loop is equipped with 26-gauge nonloaded cable from the switch out to the 15,000-ft point. Beyond 15,000 ft 22-gauge cable is used with H-88 loading. As with all loops more than 15,000 ft long, a range extender–amplifier is switched in when the loop is in use. The loop length limit for this combination is 52,000 ft. Loops more than 52,000 ft may also be installed by using a gauge of diameter larger than 22.

It should be noted that the Bell System replaces its line relays with others that are sensitive up to 2500 Ω of loop resistance with 48-V battery. Such a modification to the switch is done on long loops only. The 72 V supplied by

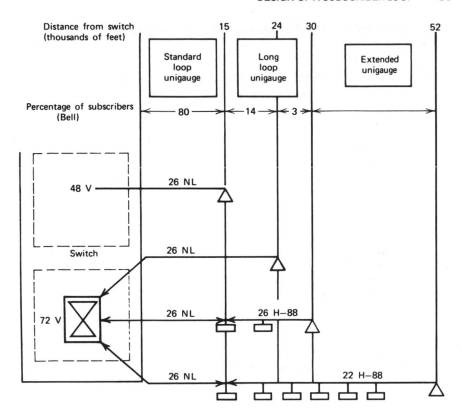

Figure 2.13 Layout of a unigauge subscriber plant.

the range extender is for pulsing, and the ringing voltage (to ring the distant telephone) is superimposed on the line with the subscriber's subset being in an on-hook condition.

Besides the savings in the expenditure on copper, unigauge displays some small improvements in transmission characteristics over older design methods of subscriber loops:

- Unigauge has a slightly lower average loss (when we look at a statistical distribution of subscribers)
- There is 15-dB average return loss on the switch side of an amplifier, compared with an average of 11 dB for older design methods

These two points echo the Bell System experience with unigauge design (Refs. 10 and 21).

2.8.5.3 Dedicated Plant The dedicated plant assigns wire pairs to subscribers or would-be subscriber locations. In the past, bridged taps were used so that subscriber loops were more versatile regarding assignment. The idea of the bridged tap is shown below:

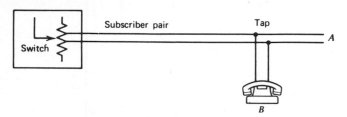

In this case station A is not connected; it is not in use but available in case station B disconnects telephone service.

The primary impact of the dedicated plant is that it eliminates the use of bridged taps. Bridged taps deteriorate the quality of transmission by notably increasing the capacitance of a loop.

2.8.5.4 Fine Gauge and Minigauge Techniques Fine gauge and minigauge techniques essentially are refinements of the unigauge concept. In each case the principal object is to reduce the amount of copper. Obviously, one method is to use still smaller gauge pairs on shorter loops. Consideration has been given to the use of gauges as small as 32. Another approach is to use aluminum as the conductor. When aluminum is used, a handy rule of thumb to follow is that ohmic and attenuation losses of aluminum may be equated to copper in that aluminum wire always is the next "standard gauge" larger than its copper counterpart if copper were to be used. Some examples are listed as follows:

Copper	Aluminum
19	17
22	20
24	22
26	24

Aluminum has some drawbacks as well. The major ones are summed up as follows:

- Aluminum is not to be used on the first 500 yd of cable where the cable has a large diameter (i.e., more loops before branching)
- It is more difficult to splice than copper

- It is more brittle than copper
- Because the equivalent conductor is larger than its copper counterpart, an equivalent aluminum cable with the same conductivity-loss characteristics as copper will have a smaller pair count in the same sheath.

2.8.5.5 Resistance Design Resistance design (RD) is a method of designing subscriber loops based on establishing a common maximum resistance limit for a switch. RD was developed for use in North America. The maximum resistance limit defines a perimeter around a switch or local central office which is called the *resistance design boundary*. For those subscribers outside of this boundary served by the same switch, long route design (LRD) procedures may be used. LRD is described in Section 2.8.5.6.

In North America, the design objective value for the maximum loop loss is 8 dB. Nearly all subscriber loops that were designed using RD procedures have losses that fall well below 8 dB.

Ref. 21 (Vol. 3) defines the following additional terms used in RD:

1. *Resistance design limit* is the maximum value of loop resistance to which the RD method is applicable. This value is set at 1300 Ω primarily to control transmission loss.
2. *Resistance design area* is that area enclosed within the resistance design boundary.
3. *Switch supervisory limit* is the conductor loop resistance beyond which the operation of the switch supervisory equipment is uncertain (i.e., the maximum loop resistance is reached or exceeded).
4. *Switch design limit*. This limit is set at 1300 Ω. However, some switches have supervisory limits *under* 1300 Ω. This lower value then becomes the switch design limit.
5. The *design loop* is the subscriber loop under study for a given distribution area to which the switch design limit is applied to determine conductor sizes (i.e., gauges or diameters). It is normally the longest loop expected for the period of fill of the cable involved.
6. The *theoretical design* is the subscriber cable makeup consisting of the two finest standard consecutive gauges necessary in the design loop to meet the switch design limit.

There are three steps or procedures that must be carried out before proceeding with resistance design: (1) determination of the resistance design boundary, (2) determination of the design loop, and (3) selection of cable gauge(s) to meet design objectives.

The resistance design boundary is applied in medium to heavy subscriber density. LRD procedures are applied in areas of sparser subscriber density.

The design loop length is based on local and forecast service requirements. The planned ultimate longest length loop for the job being considered is the design loop, and the theoretical design and gauge selection are based on it.

Ref. 21 points out that on some cables, pairs may have lengths extending beyond the resistance design area. These pairs are handled using LRD.

The theoretical design is used to determine the wire gauge or combinations of gauges for any loop. If more than one gauge is required, Ref. 21 states that the most economical approach, neglecting existing plant, is the use of the two finest consecutive standard gauges that meet a particular switch design limit. The smaller of the two gauges is usually placed outward from the serving switch because it usually has a larger cross section of pairs. Since the design loop length has been determined, the resistance per kft (or km) for each gauge may be determined from Tables 2.9, 2.10, 2.11 and 2.13. The theoretical design can now be calculated from the solution of two simultaneous equations.

The following example was taken from Ref. 21. Suppose we wished to design a 32-kft loop with a maximum loop resistance of 1300 Ω. If we were to use 24-gauge copper pair, Table 2.9 shows that we exceed the 1300-Ω limit; if we use 22 gauge, we are under the limit by some amount. Therefore, what combination of the two gauges in series would just give us 1300 Ω? The loop requires five H66 load coils, each of which has a 9-Ω resistance. It should be noted that the 1300-Ω limit value does not include the resistance of the telephone subset.

Let X = the kilofeet value of the length of the 24-gauge pair and Y = the kilofeet value of the length of the 22-gauge pair. Now we can write the first equation:

$$X + Y = 32 \text{ kft}$$

Table 2.9 shows the resistance of a 24-gauge pair as 51.9 Ω/kft, and for the 22-gauge pair, 32.4 Ω/kft. We can now write the second simultaneous equation:

$$51.9X + 32.4Y + 5(9) = 1300 \ \Omega$$
$$X = 11.2 \text{ kft of 24-gauge cable}$$
$$Y = 32 - X = 20.8 \text{ kft of 22-gauge cable}$$

We stated earlier that if the resistance design rules are followed, the North American 8-dB objective loss requirement will be met for all loops. However, to ensure that this is the case, the following additional rules should be adhered to:

- Inductive load all loops over 18 kft long
- Limit the cumulative length of all bridged taps on nonloaded loops to 6 kft or less

Ref. 21 recommends H88 loading where we know the spacing between load coils is 6000 ft with a spacing tolerance of ± 120 ft. Wherever possible it is desirable to take deviations greater than ± 120 ft on the short side so that correction may later be applied by normal build-out procedures.

The first load section out from the serving switch is 3000 ft for H66/H88 loading. In the measurement of this length, due consideration should be given to switch wiring so that the combination is equivalent to 3000 ft. It should be remembered that the spacing of this first coil is most critical to achieve acceptable return loss and must be placed as close to the recommended location as physically and economically possible.

2.8.5.6 Long Route Design (LRD)

The long route design procedure uses several zones corresponding to ranges of resistance in excess of 1300 Ω. Of course, each loop must be able to carry out the supervisory signaling function and meet the 8-dB maximum loop attenuation rule (North America). LRD provides for a specific combination of electronic range extenders and/or fixed-gain devices (VF repeaters/amplifiers) to meet the supervision and loss criteria.

On most long loops, a range extender with gain is employed at the switch. The range (loop) extender boosts the standard -48 V by an additional 30 V, and a gain of 3 to 6 dB is provided.

Table 2.14   Previous applicable loop design rules

Design Parameters	RD	LRD	Unigauge Design
Loop resistance (Ω^a)	0–1300	1301–3600	0–2500
Load coils	Full H88 > 18 kft (not including BT)	Full H88	None to 24 kft Partial > 24 kft

End sections (ES) and Bridged tap (BT)	Nonloaded:
	  BT = 6 kft
	  (Max.)
	Loaded:
	  ES + BT = 15 kft
	  (max.)
	  3 kft min.
	  12 kft recommended
		ES + BT = 12 kft	Nonloaded:
			  BT = 6 kft
			
			Loaded:
			  ES + BT = 12 kft

Transmission limitations	None loop required	Gain in the to loop for loop resistances greater than 1600 Ω	Gain applied to loop lengths > 15 kft
Cable gauging	Any combination[b]	Any combination[b]	Buffer cable needed[c]

Source: Ref. 24. Courtesy of BELLCORE. Copyright © 1986 Bell Communications Research.

[a]Includes only the resistance of the cable and loading coils.

[b]19-, 22-, 24-, or 26-gauge cable.

[c]Requires specified amounts of 26-gauge cable adjacent to the end office.

2.8.5.7 Present Loop Design Rules for North America

Table 2.14 reviews the basic North American loop design methods, some of which have been in use for over the last 30 years. In the mid-1980s a new or modified set of subscriber design rules were established. These are shown in Table 2.15.

The RD design procedures were modified and are called revised resistance design (RRD). RRD is applicable for loops with 1500 Ω or less of resistance and that are 24 kft or less in length. Loops longer than 1500 Ω are implemented using DLC (digital loop carrier) as a first choice or CREG (concentrated range extension with gain) or modified LRD (MLRD) per line for range-extended loops. CREG and MLRD are modifications of unigauge (UG) design and LRD, respectively.

A major modification to the RD plan in the RRD plan allows a higher loop resistance, 1500 Ω rather than 1300 Ω, for switches that provide for the increased resistance range and reduces the number of bridged taps allowed. The maximum length of 18 kft for nonloaded loops is the same for both plans, except in the RRD plan the maximum length includes bridged taps, whereas in RD it does not. These rule changes result in improved transmission performance and outside plant savings.

The major modifications to the LRD plan are evident when comparing Table 2.14 with Table 2.15. The range 0–1500 Ω is served by the RRD plan, and the range beyond 1500 Ω by either the MLRD or DLC, with the DLC being the first choice. Under the MLRD the 1500–2000-Ω range is designated Resistance Zone 18 (RZ18) and requires 3-dB gain. The 2000–2800-Ω

Table 2.15 Present loop design rules, North America

Design Parameter	RRD	MLRD	CREG
Loop resistance (Ω^a)	0–18 kft: 1300 Ω max. 18–24 kft: 1500 Ω max. > 24 kft: DLC method[b]	1501–2800	0–2800
Load coils End sections (ES) and Bridged tap (BT) (max.)	Full H88 > 18 kft Nonloaded: total cable + BT = 18 kft Max. BT = 6 kft	Full H88 ES + BT = 3–12 kft	Full H88 > 15 kft Nonloaded: cable + BT = 15 kft Max. BT = 6 kft
	Loaded: ES + BT = 3–12 kft		Loaded: ES + BT = 3–12 kft
Transmission limitations	None	Range extension with gain required for loop resistances greater than 1500 Ω	Gain range extention required for loop resistances greater than 1500 Ω
Cable gauging	Two gauge combinations preferred (22-, 24-, 26-gauge)	Two gauge combinations preferred (22-, 24-, 26-gauge)	Two gauge combinations preferred (22-, 24-, 26-gauge)

Source: Ref. 24. Courtesy of BELLCORE. Copyright 1986 Bell Communications Research.
[a]Includes only the resistance of the cable and loading coils.
[b]DLC method is preferred over MLRD and CREG for new growth economics control.

range (RZ28) requires 6-dB gain. When a serving switch is only equipped with the newer range extenders with gain that automatically switch their gain setting to provide from 3- to 6-dB net gain as required, the need to maintain and administer separate transmission zones in MLRD plan is not necessary. From this standpoint, all MLRD can be considered a single range-extended zone. MLRD, unlike LRD, is not limited to long rural loops and can be used anywhere in the serving area where economically justified.

The CREG plan, as shown in Table 2.15, is the replacement for UG design plan. The CREG plan, like its UG counterpart, permits increased use of finer-gauge pair cable in the subscriber plant by providing a repeater (VF amplifier) behind a stage of switching concentration. Unlike the UG plan, the CREG outside plant design is compatible with the loading employed with the RRD and MLRD plans. The CREG plan coupled with full H88 loading gives the CREG design better overall performance than the UG design (Ref. 21).

Digital Loop Carrier (DLC) Systems. DLC is an alternative to providing long route loops by wire pair VF facilities. One such system uses T1 (1.544 Mbps, see Chapter 3) with specially designed terminals. It can serve up to 40 loops over a single repeatered line consisting of two pairs, one for each direction of transmission. Such a system is typically used on long routes when relatively high subscriber line growth is forecast for the planning period and when such routes would require relief in the feeder portions.

Another advantage of DLC is that it can provide improved transmission loss distributions. One such system displays a 1000-Hz insertion loss of 2.0 dB between a serving switch and remote terminals regardless of the length of the digital section. Of course, additional losses are incurred between the remote terminals and the subscriber premise subsets. The low insertion loss of the digital portion of such loops allows up to 5.0-dB additional loss in the distribution cables.

2.9 DESIGN OF LOCAL AREA TRUNKS (JUNCTIONS)

2.9.1 Introduction

Exchanges in a common local area often are connected on a full mesh basis (see Ref. 22, chapt. 2 for definition of mesh connection). Depending on distance and certain other economic factors, these trunk circuits use VF transmission over cable. In view of the relatively small number of these trunk circuits when compared to the number of subscriber lines,* it is generally economically desirable to minimize attenuation in this portion of the network.

One approach used by some telephone companies (administrations) is to allot one-third of the total end-to-end reference equivalent to each sub-

---

*Due to the inherent concentration in local switches (e.g., 1 trunk for 3–25 subscribers, depending on design).

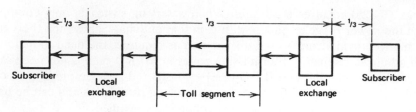

Figure 2.14   One approach to loss assignment.

Table 2.16   Inserted connection loss objectives for local trunks

	Inserted Connection Loss (dB)	
Trunk Types	Nongain	Gain
Direct trunks[a]	0–5.0	3.0 (5.0 max.)
Tandem trunks	0–4.0	3.0 (4.0 max.)
Intertandem trunks		
Terminated at sector tandems at		
both ends		1.5
Terminated in a toll center or		
sector tandem that meets		
terminal balance objectives		
at one or both ends		0.5

Source: Ref. 21.
[a]These ICLs apply to direct trunks less than 200 mi in length. Direct trunks more than 200 mi long are designed in accordance with the VNL plan.

scriber's loop and one-third to the trunk network. Figure 2.14 illustrates this concept. For instance, if the transmission plan called for a 24-dB overall reference equivalent (ORE), then one-third of 24 dB, or 8 dB, would be assigned to the trunk plant. Of this we may assign 4 dB to the four-wire portion or toll segment of the network, leaving 4 dB for local VF trunks or 2 dB at each end. The example has been highly simplified, of course.

The Bell System* uses the design objectives given in Table 2.16 for inserted connection loss assignment for local trunks.

From this it can be seen that the approach to the design of VF trunks varies considerably from that used for subscriber loop design. Although we must ensure that signaling limits are not exceeded, almost always the transmission limit will be exceeded well before the signaling limit. The tendency to use larger diameter cable on long routes is also evident.

We follow a similar approach as on subscriber loops for inductive loading. If loading is to be used, the first load coil is installed at distance $D/2$, where

*Prior to divestiture, the guideline is still valid for analog networks.

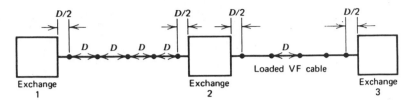

Figure 2.15   Loading of VF trunks (junctions).

D is the normal separation distance between load points. Take the case of H loading, for instance. The distance between load points is 1830 m, but the first load coil from the exchange is placed at $D/2$ or 915 m from the exchange. Then if an exchange is bypassed, a full-load section exists. This concept is illustrated in Figure 2.15.

Now consider this example. A loaded 500-pair VF trunk cable extends across town. A new switching center is to be installed along the route where 50 pairs are to be dropped and 50 inserted. It would be desirable to establish the new switch midway between load points. At the switch 450 circuits will bypass the office (switch). Using this $D/2$ technique, these circuits need no conditioning; they are full-load sections (i.e., $D/2 + D/2 = 1D$, a full-load section). Meanwhile, the 50 circuits entering from each direction are terminated for switching and need conditioning so that each looks electrically like a full-load section. However, the physical distance from the switch out to the first load point is $D/2$ or, in the case of H loading, 915 m. To make the load distance electrically equivalent to 1830 m, line build-out (LBO) is used. This is done simply by adding capacitance to the line.

Suppose that the location of the new switching center was such that it was not halfway, but at some other fractional distance. For the section comprising the shorter distance, LBO is used. For the other, longer run, often a half-load coil is installed at the switching center and LBO is added to trim up the remaining electrical distance.

2.9.2   Local Trunk (Junction) Design Considerations

The basic considerations in the design of local trunks (junctions) are loss, stability, signaling, noise, and cost. Each are interrelated such that a change in value of one may affect the others. This forces considerable reiteration in the design process, and such designs are often a compromise.

One major goal is to optimize return loss on trunk facilities. This turns out to be a more manageable task than that required in the subscriber distribution plant. In North America the characteristic impedance of local wire trunks in most cases is 900 Ω in series with a 2.16-μF capacitor to match the impedance of the local (end offices) exchanges. It should be pointed out that some tandem and intertandem trunks connect to 600-Ω tandem switches. Most local analog trunks are two wire.

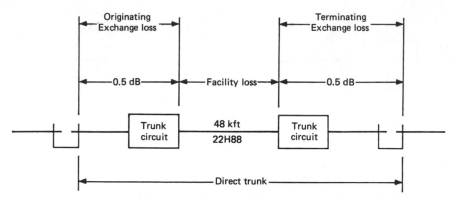

Figure 2.16 Elements of direct trunk loss.

In North American practice, the term *ICL* (inserted connection loss) is used. It is defined for a trunk as the net loss at 1000 Hz inserted between outgoing switch appearances by switching a trunk into an actual operating connection. The ICL on a switched connection is the loss from the outgoing side of a trunk in an originating switch *through* the outgoing switch of the terminating exchange.

Ref. 21 provides a good example of ICL application. Consult Figure 2.16. When dealing with VF facilities including passive device losses, the difference between the line loss and the ICL is the amount of gain that must be inserted in the circuits. The direct trunk in Figure 2.16 has a line loss of 0.5 dB + facility loss + 0.5 dB. The 1000-Hz insertion loss of 48 kft of 22-gauge H88 loaded cable between 900-Ω impedances is 7.2 dB. This exceeds the maximum objective of trunks without gain. Thus gain must be added. The ICL objective for a direct trunk with gain is 3 dB. Therefore the amount of gain to be added is $(0.5 + 7.2 + 0.5) - 3.0 = 5.2$ dB. The 7.2 dB is the facility loss and the 0.5 dB is the nominal planning insertion loss of an analog switch. Because the facility is two wire, the use of negative-impedance VF repeaters would be recommended.

For the case of carrier facilities (i.e., FDM or PCM), gain is reduced or padding (attenuation) added so that the overall trunk loss equals the ICL.

When applying standard North American practice, a first step in trunk design is to meet the ICL objective. Other considerations mentioned above are signaling limits, stability criteria, repeater gain, and the effects of the insertion of other miscellaneous circuit devices. The interrelationship of these factors should be reviewed. Some typical ICL values used in North America are shown in Table 2.16.

Local trunks commonly use dc signaling (supervision), and similar procedures are carried out as we did for subscriber loops, namely, dealing with the resistance limit. Signaling limits can be extended by use of range extenders. If carrier facilities are used, an ac signaling scheme must be employed such as

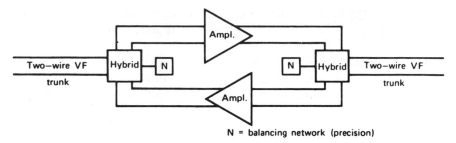

Figure 2.17   Simplified block diagram of a VF repeater.

2600-Hz SF signaling. PCM trunks would use robbed-bit signaling as described in Chapter 3.

2.10   VF REPEATERS

VF repeaters in telephone terminology imply the use of *uni*directional amplifiers at VF on VF trunks. On a two-wire trunk two amplifiers must be used on each pair with a hybrid in and a hybrid out. A simplified block diagram is shown in Figure 2.17.

The gain of a VF repeater can be run up as high as 20 or 25 dB, and originally they were used at 50-mi intervals on 19-gauge loaded cable in the long-distance (toll) plant. Today they are seldom found on long-distance circuits, but they do have application on local trunk circuits where the gain requirements are considerably less. Trunks using VF repeaters have the repeater's gain adjusted to the equivalent loss of the circuit minus the 4-dB loss to provide the necessary singing margin. In practice a repeater is installed at each end of the trunk circuit to simplify maintenance and power feeding. Gains may be as high as 6–8 dB.

An important consideration with VF repeaters is the balance at the hybrids. Here precision balancing networks may be used instead of the compromise networks employed at the two-wire–four-wire interface (Section 2.6.3). It is common to achieve a 21-dB return loss, 27 dB is also possible, and, theoretically, 35 dB can be reached.

Another repeater commonly used on two-wire trunks is the negative-impedance repeater. This repeater can provide a gain as high as 12 dB, but 7 or 8 dB is more common in practice. The negative-impedance repeater requires an LBO at each port and is a true two-way, two-wire repeater. The repeater action is based on regenerative feedback of two amplifiers. The advantage of negative-impedance repeaters is that they are transparent to dc signaling. On the other hand, VF repeaters require a composite arrangement to pass dc signaling. This consists of a transformer bypass.

2.11 TRANSMISSION CONSIDERATIONS OF TELEPHONE SWITCHES IN THE LONG-DISTANCE NETWORK (FOUR WIRE)

Any device placed in an analog transmission tends to degrade the quality of transmission. Telephone switches, unless properly designed, well selected from a system's engineering point of view, and properly installed, may

Table 2.17   Transmission characteristics of analog switches (four-wire switching)

Item	1 (Q.45)[a]	2 (Ref. 19 Criterion)[a]
Loss	0.5 dB	0.5 dB
Loss, dispersion[b]	< 0.2 dB	< 0.2 dB
Attenuation–frequency response	300–400 Hz: $-0.2/+0.5$ dB	$-0.1/+0.2$ dB
	400–2400 Hz: $-0.2/+0.3$ dB	$-0.1/+0.2$ dB
	2400–3400 Hz: $-0.2/+0.5$ dB	$-0.1/+0.3$ dB
Impulse noise	5 in 5 min above -35 dBm0	5 in 5 min, 12 dB above floor of random noise[c]
Noise		
Weighted	200 pWp	25 pWp
Unweighted	1000,000 pW	3000 pW
Unbalanced against ground	300–600 Hz: 40 dB 600–3400 Hz: 46 dB	300–3000 Hz: 55 dB 3000–3400 Hz: 53 dB
Crosstalk		
Between go and return paths	60 dB	65 dB ⎱ in the band
Between any two paths	70 dB	80 dB ⎰ 200–3200 Hz
Harmonic distortion		50 dB down with a -10-dBm signal for 2nd harmonic, 60 dB down for 3rd harmonic
Impedance variation with frequency[d] (or unbalance to ground)	200 Hz: 15 dB 300–600 Hz: 15 dB 600–3400 Hz: 20 dB	300 Hz: 18 dB 500–2500 Hz: 20 dB 3000 Hz: 18 dB 3400 Hz: 15 dB

[a]Reference frequency, where required, is 800 Hz for column 1 and 1000 Hz for column 2.
[b]Dispersion loss is the variation in loss from calls with the highest loss to those with the lowest loss. This important parameter affects circuit stability.
[c]Ref. 23.
[d]Expressed as return loss.

seriously degrade a transmission network. Transmission specifications of a switch must be set forth clearly when switches are to be purchased. One reference that sets forth a series of specifications for toll switches is CCITT Rec. Q.45. The reader should also consult the G.100 series of CCITT recommendations.

In practice it has been found that much of the criterion specified in CCITT Rec. Q.45 is rather loose. Table 2.17 compares Q.45 recommendations with a stricter criterion (based on Ref. 19).

2.12 CCITT INTERFACE

2.12.1 Introduction

To facilitate satisfactory communications between telephone subscribers in different countries, the CCITT has established certain transmission criteria in the form of recommendations. According to CCITT a connection is satisfactory if it meets certain criteria for the following:

- Corrected reference equivalent or a loudness rating
- Noise
- Echo
- Singing

The following paragraphs outline and highlight these criteria from the point of view of a telephone company or administration's international interface. This is the point where the national network meets the international system.

2.12.2 Maximum Number of Circuits in Tandem

Consulting Figure 2.18 we count four links for the international portion of a telephone call and four links each for national extensions. Hence there are a total of 12 links in tandem, which represents the maximum recommended by CCITT (CCITT Recs. G.101 and E.171) under ordinary circumstances. The number of links is limited to assure that the limits of noise (or bit errors), stability, and overall loudness ratings are maintained.

2.12.3 Noise

CCITT Rec. G.103 treats noise. It equates noise to the length of a circuit. This is equated at the rate of 4 pWp/km. Assume that a national connection is 2500 km long, or must extend 2500 km to reach the international interface. Then we would expect to measure no more than 10,000 pWp at that point. CCITT Recs. G.152 and G.153 contain further information. Actual design

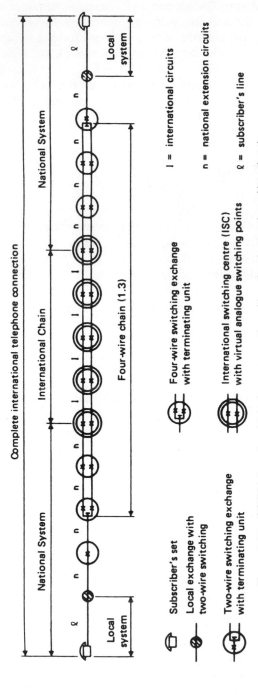

Figure 2.18 CCITT international telephone connection with national extensions. Note that there are a total of 12 links from originating local exchange to terminating local exchange in another country. This represents the maximum allowable number of links for an international call for countries of average size. From CCITT Rec. G.101. Blue Books, Figure 3 / 6.101, Fascicle III.1, Page 12, IXth Plenary Assy., Melbourne 1988.

objectives should improve on these specifications by proper choice of equipment and system layout and engineering.

Absolute noise maxima on the receive side of an international connection should not exceed 50,000 pWp referred to a zero relative level point of the first circuit in the chain (CCITT Rec. G.143). This maximum noise level is permissible if there are six international circuits in tandem.

2.12.4 Variation of Transmission Loss with Time

This important parameter effects stability, CCITT Rec. G.151C states that

> The standard deviation of the variation in transmission loss of a circuit should not exceed 1 dB.... The difference between the mean value and the nominal value of the transmission loss for each circuit should not exceed 0.5 dB.

2.12.5 Crosstalk (CCITT Rec. G.151D)

The near-end and far-end crosstalk (intelligible crosstalk only) measured at audio frequencies at trunk exchanges between two complete circuits in terminal service position should not numerically be less than 58 dB.

Between go and return paths of the same circuit in a four-wire long-distance (toll) exchange, intelligible crosstalk should be at least 43 dB down.

2.12.6 Attenuation Distortion

The worst condition for attenuation distortion is shown in Figure 2.19 (taken from CCITT Rec. G.131). It assumes that the nominal 4-kHz voice channel is used straight through. The attenuation distortion as shown in the figure is a result of 12 circuits of a four-wire chain in tandem. Note that the slope from 300 to 600 Hz and from 2400 to 3400 Hz is approximately 6.5 dB. The slope for one link, therefore, would be 6.5/12 or approximately 0.5 dB. From this we can derive the attenuation distortion permissible for each link to the international interface. For further information consult Chapter 3, which gives a discussion of attenuation distortion of FDM carrier equipment back to back.

2.12.7 Corrected Reference Equivalent and Loudness Rating

Subjective tests have shown that the optimum range of overall CREs (corrected reference equivalents) for telephone connections is approximately 5–16 dB, with the optimum value between 7 and 11 dB. Similarly, the optimum range of overall LRs (loudness ratings) is about 2–11 dB, with the optimum in the vicinity of 5 dB.

Long international telephone connections ordinarily require sufficient transmission loss to control echo and stability. Thus it has been provisionally

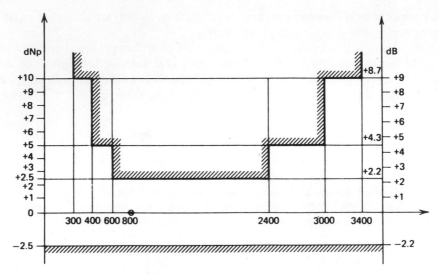

Figure 2.19 Permissible attenuation variation with respect to its value measured at 800 Hz (objective for worldwide four-wire chain of 12 circuits in terminal service). (From CCITT Rec. G.132; courtesy of ITU–CCITT.)

agreed that the long-term objective for the traffic-weighted mean value of the distribution of the planning values over overall CREs for international connections should lie in the range of 13–16 dB and for overall LRs in the range of 8–11 dB.

2.12.8 One-Way Propagation Time

CCITT recommends the following limitations on mean, one-way propagation time when echo sources exist and appropriate echo control devices are used such as echo suppressors or cancellers (from CCITT Rec. G.114, Red Books):

a) 0 to 150 ms acceptable

b) 150 to 400 ms, acceptable, provided that care is exercised on connections when the mean, one-way propagation time exceeds about 300 ms, and provided that echo control devices designed for long-delay circuits are used

c) above 400 ms, unacceptable

We can estimate propagation delay (one-way) on analog circuit with the following formula:

$$\text{Delay (ms) (one-way)} = 12 + (0.004 \times \text{distance in km})$$

For purely digital networks:

$$\text{Delay (ms) (one-way)} = 3 + (0.004 \times \text{distance in km})$$

On the first equation the value 12 ms takes into account local plant wire extensions with loading or about 160 km of H88 loaded wire pair cable. For the second equation the value 3 ms takes into account the delay in one PCM coder and decoder plus five digital exchanges in tandem.

The constant 0.004 ms is for coaxial cable or radio relay transmission. If PCM on fiber optic cable is used, the value should be 0.005 ms.

2.12.9 Echo Suppressors

Echo suppressors should be used on all connections where the *round-trip* delay (propagation time) exceeds 50 ms. Echo suppressors are covered in CCITT Rec. G.161.

REVIEW EXERCISES

1. Where does the frequency response peak for a carbon subset transmitter (e.g., what frequency or frequencies)?

2. A subscriber loop provides a metallic electrical path for four purposes. Name the four.

3. What singular parameter does reference equivalent measure at the subscriber subset?

4. In our present analog network, why is it difficult to achieve ideal reference equivalent or loudness loss values? (This is a conceptual question.)

5. Why has corrected reference equivalent been introduced to replace the well-known reference equivalent?

6. What are the three basic components of overall loudness rating (OLR)? (That is, overall loudness rating $= x + y + z$.)

7. At generally what level of psophometric weighted circuit noise (dBmp) does such noise begin to impact speech transmission quality?

8. Wire pair transmission has notable limitations when used on a geographically extended network. Name five basic transmission methods that can be used to extend the network utilizing the same wire pair.

9. Define two-wire and four-wire transmission.

10. Why is it conceptually important to achieve as high as possible a return loss between the two-wire side of a hybrid and its associated balancing network?

11. In question 10, give a one-sentence reason why it is difficult to achieve such high return loss values. What is the generally accepted median value of such return loss?

12. What is liable to occur when a hybrid is connected to a two-wire connection with a poor match?

13. Define echo and singing.

14. Why is stability return loss important? Differentiate stability return loss from echo return loss.

15. What is the total *design* loss of a hybrid? This is made up of two components. What are they?

16. When a hybrid demonstrates a poor return loss for a particular two-wire connection, what two impairments may occur?

17. How do we prevent singing?

18. How does delay affect echo as an impairment to a speech subscriber?

19. Echo can be controlled by adding loss. Loss is added as a function of delay. If we continue to add loss, how does it affect overall subscriber satisfaction? At what value(s) of round-trip delay must we use echo suppressors? Discuss this situation, including the effects on overall reference equivalent (ORE), corrected reference equivalent (CRE), or overall loudness rating (OLR).

20. When we examine the voice channel, what frequency bands are most susceptible as a cause of singing?

21. What makes via net loss (VNL) different from fixed-loss plans?

22. What is the loss value assigned across the digital toll network for all connections no matter the length? We refer here to the North American transmission plan.

23. In the design of a subscriber loop with inductive loading, besides the wire loop itself, what are the additional resistance values that must be taken into account? Give a budgetary value for each in ohms.

24. Given a cable with 26-gauge wire pairs displaying a loss of 0.51 dB/1000 ft of loop and a resistance of 83.5 Ω per 1000 ft of loop, an exchange that handles up to 1500 Ω of loop resistance, and a transmission plan that allows up to 8.0-dB loop loss, what is the maximum length of loop permitted without conditioning of any sort?

25. Give four ways of extending a loop to rural subscribers (generic, not specific).

26. 24H88 load coils are used on a loop. Explain the three elements of this load coil coding.

27. Resistance design does not treat loop loss per se. With revised resistance design (RRD), a resistance of 1500 Ω is accommodated. How do we know that we do not exceed the maximum loop loss limits?

28. How do standard long route design (LRD) procedures meet the additional loss and resistance ($> 1300\ \Omega$) requirements found on such loops?

29. What is the one major overriding advantage of using digital loop carrier (DLC) on long route design?

30. Define inserted connection loss (ICL) as applied to North American practice regarding local trunk design.

31. Name at least four important parameters one has to consider in the design of local trunks.

32. What is the range of *applied* gain values in dB we would expect to find used for VF repeaters?

33. What type of VF repeater lends itself well to two-wire metallic VF circuits? For what reason?

34. Why do we limit the number of links in tandem on an international connection? What are the ramifications of this limitation on internal national connections?

REFERENCES AND BIBLIOGRAPHY

1. *Reference Data for Radio Engineers*, 6th ed., Howard W. Sams, Indianapolis, IN, 1977.
2. *Outside Plant*, U.S. Army Tech. Manual TM-486-5.
3. M. A. Clement, "Transmission," *Telephony*. 1970.
4. CCITT, Blue Books, XIXth Plenary Assembly, Melbourne, 1988, vol. III, G recommendations.
5. *National Networks for the Automatic Service*, ITU, Geneva, chap. V.
6. *Overall Communications System Planning*, vols. I–III, IEEE New Jersey Section Seminar, 1964.
7. *Transmission Systems for Communications*, 5th ed., Bell Telephone Laboratories, American Telephone and Telegraph Co., Holmdel, NJ 1982.
8. F. T. Andrew and R. W. Hatch, "National Telephone Network Planning in the ATT," *IEEE Trans. Commun. Technol.*, June 1971.
9. "Terminal Balance, Description and Test Methods," *Autom. Electr. Tech. Bull.*, 305–351, June 1961.

10. P. A. Gresh et al., "A Unigauge Design Concept for Telephone Customer Loop Plant," *IEEE Trans. Commun. Technol.*, Apr. 1968.

11. *Principles of Electricity Applied to Telephone and Telegraph Work*, American Telephone and Telegraph Co., New York, 1961.

12. *Rural Electrification Administration Telephone Engineering and Construction Manual*, Sec. 400 series.

13. *ITT Pentaconta Manual*, PC-1000 A/B/B1.

14. H. R. Huntley, "Transmission Design of Intertoll Telephone Trunks," *Bell Syst. Tech. J.*, Sept. 1953.

15. MIL-STD-188C with Notice 1, U.S. Department of Defense, Washington, DC.

16. *DCS Engineering Installation Manual*, DCAC 330-175-1, through Change 9, U.S. Department of Defense, Washington, DC.

17. *Lenkurt Demodulator*, Lenkurt Electric Corp., San Carlos, CA, July 1960, Oct. 1962, Mar. 1964, Jan. 1964, Aug. 1966, June 1968, July 1966, Aug. 1973.

18. *Local Telephone Networks*, ITU–CCITT, Geneva, 1968.

19. USITA Symposium, Apr. 1970, Open Questions 18–37.

20. *Outside Plant*, Telecommunication Planning Documents, ITT Laboratories, Spain, 1973.

21. *Telecommunication Transmission Engineering*, 2nd ed., vols. 1–3, American Telephone and Telegraph Co., New York, 1977.

22. R. L. Freeman, *Telecommunication System Engineering*, 2nd ed., Wiley, New York, 1989.

23. Bell System Technical Reference, *Transmission Parameters Affecting Voiceband Data Transmission—Description of Parameters*, Publ. 41008, American Telephone and Telegraph Co., New York, July 1964.

24. *Notes on the BOC Intra-LATA Networks—1986,* TR-NPL-000275, Bell Communications Research, Livingston, NJ, 1986.

25. "General Characteristics for International Telephone Connections and International Telephone Circuits," CCITT Rec. G.101, Red Books, VIIIth Plenary Assembly, Malaga-Torremolinos, 1984.

26. "Mean One-Way Propagation Time," CCITT Rec. G.114, Red Books, vol. 3, VIIIth Plenary Assembly, Malaga-Torremolinos, 1984.

27. Gerald Ash, "Design and Control of Networks with Dynamic Nonhierarchical Routing," *IEEE Communications Magazine*, Oct. 1990.

3

MULTIPLEXING
TECHNIQUES

3.1 DEFINITION AND SCOPE

Multiplexing deals with the transmission of two or more signals over a common transmission facility such as a wire pair or a radio carrier. When multiplexing is carried out in the frequency domain, we call it frequency division multiplex (FDM). It may also be carried out in the time domain, and this is called time division multiplex (TDM).

With a FDM system each information-bearing channel is assigned a distinct frequency slot in a band of frequencies. Commonly, with telephony, each voice channel modulates a different carrier signal which permits translation of that voice channel to its own frequency slot or segment of a broadband spectrum that is different from all other modulating channels sharing the same spectrum. Such derived signals are combined in an electrical network for transmission to the line at one end of a circuit and then separated electrically at the other end. Because carrier frequencies are used in the multiplexing process, the term *carrier techniques* (or *carrier transmission*) evolved.

With TDM, each information-bearing channel, such as the voice channel, is assigned a time slot. Each channel then occupies the entire frequency spectrum for a very short period of time.

The bulk of this chapter discusses these two multiplexing methods and their associated waveforms. FDM is first treated conceptually and then standard CCITT and North American practices are introduced.

The subsequent TDM discussion is based on common practice of pulse code modulation (PCM). Both North American and CCITT standards are discussed. We then give short descriptions of other time division techniques such as delta modulation as illustrated by its practical application with CVSD (continuous variable slope delta modulation).

Before we launch into multiplexing, keep in mind that all multiplex systems work on a four-wire basis. The transmit and receive paths are separate. Two-wire and four-wire transmission and the conversion from two-wire to four-wire systems are covered in Section 2.6.3.

3.2 FREQUENCY DIVISION MULTIPLEX

3.2.1 Introduction

In FDM an available channel bandwidth is divided into a number of nonoverlapping frequency slots. Each slot carries a single information-bearing signal such as a voice channel. We can consider an FDM multiplexer as a frequency translator. At the opposite end of the circuit a demultiplexer filters and translates the channel frequency slots into the original individual information-bearing channels. This concept is shown in Figure 3.1.

In practice, the "frequency translator" (multiplexer) uses single sideband modulation of RF carriers. A different RF carrier is used for each channel to be multiplexed. This technique is based on mixing or heterodyning a signal to be multiplexed, typically a voice channel, with an RF carrier. A simplified block diagram of an FDM link is shown in Figure 3.2.

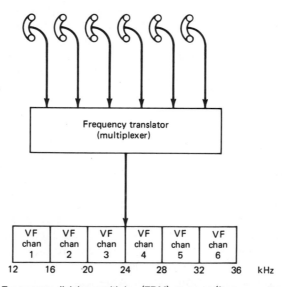

Figure 3.1 Frequency division multiplex (FDM) concept (frequency translation).

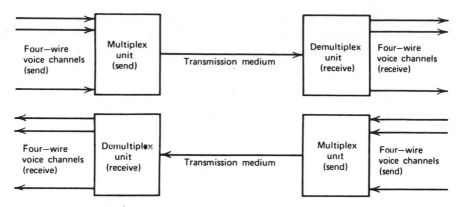

Figure 3.2   Simplified block diagram of an FDM link.

3.2.2   Mixing

The heterodyning or mixing of signals of frequencies A and B is shown below. What frequencies may be found at the output of the mixer?

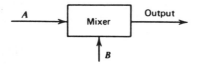

Both the original signals will be present as well as the signals representing their sum and their difference in the frequency domain. Thus at the output of the above mixer we will have present the signals of frequency $A, B, A + B$, and $A - B$. Such a mixing process is repeated many times in FDM equipment.

Let us now look at the boundaries of the nominal 4-kHz voice channel. These are 300 and 3400 Hz. Let us further consider these frequencies as simple tones of 300 and 3400 Hz. Now consider the mixer below and examine the possibilities at its output:

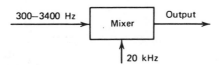

First, the output may be the sum or

$$\begin{array}{r} 20{,}000 \text{ Hz} \\ +\quad 300 \text{ Hz} \\ \hline 20{,}300 \text{ Hz} \end{array} \qquad \begin{array}{r} 20{,}000 \text{ Hz} \\ +\ 3{,}400 \text{ Hz} \\ \hline 23{,}400 \text{ Hz} \end{array}$$

A simple low-pass filter could filter out all frequencies below 20,300 Hz.

Now imagine that instead of two frequencies, we have a continuous spectrum of frequencies between 300 and 3400 Hz (i.e., we have the voice channel). We represent the spectrum as a triangle:

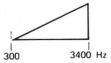

As a result of the mixing process (translation) we have another triangle as follows:

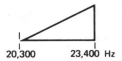

When we take the sum, as we did above, and filter out all other frequencies, we say we have selected the upper sideband. Therefore we have a triangle facing to the right, and we call this an upright or erect sideband.

We can also take the difference, such that

$$
\begin{array}{r}
20{,}000 \text{ Hz} \\
-\ \ \ \ 300 \text{ Hz} \\
\hline
19{,}700 \text{ Hz}
\end{array}
\qquad\qquad
\begin{array}{r}
20{,}000 \text{ Hz} \\
-\ 3{,}400 \text{ Hz} \\
\hline
16{,}600 \text{ Hz}
\end{array}
$$

and we see that in the translation (mixing process) we have had an inversion of frequencies. The higher frequencies of the voice channel become the lower frequencies of the translated spectrum, and the lower frequencies of the voice channel become the higher when the difference is taken. We represent this by a right triangle facing the other direction:

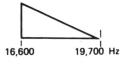

This is called an inverted sideband. To review, when we take the sum, we get an erect sideband. When we take the difference, frequencies invert and we have an inverted sideband represented by a triangle facing left.

Now let us complicate the process a little by translating three voice channels into the radio electric spectrum for simultaneous transmission on a specific medium, a pair of wire lines, for example. Let the local oscillator (mixing) frequency in each case be 20, 16, and 12 kHz. The mixing process is shown in Figure 3.3.

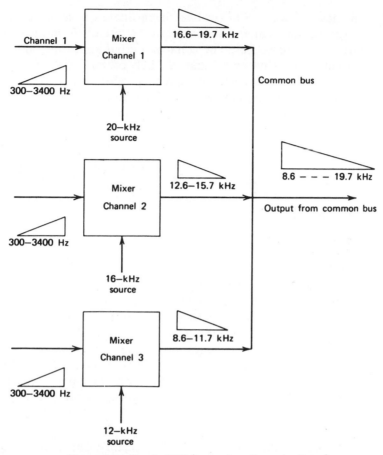

Figure 3.3 Simple FDM (transmit portion only shown).

From Figure 3.3 the difference frequencies are selected in each case as follows:

For channel 1	20,000 Hz	20,000 Hz
	− 300 Hz	− 3,400 Hz
	19,700 Hz	16,600 Hz
For channel 2	16,000 Hz	16,000 Hz
	− 300 Hz	− 3,400 Hz
	15,700 Hz	12,600 Hz
For channel 3	12,000 Hz	12,000 Hz
	− 300 Hz	− 3,400 Hz
	11,700 Hz	8,600 Hz

In each case the lower sidebands have been selected as mentioned above, and all frequencies above 19,700 Hz have been filtered from the output as well as the local oscillator carriers themselves. The outputs from the modulators terminate on a common bus. The common output appearing on this bus is a band of frequencies between 8.6 and 19.7 kHz containing the three voice channels that have been translated in frequency. They now appear on one two-wire circuit ready for transmission. They may be represented by a single inverted triangle as shown:

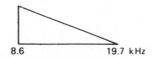

8.6   19.7 kHz

3.2.3 The CCITT Modulation Plan

3.2.3.1 Introduction A modulation plan sets forth the development of a band of frequencies called the line frequency (i.e., ready for transmission on the line or transmission medium). The modulation plan usually is a diagram showing the necessary mixing, local oscillator mixing frequencies, and the sidebands selected by means of the triangles described previously in a step-by-step process from voice channel input to line frequency output. The CCITT has recommended a standardized modulation plan with a common terminology. This allows large telephone networks, on both national and multinational systems, to interconnect. In the following paragraphs the reader is advised to be careful with terminology.

3.2.3.2 Formation of the Standard CCITT Group The standard *group* as defined by the CCITT occupies the frequency band of 60–108 kHz and contains 12 voice channels. Each voice channel is the nominal 4-kHz channel occupying the 300–3400-Hz spectrum. The group is formed by mixing each of the 12 voice channels, with a particular carrier frequency associated with each channel. Lower sidebands are then selected. Figure 3.4 shows the preferred approach to the formation of the standard CCITT group. It should be noted that in the 60–108-kHz band, voice channel 1 occupies the highest frequency segment by convention, between 104 and 108 kHz. The layout of the standard group is shown in Figure 3.5. The applicable CCITT recommendation is G.232.

Single-sideband suppressed carrier (SSBSC) modulation techniques are universally utilized. CCITT recommends that carrier leak be down to at least −26 dBm0 referred to a 0 relative level point (see Section 1.9.5).

3.2.3.3 Alternative Method of Formation of the Standard CCITT Group
The economy of filter design has caused some manufacturers to use an alternative method to form a group. This is done by an intermediate modula-

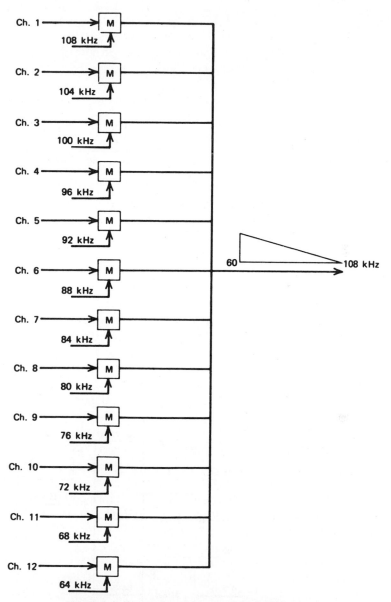

Figure 3.4 Formation of the standard CCITT group.

tion step forming four pregroups. Each pregroup translates three voice channels in the intermediate modulation step. The translation process for this alternative method is shown in Figure 3.6. For each pregroup the first voice channel modulates a 12-kHz carrier, the second a 16-kHz carrier, and the third a 20-kHz carrier. The upper sidebands are selected in this case and carriers are suppressed.

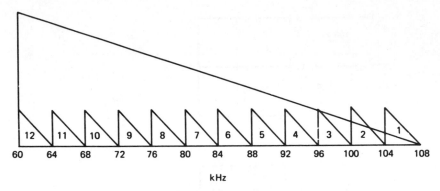

Figure 3.5   Standard CCITT group.

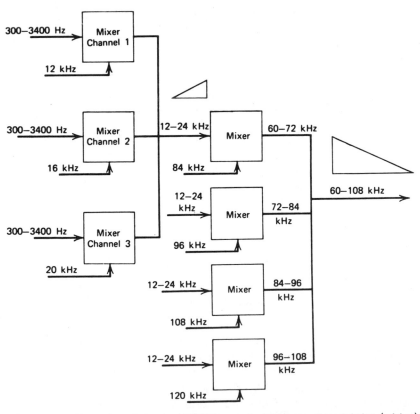

Figure 3.6   Formation of standard CCITT group by two steps of modulation (mixing).

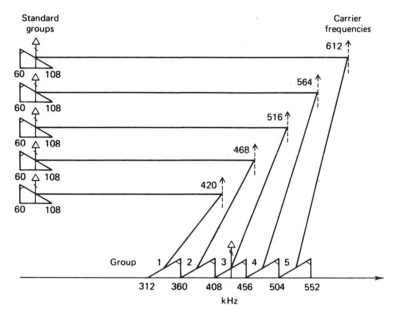

Figure 3.7 Formation of the standard CCITT supergroup. *Note*: Vertical arrows show group level regulating pilot tones (see Section 3.2.5). (From CCITT G.233; courtesy of ITU – CCITT.)

The result is a subgroup occupying a band of frequencies from 12 to 24 kHz. The second modulation step is to take four of these pregroups so formed and translate them, each to their own frequency segment, in the band of 60–108 kHz. To achieve this the pregroups are modulated by carrier frequencies of 84, 96, 108, and 120 kHz, and the lower sidebands are selected, properly inverting the voice channels. This dual modulation process is shown in Figure 3.6.

The choice of the one-step or two-step modulation is an economic trade-off. Adding a modulation stage adds noise to the system.

3.2.3.4 *Formation of the Standard CCITT Supergroup* A supergroup contains five standard CCITT groups, equivalent to 60 voice channels. The standard supergroup before translation occupies the frequency band of 312–552 kHz. Each of the five groups making up the supergroup is translated in frequency to the supergroup band by mixing with the proper carrier frequencies. The carrier frequencies are 420 kHz for group 1, 468 kHz for group 2, 516 kHz for group 3, 564 kHz for group 4, and 612 kHz for group 5. In the mixing process the difference is taken (lower sidebands are selected). This translation process is shown in Figure 3.7.

3.2.3.5 *Formation of the Standard CCITT Basic Mastergroup and Supermastergroup* The basic mastergroup contains five supergroups, 300 voice channels, and occupies the spectrum of 812–2044 kHz. It is formed by

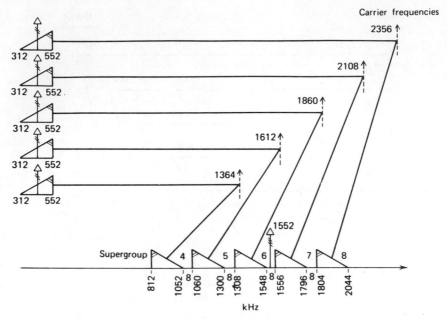

Figure 3.8 Formation of the standard CCITT mastergroup. (From CCITT Rec. G.233; courtesy of ITU – CCITT.)

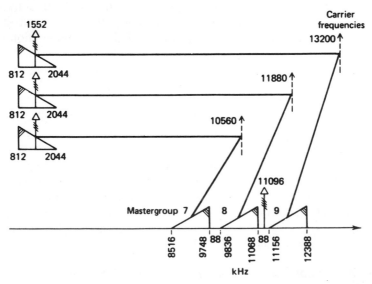

Figure 3.9 Formation of the standard CCITT supermastergroup. (From CCITT Rec. G.233; courtesy of ITU – CCITT.)

translating the five standard super groups, each occupying the 312–552 kHz band, by a process similar to that used to form the supergroup from five standard CCITT groups. This process is shown in Figure 3.8.

The basic supermastergroup contains three mastergroups and occupies the band of 8516–12,388 kHz. The formation of the supermastergroup is shown in Figure 3.9.

3.2.3.6 The "Line" Frequency

The band of frequencies that the multiplex applies to the line, whether the line is a radiolink, coaxial cable, wire pair, or open-wire line, is called the line frequency. Another expression often used is HF (or high frequency), not to be confused with high-frequency radio, discussed in Chapter 8.

The line frequency in this case may be direct application of a group or supergroup to the line. However, more commonly a final translation stage occurs, particularly on high-density systems. Several of these line configurations are shown below.

Figure 3.10 shows the makeup of the basic 15-supergroup assembly. Figure 3.11 shows the makeup of the standard 15-supergroup assembly No. 3 as derived from the basic 15-supergroup assembly shown in Figure 3.10. Figure 3.12 shows the development of a 600-channel standard CCITT line frequency.

3.2.4 Loading of Multichannel FDM Systems

3.2.4.1 Introduction

Most of the FDM (carrier) equipment in use today carries speech traffic, sometimes misnamed "message traffic" in North America. In this context we refer to full-duplex conversations by telephone between two "talkers." However, the reader should not lose sight of the fact that there is a marked increase in the use of these same intervening talker facilities for data transmission.

For this discussion the problem essentially boils down to that of human speech and how multiple telephone users may load a carrier system. If we load a carrier system too heavily, meaning here that the input levels are too high, intermodulation noise and crosstalk will become intolerable. If we do not load the system sufficiently, the signal-to-noise ratio will suffer. The problem is fairly complex because speech amplitude varies:

- With talker volume
- At a syllabic rate
- At an audio rate
- With varying circuit losses as different loops and trunks are switched into the same channel bank voice channel input

Also the loading of a particular system varies with the busy hour.*

*The *busy hour* is a term used in traffic engineering and is defined by the CCITT as "the uninterrupted period of 60 minutes during which the average traffic flow is maximum."

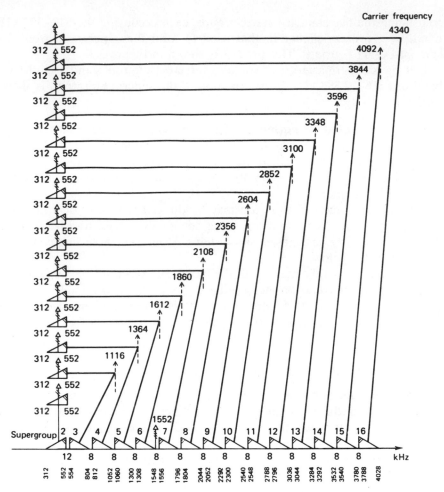

Figure 3.10 Makeup of basic CCITT 15-supergroup assembly. (From CCITT Rec. G. 233; courtesy of ITU – CCITT.)

3.2.4.2 Speech Measurement
The average power measured in dBm of a typical single talker is

$$P_{dBm} = V_{VU} - 1.4 \qquad (3.1)$$

where V_{VU} = reading of a Volume Unit (VU) meter (see Section 1.9.5). In other words a 0 VU talker has an average power of −1.4 dBm.

Empirically, for a typical talker the peak power is about 18.6 dB higher than the average power. Peakiness of the speech level means that the carrier equipment must be operated at a low average power to withstand voice peaks so as not to overload and cause distortion. Thus the primary concern is that

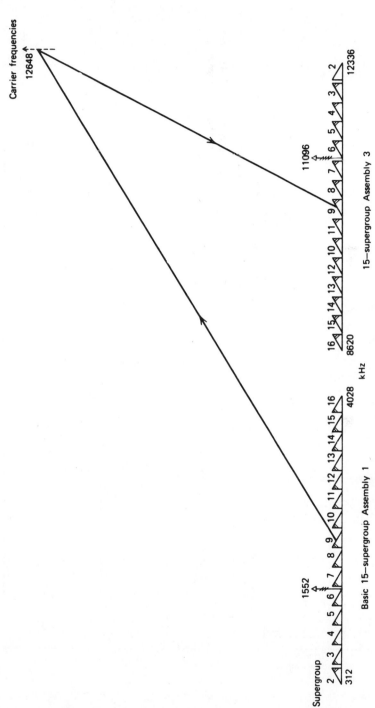

Figure 3.11 Makeup of standard 15-supergroup assembly 3 as derived from basic 15-supergroup assembly. (From CCITT Rec. G.233; courtesy of ITU – CCITT.)

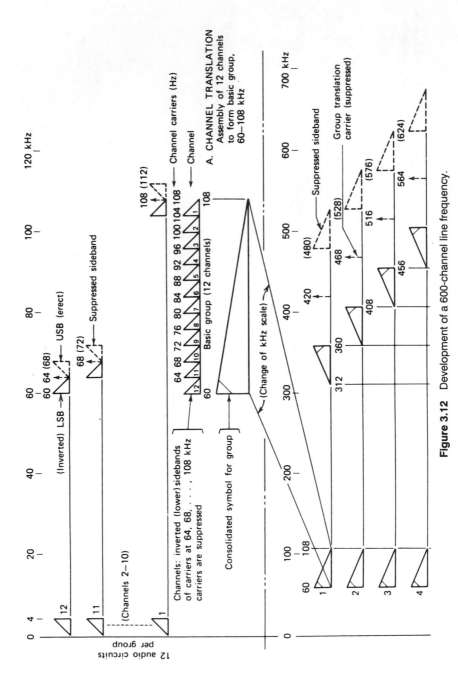

Figure 3.12 Development of a 600-channel line frequency.

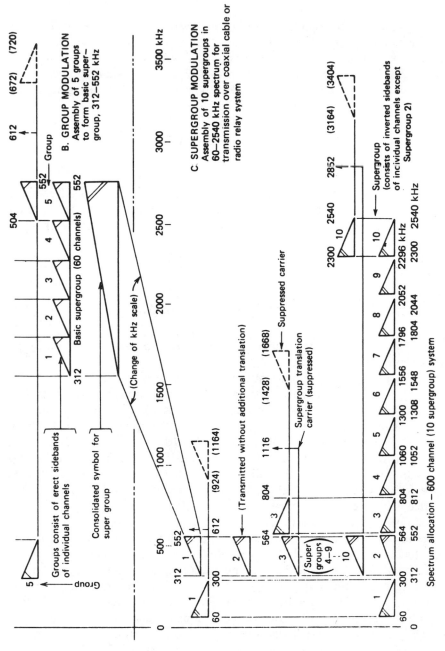

B. GROUP MODULATION Assembly of 5 groups to form basic super-group, 312–552 kHz

C. SUPERGROUP MODULATION Assembly of 10 supergroups in 60–2540 kHz spectrum for transmission over coaxial cable or radio relay system

Spectrum allocation – 600 channel (10 supergroup) system

Figure 3.12 (*Continued*)

119

of voice peaks or spurts. These can be related to an activity factor T_a. T_a is defined as that proportion of the time that the rectified speech envelope exceeds some threshold. If the threshold is about 20 dB below the average power, the activity dependence on threshold is fairly weak. We can now rewrite our equation for average talker power in dBm, relating it to the activity factor T_a as follows:

$$P_{dBm} = V_{VU} + 10 \log T_a$$

If $T_a = 0.725$, the results will be the same as for equation 3.1.

Consider now adding a second talker operating on a different frequency segment on the same equipment, but independent of the first talker. Translations to separate frequency segments are described earlier in the chapter. With the second talker added, the system average power will increase by 3 dB. If we have N talkers, each on a different frequency segment, the average power developed will be:

$$P_{dBm} = V_{VU} - 1.4 + 10 \log N \tag{3.2}$$

where P_{dBm} = power developed across the frequency band occupied by all the talkers.

Empirically, we have found that the peakiness or peak factor of multitalkers over a multichannel analog system reaches the characteristics of random noise peaks when the number of talkers N exceeds 64. When $N = 2$, the peaking factor is 18 dB; for 10 talkers it is 16 dB, for 50 talkers 14 dB, and so forth. Above 64 talkers the peak factor is 12.5 dB.

An activity factor of 1, which we have been using above, is unrealistic. This means that someone is talking all the time on the circuit. The traditional figure for activity factor accepted by CCITT and used in North American practice is 0.25. Let us see how we reach this lower figure.

For one thing, the multichannel equipment cannot be designed for N callers and no more. If this were true, a new call would have to be initiated every time a call terminated or calls would have to be turned away for an "all trunks busy" condition. In the real-life situation, particularly for automatic service, carrier equipment, like switches, must have a certain margin by being overdimensioned for busy-hour service. For this overdimensioning we drop the activity factor to 0.70. Other causes reduce the figure even more. For instance, circuits are essentially inactive during call setup as well as during pauses for thinking during a conversation. The factor of 0.70 now reduces to 0.50. This latter figure is divided in half owing to the talk–listen effect. If we disregard isolated cases of "double-talking," it is obvious that on a full-duplex telephone circuit, while one end is talking, the other is listening. Hence a circuit (in one direction) is idle half the time during the "listen" period. The resulting activity factor is 0.25.

3.2.4.3 *Overload*   In Section 1.9.6, where we discussed intermodulation (IM) products, we showed that one cause of IM products is overload. One definition of overload follows (Ref. 4):

> The overload point, or overload level, of a telephone transmission system is 6 dB higher than the average power in dBm0 of each of two applied sinusoids of equal amplitude and of frequencies a and b, when these input levels are so adjusted that an increase of 1 dB in both their separate levels causes an increase, at the output, of 20 dB in the intermodulation product of frequency $2a - b$.

Up to this point we have been talking about average power. Overload usually occurs when instantaneous signal peaks exceed some threshold. Consider that peak instantaneous power exceeds average power of a simple sinusoid by 3 dB. For multichannel systems the peak factor may exceed that of a sinusoid by 10 dB.

White noise is often used to simulate multitalker situations for systems with more than 64 operative channels.

3.2.4.4 *Loading*   For loading multichannel FDM systems, CCITT recommends (CCITT Rec. G.223):

> It will be assumed for the calculation of intermodulation below the overload point that the multiplex signal during the busy hour can be represented by a uniform spectrum [of] random noise signal, the mean absolute power level of which, at a zero relative level point—(in dBm0)

$$P_{av} = -15 + 10 \log N \quad \text{when } N \geq 240 \qquad (3.3)$$

and

$$P_{av} = -1 + 4 \log N \quad \text{when } 12 \leq N < 240 \cdots . \qquad (3.4)$$

where N = numbers of 4-kHz voice channels.

All logs are to the base 10. *Note*: These equations apply only to systems without preemphasis and using independent amplifiers in both directions. Preemphasis is discussed in Section 4.5.2.2. An activity factor of 0.25 is assumed. See Figure 4.14 and the discussion therewith. Examples of the application of these formulas are discussed in Section 4.5.2.5.2. It should also be noted that the formulas above include a small margin for loads caused by signaling tones, pilot tones, and carrier leaks.

Example 1.   What is the average power of the composite signal for a 600-voice channel system using CCITT loading? N is greater than 240; thus equation 3.3 is valid.

$$\begin{aligned} P_{av} &= -15 + 10 \log 600 \\ &= -15 + 10 \times 2.7782 \\ &= +12.782 \text{ dBm0} \end{aligned}$$

Example 2. What is the average power of the composite signal for a 24-voice channel system using CCITT loading? N is less than 240, thus equation 3.4 is valid.

$$P_{av} = -1 + 4\log 24$$

$$= -1 + 4 \times 1.3802$$

$$= +4.5208 \text{ dBm0}$$

3.2.4.5 *Single-Channel Loading*   A number of telephone administrations have attempted to standardize on -16 dBm0 for single-channel speech input to multichannel FDM equipment. With this input, peaks in speech level may reach -3 dBm0. Tests indicated that such peaks will not be exceeded more than 1% of the time. However, the conventional value of average power per voice channel allowed by the CCITT is -15 dBm0. (Refer to equation 3.3 and CCITT Rec. G.223.) This assumes a standard deviation of 5.8 dB and the traditional activity factor of 0.25. Average talker level is assumed to be at -11.5 VU. We must turn to the use of standard deviation because we are dealing with talker levels that vary with each talker, and consequently with the mean or average.

3.2.4.6 *Loading with Constant-Amplitude Signals*   Speech on multichannel systems has a low duty cycle or activity factor. We established the traditional figure of 0.25. Certain other types of signals transmitted over the multichannel equipment have an activity factor of 1. This means that they are transmitted continuously, or continuously over fixed time frames. They are also characterized by constant amplitude. Examples of these types of signals follow:

- Telegraph tone or tones
- Signaling tone or tones
- Pilot tones
- Data signals (particularly FSK and PSK; see Chapter 12)

Here again, if we reduce the level too much to ensure against overload, the signal-to-noise ratio will suffer, and hence the error rate will suffer.

For typical constant-amplitude signals, traditional* transmit levels (input to the channel modulator on the carrier [FDM] equipment) are as follows:

- Data: -13 dBm0
- Signaling (SF supervision), tone-on when idle: -20 dBm0
- Composite telegraph: -8.7 dBm0

---

*As taken from CCITT.

Table 3.1 Voice channel loading of data/telegraph signals

Signal Type	CCITT	North American
High-speed data	− 10 dBm0 simlex	− 10 dBm0 switched network
	− 13 dBm0 duplex	− 8 dBm0 leased line
		− 5 dBm0 occasionally
Medium-speed data		− 8 dBm0 total power
Telegraph (multichannel)		
≤ 12 channels	− 19.5 dBm0/channel	
	− 8.7 dBm0 total	
≤ 18 channels	− 21.25 dBm0/channel	
	− 8.7 dBm0 total	
≤ 24 channels	− 22.25 dBm0/channel	
	− 8.7 dBm0 total	

Source: Lenkurt Demodulator, July 1968. Courtesy of Lenkurt Electric Corp.

For one FDM system now on the market with 75% speech loading and 25% data/telegraph loading with more than 240 voice channels, the manufacturer recommends the following:

$$P_{\text{rms}} = -11 + 10 \log N \tag{3.5}$$

using − 5 dBm0 per channel for the data input levels* and − 8 dBm0 for the composite telegraph level.

Table 3.1 shows some of the standard practice for data/telegraph loading on a per-channel basis. Data and telegraph should be loaded uniformly. For instance, if equipment is designed for 25% data and telegraph loading, then voice channel assignment should, whenever possible, load each group and supergroup uniformly. For instance, it is bad practice to load one group with 75% data, while another group carries no data traffic at all. Data should also be assigned to voice channels that will not be near group band edge. Avoid channels 1 and 12 on each group for the transmission of data, particularly medium- and high-speed data. It is precisely these channels that display the poorest attenuation distortion and group delay due to the sharp roll-off of group filters (see Chapter 12).

3.2.5 Pilot Tones

3.2.5.1 Introduction Pilot tones in FDM carrier equipment have essentially two purposes:

- Control of level
- Actuation of alarms when levels are out of tolerance

---

*All VF channels may be loaded at − 8 dBm0 level, whether data or telegraph with this equipment, but for − 5 dBm0 level data, only two channels per group may be assigned this level, the remainder voice, or the group must be "deloaded" (i.e., idle channels).

3.2.5.2 *Level Regulating Pilots* The nature of speech, in particular its varying amplitude, makes it a poor prospect as a reference for level control. Ideally, simple single-sinusoid constant-amplitude signals with 100% duty cycles provide simple control information for level regulating equipment. Multiplex level regulators operate in the same manner as automatic gain control circuits on radio systems, except that their dynamic range is considerably smaller.

Modern carrier systems initiate a level regulating pilot tone on each group at the transmit end. Individual level regulating pilots are also initiated on all supergroups and mastergroups. The intent is to regulate the system level within ± 0.5 dB.

Pilots are assigned frequencies that are part of the transmitted spectrum yet do not interfere with voice channel operation. They usually are assigned a frequency appearing in the guard band between voice channels or are residual carriers (i.e., partially suppressed carriers). CCITT has assigned the following as group regulation pilots:

- 84.080 kHz (at a level of -20 dBm0)
- 84.140 kHz (at a level of -25 dBm0)

The Defense Communications Agency of the U.S. Department of Defense recommends 104.08 kHz ± 1 Hz for group regulation and alarm.

For CCITT group pilots, the maximum level of interference permissible in the voice channel is -73 dBm0p. CCITT pilot filters have essentially a bandwidth at the 3-dB points of 50 Hz (refer to CCITT Rec. G.232).

Table 3.2 presents other CCITT pilot tone frequencies as well as those standard for group regulation. Respective levels are also shown. The operat-

Table 3.2 Frequency and level of CCITT recommended pilots

Pilot for	Frequency (kHz)	Absolute Power Level at a Zero Relative Level Point (dB) (Np)
Basic group B	84.080	$-20\,(-2.3)$
	84.140	$-25\,(-2.9)$
	104.080	$-20\,(-2.3)$
Basic supergroup	411.860	$-25\,(-2.9)$
	411.920	$-20\,(-2.3)$
	547.920	$-20\,(-2.3)$
Basic mastergroup	1552	$-20\,(-2.3)$
Basic supermastergroup	11096	$-20\,(-2.3)$
Basic 15-supergroup assembly (No. 1)	1552	$-20\,(-2.3)$

Source: CCITT Rec. G.241. Courtesy of ITU–CCITT.

ing range of level control equipment activated by pilot tones is usually about ±4 or 5 dB. If the incoming level of a pilot tone in the multiplex receive equipment drops outside the level-regulating range, then an alarm will be indicated (if such an alarm is included in the system design). CCITT recommends such an alarm when the incoming level varies 4 dB up or down from the nominal (CCITT Rec. G.241).

3.2.6 Frequency Generation

In modern FDM carrier equipment a redundant master frequency generator serves as the prime frequency source from which all carriers are derived or to which they are phase locked. Providing redundant oscillators with fail-safe circuitry gives markedly improved reliability figures.

One such piece of FDM carrier equipment on the market has a master frequency generator with three outputs: 4, 12, and 124 kHz. Automatic frequency synchronization is available as an option. This enables the slave terminal to stay in exact frequency synchronization with the master terminal, providing drop-to-drop frequency stability.

The 4-kHz output of the master supply drives a harmonic generator in the channel-group carrier supply. Harmonics of the 4-kHz signal falling between 64 and 108 kHz are selected for use as channel carrier frequencies. The 12-kHz output is used in a similar manner to derive translation frequencies to form the basic CCITT supergroup (420, 468, 516, 564, and 612 kHz). The 124-kHz output drives a similar harmonic generator providing the necessary carriers to translate standard supergroups to the line frequency.

These same carrier frequencies are also used on demultiplex at a slave terminal, or at the demultiplex at a master terminal if that demultiplex is not slaved to a distant terminal.

A simplified block diagram of a typical single-sideband suppressed carrier (SSBSC) multiplex–demultiplex terminal is shown in Figure 3.13.

3.2.7 Noise and Noise Calculations

3.2.7.1 *General* Carrier equipment is the principal contributor of noise on coaxial cable systems and other metallic transmission media. On radio links it makes up about one-quarter of the total noise. The traditional approach is to consider noise from the point of view of a hypothetical reference circuit. Two methods are possible, depending on the application. The first is the CCITT method, which is based on a 2500-km hypothetical reference circuit. The second is used by the U.S. Department of Defense in specifying communications systems. Such military systems are based on a 12,000-nautical mi reference circuit with 1000-mi links and 333-mi sections.

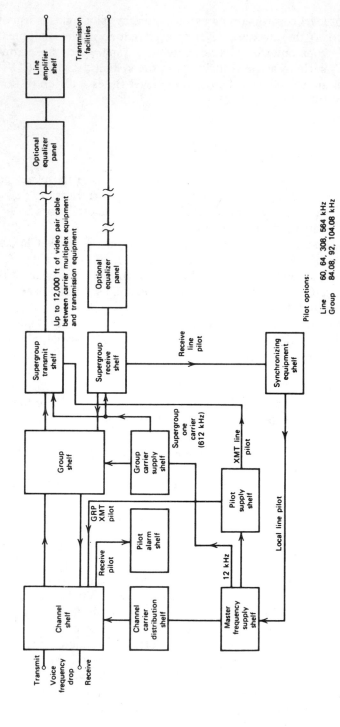

Figure 3.13 Simplified block diagrams of a typical 120-channel terminal arrangement. (Courtesy of GTE Lenkurt Inc., San Carlos, CA.)

3.2.7.2 *CCITT Approach*   CCITT Rec. G.222 states:

... The mean psophometric power, which corresponds to the noise produced by all modulating (multiplex) equipment ... shall not exceed 2500 pW at a zero relative level point. This value of power refers to the whole of the noise due to various causes (thermal, intermodulation, crosstalk, power supplies, etc.). Its allocation between various equipments can be to a certain extent left to the discretion of design engineers. However, to ensure a measurement agreement in the allocation chosen by different administrations, the following values are given as a guide to the target values:

Equipment	Maximum Value Contributed by the Send and Receive Side Together	Assumptions About Loading		
Channel modulators	200 pW0p[a]	Adjacent channels loaded with:	-15 dBm0	(Signal corresponding
		Other channels loaded with:	-6.4 dBm0	to Rec. G.227)
Group modulators	80 pW0p	Load in group to be measured:	$+3.3$ dBm0	
		Load in other groups:	-3.1 dBm0 (each)	
Supergroup modulators	60 pW0p	Load in supergroup to be measured:	$+6.1$ dBm0	
		Load in other supergroups:	$+2.3$ dBm0 (each)	
Mastergroup modulators	60 pW0p	Load in each mastergroup:	$+9.8$ dBm0	
Supermastergroup modulators	60 pW0p	Load in each supermastergroup:	$+14.5$ dBm0	
Basic 15-supergroup assembly modulators	60 pW0p	Load in each 15-supergroup assembly:		
			$+14.5$ dBm0	

[a]No account is taken of the values attributed to pilot frequencies and carrier leaks.

Experience has shown that often these target figures can be improved upon considerably. The CCITT notes that they purposely loosened the value for channel modulators. This permits the use of the subgroup modulation scheme shown above as the alternative method for forming the basic 12-channel group.

For instance, one piece of solid-state equipment now on the market, when operated with CCITT loading, has the following characteristics:

1 pair of channel modulators	224 pWp
1 pair of group modulators	62 pWp
1 pair of supergroup modulators	25 pWp
1 (single) line amplifier	30 pWp

If out-of-band signaling is used, the noise in a pair of channel modulators reduces to 75 pWp. Out-of-band signaling is discussed in Section 3.2.13.

Using the same solid-state equipment mentioned above and increasing the loading to 75% voice, 17% telegraph tones, and 8% data, the following noise

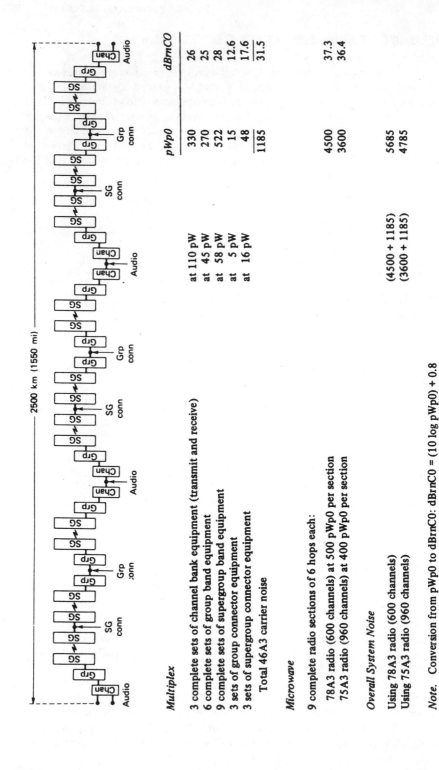

2500 km (1550 mi)

	pWp0	dBmCO
Multiplex		
3 complete sets of channel bank equipment (transmit and receive) at 110 pW	330	26
6 complete sets of group band equipment at 45 pW	270	25
9 complete sets of supergroup band equipment at 58 pW	522	28
3 sets of group connector equipment at 5 pW	15	12.6
3 sets of supergroup connector equipment at 16 pW	48	17.6
Total 46A3 carrier noise	1185	31.5
Microwave		
9 complete radio sections of 6 hops each:		
78A3 radio (600 channels) at 500 pWp0 per section	4500	37.3
75A3 radio (960 channels) at 400 pWp0 per section	3600	36.4
Overall System Noise		
Using 78A3 radio (600 channels) (4500 + 1185)	5685	
Using 75A3 radio (960 channels) (3600 + 1185)	4785	

Note.   Conversion from pWp0 to dBrnC0: dBrnC0 = (10 log pWp0) + 0.8

Figure 3.14 Typical CCITT reference system noise calculations. (Courtesy of GTE Lenkurt Inc., San Carlos, CA.)

information is applicable:

1 pair of channel modulators	322 pWp
1 pair of group modulators	100 pWp
1 pair of supergroup modulators	63 pWp
1 (single) line amplifier	51 pWp

If this equipment is used on a real circuit with the heavier loading, the sum for noise for channel modulators, group modulators, and supergroup modulator pairs is 485 pWp. Accordingly, a system would be permitted to demodulate to voice only five times over a 2500-km route (i.e., 5 × 485 = 2425 pWp). This leads to the use of through-group and through-supergroup techniques discussed in Section 3.2.9. Figure 3.14 shows a typical application of this same equipment using CCITT loading.

3.2.7.3 U.S. Military Approach   The following is excerpted from U.S. MIL-STD-188-100 under "Multiplex Noise for Voice Bandwidth Links":

The multiplex idle channel noise and the loaded noise of a long haul or of a tactical less maneuverable FDM reference voice bandwidth link when referenced to a 0 TLP shall not exceed the values given:

	FDM Noise Allocation	
Configuration	Idle Noise (pWp0)	Loaded Noise (pWp0)
Pair, channel translation sets	10	31
Pair, group translation sets	40	50
Pair, supergroup translation sets	25	50
Through-group equipment	N/A*	10
Through-supergroup equipment	25	50

3.2.8 Other Characteristics of Carrier Equipment

3.2.8.1 Attenuation Distortion   Our interest in this discussion is centered on the attenuation distortion of the voice channel (i.e., not the group, supergroup, etc.) (see CCITT Rec. G.232A). Figure 3.15a shows limits of attenuation distortion (amplitude–frequency response) as set forth in this CCITT recommendation. Figure 3.15b shows insertion loss versus frequency characteristic for single long-haul FDM reference voice bandwidth link according to MIL-STD-188-100.

3.2.8.2 Envelope Delay Distortion   The causes and effects of envelope and group delay are similar. CCITT Rec. G.232C provides guidance on group

*Not applicable.

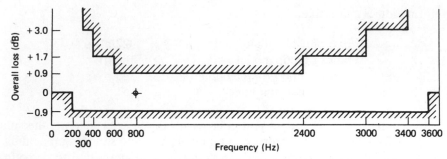

Figure 3.15a Permissible limits for the variation with frequency of the overall loss for any pair of channel transmitting and receiving equipments of one 12-channel terminal equipment. (From CCITT Rec. G.232; courtesy of ITU – CCITT.)

delay for a pair of channel modulators (back-to-back). Typical values are

Frequency Band (Hz)	Group-Delay Distortion (Relative to Minimum Delay) (ms)
400–500	5
500–600	3
600–1000	1.5
1000–2600	0.5
2600–3000	2.5

On U.S. military circuits in accordance with MIL-STD-188-100, the envelope delay distortion of a long-haul or tactical less-maneuverable FDM voice bandwidth link shall not exceed, for a pair of channel translation sets, 120 μs in the band of 600–3200 Hz, except for the band of 1000–2500 Hz, where it shall not exceed 80 μs.

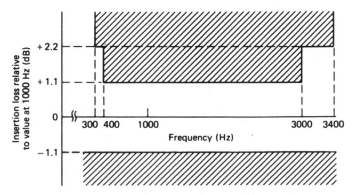

Figure 3.15b Insertion loss versus frequency characteristic for single long-haul reference voice band width link. (From MIL-STD-188-100.)

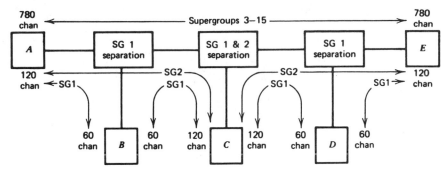

Figure 3.16 Typical drop and insert of supergroups.

3.2.9 Through-Group and Through-Supergroup Techniques

To avoid excessive noise accumulation, modulation/translation steps in long-haul carrier equipment systems must be limited. One method widely used is that of employing group connectors and through-supergroup devices. A simple application of supergroup connectors (through-supergroup devices) is shown in Figure 3.16. Here supergroup 1 is passed directly from point A to B, supergroup 2 is dropped at C, a new supergroup is inserted for onward transmission to E, and so forth. At the same time supergroups 3–15 are passed directly from A to E on the same line frequency (baseband).

The expression "drop and insert" is terminology used in carrier systems to indicate that at some point a number of channels are "dropped" to voice (if you will) and an equal number are "inserted" for transmission back in the opposite direction. If channels are dropped at B from A, B necessarily must insert channels going back to A again.

Through-group and through-supergroup techniques are used much more on long trunk routes where excessive noise accumulation can be a problem. Such route plans can be very complex. However, the savings on equipment and the reduction of noise accumulation are obvious.

When through-supergroup techniques are used, the supergroup pilot may be picked off and used for level regulation. Nearly all carrier equipment manufacturers include level regulators as an option on through-supergroup equipment, whereas through-group equipment does not usually have the option. Figure 3.17 is a simplified block diagram showing how a supergroup may be dropped (separated and inserted). CCITT Recs. G.242 and G.243 apply.

3.2.10 Line Frequency Configurations

3.2.10.1 *General* When applying carrier techniques to a specific medium such as wire pair or radio link, consideration must be given to some of the

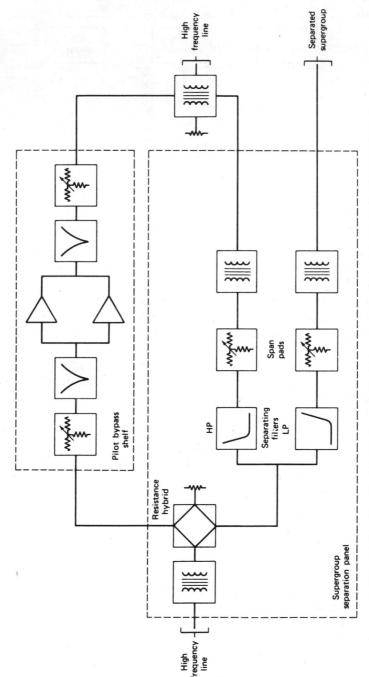

Figure 3.17 Simplified block diagram of typical supergroup drop equipment with pilot tone bypass. (Courtesy of GTE Lenkurt Inc., San Carlos, CA.)

following:

- Type of medium—metallic, optical fiber, or radio?
- If metallic, is it pair, quad, open wire, or coaxial?
- If metallic, what are amplitude distortion, envelope delay, and loss limitations?
- If optical fiber (Chapter 11), single mode or multimode, power limited or dispersion limited?
- If radio, what are the bandwidth limitations?
- What are considerations of international regulations, CCITT recommendations?
- What are we looking into at the other end(s)?

The following paragraphs review some of the more standard line frequency configurations and their applications. The idea is to answer the questions posed above.

3.2.10.2 12-Channel Open-Wire Carrier This is a line frequency configuration that permits transmission of 12 full-duplex voice channels on open-wire pole lines using single-sideband suppressed carrier (SSBSC) multiplex techniques. The industry usually refers to the "go" channels as west–east and the "return" channels as east–west. The standard modulation approach is to develop the standard CCITT 12-channel basic group and through an intermediate modulation process translate the group to one of the following (CCITT terminology):

System Type	West–East (kHz)	East–West (kHz)
SOJ-A-12	36–84	92–140
SOJ-B-12	36–84	95–143
SOJ-C-12	36–84	93–141
SOJ-D-12	36–84	94–142

It should be noted that this type of carrier arrangement replaces a single voice pair on open wire with 12 full-duplex channels. The go and return segments on the same pair are separated by directional filters. Here CCITT Rec. G.311 applies with regard to modulation plans.

Baseband or line frequency repeaters have gains on the order of 43–64 dB. Loss in an open-wire section should not exceed 34 dB under worst conditions at the highest transmitted frequency. This follows the intent of CCITT Rec. G.313. The regional Bell System companies use a similar configuration called J-carrier. Here average repeater spacing is on the order of 50 mi. Two pilot tones are used. One is for flat gain at 80 kHz for west–east and 92 kHz for east–west directions. The other pilot is called a slope pilot. Such slope pilots regulate the lossier part of the transmitted band as weather conditions change (i.e., dry to wet conditions or the reverse) as these losses increase and decrease.

3.2.10.3 Carrier Transmission over Nonloaded Cable Pairs (K-Carrier)
This technique increases the capacity of nonloaded cable facilities. Typical is the North American cable carrier called K-carrier. Separate cable pairs are used in each direction. Each pair carries 12 voice channels in the band of 12–60 kHz. This is the standard CCITT subgroup A, which is the standard CCITT basic group, 60–108 kHz, translated in frequency. On 19-gauge cable repeaters are required about every 17 mi. Pilots are used for automatic regulation on one of the following frequencies: 12, 28, 56, or 60 kHz.

3.2.10.4 Type N Carrier for Transmission over Nonloaded Cable Pairs
The N-Carrier is designed to provide 12 full-duplex voice channels on nonloaded cable pairs over distances of 20–200 mi. The modulation plan is nonstandard. The N-carrier uses double sideband emitted carrier with carrier spacings every 8 kHz. Nominal voice channel bandwidth is 250–3100 Hz. The 12 channels for one direction of transmission are contained in a band of frequencies from 44 to 140 kHz called the low band. The other direction of transmission is in a high band, 164–260 kHz. The emitted carriers serve, as pilot tones do in other systems, as level references in level-regulating equipment.

A technique known as "frogging" or "frequency frogging" is used with the N-carrier whereby the frequency groups in each direction of transmission are transposed and reversed at each repeater so that all repeater outputs are always in one frequency band and all repeater inputs are always in the other. This minimizes the possibility of "interaction crosstalk" around the repeaters through paralleling VF cables. This reversal or transposition of channel groups at each repeater provides automatic self-equalization. A 304-kHz oscillator is basic to every repeater.

N3 is the latest in the N-carrier series of the regional Bell System companies of North America. It is a 24-channel system using conventional FDM carrier techniques. It finds application on wire pair routes from about 10 to over 200 mi long. Supervisory signaling is in-band, and as many as 26 N3 terminals can share a common carrier supply.

The 24-channel line frequency can be a high- or low-group signal and is composed of two identical 12-channel groups. For each of these channel groups 12 carriers spaced at 4-kHz intervals from 148 to 192 kHz are modulated by 12 individual voice channels. After modulation the upper sideband is selected in each case and the lower sideband and carrier are suppressed. Six of the 12 carriers are reinserted as pilots for repeater regulation and at receiver terminals for regulation, frequency correction, and demodulation.

Each of the identical derived 12-channel groups is modulated by a different carrier frequency to form a 24-channel configuration occupying the band 36–132 kHz. These modulating carrier frequencies are 232 and 280 kHz. The line frequency may be a high group or a low group. To form the high group the basic 24-channel configuration is modulated again by a 304-kHz carrier and the lower sideband is selected, deriving a line frequency of 172–268 kHz. If the low group is desired, the basic 24-channel configuration (36–132 kHz) is placed directly on line.

The N3 repeater is similar in its modulation scheme to the other N-carrier repeaters described above, utilizing frequency frogging by modulating the incoming carrier with a 304-kHz signal inverting the incoming line signals.

3.2.10.5 Carrier Transmission on Star- or Quad-Type Cables

Standard carrier systems recommended by the CCITT for transmission over star or quad cables allow transmission of 12, 24, 36, 48, 60, or 120 full-duplex VF channels. The cables in all cases are nonloaded or deloaded. Regarding repeater sections, CCITT Rec. G.322 recommends that these sections have no more than a 41-dB loss at the highest modulating frequency for systems with one, two, or three groups, and 36 dB for those with four or five groups and up to two supergroups. We refer here to solid state repeaters. Figure 3.18 shows CCITT line frequency configurations recommended for star or quad cables.

3.2.10.6 Coaxial Cable Carrier Transmission Systems

Coaxial cable transmission systems using FDM carrier configurations are among the highest density transmission media in common use today. Nominal cable impedance is 75 Ω. Repeater spacing is a function of the highest modulating frequency.

A number of line frequency arrangements are recommended by the CCITT. Only several of these are discussed here. Figures 3.19 and 3.20 show several configurations for 12-MHz cables. Here repeater spacing is on the order of 3 mi (4.5 km). CCITT Rec. G.332 applies.

CCITT recommends 12,435 kHz to be used for the main-line pilot on 12-MHz line frequency configurations. This is the main level regulating pilot, and it should maintain a frequency accuracy of $\pm 1 \times 10^{-5}$. For auxiliary-line pilots, CCITT recommends 308 and/or 4287 kHz.

The CCITT 12-MHz hypothetical reference circuit for coaxial cable systems is shown in Figure 3.21. The circuit is 2500 km (1550 mi) long, consisting of nine homogeneous sections. Such an imaginary circuit is used as guidance in real circuit design for the allocation of noise. This particular circuit has in each direction of transmission

- 3 pairs of channel modulators
- 3 pairs of group modulators
- 6 pairs of supergroup modulators
- 9 pairs of mastergroup modulators

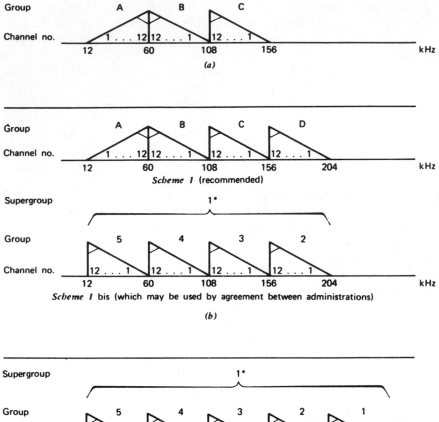

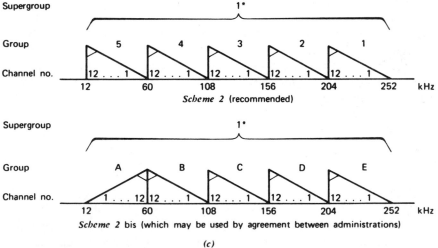

Figure 3.18 Line frequency configurations for star- or quad-type cables. (a) Systems providing one, two, or three groups. (b) Systems providing four groups. (c) Systems providing five groups. (From CCITT Rec. G.322; courtesy of ITU – CCITT.)

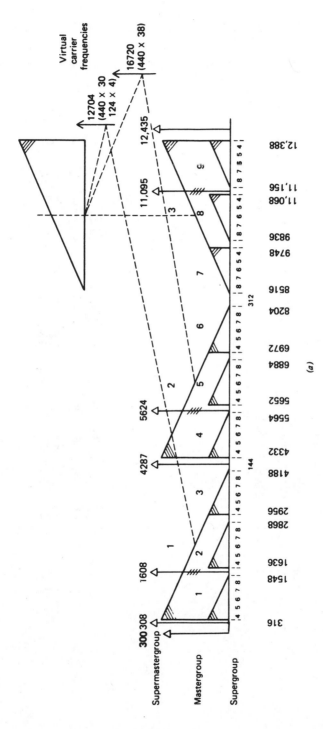

Figure 3.19 (*a*) Plan 1A frequency arrangement fro 12-MHz systems. (*b*) Plan 1B frequency arrangement for 12-MHz systems. (*c*) Plan 1B frequency arrangement for 12-MHz systems, frequencies below 4287 kHz. (From CCITT Rec. G.332; courtesy of ITU–CCITT.)

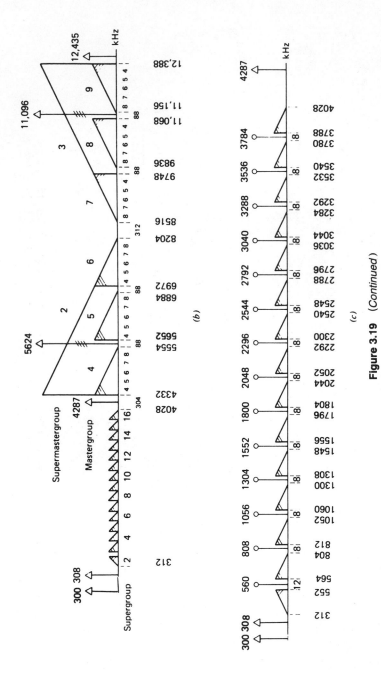

Figure 3.19 (*Continued*)

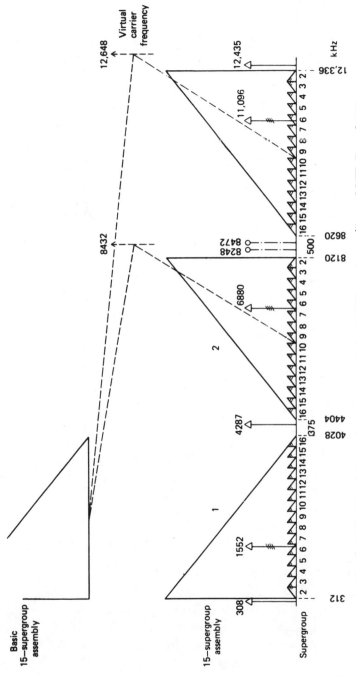

Figure 3.20 Plan 2 frequency arrangement for 12-MHz systems. (From CCITT Rec. G.332; courtesy of ITU–CCITT.)

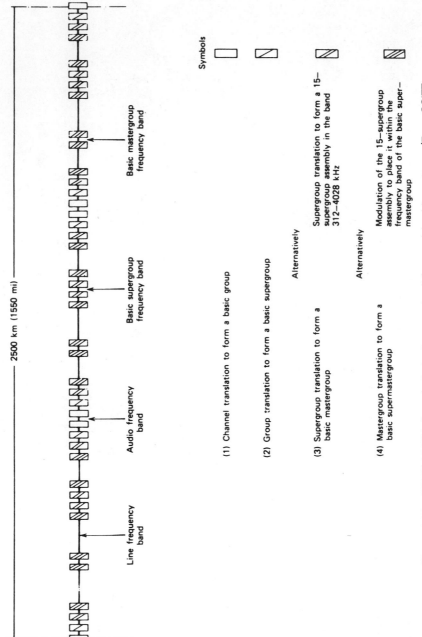

Figure 3.21 CCITT 12-MHz hypothetical reference circuit for coaxial cable systems. (From CCITT Rec. G.322; courtesy of ITU – CCITT.)

Noise allocation for each of these may be found in Section 3.2.7. When referring to pairs of modulators, the intent is that a pair consists of one modulator and a companion demodulator.

Each telephone channel at the end of the 12-MHz (2500-km) coaxial cable carrier system should not exceed 10,000 pWp of noise during any period of 1 h. 10,000 pWp is the sum of intermodulation and thermal noise. No specific recommendation regarding allocation to either type of noise is made.

Of the 10,000 pWp total noise, 2500 pWp is assigned to terminal equipment and 7500 to line equipment. Refer to CCITT Rec. G.332.

For line frequency allocations other than 12 MHz, consult the following:

2.6 MHz: repeater spacing 6 mi (9 km), line frequency 60–2540 kHz

4 MHz: repeater spacing 6 mi (9 km), 15 supergroups with line frequency 60–4028 kHz

3.2.10.7   *The L-Carrier Configuration*

L-carrier is the generic name given by AT&T of North America to their long-haul single-sideband (SSB) carrier system. Its development of the basic group and supergroup assemblies is essentially the same as that in the CCITT recommended modulation plan (Section 3.2.3).

The basic mastergroup differs, however. It consists of 600 VF channels (i.e., 10 standard supergroups). The L600 configuration occupies the 60–2788-kHz band. The U600 configuration occupies the 564–2084-kHz band. The mastergroup assemblies are shown in Figure 3.22. AT&T identifies specific long-haul line frequency configurations by adding a simple number after the letter L. For example, the L3-carrier, which is used on coaxial cable and the TH microwave,* has three mastergroups plus one supergroup comprising 1860 VF channels occupying the band of 312–8284 kHz (see Figure 3.23). L4 consists of six U600 mastergroups in a 3600 VF channel configuration. L5 is discussed in Chapter 10.

Table 3.3 compares some basic L-carrier and CCITT system parameters.

A number of multiplex arrangements are available to translate and combine U600 mastergroup signal spectra for transmission over broadband coaxial cable and microwave radio systems. One such system is the MMX-1, manufactured by Western Electric Company. This derives the L1860 line configuration, originally designed for the L3-carrier, which is shown in Figure 3.23. The three-digit numbering system shown in the figure was adopted to identify each supergroup. The first digit represents the mastergroup number and the second digit the submastergroup number. The third digit represents the supergroup number in the L600 spectrum. Elements of this numbering system are retained in later AT&T/Western Electric designs.

*An AT&T type microwave.

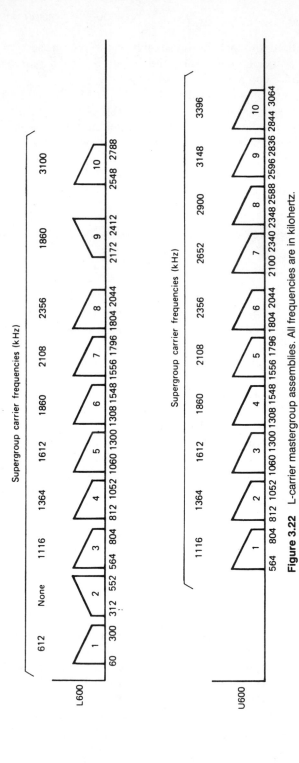

Figure 3.22 L-carrier mastergroup assemblies. All frequencies are in kilohertz.

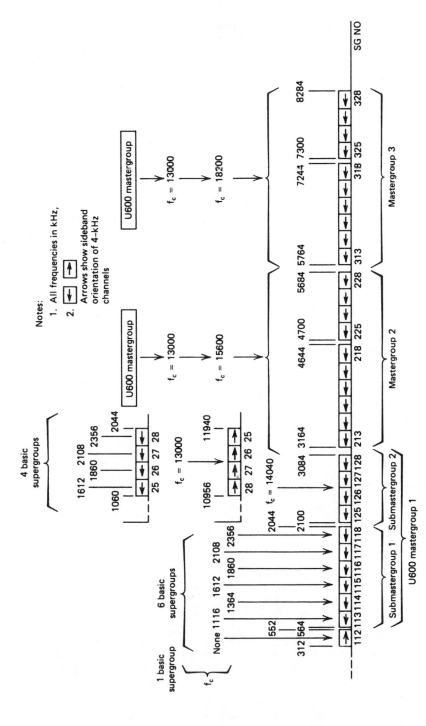

Figure 3.23 Derivation of the L1860 spectrum. (Copyright © 1977 by American Telephone and Telegraph Company.) (Ref. 13.)

Table 3.3   L-carrier and CCITT comparison table

Item	AT&T L-Carrier	CCITT
Level		
Group		
Transmit	−42 dBm	−37 dBm
Receive	−5 dBm	−8 dBm
Supergroup		
Transmit	−25 dBm	−35 dBm
Receive	−28 dBm	−30 dBm
Impedance		
Group	130 Ω balanced	75 Ω unbalanced
Supergroup	75 Ω unbalanced	75 Ω unbalanced
VF channel	200–3350 Hz	300–3400 Hz
Response	+1.0 to −1.0 dB	+0.9 to −3.5 dB
Channel carrier		
Levels	0 dBm	Not specified
Impedances	130 Ω balanced	Not specified
Signaling	2600 Hz in band	3825 Hz out of band
Group pilot		
Frequencies	104.08 kHz	84.08 kHz
Relative levels	−20 dBm0	−20 dBm0
Supergroup carrier		
Levels	+19.0 dBm per demod or mod	Not specified
Impedances	75 Ω unbalanced	Not specified
Supergroup pilot frequency	315.92 kHz	411.92 kHz
Relative supergroup pilot levels	−20 dBm0	−20 dBm0
Frequency synchronization	Yes, 64 kHz	Not specified
Line pilot frequency	64 kHz	60/308 kHz
Relative line pilot level	−14 dBm0	−10 dBm0
Regulation		
Group	Yes	Yes
Supergroup	Yes	Yes

One of the later equipments that translate mastergroups used in North America is the MMX-2 terminal, shown in Figure 3.24, which provides a line frequency spectrum from 564 to 17,548 kHz.

3.2.10.8 Direct-to-Line (DTL) Multiplex   DTL FDM equipment is an economic alternative for main route spurs and other light route applications where 600 or less VF channels are required. Each individual channel unit translates the nominal 4-kHz VF channel to its proper line frequency slot, eliminating group and supergroup translations. Further, all channel units are identical and are equipped with a field-programmable channel selection capability. Line frequencies through supergroup 10 are CCITT compatible in

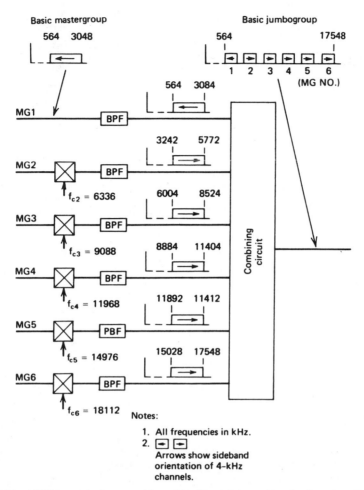

Figure 3.24 MMX-2 terminal, transmitting side. (Copyright © 1977 by American Telephone and Telegraph Company.) (Ref. 13.)

their modulation plan. A major advantage of this equipment is that all common equipment is eliminated, each channel unit being equipped with a standard frequency generator/synthesizer. The synthesizer is adjustable in the field to set the channel unit output/input to the required line frequency slot.

3.2.11 Subscriber Carrier / Station Carrier

In the subscriber distribution plant, when an additional subscriber line is required, it is often more economical to superimpose a carrier signal above the VF signal on a particular subscriber pair. This is especially true when a

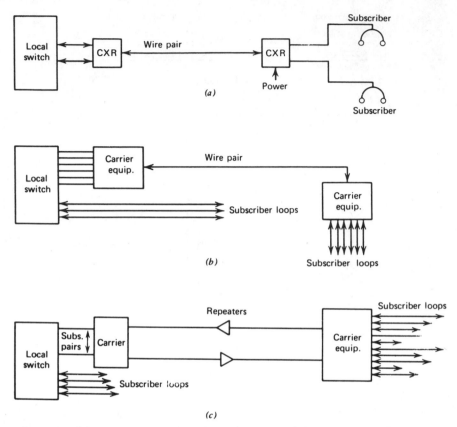

Figure 3.25 (*a*) Typical 1 + 1 subscriber carrier system. (*b*) Typical station carrier system serving six subscribers from a single wire pair. (*c*) Typical station carrier system using a wire pair for each direction of transmission to the remote carrier equipment.

cable has reached or is near exhaustion (i.e., all the subscriber voice pairs are assigned). This form of subscriber carrier is often referred to as a 1 + 1 system and is the most commonly encountered today.

The terms *subscriber carrier* and *station carrier* often are used synonymously because both systems perform the same function in much the same way. There is a difference, however. By convention we say that subscriber carrier systems are powered locally at remote distribution points. Station carrier systems are powered at the serving local switch (central office). Figures 3.25 and 3.26 outline general applications of these types of systems.

Figure 3.25*c* shows one type of station carrier system. Several are available on the market today. Some allow a local switch to serve subscribers more than 30 mi (50 km) away. This particular system carries 20 voice channels on two pairs of cable to a distant distribution point. From the distribution point standard voice pair subscriber loops can be installed with loop resistance up

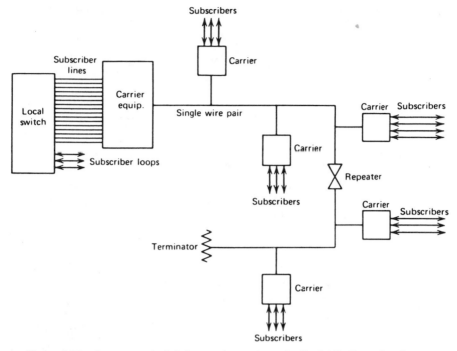

Figure 3.26   Arrangement of station carrier equipment with distribution subscribers.

to 1000 Ω (see Section 2.8.2). Another system provides for the addition of six subscribers to a single wire pair (Figure 3.25*b*). In this case transmission from the serving switch to the subscriber is in the band of 72–140 kHz. From the subscribers to the switch, frequency assignment is in the band of 8–56 kHz.

An important point when applying subscriber carrier to an existing subscriber loop is that load coils and bridged taps must be removed. The taps may act as tuned stubs, and the load coils produce a sharp cutoff, usually right in the band of interest. The load points are candidate locations for line repeaters if required.

3.2.12 Compandors

The word *compandor* is derived from two words that describe it functions: compressor–expandor, to compress and to expand. A compandor does just that. It compresses a signal on one end of a circuit and expands it on the other. A simplified functional diagram and its analogy are shown in Figure 3.27.

The compressor compresses the intensity range of speech signals at the input circuit of a communication channel by imparting more gain to weak signals than to strong signals. At the far-end output of the communication

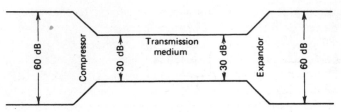

Figure 3.27 Functional analogy of a compandor.

circuit, the expandor performs the reverse function. It restores the intensity of the signal to its original dynamic range. We cover only syllabic compandors in this discussion.

The three advantages of compandors are that they

1. Tend to improve the signal-to-noise ratio on noisy speech circuits
2. Limit the dynamic power range of voice signals, reducing the chances of overload of carrier systems
3. Reduce the possibility of crosstalk

A basic problem in telephony stems from the dynamic range of talker levels. This intensity range can vary 60 dB for the weakest syllables of the weakest talker to the loudest syllables of the loudest talker. The compandor brings this range down to more manageable proportions.

An important parameter of a compandor is its compression–expansion ratio, the degree to which speech energy is compressed and expanded. It is expressed by the ratio of the *input* to the *output* power (dB) in the compressor and expandor, respectively. Compression ratios are always greater than 1 and expansion ratios are less than 1. The most common compression ratio is 2 (2 : 1). The corresponding expansion ratio is thus $\frac{1}{2}$. The meaning of a compression ratio of 2 is that the dynamic range of the speech volume has been cut in half from the input of the compressor to its output. Figure 3.28 is a simplified functional block diagram of a compressor and an expandor.

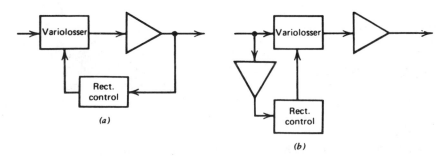

Figure 3.28 Simplified functional block diagram. (*a*) Compressor. (*b*) Expandor.

Another important criterion for a compandor is its companding range. This is the range of intensity levels a compressor can handle at its input. Usually 50–60 dB is sufficient to provide the expected signal-to-noise ratio and reduce the possibility of distortion. High-level signals appearing outside this range are limited without markedly affecting intelligibility.

But just what are the high and the low levels? Such a high or low level is referred to as an "unaffected level" or focal point. CCITT Rec. G.162 defines the unaffected level as

> . . . the absolute level, at a point of zero relative level on the line between the compressor and expander of a signal at 800 Hz, which remains unchanged whether the circuit is operated with the compressor or not.

CCITT goes on to comment:

> The unaffected level should be, in principle, 0 dBm0. Nevertheless, to make allowances for the increase in mean power introduced by the compressor, and to avoid the risk of increasing the intermodulation noise and overload which might result, the unaffected level may, in some cases, be reduced as much as 5 dB. However, this reduction of unaffected level entails a diminution of improvement in signal-to-noise ratio provided by the compandor . . . . No reduction is necessary, in general, for systems with less than 60 channels.

CCITT recommends a range of level from +5 to −45 dBm0 at the compressor input and +5 to −50 dBm0 at the nominal output of the expander. Figure 3.29 shows diagrammatically a typical compandor range of +10 to −50 dBm.

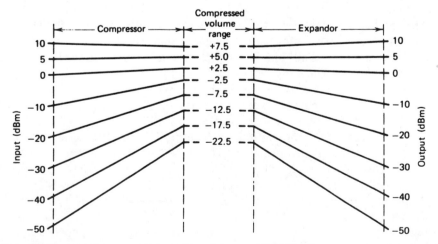

Figure 3.29 Input–output characteristics of a compandor. (Copyright © 1970 by Bell Telephone Laboratories.)

A syllabic compandor operates much in the same fashion of any level-control device where the output level acts as a source for controlling input to the device (see Figure 3.28). Automatic gain control (AGC) on radio receivers operates in the same manner. This brings up the third important design parameter of syllabic compandors: attack and recovery times. These are the response to suddenly applied signals such as a loud speech syllable or burst of syllables. Attack and recovery times are a function of design time constants and are adjusted by the designer to operate as a function of the speech envelope (syllabic variations) and *not* with instantaneous amplitude changes (such as used in PCM). If the operation time is too fast, wide bandwidths would be required for faithful transmission. When attack times are too slow, the system may be prone to overload.

CCITT Rec. G.162 specifies an attack time equal to or less than 5 ms and a recovery time equal to or less than 22.5 ms.

The signal-to-noise ratio advantage of a compandor varies with the multichannel loading factor of FDM equipment and thus depends on the voice level into the FDM channel modulation equipment. At best an advantage of 20 dB may be attained on low-level signals.

3.2.13 Signaling on Carrier Systems

In any textual discussion of carrier equipment, some space must be dedicated to signaling over carrier channels. Signaling may be broken down into two categories, supervisory and address. Supervisory signaling conveys information regarding on-hook and off-hook conditions. Address signaling routes the calls through the switching equipment. It is that signaling that contains the dialing information. Here we are concerned with only the former, supervisory signaling.

The problem stems from the fact that it would be desirable to have continuous supervisory information being exchanged during an entire telephone conversation. This may be done by one of two methods, in-band signaling and out-of-band signaling.

In-band signaling accomplishes both supervisory and address signaling inside the operative voice band spectrum (i.e., in the band of 300–3400 Hz). The supervisory function is carried out when a call is set up and when it is terminated. Hence it is not continuous. The most common type in use today is called SF signaling. SF means single frequency and is a tone, usually at 2600 Hz. Other frequencies also may be used, and most often are selected purposely in the higher end of the voice band.

A major problem with in-band signaling is the possibility of "talk-down." Talk-down refers to the activation or deactivation of supervisory equipment by an inadvertent sequence of voice tones through normal speech usage of the channel. One approach is to use slot filters to bypass the tones as well as a time-delay protection circuit to avoid the possibility of talk-down.

With out-of-band signaling, supervisory information is transmitted out of band. Here we mean above 3400 Hz (i.e., outside the speech channel).

Supervisory information is binary, either on-hook or off-hook. Some systems use tone-on to indicate on-hook and others use tone-off. One expression used in the industry is "tone-on when idle." When a circuit is idle, it is on-hook. The advantage of out-of-band signaling is that either system may be used (i.e., tone-on or tone-off when idle). There is no possibility of talk-down occurring because all supervisory information is passed out of band, away from the voice.

The most common out-of-band signaling frequencies are 3700 and 3825 Hz. 3700 Hz finds more application in North America; 3825 Hz is that recommended by CCITT (Rec. Q.21).

In the long run, out-of-band signaling is attractive from an economic and design standpoint. One drawback is that when patching is required, signaling leads have to be patched as well. In the long term, signaling equipment required may indeed make out-of-band signaling even more costly owing to the extra supervisory signaling equipment and signaling lead extensions required at each end and each time the FDM equipment demodulates to voice. The advantage is that continuous supervision is provided, whether tone-on or tone-off, during the entire telephone conversation.

3.3   TIME DIVISION MULTIPLEX — PCM

3.3.1   What is PCM?

Pulse code modulation (PCM) is a method of modulating in which a continuous analog wave is transmitted in an equivalent digital mode. The cornerstone of an explanation of the functioning of PCM is the Nyquist sampling theorem, which states (Ref. 1, Section 23):

> If a band-limited signal is sampled at regular intervals of time and at a rate equal to or higher than twice the highest significant signal frequency, then the sample contains all the information of the original signal. The original signal may then be reconstructed by use of a low-pass filter.

As an example of the sampling theorem, the nominal 4-kHz channel would be sampled at a rate of 8000 samples per second (i.e., 4000×2).

To develop a PCM signal from one or several analog signals, three processing steps are required: sampling, quantization, and coding. The result is a serial binary signal or bit stream, which may or may not be applied to the line without additional modulation steps.

One major advantage of digital transmission is that signals may be regenerated at intermediate points on links involved in transmission. The price for this advantage is the increased bandwidth required for PCM. Practical

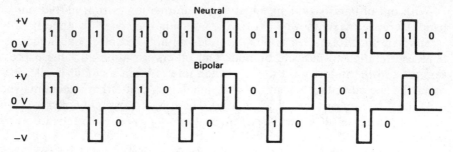

Figure 3.30 Neutral versus bipolar bit streams. Top: Alternative 1's and 0's transmitted in a neutral mode. Bottom: Equivalent in a bipolar mode.

systems require 16 times the bandwidth of their analog counterpart (e.g., a 4-kHz analog voice channel requires 16 × 4 or 64 kHz when transmitted by PCM). Regeneration of a digital signal is simplified and particularly effective when the transmitted line signal is binary, whether neutral, polar, or bipolar. An example of bipolar transmission is shown in Figure 3.30.

Binary transmission tolerates considerably higher noise levels (i.e., degraded signal-to-noise ratios) when compared to its analog counterpart (i.e., FDM, Section 3.2). This plus the regeneration capability is a great step forward in transmission engineering. The regeneration that takes place at each repeater by definition recreates a new digital signal. Therefore noise, as we know it, does not accumulate. However, there is an equivalent to noise in PCM systems that is generated in the modulation–demodulation process. This is called quantizing distortion and can be equated in annoyance to the listener with thermal noise. Regarding thermal/IM noise, let us compare a 2500-km conventional analog circuit using FDM multiplex over cable or radio with an equivalent PCM system over either medium

	FDM/Radio/Cable	PCM/Radio/Cable
Multiplex	2,500 pWp	430 pWp equivalent
Radio/cable	7,500 pWp	0 pWp
Total	10,000 pWp	430 pWp equivalent*

Error rate is another important factor (see Chapter 12). If we can maintain an end-to-end error rate on the digital portion of the system of 1×10^{-5}, intelligibility will not be degraded. A third factor is important in PCM cable applications. This is crosstalk spilling from one PCM system to another or from the send path to the receive path inside the same cable sheath.

The purpose of the discussion of PCM in this chapter is to provide a background of the problems involved with it and its transmission, including

---

*Not dependent on system length, see Section 3.3.2.7. (Ref. CCITT Rec. 713).

the several PCM formats now in use. Practical aspects, such as the design of interexchange trunks (junctions) and the prove-in distance,* compared with other forms of multiplex or VF cable media, are stressed later in this chapter. Long-distance (toll) transmission via PCM is also discussed, and reference should then be made to Chapters 4, 6, and 11. Finally, a second method of digital modulation is briefly described, namely, delta modulation.

3.3.2   Development of a PCM Signal

3.3.2.1   Sampling   Consider the sampling theorem given above. If we now sample the standard CCITT voice channel, 300–3400 Hz (a bandwidth of 3100 Hz), at a rate of 8000 samples per second, we will have complied with the theorem and we can expect to recover all the information in the original analog signal. Therefore a sample is taken every 1/8000 s, or every 125 μs. These are key parameters for our future argument.

Another example may be a 15-kHz program channel. Here the lowest sampling rate would be 30,000 times per second. Samples would be taken at 1/30,000-s intervals or every 33.3 μs.

3.3.2.2   The PAM Wave   With a couple of exceptions (e.g., SPADE, Section 6.10.2.8) practical PCM systems involve TDM. Sampling in these cases does not involve just one voice channel, but several. In practice, one system to be discussed samples 24 voice channels in sequence; another one samples 32 channels. The result of the multiple sampling is a pulse amplitude modulation (PAM) wave. A simplified PAM wave is shown in Figure 3.31, in this case a single sinusoid. A simplified diagram of the processing involved to derive a multiplexed PAM wave is shown in Figure 3.32.

If the nominal 4-kHz voice channel must be sampled 8000 times per second and a group of 24 such voice channels are to be sampled sequentially to interleave them, forming a PAM multiplexed wave, this could be done by gating. Open the gate for a 5.2-μs (125/24) period for each voice channel to be sampled successively from channel 1 through channel 24. This full sequence must be done in a 125-μs period ($1 \times 10^6/8000$). We call this 125-μs period a *frame*, and inside the frame all 24 channels are successively sampled once.

3.3.2.3   Quantization   The next step, in the process of forming a PCM serial bit stream, is to assign a binary coded sequence to each sample as it is presented to the coder.

The number of bits required to represent a character is called a code length, or, more properly, a coding *level*. For instance, a binary code with four discrete elements (a 4-level code) could code 2^4 separate and distinct

---

*The point at which a PCM transmission system becomes an economically viable alternative when applied to an analog telephone network.

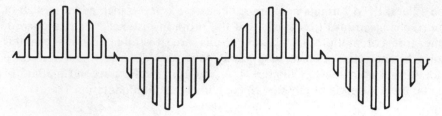

Figure 3.31 PAM wave as a result of sampling a single sinusoid.

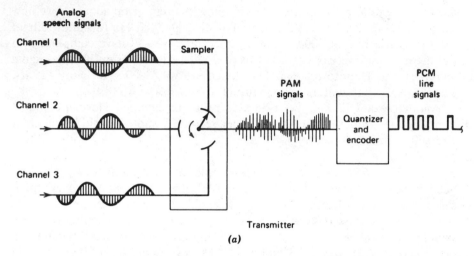

(a)

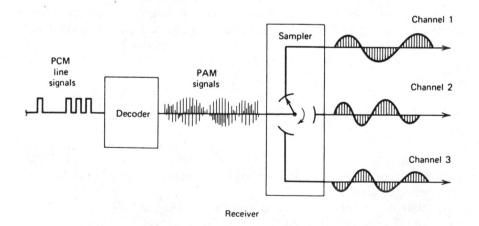

(b)

Figure 3.32 Simplified analogy of the formation of a PAM wave and PCM bit stream. (*a*) Transmitter. (*b*) Receiver. (Courtesy of GTE Lenkurt Inc., San Carlos, CA.)

meanings or 16 characters, not enough for the 26 letters in our alphabet; a 5-level code would provide 2^5 or 32 characters or meanings. The ASCII, the commonly used source code for data transmission, which is described in Chapter 12, is basically a 7-level code allowing 128 discrete meanings for each code combination ($2^7 = 128$). An 8-level code would yield 256 possibilities (see Chapter 12).

Another concept that must be kept in mind as the discussion leads into coding is that bandwidth is related to information rate (more exactly, to modulation rate) or, for this discussion, to the number of bits per second transmitted. The goal is to keep some control over the amount of bandwidth necessary. It follows, then, that the coding length (number of levels) must be limited.

As it stands, an infinite number of amplitude levels are being presented to the coder on the PAM highway. If the excursion of the PAM wave is between 0 and $+1$ V, the reader should ask how many discrete values there are between 0 and 1. All values must be considered, even 0.0176487892 V.

The intensity range of voice signals over an analog telephone channel is on the order of 50 dB (see Section 3.2.12). The 0–1-V range of the PAM highway at the coder input may represent that 50-dB range. Further, it is obvious that the coder cannot provide a code of infinite length (e.g., an infinite number of coded levels) to satisfy every level in the 50-dB range (or a range from -1 to $+1$ V). The key is to assign discrete levels from -1 V through 0 to $+1$ V (50-dB range).

The assignment of discrete values to the PAM samples is called *quantization*. To cite an example, consider Figure 3.33. Between -1 and $+1$ V, 16 quantum steps exist and are coded as follows:

Step	Code	Step	Code
0	0000	8	1000
1	0001	9	1001
2	0010	10	1010
3	0011	11	1011
4	0100	12	1100
5	0101	13	1101
6	0110	14	1110
7	0111	15	1111

Examination of Figure 3.33 shows that step 12 is used twice. Neither time it is used is it the true value of the impinging sinusoid. It is a rounded-off value. These rounded-off values are shown with a dashed line that follows the general outline of the sinusoid. The horizontal dashed lines show the points where the quantum changes to the next higher or next lower level if the sinusoid curve is above or below that value. Take step 14, for example. The curve, dropping from its maximum, is given two values of 14 consecutively.

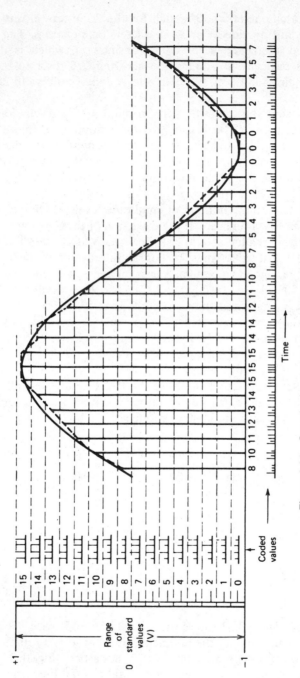

Figure 3.33 Quantization and resulting coding using 16 quantizing steps.

156

For the first, the curve is above 14, and for the second, below. That error, in the case of 14, for instance from the quantum value to the true value is called *quantizing distortion*. This distortion is the major source of imperfection in PCM systems.

In Figure 3.33, maintaining the $-1-0-+1$-V relationship, let us double the number of quantum steps from 16 to 32. What improvement would we achieve in quantization distortion? First determine the step increment in millivolts in each case. In the first case the total range of 2000 mV would be divided into 16 steps, or 125 mV per step. The second case would have 2000/32, or 62.5 mV per step. For the 16-step case, the worst quantizing error (distortion) would occur when the input to be quantized was at the half-step level, or, in this case, 125/2 or 62.5 mV, above or below the nearest quantizing step. For the 32-step case the worst quantizing error (distortion) would again be at the half-step level, or 62.5/2 or 31.25 mV. Thus the improvement in decibels for doubling the number of quantizing steps is

$$20 \log\left(\frac{62.5}{31.25} \right) = 20 \log 2$$

$$\approx 6 \text{ dB}$$

This is valid for linear quantization only (see Section 3.3.2.7). Thus increasing the number of quantizing steps for a fixed range of input values reduces quantizing distortion accordingly. Experiments have shown that if 2048 uniform quantizing steps are provided, sufficient voice signal quality is achieved.

For 2048 quantizing steps a coder will be required to code the 2048 discrete meanings (steps). We find that a binary code with 2048 separate characters or meanings (one for each quantum step) requires an 11-element code, or $2^n = 2048$. Hence $n = 11$.

With a sampling rate of 8000 samples per second per voice channel, the binary information rate per voice channel will be 88,000 bps. Consider that equivalent bandwidth is a function of information rate; the desirability of reducing this figure is therefore obvious.

3.3.2.4 *Coding* Practical PCM systems use 7- and 8-level binary codes, or

$$2^7 = 128 \text{ quantum steps}$$

$$2^8 = 256 \text{ quantum steps}$$

Two methods are used to reduce the quantum steps to 128 or 256 without sacrificing fidelity. These are nonuniform quantizing steps and companding prior to quantizing, followed by uniform quantizing. Keep in mind that the primary concern of digital transmission using PCM techniques is to transmit speech, as distinct from digital transmission covered in Chapter 12, which deals with the transmission of data and message information. Unlike data transmission, in speech transmission there is a much greater likelihood of encountering signals of small amplitudes than those of large amplitudes.

A secondary, but equally important, aspect is that coded signals are designed to convey maximum information considering that all quantum steps (meanings, characters) will have an equally probably occurrence. (We obliquely refer to this inefficiency in Chapter 12 because practical data codes assume equiprobability. When dealing with a pure number system with complete random selection, this equiprobability does hold true. Elsewhere, particularly in practical application, it does not. One of the worst offenders is our written language. Compare the probability of occurrence of the letter "e" in written text with that of "y" or "q.") To get around this problem larger quantum steps are used for the larger amplitude portion of the signal, and finer steps for signals with low amplitudes.

The two methods of reducing the total number of quantum steps can now be labeled more precisely:

- Nonuniform quantizing performed in the coding process.
- Companding (compression) before the signals enter the coder, which now performs uniform quantizing on the resulting signal before coding. At the receive end, expansion is carried out after decoding.

An example of nonuniform quantizing could be derived from Figure 3.33 by changing the step assignment. For instance, 20 steps may be assigned between 0.0 and +0.1 V (another 20 between 0.0 and −0.1 V, etc.), 15 between 0.1 and 0.2 V, 10 between 0.2 and 0.35 V, 8 between 0.35 and 0.5 V, 7 between 0.5 and 0.75 V, and 4 between 0.75 and 1.0 V.

Most practical PCM systems use companding to give finer granularity (more steps) to the smaller amplitude signals. This is instantaneous companding compared to syllabic companding described in Section 3.2.12. Compression imparts more gain to lower amplitude signals. The compression and later expansion functions are logarithmic and follow one of two laws, the A-law or the μ-law. The curve for the A-law may be plotted from the following formula:

$$Y = \frac{AX}{1 + \log A}, \qquad 0 \le v \le \frac{V}{A} \qquad (3.6a)$$

$$= \frac{1 + \log(AX)}{1 + \log A}, \qquad \frac{V}{A} \le v \le V \qquad (3.6b)$$

where $A = 87.6$. The curve for the μ-law may be plotted from the following formula:*

$$|Y| = \frac{\log{(1 + \mu|X|)}}{\log{(1 + \mu)}} \qquad (3.7)$$

where $\mu = 100$ for the original North American T1 system and 255 for later North American (DS1) systems and the CCITT 24-channel system (CCITT Rec. G.733). In these formulas:

$$X = \frac{v}{V}$$

$$Y = \frac{i}{B}$$

where v = instantaneous input voltage

V = maximum input voltage for which peak limitation is absent

i = number of the quantization step starting from the center of the range

B = number of quantization steps on each side of the center of the range (CCITT Rec. G.711, Geneva, 1976).

A common expression used in dealing with the "quality" of a PCM signal is the signal-to-distortion ratio, expressed in decibels. Parameters A and μ determine the range over which the signal-to-distortion ratio is comparatively constant. This is the dynamic range. Using a μ of 100 can provide a dynamic range of 40 dB of relative linearity in the signal-to-distortion ratio. Ref. 33 states that an 8-bit ($\mu = 255$) codec provides a theoretical signal-to-distortion (S/D) ratio greater than 30 dB across a dynamic range of 48 dB. The 8-bit A-law codec provides similar S/D for about a 42-dB dynamic range.

In actual PCM systems the companding circuitry does not provide an exact replica of the logarithmic curves shown. The circuitry produces approximate equivalents using a segmented curve, each segment being linear. The more segments the curve has, the more it approaches the true logarithmic curve desired. Such a segmented curve is shown in Figure 3.34.

If the μ-law were implemented using a segment linear approximate equivalent, it would appear as shown in Figure 3.34. Thus upon coding the first three coded digits would indicate the segment number (e.g., $2^3 = 8$). Of the seven-digit code, the remaining four digits would divide each segment in 16 equal parts to further identify the exact quantum step (e.g., $2^4 = 16$).

*For A-law and μ-law, formulas use natural logarithms.

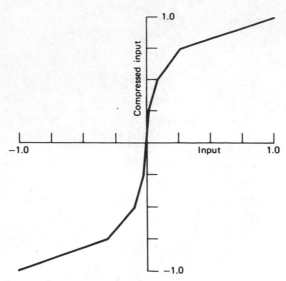

Figure 3.34 Seven-segment linear approximation of the logarithmic curve for the μ-law ($\mu = 100$). (From Ref. 4; copyright © 1970 by Bell Telephone Laboratories, from 1970 edition.) Reprinted with permission.

For small signals* the companding improvement is approximately

A-law   24 dB
μ-law   30 dB

using a seven-level code.

Coding in PCM systems utilizes a straightforward binary coding. Two good examples of this coding are shown in Figures 3.36 and 3.39.

The coding process is closely connected to quantizing. In practical systems, whether using A-law or μ-law, quantizing uses segmented equivalents as discussed above and shown in Figure 3.34. Such segmenting is a handy aid to coding. Consider Figure 3.35, which shows the segmenting used on a 24-channel PCM system (A-law) developed by Standard Telephone and Cables, U.K. (STC). Here there are seven linear approximations (segments) above the origin and seven below, providing a 13-segment equivalent of the A-law. It is 13, not 14 (i.e., 7 + 7), because the segments passing through the origin are collinear and are counted as one segment, not two segments. In this system 6 bits identify the specific quantum level, and a seventh bit identifies whether it is positive (above the origin) or negative (below the origin). The maximum negative step is assigned 0000000, the maximum positive, 1111111. Obviously we are dealing with a seven-level code providing identification of 128 quantum steps, 64 above the origin and 64 below.

*Low-level signals.

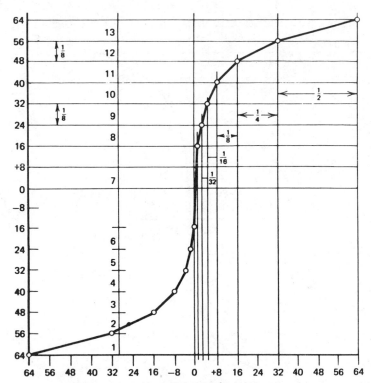

Figure 3.35 A 13-segment approximation of the *A*-law curve as used on the 24-channel STC system (PSC-24B). The abscissa represents quantized signal levels. Note that there are many more companding values at the lower signal levels than at the higher signal levels.

The CEPT 30 + 2 PCM system also uses a 13-segment approximation of the *A*-law, where $A = 87.6$. However, the segment passing through the origin contains four steps, two above and two below the *x*-axis; thus a 16-segment representation is used, which leads us to an eight-level code. The coding for this system is shown in Figure 3.36. Again, if the first code element (bit) is 1, it indicates a positive value (e.g., the quantum step is located above the origin). The following three elements (bits) identify the segment, there being seven segments above the seven segments below the origin (horizontal axis).

As an example consider the fourth positive segment, given as 1101XXXX in Figure 3.36. The first 1 indicates that it is above the horizontal axis (e.g., it is positive). The next three elements indicate the fourth step, or

$$
\begin{aligned}
0 &\text{—1000 and 1001}\\
1 &\text{—1010}\\
2 &\text{—1011}\\
3 &\text{—1100}\\
\rightarrow 4 &\text{—1101}\\
5 &\text{—1110 etc.}
\end{aligned}
$$

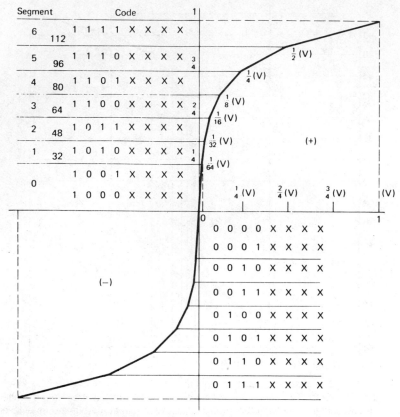

Figure 3.36 Quantization and coding used in the CEPT 30 + 2 PCM system.

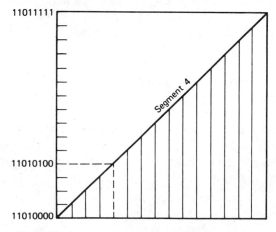

Figure 3.37 CEPT 30 + 2 PCM system, coding of segment 4 (positive).

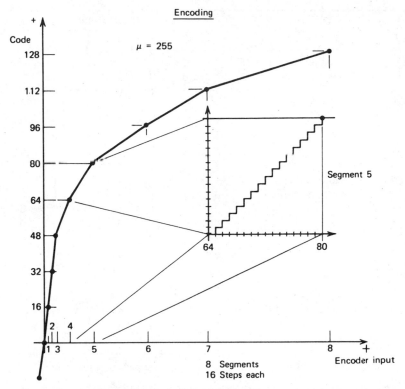

Figure 3.38   Positive portion of the segmented approximation of the μ-law quantizing curve used in the North American DS1 PCM channelizing equipment. (Courtesy of ITT Telecommunications, Raleigh, NC.)

Figure 3.37 shows a "blowup" of the uniform quantizing and subsequent straightforward binary coding of step 4; it illustrates final segment coding, which is uniform, providing 16 ($2^4 = 16$) coded quantum steps.

The North American DS1 PCM system uses a 15-segment approximation of the logarithmic μ-law. Again, there are actually 16 segments. The segments cutting the origin are collinear and are counted as 1 but are actually two, one above and one below the x-axis. This leads to the 16 value. The quantization in the DS1 system is shown in Figure 3.38 for the positive portion of the curve. Segment 5 representing quantizing steps 64–80 is shown blown up in the figure. Figure 3.39 shows the DS1 coding. As can be seen again, the first code element, whether a 1 or a 0, indicates if the quantum step is above or below the horizontal axis. The next three elements identify the segment, and the last four elements (bits) identify the actual quantum level inside that segment.

Code Level		Digit Number							
		1	2	3	4	5	6	7	8
255	(Peak positive level)	1	0	0	0	0	0	0	0
239		1	0	0	1	0	0	0	0
223		1	0	1	0	0	0	0	0
207		1	0	1	1	0	0	0	0
191		1	1	0	0	0	0	0	0
175		1	1	0	1	0	0	0	0
159		1	1	1	0	0	0	0	0
143		1	1	1	1	0	0	0	0
127	(Center levels)	1	1	1	1	1	1	1	1
126	(Nominal zero)	0	1	1	1	1	1	1	1
111		0	1	1	1	0	0	0	0
95		0	1	1	0	0	0	0	0
79		0	1	0	1	0	0	0	0
63		0	1	0	0	0	0	0	0
47		0	0	1	1	0	0	0	0
31		0	0	1	0	0	0	0	0
15		0	0	0	1	0	0	0	0
2		0	0	0	0	0	0	1	1
1		0	0	0	0	0	0	1	0
0	(Peak negative level)	0	0	0	0	0	0	1*	0

*One digit added to ensure that timing content of transmitted pattern is maintained.

Figure 3.39 Eight-level coding of the North American (AT&T) DS1 PCM system. Note that actually there are really only 255 quantizing steps because steps 0 and 1 use the same bit sequence, thus avoiding a code sequence with no transitions (i.e., 0's only). (From Ref. 16; courtesy of ITT Telecommunications, Raleigh, NC.)

3.3.2.5 The Concept of Frame

As shown in Figure 3.32, PCM multiplexing is carried out in the sampling process, sampling several sources sequentially. These sources may be nominal 4-kHz voice channels or other information sources, possibly data or video. The final result of the sampling and subsequent quantization and coding is a series of pulses, a serial bit stream that requires some indication or identification of the beginning of a scanning sequence. This identification tells the far-end receiver when each full sampling sequence starts and ends; it times the receiver. Such identification is called *framing*. A full sequence or cycle of samples is called a frame in PCM terminology.

CCITT Rec. G.702 defines a frame as

A set of consecutive digit time slots in which the position of each digit time slot can be identified by reference to a frame alignment signal. The frame alignment signal does not necessarily occur, in whole or in part, in each frame.

Consider the framing structure of several practical PCM systems. The AT&T D1* system is a 24-channel PCM system using a 7-level code (e.g., $2^7 = 128$ quantizing steps). To every 7 bits representing a coded quantum step, 1 bit is added for signaling. To the full sequence 1 bit is added, called a framing bit. Therefore a D1-frame consists of

$$(7 + 1) \times 24 + 1 = 193 \text{ bits}$$

making up a full sequence of frame. By definition 8000 frames are transmitted, so the bit rate is

$$193 \times 8000 = 1,544,000 \text{ bps}$$

The CEPT† 30 + 2 system is a 32-channel system where 30 channels transmit speech derived from incoming telephone trunks and the remaining two channels transmit signaling and synchronization information. Each channel is allotted a time slot, and we can speak of time slots 0–31 as follows:

Time Slot	Type of Information
0	Synchronizing (framing)
1–15	Speech
16	Signaling
17–31	Speech

In time slot 0 a synchronizing code or word is transmitted every second frame, occupying digits 2–8 as follows:

$$0011011$$

In those frames without the synchronizing word, the second bit of time slot 0 is frozen at a 1 so that in these frames the synchronizing word cannot be imitated. The remaining bits of time slot 0 can be used for the transmission of supervisory information signals.

The current North American basic 24-channel PCM system, typified by the AT&T D1D channel bank, varies compared with the older D1A system described above in that all 8 bits of a channel word are used in five out of six

---

*Now called D1A.
†CEPT = Conférénce Européenne des Postes et Télécommunications.

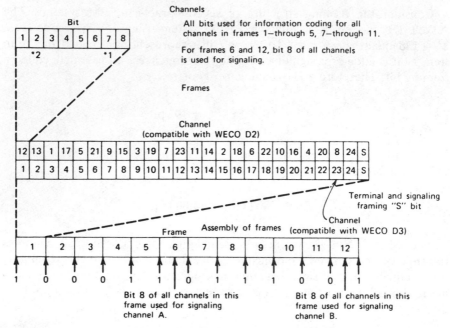

Figure 3.40 Frame structure of the North American (AT&T) D1D (DS1) PCM system for the channel bank. Note the bit "robbing" technique used on each sixth frame to provide signaling information. *Notes*: (1) If bits 1 – 6 and 8 are 0, then bit 7 is transmitted as 1. (2) Bit 2 is transmitted as 0 on all channels for transmission if end-to-end alarm. (3) Composite pattern 000110111001, etc. (Ref. 16, courtesy of ITT Telecommunications, Raleigh, NC.)

frames. In the remaining frame digit 8 is used for signaling. To accommodate these changes, it is necessary to change the framing format so that the specific frames containing signaling information can be identified. By using 8 bits instead of 7 for each channel word, allowing 256 amplitude values to be represented instead of 128 (in five out of six frames), quantizing noise is reduced. The companding characteristic is also different from its older D1A counterpart. D1D uses $\mu = 255$, whereas with D1A $\mu = 100$. This change made a significant difference in the signal-to-noise ratio over a wide range of input signals.

The D1D frame has a similar makeup as the D1A frame in that

$$8 \times 24 + 1 = 193 \text{ bits per frame}$$

producing a line data rate of $193 \times 8000 = 1.544$ Mbps. The frame structure is shown in Figure 3.40. Note that signaling is provided by "robbing" bit 8 from every channel in every sixth frame. For all other frames all bits are used to transmit information coding.

Framing and basic timing should be distinguished. Framing ensures that the PCM receiver is aligned regarding the beginning (and end) of a sequence or frame; timing refers to the synchronization of the receiver clock, specifically, that it is in step with its companion (far-end) transmit clock. Timing at the receiver is corrected via the incoming mark–space (and space–mark) transitions. It is important, then, that long periods without transitions do not occur. This point is discussed later in reference to line codes and digit inversion.

3.3.2.6 Details of and Enhancements to the DS1 Frame

A TDM demultiplexer has to identify individual time slot positions so that they can relate to the particular channel inputs of the associated multiplexer or channel bank. A typical demultiplexer uses a counter synchronized to the frame format of its companion multiplexer. There are at least five regimes available to carry out PCM frame alignment so that time slots can be appropriately identified. Two of these regimes are used in common PCM practice today. CEPT 30 + 2 uses *added channel framing*, as described in Section 3.3.2.5. The North American DS1 PCM system uses *added-bit framing* by means of a framing bit or *F-bit*. This is the 193rd pulse of a DS1 frame (i.e., the last pulse or bit). When superframe implementations are used, there are 12 frames in the conventional superframe and thus 12 framing bits that can be considered as forming a bit sequence 12 bits long (Ref. 28).

CCITT Rec. G.704 (Ref. 27) discusses allocation of *F*-bits using two distinct methods. The first uses a 24-frame multiframe (note the different semantics here), and the second, a 12-frame multiframe. The structure of these multiframes are shown in Tables 3.4 and 3.5, respectively. The reader will note that we identified three different frame alignment signals for the DS1 system. One was taken from Ref. 28, using "superframe" terminology; one was used for the 24-frame CCITT multiframe (001001 repeating); and one for the CCITT 12-frame multiframe using alternating 10's.

The structure of the 24-frame multiframe offers several enhancements. In the *F*-bit column, Table 3.4, under assignments, there are three headings: FAS, DL, and CRC. The FAS is the familiar frame alignment signal, in this case 001011. The DL means data link. Beginning with frame 1 of the multiframe, every other 193rd bit is part of a 4-kbps data link. The data link provides a communication path between primary hierarchical level terminals and contains data and idle data link sequence or a loss of frame alignment sequence. The loss of frame alignment sequence is 16 bits long, consisting of eight 1's and eight 0's.

The CRC (cyclic redundancy check) is a sequence transmitted at 2 kbps (see Chapter 12) for discussion of CRC error detection). The CRC generating polynomial is $X^6 + X + 1$. The CRC-6 message block (CMB) is a sequence of 4632 bits that is coincident with the multiframe.

Table 3.4 Multiframe structures for 24-frame multiframe

Multiframe Number	Multiframe Bit Number	F Bits			Bit Number(s) in each Channel Time Slot		Signaling Channel Designation
		Assignments					
		FAS[a]	DL[b]	CRC[c]	For Character Signal	For Signaling	
1	0	—	n	—	1–8	—	
2	193	—	—	e_1	1–8	—	
3	386	—	n	—	1–8	—	
4	579	0	—	—	1–8	—	
5	772	—	m	—	1–8	—	
6	965	—	—	e_2	1–7	8	A
7	1158	—	m	—	1–8	—	
8	1351	0	—	—	1–8	—	
9	1544	—	m	—	1–8	—	
10	1737	—	—	e_3	1–8	—	
11	1930	—	m	—	1–8	—	
12	2123	1	—	—	1–7	8	B
13	2316	—	m	—	1–8	—	
14	2509	—	—	e_4	1–8	—	
15	2702	—	m	—	1–8	—	
16	2895	0	—	—	1–8	—	
17	3088	—	m	—	1–8	—	
18	3231	—	—	e_5	1–7	8	C
19	3474	—	m	—	1–8	—	
20	3667	1	—	—	1–8	—	
21	3860	—	m	—	1–8	—	
22	4053	—	—	e_6	1–8	—	
23	4246	—	m	—	1–8	—	
24	4439	1	—	—	1–7	8	D

Source: From CCITT Rec. G.704, Table 1/G.704, Page 77, Blue Books, Vol. III, Fascicle III.4, IXth Plenary Assy, Melbourne, 1988.
[a]FAS: Frame alignment signal (. . . 001011 . . .).
[b]DL: 4 kbit/s data link (message bits m).
[c]CRC: CRC-6 block check field (check bits e_1 to e_6).

AT&T calls the 24-frame multiframe the *extended superframe format* (ESF). This format extends the basic superframe structure from 12 to 24 frames and divides the 8-kbps framing bit position pattern previously used for basic frame and robbed-bit signaling synchronization into a 2-kbps channel for CRC and a 4-kbps channel for terminal-to-terminal data link.

Reduction of the basic frame and robbed-bit synchronization rate to 2 kbps is made practical by improvements in integrated circuit technology. It is now feasible to use much more sophisticated frame-search strategies than were employed in the past, thereby permitting faster synchronization acquisition with fewer framing bits. Extension of the superframe to 24 frames also allows more signaling states to be defined.

Table 3.5 Multiframe structure for 12-frame multiframe

Frame Number	Frame Alignment Signal (See Note 1)	Multiframe Alignment Signal (S bit)	Bit Number(s) in Each Channel Time Slot		Signaling Channel Designation (See Note 2)
			For Character Signal	For Signaling	
1	1	—	1–8	—	
2	—	0	1–8	—	
3	0	—	1–8	—	
4	—	0	1–8	—	
5	1	—	1–8	—	
6	—	1	1–7	8	A
7	0	—	1–8	—	
8	—	1	1–8	—	
9	1	—	1–8	—	
10	—	1	1–8	—	
11	0	—	1–8	—	
12	—	0	1–7	8	B

Source: From CCITT Rec. G.704, Table 5/G.704, page 84, Blue Books, Vol. III, Fascicle II.4, IXth Plenary Assy., Melbourne, 1988.

Note 1: When the *S*-bit is modified to signal the alarm indications to the remote end, the *S*-bit in frame 12 is changed from 0 to 1.

Note 2: Channel associated signaling provides two independent 667-bit/s signaling channels designated A and B or one 1333 bit/s signaling channel.

The CRC-6 code has the ability to detect most errors that occur on the DS1 signal and can be used in various applications such as false-framing protection, protection switching, terminal-to-terminal performance monitoring, automatic restoration after alarms, line verification before, during, and after maintenance, and detection of errored seconds (Ref. 28).

The CEPT 30 + 2 system utilizes a similar approach with the 16-frame multiframe and a 4-kbps CRC procedure using a CRC-6 with a generating polynomial $X^4 + X + 1$.

Table 3.4 shows four different signaling bits, A, B, C, and D, and Table 3.5 shows two, A and B. This allows some latitude in supervisory signaling implementation. For example, if A and B are operated together, we derive a 1333 bps signaling channel and with A and B operating separately we derive two independent signaling channels each operating at 667 bps. In the case of the extended superframe (24-frame case), channels A, B, C, and D are available so more versatile signaling options can be offered. However, in most cases each signaling bit is associated with the channel from which it is robbed, giving that channel's idle or busy state.

3.3.2.7 Quantizing Distortion

Quantizing distortion has been defined as the difference between the signal waveform as presented to the PCM multiplex (codec*) and its equivalent quantized value. Quantizing distortion produces a signal-to-distortion ratio S/D given by (Ref. 4, equation 28-10, p. 615)

$$\frac{S}{D} = 6n + 1.8 \text{ dB (for uniform quantizing)}$$

where n = number of bits used to express a quantizing level. This bit grouping is often referred to as a PCM word. For instance, the AT&T D1A system uses a 7-bit code word to express a level, and the 30 + 2 and D1D systems use essentially 8 bits.

With a 7-bit code word (uniform quantizing),

$$\frac{S}{D} = 6 \times 7 + 1.8 = 43.8 \text{ dB}$$

This demonstrates the linear relationship between the number of digits per sample and the signal-to-distortion ratio in decibels. Each added digit increases the signal-to-distortion ratio by 6 dB. Practical signal-to-distortion values range on the order of 33–38 dB[†] for average talker levels.

3.3.2.8 Idle Channel Noise An idle PCM channel can be excited by the idle Gaussian noise, ac hum (50 or 60 Hz) and crosswalk present on the input channel. A decision threshold may be set that would control idle noise if it remains constant. With a constant-level input there will be no change in code word output, but any change of amplitude will cause a corresponding change in code word, and the effect of such noise may be an annoyance to the telephone listener, particularly during pauses in conversation.

One important overall PCM design issue to control idle channel noise is the selection of small signal quantizing values. Of course, these are the values near the origin on the segmented logarithmic companding curve. Typical curves are shown in Figures 3.34–3.36. The question arises as to how to bias these values to minimize idle channel noise. Biasing in the y-axis (up and down) is called mid-riser biasing; biasing in the x-axis (left and right) is called midtread biasing.

The length of quantizing segments is related to the power of 2. The 15-segment ($\mu = 255$ law) and the 13-segment ($A = 87.6$ law) both use the power of 2 relationship. For the μ-law we have a midtread codec with 8 segments of each polarity with a segment size increasing by a factor of 2 in successive segments. Step size at segment boundaries is adjusted so that the output levels always are halfway between decision thresholds.

---

*The term *codec*, meaning coder–decoder, is analogous to *modem* in analog circuits.
[†]Using eight-level coding.

In the case of the A-law codec we consider it to have 8 segments of each polarity where the two innermost segments of each are collinear, which leads to its name, a 13-segment codec. Again, the output levels are always midway between decision thresholds. But in this case small signal biasing values near the origin are twice as large as those for the 15-segment μ-law codec. These characteristics cause the 13-segment A-law codec to have poorer idle channel noise characteristics than its μ-law counterpart. On the other hand, the A-law codec exhibits flat signal-to-distortion performance over a somewhat wider range of signal levels.

Midtread quantization characteristics can give better quantization characteristics than midriser (Ref. 33).

CCITT Rec. G.712 states that with the input and output ports of the channel terminated in the nominal impedance, idle channel noise should not exceed -65 dBm0p.

3.3.3 Operation of a PCM Codec (Channel Bank)

PCM is four-wire. Voice channel inputs and outputs to and from a PCM multiplex are on a four-wire basis. The term *codec* is used to describe a unit of equipment carrying out the function of PCM multiplex and demultiplex and stands for coder–decoder, even though the equipment carries out more functions than just coding and decoding. A block diagram of a codec is shown in Figure 3.41.

A codec accepts 24 or 30 voice channels, depending on the system used, and digitizes and multiplexes the information. It delivers 1.544 Mbps to the line for the AT&T DS1 channelizing equipment and 2.048 Mbps for the CEPT 30 + 2 (European) channelizing equipment. On the decoder side it accepts a serial bit stream at one or the other line modulation rate, demultiplexes the digital information, and performs digital-to-analog conversion. Output to the analog telephone network is 24 or 30 nominal 4-kHz voice channels. Figure 3.41 illustrates the processing of a single analog voice channel through a codec. The voice channel to be transmitted is passed through a 3.4-kHz low-pass filter. The output of the filter is fed to a sampling circuit. The sample of each channel of a set of n channels (n usually equals 24 or 30) is released in turn to the pulse amplitude modulation (PAM) highway. The release of samples is under control of a channel gating pulse derived from the transmit clock. The input to the actual coder is the PAM highway. The coder accepts a sample of each channel in sequence and then generates the appropriate signal character (channel word) corresponding to each channel presented. The coder output is the basic PCM signal that is fed to the digit combiner, where framing alignment signals are inserted in the appropriate time slots, as well as the necessary supervisory signaling digits corresponding to each channel (European approach). The output of the digit combiner is the serial PCM bit stream fed to the line. As mentioned above,

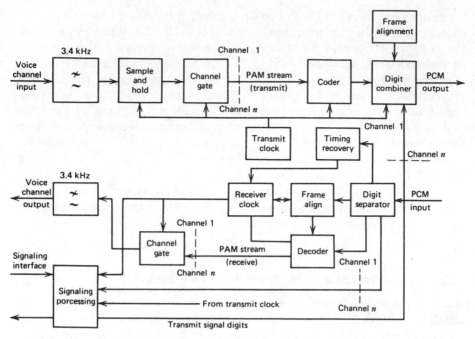

Figure 3.41   Simplified functional block diagram of a typical PCM codec.

supervisory signaling is carried out somewhat differently in the North American (AT&T) approach by robbing 1 bit in frame 6 and in frame 12 for this purpose.

On the receive side the codec accepts the serial PCM bit stream at the line rate at the digit separator. Here the signal is regenerated and split four ways to carry out the following processing functions: (1) timing recovery, (2) decoding, (3) frame alignment, and (4) signaling (supervisory). The timing recovery keeps the receive clock in synchronism with the far-end transmit clock. The receive clock provides the necessary gating pulses for the receive side of the PCM codec. The frame alignment circuit senses the presence of the frame alignment signal at the correct time interval, thus providing the receive terminal with frame alignment. The decoder, under control of the receive clock, decodes the code character (channel word) signals corresponding to each channel. The output of the decoder is comprised of the reconstituted pulses making up a PAM highway. The channel gate accepts the PAM highway, gating the n-channel PAM highway in sequence under control of the receive clock. The output of the channel gate is fed in turn to each channel filter, therefore enabling the reconstituted analog voice signal to reach each appropriate voice path. Gating pulses extract signaling information in the signaling processor and apply this information to each of the reconstituted voice channels with the supervisory signaling interface as required by the analog telephone system in question.

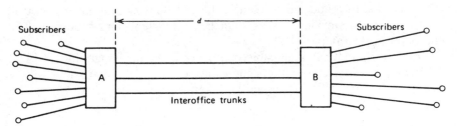

Figure 3.42 Simplified application diagram of PCM as applied to interoffice (interexchange) plant. A and B are switching centers.

3.3.4 Practical Application

3.3.4.1 General PCM found early application in expanding interoffice trunks (junctions) that have reached exhaust* or will reach exhaust in the near future. An interoffice trunk is one pair of a circuit group connecting two switching points (exchanges). Figure 3.42 sketches the interoffice trunk concept. Depending on the particular application, at some point where distance d is exceeded, it will be more economical to install PCM on the existing VF cable plant than to rip up streets and add more VF cable pairs. For the planning engineer, the distance d, where PCM becomes an economic alternative, is called the prove-in distance. d may vary from 8 to 16 km (5–10 mi), depending on the location and other circumstances. For distances less than d, additional VF cables pairs should be used for expanding the plant.

The general rule for measuring the expansion capacity of a given VF cable is as follows:

- For AT&T DS1 channelizing equipment, two VF pairs will carry 24 PCM channels.
- For CEPT 30 + 2 system as configured by ITT, two VF pairs plus a phantom pair will carry 30 PCM speech channels.

All pairs in a VF cable may not necessarily be usable for PCM transmission. One restriction is brought about by the possibility of excessive crosstalk between PCM carrying pairs. The effect of high crosstalk levels is to introduce digital errors in the PCM bit stream. The error rate may be related on a statistical basis to crosstalk, which in turn is dependent on the characteristics of the cable and the number of PCM carrying pairs.

Exhaust is an outside plant term meaning that the useful pairs of a cable have been used up (assigned) from a planning point of view.

One method to reduce crosstalk and thereby increase VF pair usage is to turn to two-cable working, rather than have the "go" and "return" PCM cable pairs in the same cable.

Another item that can limit cable pair usage is the incompatibility of FDM and PCM carrier systems in the same cable. On the cable pairs that will be used for PCM, the following should be taken into consideration:

- All load coils must be removed.
- Build-out networks and bridged taps must also be removed.
- No crosses, grounds, splits, high-resistance splices, nor moisture is permitted.

The frequency response of the pair should be measured out to 1 MHz and taken into consideration as far out as 2.5 MHz. Insulation should be checked with a megger. A pulse reflection test using a radar test set is also recommended. Such a test will indicate open circuits, short circuits, and high-impedance mismatches. A resistance test and balance test using a Wheatstone bridge may also be in order. Some special PCM test sets are available, such as the Lenkurt Electric 91100 PCM cable test set using pseudorandom PCM test signals and the conventional digital test eye pattern.

3.3.4.2 *Practical System Block Diagram*

A block diagram showing the elemental blocks of a PCM transmission link used to expand installed VF cable capacity is shown in Figure 3.43. Most telephone administrations (companies) distinguish between the terminal area of a PCM system and the repeatered line. The term *span* comes into play here. A span line is composed of a number of repeater sections permanently connected in tandem at repeater apparatus cases mounted in manholes or on pole lines along the span. A span is defined as the group of span lines that extends between two office (switching center) repeater points.

A typical span is shown in Figure 3.43. The spacing between regenerative repeaters is important. Section 3.3.4.1 mentioned the necessity of removing load coils from those trunk (junction) cable pairs that are to be used for PCM transmission. It is at these load points that the PCM regenerative repeaters are installed. On a VF line with H-type loading (see Section 2.8.4), spacing between load points is normally about 6000 ft (1830 m). It will be remembered from Chapter 2 that the first load coil out from the exchange on a trunk pairs is at half-distance, or 3000 ft (915 m). This is provident, for a regenerative repeater must also be installed at this point. This spacing is shown in Figure 3.43 (1 space = 1000 ft). The purpose of installing a repeater at this location is to increase the pulse level before entering the environment of an exchange area, where the levels of impulse noise may be quite high. High levels of impulse noise induced into the system may cause significant

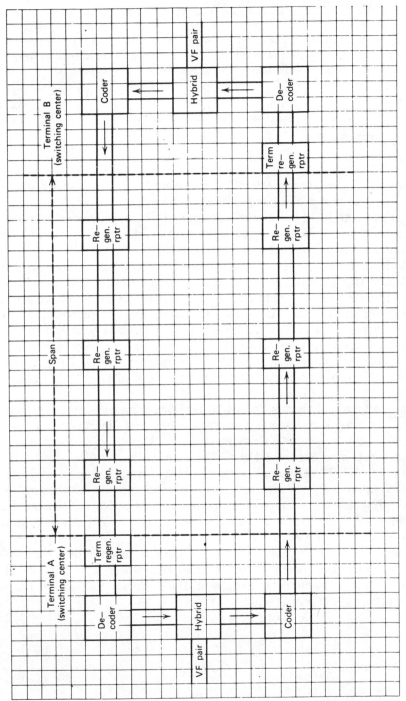

Figure 3.43 Simplified functional block diagram of a PCM link used to expand capacity of an existing VF cable. For simplicity, interface with only one VF pair is shown. Note the spacing between repeaters in the span line.

Table 3.6 Line parameters for ITT – BTM PCM

Pair Diameter (mm)	Loop Attenuation at 1 MHz (dB/km)	Loop Resistance (Ω/km)	Voltage Drop (V/km)	Maximum Distance[a] (km)	Total Number of Repeaters	Maximum System Distance (km)
0.9	12	60	1.5	3	18	54
0.6	16	100	2.6	2.25	16	36

[a]Between adjacent repeaters.

increases in the digital error rate of the incoming PCM bit streams, especially when the bit stream is of a comparatively low level.

Commonly the PCM pulse amplitude output of a regenerative repeater is on the order of 3 V. Likewise, 3 V is the voltage on the PCM line crossconnect field at the exchange (terminal area).

A guideline used by Bell Telephone Manufacturing (BTM) Company (Belgium) is that the maximum distance separating regenerative repeaters is that corresponding to a cable pair attenuation of 36 dB at the maximum expected temperature at 1024 kHz. This frequency is equivalent to the half-bit rate for the CEPT systems (e.g., 2048 kbs). Actually, repeater design permits operation on lines with attenuations anywhere from 4 to 36 dB, allowing considerable leeway in placing repeater points. Table 3.6 gives some other practical repeater spacing parameters for the CEPT–ITT–BTM 30 + 2 system.

The maximum distance is limited by the maximum number of repeaters, which in this case is a function of power feeding and supervisory considerations. For instance, the fault location system can handle up to a maximum of 18 tandem repeaters for the BTM (ITT) configuration.

Power for the BTM system is via a constant-current feeding arrangement over a phantom pair serving both the "go" and the related "return" repeaters, providing up to 150-V dc at the power feed point. The voltage drop per regenerative repeater is 5.1 V. Thus for a "go" and "return" repeater configuration, the drop is 10.2 V.

As an example, let us determine the maximum number of regenerative repeaters in tandem that may be fed from one power feed point by this system using 0.8-mm diameter pairs with a 3-V voltage drop in an 1830-m spacing between adjacent repeaters:

$$\frac{150}{(10.2 + 3)} = 11$$

Assuming power fed from both ends and an 1800-m "dead" section in the middle, the maximum distance between power feed points is approximately

$$(2 \times 11 + 1)1.8 \text{ km} = 41.4 \text{ km}$$

Fault tracing for the North American (AT&T) T1 system is carried out by means of monitoring the framing signal, the 193rd bit (Section 3.3.2.5). The framing signal (amplified) normally holds a relay closed when the system is operative. With loss of the framing signal, the relay opens, actuating alarms. By this means a faulty system is identified, isolated, and dropped from "traffic."

To locate a defective regenerator on the BTM(Belgium)–CEPT system, traffic is removed from the system, and a special pattern generator is connected to the line. The pattern generator transmits a digital pattern with the same bit rate as the 30 + 2 PCM signal, but the test pattern can be varied to contain selected low-frequency spectral elements. Each regenerator on the repeatered line is equipped with a special audio filter, each with a distinctive passband. Up to 18 different filters may be provided in a system. The filter is bridged across the output of the regenerator, sampling the output pattern. The output of the filter is amplified and transformer-coupled to a fault transmission pair, which is normally common to all PCM systems on the route, span, or section.

To determine which regenerator is faulty, the special test pattern is tuned over the spectrum of interest. As the pattern is tuned through the frequency of the distinct filter of each operative repeater, a return signal will derive from the fault transmission pair at a minimum specified level. Defective repeaters will be identified by the absence of a return signal or a return level under specification. The distinctive spectral content of the return signal is indicative of the regenerator undergoing test.

3.3.4.3 The Line Code

PCM signals as transmitted to the cable are in the bipolar mode (biternary), as shown in Figure 3.30. The marks or 1's have only a 50% duty cycle. There are several advantages to this mode of transmission:

- No dc return is required; hence transformer coupling can be used on the line.
- The power spectrum of the transmitted signal is centered at a frequency equivalent to half the bit rate.

It will be noted in bipolar transmission that the 0's are coded as absence of pulses, and the 1's are alternately coded as positive and negative pulses, with the alternation taking place at every occurrence of a 1. This mode of transmission is also called alternate mark inversion (AMI).

One drawback to straightforward AMI transmission is that when a long string of 0's is transmitted (e.g., no transitions), a timing problem may come

Table 3.7   Error rate of a binary transmission system versus signal-to-RMS noise ratio

Error Rate	S/N (dB)	Error Rate	S/N (dB)
10^{-2}	13.5	10^{-7}	20.3
10^{-3}	16	10^{-8}	21
10^{-4}	17.5	10^{-9}	21.6
10^{-5}	18.7	10^{-10}	22
10^{-6}	19.6	10^{-11}	22.2

about because repeaters and decoders have no way of extracting timing without transitions. The problem can be alleviated by forbidding long strings of 0's. Codes have been developed that are bipolar but with N zeros substitution; they are called BNZS codes. For, instance, a B6ZS code substitutes a particular signal for a string of six 0's.

Another such code is the HDB3 code (high-density binary 3), where the 3 indicates that it substitutes for binary formations with more than three consecutive 0's. With HDB3 the second and third zeros of the string are transmitted unchanged. The fourth 0 is transmitted to the line with the same polarity as the previous mark sent, which is a "violation" of the AMI concept. The first 0 may or may not be modified to a 1 to assure that the successive violations are of opposite polarity.

3.3.4.4   Signal-to-Gaussian-Noise Ratio on PCM Repeatered Lines

As we mentioned earlier, noise accumulation on PCM systems is not an important consideration. This does not mean that Gaussian noise (nor crosstalk, impulse noise) is not important. Indeed, it does affect the error performance, expressed as error rate (see Chapter 12). The error rate, from one point of view, is cumulative. A decision in error, whether 1 or 0, made anywhere in the digital system, is not recoverable.* Thus, such an incorrect decision made by one regenerative repeater adds to the existing error rate on the line, and errors taking place in subsequent repeaters further down the line add in a cumulative manner, tending to deteriorate the received signal.

In a purely binary transmission system, if a 20-dB signal-to-noise ratio is maintained, the system operates nearly error free. In this respect, consider Table 3.7.

As discussed in Section 3.3.4.3, PCM, in practice, is transmitted online with alternate mark inversion. The marks have a 50% duty cycle, permitting energy concentration at a frequency of half the transmitted bit rate. Consequently, it is advisable to add 1 or 2 dB to the values shown in Table 3.7 to achieve a desired error rate on a practical system.

---

*Unless some special form of coding is used to correct the errors (see Chapters 6 and 12).

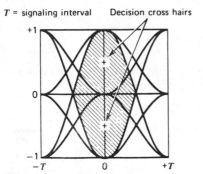

_T_ = signaling interval   Decision cross hairs

Figure 3.44 Sketch of an eye pattern.

3.3.4.5 The Eye Pattern The "eye" pattern provides a convenient method of checking the quality of a digital transmission line. A sketch of a typical eye pattern is shown in Figure 3.44. Any oscilloscope can produce a suitable eye pattern provided it has the proper rise time, which most quality oscilloscopes now available on the market have. The oscilloscope should either terminate or bridge the repeatered line or output of a terminal repeater. The display on the oscilloscope contains all the incoming bipolar pulses superimposed on one another.

Eye patterns are indicative of decision levels. The wider the eye opening vertically the better defined is the decision (whether 1 or 0 in the case of PCM). The opening is often referred to as the decision area (crosshatched in Figure 3.44). Degradations reduce the area. Eye patterns are often measured off in the vertical, giving a relative measure of the margin of decision.

Amplitude degradations shrink the eye in the vertical. Among amplitude degradations can be included echos, intersymbol interference, and decision threshold uncertainties.

Horizontal shrinkage of the eye pattern is indicative of timing degradations (i.e., jitter and decision time misalignment).

Noise is the other degradation to be considered. Usually noise may be expressed in terms of some improvement in the signal-to-noise ratio to bring the operating system into the bounds of some desired objective (see Table 3.7, for example). This ratio may be expressed as $20 \times \log^*$ of the ideal eye opening (in the vertical as read on the oscilloscope's vertical scale) to the degraded reading.

3.3.4.6 Regenerative Repeaters As we probably know, pulses passing down a digital transmission line suffer attenuation and are badly distorted by the frequency characteristic of the line. A regenerative repeater amplifies and reconstructs such a badly distorted digital signal and develops a nearly

---

*Oscilloscopes are commonly used to measure voltage. Thus we can measure the degraded opening in voltage units and compare it to the full-scale perfect opening in the same units. A ratio is developed, and we take 20 log that ratio to determine the signal-to-noise ratio.

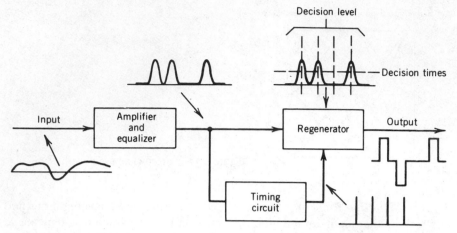

Figure 3.45 Simplified functional block diagram of a regenerative repeater for use on PCM cable systems.

perfect replica of the original at its output. Regenerative repeaters are an essential key to digital transmission in that we could say that the "noise stops at the repeater."

Figure 3.45 is a simplified block diagram of a regenerative repeater and shows typical waveforms corresponding to each functional stage of signal processing. As shown in the figure, the first stage of signal processing is amplification and equalization. Equalization is often a two-step process. The first is a fixed equalizer that compensates for the attenuation–frequency characteristic of the nominal section, which is the standard length of transmission line between repeaters (often 6000 ft). The second equalizer is variable and compensates for departures between nominal repeater section length and the actual length and loss variations due to temperature. The adjustable equalizer uses automatic line build-out (ALBO) networks that are automatically adjusted according to characteristics of the received signal.

The signal output of the repeater must be accurately timed to maintain accurate pulsewidth and space between the pulses. The timing is derived from the incoming bit stream. The incoming signal is rectified and clipped, producing square waves that are applied to the timing extractor, which is a circuit tuned to the timing frequency. The output of the circuit controls a clock-pulse generator that produces an output of narrow pulses that are alternately positive and negative at the zero crossings of the square wave input.

The narrow positive clock pulses gate the incoming pulses of the regenerator, and the negative pulses are used to run off the regenerator. Thus the combination is used to control the width of the regenerated pulses.

Regenerative repeaters are the major source of timing jitter in a digital transmission system. Jitter is one of the principal impairments in a digital network, giving rise to pulse distortion and intersymbol interference. Jitter is discussed in more detail in Section 3.3.7.2.

Most regenerative repeaters transmit a bipolar (AMI) waveform (see Figure 3.30). Such signals can have one of three possible states in any instant in time, positive, zero, or negative, and are often designated +, 0, −. The threshold circuits are gated to admit the signal at the middle of the pulse interval. For example, if the signal is positive and exceeds a positive threshold, it is recognized as a positive pulse. If it is negative and exceeds a negative threshold, it is recognized as a negative pulse. If it has a value between the positive and negative thresholds, it is recognized as a 0 (no pulse).

3.3.5 Higher Order PCM Multiplex Systems

In Section 3.2 an FDM multiplex hierarchy was developed based on the 12-channel group. Five such groups were formed into a supergroup, thence the mastergroup and the supermastergroup. Likewise, in PCM a hierarchy of multiplex is developed based on the 24- or 30-channel group, which is called level 1. Subsequent levels are then developed (i.e., levels 2, 3, 4, and in one system, level 5). Table 3.8 summarizes and compares these multiplex levels for the North American system, Japan, and Europe (based on CCITT). The North American PCM hierarchy is shown in Figure 3.46, giving respective DS line rates and multiplex nomenclature. Regarding this nomenclature, we see from the figure that M12 accepts 1-level input, delivering 2-level to the line. It actually accepts four DS1 inputs, deriving a DS2 output (6.312 Mbps). M13 accepts 1-level inputs, delivering 3-level to the line. In this case 28 DS1 inputs form one DS3 output (44.736 Mbps); the M34 takes six DS3 inputs (level 3) to form one DS4 line rate (274.176 Mbps). DS1C is a special case where two 1.544-Mbps DS1 rates are multiplexed to form a 48-channel group with a line rate of 3.152 Mbps.

Table 3.8 PCM multiplex hierarchy comparison

	Level				
System Type	1	2	3	4	5
North American AT&T type	1	2	3	4	
Number of voice channels	24	96	672	4032	
Line bit rate (Mbps)	1.544	6.312	44.736	274.176	
Japan					
Number of voice channels	24	96	480	1440	5760
Line bit rate (Mbps)	1.544	6.312	32.064	97.728	400.352
Europe					
Number of voice channels	30	120	480	1920	7680
Line bit rate (Mbps)	2.048	8.448	34.368	139.264	564.992

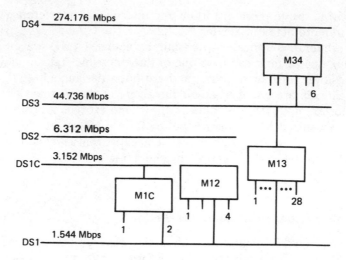

Figure 3.46 North American (AT&T) PCM hierarchy and multiplexing plan.

By simple multiplication we can see that the higher order line rate is a multiple of the lower input rate plus some number of bits. The DS1C is an example. Here the line rate is 3152 kbps, which is 2×1544 kbps + 64 kbps. The additional 64 kbps are used for multiplex synchronization and framing. Multiplex (and demultiplex) timing is very important, as one might imagine. The two DS1 signal inputs are each 1.544 Mbps plus and minus some tolerance (actually specified as ± 130 ppm). The two input signals must be made alike in repetition rate and a rate suitable for multiplexing. This is done by *bit stuffing*. In this process time slots are added to each signal in sufficient quantity to make the signal operate at a precise rate controlled by a common clock circuit in the multiplex. Pulses are inserted (or stuffed) into these time slots but carry no information. Thus it is necessary to code the signal in such a manner that these noninformation bits can be recognized and removed at the receiving terminal (demultiplex).

Of course, in the above example we are dealing with two DS1 sources that are physically separate and controlled by different clocks. If the sources were colocated (no difference in arrival due to delay) and controlled by a common clock, bit stuffing would not be necessary. However, codecs usually operate with free-running clocks (see Table 3.10, tolerance). Suppose we had two DS1 tributary codecs multiplexed by a M1C multiplexer (Figure 3.46) and each codec operated at maximum tolerance on the high side (i.e., 1.544×130 ppm $\times 2$), which is 401 bps in excess of 2×1.544 bps. If the output of the M1C were just 2×1.544 Mbps, its buffers would be accumulating 401 bps more than would be released to the multiplexed line. In some time period, depending on buffer size, the buffers would overflow and a frame would have to be dropped or "slipped" periodically. Such loss of data called a "slip," which is a highly undesirable impairment. This then is one cogent reason why PCM multiplex line rates must be greater than the aggregate nominal input rates. These added bits to compensate for input tributary variation are called "stuff" bits.

Consider the more general case using CCITT terminology. If we wish to multiplex several lower level PCM bit streams deriving from separate tributaries into a single PCM bit stream at a higher level, a process of *justification* is required (called *bit stuffing* above). CCITT defines justification (Rec. G.702)

> as a process of changing the rate of a digital signal in a controlled manner so that it can accord with a rate different from its own inherent rate usually without loss of information.

Positive justification (as above) adds or stuffs digits; negative justification deletes or slips digits.

In the case of positive justification, normally each separate tributary bit stream is read into a store at its own data rate t, but the store is read out at a rate corresponding to T/n, where T = rate of the multiplex equipment and n = number of tributary signals being multiplexed. T/n is selected relative to t so that $T/n > t$ with a sufficient margin to accommodate the difference in relative data rates of the multiplex and input tributary signals and also to allow for the addition of frame alignment and other service digits.

Under normal operational conditions there will be variations between T/n and t. To provide for these variations, the sequence of time slots at the output of each tributary store has available in it certain designated time slots known as *justifiable digit time slots*. These occur at fixed intervals, and the state of the store fill determines whether or not the justifiable digit time slot has information written into it from the store. If the store is filling, the time slot is used; if not, the time slot is ignored. By this means, a degree of "elasticity" is acquired that enables the relative timing difference to be absorbed.

Figure 3.47*a* is a simplified functional block diagram of the AT&T M12 multiplex, and Figure 3.47*b* of a higher level multiplex scheme.

3.3.5.1 Second-Level Frame Structures

The organization of the DS2 bit stream is shown in Figure 3.48. DS2 is transmitted on line at 6.312 Mbps (equivalent of 96 VF channels) and consists of four DS1 bit streams at 1.544 Mbps multiplexed, plus synchronization, framing, and stuff (justification) bits. All of the control information for the far-end demultiplexer is carried within an 1176-bit frame which is divided into four 294-bit subframes. The control-bit word, disbursed throughout the frame, begins with an M bit, as shown in Figure 3.48. The four M bits are transmitted as $011X$; the fourth bit X may be a 1 or a 0. This bit may be used as an alarm where 0 indicates an alarm

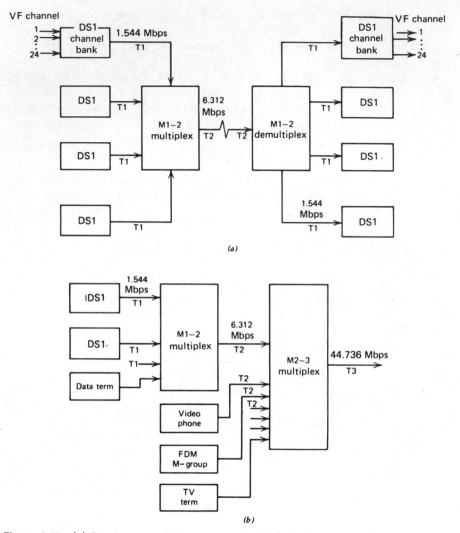

Figure 3.47 (a) Development of the 96-channel DS2 (AT&T) system by multiplexing four 24-channel DS1 channel bank outputs. (b) Development of higher order PCM (AT&T plan). (From Ref. 4; copyright © 1970 by Bell Telephone Laboratories from 1970 edition reprinted with permission.)

condition and 1 no alarm. The 011 sequence for the first three M bits identifies (formats) the frame.

Within each of the four subframes two other control sequences are used. Each control bit is followed by a 48-bit block of information of which 12 bits are taken from each of the four DS1 input signals. These are interleaved sequentially in the 48-bit block. The first bit in the third and sixth blocks is designated an F bit. The F bits are the sequence 0101 and are used to

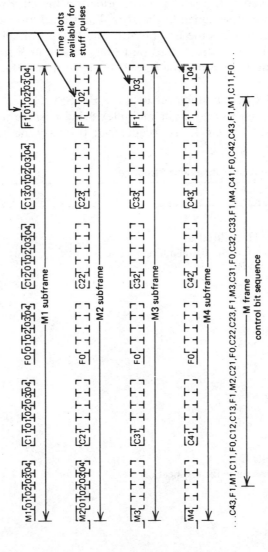

Figure 3.48 Organization of the DS2 signal bit stream. (From Ref. 13, vol. 2; copyright © 1977 by American Telephone and Telegraph Company.)

identify the location of the control bit sequences and the start of each information block.

The stuff (justification) control bits are transmitted at the beginning of each of the 48-bit blocks numbered 2, 4, and 5 within each subframe. These control bits are designated C in Figure 3.48. When a sequence is 000, no stuff pulse is present; when 111, a stuff pulse is added in the stuff position. The stuff bit positions are all assigned to the sixth 48-bit block of each subframe. In subframe 1 the stuff bit is the first bit after the F1 bit; for subframe 2 it is the second bit after the F1 bit, and so on through the fourth subframe. The nominal stuffing rate is 1796 bps for each DS1 signal input; the maximum is 5367 bps.

The output of the M12 multiplexer is in the B6ZS format.

The basic European second-order multiplex using positive justification is described in CCITT Rec. G.742. It accepts four tributary inputs, each from

Table 3.9   8448-kbps Multiplexing Frame Structure

Frame Structure	Bit Number
	Set I
Frame alignment signal (1111010000)	1–10
Alarm indication to the remote digital multiplex equipment	11
Bit reserved for national use	12
Bits from tributaries	13–212
	Set II
Justification control bits $C_{j1}{}^a$	1–4
Bits from tributaries	5–212
	Set III
Justification control bits $C_{j2}{}^a$	1–4
Bits from tributaries	5–212
	Set IV
Justification control bits $C_{j3}{}^a$	1–4
Bits from tributaries available for justification	5–8
Bits from tributaries	9–212
Tributary bit rate	2048 kbps
Number of tributaries	4
Frame length	848 bits
Bits per tributary	206 bits
Maximum justification rate per tributary	10 kbps
Nominal justification ratio	0.424

Source: CCITT Rec. G.742. Courtesy of ITU–CCITT.

[a] C_{ji} indicates the ith justification control bit of the jth tributary.

Table 3.10   AT&T DS series line rates, tolerances, and line codes (Format)

Signal	Repetition Rate (Mbps)	Tolerance (ppm)[a]	Format	Duty Cycle (%)
DS0	0.064	[b]	Bipolar	100
DS1	1.544	± 130	Bipolar	50
DS1C	3.152	± 30	Bipolar	50
DS2	6.312	± 30	B6ZS	50
DS3	44.736	± 20	B3ZS	50
DS4	274.176	± 10	Polar	100

[a]Parts per million.
[b]Expressed in terms of slip rate.

the standard CEPT30 + 2 PCM channel bank which has a nominal line data rate of 2.048 Mbps. The output of the multiplex is a nominal 8.448 Mbps on line, an equivalent of 4×30 or 120 VF channels. The multiplex frame structure is described in Table 3.9. CCITT Rec. G.742 recommends cyclic bit interleaving in the tributary numbering order and positive justification. Justification should be distributed using C_{jn} bits ($n = 1, 2, 3$). Positive justification is indicated by the signal 111 (stuffing), no stuffing by 000. Two bits per frame are available as service digits. Bit 11 of set I is used to transmit an alarm indication to the remote multiplex equipment. Faults indicated by service digits may be power supply failure, loss of tributary input (2.048 Mbps), loss of incoming 8.448-Mbps line signal, and loss of frame alignment.

3.3.5.2   Line Rates and Codes

Table 3.10 summarizes the North American DS series of PCM multiplex line rates, tolerances of these rates, and line codes. CCITT Rec. G.703 deals with line interfaces. For the CEPT30 + 2 PCM and higher order multiplex derived therefrom, Table 3.11 provides a similar summary as Table 3.10.

Table 3.11   Summary of CEPT 30 + 2 related line rates and codes

Level	Line Data Rate (Mbps)	Tolerance (ppm)	Code	Mark Peak Voltage (V)
1	2.048	± 50	HDB3	2.37 or 3[a]
2	8.448	± 30	HDB3	2.37
3	34.368	± 20	HDB3	1.0
4	139.264	± 15	CMI[b]	1 ± 0.1 V[c]

[a]2.37 V on coaxial pair; 3 V on symmetric wire pair.
[b]Coded mark inversion.
[c]Peak-to-peak voltage.

3.3.6 Transmission of Data Using PCM, as Exemplified by the AT&T DDS

The AT&T digital data system (DDS) provides duplex point-to-point and multipoint private line digital data transmission at a number of synchronous data signal rates. This system is based on the standard 1.544-Mbps DS1 PCM line rate, where individual bit streams have data rates that are submultiples of that line rate (i.e., based on 64 kbps). However, pulse slots are reserved for identification in the demultiplexing of individual user bit streams as well as for certain status and control signals and to ensure that sufficient line pulses are transmitted for receive clock recovery and pulse regeneration. The maximum data rate available to a subscriber to the system is 56 kbps, some 87.5% of the 64-kbps theoretical maximum.

The 1.544-Mbps line signal as applied to DDS service consists of 24 sequential 8-bit words (i.e., channel time slots) plus one additional framing bit. This entire sequence is repeated 8000 times a second. Note that again we have $(192 + 1)8000 = 1.544$ Mbps, where the value 192 is 8×24 (see Section 3.3.2.5). Thus the line rate of a DDS facility is compatible with the DS1 (T1) PCM line rate and offers the advantage of allowing a mix of voice (PCM) and data where the full dedication of a DS1 facility to data transmission would be inefficient.

AT&T calls the basic 8-bit word a *byte*. One bit of each 8-bit word is reserved for network control and for stuffing to meet nominal line bit rate requirements. This control bit is called a *C*-bit. With the *C*-bit removed we

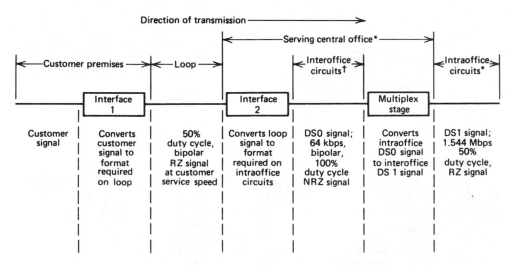

Figure 3.49 Subhierarchy of DDS signals. *Note*: Inverse processing must be provided for the opposite direction of transmission. Four-wire transmission is used throughout. *Exchange; †PCM trunk. (Copyright © 1977 by American Telephone and Telegraph Company.)

see where the standard channel bit rate is derived, namely, 56 kbps or
8000 × 7. Three subrates or submultiple data rates are also available: 2.4,
4.8, and 9.6 kbps. However, when these rates are implemented, an additional
bit must be robbed from the basic byte to establish flag patterns to route each
subrate channel to its proper demultiplexer port. This allows only 48 kbps out
of the original 64 kbps for the transmission of user data. The 48-kbps
composite total may be divided down to five 9.6-kbps channels, or ten
4.8-kbps channels, or twenty 2.4-kbps channels. The subhierarchy of DDS
signals is shown in Figure 3.49.

3.3.7 Long-Distance (Toll) Transmission by PCM

3.3.7.1 General   PCM, with its capability of regeneration, essentially eliminating the accumulation of noise as a signal traverses its transmission media,
would appear to be the choice for toll transmission or backbone long-haul
routes. One must consider the disadvantages of PCM as well. Most important
is the competition with FDM systems, the L5 system, for instance (Table 3.3).
AT&T's L5 system provides 10,800 VF channel capacity over long-haul
coaxial cable media. The required bandwidth for this capacity on the cable is
60 MHz. To transmit the same number of channels by PCM would require on
the order of 16 times the bandwidth.

Keep in mind the relationship briefly covered in Section 3.3.1 wherein this
16-multiple concept is shown: a 4-kHz voice channel requires an equivalent
PCM bandwidth of 64 kHz, assuming 1-Hz bandwidth per PCM bit transmitted. Therefore a 10,800 VF (4-kHz) channel configurations would require, if
transmitted by PCM, about 691.2 MHz. The available bandwidth is still at a
premium, whether by wire cable or radio.

PCM at the DS1 line rate in the T1 repeatered line is capable of up to
200-mi (320-km) routes. In this case regenerative repeaters are spaced about
1 statute mi (1.6 km) apart. The transmission medium is a wire pair for each
direction in standard multipair telephone cable. Powering points are at 36-mi
(58-km) intervals. The AT&T T2 system designed for the DS2 line rate
(equivalent to 96 VF channels) may be used for distances up to 500 mi (800
km) in length. T2 requires special low-capacitance wire pairs in separate
cables for opposite directions of transmission, crosstalk being a major system
design consideration. Nominal repeater spacing is 15,000 ft (4570 m).

The AT&T T4M system is designed for the DS4 line rate, 274.176 Mbps,
equivalent to 4032 VF channels, over 0.375-in. (9.5-mm) coaxial cable. T4M
repeaters are spaced up to 5700 ft (1738 m) apart, and systems can work up
to 500 mi (800 km) in length.

PCM on fiber optic cables, on both local trunk and toll (long-distance)
routes, is continuing to be implemented. With fiber optics, equivalent bandwidth is provided more economically, this being the major restraint covered
above. The implementation of digital transmission on glass fiber cable is

being spurred forward by the ever-increasing number of digital exchanges being installed requiring good-quality wideband landline trunks. One such high-capacity long-distance route in the United States is AT&T eastern corridor route from Boston via New York City to Washington, DC. Many transcontinential PCM optical fiber systems are now operational. For one thing, optical fiber can provide the bandwidth relatively cheaply; another advantage is that a fraction of the repeaters are required per 100 km (60 mi), easing the jitter buildup problem.

3.3.7.2 *Jitter* There is one other important limitation of present-day technology on using PCM as a vehicle for long-haul transmission. This is jitter, more particularly, timing jitter.

A general definition of jitter is "the movement of zero crossings of a signal (digital or analog) from their expected time of occurrence." In Chapter 12 it is called unwanted phase modulation or incidental FM. Such jitter or phase jitter affects the decision process or the zero crossing in a digital data modem. Much of this sort of jitter can be traced to the intervening FDM equipment between one end of a data circuit and the other.

PCM has no intervening FDM equipment, and jitter in PCM systems takes on different characteristics. However, essentially the effect is the same—uncertainty in a decision circuit as to when a zero crossing (transition) takes place, or the shifting of a zero crossing from its proper location. In PCM it is more proper to refer to jitter as timing jitter.

The primary source of timing jitter is the regenerative repeater. In the repeatered line, jitter may be systematic or nonsystematic. Systematic jitter may be caused by offset pulses (i.e., where the pulse peak does not coincide with regenerator timing peaks, or transitions are offset), intersymbol interferences (dependent on specific pulse patterns), and local clock threshold offset. Nonsystematic jitter may be traced to timing variations from repeater to repeater and to crosstalk.

In long chains of regenerative repeaters, systematic jitter is predominant and cumulative, increasing in rms value as $N^{1/2}$, where N = number of repeaters in the chain. Jitter is also proportional to a repeater's timing filter bandwidth. Increasing the Q of these filters tends to reduce the jitter of the regenerated signal, but it also increases the error rate due to sampling the incoming signal at nonoptimum times.

The principle effect of jitter on the resulting analog signal after decoding is to distort the signal. The analog signal derives from a PAM pulse train which is then passed through a low-pass filter. Jitter displaces the PAM pulses from their proper location, showing up as undesired pulse position modulation (PPM).

Because jitter varies with the number of repeaters in tandem, it is one of the major restricting parameters of long-haul high-bit-rate PCM systems on metallic media. Jitter can be reduced in future systems by using elastic store at each regenerative repeater (costly) and high-Q phase-locked loops.

3.3.7.3 Error Performance

3.3.7.3.1 North American Perspective. The error rate on a digital connection is the fraction of bits received that differ in binary value from the corresponding bits in the transmitted bit stream. Bit errors accumulate on a digital network from the transmitter (source) analog-to-digital (A/D) conversion point to the receiver (destination or sink) digital-to-analog (D/A) conversion point. An objective for end-to-end connections through the switched digital network (SDN) is that the error rate should be less than 1×10^{-6} on at least 95% of the connections (Ref. 35).

Most digital facilities and connections have error rates lower than 1×10^{-6}. Where a higher error rate occurs on a connection, it is usually caused by a high error rate at one facility. The error rate objectives for digital facilities are therefore designed to control the probability that one or more facilities in a connection will have an error rate greater than 1×10^{-6}. The allocated objective of each digital facility specifies the percent of such facilities or the percentage of time for each facility for which the error rate should be less than 1×10^{-6}. The specified percentage is chosen so that on representative connections the total probability of having an error rate greater than 1×10^{-6} will be about 5% (Ref. 35).

If we consider transmission of speech telephony only on a connection, PCM can withstand much greater error rates. Intelligibility is actually lost when the bit error rate (BER) reaches about 1×10^{-2}. This error rate is unacceptable in the PSTN not because of intelligibility but because of supervisory signaling. Studies carried out by British Telecom indicate that a BER of 1×10^{-3} or better is required to maintain the supervisory signaling channel in its desired condition throughout the connection (i.e., idle or busy).

3.3.7.3.2 CCITT Perspective. CCITT Rec. G.821 (Ref. 36) error performance objectives are based on a 64-kbps circuit-switched connection used for voice traffic or as a "bearer channel" for data traffic.

The CCITT error performance parameters are defined as follows (CCITT Rec. G.821):

> The percentage of averaging periods each of time interval $T(0)$ during which the bit error rate (BER) exceeds a threshold value. The percentage is assessed over a much longer time interval $T(L)$.

A suggested interval for $T(L)$ is 1 mo.

It should be noted that total time $T(L)$ is broken down into two parts:

- Time that the connection is available
- Time that the connection is unavailable

The following BERs and intervals are used in CCITT Rec. G.821 in the

Table 3.12 CCITT error performance objectives for international ISDN connections

Performance Classification	Objective (Note 3)
(a) (Degraded minutes) (Notes 1, 2)	Fewer than 10% of one-minute intervals to have a bit error ratio worse than $1 \cdot 10^{-6}$ (Note 4)
(b) (Severely errored seconds) (Note 1)	Fewer than 0.2% of one-second intervals to have a bit error ratio worse than $1 \cdot 10^{-3}$
(c) (Errored seconds) (Note 1)	Fewer than 8% of one-second intervals to have any errors (equivalent to 92% error-free seconds)

Source: From CCITT Rec. G.821, Fascicle III.5, Table 1/G.821, page 29, CCITT Blue Books, XVith Plenary Assy, Melbourne, Nov. 1988.

Note 1. The terms "degraded minutes," "severely errored seconds" and "errored seconds" are used as a convenient and concise performance objective "identifier." Their usage is not intended to imply the acceptability, or otherwise, of this level of performance.

Note 2. The one-minute intervals mentioned in Table 1/G.821 and in the notes (i.e. the periods for $M > 4$ in Annex B) are derived by removing unavailable time and severely errored seconds from the total time and then consecutively grouping the remaining seconds into blocks of 60. The basic one-second intervals are derived from a fixed period.

Note 3. The time interval T_L, over which the percentages are to be assessed has not been specified since the period may depend upon the application. A period of the order of any one moment is suggested as a reference.

Note 4. For practical reasons, at 64 kbit/s, a minute containing four errors (equivalent to an error ratio of 1.04×10^{-6}) is not considered degraded. However, this does not imply relaxation of the error ratio objective of $1 \cdot 10^{-6}$.
See Table 3.13 for interpretation.

statement of objectives (Ref. 36):

- A BER of less than 1×10^{-6} for $T(0) = 1$ min
- A BER of less than 1×10^{-3} for $T(0) = 1$ s
- Zero errors for $T(0) = 1$ s

Table 3.12 gives the CCITT error performance objectives. Table 3.13 gives some guidelines for interpreting Table 3.12.

3.3.7.4 Slips

3.3.7.4.1 Definition. A *slip* is a synchronization impairment of a digital network. No two free-running digital network clocks are exactly synchronized. Consider the simple case where two switches are connected by digital links. An incoming link to a switch terminates in an elastic store to remove

Table 3.13   Guidelines for the interpretation of table 3.12

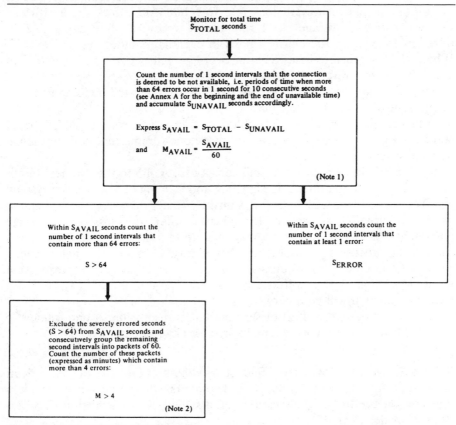

Note 1 — The result is rounded off to the next higher integer.
Note 2 — The last packet which may be incomplete is treated as if it were a complete packet with the same rules being applied.

Performance classification (see Table 1/G.821)	Objective
(a)	$\dfrac{M > 4}{M_{AVAIL}} < 10\%$
(b)	$\dfrac{S > 64}{S_{AVAIL}} < 0.2\%$
(c)	$\dfrac{S_{ERROR}}{S_{AVAIL}} < 8\%$

Source: From CCITT Rec. G.121, Annex B, Fascicle III.5, XVith Plenary Assy., Melbourne, 1988.

transmission timing jitter. The elastic store at the incoming interface is *written into* at the recovered clock rate $R(2)$ but *read out* by the local clock rate $R(1)$. If the average clock rate of the recovered line clock $R(2)$ is different from $R(1)$, the elastic store will eventually underflow or overflow. When $R(2)$ is greater than $R(1)$, an overflow will occur, causing a loss of data. When the reverse happens, that is, $R(1)$ is greater than $R(2)$, an underflow occurs, causing extraneous data inserted into the bit stream. These disruptions of data are called *slips*.

Slips must be controlled to avoid loss of frame synchronization. The most common approach is to purposely add or delete a frame that does not affect framing synchronization.

Slips do not impair voice transmission per se. A slip to the listener sounds like a click. The impairments to signaling and framing depend on the system, whether it is DS1 or CEPT 30 + 2. Consider CEPT 30 + 2. It uses channel-associated signaling in channel 16 and may suffer from loss of multiframe alignment due to slip. Alignment may take up to 5 ms to be reacquired, and calls in the process of being set up may be lost. Common channel signaling such as CCITT SSN 7 is equipped with error detection and retransmission features so that a slip will cause increased retransmit activity, but the signaling function will be otherwise unaffected.

One can imagine the effect of slips on data transmission. A one-frame slip causes a byte of extraneous data or a byte of errored data.

3.3.7.4.2 North American Slip Objectives.

The North American end-to-end slip objective is one slip in five hours, where one slip in 10 hours (half of the end-to-end allocation) is assigned to transmission facilities between nodes (Ref. 35).

The worst case, in which all reference clocks are lost, is 255 slips per day. However, Ref. 37 (AT&T) states that during normal operation the network operates without slips and what slips do occur are traced to uncontrollable impairments, such as reference clock failures. But most clocks are redundant, so the probability of clock failure is very low (Ref. 37).

3.3.7.4.3 CCITT Slip Criteria.

The CCITT slip objective is one slip per 70 days per plesiochronous interexchange link.

The CCITT HRX (hypothetical reference connection) includes 13 nodes operating in the plesiochronous mode connected by 12 links. Hence the nominal slip performance of a connection is 70/12 or one slip in 5.8 days. CCITT Rec. G.822 (Ref. 38) states that acceptable controlled slip rate performance on a 64-kbps international connection is less than five slips in 24 h for more than 98.9% of the time. Forty percent of the slip rate is allocated to the local network on each end of the connection, 6% for each national transit (toll) exchange, and 8% for the international transit portion (see CCITT Rec. G.822).

3.3.8 Delta Modulation (DM)

3.3.8.1 Basic DM   DM is another method of transmitting an audio (analog) signal in a digital format. It is quite different from PCM in that coding is carried out before multiplexing and the code is far more elemental, actually coding at only 1 bit at a time.

The DM code is a one-element code and differential in nature. Of course we mean here that comparison is always made to the prior condition. A 1 is transmitted to the line if the incoming signal at the sampling instant is greater than the immediate previous sampling instant; it is a 0 if it is of smaller amplitude. With DM the derivative of the analog input is transmitted rather than the instantaneous amplitude as in PCM. This is achieved by integrating the digitally encoded signal and comparing it with the analog input to decide which of the two has the larger amplitude. The polarity of the next binary digit placed on line is either plus or minus, to reduce the amplitude of the two waveforms (i.e., analog input and integrated digital output [previous digit]). We thus see the delta encoder basically as a feedback circuit, as shown in Figure 3.50.

Let's see how this feedback concept is applied to the delta encoder. Figure 3.51 illustrates the application. The switch is the double NAND gate and flip-flop. The comparator is the amplifier in Figure 3.50, and the feedback network is the integrating network.

The basic delta decoder consists of a current source, integrating network, amplifier, and low-pass filter. Figure 3.52 illustrates a simplified delta decoder.

We have seen that the digital output signal of the delta coder is indicative of the slope of the analog input signal (its derivative is the slope)—a 1 for positive slope and 0 for negative slope. But the 1 and 0 give no idea of an instantaneous or even semi-instantaneous slope. This leads to the basic weakness found in the development of the DM system, namely, poor dynamic range or poor dynamic response, given a satisfactory signal-to-quantizing-noise ratio. For delta circuits this limit is about 26 dB. A number greater than 26 dB is generally satisfactory, and a number numerically less than 26

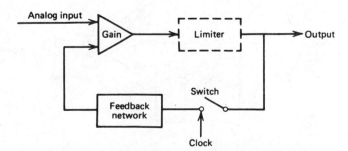

Figure 3.50 Basic electronic feedback circuit used in DM.

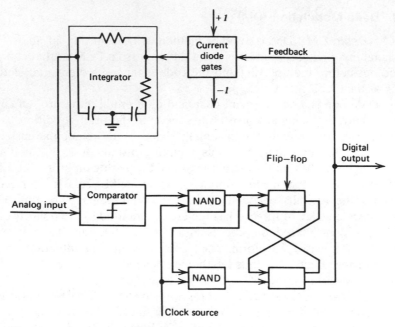

Figure 3.51   Delta encoder.

dB is unsatisfactory. The reader is cautioned not to numerically equate quantizing noise in PCM to quantizing noise in DM, although the concept is the same.

One method used to improve the dynamic range of a DM system is to utilize two integrator circuits (we showed just one in Figure 3.51). This is called *double integration*. Companding provides further improvement, Table 3.14 compares several 56-kbps digital systems with a 3-kHz bandwidth input analog signal regarding dynamic range.

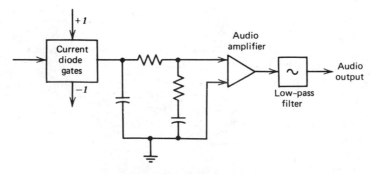

Figure 3.52   Delta decoder.

Table 3.14   Dynamic range—digital modulation systems compared

System	Maximum Signal-to-Quantizing-Noise Ratio (dB)	Dynamic Range for Minimum Signal-to-Quantizing-Noise Ratio of 26 dB (dB)
Basic DM, single integration	34	8
Basic DM, double integration	44	18
Companded DM, single integration	34	15
Companded DM, double integration	44	31
Seven-digit companded PCM	30-dB UVR[a]	At S/N of 31 dB

[a]Useful volume range.

As seen from Table 3.14, companded DM has properties equal to PCM. One reason we can take advantage of the good coding nature of companded DM is that voice signals are predictable, unlike band-limited random signals. The predictability can be based on knowledge of the speech spectrum or on the autocorrelation function. Therefore delta coders can be designed on the principle of prediction.

The advantages of DM over PCM are as follows:

- Multiplexing is carried out by simple digital multiplexers, whereas PCM interleaves analog samples
- It is essentially more economical for small numbers of channels
- It has few varieties of building blocks
- It is less complex, consequently giving improved reliability
- Intelligibility is maintained with a BER down to 10^{-2}

The disadvantages remain for the dynamic range and for the multiplexing of many channels.

DM has found wide application in military communications where low-bit-rate digital systems are required. Both the U.S. and NATO forces have fielded large quantities of delta multiplexers and switches based on 16 and 32 kbps. DM is also used in the commercial telephony world on thin-line telephone systems, for rural subscribers, and in certain satellite DAMA applications, such as Canada's TeleSat.

3.3.8.2   CVSD — A Subset of DM   Continuous variable-slope DM (CVSD) is a DM scheme that improves the basic weakness of DM, namely, small

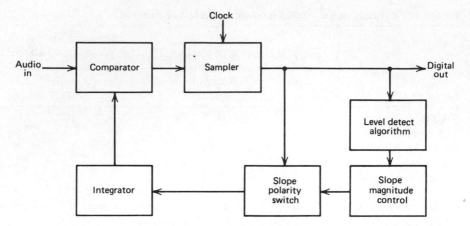

Figure 3.53 A CVSD encoder.

dynamic range over which the noise level is constant. CVSD is used by the U.S. military forces Tri-Tac communications system and by NATO's Eurocom with 16- and 32-kbps digital data line rates.

A CVSD coder is shown in Figure 3.53, a decoder in Figure 3.54, and the CVSD waveforms in Figure 3.55. CVSD circuitry provides increased dynamic range capability by adjusting the gain of the integrator. For a given clock frequency and input bandwidth, the additional circuitry increases the delta modulator's dynamic range (i.e., up to 50 dB of range). External to the basic delta modulator is an algorithm that monitors the past few outputs of the delta modulator in a simple shift register. The register is usually 3–4 bits long, depending on the application. A CVSD algorithm monitors the contents of the shift register and indicates whether it contains all 1's or all 0's. This condition is called coincidence. When it occurs, it indicates that the gain of the integrator is too small. In this particular design, the coincidence

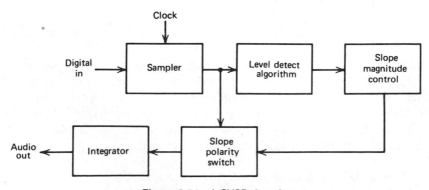

Figure 3.54 A CVSD decoder.

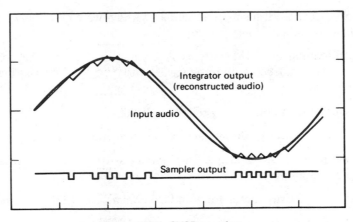

Figure 3.55   CVSD waveforms.

changes a single-pole low-pass filter. The voltage output of this syllabic filter controls the integrator gain through a pulse amplitude modulator whose other input is the sign bit or up/down control.

The algorithm provides a means of measuring the average power or level of the input signal. The purpose of the algorithm is to control the gain of the integrator and to increase the dynamic range. The algorithm is repeated in the receiver to recover level data in the decoding process. Because the algorithm only operates on past serial data, it changes the nature of the bit stream without changing the channel bit rate.

The effect of the algorithm is to compand the input signal. If a CVSD encoder is fed into a basic delta demodulator, the output of that demodulator will reflect the shape of the input signal, but all of the output will be at an equal level. Thus the algorithm at the output is needed to restore the level variations.

REVIEW EXERCISES

1. Define the two basic generic types of multiplexing covered in the text.

2. The input to a mixer is the standard CCITT voice channel, and a local oscillator injects a 20-kHz signal into the mixer. Define the lower sideband output of the mixer (frequency domain).

3. Define the standard CCITT group regarding voice channel capacity and the frequency band that it occupies.

4. Define the standard CCITT supergroup regarding voice channel capacity, number of groups, and the frequency band that it occupies.

5. List at least three transmission media to which we could apply FDM line frequency configurations.

6. Speech loading of FDM systems is a complex matter. The complexity arises because speech amplitude varies with _____? List at least three parameters.

7. How does volume unit (VU) relate to dBm?

8. Relate peak power of an average talker to average power of a talker.

9. If we load an FDM multiplex equipment with −16 dBm, what is the composite loading of 500 channels each at −16 dBm?

10. What is the activity factor for standard commercial FDM equipment?

11. Name one deleterious effect of overload of FDM equipment on network derived connections from that equipment.

12. Give the CCITT loading formula for FDM equipment operating with more than 240 channels.

13. Name four types of constant-amplitude signals that might be transmitted on FDM equipment.

14. Why must system operators control level and channel assignment of constant amplitude signals transmitted on FDM equipment?

15. What is the principal purpose of an FDM pilot tone? A secondary purpose?

16. On long routes why is it necessary to limit the number of modulation/demodulation steps of FDM systems?

17. What is the primary reason through-group and through-supergroup techniques are used, particularly on long routes?

18. Differentiate the terms *subscriber carrier* and *station carrier*.

19. Give at least two applications of subscriber/station carrier.

20. Give the three principal advantages of compandors.

21. Regarding FDM implementations, compare in-band and out-of-band signaling.

22. What is the main advantage of pulse code modulation (PCM) over its analog counterpart? Give two disadvantages.

23. What are the three basic steps in the development of a PCM signal from an analog input?

24. In accordance with the Nyquist sampling theorem, what is the sampling rate for the standard nominal 4-kHz voice channel? For a 15-kHz program channel? For a 4.2-MHz TV video channel?

25. A 7-level PCM code word has how many distinct code word possibilities? An 8-level code? An 11-level code?

26. What is the cause of quantization noise developed in a PCM coder–decoder combination?

27. If we double the number of quantizing steps in a PCM system using linear quantization, what improvement in dB would we achieve in quantizing noise?

28. What does the first significant bit in a PCM code word indicate? What do bits 2, 3, and 4 of that code word indicate to the decoder? (Assume 8-bit PCM code words.)

29. Approximately what signal-to-distortion (S/D) ratio may we expect in an 8-bit PCM system, whether μ-law of A-law?

30. Show how we numerically derive the two basic PCM bit rates (i.e., 1.544 Mbps and 2.048 Mbps).

31. How many bits are there in the North American DS1 frame?

32. There are two basic framing architectures used in today's PCM systems. Define and discuss these framing strategies (DS1 1.544 Mbps and CEPT 30 + 2 2.048 Mbps).

33. Describe the two DS1 multiframe strategies. What are some of the applications of the derived data channel? How will the derived CRC improve performance?

34. Discuss and compare how supervisory signaling is carried out with DS1 and CEPT 30 + 2 PCM systems.

35. What are the causes of idle channel noise in a PCM system? How is such noise mitigated?

36. When PCM is used on local metallic pair trunks, compare single- and two-cable working.

37. What is a *violation* in an AMI (bipolar) line code?

38. What are the meanings and implications of BNZS and HDB3? Why are they necessary?

39. Key to the principal advantage of digital transmission is the regenerative repeater. Explain why.

40. Why does a PCM system operating on metallic media use a waveform such as AMI rather than just a simple serial binary bit stream?

41. An M12 multiplexer provides a DS2 line bit rate of 6.312 Mbps at its output and has four 1.544-Mbps DS1 inputs. If we multiply 1.544 Mbps by four, we get 6.176 Mbps. Hence the M12 output is greater than the

aggregate of the four inputs by 136 kbps. Why? What are the extra bits used for?

42. Explain the rationale of justification/stuffing.

43. How does the AT&T DDS system transmit computer data on a DS1 bit stream? Include the use of DS0 and overhead bits in the discussion.

44. What is the basic measure of quality of a PCM link.

45. Relate jitter accumulation on a long repeatered line to the number of regenerative repeaters in tandem on the line.

46. Give the two basic transmission impairments of a digital transmission system that (in general) do not affect analog transmission systems.

47. In the text we described three BER threshold points for a PCM system. What are they? Discuss each regarding rationale.

48. What causes *slips*?

49. Continuous variable-slope DM (CVSD) is commonly used on many military digital networks. Describe how it works, and discuss its drawbacks and advantages.

50. Describe signal-to-quantizing-distortion ratio (dB) for 16- and 32-kbps CVSD systems and compare these values to a typical 8-bit PCM system in terms of quality. Give reasons for your answers.

REFERENCES AND BIBLIOGRAPHY

1. *Reference Data for Radio Engineers*, 6th ed., Howard W. Sams, Indianapolis, IN, 1977.
2. MIL-STD-188-100, U.S. Department of Defense, Washington, DC, Dec. 1972.
3. CCITT, Blue Books, XIXth Plenary Assembly, Melbourne, 1988, vol. III, G recommendations.
4. *Transmission Systems for Communications*, 5th ed., Bell Laboratories, Holmdel, NJ, 1982.
5. MIL-STD-188C with Notice 1, U.S. Department of Defense, Washington, 1970.
6. *Principles of Electricity as Applied to Telephone and Telegraph Work*, American Telephone and Telegraph Co., New York, 1961.
7. D. M. Hamsher, *Communication System Engineering Handbook*, McGraw-Hill, New York, 1967.
8. MIL-STD-188-311, U.S. Department of Defense, Washington, DC, 1971.
9. R. L. Freeman, *Reference Manual for Telecommunications Engineering*, Wiley, New York, 1985.
10. B. D. Holbrook and J. T. Dixon, "Load Rating Theory for Multichannel Amplifiers," *Bell Syst. Tech. J.*, vol. 18, 624–644, Oct. 1939.

11. *Engineering and Equipment Considerations—46A*, Issue 1, Lenkurt Electric Corp., San Carlos, CA, 1970.

12. R. L. Freeman, *Telecommunications System Engineering*, 2nd ed., Wiley, New York, 1989.

13. *Telecommunications Transmission Engineering*, 2nd ed., vols. 1–3, American Telephone and Telegraph Co., New York, 1977.

14. Seminar on Pulse Code Modulation Transmission, Standard Telephone and Cables Ltd., Basildon, UK, June 1967.

15. *PCM System Applications 30 + 2TS*, ITT/BTM, Antwerp, Belgium, 1973.

16. *Operations and Maintenance Manual for T324 PCM Cable Carrier System*, ITT Telecommunications, Raleigh, NC, Apr. 1973.

17. K. W. Catermole, *Principles of Pulse Code Modulation*, Illife, London, 1969.

18. W. C. Sain, "Pulse Code Modulation Systems in North America," *Electr. Commun.* (*ITT*), vol. 48, no. 1/2, 1973.

19. J. V. Marten and E. Brading, "30-Channel Pulse Code Modulation System," *Electr. Commun.* (*ITT*), vol. 48, no. 1/2, 1973.

20. R. B. Moore, "T2 Digital Line System," *Proceedings of the IEEE International Conference on Communications*, Seattle, WA, June 1973.

21. J. R. Davis, "T2 Repeaters and Equalization," *Proceedings of the IEEE International Conference on Communications*, Seattle, WA, June 1973.

22. Bell System Technical Reference, *Digital Channel Bank Requirements and Objectives*, PUB 43801, AT&T, New York, Nov. 1982.

23. *T1G Digital Carrier System Planner's Guide*, AT&T, New York, 1984.

24. G. C. Hartley et al., "Techniques of Pulse-Code Modulation in Communication Networks," *IEE Monograph Series*, Cambridge University Press, Cambridge, 1967.

25. P. Bylanski and D. G. W. Ingram, "Digital Transmission Systems," *IEE Telecommunication Series No. 4*, Peter Peregrinus Ltd., Stevenage, Herts, UK, 1967.

26. H. Nyquist, "Certain Topics in Telegraph Transmission Theory," *Trans. AIEE*, vol. 47, 617–644, Apr. 1928.

27. "Functional Characteristics of Interfaces Associated with Network Nodes," CCITT Rec. G.704, Blue Books, XIXth Plenary Assembly, Melbourne, 1988, vol. III.

28. *The Extended Superframe Format Interface Specification*, Compatibility Bulletin no. 142, AT&T, New York, 1983.

29. C. E. Shannon, "A Mathematical Theory of Communication," *Bell Syst. Tech. J.* vol. 27, 623–656, 1948.

30. Gregory R. Werth, *The Fundamentals of DS2: Parts 1 and 2*, Telecommunications, Norwood, MA, Aug. and Sept. 1989.

31. H. Akima, *Noise Power Due to Digital Errors in a PCM Channel*, OT Rep. 78-139, NTIS PB 277-477, U.S. Department of Commerce, Office of Telecommunications, Washington, DC, Jan. 1978.

32. H. R. Schindler, "Delta Modulation," *IEEE Spectrum*, Oct. 1970.

33. John C. Bellamy, *Digital Telephony*, 2nd ed., Wiley, New York, 1990.

34. "Performance Characteristics of PCM Channels between 4-Wire Interfaces at Voice Frequency," CCITT Rec. G.712, Blue Books, XIXth Plenary Assembly, Melbourne, 1988, vol. III.

35. "Notes on the BOC INTER-LATA Networks–1986," BELLCORE Tech. Ref. TR-NPL-000275, Issue 1, Apr. 1986 BELLCORE, Livingston, NJ 1986.

36. "Error Performance of an International Digital Connection Forming Part of an Integrated Services Digital Network," CCITT Rec. G.821, Blue Books, Fascicle III.5, IXth Plenary Assembly, Melbourne, 1988.

37. "Digital Synchronization Network Plan," AT&T Pub. 60110, AT&T, Florham Park, NJ, 1983.

38. "Controlled Slip Rate Objectives on an International Digital Connection," CCITT Rec. G.822, Fascicle III.5 IXth Plenary Assembly, Melbourne, 1988.

4

LINE-OF-SIGHT MICROWAVE (RADIOLINK) SYSTEMS

4.1 INTRODUCTION

In this chapter we discuss line-of-sight microwave, called radiolinks in Europe. Line-of-sight (LOS) microwave is a widely employed broadband transmission medium commonly used to transport the analog FDM or digital PCM/CVSD baseband described in Chapter 3.

Among the many applications of LOS microwave are

1. Point-to-point links as a backbone or tails of large networks for common carriers, specialized common carriers, and private and government entities.
2. Point-to-multipoint systems for TV, telephony, data, or various mixes thereof.
3. Transport of TV or other video signals such as community antenna television (CATV) head-end extension, broadcast transport, and studio-to-transmitter links.
4. Specialized digital and digital data networks.
5. Power and pipeline companies for the transport of telemetry, command, and control information.
6. Air traffic control center interconnectivity.
7. Short-haul applications such as linking offices and buildings in congested urban areas; final connectivities for common carrier/specialized common carriers; tails off fiber optic trunks.
8. Military applications: fixed point-to-point, point-to-multipoint, and transportable point-to-point.

Table 4.1a   Letter designations for microwave bands

Subband	Frequency (GHz)	Wavelength (cm)
	P Band	
	0.225	133.3
	0.390	76.9
	L Band	
p	0.390	76.9
c	0.465	64.5
l	0.510	58.8
y	0.725	41.4
t	0.780	38.4
s	0.900	33.3
x	0.950	31.6
k	1.150	26.1
f	1.350	22.2
z	1.450	20.7
	1.550	19.3
	S Band	
e	1.55	19.3
f	1.65	18.3
t	1.85	16.2
c	2.00	15.0
q	2.40	12.5
y	2.60	11.5
g	2.70	11.1
s	2.90	10.3
a	3.10	9.67
w	3.40	8.32
h	3.70	8.10
z[a]	3.90	7.69
d	4.20	7.14
	5.20	5.77
	X Band	
a	5.20	5.77
q	5.50	5.45
y[a]	5.75	5.22
d	6.20	4.84
b	6.25	4.80
r	6.90	4.35
c	7.00	4.29
l	8.50	3.53
	9.00	3.33

Table 4.1a (*Continued*)

Subband	Frequency (GHz)	Wavelength (cm)
	X Band	
s	9.60	3.13
x	10.00	3.00
f	10.25	2.93
k	10.90	2.75
	K Band	
	10.90	2.75
p	12.25	2.45
s	13.25	2.26
e	14.25	2.10
c	15.35	1.95
u^b	17.25	1.74
t	20.50	1.46
q^b	24.50	1.22
r	26.50	1.13
m	28.50	1.05
n	30.70	0.977
l	33.00	0.909
a	36.00	0.834
	Q Band	
a	36.0	0.834
b	38.0	0.790
c	40.0	0.750
d	42.0	0.715
e	44.0	0.682
	46.0	0.652
	V Band	
a	46.0	0.652
b	48.0	0.625
c	50.0	0.600
d	52.0	0.577
e	54.0	0.556
	56.0	0.536
	W Band	
	56.0	0.536
	100.0	0.300

Source: Ref. 1.

[a]C band includes S_z through X_y (3.90–6.20 GHz).
[b]K_1 band includes K_u through K_q (15.35–24.50 GHz).

Many of these application implementations are now being tempered by fiber optic links. Under certain circumstances fiber optics is more cost-effective and provides considerably greater information bandwidth. Fiber, however, is hampered by right-of-way requirements and by cables being severed by construction activities.

4.1.1 Definition

Let us define radiolink systems as those that fulfill the following require-ments:

1. Signals follow a straight line or LOS path.
2. Signal propagation is affected by free-space attenuation and precipitation.

Table 4.1b   Microwave radiolink frequency assignments for fixed service[a]

General Frequency Assignments

450–470 MHz	5,925–6,425 MHz
890–960 MHz	(7,250)7,300–8,400 MHz
1,710–2,290 MHz	10,550–12,700 MHz
2,550–2,960 MHz	14,400–15,250 MHz
3,700–4,200 MHz	17,700–19,700 MHz

Specific Frequency Assignments, United States

Service	GHz	Service	GHz
Military	1.710–1.850	Common carrier (space)	5.925–6.425
Operational fixed	1.850–1.990	Operational fixed	6.575–6.875
Studio transmitter link	1.990–2.110	Studio transmitter link	6.875–7.125
Common carrier	2.110–2.130	Military	7.125–7.750
Operational fixed	2.130–2.150	Military	7.750–8.400
Common carrier	2.160–2.180	Common carrier	10.7–11.7
Operational fixed	2.180–2.200	Operational fixed	12.2–12.7
Operational fixed (TV only)	2.500–2.690	CATV studio transmitter link (CARS)[b]	12.7–12.95
Common carrier (space)	3.700–4.200	STL	12.95–13.2
Military	4.400–5.000	Military	14.4–15.25
		Common carrier	17.7–19.7
		Common carrier	21.2–23.6
		Common carrier	23.5–29.5

[a]Point-to-point communications and some other nonmobile applications.
[b]CATV = community antenna television; CARS = community antenna radio service.

3. Use of frequencies greater than 150 MHz, thereby permitting transmission of more information per RF carrier by use of a wider information baseband.
4. Use of angle modulation (i.e., FM or PM), digital modulation, or spread-spectrum and time-sharing techniques.

Table 4.1*a* gives conventional letter designations for microwave bands and Table 4.1*b* shows typical assignments for the frequency region through 19 GHz.

A valuable characteristic of LOS transmission is that we can predict the level of a signal arriving at a distant receiver with known accuracy.

4.2   LINK ENGINEERING

Engineering a radiolink system involves the following steps:

1. Selection of sites (radio equipment plus tower locations) that are in line-of-sight of each other.
2. Selection of an operational frequency band from those set forth in Table 4.1*b*, considering RF interference environment and legal restraints.
3. Development of path profiles to determine radio tower heights. If tower heights exceed a certain economic limit, then step 1 must be repeated, bringing the sites closer together or reconfiguring the path, usually along another route. In making a profile, it must be taken into consideration that microwave energy is
   - Attenuated or absorbed by solid objects
   - Reflected from flat conductive surfaces such as water and sides of metal buildings
   - Diffracted around solid objects
   - Refracted or bent by the atmosphere; often the bending is such that the beam may be extended beyond the optical horizon
4. Path calculations. After setting a propagation reliability expressed as a percentage of time, the received signal will be above threshold level. Often this level is the FM improvement threshold of the FM receiver. To this level a margin is set for signal fading under all anticipated climatic conditions.
5. Making a path survey to ensure correctness of steps 1–4. It also provides certain additional planning information vital to the installation project or bid.
6. Establishment of a frequency plan and necessary operational parameters.

7. Equipment configuration to achieve the fade margins set in step 4 most economically.
8. Installation.
9. Beam alignment, equipment lineup, checkout, and acceptance by a customer.

Reference will be made, where applicable, to these steps so that the reader will be exposed to practical radiolink problems.

4.3 PROPAGATION

4.3.1 Free-Space Loss

Consider a signal traveling between a transmitter at A and a receiver at B:

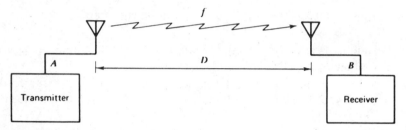

The distance between antennas is D and the frequency of transmission is f. Let D be in kilometers and f in megahertz; then the free-space loss in decibels may be calculated with the following formula:

$$L_{dB} = 32.44 + 20 \log D + 20 \log f \tag{4.1a}$$

If D is in statute miles, then

$$L_{dB} = 36.58 + 20 \log D + 20 \log f \tag{4.1b}$$

(all logarithms are to the base 10).

Suppose that the distance separating A and B were 40 km ($D = 40$). What would the free-space path loss be at 6 GHz ($f = 6000$)?

$$
\begin{aligned}
L &= 32.44 + 20 \log 40 + 20 \log 6 \times 10^3 \\
&= 32.44 + 20 \times 1.6021 + 20 \times 3.7782 \\
&= 32.44 + 32.042 + 75.564 \\
&= 140.046 \text{ dB}
\end{aligned}
$$

Let us look at it another way. Consider a signal leaving an isotropic antenna.*

---

*An isotropic antenna is an ideal antenna with a reference gain of 1 (0 dB). It radiates uniformly in all directions (i.e., is perfectly omnidirectional).

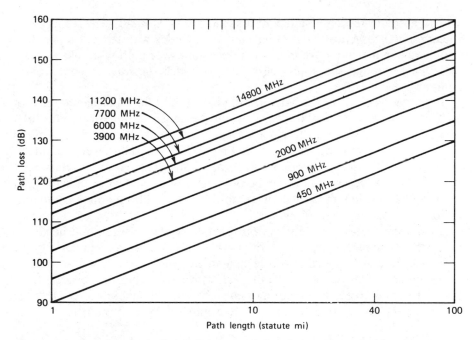

Figure 4.1   Path loss in decibels related to path length in statute miles for eight discrete radiolink frequencies.

At one wavelength (1λ) away from the antenna, the free-space attenuation is 22 dB. At two wavelengths (2λ) it is 28 dB; at four wavelengths it is 34 dB. Every time we double the distance, the free-space attenuation (or loss) is 6 dB greater. Likewise, if we halve the distance, the attenuation (or loss) decreases by 6 dB.

Figure 4.1 relates free-space path loss to distance at eight discrete frequencies between 450 and 14,800 MHz.

4.3.2   Bending of Radio Waves above 100 MHz from Straight-Line Propagation

Radio waves traveling through the atmosphere do not follow true straight lines. They are refracted or bent. They may also be diffracted.

The velocity of an electromagnetic wave is a function of the density of the media through which it travels. This is treated by Snell's law, which provides a valuable relationship for an electromagnetic wave passing from one medium to another (i.e., from an air mass with one density to an air mass with another density). It states that the ratio of the sine of the angle of incidence to the sine of the angle of refraction is equal to the ratio of the respective velocities in the media. This is equal to a constant that is the refractive index of the second medium relative to the first medium.

The absolute refractive index of a substance is its index with respect to a vacuum and is practically the same value as its index with respect to air. It is the change in the refractive index that determines the path of an electromagnetic wave through the atmosphere, or how much the wave is bent from a straight line.

If radiowaves above 100 MHz traveled a straight line, the engineering of LOS microwave (radiolink) systems would be much easier. We could then accurately predict the height of the towers required at repeater and terminal stations and exactly where the radiating device on the tower should be located (steps 1 and 3, Section 4.2). Essentially what we are dealing with here, then, is a method to determine the height of a microwave radiator (i.e., an antenna or other radiating device) to permit reliable radiolink communication from one location to another.

To determine tower height, we must establish the position and height of obstacles in the path between stations with which we want to communicate by radiolink systems. To each obstacle height, we will add earth bulge. This is the number of feet or meters an obstacle is raised higher in elevation (into the path) owing to earth curvature or earth bulge. The amount of earth bulge in feet at any point in a path may be determined by the formula

$$h = 0.667d_1d_2 \tag{4.2}$$

where d_1 = distance from the near end of the link to the point (obstacle location), and d_2 = distance from the far end of the link to the obstacle location.

The equation will become more useful if it is made directly applicable to the problem of ray bending. As the equation is presented above, the ray is unbent or a straight line.

Atmospheric refraction may cause the ray beam to be bent toward the earth or away from the earth. If it is bent toward the earth, it is as if we shrank the earth bulge or lowered it from its true location. If the beam is bent away from the earth, it is as if we expanded the earth bulge or raised it up toward the beam above its true value. This lowering or raising is handled mathematically by adding a factor K to the earth bulge equation. It now becomes

$$h_{\text{ft}} = \frac{0.667d_1d_2}{K} \quad (d_1 \text{ and } d_2 \text{ in mi}) \tag{4.3a}$$

$$h_{\text{m}} = \frac{0.078d_1d_2}{K} \quad (d_1 \text{ and } d_2 \text{ in km}) \tag{4.3b}$$

The K-factor can be calculated from the formula

$$K = \frac{\text{effective earth radius}}{\text{true earth radius}} = \frac{r}{r_0} \tag{4.4}$$

The value commonly used for r_0 is 6370 km. We calculate the effective earth radius, r, from the formula

$$r = r_0\{1 - 0.04665 \exp(0.005577N_s)\}^{-1} \tag{4.5}$$

Note that the use of the notation "exp" means e, the natural number raised to the indicated power, in this case $(0.005577N_s)$.

For example, if the surface refractivity N_s is 301, we substitute that value in equation 4.5 and

$$r = 6370[1 - 0.04665e^{(0.005577 \times 301)}]^{-1}$$

$$= 6370[1 - 0.04665 \times 5.3585]^{-1}$$

$$= 6370(1 - 0.24997)^{-1}$$

$$= 6370 \times 1.333$$

$$= 8493 \text{ km}$$

Note that the K factor comes out to be 1.33 or $\frac{4}{3}$.

Of course we now need to find a value for surface refractivity N_s. This is the refractivity at the altitude of the LOS microwave site that we selected or the average refractivity of the path. The sea level refractivity can be obtained for the area from a nearby weather bureau or from a chart such as in Figure 4.4. To calculate N_s when we are given N_0, the mean sea level refractivity, from the following formula:

$$N_s = N_0 \exp(-0.1057h_s) \tag{4.6}$$

where h_s is the altitude above mean sea level (in kilometers) of the LOS radio site.

For example, the sea level refractivity is about 312 around the Boston, Massachusetts, area of the United States. A potential site near Boston is 220 m above sea level (i.e., 0.22 km). Using equation 4.6 we have

$$N_s = N_0 X e^{(-0.1057 \times 0.22)}$$

$$N_s = 304.8$$

If the K factor is greater than 1, the ray beam is bent toward the earth, which essentially allows us to shorten radiolink towers. If K is less than 1, the earth bulge effectively is increased, and the path is shortened or the tower height must be increased. Figure 4.2 gives earth curvature, in feet, for various values of K.

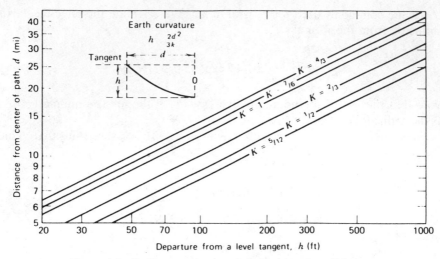

Figure 4.2 Earth curvature or earth bulge for various K factors.

Many texts on radiolinks refer to normal refraction, which is equivalent to a K factor of $\frac{4}{3}$ or 1.33. It follows a rule of thumb that applies to refraction in that a propagated wave front (or beam) bends toward the region of higher density, that is, toward the region having the higher index of refraction. Older texts insisted that the K factor should nearly always be $\frac{4}{3}$. Care should be taken when engineering radiolinks that the $\frac{4}{3}$ theory (standard refraction or normal refraction) not be accepted "carte blanche" on many of the paths likely to be encountered. However, $K = \frac{4}{3}$ may be used for gross planning of radiolink systems. Figure 4.3 may be used to estimate tower heights with $K = \frac{4}{3}$ for smooth earth paths (no obstacles besides midpath earth bulge). We also can calculate the smooth earth distance to the horizon ($K = \frac{4}{3}$) using the formula $d = \sqrt{2h}$ where d is in statute miles and h is in feet.

Another factor must be added to the obstacle height to obtain an effective obstacle height. This is the Fresnel clearance. It derives from electromagnetic wave theory that a wave front (which our ray beam is) has expanding properties as it travels through space. These expanding properties result in reflections and phase transitions as the wave passes over an obstacle. The result is an increase or decrease in the received signal level. The amount of additional clearance that must be allowed to avoid problems with the Fresnel phenomenon (diffraction) is expressed in Fresnel zones. Optimum clearance of an obstacle is accepted as 0.6 of the first Fresnel zone radius. The first Fresnel zone radius, in feet, may be calculated with the following formula:

$$R_{ft} = 72.1\sqrt{\frac{d_1 d_2}{FD}} \tag{4.7a}$$

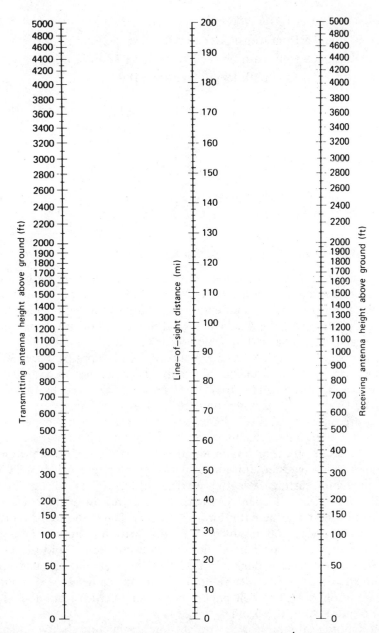

Figure 4.3 LOS distance for smooth spherical earth with $K = \frac{4}{3}$; distance in statute miles, tower heights in feet.

where F = frequency of signal (GHz)

d_1 = distance from transmitter to path obstacle (statute mi)

d_2 = distance from path obstacle to receiver (statute mi)

$D = d_1 + d_2$, total path length (statute mi)

To determine the first Fresnel zone radius when using metric units, use

$$R_{\mathrm{m}} = 17.3\sqrt{\frac{d_1 d_2}{FD}} \qquad (4.7b)$$

where F = frequency of signal (GHz), d_1, d_2, and D are the same as in equation 4.7a but in kilometers, and R_{m} is in meters.

4.3.3 Path-Profiling—Practical Application

After tentative terminal or repeater sites have been selected, path profiles are plotted on rectangular graph paper. Obstacle information is taken from topographical maps. For the continental United States the best topographical maps available are from the U.S. Geological Survey. They are $7\frac{1}{2}$-min maps of latitude and longitude with a scale 1:24,000, where 1 in. = 2000 ft, and 15-min maps with a scale of 1:62,5000, where 1 in. = approx. 1 mi. 30-min and 1-deg maps are also available, but their scales are not fine enough for path profile application. Many areas of Canada are covered by maps with scales of 1:50,000 (1 in. = 0.79 mi) and 1:63,360 (1 in. = 1 mi).

Profiles are made on available linear graph paper; 10 divisions per inch is suggested. Any convenient combination of vertical and horizontal scales may be used. For paths 30 mi or less in length, 2 mi to the inch plotted on the horizontal scale is suggested. For longer paths, graph paper may be extended by trimming and pasting. For the vertical scale, 100 ft to the inch is satisfactory for fairly flat country, where there is no more than an 800-ft change in altitude along the path; for hilly country, 200 ft to the inch and for mountainous country, 500 or 1000 ft to the inch are appropriate. One suggestion to preserve a proper relationship between height and distance is that if the distance scale is doubled, the height scale should be quadrupled. All heights should be shown with reference to mean sea level (MSL), but the base (0 elevation on the graph paper) need not be MSL, but the lowest elevation of interest on the profile.

On the topological map draw a straight line connecting the two adjacent radiolink sites. Carefully trace with your eye or thumb down the line from one site to the other, marking all obstacles or obstructions and possible points of reflection, such as bodies of water, marshes, or desert areas, assigning consecutive letters to each obstacle.

Plot the horizontal location of each point on the graph paper. Mark the path midpoint, which is the point of maximum earth bulge and should be

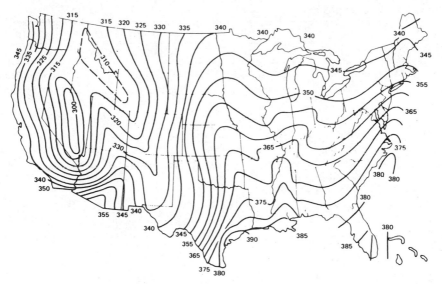

Figure 4.4 Sea level refractivity index N_0 for the continental United States: maximum for worst month (August).

marked as an obstacle. Determine the K factor by one of the following methods:

1. Refer to a sea level refractivity profile chart such as in Figure 4.4. Select the appropriate N_0 for the area of interest; correct for the altitude of midpath (equation 4.6). Calculate the K-factor from equations 4.4 and 4.5.
2. Lacking refractivity information, plot path profile using three K factors: 1.33, 1.0, and 0.5. A later field survey will help to decide the appropriate K factor. Table 4.2 will also be a useful guide.

For each obstacle point compute d_1, the distance to one repeater site, and d_2, the distance to the other. Compute the equivalent earth curvature EC for each point with equation 4.3:

$$EC = \frac{0.667 d_1 d_2}{K} \qquad (4.8)$$

Compute the Fresnel zone clearance by equation 4.7 or by using Figures 4.5 and 4.6. Figure 4.5 gives the Fresnel clearance for midpath for various bands of frequencies. For other obstacle points compute the percentage of total path length. For instance, on a 30-mi path, midpoint is 15 mi (50%). However, a point 5 mi from one site is 25 mi from the other, or represents $\frac{5}{30}$ and $\frac{25}{30}$, or $\frac{1}{6}$ and $\frac{5}{6}$. Converted to percentage, $\frac{1}{6} = 16.6\%$ and $\frac{5}{6} = 83\%$. Apply

Table 4.2   K factor guide[a]

	Propagation Conditions				
	Perfect	Ideal	Average	Difficult	Bad
Weather	Standard atmosphere	No surface layers or fog	Substandard, light fog	Surface layers, ground fog	Fog moisture over water
Typical	Temperate zone, no fog, no ducting, good atmospheric mix day and night	Dry, mountainous no fog	Flat, temperate, some fog	Coastal water tropical	Coastal,
K factor	1.33	1–1.33	0.66–1.0	0.66–0.5	0.5–0.4

[a]For 99.9–99.99% path reliability.

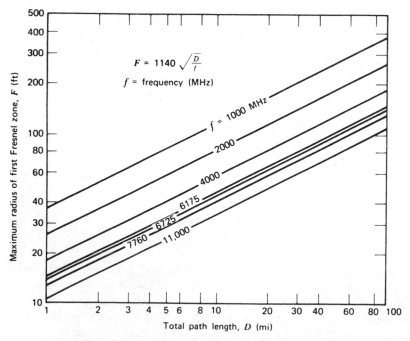

Figure 4.5   Midpath Fresnel clearance for first Fresnel zone. 0.6 of this value is used in the calculations or the value is calculated in conjunction with Figure 4.6. (Courtesy of GTE Lenkurt Inc., San Carlos, CA.)

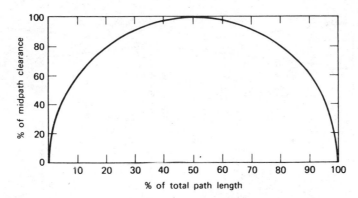

Figure 4.6 Conversion from midpath Fresnel zone clearance for other than midpath obstacle points.

this percentage on the *x* axis of Figure 4.6. In this case 16.6% or 83% is equivalent to 76% of midpath clearance. If midpath Fresnel clearance were 40 ft, only 30.4 ft of Fresnel clearance would be required at the 5-mi point. Remember, we use 0.6 of R, the first Fresnel zone radius, for the clearance value.

Set up a table on the profile chart as follows:

Obstacle Point	d_1	d_2	EC	F (Fresnel)
A				
B				
C				
D				
E				

On the graph paper plot the height above sea level of each obstacle point. To this height add EC (earth curvature), the F (Fresnel zone clearance), and if tree growth exists, add 40 ft for trees and 10 ft for additional growth. If undergrowth exists, assign 10 ft for vegetation. A path survey will confirm or deny these figures or will permit adjustment.

Minimum tower heights may now be determined by drawing a straight line from site to site through the highest obstacle points. These often cluster around midpath. Figure 4.7 is a hypothetical profile exercise.

4.3.4 Reflection Point

From the profile, possible reflection points may be obtained. The objective is to adjust the tower heights such that the reflection point is adjusted to fall on land area where the reflected energy will be broken up and scattered. Bodies

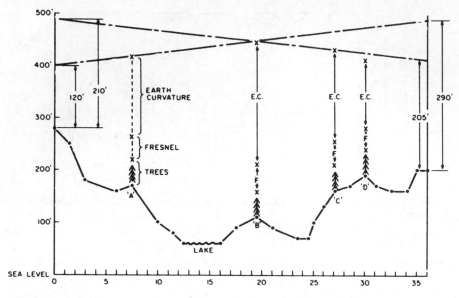

Path profile data base

			Total Height
Obstacle	d_1 (mi)	d_2 (mi)	0.6 Fresnel (ft)	EC (ft)[b]	Vegetation	Extend. (ft)
---	---	---	---	---	---	---
A	7.5	28.5	43	152	50	245
B	19.4	16.6	53	233	50	336
C	27.0	9.0	46	176	50	272
D	30.0	6.0	39	130	50	219

[a]Frequency band; 6 GHz; K factor 0.92; $D = d_1 + d_2 = 36$ mi; vegetation/tree conditions 40 ft plus 10 ft growth.

Figure 4.7 Example path profile.

of water and other smooth surfaces cause reflections that are undesirable. Figure 4.8 will assist in adjusting the reflection point. It uses a ratio of tower heights h_1/h_2, and the shorter tower height is always h_1. The reflection area lies between a K factor of grazing ($K = 1$) and a K factor of infinity. The distance expressed is always from h_1, the shorter tower. By adjusting the ratio h_1/h_2, the reflection point can be moved. The objective is to ensure that the reflection point does not fall on an area of smooth terrain or on water, but rather on a land area where the reflected energy will be broken up or scattered (e.g., wooded areas).

For a highly reflective path, space diversity operation may be desirable to minimize the effects of multipath reception (see Section 4.5.2.8).

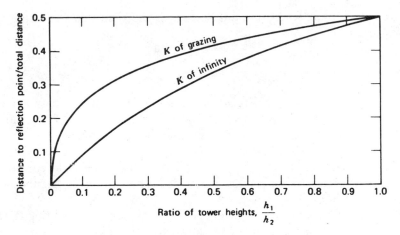

Figure 4.8 Calculation of reflection points.

4.4 PATH CALCULATIONS

4.4.1 General

The next step in path engineering is to carry out path calculations. Essentially this entails the determination of equipment parameters and configurations to meet a minimum performance requirement. Such performance requirements are usually related to noise in an equivalent voice channel or they are related to the signal-to-noise ratio, or both. Either way, the requirement is given for a percentage of time. For a single path this may be stated as 99, 99.9, or 99.99% of the time said performance exceeds a certain minimum. This is often called propagation reliability. Table 4.3 states reliability percentages

Table 4.3 Reliability versus outage time

Reliability (%)	Outage Time(%)	Outage Time		
		Per Year	Per Month (avg.)	Per Day (avg.)
0	100	8760 h	720 h	24 h
50	50	4380 h	360 h	12 h
80	20	1752 h	144 h	4.8 h
90	10	876 h	72 h	2.4 h
95	5	438 h	36 h	1.2 h
98	2	175 h	14 h	29 min
99	1	88 h	7 h	14 min
99.9	0.1	8.8 h	43 min	1.4 min
99.99	0.01	53 min	4.3 min	88.6 s
99.999	0.001	5.3 min	26 s	0.86 s

versus outage time, where the outage time is the time that the requirement will not be met.

4.4.2 Basic Path Calculations

4.4.2.1 Introduction Whereas the path profile provides data on tower heights to achieve the necessary ray beam path clearance, from the path calculation we derive parameters for dimensioning radio equipment. This includes antenna size or aperture, transmitter power output, receiver noise figure, required bandwidth, whether diversity is to be used, and performance measured in signal-to-noise ratio, noise in the derived voice channel, or bit error rate (BER).

The example we use here is a system using frequency modulation (FM). In Section 4.6 we treat digital line-of-sight microwave links.

4.4.2.2 The FM Receiver The receiver is the starting point on a path calculation. In this case it is an FM receiver. The following question must be answered: What signal level entering the receiver will give the desired link performance?

For the basic approach the case must be simplified. Figure 4.9 is a curve comparing input power level to some typical FM receiver to output signal-to-noise (S/N) ratio in a derived voice channel. Let us assume that the

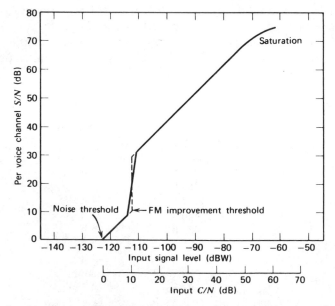

Figure 4.9 Input carrier-to-noise C/N in decibels versus output signal-to-noise ratio S/N (per voice channel) for a typical FM radiolink receiver.

thermal noise threshold for this receiver is -122.5 dBW. If the input RF carrier level was exactly this value the carrier-to-noise (C/N) ratio would be zero. When the input carrier level, which we call receive signal level (RSL), increases to -112.5 dBW, the carrier-to-noise ratio is 10 dB. The carrier-to-noise ratio in decibels is shown below the input carrier level in Figure 4.9.

The relationship between the input carrier-to-noise ratio of an FM receiver and the output signal-to-noise ratio is as follows. At noise threshold, for every 1-dB increase of carrier-to-noise ratio, the signal-to-noise ratio of the output increases approximately 1 dB. This occurs up to a carrier-to-noise ratio of 10 dB or a little greater (11 or 12 dB), when a "capture effect" takes place and the output signal-to-noise ratio suddenly jumps to 30 dB. This point of "capture" is called the FM improvement threshold and is shown in Figure 4.9. Beyond this point the signal-to-noise ratio at the receiver output improves again by 1 dB for every increase of 1 dB of input carrier-to-noise ratio, up to a point where compression starts to take effect (saturation). For instance, if the input carrier-to-noise ratio were 15 dB, we might expect the output signal-to-noise ratio to be approximately 35 dB. This assumes that the FM system has been "adjusted" properly (e.g., there is sufficient deviation to effect FM improvement; see Section 4.5.2.3).

For many radiolink systems an input carrier-to-noise ratio of 10 dB is used as a starting point for path calculations. The reason is obvious, for it is at this point that we start to get the improved signal-to-noise ratio. For future discussion in this chapter we will use the same point of departure. We can calculate this point. It is the noise threshold level, calculated in Chapter 1, plus 10 dB. The FM improvement threshold is

$$\text{FM improvement threshold (dBW)} = 10 \log kTB_{IF} + \text{NF}_{dB} + 10 \text{ dB} \quad (4.9)$$

where T = noise temperature (K)

B_{IF} = noise bandwidth (bandwidth of the intermediate frequency) (Hz)

k = Boltzmann's constant (1.38×10^{-23} W-s/deg)

NF = receiver noise figure (dB)

Simplified for an uncooled receiver (i.e., $T = 290$ K or 62°F),

$$\text{FM improvement threshold (dBW)}$$
$$= -204 \text{ dBW} + \text{NF} + 10 \text{ dB} + 10 \log B_{IF} \quad (4.10)$$

Figure 4.10 gives threshold values for several receiver NFs. However, the threshold may be computed rapidly by directly subtracting the receiver NF, the 10 dB, and then working with the bandwidth. If the bandwidth were only 10 Hz, we would subtract (algebraically add) 10 dB; if it were 20 Hz, we would subtract 13 dB; 100 Hz, 20 dB; 1000 Hz, 30 dB; 10 MHz, 70 dB; 20 MHz, 73 dB; and so on.

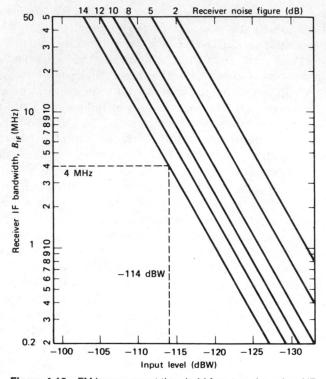

Figure 4.10 FM improvement threshold for several receiver NFs.

Consider an uncooled receiver with a 14-dB NF and 4-MHz IF bandwidth. What would the FM improvement threshold be?

$$\text{Threshold}_{dBW} = -204 + 14 + 10 + 66$$
$$= -114 \text{ dBW}$$

This is the necessary input level to reach an FM improvement threshold. The required IF bandwidth may be estimated by Carson's rule, which states

$$B_{IF} = 2 \,(\text{highest modulating baseband frequency} + \text{peak FM deviation*}).$$
$$(4.11)$$

Example. A video signal with a 4.2-MHz baseband modulates an FM transmitter with a peak deviation of 5 MHz. What is B_{IF}?

$$B_{IF} = 2(4.2 \text{ MHz} + 5 \text{ MHz})$$
$$= 18.4 \text{ MHz}$$

*See Section 4.5.2.3 for calculating peak FM deviation.

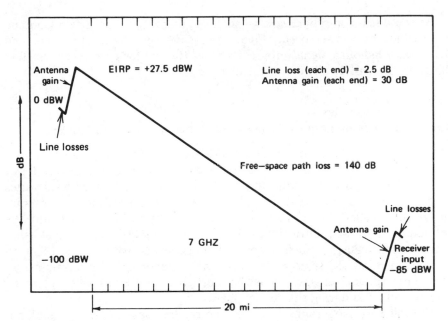

Figure 4.11 Radiolink gains and losses (simplified). Transmitter output = 0 dBW.

The FM improvement threshold may be calculated when given the peak deviation and the bandwidth of the information baseband as well as the receiver NF (or noise temperature; see Section 1.12).

For the most basic path calculation, consider a receiver with an FM improvement threshold of − 114 dBW, a free-space path loss of 140 dB, an isotropic antenna* at both ends, and lossless transmission lines. What would the transmitter output have to be to provide a − 114 dBW input level to the receiver?

$$\text{Output (dBW)} = -114 + 140 = +26 \text{ dBW}$$
$$+ 26 \text{ dBW} \simeq 400 \text{ W}$$

For this case both antennas have unity gain and the transmission lines are lossless. Now extend the example for 2.0-dB line losses and 20-dB antenna gain at each end:

$$\text{Output (dBW)} = -114 \text{ dBW} + 2.0 - 20 \text{ dB} + 140 - 20 \text{ dB} + 2.0$$
$$= -10 \text{ dBW} = 0.1 \text{ W}.$$

Path calculations, therefore, deal with adding gains and losses to arrive at a specified system performance.

*Unrealistic, but useful for discussion.

As the discussion progresses, we shall see many means to attain gain and the various ways losses occur. Figure 4.11 shows graphically the gains and losses as a radiolink signal progresses from transmitter to receiver. Section 4.5.2.6 deals with detailed path calculations.

4.4.3 The Mechanism of Fading—An Introductory Discussion

General Up to this point we have considered that the received signal level remains constant. On most paths, particularly on shorter ones, at lower frequencies, much of the time this holds true. When a receive level varies from the free-space calculated level for a given far-end transmitter output, the result is called fading. Fading due to propagation mechanisms involves refraction, reflection, diffraction, scattering, focusing attenuation, and other miscellaneous causes.* Such factors, when associated with fading, relate to certain conditions classified as meteorological phenomena and terrain geometry. The radiolink engineer must be alert to these factors when planning specific links and during site survey phases.

The following paragraphs cover the two general types of fading, multipath and power.

Multipath Fading This type of fading is due to interference between a direct wave and another wave, usually a reflected wave. The reflection may be from the ground or from atmospheric sheets or layers. Direct path interference may also occur. It may be caused by surface layers of strong refractive index gradients or horizontally distributed changes in the refractive index.

Multipath fading may display fades in excess of 30 dB for periods of seconds or minutes. Typically this form of fading will be observed during quiet, windless, and foggy nights, when temperature inversion near the ground occurs and there is not enough turbulence to mix the air. Thus stratified layers, either elevated or ground based, are formed. Two-path propagation may also be due to specular reflections from a body of water, salt beds, or flat desert between transmitting and receiving antennas. Deep fading of this latter type usually occurs in daytime on over-water paths or other such paths with high ground reflection. Vegetation or other "roughness" found on most radiolink paths breaks up the reflected components, rendering them rather harmless. Multipath fading at its worst is independent of obstruction clearance, and its extreme condition approaches a Rayleigh distribution.

---

*Above 10 GHz, rainfall is also an important factor; see Chapter 7.

Power Fading   Dougherty (Ref. 26) defines power fading as a

... partial isolation of the transmitting and receiving antennas because of

- Intrusion of the earth's surface or atmospheric layers into the propagation path (earth bulge or diffraction fading)
- Antenna decoupling due to variation of the refractive index gradient
- Partial reflection from elevated layers interpositioned between terminal antenna elevations
- "Ducting" formations containing only one of the terminal antennas
- Precipitation along the propagation path...

Power fading is characterized by marked decreases in the free-space signal level for extended time periods. Diffraction may persist for several hours with fade depths of 20–30 dB.

4.5   SPECIFIC LINE-OF-SIGHT (RADIOLINK) TRANSMISSION TECHNIQUES

4.5.1   Introduction

There are two generic LOS microwave transmission techniques in use today:

- Analog systems typically using frequency modulation (FM)
- Digital systems

Historically, FM microwave has been the workhorse of the industry; however, in the late 1970s digital microwave began to be favored. This trend has been accelerating, and the network is evolving to all-digital. FM links still maintain a niche, though, particularly for the transport of video signals.

This section treats FM LOS systems with sufficient detail for a first-cut link design effort. Even though FM systems are being phased out in favor of digital systems, a good understanding of this mature analog technology will prepare the reader to easily transition to the digital world. It also is an excellent lead into Chapters 5–7, on troposcatter, satellite communications, and millimeter wave transmission, respectively.

4.5.2   FM Line-of-Sight Microwave

4.5.2.1   General   Figure 4.12*a* and *b* contains simplified block diagrams of an FM radiolink transmitter and a receiver, respectively. The input information to the transmitter (baseband) and the output from the receiver are considered to be groupings of nominal 4-kHz telephone channels in an

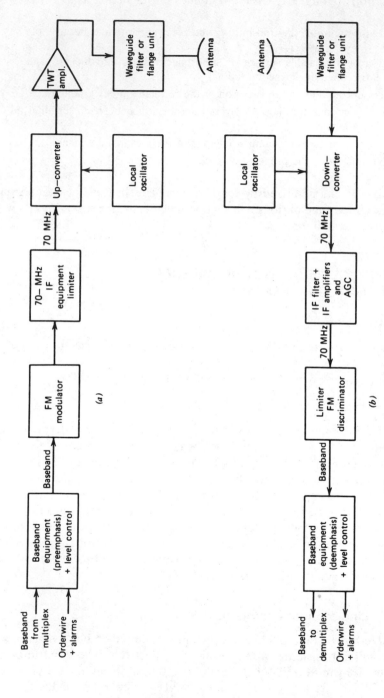

Figure 4.12 (a) Typical radiolink terminal transmitter. (b) Typical radiolink terminal receiver.

SSBSC FDM configuration. This type of multiplexing is discussed in Chapter 3. Orderwire plus alarm information is inserted from 300 Hz to 12 kHz for many installations accepting an FDM multitelephone channel input. Therefore the composite baseband will be made up of the following:

300 Hz–12 kHz	Orderwire/alarms
12 kHz–60 kHz	A 12-channel group of FDM multichannel information

or

300 Hz–12 kHz	Orderwire/alarms
60 kHz–*	Supergroups/mastergroups of multichannel information

or

300 Hz–12 kHz	Orderwire/alarms
312 kHz–*	Supergroups/mastergroups of multichannel information

Other systems use a 12-kHz spectrum with an SSB signal at 3.25 MHz for alarms and orderwire. Video plus program channel subcarriers is another information input that will be discussed in Chapter 13. The orderwire/alarms input for video is necessarily inserted above the video in the baseband.

4.5.2.2 *Preemphasis–Deemphasis*

The output characteristics of an FM receiver without system preemphasis–deemphasis, with an input FM signal of a modulation of uniform amplitude, are such that it has linearly increasing amplitude with increasing baseband frequency. This is shown in Figure 4.13. Note the ramplike or triangular noise of the higher frequency baseband components. The result is decreasing signal-to-noise ratio with increasing baseband frequency. The desired receiver output is a constant signal-to-noise ratio across the baseband. Preemphasis at the transmitter and deemphasis at the receiver achieve this end.

Preemphasis is accomplished by increasing the peak deviation during the FM modulation process for higher baseband frequencies. This increase of peak frequency deviation is done in accordance with a curve designed to effect compensation for the ramplike noise at the FM receiver output. CCIR[†]

---

*See Chapter 3 for specific baseband makeup.

[†]International Consultive Committee for Radio.

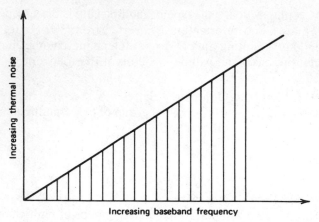

Figure 4.13 Sketch of increasing thermal noise from the output of an FM receiver in a system without preemphasis–deemphasis.

has recommended standardization on the curve shown in Figure 4.14 for multichannel telephony (CCIR Rec. 275-2), and for video transmission a preemphasis curve may be found in CCIR Rec. 405-1 (Figure 4.15).

The preemphasis characteristic is achieved by applying the modulating baseband to a passive network that "forms" the input signal. At the receive end after modulation, and baseband signal is applied to a deemphasis network which restores the baseband to its original amplitude configuration. This is shown diagrammatically in Figure 4.16.

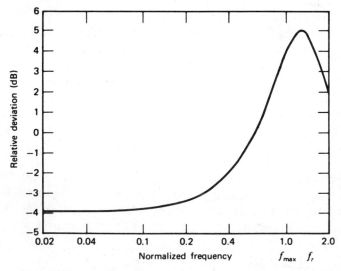

Figure 4.14 Preemphasis characteristic for multichannel telephony. (From CCIR Rec. 275; courtesy of ITU – CCIR.)

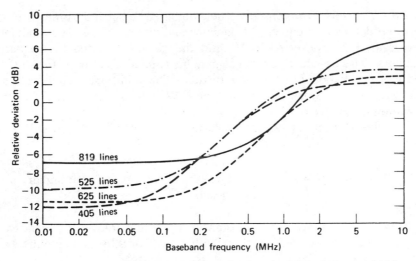

Figure 4.15 Television preemphasis characteristics for 405-, 525-, 625-, and 819-line systems. 0 dB corresponds to a deviation of 8 MHz for a 1-V peak-to-peak signal. (From CCIR Rec. 405; courtesy of ITU – CCIR.)

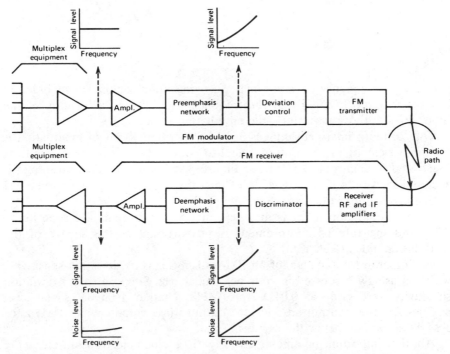

Figure 4.16 Simplified block diagram of an FM radiolink system, showing the effects of preemphasis on the signal and thermal noise.

4.5.2.3 *The FM Transmitter* The combined baseband (orderwire and FDM multiplex line frequency) frequency modulates a wave. The modulated wave is at IF or converted to IF, and thence up-converted to the output frequency. The up-converted signal may be applied directly to the antenna system for radiation or amplified further. The additional amplification is usually carried out by traveling wave tube (TWT) or solid-state amplifier(SSA) for radiolink systems.

Radiolink transmitter outputs have fairly well been standardized as follows:

0.1 W	− 10 dBW	(usually for transmitters operating at higher frequencies)
1.0 W	0 dBW	
10 W	+ 10 dBW	(for transmitters operating at lower frequencies or those with a TWT or SSA final amplifier)

The percentage of modulation or the modulation index is a most important parameter for radio transmitters. The percentage of modulation is used to describe the modulation in AM transmitters. The modulation index is the equivalent parameter for FM transmitters. The modulation index for frequency modulation is

$$M = \frac{F_d}{F_m} \tag{4.12}$$

where F_d = frequency deviation and F_m = maximum significant modulation frequency. Keep in mind that the frequency deviation is a function of the input level to the transmitter—the modulation input.

One of the primary advantages of FM over other forms of modulation is the improvement of the output signal-to-noise ratio for a given input, as illustrated in Figure 4.17 (also see Figure 4.9). However, this advantage is accomplished only if the signal level into the receiver is greater than FM improvement threshold (see Section 4.4.2).

If the modulation index is in excess of 0.6, FM systems are superior to AM, and the rate of improvement is proportional to the square of the modulation index (Ref. 12).

An important parameter for an FM transmitter is its deviation sensitivity, which is usually expressed in volts per megahertz. Suppose that a deviation sensitivity were given as 0.05 V rms/MHz. Then if a 0.05-V signal level appeared at the transmitter input, the output wave would deviate 1 MHz (or 1 MHz above and below the carrier).

Another important parameter given for FM wideband transmitters is the rms deviation per channel at the test tone level. Table 4.4 relates channel capacity and rms deviation (without preemphasis).

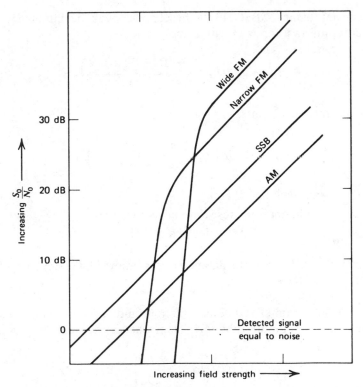

Figure 4.17 Advantage of FM over AM. Relative signal-to-noise ratios as a function of field strength. (Courtesy of the Institute of Telecommunication Sciences, Office of Telecommunications, U.S. Department of Commerce NBS Tech. Note 103.)

Table 4.4   Frequency deviation without preemphasis

Maximum Number of Channels	Rms Deviation per Channel[a] (kHz)
12	35
24	35
60	50, 100, 200
120	50, 100, 200
300	200
600	200
960	200
1260	140, 200
1800	140
2700	140

Source: CCIR Rec. 404-2.

[a]For 1 mW, 800-Hz tone at a point of 0 reference level.

For systems using SSBSC FDM basebands, peak deviation D may be calculated as follows, for $N = 240$ or more,

$$D = 4.47d\left(\log^{-1}\frac{-15 + 10\log N}{20}\right) \tag{4.13a}$$

or for $N =$ between 12 and 240,

$$D = 4.47d\left(\log^{-1}\frac{-1 + 4\log N}{20}\right) \tag{4.13b}$$

where $D =$ peak deviation (kHz)

$\quad\quad\; d =$ per channel rms test tone deviation (kHz)

$\quad\quad\; N =$ number of SSBSC voice channels in the system

(See Section 3.2.4.4 for more detailed explanation of the loading of FDM systems.)

Example. A 300-channel radiolink system would have a 200-kHz/channel rms deviation according to Table 4.4. Then what is the peak deviation?

$$D = (4.47)(200)\left(\log^{-1}\frac{9.77}{20}\right) = 2753 \text{ kHz}$$

The question arises, how much deviation will optimize FM transmitter operation? We know that as the input signal to the transmitter is increased, the deviation is increased. Again, as the deviation increases, the FM improvement threshold becomes more apparent. In effect we are trading off bandwidth for thermal noise improvement. However, with increasing input levels, the IM noise of the system begins to increase. There is some point of optimum input to a wideband FM transmitter where the thermal noise improvement in the system has been optimized and where the IM noise is not excessive. This concept is shown in Figure 4.18.

Frequency stability requirements for FM systems are far less severe than for equivalent SSB systems (see Chapter 3); $\pm 0.001\%$/mo is usually sufficient.

4.5.2.4 The FM Receiver In most applications the radiolink receiver shares the same antenna and waveguide as its companion transmitter to or from a common distant end. Figure 4.12b is a simplified block diagram of a typical radiolink receiver (ideal configuration). The receiver may or may not be connected to the common waveguide manifold by means of a circulator. It also may use a bandpass or preselector filter. Both the circulator and the preselector filter reduce the effects of adjacent transmitter energy to a

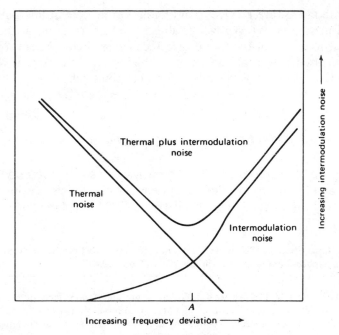

Figure 4.18 Optimum setting for deviation is shown by letter A.

negligible amount in the receiver front end. Radiolink receivers operating at lower frequencies use coaxial cable transmission lines instead of waveguide.

From the manifold or via a circulator and/or preselector, the incoming FM signal next looks into a mixer or down-converter. This unit heterodynes the received signal with the local oscillator signal to produce an IF. Most installations have standardized on a 70-MHz IF. However, CCIR discusses a 140-MHz IF for systems designed to carry voice channels in excess of 1800 in a standard CCITT FDM configuration.

From the mixer or down-converter output, the IF is fed through several amplification stages, often through a phase equalizer to correct delay distortion introduced by IF (and RF) filters. IF gains commonly are on the order of 80–90 dB.

The output of the IF is fed to a limiter–discriminator which is the FM detector or demodulator. The output of the discriminator is the composite baseband made up of the information baseband plus orderwire and alarm signals.

After demodulation the composite baseband is passed through a deemphasis network (see Figure 4.16). Thence the composite signal is split, the information baseband being directed to the demultiplex equipment and the orderwire (service channel) and alarms to the orderwire equipment and alarm display.

The LOS microwave receiver plays an important role in link path calculations and noise analysis. We will establish a reference point at the input of the first active stage of the microwave receiving system. Where a low noise amplifier is used, it will be the input to this device; otherwise it will be the input to the mixer. It is at this reference point where we measure the important receive signal level (RSL). The RSL is the level of the received RF carrier measured in either dBm or dBW.

4.5.2.5 Wideband FM Radio Noise Considerations

4.5.2.5.1 Noise Planning for FM Microwave Links—Requirements. In the design of an LOS microwave link or system of links, a noise requirement will be specified. In the case of a system carrying FDM telephony, noise power in

Table 4.5   Noise accumulation in an FDM voice channel due to radio portion on 2500-km hypothetical reference circuit – CCIR Rec. 395-2

1. That, in circuits established over real links that do not differ appreciably from the hypothetical reference circuit, the psophometrically weighted noise power at a point of zero relative level in the telephone channels of frequency-division multiplex radio-relay systems of length L, where L is between 280 and 2500 km, should not exceed:
    1.1 $3L$ pW 1-min mean power for more than 20% of any month;
    1.2   47,500 pW 1-min mean power for more than $(L/2500) \times 0.1\%$ of any month; it is recognized that the performance achieved for very short periods of time is very difficult to measure precisely and that in a circuit carried over a real link, it may, after installation, differ from the planning objective;
2. That circuits to be established over real links, the composition of which, for planning reasons, differs substantially from the hypothetical reference circuit, should be planned in such a way that the psophometrically weighted noise power at a point of zero relative level in a telephone channel of Length L, where L is between 50 and 2500 km, carried in one or more baseband sections of frequency-division multiplex radio links, should not exceed:
    2.1   For $50 \leq L \leq 840$ km:
        2.1.1 $3L$ pW + 200 pW 1-min mean power for more than 20% of any month,
        2.1.2   47,500 pW 1-min mean power for more than $(280/2500) \times 0.1\%$ of any month when L is less than 280 km, or more than $(L/2500) \times 0.1\%$ of any month when L is greater than 280 km;
    2.2   For $840 < L \leq 1670$ km:
        2.2.1 $3L$ pW + 400 pW 1-min mean power for more than 20% of any month.
        2.2.2   47,500 pW 1-min mean power for more than $(L/2500) \times 0.1\%$ of any month;
    2.3   For $1670 < L \leq 2500$ km:
        2.3.1 $3L$ pW + 600 pW 1-min mean power for more than 20% of any month
        2.3.2   47,500 pW 1-min mean power for more than $(L/2500) \times 0.1\%$ of any month;
3. That the following Note should be regarded as part of the Recommendation: NOTE 1. Noise in the frequency-division multiplex equipment is excluded. On a 2500 km hypothetical reference circuit the CCITT allows 2500 pW mean value for this noise in any hour.

Source: CCIR Rec. 395-2 (Ref. 3). Courtesy of ITU–CCIR.
[a]The level of uniform-spectrum noise power in a 3.1-kHz band must be reduced by 2.5 dB to obtain the psophometrically weighted noise power.

the derived voice channel will be specified. For a system carrying video, the requirement will probably be stated as a weighted signal-to-noise ratio.

Because the carrier-to-noise ratio at the receiver input varies with time, noise power is specified statistically. Specifications are typically based on Table 4.5 and derive from the following excerpt from CCIR Rec. 393-4 (Ref. 3):

> The noise power at a point of zero relative level in any telephone channel on a 2500 km hypothetical reference circuit for frequency division multiplex radio relay systems should not exceed the values given below, which have been chosen to take account of fading:
>
> - 7500 pW0p, psophometrically weighted one minute mean power for more than 20% of any month;
> - 47,500 pW0p, psophometrically weighted mean power for more than 0.1% of any month;
> - 1,000,000 pW0, unweighted (with an integrating time of 5 ms) for more than 0.01% of any month.

If we were to connect a psophometer to the end of a 2500-km (reference) circuit made up of homogeneous radio-relay sections, in a derived voice channel we should read values no greater than those shown above plus 2500 pW0p mean noise power due to the FDM equipment contribution. For the first entry above, we would add the 2500 pW0p to the 7500 pW0p value for a total of 10,000 pW0p which would not be exceeded for more than 20% of any month; and so on for the other two entries.

The problem addressed in this section is to ensure that the radiolink design can meet this or other similar criteria.

4.5.2.5.2 Noise in the Derived FDM Voice Channel

INTRODUCTION. On an FM LOS microwave link, noise is examined during conditions of fading and for the unfaded or median noise condition. For the low RSL or faded condition, Ref. 2 states that

> the noise in a derived (FDM) voice channel at FM threshold, falls approximately at, or slightly higher than the level considered to be the maximum tolerable noise for a telephone channel in the public network. By present standards, this maximum is considered to be 55 dBrnc0 (316,200 pWp0). In industrial systems, a value of 59 dBrnc0 (631,000 pWp0) is commonly used as the maximum acceptable noise level.

FM threshold, therefore, is the usual point where fade margin is added (Section 4.5.2.7) to achieve the unfaded RSL. In many systems, however, the CCIR guideline of 3 pWp/km may not be achieved at the high-signal-level condition, and the unfaded RSL may have to be increased still further.

At low RSL, the primary contributor to noise is thermal noise. At high-signal-level conditions (unfaded RSL), there are three contributors:

- Thermal noise
- Radio equipment IM noise
- IM noise due to antenna feeder distortion

Each are calculated for a particular link for unfaded RSL and each value is converted to noise power in mW, pW, or pWp and summed. The sum is then compared to the noise apportionment for the link from Table 4.5.

CALCULATION OF THERMAL NOISE. The conventional formula (Ref. 19) to calculate signal-to-noise power ratio on FM radiolink (test tone to flat weighted noise for thermal noise) is

$$\frac{S}{N} = \frac{RSL}{(2ktbF)[\Delta f/f_c]^2} \tag{4.14}$$

where b = channel bandwidth (3.1 kHz)

Δf = channel test tone peak deviation (as adjusted by preemphasis)

f_c = highest channel (center) frequency in baseband (KHz)

F = the noise factor of the receiver (i.e., the numeric equivalent of the noise figure, dB)

$kt = 4 \times 10^{-18}$ mW/Hz

RSL = receive signal level (mW)

S/N = test tone-to-noise ratio (numeric equivalent of S/N, dB)

In the more useful decibel form, equation 4.14 is

$$\left(\frac{S}{N}\right)_{dB} = RSL_{dBm} + 136.1 - NF_{dB} + 20\log\left(\frac{\Delta f}{f_c}\right) \quad \text{(flat)} \tag{4.15}$$

In a similar fashion (Ref. 2) the noise power in the derived voice channel can be calculated as follows:

$$P_{dBrnc0} = -RSL_{dBm} - 48.1 + NF_{dB} - 20\log\left(\frac{\Delta f}{f_c}\right) \tag{4.16}$$

and

$$P_{pWp0} = \log_{10}^{-1}\left[\frac{-RSL_{dBm} - 48.6 + NF_{dB} - 20\log(\Delta f/f_c)}{10}\right] \tag{4.17}$$

Note: Values for Δf, the *channel* test tone *peak* deviation (Table 4.4): 200-kHz rms deviation for 60–960 VF channel loading and 140-kHz rms deviation for transmitters loaded with more than 1200 voice channels. Per channel peak deviation is 282.8 and 200 kHz, respectively (e.g., use sinusoid peaking factor of 1.414 from rms).

Example. A 50-hop microwave system is designed to meet CCIR noise requirements. The system is 2500 km long, so each hop contributes 150 pWp of noise. The system carries 300 FDM channels and the highest channel center frequency is 1248 kHz. The receiver noise figure is 10 dB. Calculate the receiver RSL at the end of one hop.

Use equation 4.16 and set 150 pWp0 equal to the value of the right-hand side of the equation:

$$150 \text{ pWp0} = \log^{-1}\left[\frac{-\text{RSL} - 48.6 + 10 \text{ dB} - 20\log(\Delta f/f_c)}{10}\right]$$

Calculate the value of

$$20\log\left(\frac{\Delta f}{f_c}\right) = 20\log\left(\frac{282.8}{1248}\right)$$

$$= -12.9 \text{ dB}$$

Then

$$10\log(150) = -\text{RSL} - 48.6 + 10 + 12.9$$

$$= -47.4 \text{ dBm}$$

Table 4.6 presents equivalent values of dBrnc, pWp, and signal-to-noise power ratio for the standard voice channel.

CALCULATION OF RADIO EQUIPMENT IM NOISE.   Up to this point we have only dealt with thermal noise in a radiolink. In an operational analog radiolink, a second type of noise can be equally important. This is intermodulation noise (IM noise).

IM noise is caused by nonlinearity when information signals in one or more channels give rise to harmonics or intermodulation products that appear as unintelligible noise in other channels. In an FDM/FM radiolink, nonlinear noise in a particular channel varies as the multiplex signal level and the position of the channel in the multiplex baseband spectrum. For a fixed multiplex signal level and for a specific FDM channel, nonlinear noise is constant.

Table 4.6 Approximate equivalents for signal-to-noise ratio (S/N) and common noise level values[a]

dBrnc0	pWp0	dBm0p	S/N	NPR (flat)	dBrnc0	pWp0	dBm0p	S/N	NPR (flat)
0	1.0	−90	88	71.6	30	1,000	−60	58	41.6
1	1.3	−89	87	70.6	31	1,259	−59	57	40.6
2	1.6	−88	86	69.6	32	1,585	−58	56	39.6
3	2.0	−87	85	68.6	33	1,995	−57	55	38.6
4	2.5	−86	84	67.6	34	2,520	−56	54	37.6
5	3.2	−85	83	66.6	35	3,162	−55	53	36.6
6	4.0	−84	82	65.6	36	3,981	−54	52	35.6
7	5.0	−83	81	64.6	37	5,012	−53	51	34.6
8	6.3	−82	80	63.6	38	6,310	−52	50	33.6
9	7.9	−81	79	62.6	39	7,943	−51	49	32.6
10	10.0	−80	78	61.6	40	10,000	−50	48	31.6
11	12.6	−79	77	60.6	41	12,590	−49	47	30.6
12	15.8	−78	76	59.6	42	15,850	−48	46	29.6
13	20.0	−77	75	58.6	43	19,950	−47	45	28.6
14	25.2	−76	74	57.6	44	25,200	−46	44	27.6
15	31.6	−75	73	56.6	45	31,620	−45	43	26.6
16	39.8	−74	72	55.6	46	39,810	−44	42	25.6
17	50.1	−73	71	54.6	47	50,120	−43	41	24.6
18	63.1	−72	70	53.6	48	63,100	−42	40	23.6
19	79.4	−71	69	52.6	49	79,430	−41	39	22.6
20	100	−70	68	51.6	50	100,000	−40	38	21.6
21	126	−69	67	50.6	51	125,900	−39	37	20.6
22	158	−68	66	49.6	52	158,500	−38	36	19.6
23	200	−67	65	48.6	53	199,500	−37	35	18.6
24	252	−66	54	47.6	54	252,000	−36	34	17.6
25	316	−65	63	46.6	55	316,200	−35	33	16.6
26	398	−64	62	45.6	56	398,100	−34	32	15.6
27	501	−63	61	44.6	57	501,200	−33	31	14.6
28	631	−62	60	43.6	58	631,000	−32	30	13.6
29	794	−61	59	42.6	59	794,300	−31	29	12.6

Source: Extracted from EIA RS-252A (Ref. 19). Courtesy of Electronics Industries Association.
[a]This table is based on the following commonly used relationships, which include some rounding off for convenience:

Correlations between columns 3 and 4 are valid for all types of noise. All other correlations are valid for white noise only. dBrnc0 = $10 \log_{10}$ pWp0 = dBm0p + 90 = 88 − S/N (flat) = 7.16 − NPR.

For low receive signal levels at the far-end FM receiver, such as during conditions of deep fades, thermal noise limits performance. During the converse condition, when there are high signal levels, IM noise may become the limiting performance factor.

In an FM radio system, nonlinear (IM) noise may be attributed to three principal factors: (1) transmitter nonlinearity, (2) multipath effects of the medium, and (3) receiver nonlinearity. Amplitude and phase nonlinearity are equally important in contribution to total noise and each should be carefully considered.

A common method of measuring total noise on an FM radiolink under maximum (traffic) loading conditions consists of applying a "white noise" signal at the baseband input port of the FM transmitter. A white noise generator produces a noise spectrum approximating that produced by the FDM equipment. The output noise level of the generator is adjusted to a desired multiplex composite baseband level (composite noise power). Then a notched filter is switched in to clear a narrow slot in the spectrum of the noise signal, and a noise analyzer is connected to the output of the system. The analyzer can be used to measure the ratio of the composite noise power to the noise power in the cleared slot. The noise power is equivalent to the total noise (e.g., thermal plus IM noise) present in the slot bandwidth. Conventionally, the slot bandwidth is made equal to that of a single FDM voice channel.

The most common unit of noise measurement in white noise testing is noise power ratio (NPR), which is defined as follows (Ref. 8):

NPR is the decibel ratio of the noise level in a measuring channel with the baseband fully loaded...to...the level in that channel with all of the baseband noise loaded except the measuring channel.

The notched (slot) filters used in white noise testing have been standardized for radiolinks by CCIR in Rec. 399-3 (Ref. 3) for 10 common FDM baseband configurations. Available measuring channel frequencies and high and low baseband cutoff frequencies are shown in Table 4.7. In an NPR test, usually three different slots are tested separately: high frequency, midband, and low frequency

When an NPR measurement is made at high RF signal levels, such as when the measurement is made in a back-to-back configuration, the dominant noise component is equipment IM noise. This parameter can be used as an approximation of the equipment IM noise contribution. This value together with stated equivalent noise loading should also be available from manufacturer's published specifications on the equipment to be used. Modern, new radiolink terminal equipment (i.e., that equipment that accepts an information baseband for modulation ad demodulates an RF signal to baseband) should display an NPR of at least 55 dB when tested back-to-back. It should be noted, however, that diversity combining can improve NPR. This is because in equal gain and maximal ratio combiners, the signal powers are added coherently, whereas the IM noise contribution, which is similar (for this discussion) to other noise, is added randomly. Reference 18 allows a 3-dB improvement in NPR when diversity combining is used on a link.

Up to this point NPR has been treated for terminal radio equipment or baseband repeaters. If the designer is concerned with heterodyne repeaters (Section 4.5.2.10), the white noise test procedure as previously described

Table 4.7 CCIR measurement frequencies for white noise testing

System Capacity (Channels)	Limits of Band Occupied by Telephone Channels (kHz)	Effective Cutoff Frequencies of Band-Limiting Filters (kHz) High Pass	Low Pass	Frequencies of Available Measuring Channels (kHz)								
				70	270	534	1,248	2,438	3,886	5,340	7,600	11,700
60	60–300	60 ± 1	300 ± 2	70	270							
120	60–552	60 ± 1	552 ± 4	70	270	534						
300	60–1,300 / 64–1,296	60 ± 1	1,296 ± 8	70	270	534	1,248					
600	60–2,540 / 54–2,660	60 ± 1	2,600 ± 20	70	270	534	1,248	2,438				
960	60–4,028 / 64–4,024	60 ± 1	4,100 ± 30	70	270	534	1,248	2,438	3,886			
900	316–4,188	316 ± 5	4,100 ± 30			534	1,248	2,438	3,886			
1,260	60–5,636 / 60–5,564	60 ± 1	5,600 ± 50	70	270	534	1,248	2,438	3,886	5,340		
1,200	316–5,564	316 ± 5	5,600 ± 50			534	1,248	2,438	3,886	5,340		
1,800	312–8,120 / 312–8,204 / 316–8,204	316 ± 5	8,160 ± 75			534	1,248	2,438	3,886	5,340	7,600	
2,700	312–12,336 / 316–12,388 / 312–12,388	316 ± 5	12,360 ± 100			534	1,248	2,438	3,886	5,340	7,600	11,700

Source: CCIR Rec. 399-3 (Ref. 3). Courtesy of ITU–CCIR.

cannot be carried out per se, and the designer should rely on manufacturer's specifications. Alternatively, about a 4-dB improvement (Ref. 18) in NPR over baseband radio equipment may be assumed, or for a new IF repeater of modern design, an NPR of at least 59 dB should be achieved.

Specifying the noise in a test channel by NPR provides a relative indication of IM noise and crosstalk. An alternative is to express the noise in decibels relative to a specified absolute signal level in a test channel. In this case we can define the signal-to-noise power ratio (S/N) as the decibel ratio of the level of the standard test tone to the noise in a standard channel bandwidth (3100 Hz) within the test channel or

$$\left(\frac{S}{N}\right)_{\text{dB}} = \text{NPR} + \text{BWR} - \text{NLR} \tag{4.18}$$

where NLR = noise load ratio and BWR = bandwidth ratio.

For FDM telephony baseband loading of an FM microwave transmitter following CCIR recommendations, NLR in decibels may be calculated using the following formulas:

$$\text{NLR}_{\text{dB}} = -1 + 4\log N \tag{4.19}$$

for FDM configurations from 12 to 240 voice channels and

$$\text{NLR}_{\text{dB}} = -15 + 10\log N \tag{4.20}$$

for FDM configurations where N is greater than 240. For U.S. military systems, the value

$$\text{NLR}_{\text{dB}} = -10 + 10\log N \tag{4.21}$$

is used.

Using AT&T practice, the following formula is used in lieu of equation 4.20:

$$\text{NLR}_{\text{dB}} = -16 + 10\log N \tag{4.22}$$

The bandwidth ratio

$$\text{BWR}_{\text{dB}} = \left(\frac{\text{occupied baseband of white noise test signal}}{\text{voice channel bandwidth}}\right) \tag{4.23}$$

The denominator in equation 4.23 can usually be taken as 3.1 kHz. The numerator can be taken from Table 4.7.

Example. A particular FM radiolink transmitter and receiver back-to-back display an NPR of 55 dB. They have been designed and adjusted for 960 VF channel operation and will use CCIR loading. What is the S/N in a voice channel?

Consult Table 4.7. The baseband occupies 60–4028 kHz.

$$\text{BWR} = 10\log\left(\frac{4028 - 60}{3.1}\right) \text{ (kHz)}$$

$$= 10\log\left(\frac{3968}{3.1}\right)$$

$$= 31.07 \text{ dB}$$

$$\text{NLR} = -15 + 10\log(960)$$

$$= 14.82 \text{ dB}$$

$$\frac{S}{N} = 55 \text{ dB} + 31.07 \text{ dB} - 14.82 \text{ dB}$$

$$= 71.25 \text{ dB}$$

IM NOISE DUE TO ANTENNA FEEDER DISTORTION.   Antenna feeder distortion or echo distortion is caused by mismatches in the transmission line connecting the radio equipment to the antenna. These mismatches cause echos or reflections of the incident wave. Similar distortion can be caused by long IF runs; however, in most cases, this can be neglected.

Echo distortion actually results from a second signal arriving at the receiver but delayed in time by some given amount. It should be noted that multipath propagation may also cause the same effect. In this case, though, the delay time is random and continuously varying, thereby making analysis difficult, if not impossible.

The level of the echo signal is an inverse function of the return loss at each end of the transmission line and its terminating device (i.e., the antenna at one end and the communication equipment at the other). An echo signal so generated will be constant, since the variables that established it are constant. Thus the distortion created by the echo will be constant but contingent on modulation. In other words, if the carrier were unmodulated, there would be no distortion due to echo. When the carrier is modulated, echo appears.

Ref. 4 provides a method of calculating IM noise due to antenna feeder distortion. Such IM noise makes a contribution to total noise at each end of a link. For noise budgeting purposes, allow about 7-pWp IM feeder noise contribution at each end of a link, for a total of 14 pWp.

TOTAL NOISE IN THE DERIVED VOICE CHANNEL.   To calculate the total noise in the voice channel from each of the three noise contributors, convert to absolute values of noise power such as mW, pW, or pWp, sum, and reconvert to a decibel unit if so desired. The thermal noise value should be taken for unfaded RSL conditions.

Example. Suppose the thermal noise contribution were 206 pWp, the IM noise contribution 37.5 pWp, and antenna IM feeder noise contribution 12.5 pWp. The total noise would be 256 pWp, or simply the sum of the three contributors.

SOME NOTES ON NOISE IN THE VOICE CHANNEL

- A voice channel with psophometric weighting has a 2.5-dB noise improvement over a flat channel.
- dBrnc (dB above reference noise with C-message weighting): The reference frequency/level is a 1000-Hz tone at -90 dBm (Ref. 32).
- pWp (picowatts of noise power with psophometric weighting): The noise power reference frequency/level is an 800-Hz tone where 1 pWp $= -90$ dBm (pWp = pW $\times$ 0.56) (Ref. 32).
- dBmp (psophometrically weighted noise power measured in dBm), where, with an 800-Hz tone 0 dBmp = 0 dBm. For flat noise in the band 300–3400 Hz, dBmp = dBm $-$ 2.5 dB (ref. 32).
- 0 dBrnc $= -88.5$ dBm. Commonly, the -88.5-dBm value is rounded off to -88 dBm, thus 0 dBrnc $= -88.0$ dBm (Ref. 32).
- dBrnc = dBmp + 90 dB: It should be noted that C-message weighting and psophometric weighting in fact vary by 1 dB (-1.5 dB and -2.5 dB) and this equivalency has an inherent error of 1 dB. However, the equivalency is commonly accepted in the industry (Ref. 32).
- dBrnc0 = 10 log pWp0 + 0.8 dB = dBmp + 90.8 dB = 88.3 (dB) $-$ $(S/N)_{dB \, (flat)}$ (Ref. 2).

To calculate flat noise in the test channel, the following expression applies:

$$P_{tcf} = \log^{-1}\left(\frac{90 - (S/N)_{dB}}{10}\right) \quad (pW0) \qquad (4.24)$$

and to calculate psophometrically weighted noise:

$$P_{tcp} = 0.56 \log^{-1}\left(\frac{90 - (S/N)_{dB}}{10}\right) \quad (pWp0) \qquad (4.25)$$

Example. If S/N in a voice channel is 71.25 dB, what is the noise level in that channel in pWp0?
Use equation 4.25:

$$P_{tcp} = 0.56 \log^{-1}\left(\frac{90 - 71.25}{10}\right)$$
$$= 0.56 \times 74.99$$
$$= 41.99 \text{ pWp0}$$

4.5.2.6 Detailed Path Calculations

4.5.2.6.1 Introduction. In Section 4.4.2 we introduced the concept of path calculations. Path calculations size or dimension radio equipment to meet noise requirements for analog links or bit error rate (BER) requirements for digital links. The path profile tells us how high the towers must be. Path calculations provide us with the necessary equipment parameters, such as

- Antenna size (gain)
- Transmitter power output
- Receiving system noise characteristics (e.g., use of a low-noise amplifier [LNA] option)
- Diversity (if required), order, and type
- Transmission line characteristics
- Resiting; increasing the number of links (hops), thus decreasing the distance between certain adjacent sites

This last item on the list certainly is the most costly. Under certain circumstances, however, it may remain the ultimate alternative to meet system noise or BER requirements.

Each item on the list above has some limit as to its ability to enhance system performance. The limit may be brought about by economic factors or by the laws of physics. For instance, antenna gain is a function of antenna aperture (size). In general, LOS microwave antennas with apertures greater than 12 ft (3.8 m) are not recommended, solely because of cost. The cost is not from the antenna itself. It is due to the sail area of the antenna, which becomes so great that special tower stiffening is required to reduce twist and sway of the tower during high wind conditions.

One major objective of a path calculation is to calculate RSL (receive signal level) for unfaded conditions. By varying one or several of the parameters associated with the list of factors above, we can adjust the RSL. In the unfaded condition our primary interest is in the thermal noise level of the derived FDM voice channel (see Section 4.5.2.5.2). It must be a certain specified value to meet system requirements. We then will calculate a fade margin by increasing the RSL still further. The fade margin is the amount we overbuild a link/system to meet specified time availability* requirement. The calculation of the amount of required fade margin is described in Section 4.5.2.7.

4.5.2.6.2 The Path Calculation Process. We introduced the basic concepts of path calculations in Section 4.4.2. The steps required to calculate the RSL are presented below. Figure 4.19 is a model of a typical microwave LOS

---

*As used in this text, time availability and propagation reliability are synonymous.

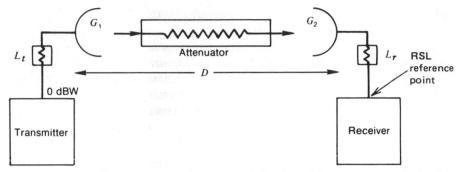

Figure 4.19 Simplified model, radiolink path analysis. L_t and L_r are the transmission line losses; G_1 and G_2 are the antenna gains.

link. The figure illustrates antennas (where we achieve a form of gain), transmission line losses, and the RSL reference point at the input of the first active stage of the receiver.

These steps are for calculating the RSL (also see Figure 4.20):

1. Calculate the effective isotropically radiated power (EIRP) of the microwave transmit installation (Section 1.14).
2. Calculate the free-space loss (FSL) (Section 4.3.1, equation 4.1).
3. Calculate the IRL (isotropic receive level). This is the RF signal level impinging on the far-end receive antenna as if it were an isotropic antenna:

$$\text{IRL}_{\text{dBW}} = \text{EIRP}_{\text{dBW}} - \text{FSL}_{\text{dB}} \qquad (4.26)$$

4. Calculate the unfaded RSL. This is done by adding the receive antenna gain to the IRL and subtracting the receive transmission line losses:

$$\text{RSL}_{\text{dBW}} = \text{IRL}_{\text{dBW}} + \text{rec. ant. gain}_{\text{(dB)}} - \text{line losses}_{\text{(dB)}} \quad (4.27)$$

5. Calculate the thermal noise threshold of the far-end receiving system (Section 4.4.2 and equation 1.20b):

$$P_{\text{n(dBW)}} = -204 \text{ dBW} + \text{NF}_{\text{bB}} + 10 \log(\text{IF bandwidth}_{\text{Hz}})$$

6. Calculate the carrier-to-noise ratio (C/N):

$$C/N_{\text{(dB)}} = \text{RSL}_{\text{dBW}} - P_{\text{n(dBW)}} \qquad (4.28)$$

Site .... To site ....
Equipment type ...... Frequency (MHz) .... Receiver noise figure.. dB
Type of traffic .... Highest baseband frequency..
Frequency deviation.. B_{IF} ....
Diversity or hot standby................. Radio pilot ....
Reference path profile .... True bearing $A-B$ ....$_\circ$
Path length (statute mi or km) .... True bearing $B-A$ ....$_\circ$

Step	Add/Subtract		Comments
1. EIRP			
A. Transmitter output (dBW)	+		
B. Connector/transition losses	−		Include directional cplr loss (if any)
C. Circulator loss	−		
D. Transmission line loss	−		
E. Transmit antenna gain	+		
EIRP		dBW	
2. Free-space loss	−		
3. Isotropic receive level		dBW	
4. Receive signal level			
A. Receive antenna gain (dB)	+		
B. Transmission line loss	−		
C. Circulator loss	−		
(RSL)		dBW	
5. Thermal noise threshold		dBW	
6. Carrier-to-noise ratio (C/N)		dB	
7. FM improvement reference (threshold)		dBW	
A. Reference threshold to meet noise requirements		dBW	If not the same as FM improvement
8. Margin		dB	Steps 4–7
9. Improvements			
Preemphasis		dB	See Figure 4.21
Diversity		dB	
Weighting		dB	
10. Noise in VF channel			
Thermal	+	pWp/dBrnc	
IM	+	pWp/dBrnc	
Antenna feeder IM	+	pWp/dBrnc	
Total Noise in VF Channel		pWp/dBrnc	

Figure 4.20

11. Allocated noise requirement   pWp/dBrnc
12. Noise Margin   dB
13. Calculated fade margin   dB   13 should equal 8

Notes:

The heading section is self-explanatory. This is a path calculation work sheet for one-hop connecting sites *A* and *B*. These sites are usually assigned geographic names.

Type of traffic: FDM or TV.

Frequency deviation (peak deviation) and B_{IF} (bandwidth of the IF): use equations 4.13 and 4.11, respectively.

Pilot tones (radio): see Section 4.5.2.9.

The true bearings should be calculated during the path profile exercise. These bearings are required for initial antenna pointing.

Step 1, items B, C, and D step 4, items B and C: see Section 4.7.

Step 1E and Step 4A, see Section 4.7.

Steps 1–6 are also applicable for digital LOS microwave links.

Figure 4.20   *(Continued)*

7. Calculate the FM improvement threshold reference point (equation 4.10):

$$\text{FM improvement threshold}_{dBW}$$
$$= -204 \text{ dBW} + \text{NF}_{dB} + 10 \log(\text{IF bandwidth}_{Hz})$$

8. Calculate the link margin by algebraically subtracting the FM improvement threshold from the RSL. *Note*: Most or all of this margin can be assigned as fade margin.

9. Improvements in the noise level in a voice channel:
   - Preemphasis improvement from Figure 4.21
   - If we use diversity, Section 4.5.2.8, with dual diversity we may take a 3-dB improvement in the unfaded condition
   - Psophometric weighting provides a noise improvement of 2.5 dB; C-message weighting, 2 dB

10. Calculate thermal, intermodulation (IM), and antenna feeder IM noise contributions (Section 4.5.2.5.2). Convert to watt-units and add. Include improvements from item 9 above.

11. Compare the sum of noise level values and improvements in item 10 above to the noise allocation requirement (Section 4.5.2.5.1).

12. Calculate the noise margin (items 11–10).

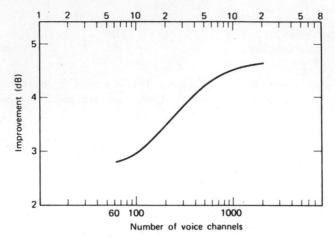

Figure 4.21 Preemphasis improvement in decibels for a given number of FDM voice channels.

13. Calculate the fade margin (Section 4.5.2.7). This value should be equal or greater than the value in item 8. If the value is less, improve one or several of the parameters starting at the top of the list in Section 4.5.2.6.1. The items are listed from least costly to most costly to implement.

The best way to illustrate the above process is by working a typical example. Once the process is well understood, such a path calculation may be carried out in a tabular fashion. It also lends itself well to a simple PC program using either BASIC or FORTRAN.

Example. An FM LOS microwave link operates at 6.150 GHz. The required receiver IF bandwidth is 20 MHz. The transmitter output power at the flange is 1 W. The receiver front end's first active stage is a mixer with a noise figure of 9 dB and is preceded by a bandpass filter with an insertion loss of 0.5 dB. The path length is 21 mi; the antennas at each end have a 35-dB gain and the transmission line losses at each end are 3 dB. If the FM improvement threshold is used as the unfaded reference, what fade margin is available?

1. EIRP (dBW) = 0 dBW − 3 dB + 25 dB
   = +32 dBW
2. FSL (dB) = 36.58 + 20 log 21 + 20 log 6150
   = 138.8 dB
3. IRL = EIRP − FSL
   = +32 dBW − 138.8 dB
   = −106.8 dBW

4. RSL = IRL + rec. ant. gain − line losses
    = − 106.8 dBW + 35 dB − 3 dB
    = − 74.8 dBW

5. Thermal noise threshold = − 204 dBW + NF (dB) + 10 log (BW)
                          = − 204 dBW + 9 dB + 73.01 dB
                          = − 122 dBW = N (dBW)

6. C/N = RSL/N = RSL (dBW) − N (dBW)
    = − 74.8 dBW − (− 122 dBW)
    = 47.2 dB

7. Margin to FM improvement reference = C/N − 10 dB
                                     = 47.2 dB − 10 dB = 37.2 dB

8. Margin = 37.2 dB

    37.2 dB is the fade margin available.

4.5.2.7 Determination of Fade Margin   Fading is a random increase in path loss during abnormal propagation conditions. During such conditions path loss may increase 10, 20, 30 dB or more for short periods. The objective of this subsection is to assist in setting a margin or overbuild in system design to minimize the effects of fading.

The factors involved in fading phenomena are many and complex. They have been discussed in some detail in Section 4.4.3. The principal types of fading below 10 GHz is multipath fading. CCIR Rep. 784 (Ref. 3) states that

> ...multipath effects due to atmosphere increase slowly with frequency, but much more rapidly with path length (the deep fading probability follows an approximate law $f \cdot d^{3.5}$, where f is the carrier frequency and d is distance).

Determining the fade margin without resorting to live path testing is not an easy matter for the radio system engineer. One approach is to simply assume what is often considered the worst fading condition on a single radiolink hop. This is the familiar Rayleigh fading, which can be summarized as follows:

Single-Hop Propagation Reliability (%)	Required Fade Margin (dB)
90	8
99	18
99.9	28
99.99	38
99.999	48

For other propagation reliability values, simple interpolation can be used. For instance, for a hop where a 99.95% reliability is desired, the fade margin required for Rayleigh would be 33 dB.

This expresses worst-case fading and does not take into account the variables mentioned in the CCIR report, namely, hop length (distance) and frequency. Adding margin costs money, and it would be desirable to reduce the margin yet maintain hop reliability.

Further refinements have been made in the methodology of estimating fade margins. Experience in the design of many radiolink systems has shown that the incidence of multipath fading varies not as only as a function of path length and frequency but also as a function of climate and terrain conditions. It has been found that hops over dry, windy, mountainous areas are the most favorable, displaying a low incidence of fading. Worst fading conditions usually occur in coastal areas that are hot and humid, and inland temperate regions are somewhere in between the extremes. Flat terrain along a radiolink path tends to increase the probability of fading, while irregular hilly terrain, especially with vegetation, tends toward lower incidence of fading (e.g., depth and frequency of events).

W. T. Barnett and A. Vigants of Bell Telephone Laboratories (Ref. 2) have developed an empirical method to further refine the estimation of fade margins. The percentage we have been using really expresses an availability A of a radiolink path or hop, sometimes called *time availability*. Subtracting the percentage from 1.000 gives the path unavailability u. As a percentage, $A = 100 (1 - u)$. Following Barnett's argument (in general), let U_{ndp} be the nondiversity annual outage probability and r the fade occurrence factor:

$$r = \frac{\text{actual fade probability}}{\text{Rayleigh fade probability}} \qquad (4.29)$$

If F is the fade margin in decibels, then

$$r = \frac{\text{actual fade probability}}{10^{-F/10}} \qquad (4.30)$$

For the worst month,

$$r_m = a \times 10^{-5} \left(\frac{f}{4}\right) D^3 \qquad (4.31)$$

where D = path length (statute mi)
$\quad\;\; f$ = frequency (GHz)
$\quad\;\; F$ = fade margin (dB)
$\quad\;\; a$ = 4 for very smooth terrain, over water, flat desert
$\quad\quad\;\;$ = 1 for average terrain with some roughness
$\quad\quad\;\;$ = 0.25 for mountains, very rough or very dry terrain

Over a year,

$$r_{yr} = br_m \tag{4.32}$$

where b = 0.5 for hot, humid coastal areas
  = 0.25 for normal, interior temperate or subarctic areas
  = 0.125 for mountainous or very dry but nonreflective areas

$$U_{ndp} = r_{yr}(10^{-F/10}) = br_m(10^{-F/10})$$
$$= 2.5abfD^3(10^{-F/10})(10^{-6}) \tag{4.33}$$

Example. Given a 25-mi path with average terrain but with some roughness in an inland temperate climate, and a link operating at a frequency of 6.7 GHz with a desired propagation reliability of 99.95%, what fade margin should be assigned to the link?

$$U_{ndp} = 1 - \text{percentage}$$
$$= 1 - 0.9995$$
$$= 0.0005$$

then

$$0.0005 = 2.5abfD^3(10^{-F/10})(10^{-6})$$
$$= 2.5(1)(0.25)(6.7)25^3(10^{-F/10})(10^{-6})$$
$$F = 31.2 \text{ dB}$$

It is good design practice to add a miscellaneous loss margin to the derived fade margin by whatever method the derivation has been made. This additional margin is required to account for minor antenna misalignment and system gain degradation (e.g., waveguide corrosion, and transmitter output and receiver NF degradations due to aging). The Defense Communications Agency of the U.S. Department of Defense recommends 6 dB (Ref. 34) for this additional margin.*

Diversity also reduces fade margin requirements as established above; it tends to reduce the depth of fades. Figure 4.22 shows how fade margins may be reduced with diversity. For instance, assume a Rayleigh distribution (random fading) with a 35-dB fade margin with no diversity. Then using frequency diversity with a frequency separation of 2%, only a 23.5-dB fade margin would be required to maintain the same path reliability (i.e., the percentage of time the level is exceeded).

4.5.2.8 Diversity Operation Various types of diversity operation are widely used on point-to-point high-frequency (HF) systems (Chapter 8),

---

*For more information on fading, consult Ref. 36.

transhorizon links (Chapter 5), and to an increasing extent on LOS microwave links (radiolinks). Diversity is attractive for the following reasons:

- It tends to reduce depth of fades on combined output
- It provides improved equipment reliability (if one diversity path is lost due to equipment failure, other path[s] remain in operation)
- Depending on the type of combiner in use, the combined output signal-to-noise ratio is improved over that of any single signal path

Diversity reception is based on the fact that radio signals arriving at a point of reception over separate paths may have noncorrelated signal levels. More simply, at one instant of time a signal on one path may be in a condition of fade while the identical signal on another path may not.

First one must consider what are separate paths and how "separate" must they be (CCIR Rep. 376-5; Ref. 3). The separation may be in

- Frequency
- Space (including angle of arrival and polarization)
- Time (a time delay of two identical signals on parallel paths)
- Path (signals arrive on geographically separate paths)

The most common forms of diversity in radiolink systems are those of frequency and space. A frequency diversity system utilizes the phenomenon that the period of fading differs for carrier frequencies separated by 2–5%. Such a system employs two transmitters and two receivers, with each pair tuned to a different frequency (usually 2–3% separation, since the frequency band allocations are limited). If the fading period at one frequency extends for a period of time, the same signal on the other frequency will be received at a higher level, with the resultant improvement in propagation reliability.

As far as equipment reliability is concerned, frequency diversity provides a separate path, complete and independent, and consequently one whole order of reliability has been added. Besides the expense of the additional equipment, the use of additional frequencies without carrying additional traffic is a severe disadvantage to the employment of frequency diversity, especially when frequency assignments are even harder to get in highly developed areas where the demand for frequencies is greatest.

One of the main attractions of space diversity is that no additional frequency assignment is required. In a space diversity system, if two or more antennas are spaced many wavelengths apart (in the vertical plane), it has been observed that multipath fading will not occur simultaneously at both antennas. Sufficient output is almost always available from one of the antennas to provide a useful signal to the receiver diversity system. The use of two antennas at different heights provides a means of compensating, to a certain degree, for changes in electrical path differences between direct and

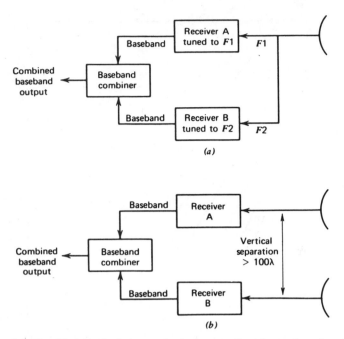

Figure 4.22 (*a*) Simplified block diagram of a frequency diversity configuration $F_1 < F_2 +$ 00.2F_2. (*b*) Simplified block diagram of a space diversity configuration.

reflected rays by favoring the stronger signal in the diversity combiner. The antenna separation required for optimum operation of a space diversity system may be calculated using the following formula:

$$S = \frac{3\lambda R}{L} \tag{4.34}$$

where S = separation (m)

   R = effective earth radius (m)

   λ = wavelength (m)

   L = path length (m)

However, any spacing between 100 and 200λ is usually found to be satisfactory. The goal in space diversity is to make the separation of diversity antennas such that the reflected wave travels a half-wavelength further than the normal path. In addition, the CCIR states that for acceptable space diversity operation the spacing should be such that the value of the space correlation coefficient does not exceed 0.6 (CCIR Rep. 376-5; Ref. 3).

Figure 4.22*a* is a simplified block diagram of a radiolink frequency diversity system and Figure 4.22*b* that of a space diversity system. Figure 4.23

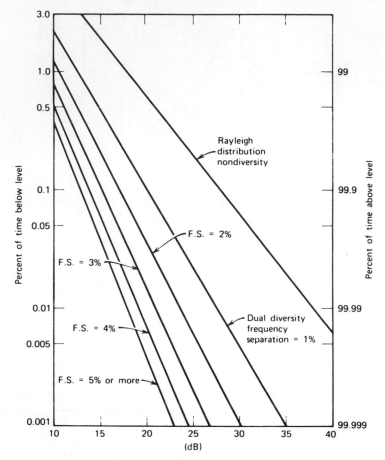

Figure 4.23 Approximate interference fading distribution for a nondiversity system with Rayleigh fading versus frequency diversity systems for various percentages of frequency separation (F.S).

shows approximate multipath fading for nondiversity versus frequency diversity systems for various percentages of frequency separation.

Diversity Combiners

GENERAL. A diversity combiner combines signals from two or more diversity paths. Combining is traditionally broken down into two major categories:

- Predetection
- Postdetection

The classification is made according to where in the reception process the combining takes place. Predetection combining takes place in the IF. How-

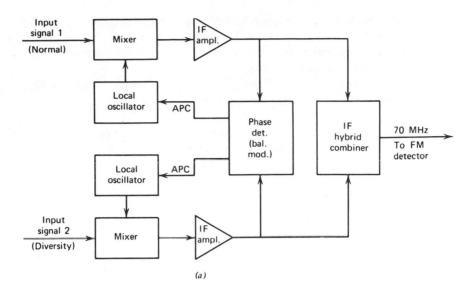

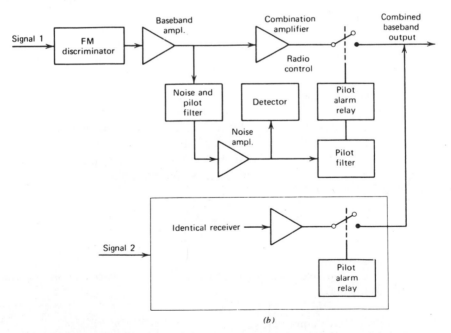

Figure 4.24 (*a*) Predetection combiner. APC = automatic phase control. (*b*) Postdetection combiner (maximum ratio squared).

ever, at least one system* performs combining at RF. With the second type, combining is carried out at baseband (i.e., after detection).

For predetection combining, phase control circuitry is required unless some form of path selection is used.

Figure 4.24a and b shows simplified functional block diagrams of radiolink receiving systems using predetection and postdetection combiners.

TYPES OF COMBINERS.   Three types of combiners find more common application in radiolink diversity systems. These are

- Selection combiner
- Equal gain combiner
- Maximal ratio combiner (ratio squared)

The selection combiner uses but one receiver at a time. The output signal-to-noise ratio is equal to the input signal-to-noise ratio from the receiver selected for use at the time.

The equal gain combiner simply adds the diversity receiver outputs, and the output signal-to-noise ratio of the combiner is

$$\frac{S_o}{N_o} = \frac{S_1 + S_2}{2N} \tag{4.35}$$

where N = receiver noise.

The maximal ratio combiner uses a relative gain change between the output signals in use. For example, let us assume that the stronger signal has unity output and the weaker signal has an output proportional to gain G. It then can be shown that $G = S_2/S_1$ such that the signal gain is adjusted to be proportional to the ratio of the input signals. We then have

$$\left(\frac{S_o}{N_o}\right)^2 = \left(\frac{S_1}{N}\right)^2 + \left(\frac{S_2}{N}\right)^2 \tag{4.36}$$

where N = receiver noise.

For the signal-to-noise ratio equation for the latter two combiners, we assume the following:

- All receivers have equal gain
- Signals add linearly; noise adds on an rms basis
- Noise is random
- All receivers have equal noise outputs N
- The output (from the combiner) signal-to-noise ratio S_o/N_o is a constant

*System manufactured by STC (U.K.).

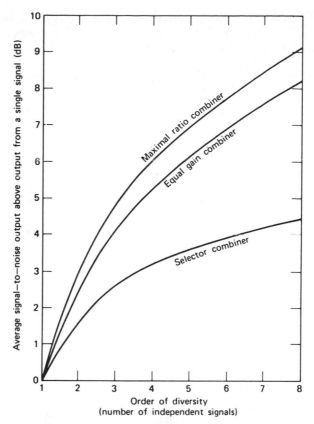

Figure 4.25 Signal-to-noise ratio improvement in a diversity system for various orders of diversity. (Ref. 13. Courtesy of the Institute of Telecommunication Sciences, Office of Telecommunications, U.S. Department of Commerce.)

Figure 4.25 shows graphically a comparison of the three types of combiners. We see the gain that we can expect in output signal-to-noise ratio for various orders of diversity for the three types of combiners discussed, assuming a Rayleigh distribution. The order of diversity refers to the number of independent diversity paths. If we were to use space *or* frequency diversity alone, we would have two orders of diversity. If we used space *and* frequency diversity, we would then have four orders of diversity.

The reader should bear in mind that the efficiency of diversity depends on the correlation of fading of the independent diversity paths. If the correlation coefficient is zero (i.e., there is no relationship in fading for one path to another), we can expect maximum diversity enhancement. The efficiency of a diversity systems drops by half with a correlation coefficient of 0.8, and nearly full efficiency can be expected with a correlation coefficient of 0.3.

4.5.2.9 Pilot Tones   A radio continuity pilot is provided between radio terminals which is independent of the multiplex pilot tones. The pilot or pilots provided here are used for

- Gain regulating
- Monitoring
- Frequency comparison
- Measurement of level stability
- Control of diversity combiners

Table 4.8   Radio continuity pilot tones

System Capacity (Channels)	Limits of Band Occupied by Telephone Channels (kHz)	Frequency Limits of Baseband (kHz)[a]	Continuity Pilot Frequency (kHz)	Deviation (rms) Produced by the Pilot (kHz)[b]
24	12–108	12–108	116 or 119	20
60	12–252 60–300	12–252 60–300	304 or 331	24, 50, 100[d]
120	12–552 60–552	12–552 60–552	607[e]	25, 50, 100[d]
300	60–1,300	60–1364	1499, 3200[f] or 8500[f]	100 or 140
600	60–2,540 64–2,660	60–2792	3200 or 8500	140
960 900	60–4,028 316–4,188	60–4287	4715 or 8500	140
1260 1200	60–5,636 60–5,564 316–5,564	60–5680	6199 8500	100 or 140 140
1800	312–8,204 316–8,204	300–8248	9203	100
2700	312–12,388 316–12,388	308–12,435	13,627	100
Television			8500 9023[g]	140 100

Source: CCIR Rec. 401–2 (Ref. 3). Courtesy of ITU–CCIR.

[a]Including pilot or other frequencies that might be transmitted to line.

[b]Other values may be used by agreement between the administrations concerned.

[c]This deviation does not depend on whether or not a preemphasis network is used in the baseband.

[d]Alternative values dependent on whether the deviation of the signal is 50, 100, or 200 kHz (CCIR Rec. 404-2; Ref. 3).

[e]Alternatively, 304 kHz may be used by agreement between the administrations concerned.

[f]For compatibility in the case of alternate use with 600-channel telephony systems and television systems.

[g]The frequency 9023 kHz is used for compatibility purposes between 1800-channel telephony systems and television systems when the establishment of multiple sound channels so indicates.

The latter application involves the simple sensing of continuity by a diversity combiner. The presence of the continuity pilot tells the combiner that the diversity path is operative. The problem is that the most commonly used postdetection combiners, the maximal ratio type, selection type, and others, use noise as the means to determine the path contribution to the combined output. The path with the least noise, as in the case of the maximal ratio combiner, provides the greatest path contribution.

If, for some reason, a path were to fail, it would be comparatively noiseless and would provide 100% contribution. Thus we would have a no-signal output. In this case the continuity pilot tells the combiner that the path is a valid one.

Pilots are inserted prior to modulation. They are stopped (eliminated) at baseband output and are reinserted anew if another radiolink hop is added. Table 4.8, taken from CCIR Ref. 401-2 (Ref. 3), provides a list of recommended radio continuity pilots.

4.5.2.10 *Repeaters for Analog Microwave LOS Systems* Repeaters are used to extend a line-of-sight microwave system for several additional miles or all the way across a continent. As a minimum a repeater receives a signal at frequency F_1, amplifies it, translates the frequency to F_2, and amplifies and radiates that signal.

There are three types of analog radio repeaters: baseband, IF repeater, and RF repeater. These are shown in Figure 4.26.

Repeater sites that have drop and insert requirements use baseband repeaters. A baseband repeater fully demodulates the incoming RF signal to baseband. In its simplest configuration, the demodulated baseband is used to modulate the transmitter used in the link section. If a switch or concentrator is associated with a repeater, the baseband may be partially or entirely demultiplexed and remultiplexed so that the switch may have access to individual voice channels. A typical baseband repeater is shown in Figure 4.26*a*.

Two other types of repeaters are also available: the IF heterodyne repeater (Figure 4.26*b*) and the RF heterodyne repeater (Figure 4.26*c*). The IF repeater is attractive for use on long backbone systems where noise and/or differential phase and gain should be minimized.

Generally, a system with fewer modulation–demodulation stages or steps is less noisy. The IF repeater eliminates two modulation steps. It simply translates the incoming signal to IF with the appropriate local oscillator and a mixer, amplifies the derived IF, and then up-converts it to a new RF frequency. The up-converted frequency may then be amplified by a traveling wave tube (TWT) or solid state amplifier.

With an RF heterodyne repeater (Figure 4.26*c*), amplification is carried out directly at RF frequencies. The incoming signal is amplified, up- or down-converted, and amplified again, usually by a TWT, and then reradiated. RF repeaters are troublesome in their design in such things as sufficient selectivity, limiting and automatic gain control, and methods to correct

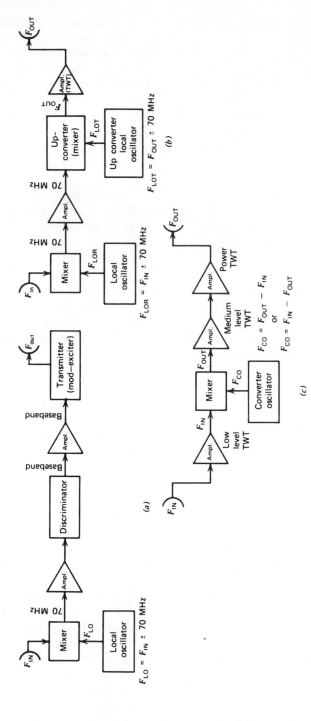

Figure 4.26 Radiolink repeaters. (a) Baseband repeater. (b) IF heterodyne repeater. (c) RF heterodyne repeater. F_{IN} = input frequency to receiver; F_{OUT} = output frequency of the transmitter; F_{CO} = output frequency of the converter local oscillator; F_{LOR} = frequency of the receiver local oscillator; F_{LOT} = frequency of the transmitter local oscillator; F_{LO} = frequency of the local oscillator; TWT = traveling wave tube.

envelope delay. However, some RF repeaters are now available, particularly for operation below 6 GHz.

4.6 DIGITAL LINE-OF-SIGHT MICROWAVE LINKS

4.6.1 Definition and Scope

Most new line-of-sight microwave systems used for duplex communications are digital systems. Many existing analog microwave radiolinks are being replaced with digital implementations. Much of the impetus toward digital is driven by the evolving all-digital telecommunications network. First, let's define digital radio.

A digital radio for this discussion is defined as a radio set in which one or more properties (amplitude, frequency, and phase) of the RF are quantized by a modulating signal. Digital implies fixed sets of discrete values. A digital radio waveform, then, can assume one of a discrete set of amplitude levels, frequencies, or phases as a result of the modulating signal. Let us assume that the modulating signal is a serial synchronous bit stream.

For a given system the information in the serial bit stream may be PCM, DM (Chapter 3), or any other form of serial data information (Chapter 12). Although we define data transmission in Chapter 12 as the digital transmission of record traffic, graphics, numbers, and so on, the industry has given *data* a wider meaning in many instances. Depending on the person one is talking to, data can mean any digital bit stream, no matter what its information content. For instance, *data rate* is a common term that is synonymous with bit rate. We shall also be using the term *symbol rate*, meaning the number of transitions or changes in state per second. In a two-state or two-level system the symbol rate measured in bauds is the same as the data rate in bits per second. If there are more than two states or levels, it is called an *M*-ary system, where *M* is the number of states. For example, 8-ary frequency shift keying (FSK) is an eight-frequency system, and at any instant one of the eight frequencies is transmitted.

We must not lose sight of the system aspects of digital radio. A digital radio transmitter does not sit out there alone; it must work with a far-end receiver making up a link. There may be several links in tandem. These links may or may not be part of a larger network. Timing, bit count integrity, justification (slips and stuffing), bit error rate (BER), and special coding may all impact the design of a digital radio link.

Let us adopt CCIR practice (CCIR Rep. 378-5; Ref. 3) to define capacity:

- Small-capacity radio-relay systems for the transmission of digital signals with gross bit rates up to and including 10 Mbps
- Medium-capacity radio-relay systems with gross bit rates ranging from 10 Mbps to about 100 Mbps
- Large-capacity radio-relay systems with gross bit rates greater than about 100 Mbps

4.6.2 Application

Digital radio-relay systems for the common carrier (telecommunication administration) or specialized common carrier find application in

- Local digital trunks and, in particular, connecting digital exchanges.
- High-usage routes connecting digital exchanges in the local area.
- To expand a VF cable route where radio proves, under certain circumstances, more economic than expanding existing cable facilities either with PCM or by adding additional metallic pair cable.
- On toll (long-distance) routes connecting digital toll exchanges or with an analog space division exchange (switch) on one end and a digital exchange on the other. On shorter routes radiolinks and on longer routes satellite links may be more applicable.
- For all-digital networks where radio proves in on an economic basis over cable, whether metallic or fiber optic, for planning periods up to 15 years in the future.

Other applications of digital radio are

- For military digital networks such as Tri-Tac, MSE and Ptarmigan. In the military environment, digital transmission, especially radio, is more attractive than analog because it is much easier to make secure by on-line encryption. Digital radio tolerates fair-quality links in tandem when compared to analog radiolinks.
- Private data networks.

4.6.3 Basic Digital Radio and Link Design Considerations

4.6.3.1 The Primary Feature Digital radio system design follows most of the basic procedures outlined in Section 4.5, but with one overriding issue that must be resolved by the system design engineer, namely, that of spectrum conservation without significant performance degradation. Consider that we can accommodate 1800 nominal 4-kHz VF channels in approximately 30 MHz of RF bandwidth using conventional FM, as discussed in Section 4.5.2. The same number of voice channels using PCM, allowing 1 bps/Hz of bandwidth, then would require 1800×64 kHz or 115 MHz of bandwidth. The RF spectrum is one of our natural resources, and we should not be wasteful of it. Therefore something has to be done to make digital radio transmission more bandwidth conservative. It is the 1 bps/Hz assumption that we will be working with in this section. The problem of achieving more bits per hertz will be treated at length further on in the chapter.

The FCC has taken up the issue in Part 21.122 of *Rules and Regulations*, stating:

Microwave transmitters employing digital modulation techniques and operating below 15 GHz shall, with appropriate multiplex equipment, comply with the following additional requirements:

1. The bit rate in bits per second shall be equal to or greater than the bandwidth specified by the emission designator in hertz (e.g., to be acceptable, equipment transmitting at a 20-Mb/s rate must not require a bandwidth greater than 20 MHz).

2. Equipment to be used for voice transmission shall be capable of satisfactory operation within the authorized bandwidth to encode at least the following number of voice channels:

Frequency Range	Number of Encoded Voice Channels
2,110–2,130 MHz	96
2,160–2,180 MHz	96
3,700–4,200 MHz	1,152
5,925–6,425 MHz	1,152
10,700–11,700 MHz	1,152

The FCC rules the following on emission limitation (part 21.106):

When using transmissions employing digital modulation techniques

(i) For operating frequencies below 15 GHz, in any 4-kHz band, the center frequency of which is removed from the assigned frequency by more than 50% up to and including 250% of the authorized bandwidth: As specified by the following equation but in no event less than 50 dB:

$$A = 35 + 0.8(P - 50) + 10 \log_{10} B \tag{4.37}$$

where A = attenuation (dB) below the mean output power level
$\quad\quad P$ = percent removed from carrier frequency
$\quad\quad B$ = authorized bandwidth (MHz)
(Attenuation greater then 80 dB is not required.)

(ii) For operating frequencies above 15 GHz, in any 1-MHz band, the center frequency of which is removed from the assigned frequency by more than 50% up to and including 250% of the authorized bandwidth: As specified by the following equation but in no event less than 11 dB:

$$A = 11 + 0.4(P - 50) + 10 \log_{10} B \tag{4.38}$$

(Attenuation greater than 56 dB is not required.)

(iii) In any 4-kHz band, the center frequency of which is removed from the assigned frequency by more than 250% of the authorized bandwidth: At least $43 + 10 \log_{10}$ (mean power output in watts) dB or 80 dB, whichever is the lesser attenuation...

CCIR also provides guidance on spectral occupancy for digital radio systems. Reference should be made to CCIR Reps. 378-5 and 610-1 (Ref. 3). However, CCIR is less specific on spectral occupancy versus bit rate and numbers of equivalent voice channels. It deals more with frequency allocations for specific operational bands.

4.6.3.2 Other Design Considerations in Contrast to Analog FM Systems

The measurement of performance of a digital radio system is the BER, often expressed in a time percentage. FM analog systems use signal-to-noise ratio in decibels or noise power weighted or unweighted (flat), which is often expressed in dBm or pW and referenced to the derived nominal 4-kHz VF channel. For radio systems this measure of performance should be related to a given time percentage of the year or worst month.

E_b/N_0 is an expression commonly used on digital radio systems, usually related to BER. We first introduced E_b/N_0 in Section 1.11. It expresses "energy per bit per hertz of noise spectral density." This is a fairly universal figure of merit for bit decisions. We will be using E_b/N_0 rather than S/N when discussing digital radio performance. If we are given a particular waveform, we can derive a plot of E_b/N_0 versus BER. Figure 4.29 is a family of such curves. Since we will be using logarithmic values for E_b/N_0, it is helpful to state the identity

$$\frac{E_b}{N_0} = E_{b\,(\mathrm{dBW/dBm})} - N_{0\,(\mathrm{dBW/dBm})} \tag{4.39}$$

In Section 4.6.5 we give a more rigorous mathematical definition of E_b/N_0.

Since we are dealing with synchronous serial data, synchronization is a very important system aspect. It would seem that this goes without saying. On radiolinks over, say 7 km (4.3 mi) long, we will encounter some sort of fading at least some of the time. Fading on radiolinks (LOS microwave) is more probably caused by multipath, or at least we can say that it is the principal offender. When fades are deep enough, a dropout occurs and synchronization (i.e., the receiver is in synchronization with its companion far-end transmitter) may be lost. A digital radio system must be designed to withstand short periods of dropout without losing synchronization. This capability is expressed as maintenance of bit count integrity (BCI) and is often measured in milliseconds. The system should also have some rapid automatic means of resynchronizing itself once BCI is exceeded.

One result of multipath on microwave radiolinks is dispersion. Dispersion means that an arriving signal at the receiver has been dispersed or "elongated." In the case of a pulse, there would be a main signal component arriving in its proper time slot, this signal having followed the direct and expected path. During multipath conditions, reflected signal energy due to multipath conditions arrives somewhat later, spilling into the time slot of the next pulse. Dispersion is a measure of elongation of smearing, usually given in nanoseconds. Such dispersion causes intersymbol interference, deteriorating link error performance. Digital radio links often can operate up to 10 Mbps without concern for dispersion. Those operating at rates greater than 10 Mbps and especially those at rates higher than 20 Mbps should take dispersion into account in the link design and equipment selection.

On digital radio systems there are basically two causes of deteriorated error performance. One is multipath dispersion just discussed. A second is due to Gaussian noise in the receiving system. Basic to any link design is to establish a value of E_b/N_0 to meet the required error performance criteria of the link expressed as BER. Deep fades will degrade this value for the period of the fade, causing an error burst. Another cause of errors is on systems designed to operate near the E_b/N_0 margin, resulting in bursts of errors and periods when the link is comparatively error free. Thus the system designer at the outset must establish whether a digital radiolink is dominated by random errors or by bursts of errors. Earth stations* would probably fall into the random error category, whereas longer LOS radiolinks and tropospheric scatter links would fall into the bursty error category. If coding[†] is used, and we mean channel coding here, to improve error performance, the code selected will much depend on whether a link will expect random errors or bursty errors. Other causes of errors are timing jitter and interference.

4.6.3.3 Error Rate Performance   CCIR Rec. 594-1 provides recommended BER performance over a 64-kbps 2500-km hypothetical reference digital path (HRDP):

- That the BER should not exceed the following values:
  - 1×10^{-6} during more than 0.4% of any month; integration time, 1 min
  - 1×10^{-3} during 0.054% of any month; integration time, 1 s
- That the total errored seconds should not exceed 0.32% of any month

The above values take into account fading, interference, and other sources of degradation.

CCIR Rec. 634 (Ref. 3) deals with error performance objectives for real digital radio-relay links (LOS microwave links) forming part of a high-grade circuit within an integrated services digital network (ISDN). It deals with an ISDN link length L of between 280 and 2500 km. The recommendation

---

*Operating below 11 GHz and with elevation angle > 7°.
[†]Coding is discussed in Chapter 6.

provides the following values of BER:

- Equal to or greater than 1×10^{-3} for no more than $(L/2500) \times 0.054\%$ of any month; integration time of 1 s
- Equal to or greater than 1×10^{-6} for no more than $(L/2500) \times 0.4\%$ of any month; integration time of 1 min
- Errored seconds of no more than $(L/2500) \times 0.32\%$ of any month

The residual bit error ratio (RBER):

$$\text{RBER} = \frac{L \times 5 \times 10^{-9}}{2500} \qquad (4.40)$$

CCIR Rep. 930-1 defines RBER as the error ratio in the absence of fading and includes allowance for system inherent errors, environmental and aging effects, and long-term interference. An RBER over a 2500-km high-grade real circuit of 5.0×10^{-9} is specified.

The rationale for selecting a bottoming-out value of BER at 1×10^{-3} is purposeful. For digital systems carrying telephone traffic, error bursts can cause call dropout due to the loss of supervisory signaling data. A United Kingdom proposal to the CCIR suggests that with an error rate threshold of 1×10^{-3}, any error burst of 0.07 s or less would not cause call dropout.

Bellcore (Ref. 37) specifies an objective end-to-end connection through switched digital network BER value of less than 1×10^{-6} on at least 95% of connections.

The requirements of computer data users are what forces such low BER values on the digital network. As we mentioned in Chapter 3, PCM speech intelligibility remains satisfactory with BERs in the range of 10^{-2}.

4.6.3.4 Modulation Techniques and Spectral Efficiency

4.6.3.4.1 Introduction. There are three generic modulation techniques available: AM, FM, and PM. Terminology of the industry often appends the letters "SK" to the first letter of the modulation type, as in ASK, meaning amplitude shift keying; FSK, meaning frequency shift keying (classified as digital FM); and PSK, meaning phase shift keying. Any of the three basic modulation techniques may be two level or multilevel. For the two-level case, one state represents the binary "1" and the other state a binary "0." For multilevel or *M*-ary systems there are more than two levels or states, usually a multiple of 2, with a few exceptions, such as partial response systems, duo-binary being an example. Four-level or 4-ary systems are in common use, such as QPSK. In this case, each level or state represents two information bits or coded symbols. For 8-level systems such as 8-ary FSK or 8-ary PSK, 3 bits are transmitted for each transition or change of state, and for a 16-level

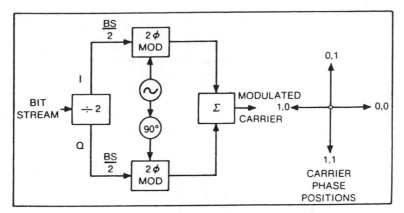

Figure 4.27 A QPSK (4-ary PSK) modulator. I = in-phase; Q = quadrature.

system (16-ary) 4 bits are transmitted for each change of state or transition. Of course, in *M*-ary systems some form of coding or combining is required prior to modulation and decoding after demodulation to recover the original bit stream. Conceptually, a typical QPSK modulator is shown in Figure 4.27.

Let us consider that each of the three basic modulation techniques may be represented by a modulated sinusoid. At the far-end receiver, some sort of detection must be carried out. Coherent detection requires a sinusoidal reference signal extremely closely matched in both frequency and phase to the received carrier. This phase frequency reference may be obtained from a transmitted pilot tone or from the modulated signal itself. Noncoherent detection, being based on waveform characteristics independent of phase (e.g., energy or frequency), does not require a phase frequency. FSK commonly uses noncoherent detection.

After detection in the receiver, there is usually some device that carries out a decision process, although in less sophisticated systems this process may be carried out in the detector itself. Some decision circuits make decisions on a baud-by-baud basis. Others obtain some advantage by examining the signal over several baud intervals before making each "baud" decision. The observation interval is the portion of the received waveform examined by the decision device.

4.6.3.4.2 Selection of a Spectrally Efficient Modulation Technique. There are three aspects to be considered in selecting a particular modulation technique:

1. To meet the spectral efficiency (confinement) requirements (i.e., a certain number of megabits for a given bandwidth, as described in Section 4.6.3.1)
2. To meet realizable performance requirements (Section 4.6.3.3)
3. Economic constraints that are a function of equipment complexity

Table 4.9   Comparison of some digital modulation types

System	Variant			W (dB)[a]	Nyquist Bandwidth
Amplitude modulation	Full-carrier binary double-sideband with envelope detection			17	B
	Double-sideband, suppressed-carrier, two binary channels in quadrature with coherent detection			10.5	$B/2$
	Double-sideband, suppressed-carrier, two binary channels in quadrature with differentially coherent detection			12.8	$B/2$
	Single-sideband binary, suppressed-carrier[b]			10.5	c
	Vestigial-sideband binary, suppressed-carrier, with coherent detection[b]			11.3	c
	Vestigial-sideband, reduced-carrier, with coherent detection[b]			11.8	c
	Vestigial-sideband binary, suppressed-carrier, 50% amplitude modulation with envelope detection[b]			17.8	c
Phase modulation with coherent detection[e]	2-state	Binary phase-shift keying		10.5	B
		Reduced bandwidth binary phase-shift keying		15.5[d]	B
	4-state	Quaternary phase-shift keying		10.5	$B/2$
		Reduced bandwidth quaternary with phase-shift keying		15.5[d]	$B/2$
	8-state			13.8	$B/3$
Phase modulation with differentially coherent detection[e]	2-state			11.2	B
	4-state			12.8	$B/2$
	8-state			16.8	$B/3$
Frequency modulation, with discriminator detection	2-state			13.4	B
	3-state (duo-binary)			15.9	B
	4-state			20.1	$B/2$
	8-state			25.5	$B/3$
Quadrature amplitude modulation (QAM)	16-QAM			17	$B/4$
	32-QAM			18.9	$B/5$
	64-QAM			22.5	$B/6$
	128-QAM			24.3	$B/7$
	256-QAM			27.8	$B/8$
Other modulation methods with coherent detection	Two 3-state class 1 partial response channels in quadrature			13.5	$B/2$

Source: CCIR Rep. 378-5, Green Books, vol. IX, 1986, Part 1, p. 242. Courtesy of ITU–CCIR.

[a] $P_e = 10^{-6}$.

[b] The maximum steady-state power depends on the shape of the modulation pulses. These figures are therefore based on average power.

[c] The Nyquist bandwidth for single- and vestigial-sideband systems is difficult to clearly define.

[d] The modulation method introduces controlled amounts of inter-symbol interference thus the figure is a quasi-W value, not related to peak value.

[e] All digital phase modulation may be obtained directly by phase modulation or indirectly by methods of amplitude or frequency modulation.

The theoretical bandwidth occupancy of a radio system using binary modulation (i.e., two-state) is 1 bit per Hz. For instance, a system transmitting 10 Mbps using binary FSK, PSK, or ASK would require a theoretical bandwidth of 10 MHz.

CCIR Rep. 378-5 (Ref. 3) gives some excellent guidance on the selection of digital modulation schemes. The recommendation states in part:

Different modulation techniques may be compared on the basis of their Nyquist bandwidth and power requirements, the latter being indicated by W the normalized carrier-to-noise power ratio, and is given by

$$W = 10 \log(W_{in}/W_n f_n) \tag{4.41}$$

where W_{in} = received maximum steady-state signal power (the highest value of the mean power during any one radio-frequency cycle)

  W_n = noise-power-density at the receiver input

  f_n = bandwidth numerically equal to the bit rate (B) of a binary signal before the modulation process

The recommendation goes on to say that the bit rate B is the gross bit rate transmitted along the radio system, and takes into account the redundancy possibly introduced for service or supervisory channels, error control, and so forth. Table 4.9 shows a comparison of various modulation implementations where W corresponds to a BER of 1×10^{-6}. Theoretical bandwidths are given in the table.

Bit packing is a term used when discussing spectrally efficient modulation techniques. The packing ratio is the ratio of bits to hertz of bandwidth. For instance, the theoretical packing ratio of QPSK is 2 bits per hertz of bandwidth; 32-QAM is 5 bits/Hz, and so forth.

Digital modulation schemes are often represented by space diagrams. Typical signal space diagrams are shown in Figure 4.28. Space diagrams are nearly always symmetrical, so often only one quadrant is shown. Figure 4.28*a*

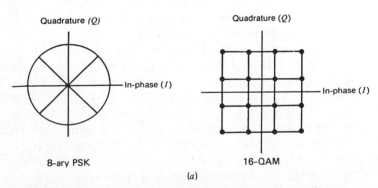

Figure 4.28*a*   Two signal space diagrams, all 4 quadrants shown.

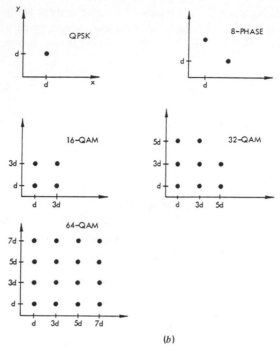

(b)

Figure 4.28b One quadrant representation of signal space diagrams, sometimes called signal state constellations. *d* is amplitude distance. (From Ref. 38; courtesy of Raytheon Company.)

shows simple signal space diagrams and Figure 4.28*b* shows one-quadrant representations of several digital modulation schemes.

The U.S. Federal Communications Commission requires 4.5 bits/Hz (Ref. 37) to meet spectral conservation requirements introduced in Section 4.6.3.1. From Table 4.9, it would appear that the 32-QAM waveform with 5 bits/Hz packing ratio would meet and exceed the 4.5-bits/Hz requirement. This may not necessarily be so. The values given in the table for W are theoretical values. It will be noted that the M-ary schemes discussed have a 2^n-relationship, where n is the theoretical packing ratio.

In practical systems, the packing ratios are less optimistic. For instance, we will find that binary FSK and PSK have packing ratios more on the order of 0.8 bits/Hz; QPSK 1.9 bits/Hz, 8-ary PSK 2.6 bits/Hz, 16-ary PSK 2.9 bits/Hz, 16-QAM 3.1 bits per Hz, 32-QAM 4 bits/Hz, and 64-QAM about 4.9 bits/Hz.

Care should be taken when dealing with E_b/N_0 values versus equivalent BERs, whether the values are theoretical or practical. Figure 4.29 shows some typical theoretical E_b/N_0 values. Of course, practical values should be used when carrying out a link path analysis (link budget). We refer to the difference between theoretical values and practical values as "modulation implementation loss." Modulation implementation loss is equipment driven.

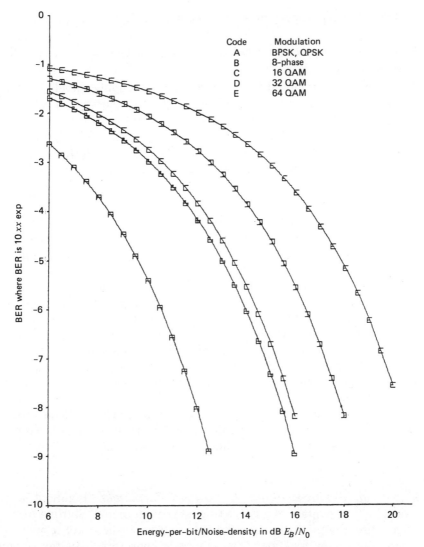

Figure 4.29 BER performance for selected modulation types (Ref. 38; courtesy of Raytheon Company.)

Therefore, for a specific equipment implementation the link design engineer should use equipment manufacturer's E_b/N_0 values.

Some of the equipment/system design constraints that force the ideal into the practical domain are (Ref. 4)

1. The effect of nonideal sharp cutoff spectral shaping filters
2. The effect of additional linear distortion created by realistic filters
3. The effect of signaling with a peak-power-limited amplifier

4. The relative efficiency between preamplifier filtering and postamplifier filtering

5. The effect of practical nonlinearities encountered with real amplifiers on system performance

6. Techniques(s) required for adapting realistic amplifier nonlinearities in order to render the amplifier linear for the purpose of supporting highly bandwidth-efficient digital data radio transmission.

7. The effect on performance of practical imperfections in an implementation of a bandwidth-efficient modem, including a baseband equalizer for counteracting linear distortion caused by realistic filter characteristics.

Many system problems arise at the receiver, which must resolve which signal point was transmitted. Outside disturbances tend to deteriorate the system by masking or otherwise confusing the correct signal point degrading error performance.

One such outside disturbance is additive Gaussian noise. The bandwidth-efficient modem consequently requires a higher signal-to-noise ratio for a given symbol rate as the number of bits per second per hertz is increased.

Another type of outside disturbance is created by cochannel or adjacent-channel interference. Here the bandwidth-efficient modem is more vulnerable to interference, since less interference power is required to push the transmitted signal point to an adjacent point, thus resulting in a hit causing errors at the receiver.

For the system engineer, one of the most perplexing outside disturbances is caused by the transmission medium itself. This is multipath distortion causing signal dispersion. This problem is dealt with, in part, in Subsection 4.6.4.

4.6.3.5 Digital Line-of-Sight Microwave Terminals

4.6.3.5 Digital Line-of-Sight Microwave Terminals Figure 4.30 is a functional block diagram of a typical digital radio terminal. Starting from the bottom of the figure on the transmit line (left side) there are a number of baseband processing functions. The line code converter takes the standard PCM line codes, which have been implemented for good baseband transmission properties, and converts the code usually to nonreturn-to-zero (NRZ) format. Then the resulting code is scrambled by means of a pseudo-random-binary sequence (PRBS). This action tends to remove internal correlation among symbols such as long strings of 1s or 0s. The resulting modulation with the PRBS provides an output with a constant power spectrum.

The signal may then be channel coded for forward error correction,* although this is not often done on terrestrial radiolinks. Differential coding/decoding is one method to remove phase ambiguity at the receive

*FEC is discussed in Chapter 6.

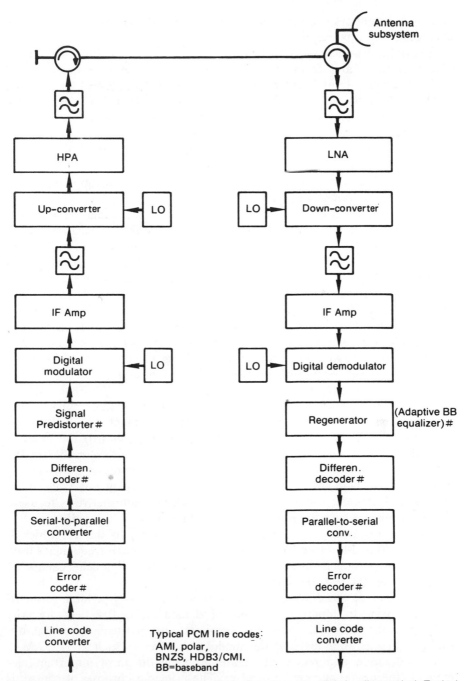

Figure 4.30 Typical functional block diagram of a microwave digital radio terminal. Typical PCM line codes: AMI, polar, BNZS, HDB3 / CMI. BB = baseband, # = optional.

end. A coherent system would not require this function. However, a phase coherence would be required.

A serial-to-parallel converter divides the serial bit stream into two components for the I and Q inputs* of a phase modulator (see Figure 4.27.) Multilevel coders (and decoders) are required for 8-ary PSK and for so-called QAM modulation schemes. (Refer to Section 4.6.3.4). In the cases of 16-QAM, 32-QAM, and 64-QAM more than two amplitude levels are used per orthogonal modulation. Therefore, the digital data have to be converted to multilevel logic. For instance, 64-QAM requires two 8-level modulation signals.

The predistorter is designed to compensate for the distortion imparted on the signal by the power amplifier.

The digital modulator, of course, carries out the modulation function. Modulation to achieve spectral efficiency was discussed in Section 4.6.3.4. The output of the modulator is then amplified, filtered, and passed to the up-converter. The up-converter translates this signal to the operating frequency of the terminal. The output of the up-converter is then fed to the high-power amplifier (HPA), which amplifies the signal to the desired output level. The HPA may be a TWT or a solid state amplifier (SSA). Commonly, power outputs of HPAs for microwave LOS operation are 0.1, 1, or, in some cases, 10 W. The output of the HPA generally incorporates a bandpass filter to reduce spurious and harmonic out-of-band signals. The signal is then fed to the antenna subsystem for radiation to the distant end.

The received signal, starting from the antenna downward in Figure 4.30, is fed from the antenna subsystem through a bandpass filter through the low-noise amplifier (LNA) to a down-converter. In many installations the LNA can be omitted. This decision is driven by economics. If the link can tolerate the additional thermal noise, some savings can be made on the installation. Common down-converters display noise figures of the 7–11 dB, whereas GaAs field-effect transistor (FET) LNAs display noise figures of from 1 to 3 dB, and above 20 GHz, 3.5–5 dB. One alternative is to use a low-noise mixer where the noise figure is on the order of 2.5–5 dB.

The down-converter translates the incoming signal to the IF, which is commonly 70 or 140 MHz. These are the more common IF frequencies used in the industry, and it does not mean that other IF frequencies cannot be used, such as 300, 600, and 700 MHz.

The output of the down-converter is then fed through a filter to an IF amplifier prior to inputting the digital demodulator. Excellent linearity throughout the receive chain is extremely important, particularly on high-bit-rate equipments using higher order modulation schemes such as 64-QAM.

Demodulation and regeneration are probably the most important elements in the digital microwave receiver. The principle purpose, of course, is to achieve a demodulator serial bit stream output that is undistorted and comparatively free of intersymbol interference.

*I and Q inputs = "in-phase" and "quadrature" inputs.

Coherent demodulation is commonly used. Reference phase has to be maintained. The vector status of the modulated signal is compared with that of the carrier. However, since the carrier is not available in the received modulation signal, it has to be reproduced from it. One method that can be used for an AM-SSB signal is to employ a squaring process where a discrete spectral component is available at twice the frequency of the carrier from which a carrier can then be derived. Another method is to use a Costas loop for carrier regeneration.

After demodulation, the data clock at the incoming symbol rate is regenerated from the baseband digital bit stream. The clock is usually derived in a similar manner as the carrier. A phase-lock loop (PLL) is synchronized to the spectral component occurring at the clock rate. Here, timing jitter is a major impairment that should be prevented or minimized. Specifications on jitter are called out in CCITT Rec. G.703 (Ref. 3).

The demodulated signal is then regenerated by means of the regenerated clock and a sample-and-hold circuit is applied to regenerate the actual signal. For modulation schemes using M-PSK, the I and Q channel sequences are regenerated separately. The regenerative repeater may also incorporate a baseband equalizer to reduce signal distortion. Methods of reducing distortion on a digital radiolink as discussed in Section 4.6.4.

For the case of I and Q bit streams, the combining of these two bit streams into a single serial bit stream is accomplished in the parallel-to-serial converter. If FEC is implemented, the signal is then error decoded and the resulting signal is then conditioned for line transmission in the line code converter.

4.6.4 Mitigation Techniques for the Effects of Multipath Fading

In analog radiolink systems, multipath fading results in an increase in thermal noise as the RSL drops. In digital radio systems, however, there is a degradation in BER during periods of fading that is usually caused by intersymbol interference due to multipath. Even rather shallow fades can cause relatively destructive amounts of intersymbol interference. This interference results from frequency-dependent amplitude and group delay changes. The degradation depends on the magnitude of in-band amplitude and delay distortion. This, in turn, is a function of fade depth and time delay between the direct and reflected signals.

Five of the most common methods to mitigate the effects of multipath in digital radiolinks are

- System configuration (i.e., adjusting antenna height to avoid ground reflection; implementation of space and/or frequency diversity)
- Use of IF combiners in diversity configurations
- Use of baseband switching combiners in diversity configuration
- Adaptive IF equalizers
- Adaptive transversal equalizers

System-configuration techniques have been described previously in this section.

An optimal IF combiner for digital radio receiving subsystems can be designed to adjust adaptively to path conditions. One such combining technique, the maximum power IF combiner, vectorially adds the two diversity paths to give maximum power output from the two input signals. This is done by conditioning the signal on one path with an endless phase shifter, which rotates the phase on this path to within a few degrees of the signal on the other diversity path prior to combining. The output of this type of combiner can display in-band distortion that is worse than the distortion on either diversity path alone, but functions well to keep the signal at an acceptable level during deep fades on one of the diversity paths.

A minimum-distortion IF combiner operation is similar in most respects to the maximum-power IF combiner but uses a different algorithm to control the endless phase shifter. The output spectrum of the combiner is monitored for flatness such that the phase of one diversity path is rotated and, when combined with the second diversity path, produces a comparatively flat spectral output. The algorithm also suppresses the polarity inversion on the group delay, which is present during nonminimum phase conditions. One disadvantage of this combiner is that it can cancel two like signals such that the signal level can be degraded below threshold.

Ref. 38 suggests a dual algorithm combiner that functions primarily as a maximum-power combiner and automatically converts to a minimum-distortion combiner when signal conditions warrant. Using space diversity followed by a dual-algorithm combiner can give improvement factors better than 150.

Adaptive IF equalizers attempt to compensate directly at IF for multipath passband distortion. Digital radio transmitters emit a transmit spectra of relatively fixed shape. Hence, various points on the spectrum can be monitored, and when distortion is present, corrective action can be taken to restore spectral fidelity. The three most common types of IF adaptive equalizers are shape-only equalizers, slope and fixed notch equalizers, and tracking notch equalizers.

Another equalizer is the adaptive transversal equalizer, which is efficient at canceling intersymbol interference due to signal dispersion caused by multipath. The signal energy dispersion can be such that energy from a digital transition or pulse arrives both before and after the main bang of the pulse. The equalizer uses a cascade of baud delay sections that are analog elements to which the symbol or baud sequence is inputted. The "present" baud or symbol is defined as the output of the Nth section. Sufficient sections are required to encompass those symbols or bauds that are producing the distortion. These transversal equalizers provide both feed forward and feedback information. There are linear and nonlinear versions. The nonlinear version is sometimes called a decision feedback equalizer. Reference 39 reports that both the IF and transversal equalizers show better than three times improvement in error rate performance over systems without such equalizers.

4.6.5 Link Calculations for Digital Line-of-Sight Microwave

The procedure for link analysis for digital LOS microwave links follows that for analog microwave links (Section 4.5.2.6) or simply

- Calculate EIRP
- Algebraically add FSL and other losses due to the medium (P_L) such as gaseous absorption loss
- Add receiving antenna gain (G_r)
- Algebraically add line losses (L_{lr})

from which we derive the RSL in dBW or dBm:

$$\text{RSL}_{dBW} = \text{EIRP}_{dBW} + \text{FSL} + P_L + G_r + L_{Lr} \tag{4.42}$$

RSL, as defined in this text, is the signal level at any given time at the input to the first active stage of a receiver chain whether an LNA or a mixer.

Whereas on an analog radiolink, the measure of quality is S/N and noise in the derived voice channel, the measure of quality on a digital link is BER. To derive a value for BER, we must first calculate E_b/N_0. E_b/N_0 expresses received signal energy per bit per hertz of thermal noise. (See Sections 1.11 and 4.6.3.2.)

One approach is to break E_b/N_0 down into E_b and N_0:

$$E_b = \frac{\text{RSL}}{\text{bit rate}}$$

or logarithmically

$$E_b = \text{RSL}_{dBW} - 10\log(\text{BR}) \tag{4.43}$$

where BR = bit rate

$$N_0 = kT \text{ or } -228.6 \text{ dBW} + 10\log T_{sys} \tag{4.44}$$

where T_{sys} is the effective noise temperature of the receiving system, or for a system operating at room temperature

$$N_0 = -204 \text{ dBW} + \text{NF}_{dB} \tag{4.45}$$

where NF_{dB} is the receiver noise figure.

For digital radiolinks we calculate C/N_0, which is the ratio of the received carrier level to thermal noise in 1-Hz bandwidth. Here it should be noted that C and RSL are synonymous. Thus we can express

$$\frac{C}{N_0} = \text{RSL} - (-204 \text{ dBW}) - \text{NF}_{dB} \tag{4.46a}$$

or

$$\frac{C}{N_0} = \text{RSL} - (228.6 \text{ dBW}) - 10 \log T_{\text{sys}} \qquad (4.46b)$$

and

$$\frac{E_b}{N_0} = \frac{C}{N_0} - 10 \log(\text{BR}) \qquad (4.47)$$

Furthermore,

$$\frac{E_b}{N_0} = \text{RSL}_{\text{dBW}} - (-204 \text{ dBW}) - \text{NF}_{\text{dB}} - 10 \log(\text{BR}) \qquad (4.48a)$$

or

$$\frac{E_b}{N_0} = \text{RSL}_{\text{dBW}} - (-228.6 \text{ dBW}) - \log T_{\text{sys}} - 10 \log(\text{BR}) \quad (4.48b)$$

During the process of carrying out a link analysis, we establish at the outset the desired E_b/N_0 derived from curves such as those in Figure 4.29. Of course, the curves in the figure are for *ideal* systems. For a real system we need *practical* curves, which can be obtained from the modem or equipment manufacturer. The difference between ideal values for E_b/N_0 and practical values is on the order of 0.5 to over 5 dB. This difference is referred to as modulation implementation loss or system degradation. Table 4.10 gives a budget for contributors to modulation implementation loss.

Table 4.10   Typical degradation budget for 90-Mbps digital radiolink

Cause[a]	Degradation (dB)
A. Modem, AWGN back-to-back	
1. Phase and amplitude errors of modulator	0.1
2. ISI caused by filters	1.0
3. Carrier recovery phase noise	0.1
4. Differential encoding/decoding	0.3
5. Jitter (imperfect sampling instants)	0.1
6. Excess noise bandwidth of receiver (demodulator)	0.5
7. Other hardware impairments	
(temperature variations, aging, etc.)	0.4
Modem total	2.5 dB
B. RF Channel imperfections	
1. AM/PM conversion of the quaslinear output stage	1.5
2. Band limitation and group delay	0.3
3. Adjacent RF channel interference	1.0
4. Feeder echo distortion	0.0 dB
Channel total	
Total degradation	5.5 dB

Source: Ref. 40. Courtesy of Prentice-Hall.
[a]AWGN = additive white Gaussian noise; ISI = intersymbol interference.

Example. Assume a digital LOS microwave link with an operating frequency of 6 GHz with a path length of 15 mi. The RF carrier transmits 90 Mbps (1344 VF channels of North American PCM). 32-QAM modulation is selected. The modulation implementation loss is 3.2 dB and the desired BER per hop is 1×10^{-8}. Turning to Figure 4.29, D curve, we find that the ideal E_b/N_0 required for the BER is 18.2 dB. We assume for the example no fading, 0-dB margin, receiver noise figure of 7 dB, and waveguide losses at each end of 1.5 dB. We will use antennas with a gain of 25 dB at each end of the link. Calculate the required transmitter output power under the given conditions.

Calculate N_0 using equation 4.45:

$$N_0 = -204 \text{ dBW} + 7 \text{ dB}$$
$$= -197 \text{ dBW}$$

Calculate the real E_b/N_0 by adding the modulation implementation loss to the idealized E_b/N_0:

$$E_b/N_0 = 18.2 \text{ dB} + 3.2 \text{ dB}$$
$$= 21.4 \text{ dB}$$

This tells us that the energy per bit (E_b) must be 21.4 dB in level greater than the N_0 level value; therefore

$$E_b = N_0 + 21.4 \text{ dB}$$
$$= -175.6 \text{ dBW}$$

We now can calculate RSL (or C). RSL will be $10 \log(\text{bit rate})$ greater in level than the E_b value just calculated or

$$\text{RSL}_{dBW} = E_{b\,(dBW)} + 10 \log(90 \times 10^6)$$
$$= -175.6 \text{ dBW} + 79.54 \text{ dB}$$
$$= -96.06 \text{ dBW}$$

This is the minimum RSL to meet the 0-dB margin (unfaded) requirement. Now we calculate the free-space loss (FSL) for the link using equation 4.1:

$$\text{FSL}_{dB} = 36.58 + 20 \log 15 + 20 \log 6000$$
$$= 135.66 \text{ dB}$$

Then, using equation 4.42:

$$\text{RSL}_{dBW} = \text{EIRP}_{dBW} + (-135.66 \text{ dB}) + 25 \text{ dB} + (-1.5 \text{ dB})$$
$$\text{RSL}_{dBW} = \text{EIRP}_{dBW} - 112.16 \text{ dB}$$

But RSL = -96.06 dBW, so

$$\text{EIRP}_{dBW} = -96.06 \text{ dBW} + 112.16 \text{ dB}$$

and

$$\text{EIRP}_{dBW} = +16.1 \text{ dBW}$$

Turn to Section 1.14 and

$$\text{EIRP}_{dBW} = \text{transmitter output (dBW)} - \text{line loss (dB)}$$
$$+ \text{ antenna gain (dB)}$$
$$16.1 \text{ dBW} = \text{transmitter output (dBW)} - 1.5 \text{ dB} + 25 \text{ dB}$$
$$\text{Transmitter output (dBW)} = +16.1 \text{ dBW} - 23.5 \text{ dB}$$
$$= -7.4 \text{ dBW}$$
$$= 182 \text{ mW}$$

4.7 ANTENNA SUBSYSTEMS

4.7.1 General

For conventional LOS microwave links, the antenna subsystem offers more room for trade-off to meet system requirements than any other subsystem. Basically, the antenna group looking outward from a transmitter must have

- Transmission line (waveguide or coaxial line)
- Antenna: a reflecting surface or device
- Antenna: a feed horn or other feeding device

In addition the antenna system may have

- Circulators
- Directional couplers
- Phasers
- Passive reflectors
- Radome

4.7.2 Antennas

Below 700 MHz, antennas used for point-to-point microwave systems are often a form of yagi and are fed with coaxial transmission lines. Above 700 MHz some form of parabolic reflector-feed arrangement is used. 700 MHz is no hard and fast dividing line. Above 2000 MHz the transmission line is usually waveguide. The same antenna is used for transmission and reception.

An important antenna parameter is its radiation efficiency. Assuming no losses, the power radiated from an antenna would be equal to the power delivered to the antenna. Such power is equal to the square of the rms current flowing on the antenna times a resistance, called the radiation resistance:

$$P = I_{rms}^2 R \qquad (4.49)$$

where P = radiated power (W)

R = radiation resistance (Ω)

I = current (A)

In practice, all the power delivered to the antenna is *not* radiated in space. The radiation efficiency is defined as the ratio of the power radiated to the total power delivered to an antenna.

To derive a more realistic equation to express power, we divide the resistance, which we shall call the terminal resistance, into two component parts, namely, R = radiation resistance and R_1 = equivalent terminal loss resistance, so that

$$P = I_{rms}^2(R + R_1) \qquad (4.50)$$

and

$$\text{Radation efficiency } (\%) = \frac{R}{R + R_1} \times 100 \qquad (4.51)$$

Antenna gain is a fundamental parameter in radiolink engineering. Gain is conventionally expressed in decibels and is an indication of the antenna's concentration of radiated power in a given direction. Antenna gain expressed anywhere in this work is gain over an isotropic.* An isotropic is a theoretical antenna with a gain of 1 (0 dB). In other words, it is an antenna that radiates equally in all directions.

For parabolic reflector-type antennas, gain is a function of the diameter of the parabola D and the frequency f. Theoretical gain is expressed by the formula

$$G_{dB} = 20 \log F_{MHz} + 20 \log_{10} D_{ft} - 52.5 \qquad (4.52)$$

for a 55% surface efficiency.

Figure 4.31 is a graph from which the gain of a parabolic reflector type of antenna may be derived for several discrete reflector diameters.

Note: In practice we assume surface efficiency of usually around 55% for radiolink systems. Chapter 6 discusses antennas with improved efficiencies.

---

*Often expressed in dBi.

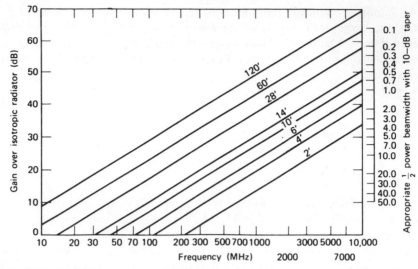

Figure 4.31   Parabolic antenna gains for discrete reflector diameters. (From Ref. 13; courtesy of Institute of Telecommunication Sciences, Office of Telecommunications, U.S. Department of Commerce.)

For uniformly illuminated parabolic reflectors, an approximate relation between half-power beamwidth and gain is expressed by

$$Q \cong \frac{142}{\sqrt{G}} \text{ (degrees)} \tag{4.53}$$

where Q = half-power beamwidth and

$$\sqrt{G} = \text{antilog}\left(\frac{G_{dB}}{20}\right)$$

In practice absolutely uniform illumination is not used so that side lobe levels may be reduced using a tapered form of illumination. Expect beamwidths to be 0.1–0.2° wider in practical applications.

Beamwidths are narrow in radiolink systems. Table 4.11 illustrates their narrowness. From this it is evident that considerable accuracy is required in pointing the antenna at the distant end.

Table 4.11   Antenna beamwidths

Gain (dB)	Half-Power Beamwidth
30	5°
35	3°
44	1°

Polarization   In radiolink systems antennas use linear polarization and, depending on the feed, may be horizontally or vertically polarized, or both. That is, an antenna radiating and receiving several frequencies at once often will radiate adjacent frequencies with opposite polarizations, or the received polarization is opposite to the transmit. Isolation of 26 dB or better may be expected between polarizations, and on well-designed installations 35 dB, allowing closer interworking at installations using multi-RF carrier operation.

4.7.3 Transmission Lines and Related Devices

Waveguide and Coaxial Cable   Waveguides may be used on installations operating above 2 GHz to carry the signal from the radio equipment to the antenna (and vice versa). For those systems operating above 4 GHz it is mandatory from a transmission efficiency point of view. Coaxial lines are used on those systems operating below 2 GHz.

From a systems engineering aspect, the concern is loss with regard to a transmission line. Figures 4.32 and 4.33 identify several types of commonly used waveguides and coaxial cable expressing loss versus frequency.

Waveguide installations are always maintained under dry air or nitrogen pressure to prevent moisture condensation within the guide. Any constant positive pressure up to 10 lb/in.^{2} (0.7 kg/cm^2) is adequate to prevent "breathing" during temperature cycles.

Waveguides may be rectangular, elliptical, or circular. Nearly all older installations used rectangular waveguide exclusively. Today its use is limited to routing in tight places, where space is limited. However, bends are troublesome and joints add 0.06-dB loss each to the system. Optimum electrical performance is achieved by using the minimum number of components. Therefore it has become the practice that wherever a single length is required, elliptical waveguide is used from the antenna to the radio equipment without the addition of miscellaneous flex-twist or rigid sections that are used for rectangular waveguide.

Circular waveguide offers generally lower loss plus dual polarized capability such that only one waveguide run up the tower is necessary for dual polarized installations. Circular waveguide is used when the run is long so that excessive loss is not introduced.

Transmission Line Devices   A ferrite load isolator is a waveguide component that provides isolation between a single source and its load, reducing ill effects of voltage standing wave ratio (VSWR) and often improving stability as a result. Most commonly, load isolators are used with transmitting sources absorbing much of the reflected energy from high VSWR. Owing to the ferrite material with its associated permanent magnetic field, ferrite load isolators have a unidirectional property. Energy traveling toward the antenna is relatively unattenuated, whereas energy traveling back from the antenna

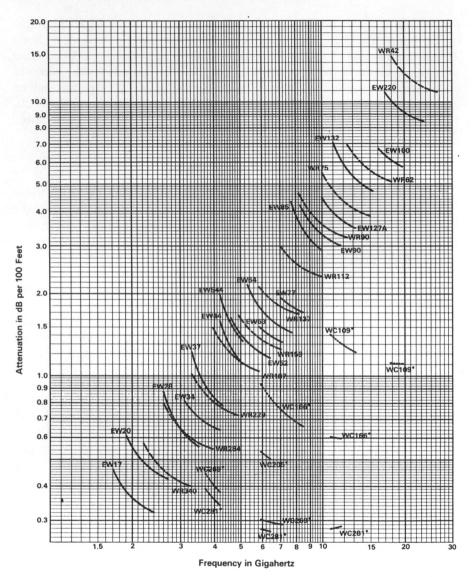

Figure 4.32 Waveguide attenuation. Add 0.3 dB to allow for top and bottom transition. (From Ref. 24, courtesy of Andrew Corporation.)

undergoes fairly severe attenuation. The forward and reverse attenuations are on the order of 1 and 40 dB, respectively.

A waveguide circulator is used to couple two or three microwave radio equipments to a single antenna. The circulator shown in Figure 4.34 is a four-port device. It consists essentially of three basic waveguide sections combined into a single assembly. The center section is a ferrite nonreciprocal phase shifter. An external permanent magnet causes the ferrite material to

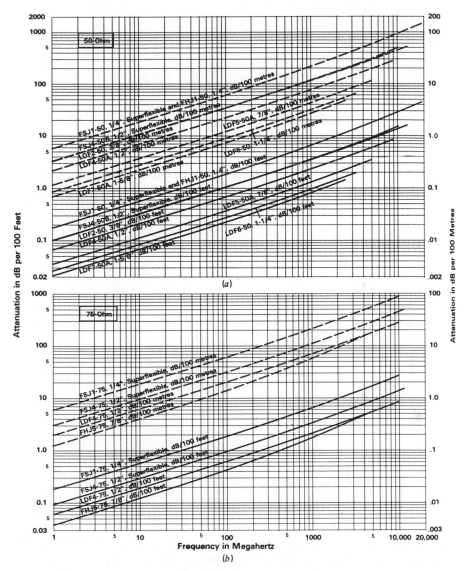

Figure 4.33 Attenuation per unit length for certain types of coaxial cable. Attenuation curves based on VSWR 1.0 : 1, 50-Ω, copper conductors. For 75-U cables multiply by 0.95. (*a*) Air dielectric. (*b*) Foam dielectric. (From Ref. 24; courtesy of Andrew Corporation).

exhibit phase shifting characteristics. Normally an antenna transmission line is connected to one arm and either three radio equipments are connected to the other three arms, or two equipments and a shorting plate. Attenuation in a clockwise direction from arm to arm is low, on the order of 0.5 dB, whereas in the counterclockwise direction from arm to arm it is high, on the order of 20 dB.

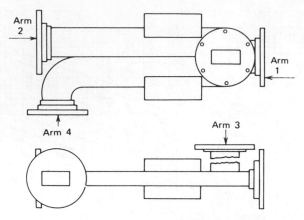

Figure 4.34 Waveguide circulator (four ports).

A power splitter is a simple waveguide device that divides the power coming from or going to the antenna. A 3-dB power split divides power in half; such a device could be used, for instance, to radiate the power from a transmitter in two directions. A 20-dB power split, or 30 dB, has an output that serves to sample the power in a transmission line. Often such a device has directional properties (therefore called a directional coupler) and is used for VSWR measurements allowing measurement of forward and reverse power.

Magic tees or hybrid tees are waveguide devices used to connect several equipments to a common waveguide run.

4.8 SYSTEM AND LINK PROPAGATION RELIABILITY (TIME AVAILABILITY)

Reliability is a principle concern of the microwave design engineer, whether a design involves an LOS microwave system of links in tandem or only just one link. There are two facets to reliability: outage due to propagation and outage due to equipment malfunction or failure. Availibility is a term we can expect to encounter. We define availability as the percentage of time a system or link meets performance requirements; unavailability is the percentage of time the system or link does not meet requirements.

Time availability is a term that has become synonymous with propagation reliability. We use the two terms interchangeably.

CCIR Rep. 445-3 gives five general causes of interruption:

1. Equipment and power failure
2. Propagation
3. Interference

4. Support facilities (e.g., collapse of a tower, etc.)
5. Human activity (human error)

Previous editions of CCIR Rep. 557 allocated about half of the unavailability time to propagation (deep fades) and half to the remaining four items on the list above. CCIR Rec. 557-1 (XVIth Plenary Assembly, 1986) does not take this liberty.

Suppose we did assume that half the outage time was due to propagation and we alot the remainder to "other." If a system availability is stated as 99.7%, then the unavailability would be 0.3% (i.e., $1.000 - 0.997 = 0.003$ or 0.3%). What time availability would we design the system for? We divide the unavailability in half and subtract that value from 1.0000. In this case 0.15% or 0.0015; $1.0000 - 0.0015 = 99.85\%$.

Suppose there were 10 hops in tandem on a certain LOS microwave system and the system time availability was 99.85%; what would the per hop time availability be? Calculate the system unavailability (time). This is 0.0015. Divide this value by 10 and subtract from 1.0000; $1.0000 - .00015 = 99.985\%$.

This method we show is extremely conservative because it assumes that fading is well correlated along the microwave route or that all paths fade nearly simultaneously. This is highly unlikely. Nevertheless, many system engineers use this conservative technique.

One guideline is that for long systems of many hops in tandem (e.g. more than 10) only one-third should be permitted to fade at once. Such assumption is fairly safe, erring on the conservative side, and can provide considerable relief for per hop time availability values.

CCIR Rep. 445-3 (Ref. 3) gives the equation shown below to calculate availability and assumes a two-way circuit:

Overall availability is defined by the following formula:

$$A = 100 - (2500 \times 100/L) \times [(T_1 + T_2 - T_b)/T_s] \qquad (4.54)$$

where A = percentage availabilty based on 2500 km
L = length of one-way radio channel under consideration (km)
T_1 = total interruption time for at least 10 consecutive seconds for one direction of transmission (s)
T_2 = total interruption time for at least 10 consecutive seconds for the other direction of transmission (s)
T_b = bidirectional interruptions for at least 10 consecutive seconds (s)
T_s = period of study (s)

Note: For short circuits, the unavailability is unlikely to add in a linear manner and it may be necessary to define a minimum value for the parameter L. For higher grade circuits, the value of 280 km quoted in Report 930 might be appropriate.

For unidirectional transmission set $T_2 = 0$ and $T_b = 0$ in equation 4.54.

"Fundamental system availability" is defined by CCIR (Rep. 445-3) as a system without built-in redundancy (e.g., hot-standby), and is taken to be a function of the mean time between failure(s) (MTBF) and of the mean time to repair (MTTR). We use the familiar formula in this case as

$$A = \frac{\text{MTBF}}{\text{MTBF} + \text{MTTR}} \qquad (4.55)$$

The unit for MTBF and MTTR is hours. CCIR uses 10 for MTTR. The number is large because it assumes no technician on duty. The U.S. Department of Defense may use 20 min because it assumes a technician on duty and spare parts are available on site.

Example. What is the availability of a system where the MTBF is 40,000 and the MTTR is 10 h?

$$A = \frac{40,000}{40,000 + 10}$$
$$= 0.99975 \text{ or } 99.975\%$$

If we change MTTR to 20 min or 0.33 h, then we get an availability of 99.999%.

4.9 HOT-STANDBY OPERATION

LOS microwave radiolinks commonly provide transport of multichannel telephone service and/or point-to-point broadcast television on high-priority backbone routes. A high order of route reliability is essential. Route reliability depends on path reliability or link time availability (propagation) and equipment/system reliability. We have discussed path reliability and equipment reliability in Section 4.8. Redundancy is one way to achieve equipment reliability to minimize downtime and maximize link availability regarding equipment degradation or failure.

One straightforward way to achieve redundancy effectively is to provide a parallel terminal/repeater system. Frequency diversity effectively does just this. With this approach all equipment is active and operated in parallel with two distinct systems carrying the same traffic. This is expensive, but necessary if a high order of link reliability is desired. Here we mean route reliability.

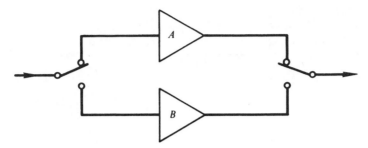

Figure 4.35 The concept of hot-standby protection.

Often the additional frequency assignments to permit operation in frequency diversity are not available. When this is the case, the equivalent equipment reliability may be achieved by the use of a hot-standby configuration. Figure 4.35 illustrates the hot-standby concept.

The equipment marked A and B in Figure 4.35 could be single modules, shelves or groups of modules, or whole equipment racks. On a complex equipment, such as a radiolink terminal, whether digital or analog, more than one set of protection can exist.

On a radiolink terminal such as shown in Figure 4.36 the sections marked "baseband" and "RF" are both hot-standby protected. The operation of the protection system of these two sections, however, is independent. The protection system is broken down further in that the protection for the transmit and receive paths operate independently, as shown in Figure 4.37. It should be noted that the switches ahead of each set of modules have been replaced by a signal splitter. This technique allows the signal to be fed into each set of modules simultaneously.

As the expression indicates, hot standby is the provision of parallel redundant equipment such that this equipment can be switched in to replace the operating on-line equipment nearly instantaneously when there is a failure in the operating equipment. The switchover can take place in the order of microseconds or less. The changeover of a transmitter and/or receiver line can be brought about by a change, over/under a preset amount,

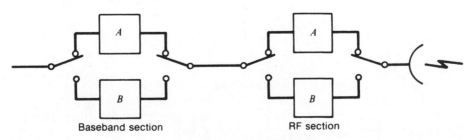

Baseband section   RF section

Figure 4.36 Hot-standby radio.

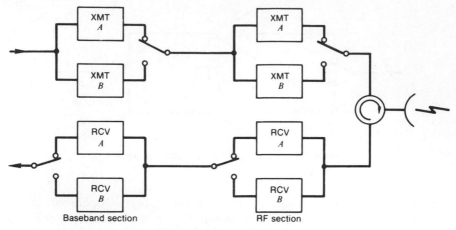

Figure 4.37 Separate transmit and receive hot-standby radio.

in one of the following values: for a transmitter,

- Frequency
- RF power
- Demodulated baseband (radio) pilot level

and for a receiver,

- Automatic gain control (AGC) voltage
- Squelch
- Received pilot level
- Degraded bit error rate for a digital system
- Frame misalignment for a digital system

(*Note*: Digital systems do not use pilot tones.)

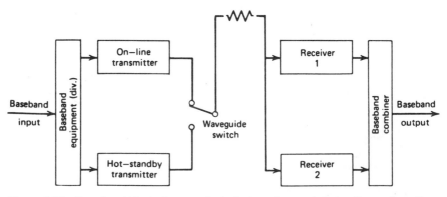

Figure 4.38 Functional block diagram of a typical one-for-one hot-standby configuration.

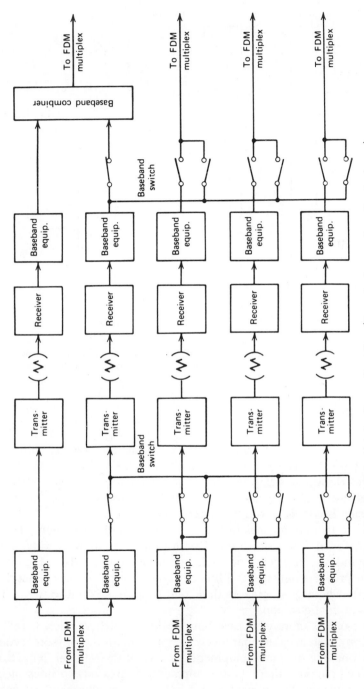

Figure 4.39 One-for-*n* hot-standby configuration. (A one-for-four configuration is shown.)

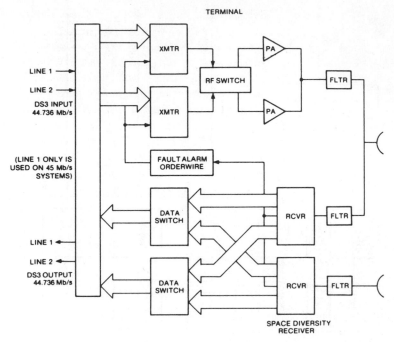

Figure 4.40 A typical hot-standby (one-for-one) digital implementation using space diversity. (From Ref. 39; courtesy of Rockwell International, Collins Transmission Division.)

Hot-standby-protection systems provide sensing and logic circuitry for the control of waveguide switches (or coaxial switches where appropriate), in some cases IF switches as well as baseband switches on transmitters, and IF and baseband output signals on receivers. The use of a combiner on the receiver side is common with both receivers on line at once.

There are two approaches to the use of protection equipment. These are called one-for-one and one-for-n. One-for-one operation provides one full line of standby equipment for each operational system (see Figure 4.38). One-for-n provides only one full line of equipment for n operational lines of equipment, where n is greater than one. Figure 4.39 illustrates a typical one-for-four configuration.

One-for-one is more expensive but provides a higher order of reliability. Its switching system is comparatively simple. One-for-n is more economic, with only one line of spare equipment for several operational lines. It is less reliable (i.e., suppose there were equipment failures in two lines of operational equipment of the n lines), and switching is considerably more complex.

Figure 4.40 shows a typical digital hot-standby configuration with space diversity.

4.10 SERVICE CHANNELS

Service channels are separate facilities from the information baseband (on analog radiolinks) but transmitted on the same carrier. Service channels operate in frequency slots (on analog microwave) below the information baseband for FDM telephony operation and above the video baseband for those links transmitting video (and associated but separated aural channels). Service channels are used for maintenance, link, and network coordination, and, in these cases, may be called "orderwire(s)." They may also be assigned to carry network status data and fault information from unmanned radio relay sites. For FDM telephony links, service channels commonly occupy the band from 300 Hz to 12 kHz of the transmitted baseband, allowing three nominal 4-kHz channel operation.

Digital microwave radiolinks utilize specific time slots in the digital bit stream for service channels. At each terminal and repeater, local digital service channels are dropped and inserted. An express orderwire can also be accommodated. This, of course, requires a reconstitution of the information bit stream at the terminal locations to permit the insertion of service channel information. Another method is to use a separate carrier, either analog or digitally modulated.

4.11 ALARM AND SUPERVISORY SUBSYSTEMS

Many LOS microwave sites are unattended, especially repeater sites. To ensure improved system availability, it is desirable to know the status of unattended sites at a central or manned location. This is accomplished by means of a fault-reporting system. Commonly, such fault alarms are called status reports. The microwave sites originating status reports are defined as reporting stations. A site that receives and displays such reports is defined as a supervisory location. This is the standard terminology of the industry. Normally, supervisory locations are those terminals that terminate an LOS microwave section. Status reports may also be required to be extended over a wire circuit to a remote location, often a maintenance center.

The following functions at a microwave site, which is a reporting location (unmanned), are candidate functions for status reports:

Equipment Alarms

Loss of receive signal
Loss of pilot (at receiver)
High noise level (at receiver)
Power supply failure

Loss of modulating signal

TWT overcurrent

Low transmitter output

Off-frequency operation

Hot-standby actuation

Site Alarms

Illegal entry

Commercial power failure

Low fuel supply

Standby power unit failure

Standby power unit on-line

Tower light status

Additional Fault Information for Digital Systems

Loss of bit count integrity (BCI)

Loss of sync

Excessive BER

Often alarms are categorized into "major" (urgent) and "minor" (nonurgent) in accordance with their importance. For instance, a major alarm would be one where the fault would cause the system or link to go down (cease operation) or seriously deteriorate performance. A major alarm may be audible as well as visible on the status panel. A minor alarm may then show only as an indication on the status panel. On military equipment, alarms are a part of built-in-test equipment (BITE).

The design intent of alarm or BITE systems is to make all faults binary: a tower light is either on or off; the RSL (receive signal level) has dropped below a specified level, -100 dBW, for example; the transmitter power output is 3 dB below its specified output; or the noise on a derived analog channel is above a certain level in picowatts. By keeping all functions binary, using relay closure (or open) or equivalent solid-state circuitry, the job of coding alarms for transmission is made much easier. Thus all alarms are of a "go/no-go" nature.

4.11.1 Transmission of Fault Information

On analog systems, common practice today is to transmit fault information in a voice channel associated with the service channel groupings of voice channels (Section 4.10). Binary information is transmitted by VF telegraph equipment using frequency shift keying (FSK) or tone-on, tone-off (see Chapter 12). Depending on the system used, 16, 18, or 24 tone channels may occupy the voice channel assigned. A tone channel is assigned to each reporting location (i.e., each reporting location will have a tone transmitter operating on the specific tone frequency assigned to it). The supervisory location will have a tone receiver for each reporting (unmanned site under its supervision) location.

At each reporting location the fault or BITE points previously listed are scanned every so many seconds, and the information from each monitor or scan point is time division multiplexed in a simple serial bit stream code. The data output from each tone receiver at the supervisory location represents a series of reporting information on each remote unmanned site. The coded sequence in each case is demultiplexed and displayed on the status panel.

A simpler method is the tone-on, tone-off method. Here the presence of a tone indicates a fault in a particular time slot; in another method it is indicated by the absence of a tone. A device called a fault-interrupter panel is used to code the faults so that different faults may be reported on the same tone frequency.

On digital microwave radiolinks, fault information in a digital format is stored and then inserted into one of the service channel time slots, a time slot is reserved for each reporting location.

4.11.2 Remote Control

Through a similar system to that previously described, which operates in the opposite direction, a supervisory station can control certain functions at reporting locations via a voice frequency (VF) telegraph tone line (for analog systems) with a tone frequency assigned to each separate reporting location on the span. If only one condition is to be controlled, such as turning on tower lights, then a mark condition could represent lights on and a space could represent lights off. If more than one condition is to be controlled, then coded sequences are used to energize or deenergize the proper function at the remote reporting location.

There is an interesting combination of fault reporting and remote control that particularly favors implementation on long spans. Here only summary status is normally passed to the supervisory location, that is, a reporting location is either in a "go" or "no-go" status. When a "no-go" is received, that reporting station in question is polled by the supervisory location, and detailed fault information is then released. Polling may also be carried out on a periodic basis to determine detailed minor alarm fault data.

On the digital radiolink equipment of one manufacturer (Ref. 5) the following operational support system maintenance functions and performance monitoring are listed below:

Maintenance Functions

Alarm Surveillance

Alarm reporting
Status reporting
Alarm conditioning
Alarm distribution
Attribute report

Control Functions

Allow–inhibit local alarms
Operate alarm cutoff
Allow–inhibit protection switching
Operate release protection switch
Remove–restore service
Restart processor
Preemptive switching–override an existing protective switch
Activate/restore lockout (lockout prevents switching)
Local/remote control (i.e., inhibits local operation)
Operate–release loopback
Command (control) verification (i.e., set status point)
Completion acknowledgement (i.e., completed the command received earlier)

Performance Monitoring

Report performance monitoring data such as BER, sync error, error second, and so on.
Inhibit–allow performance monitoring data (i.e., collect but don't send unless asked)
Start–stop performance monitoring data
Initialize (reset) performance monitoring data storage registers

4.12 FREQUENCY PLANNING AND THE INTERFERENCE PROBLEM

4.12.1 General

To derive maximum performance from a microwave radiolink system, the systems engineer must set out a frequency usage plan that may or may not have to be approved by the local administration.

The problem has many aspects. First, the useful RF spectrum is limited, from above direct current to about 150 GHz. The frequency range of discussion for microwave radiolinks is essentially from the VHF band at 150 MHz (overlapping) to the millimeter region of 23 GHz. Second, the spectrum from 150 MHz to 23 GHz must be shared with other services such as radar, navigational aids, research (i.e., space), meteorological, and broadcast. For point-to-point communications, we are limited by international agreement to those bands shown in Table 4.1b.

Although many of the allocated bands are wide, some up to 500 MHz in width, FM by its very nature is a wideband form of emission. It is not uncommon to have an RF bandwidth $B_{RF} = 25$ or 30 MHz for just one emission. Guard bands must also be provided. These are a function of the frequency drift of transmitters as well as "splatter" or out-of-band emission, which in some areas is not well specified.

Occupied bandwidth is discussed in Sections 4.4.2 and 4.6.3.1.

4.12.2 Spurious Emission

For a digital LOS microwave transmit terminal, we discussed spectral confinement of the digital waveform in accordance with FCC Regulations, Part 21 (21.106). This FCC paragraph is titled "emission limitations." The FCC is very specific of signal attenuation as a function of frequencies removed from the carrier center frequency. Portions of the FCC regulation is quoted in Section 4.6.3.1.

The CCIR discusses the issue of spurious emission in Rep. 937-1. Both the FCC and CCIR give power density values in a 4-kHz bandwidth. CCIR references spurious signal levels at the antenna feeder input:

For an analog signal, the maximum spurious power density is -70 dB (W/4 kHz)

For digital transmission, spurious is broken down into two parts:
- Equal or less than -40 dB (W/4 kHz) for CW* emissions
- Equal or less than -70 dB(W/4 kHz) for noise spectrum like emissions

The reference point again is at the antenna feeder input.

4.12.3 Radio Frequency Interference (RFI)

On planning a new radiolink system or on adding RF carriers to an existing installation, careful consideration must be given to the RFI of the existing (or planned) emitters in the area. Usually the governmental authorizing agency has information on these and their stated radiation limits. Typical limits have

---

*CW = carrier wave.

been given above. Equally important as those limits are those of antenna directivity and side lobe radiation. Not only must the radiation of other emitters be examined from this point of view, but also the capability of the planned antenna to reject unwanted signals. The radiation pattern of all licensed emitters should be known. Convert the lobe level in the direction of the planned installation to EIRP in dBW. This should be done for all interference candidates within interference frequency range. For each emitter's EIRP, compute a path loss to the planned installation to determine interference. Such a study could well affect a frequency plan or antenna design.

Nonlicensed emitters should also be looked into. Many such emitters may be classified as industrial noise sources such as heating devices, electronic ovens, electric motors, or unwanted radiation from your own and other microwave installations (i.e., radar harmonics). In the 6-GHz band a coordination contour should be carried out to verify interference from earth stations (see CCIR Rep. 448 and Rec. 359-5). For a general discussion on the techniques for calculating interference noise in radiolink systems, see CCIR Rep. 388-3.

Another RFI consideration is the interference that can be caused by radiolink transmitting systems with the fixed satellite service in those frequency bands that are shared with that service. Guidance on this potential problem is provided by CCIR Rec. 406-5. Some of the major points of the CCIR recommendation are summarized below:

1. In those shared bands between 1 and 10 GHz the maximum EIRP of a radiolink transmitting system should not exceed +55 dBW, and the input power to the transmitting antenna should not exceed +13 dBW. As far as practicable, sites for new radiolink transmitting stations where the EIRP exceeds +35 dBW should be selected so that the direction of maximum radiation of any antenna will be at least 2° away from a geostationary satellite orbit. In special situations the EIRP should not exceed +47 dBW for any antenna directed within 0.5° of a geostationary orbit and +47 to +55 dBW on a linear dB scale (8 dB per angular degree) for any antenna beam directed between 0.5° and 1.5° of a geostationary satellite orbit.

2. In those shared bands between 10 and 15 GHz, the maximum EIRP of a radiolink transmitting system should not exceed +55 dBW, and the input power to the antenna system should not exceed +10 dBW, as far as practicable, transmitting stations where the EIRP exceeds +45 dBW should be selected so that the direction of maximum radiation of any antenna will be at least 1.5° away from the geostationary satellite orbit.

4.12.4 Overshoot

Overshoot interference may occur when radiolink hops in tandem are in a straight line. Consider stations *A*, *B*, *C*, and *D* in a straight line, or that a straight line on a map drawn between A and C also passes through B and D. Link $A-B$ has frequency F_1 from A to B. F_1 is reused in the direction C to D. Care must be taken that some of the emission F_1 on the $A-B$ hop does not spill over into the receiver at D. Reuse may even occur on an A, B, and C combination, so F_1 at A to B may spill into a receiver at C tuned to F_1. This can be avoided, provided stations are removed from the straight line. In this case the station at B should be moved to the north of a line A to C, for example.

4.12.5 Transmit – Receive Separation

If a transmitter and receiver are operated in the same frequency band at a radiolink station, the loss between them must be at least 120 dB. One way to assure the 120 figure is to place all "go" channels in one-half of an assigned band and all "return" channels in the other. The terms *go* and *return* are used to distinguish between the two directions of transmission.

4.12.6 Basis of Frequency Assignment

"Go" and "return" channels are assigned as in the preceding section. For adjacent RF channels in the same half of the band, horizontal and vertical polarizations are used alternately. To carry this out we may assign, as an example, horizontal polarization H to the odd-numbered channels in both directions on a given section and vertical polarization V to the even-numbered channels. The order of isolation between polarizations is on the order of 26 dB, but often specified as 35 dB or more.

In order to prevent interference between antennas at repeaters between receivers on one side and transmitters in the same chain on the other side of the station, each channel shall be shifted in frequency (called "frequency frogging") as it passes through the repeater station. Recommended separations or shifts of frequency are shown in Table 4.12.

4.12.7 IF Interference

Care must be taken when assigning frequencies of transmitter and receiver local oscillators as to whether these are placed above or below the desired

Table 4.12 Frequency frogging at radiolink repeaters

Number of Voice Channels	Minimum Separation (MHz)	
	2000–4000 MHz	6000–8000 MHz
120 or less	120	161
300 or more	213	252

Source: USAF Tech. Order TO 3IR5-1-9 (Ref. 7).

operating frequency. Avoid frequencies that emit the received channel frequency F_R and check those combinations of $F_R \pm 70$ MHz for equipment with 70-MHz IFs or $F_R \pm 140$ MHz when 140-MHz IFs are used. Often plots of all station frequencies are made on graph paper to assure that forbidden combinations do not exist. When close frequency stacking is desired, and/or nonstandard IFs are to be used, the system designer must establish rules as

Table 4.13 CCIR Recommendations for preferred radio-frequency channel arrangements for radio-relay systems, used for international connections

Recommendation	Frequency Band (GHz)	Maximum Capacity in Analogue Operation of Each Radio Carrier (Telephone Channels of the Equivalent)	Capacity of Each Digital Channel[a]	Preferred Center Frequency[b] f_0 (MHz)
283	2	60/120/300/960[c]	Small, medium	1,808 2,000 2,203 2,586
382	2, 4	600/1800	Medium	1,903 2,101 4,005.5[d]
635	4		Medium, high	4,200[e]
383	6	600/1800	High	6,175
384	6	1260/2700	High	6,770
385	7	60/120/300		7,575
386	8	300/960[f]		8,350[f]
387	11	600/1800	Small, medium, high	11,200
497	13	960	Medium	12,996[e]
636	15		Small, medium	11,701[e]
595	18		Small, medium, high	18,700
637	23	Various services	Small, medium, high	21,196[e]

Source: CCIR Op. 14-5. (Ref. 3; Courtesy of ITU–CCIR.)

[a]The definition of the terms *small-*, *medium-*, and *high-capacity* digital systems is given in Rep. 378.
[b]Other center frequencies may be used by agreement between administrations concerned.
[c]The 960-channel capacity can only be used with the center frequency 2586 MHz.
[d]In some countries, mostly in a large part of region 2 and in certain other areas, a reference frequency $f_r = 3700$ MHz is used at the lower edge of a band 500 MHz wide (see Annex I to Rec. 382).
[e]Reference frequency.
[f]In some countries, a maximum capacity of 1800 telephone channels or the equivalent on each radio-frequency carrier may be used with a preferred center frequency of 8000 MHz. The width of the radio-frequency band occupied is 500 MHz.

to minimum adjacent channel spacing and receive–transmit channel spacing. A listing of CCIR recommendations regarding channel spacing is given in the next section.

4.12.8 CCIR Recommendations

Regarding frequency assignments, Table 4.13 is presented as a guide to the relevant CCIR recommendations for radiolink systems.

4.13 ANTENNA TOWERS AND MASTS

4.13.1 General

Two types of towers are used for LOS microwave systems: guyed and self-supporting. However, other natural or man-made structures should also be considered or at least taken advantage of. Radiolink engineers should consider mountains, hills, and ridges so that tower heights may be reduced. They should also consider office buildings, hotels, grain elevators, high-rise apartment houses, and other steel structures (e.g., the sharing of a TV broadcast tower) for direct antenna mounting. For tower heights of 30–60 ft, wooden masts are often used.

One of the most desirable construction materials for a tower is hot-dipped galvanized steel. Guyed towers are usually preferred because of overall economy and versatility. Although guyed towers have the advantage that they can be placed closer to a shelter or building than self-supporting types, the fact that they need a larger site may be a disadvantage where land values are high. The larger site is needed because additional space is required for installing guy anchors. Table 4.14 shows approximate land areas needed for several tower heights.

Tower foundations should be reinforced concrete with anchor bolts firmly embedded. Economy or cost versus height trade-offs usually limit tower heights to no more than 300 ft (188 m). Soil bearing pressure is a major consideration in tower construction. Increasing the foundation area increases soil bearing capability or equivalent design pressure. Wind loading under no-ice (i.e., normal) conditions is usually taken as 30 lb/ft^2 for flat surfaces. A design guide (EIA RS-222-B, Ref. 19) indicates that standard tower foundations and anchors for self-supporting and guyed towers should be designed for a soil pressure of 4000 lb/ft^2 acting normal to any bearing area under specified loading.

4.13.2 Tower Twist and Sway

As any other structure, a radiolink tower tends to twist and sway due to wind loads and other natural forces. Considering the narrow beamwidths referred

Table 4.14 Minimum land area required for guyed towers

Tower Height (ft)	Area Required[a] (ft)				
	80% Guyed	75% Guyed	70% Guyed	65% Guyed	60% Guyed
60	87 × 100	83 × 96	78 × 90	74 × 86	69 × 80
80	111 × 128	105 × 122	99 × 114	93 × 108	87 × 102
100	135 × 156	128 × 148	120 × 140	113 × 130	105 × 122
120	159 × 184	150 × 174	141 × 164	132 × 154	123 × 142
140	183 × 212	178 × 200	162 × 188	152 × 176	141 × 164
160	207 × 240	195 × 226	183 × 212	171 × 198	159 × 184
180	231 × 268	218 × 252	204 × 236	191 × 220	177 × 204
200	255 × 296	240 × 278	225 × 260	210 × 244	195 × 226
210	267 × 304	252 × 291	236 × 272	220 × 264	204 × 236
220	279 × 322	263 × 304	246 × 284	230 × 266	213 × 246
240	303 × 350	285 × 330	267 × 308	249 × 288	231 × 268
250	315 × 364	296 × 342	278 × 320	254 × 282	240 × 277
260	327 × 378	308 × 356	288 × 334	269 × 310	249 × 288
280	351 × 406	330 × 382	309 × 358	288 × 332	267 × 308
300	375 × 434	353 × 408	330 × 382	308 × 356	285 × 330
320	399 × 462	375 × 434	351 × 406	327 × 376	303 × 350
340	423 × 488	398 × 460	372 × 430	347 × 400	321 × 372
350	435 × 502	409 × 472	383 × 442	356 × 411	330 × 381
360	447 × 516	420 × 486	393 × 454	366 × 424	339 × 392
380	471 × 544	443 × 512	414 × 478	386 × 446	357 × 412
400	495 × 572	465 × 536	425 × 502	405 × 468	375 × 434
420	519 × 599	488 × 563	456 × 527	425 × 490	393 × 454
440	543 × 627	510 × 589	477 × 551	444 × 513	411 × 475

Tower Height (ft)	Area Required[a] (acre)				
	80% Guyed	75% Guyed	70% Guyed	65% Guyed	60% Guyed
60	0.23	0.21	0.19	0.17	0.15
80	0.38	0.34	0.30	0.26	0.23
100	0.56	0.50	0.44	0.39	0.34
120	0.77	0.69	0.61	0.53	0.46
140	1.03	0.91	0.80	0.70	0.61
160	1.31	1.16	1.03	0.90	0.77
180	1.63	1.45	1.27	1.11	0.96
200	1.99	1.76	1.55	1.35	1.16
210	2.18	1.93	1.70	1.48	1.27
220	2.38	2.11	1.85	1.61	1.39
240	2.81	2.49	2.18	1.90	1.63
250	3.04	2.69	2.36	2.05	1.76
260	3.27	2.89	2.54	2.21	1.90

Tower Height (ft)	Area Required[a] (acre)				
	80% Guyed	75% Guyed	70% Guyed	65% Guyed	60% Guyed
280	3.77	3.33	2.92	2.54	2.18
300	4.30	3.80	3.33	2.89	2.49
320	4.87	4.30	3.77	3.27	2.81
340	5.48	4.84	4.24	3.65	3.15
350	5.79	5.11	4.48	4.88	3.33
360	6.12	5.40	4.73	4.10	3.52
380	6.79	5.99	5.25	4.55	3.90
400	7.50	6.62	5.79	5.02	4.30
420	8.24	7.27	6.36	5.52	4.73
440	9.03	7.96	6.96	6.03	5.17

[a]Preferred area is a square using the larger dimension of minimum area. This will permit orienting tower in any desired position.

Table 4.15 Nominal twist and sway values for microwave tower – antenna – reflector systems[a]

	Tower-Mounted Antenna		Tower-Mounted Passive Reflector		
A Total Beamwidth of Antenna or Passive Reflector between Half-Power Points (°)	*B* Limits Limits of Movement of Antenna Beam with Respect to Tower ($\pm°$)	*C* Limits Limits of Tower Twist or Sway at Antenna Mounting Point ($\pm°$)	*D* Limits Limits of Movements of Passive Reflector with Respect to Tower ($\pm°$)	*E* Limits Limits of Tower Twist at Passive Reflector Mounting Point ($\pm°$)	*F* Limits Limits of Tower Sway at Passive Reflector Mounting Point ($\pm°$)
14	0.75	4.5	0.2	4.5	4.5
13	0.75	4.5	0.2	4.5	4.3
12	0.75	4.5	0.2	4.5	3.9
11	0.75	4.5	0.2	4.5	4.6
10	0.75	4.5	0.2	4.5	3.3
9	0.75	4.5	0.2	4.5	3.9
8	0.75	4.2	0.2	4.5	2.6
7	0.6	4.1	0.2	4.5	2.3
6	0.5	4.0	0.2	4.3	2.1
5	0.4	3.4	0.2	3.7	1.8
4	0.3	3.1	0.2	3.3	1.6
3.5	0.3	2.9	0.2	2.9	1.4
3.0	0.3	2.3	0.1	2.5	1.2
2.5	0.2	1.9	0.1	2.1	1.0
2.0	0.2	1.5	0.1	1.7	0.9
1.5	0.2	1.1	0.1	1.2	0.6
1.0	0.1	0.9	0.1	0.9	0.5
0.75[b]	0.1	0.7	0.1	0.7	0.4
0.5[b]	0.1	0.4	0.1	0.4	0.2

Source: EIA RS-222B. (Ref. 19; courtesy of the Electronics Industry Association.)

[a]The values are tabulated as a guide for systems design and are based on the values that have been found satisfactory in the operational experience of the industry. These data are listed for reference only.
[b]These deflections are extrapolated and are not based on experience of the industry.

Notes:

1. Half-power beamwidth of the antenna to be provided by the purchaser of the tower.
2. A. The limits of beam movement resulting from an antenna mounting on the tower are the sum of the appropriate figures in columns *B* and *C*.
   B. The limits of beam movement resulting from twist when passive reflectors are employed are the sum of the appropriate figures in columns *D* and *E*.
   C. The limits of beam movement resulting from sway when passive reflectors are employed are twice the sum of the appropriate figures in columns *D* and *F*.
   D. The tabulated values in columns *D*, *E*, and *F* are based on a vertical orientation of the antenna beam.
3. The maximum tower movement shown above (4.5°) will generally be in excess of that actually experienced under conditions of 20 lb/ft^2 wind loading.
4. The problem of linear horizontal movement of a reflector–parabola combination had been considered. It is felt that in a large majority of cases, this will present no problem. According to tower manufacturers, no tower will be displaced horizontally at any point on its structure more than 0.5 ft/100 ft of height under its designed wind load.
5. The values correspond to 10-dB gain degradation under the worst combination of wind forces at 20 lb/ft^2. This table is meant for use with standard antenna–reflector configurations.
6. Twist and sway limits apply to 20 lb/ft^2 wind load only, regardless of survival or operating specifications. If there is a requirement for these limits to be met under wind loads greater than 20 lb/ft^2, such requirements must be specified by the user.

to in Section 4.7.2 (Table 4.11), with only a little imagination we can see that only a very small deflection of a tower or antenna will cause a radio ray beam to fall out of the reflection face of an antenna on the receive side or move the beam out on the far-end transmit side of a link.

Twist and sway, therefore, must be limited. Table 4.15 sets certain limits. The table has been taken from EIA RS-222B. From the table we can see that angular deflection and tower movement are functions of wind velocity. It should also be noted that the larger the antenna, the smaller the beamwidth, besides the fact that the sail area is larger. Hence the larger the antenna (and the higher the frequency of operation), the more we must limit the deflection.

To reduce twist and sway, tower rigidity must be improved. One generality we can make is that towers that are designed to meet required wind load or ice load specifications are sufficiently rigid to meet twist and sway tolerances. One way to increase rigidity is to increase the number of guys, particularly at the top of the tower. This is often done by doubling the number of guys from three to six.

4.14 PLANE REFLECTORS AS PASSIVE REPEATERS

A plane reflector as a passive repeater offers some unique advantages. Suppose that we wish to provide multichannel service to a town in a valley, and a mountain is nearby with poor access to its top. The radiolink engineer should consider the use of a passive repeater as an economic alternative. A prime requirement is that the plane reflector be within line-of-sight of the terminal antenna in town as well as line-of-sight of the distant microwave radiolink station. Such a passive repeater installation may look like the following example, where a = the net path loss in decibels:

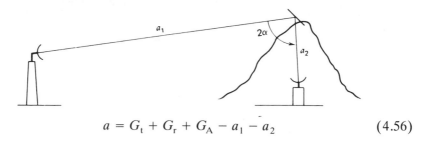

$$a = G_t + G_r + G_A - a_1 - a_2 \tag{4.56}$$

where a_1 = path loss on path 1 (dB)
   a_2 = path loss on path 2 (dB)
   G_t = transmitting antenna gain
   G_r = receiving antenna gain
   G_A = passive reflector gain

Let us concern ourselves with G_A for the moment. The gain of a passive reflector results from the capture of a ray beam of RF energy from a distant antenna emitter and the redirection of it toward a distant receiving antenna. The gain of the passive reflector is divided into two parts: (1) incoming energy and (2) redirected or reflected energy. The gain for (1) and (2) is

$$G_A = 20 \log \frac{4\pi A \cos \alpha}{\lambda^2} \qquad (4.57)$$

where α = one-half the included horizontal angle between incident and reflected wave and A = the surface area (ft^2 or m^2) of the passive reflector. If A is in square feet, then λ must also be in feet.

The reader will find the following relationship useful:

$$\text{Wavelength (ft)} = \frac{985}{F} \qquad (4.58)$$

where F is a megahertz. Passive reflector path calculations are not difficult. The first step is to determine if the shorter path (a_2 path in the figure above) places the passive reflector in the near field of the nearer parabolic antenna. To determine whether near field or far field, solve the following formula:

$$\frac{1}{K} = \frac{\pi \lambda d'}{4A} \qquad (4.59)$$

If the ratio $1/K$ is less than 2.5, a near-field condition exists; if $1/K$ is greater than 2.5, a far-field condition exists. d' = length of the path in question (i.e., the shorter distance).

For the far-field condition, consider paths 1 and 2 as separate paths and sum their free-space path losses. Determine the gain of the passive plane reflector using equation 4.57. Sum this gain with the two free-space path losses algebraically to obtain the net path loss.

For the near-field condition, where $1/K$ is less than 2.5, the free-space path loss will be that of the longer hop (e.g., a_1 above). Algebraically add the repeater gain (or loss) which is determined as follows. Compute the parabola/reflector coupling factor l:

$$l = D'\sqrt{\frac{\pi}{4A}} \qquad (4.60)$$

where D' = diameter of parabolic antenna (ft)
A = effective area of passive reflector (ft^2)

Figure 4.41 is now used to determine near-field gain or loss. The $1/K$ value is on the abscissa, and l is the family of curves shown.

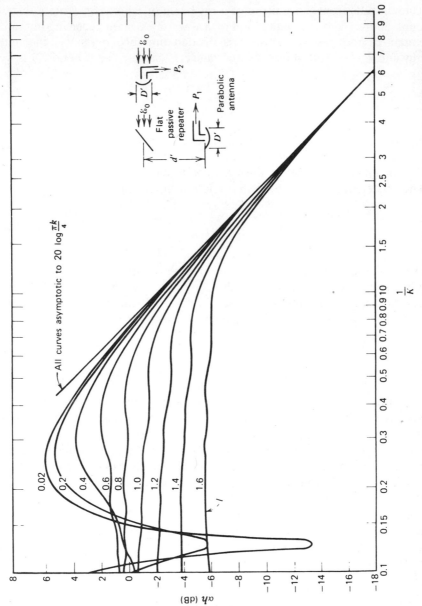

Figure 4.41 Antenna – reflector efficiency curves. (From Ref. 17; courtesy of Microflect, Inc., Salem, OR.)

The examples below are from Ref. 17.

Example 1. *Far-field:* A plane passive reflector 10×16 ft, or 160 ft^2, is erected 21 mi from one active site and only 1 mi from the other. $2\alpha = 100°$ and $\alpha = 50°$. The operating frequency is 2000 MHz. By formula the free-space loss for the longer path is 129.5 dB, and for the shorter path it is 103 dB.

Calculate the gain G_A of the passive plane reflector, equation 4.57:

$$G_A = \frac{20 \log 4 \times 160 \cos 50°}{(985/2000)^2} = 20 \log 5.340$$

$$= 74.6 \text{ dB}$$

Net path loss $= -129.5 - 103 + 74.6 = -157.9$ dB

Example 2. *Near-field:* The passive reflector selected in this case is 24×30 ft. The operating frequency is 6000 MHz. The long leg is 30 mi and the short leg 4000 ft. A 10-ft parabolic antenna is associated with the active site on the short leg. (6 GHz is approximately equivalent to 0.164 ft.) Determine $1/K$:

$$\frac{1}{K} = \frac{\pi \lambda d}{4A} = \frac{\pi(0.164)(4000)}{4 \times 720} = 0.717$$

Note that this figure is less than 2.5, indicating the near-field condition. Calculate l:

$$l = D'\sqrt{\pi/4A} = 10\sqrt{\pi/4 \times 720} = 0.33$$

Using these two inputs, the values of l and $1/K$, we go to Figure 4.41 and find the net gain of the system to be $+0.2$ dB. The net free-space loss is then $+0.2 - 142.3 = -142.1$ dB. The free-space loss of the 30-mi leg is 142.3 dB.

REVIEW EXERCISES

1. Give at least five applications of LOS microwave systems for telecommunications.

2. What are the principal competing transmission mediums for LOS microwave? Give some advantages and disadvantages of each.

3. Give the four most important steps in the design of an LOS microwave link.

4. Calculate the free-space loss (FSL) in dB for the following LOS microwave paths: 1 GHz, 22 mi; 3800 MHz, 45 mi; 12.27 GHz, 7.3 mi; and an Andean path 127 mi long at 2.1 GHz.

5. What is the smooth earth distance to the horizon ($K = \frac{4}{3}$) for an antenna mounted on the top of a tower 50 ft high; 450 ft high?

6. In a particular geographic area and altitude, the effective earth radius is found to be 7800 km. Calculate the equivalent K factor.

7. What do we derive from a path profile? (What does it tell us?)

8. When drawing a path profile, we calculated the appropriate K factor as 0.85. How will such a K factor impact tower height compared to a K factor of 1.0?

9. What is the standard Fresnel zone clearance for a microwave ray beam to clear an obstacle?

10. A Fresnel zone clearance is really a function of only two variables. What are they? Where in an LOS microwave path is the Fresnel zone clearance maximum?

11. Calculate the FM improvement threshold for a receiver operating at room temperature with a noise figure of 9 dB and an IF bandwidth of 20 MHz.

12. Using Carson's rule, calculate the IF bandwidth of a receiver operating at 6 GHz where the highest modulating frequency is 1.248 MHz and the peak deviation is 3.5 MHz.

13. What is the primary purpose of the path calculation exercise?

14. Name the two basic types of fading one is most likely to encounter on an LOS microwave link.

15. Why is preemphasis–deemphasis needed on broadband FM microwave links?

16. Give three fairly standardized LOS microwave transmitter power outputs in dBW.

17. Determine the peak deviation of a 960 VF channel FDM/FM transmitter using standard CCIR deviation per channel values (Table 4.4).

18. Using a rough rule of thumb based on CCIR recommendations, what would we expect the thermal noise in an analog voice channel to be at the end of a 1000-km multihop LOS microwave system (considering the radio equipment contribution to noise only)?

19. Name the three noise sources of an FM LOS microwave radio terminal that must be summed to arrive at the total noise in a derived FDM/FM voice channel.

20. Calculate the RSL value for an FDM/FM microwave link where the thermal noise in the voice channel (derived from the radio equipment) is

200 pWp (do not include preemphasis improvement). The link carries 960 VF channels and the receiver noise figure is 8 dB.

21. What is an acceptable NPR for an FDM/FM microwave *terminal*?

22. What does noise power ratio (NPR) tell us about a microwave radio terminal or repeater?

23. What is the cause of IM noise due to antenna feeder distortion? How can it be reduced?

24. What is the psophometric weighting improvement value in dB over flat noise in the voice channel?

25. What is the commonly accepted value in dBm of 0 dBrnc?

26. The S/N in a voice channel is 63 dB. What is the equivalent value of thermal noise measured in pWp?

27. We designed an LOS microwave radio hop that we find, late in the acceptance test period, does not meet noise or BER requirements. If it is a digital link, we assume measures have been taken to mitigate the effects of dispersion. Now list in order of economic impact (least to most), measures that we can take to bring the performance up to specifications.

28. Calculate the EIRP in dBW of a microwave transmitting installation where the transmitter output is 0.5 W, the total transmission line loss is 5 dB, and the antenna gain is 33 dB.

29. Determine the IRL at the receive end of a 4-GHz link 25 mi long where the transmitter has 1-W output to the flange, the transmission line loss is 4 dB, and the transmit antenna has a gain of 37 dB. There is a radome at each end with a transmission loss of 1 dB (each).

30. Calculate the RSL in dBW from question 29 above where the receive installation has the same antenna type but the transmission line loss is 5.5 dB.

31. Determine the total noise in a derived voice channel (radio portion only) where the IM antenna feeder noise sum of both ends is 10 dBrnc0, the thermal noise is 21 dBrnc0, and the IM noise is 16 dBrnc0. Give the answer in dBrnc0.

32. What fade margin is required for a microwave LOS link with a time availability requirement of 99.997%? Consider the worst-case situation.

33. When we calculate fade margin using the Barnett method, fade margin is a function of what five factors?

34. What are the four types of diversity listed in the text?

35. In the United States, if diversity is required on a LOS microwave LOS hop, in nearly all situations space diversity is mandatory. Why?

36. What are the three generic types of repeaters used on analog microwave radio systems? Discuss where each is applied or not applied.

37. Discuss bandwidth occupancy of digital microwave LOS radios that carry standard 8-bit PCM telephone channels versus FDM/FM and FDM/SSB. What measures are used to conserve bandwidth in the digital case?

38. Calculate N_0 for a digital radio receiver with a noise figure of 10 dB; of 4 dB?

39. What is the meaning of bit count integrity (BCI)?

40. Give at least four causes of error on a digital LOS microwave link.

41. Relate E_b/N_0 to RSL, bit rate, and receiver noise figure.

42. What is the gating BER value at the terminal end of a series of digital radio hops (terminating at a switch)? Why? (We refer here to digital speech telephony/PCM.) What drives this high value to a far improved BER (i.e., what type of use)?

43. If using binary phase shift keying (BPSK) for a 10-Mbps radio requires 10 MHz bandwidth (theoretical), what theoretical bandwidth would be needed for 32-QAM for the same bit rate?

44. As M increases on an M-QAM system, what is the effect on E_b/N_0 for a fixed BER?

45. If an RSL is -76 dBW and the bit rate is 20 Mbps, what is E_b in dBW?

46. An LOS microwave parabolic dish antenna has an aperture of 6 ft and operates at 12 GHz. What is its gain in dB? Assume 55% aperture efficiency.

47. A certain antenna has a 38-dB gain. What is the equivalent 3-dB beamwidth?

48. Identify at least four causes of link reliability degradation.

49. Give two applications of a power splitter.

50. There are two generic types of hot-standby operation. Argue the pros and cons of each.

51. The availability of a certain microwave link is 99.97%. What is the equivalent unavailability?

52. What are service channels used for? Name two applications.

53. Most microwave repeater sites are unmanned. We wish to receive fault information from each repeater site at a manned location. List at least seven equipment alarm malfunctions and at least four site alarms to be monitored. List three fault alarm functions to be monitored that are peculiar to digital operation.

54. We have 10 links in tandem, and at the output of a far-end receiver we wish to have 99.997% availability. We assign half the unavailability to propagation outage and half to other outages including equipment outage. Using Rayleigh fading criteria only, what fade margin will we need at the receive side of each site?

55. Tower twist and sway limits are a function of what? Argue why large dish antennas on high towers should be avoided.

56. Under what circumstances would we use a plane reflector as a passive repeater?

REFERENCES AND BIBLIOGRAPHY

1. *Reference Data for Radio Engineers*, 6th ed., Howard W. Sams, Indianapolis, IN, 1977.
2. *Engineering Considerations for Microwave Communication Systems*, Lenkurt Electric Corp., San Carlos, CA, 1975.
3. *Recommendations and Reports of the CCIR*, XVIth Plenary Assembly, Dubrovnik, 1986, vols. IV and IX.
4. Roger L. Freeman, *Reference Manual for Telecommunications Engineering*, Wiley, New York, 1985.
5. T. K. Fitzsimmons, *VHF/UHF/Microwave LOS Terrestrial Propagation and Systems Design*, AGARD Report No. 744, NATO Headquarters, Brussels, Oct. 1986.
6. Robert M. Gagliardi, *Introduction to Communications Engineering*, 2nd ed., Wiley, New York, 1988.
7. *Microwave Radio Relay Systems*, USAF Tech. Order TO 31R5-1-9, U.S. Department of Defense, Washington, DC, Apr. 1965.
8. W. Oliver, *White Noise Loading of Multichannel Communication Systems*, Marconi Instruments Ltd., St. Albans, UK, Sept. 1964.
9. *Electrical Communications Systems Engineering—Radio*, U.S. Army Tech. Manual TM-11-486-6, U.S. Department of Defense, Washington, DC, 1956.
10. *Transmission Systems for Communications*, 5th ed., Bell Telephone Laboratories (AT&T), New York, 1982.
11. B. D. Holbrook and J. T. Dixon, "Load Rating Theory for Multichannel Amplifiers," *Bell Syst. Tech. J.*, vol. 18, 624–644, Oct. 1939.
12. E. F. Plarman and J. J. Tary, *Required Signal-to-Noise Ratios, RF Signal Power and Bandwidth for Multichannel Radio Communication Systems*, Tech. Note 100, National Bureau of Standards, Boulder, CO, Jan. 1962.

13. A. P. Barkhausen et al., *Equipment Characteristics and Their Relationship to Performance for Tropospheric Scatter Communication Circuits*, Tech. Note 103, National Bureau of Standards, Boulder, CO, Jan. 1963.

14. *Design Handbook for Line-of-Sight Microwave Communication Systems*, MIL-HDBK-416, U.S. Department of Defense, Washington, DC, 1977.

15. F. E. Terman, *Radio Engineering*, 4th ed., McGraw-Hill, New York, 1954.

16. D. C. Livingston, *The Physics of Microwave Propagation*, GTE Monograph, General Telephone and Electronics, Bayside, NY, 1967.

17. *Microflect Passive Repeater Engineering Manual No. 161*, Microflect Inc., Salem, OR, 1962.

18. Warren L. Stutzman and Gary A. Thiele, *Antenna Theory and Design*, Wiley, New York, 1981.

19. EIA Standards RS-195B, RS-203, RS-222B, RS-250B and RS252A, Electronics Industries Association, Washington, DC.

20. K. Bullington, "Radio Propagation Fundamentals," *Bell Syst. Tech. J.*, May 1957.

21. J. Jasik, *Antenna Engineering Handbook*, 2nd ed., McGraw-Hill, New York, 1986.

22. R. L. Marks et al., *Some Aspects of the Design of FM Line-of-Sight Microwave and Troposcatter Systems*, USAF Rome Air Development Center, Rome, NY, U.S. National Technical Information Service, AD 617-686, Springfield, VA, Apr. 1965.

23. J. Fagot and P. Magne, *Frequency Modulation Theory*, Pergamon Press, London, 1961.

24. *Andrew Catalog 34*, Andrew Corp., Orland Park, IL, 1988.

25. *Collins Telecommunications Equipment Catalog 2*, Rockwell International, Collins Radio, Dallas, TX, 1982.

26. H. T. Dougherty, *A Survey of Microwave Fading Mechanisms, Remedies and Applications*, ESSA Tech. Report ERL 69-WPL-4, Boulder, CO, Mar. 1968.

27. P. F. Panter, *Communication Systems Design—Line-of-Sight Microwave and Troposcatter Systems*, McGraw-Hill, New York, 1962.

28. R. F. White, *Reliability in Microwave Communication Systems—Prediction and Practice*, Lenkurt Electric Corp., San Carlos, CA, 1970.

29. A. P. Barsis et al., *Analysis of Propagation Measurements over Irregular Terrain in the 76 to 9200 MHz Range*, ESSA Tech. Report ERL 114-ITS-82, Institute of Telecommunication Sciences (NBS), Boulder, CO, Mar. 1969.

30. Roger L. Freeman, *Telecommunication System Engineering*, 2nd ed., Wiley, New York, 1989.

31. M. J. Tant, *The White Noise Book*, Marconi Instruments, Ltd., St. Albans, UK, 1969.

32. H. Brodhage and W. Hormuth, *Planning and Engineering Radio Links*, Siemens/Heyden & Sons, London, 1977.

33. Topographic Maps, U.S. Geological Survey, Arlington, VA.

34. *Design Objectives of DCEC LOS Digital Radio Links*, Eng. Pub DCEC EP 27-77, Defense Communications Engineering Center, Washington, DC, 1977.

35. A. Longley and R. K. Reasoner, *Comparison of Propagation Measurements with Predicted Values in the 20–10,000 MHz Range*, ESSA Tech. Report ERL 148-ITS-97, Institute of Telecommunication Sciences (NBS), Boulder, CO, 1970.

36. *Notes on the BOC Intra-LATA Networks—1986*, TR-NPL-000275, Bell Communications Research, Holmdel, NJ, 1986.

37. *Principles of Digital Transmission*, ER79-4307, Raytheon Co. under U.S. government contract MDA-904-79-C-0470, Sudbury, MA, 1979 (limited circulation).

38. E. W. Allen, "The Multipath Phenomenon in Line-of-Sight Digital Transmission Systems," *Microwave J.*, 1984.

39. "Transmission Systems Engineering Symposium," Rockwell International, Collins Transmission Systems Division, Dallas, TX, Sept. 1985.

40. K. Feher, "Digital Communications Microwave Applications," Prentice Hall, Englewood Cliffs, NJ, 1981.

5

BEYOND LINE-OF-SIGHT: TROPOSPHERIC SCATTER AND DIFFRACTION LINKS

5.1 INTRODUCTION

Tropospheric scatter and diffraction are methods of propagating microwave energy beyond LOS or "over the horizon." Communication systems utilizing the tropospheric scatter/diffraction phenomena handle from 12 to 240 multiplexed telephone channels. Well-planned tropospheric scatter diffraction links may have propagation reliabilities on the order of 99.9% or better. These reliabilities are comparable to those of radiolink (LOS microwave) discussed in the preceding chapter. In fact, the discussion of tropospheric scatter is a natural extension of Chapter 4.

Tropospheric scatter takes advantage of the refraction and reflection phenomena in a section of the earth's atmosphere called the troposphere. This is the lower portion of the atmosphere from sea level to a height of about 11 km (35,000 ft). UHF/SHF signals are scattered in such a way as to follow reliable communications on hops up to 640 km (400 mi). Long distances of many thousands of kilometers may be covered by operating a number of hops in tandem. The North Atlantic Radio System (NARS) of the U.S. Air Force is an example of a lengthy tandem system. It extends from Canada to Great Britain via Greenland, Iceland, the Faeroes, and Scotland. A mix of radiolinks (LOS microwave) and tropospheric scatter is becoming fairly common. The Canadian National Telephone Company (CNT) operates such a system in the Northwest Territories. The Bahama Islands are interconnected for communications by a mix of LOS microwave, tropospheric scatter, and HF.

Tropospheric scatter systems generally use transmitter power outputs of 1 or 10 kW, parabolic-type antennas with diameters of 4.5 m (15 ft), 9 m (30 ft), or 18 m (60 ft), and sensitive (uncooled) broadband FM receivers with front-end NFs on the order of 1.0–4.0 dB. A tropospheric scatter installation is obviously a bigger financial investment than an LOS microwave installation. Tropospheric scatter/diffraction, however, has many advantages for commercial application that could well outweigh the issue of high cost. These advantages are summarized as follows. It

1. Reduces the number of stations required to cover a given large distance when compared to radiolinks. Tropospheric scatter may require from one-third to one-tenth the number of stations as a radiolink system over the same path.
2. Provides reliable multichannel communication across large stretches of water (e.g., over inland lakes, to offshore islands, between islands) or between areas separated by inaccessible terrain.
3. May be ideally suited to meet toll-connecting requirements of areas of low population density.
4. Useful when radio waves must cross territories of another political administration.
5. Requires less maintenance staff per route-kilometer than conventional radiolink systems over the same route.
6. Allows multichannel communication with isolated areas, especially when intervening territory limits or prevents the use of repeaters.
7. Desirable for multichannel communications in the tactical military field environment for links from 30 to 200 mi long (50–340 km).
8. Thin-line (e.g., $\leq$ 16-kbps) military systems with links up to 800 mi long (1480 km).

5.2 THE PHENOMENON OF TROPOSPHERIC SCATTER

There are a number of theories explaining over-the-horizon communications by tropospheric scatter. One theory postulates atmospheric air turbulence, irregularities in the refractive index, or similar homogeneous discontinuities capable of diverting a small fraction of the transmitted radio energy toward a receiving station. This theory accounts for the scattering of radio energy in a way much as fog or moisture seems to scatter a searchlight on a dark night. Another theory is that the air is stratified into discrete layers of varying thickness in the troposphere. The boundaries between these layers become partially reflecting surfaces for radio waves and thereby scatter the waves downward over the horizon.

Figure 5.1 is a simple diagram of a tropospheric scatter link showing two important propagation concepts. These are

- Scatter angle, which may be defined as either of two acute angles formed by the intersection of the two portions of the tropospheric scatter beam

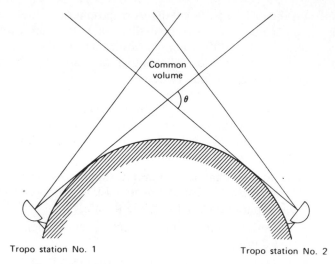

Figure 5.1 Tropospheric scatter model. θ = scatter angle.

(lower boundaries) tangent to the earth's surface. Keeping the angle small effectively reduces the overall path attenuation (see Figure 5.1).

• Scatter volume or "common volume" is the common enclosed area where the two beams intercept.

5.3 TROPOSPHERIC SCATTER FADING

Fading is characteristic of tropospheric scatter. It is handy to break fading in tropospheric scatter systems into two types, slow and fast fading. Expressed another way, these are long-term (slow) and short-term variations in the received signal level.

When referring to tropospheric scatter received signal level, we usually use the median received level as reference. In general the hourly median and minute median are the same. In Chapter 4, the reference level was the unfaded signal level, which turned out to approximate sufficiently the calculated level under no-fade conditions. Such a straightforward reference signal level is impossible in tropospheric scatter because a tropospheric scatter signal is in a constant condition of fade. Thus for path calculations and path loss, we refer to the long-term median, usually extended over the whole year.

At any one moment a received tropospheric scatter signal will be affected by both slow and fast fading. It is believed that fast fading is due to the effects of multipath (i.e., due to a phase incoherence at various scatter angles). Fast fading is treated statistically "within the hour" and has a Rayleigh distribution with a sampling time of 1–7 min, although in some

Table 5.1   Time block assignments

Time Block	Month	Hours
1	Nov.–Apr.	0600–1300
2	Nov.–Apr.	1300–1800
3	Nov.–Apr.	1800–2400
4	May–Oct.	0600–1300
5	May–Oct.	1300–1800
6	May–Oct.	1800–2400
7	May–Oct.	0000–0600
8	Nov.–Apr.	0000–0600

circumstances it has been noted up to 1 h. The fading rate depends on both frequency and distance or length of hop.

The U.S. National Bureau of Standards (NBS) describes long-term variations in signal level as variations of *hourly median* values of transmission loss. This is the level of transmission loss that is exceeded for a total of one half of a given hour. A distribution of hourly medians gives a measure of long-term fading. Where these hourly medians are considered over a period of 1 month or more, the distribution is log normal.

In studying variations in tropospheric scatter transmission path loss (fading), we have had to depend on empirical information. The signal level varies with the time or day, the season of year, and the latitude, among other variables. To assist in the analysis and prediction of long-term signal variation, the hours of the year have been broken down into eight time blocks, given in Table 5.1.

Most commonly we refer to time block 2 for a specific median path loss. Time block 2 may be thought of as an average winter afternoon in the temperate zone of the Northern Hemisphere.

It should be noted that signal levels average 10 dB lower in winter than in summer, and that morning or evening signals are at least 5 dB higher than midafternoon signals. Slow fading is believed due to changes in path conditions such as atmospheric changes (e.g., a change in the index of refraction of the atmosphere).

5.4   PATH LOSS CALCULATIONS

Tropospheric scatter paths typically display considerably larger losses when compared to radiolink (LOS microwave) paths. Losses up to 260 dB are not uncommon. There are a number of acceptable methods of estimating path losses for tropospheric scatter systems. One such method is outlined in CCIR Rep. 238-3 (Ref. 17), and another is described in NBS Tech. Note 101 (Ref. 3). These are more commonly known as the CCIR method and the NBS method. Their approach to the problem is somewhat similar. In the following paragraphs we will summarize some of the more important aspects of the NBS method as described in Ref. 5. The procedure has been considerably simplified and abbreviated for this discussion.

The objective is to predict a long-term path loss that will not be exceeded for specified time availabilities, such as 50, 90, 99, 99.9, or 99.99% of the time. In the previous chapter on LOS microwave we referred to this as propagation reliability.

The NBS method describes how to calculate these losses with a 50% probability (confidence level) of being the correct prediction for a path in question. Then it shows how to systematically add margin to assure improved probability that the prediction will be correct or more than the minimum necessary on a particular path. This probability of prediction is called *service probability*. It is common with tropospheric scatter path calculations to show a 50 and a 95% service probability. Service probability indicates the confidence level of the prediction.

5.4.1   Mode of Propagation

An over-the-horizon microwave path can and often does display two modes of propagation, diffraction and tropospheric scatter. In most cases either one or the other will predominate. Particularly on shorter paths the possibility of the diffraction mode should be investigated. Experienced engineers in over-the-horizon systems can often identify which propagation mode can be expected during the path profile and path survey phases of the link engineering effort. In the absence of this expertise, the following criteria may be used as an aid to identify the principal propagation mode (Ref. 5):

1. The distance at which diffraction and tropospheric scatter losses are approximately equal is $65(100/f)^{1/3}$ km, where f = radio frequency in MHz. For path lengths less than this value, diffraction will be the predominant mode, and vice versa.
2. For paths having angular distances of 20 mrad or more, the diffraction mode may be neglected and the path can be considered to be operating in the troposcatter mode. (Angular distance is explained below.)

5.4.2   Basic Long-Term Tropospheric Scatter Transmission Loss

Following the NBS method, we determine the basic long-term tropospheric scatter transmission loss L_{bsr}:

$$L_{bsr} = 30 \log f - 20 \log d + F(\theta d) - F_0 + H_0 + A_a \qquad (5.1)$$

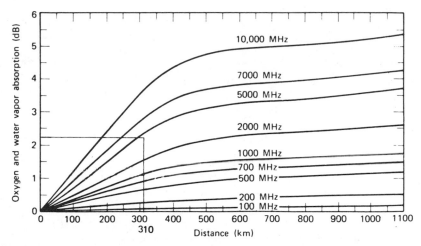

Figure 5.2 Determination of median oxygen and water vapor absorption in decibels for various operating frequencies when path length is given (for August, Washington, DC). (Courtesy of the Institute of Telecommunication Sciences, Office of Telecommunications, U.S. Department of Commerce.)

where f = operating frequency (MHz)
            d = great-circle path length (km)
     $F(\theta d)$ = attenuation function (dB)
         F_0 = scattering efficiency correction factor
         H_0 = frequency gain function
         A_a = atmospheric absorption factor from Figure 5.2

We have simplified our procedure by neglecting the frequency gain function and the scattering efficiency correction factor.

The numerical value of the first two terms of equation 5.1 are determined by substituting the assigned frequency in megahertz of the radio system to be installed (see Section 5.9.2) and the great-circle distance in kilometers. The third term requires some detailed discussion.

5.4.3 Attenuation Function

The attenuation function $F(\theta d)$, is derived from Figure 5.3. θd is the product of the angular distance (scatter angle) in radians and the great-circle path length in kilometers. The following is an abbreviated method of approximating the scatter angle θ. We assume that a path profile has been carried out (see Section 4.3.3). Arbitrarily, one site is denoted the transmitter site t and the other site the receiver site r. From the profile the horizon location in the direction of the distant site and its altitude above mean sea level (MSL) are determined as well as its distance from its corresponding site. For all

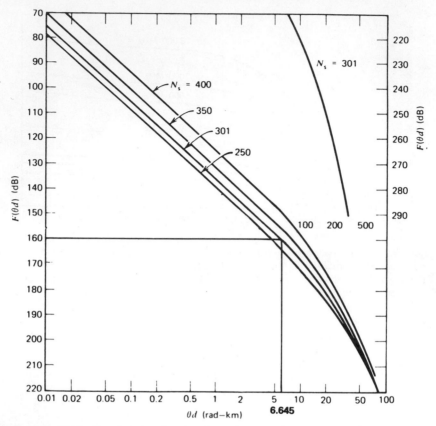

Figure 5.3 Attenuation function for the determination of scatter loss. (Courtesy of the Institute of Telecommunication Sciences, Office of Telecommunications, U.S. Department of Commerce.)

further calculations distances and altitudes (elevations) are measured in kilometers and angles in radians. It is important to use only these units throughout. Figure 5.4 will assist in identifying the following distances, elevations, and angles. Let

d = great-circle distance between transmitter and receiver sites

d_{Lt} = distances from transmitter site to transmitter horizon

d_{Lr} = distance from receiver to receiver horizon

h_{ts} = elevation above MSL to center of transmitting antenna (km)

h_{rs} = elevation above MSL to center of receiving antenna (km)

h_{Lt} = elevation above MSL of transmitter horizon point (km)

h_{Lr} = elevation above MSL of receiver horizon point (km)

N_0 = surface refractivity corrected for MSL (For the continental United States use Figure 4.4, and for other locations use Ref. 23, Figures 1.3 and 1.4)

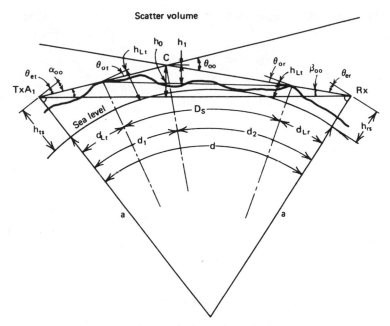

Figure 5.4 Tropospheric scatter path geometry.

Adjust the surface refractivity N_0 for the elevation of each site by the following formula:

$$N_{ts, rs} = N_0 \exp(-0.1057h_{ts}, h_{rs})$$ (5.2)

Compute N_s:

$$N_s = \frac{N_{ts} + N_{rs}}{2}$$ (5.3)

Calculate the effective earth radius by the following formula:

$$a = a_0(1 - 0.04665 \exp^{0.005577N_s})^{-1}$$ (5.4)

where a_0 = true earth radius, 6370 km. If $N_s = 301$ then $a = 8500$ km, which is the familiar $\frac{4}{3}$ earth radius case alluded to in Chapter 4.

Calculate the antenna take-off angles at each site by the following equations:

$$\theta_{et} = \frac{h_{Lt} - h_{ts}}{d_{Lt}} - \frac{d_{Lt}}{2a}$$ (5.5a)

$$\theta_{er} = \frac{h_{Lr} - h_{rs}}{d_{Lr}} - \frac{d_{Lr}}{2a}$$ (5.5b)

Calculate the scatter angle components α_0 and β_0 by the following formulas:

$$\alpha_0 = \frac{d}{2a} + \theta_{et} + \frac{h_{ts} - h_{rs}}{d} \tag{5.6a}$$

$$\beta_0 = \frac{d}{2a} + \theta_{er} + \frac{h_{rs} - h_{ts}}{d} \tag{5.6b}$$

Calculate the scatter angle (often called *angular distance*) θ_0 by the following equation:

$$\theta_0 = \alpha_0 + \beta_0 \, (\text{rad}) \tag{5.7}$$

Multiply θ_0 in radians by the path length in kilometers. This is θd.

Determine $F(\theta d)$ from Figure 5.3 using the product θd calculated above, and interpolate, if necessary, for the value of N_s taken from equation 5.3. L_{bsr} is now calculated by equation 5.1, neglecting terms F_0 and H_0.

5.4.4 Basic Median Transmission Loss

The predicted median long-term transmission loss $L_n(0.5, 50)$, abbreviated $L_n(0.5)$, for the appropriate climatic region n is related to L_{bsr} by the following formula:

$$L_n(0.5) = L_{bsr} - V_n(0.5, d_e) \tag{5.8}$$

where $L_n(0.5)$ = predicted transmission loss (in dB) exceeded by half of all hourly medians, and hence the yearly median value, and $V_n(0.5, d_e)$ = variability of the median value about the basic long-term transmission loss L_{bsr} for the appropriate climatic region n and the effective distance d_e. NBS has established eight climatic regions for the world as follows (Ref. 5):

1. Continental temperature. Large land mass, 30–60°N latitude, 30–60°S latitude.
2. Maritime temperature overland. In this region, prevailing winds, unobstructed by mountains, carry moist maritime air inland. 20–50°N, 20–50°S latitudes, typified by United Kingdom, the west coasts of North America and Europe, and northwestern coastal areas of Africa.
3. Maritime temperature oversea. Fully over-water paths in temperate regions.
4. Maritime subtropical overland. 10–30°N, 10–30°S latitudes, near the sea with defined rainy and dry seasons.
5. Maritime subtropical. Latitudes same as region 4. Over-water paths. However, valid curves are not available due to lack of empirical data for this region. Use region 3 or region 4, whichever is more applicable.

6. Desert, Sahara. Regions with year-round semiarid conditions.

7. Equatorial. $\pm 20°$ latitude from the equator, characterized by monotonous heavy rains and high average summer temperatures.

8. Continental subtropical. Usually 20–40°N latitude, an area of monsoons with seasonal extremes of summer rainfall and winter drought.

Select the most appropriate region (n) for the path in question, then compute the effective distance d_e. To calculate d_e, effective antenna heights are required, namely, h_{te} and h_{re}. These heights are functions of the average elevation of the terrain between each antenna and its respective radio horizon in the direction of the distant end of the path. For smooth earth condition (i.e., typically an over-water path or a hypothetical path with no obstacles except central earth bulge), h_{te} and h_{re} are the effective elevations of each site above MSL. Under real overland conditions, the effective height is the average height above MSL of the central 80% between the antenna and its radio horizon.

Calculate d_L by the following formula:

$$d_L = 3\sqrt{2h_{te} \times 10^3} + 3\sqrt{2h_{re} \times 10^3} \qquad (5.9)$$

Determine d_{sl} by the following formula:

$$d_{sl} = 65\left(\frac{100}{f}\right)^{1/3} \qquad (5.10)$$

where f = frequency (MHz).

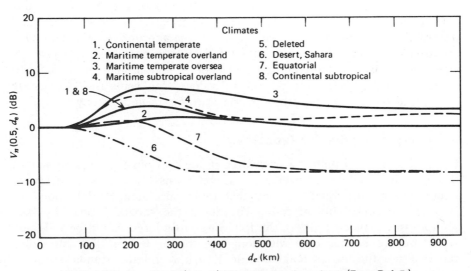

Figure 5.5 Function $V_n(0.5, d_e)$ for eight climatic regions. (From Ref. 5.)

There are two cases to calculate d_e. Use whichever is applicable.

1. If $d \leq d_L + d_{sl}$, then

$$d_e = \frac{130d}{(d_L + d_{sl})} \qquad (5.11a)$$

2. If $d > d_L + d_{sl}$, then

$$d_e = 130 + d - (d_L + d_{sl}) \qquad (5.11b)$$

With the climatic region n and the effective distance d_e determined, derive V_n in decibels from Figure 5.5. Calculate $L_n(0.5)$ using equation 5.8. The value $L_n(0.5, 50)$ represents the long-term median path loss for a 50% time availability and 50% service probability.

5.4.5 The 50% Service Probability Case

The next step is to extend the time availability to the specified or desired value for the tropospheric scatter path in question. Often it is convenient to state time availability for a number of percentages as follows:

Time Availability q (%)	Path Loss
50	$L_n(0.5, 50)$
90	A
99	B
99.9	C
99.99	D

Values for A, B, C, and D in the path loss column are determined by adding a factor to $L_n(0.5, 50)$ called $(Y_n(q, 50, d_e)$, where $q = 0.9$, 0.99, 0.999, and 0.9999. Y_n values are derived from curves for the appropriate climatic region and frequency band. One example family of curves is presented in Figure 5.6, where n is region 1, the continental temperate region, and for frequencies above 1 GHz. Y_n is derived for several values of q using the appropriate effective distance d_e of the tropospheric scatter path under study.

5.4.6 Improving Service Probability

Under the values of path loss calculated in the previous section, only half of the paths installed for a specific set of conditions would have a measured long-term path loss equal to or less than those calculated. By definition, this is a service probability of 50%. To extend the service probability (i.e., improve the confidence level), the following procedures should be followed. Again we are dealing only with long-term power fading. The basic data required are the values obtained for $Y_n(q, d_e)$ and the standard normal deviate Z_{mo} for the service probability desired. Several standard normal

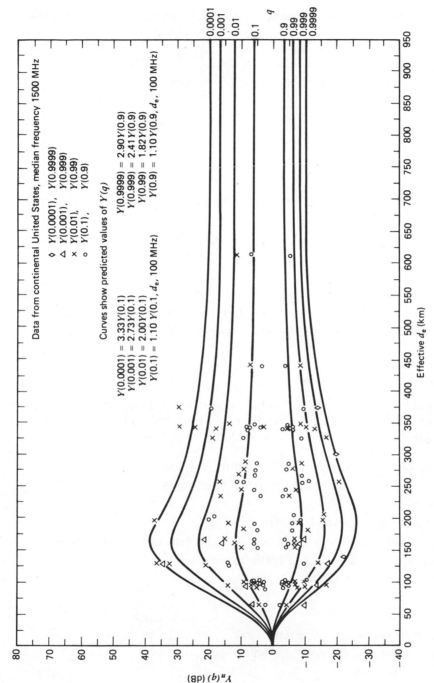

Figure 5.6 Long-term power fading, continental temperature climate, for frequencies greater than 1000 MHz. (From Ref. 5.)

Data from continental United States, median frequency 1500 MHz

◇ $Y(0.0001)$, $Y(0.9999)$
△ $Y(0.001)$, $Y(0.999)$
× $Y(0.01)$, $Y(0.99)$
○ $Y(0.1)$, $Y(0.9)$

Curves show predicted values of $Y'(q)$

$Y(0.0001) = 3.33Y(0.1)$
$Y(0.001) = 2.73Y(0.1)$
$Y(0.01) = 2.00Y(0.1)$
$Y(0.1) = 1.10\,Y(0.1, d_e, 100\ \text{MHz})$

$Y(0.9999) = 2.90Y(0.9)$
$Y(0.999) = 2.41Y(0.9)$
$Y(0.99) = 1.82Y(0.9)$
$Y(0.9) = 1.10\,Y(0.9, d_e, 100\ \text{MHz})$

deviates and their corresponding service probabilities are provided below:

Service Probability (%)	Standard Normal Deviate Z_{mo}
50	0
60	0.25
75	0.75
80	0.85
90	1.28
95	1.65
99	2.35

Calculate the path-to-path variance $\sigma_c^2(q)$ where q is the corresponding (or desired) time availability:

$$\sigma_c^2(q) = 12.73 + 0.12Y_n^2(q, d_e) \qquad (5.12)$$

Determine the prediction error $\sigma_{rc}(q)$ by the following equation:

$$\sigma_{rc}(q) = \sqrt{\sigma_c^2(q) + 4} \qquad (5.13)$$

Calculate the product of Z_{mo} and $\sigma_{rc}(q)$. This value is now added to the path loss value for the corresponding time availability q given in the previous section.

5.4.7 Example Problem

Assume smooth earth condition (i.e., no intervening obstacles besides earth bulge) and calculate the path loss from Newport, NY, to Bedford, MA (U.S.). The great-circle distance between the sites is 310.5 km.

Site elevation, Newport, $h_{ts} = 2000$ ft (0.61 km)
Site elevation, Bedford, $h_{rs} = 100$ ft (0.031 km)
$N_0 = 310$
Operating frequency 4700 MHz
$d_{Lt} = 102$ km (smooth earth); $d_{Lr} = 23$ km (smooth earth) (see Figure 4.3)
$h_{Lt}, h_{Lr} = 0.0$ km (smooth earth, by definition)
$N_{st} = 291$; $N_{sr} = 309$
$N_s = 300$
$a = 8500$ km
$\theta_{et} = 0.0119$ rad; $\theta_{er} = -0.0027$ rad
$\alpha_0 = 0.008$ rad; $\beta_0 = 0.0137$ rad
$\theta = 0.0217$ rad
$\theta d = 0.0217 \times 310.5 = 6.784$ km-rad

From Figure 5.3, $F(\theta d) = 160$ dB.

Determine L_{bsr}:

$$30 \log f = 110.16 \text{ dB} \qquad \text{where } f = 4700 \text{ MHz}$$
$$-20 \log d = -49.84 \text{ dB}$$
$$F(\theta d) = 160 \text{ dB}$$
$$\underline{A_a = 2.2 \text{ dB}} \qquad \text{(from Figure 5.2)}$$

$$L_{bsr} = 222.52 \text{ dB}$$

Calculate d_L and d_{sL}:

$$h_{te} = 0.609 \text{ km} = h_{ts}; \qquad h_{re} = 0.031 \text{ km} = h_{rs}$$

$$f = 4700 \text{ MHz}; \qquad d_{sL} = 65(100/f)^{1/3} = 18 \text{ km}$$

$$\text{(from equation 5.10)}$$

$$d_L = 3(2h_{te} \times 10^3)^{1/2} + 3(2h_{re} \times 10^3)^{1/2} = 128 \text{ km}$$

$$\text{(from equation 5.9)}$$

$$d_e = 130 + 310.5 - 128 - 18 \text{ km} = 294.5 \text{ km} \qquad \text{(from equation 5.11b)}$$

Determine V_n from Figure 5.5:

$$V_n = 3.5 \text{ dB} \qquad n = \text{region 1}$$

Then

$$L_n(0.5) = 222.52 - 3.5 = 219.02 \text{ dB}$$

This is the predicted path loss for a 50% time availability and 50% service probability. All further path loss calculations are based on this value. We now make up a path loss/time distribution table similar to that shown in Section 5.4.5 for the 50% service probability case:

Time Availability q (%)	$Y_1(q)$ (dB)	Transmission Loss (dB)
50	0	219
90	7	226
99	13	232
99.9	17	236
99.99	20	239

The values for $Y_1(q)$ were taken from Figure 5.6.

We now prepare a similar table for the 95% service probability case:

Time Availability q (%)	$Z_{mo}\sigma_{rc}(q)$ (dB)	Transmission Loss (dB)
50	6.7	226
90	7.8	234
99	10.0	242
99.9	11.8	245
99.99	13.3	252

5.5   APERTURE-TO-MEDIUM COUPLING LOSS

Some tropospheric scatter link designers include aperture-to-medium coupling loss as another factor in the path loss equation, and others subtract the values from the antenna gains. In any event the loss must be included somewhere.

Aperture-to-medium coupling loss has sometimes been called the *antenna gain degradation*. It occurs because of the very nature of tropospheric scatter in that the antennas used are not doing the job we would expect them to do. This is evident if we use the same antenna on an LOS (or radiolink) path. The problem stems from the concept of the common volume. High-gain parabolic antennas used on tropospheric scatter paths have very narrow beamwidths (see Section 5.9.3). The tropospheric scatter loss calculations consider a larger common volume than would be formed by these beamwidths. As the beam becomes more narrow due to the higher gain antennas, the received signal level does not increase in the same proportion as it would

Table 5.2   Aperture-to-median coupling loss

Antenna Beam width Ratio, θ/Ω	Coupling Loss (dB)	Antenna Beam width Ratio, θ/Ω	Coupling Loss (dB)
0.3	0.18	1.4	2.95
0.4	0.40	1.5	3.22
0.5	0.60	1.6	3.55
0.6	0.90	1.7	3.80
0.7	1.10	1.8	4.10
0.75	1.20	1.9	4.25
0.8	1.40	2.0	4.63
0.9	1.70	2.1	4.90
1.0	1.95	2.2	5.20
1.1	2.2	2.3	5.48
1.2	2.42	2.4	5.70
1.3	2.75	2.5	6.00

Source: Ref. 11.

under free-space (LOS) propagation conditions. The difference between the free-space expected gain and its measured gain on a tropospheric scatter hop is called the aperture-to-medium coupling loss. This loss is proportional to the scatter angle θ and the beam width Ω. The beamwidth may be calculated from

$$\Omega = \frac{7.3 \times 10^4}{F \times D_r} \tag{5.14}$$

where F = carrier frequency (MHz) and D_r = antenna reflector diameter (ft).

The ratio θ/Ω is computed, and from this ratio the aperture-to-medium coupling loss may be derived from Table 5.2.

Example. Calculate the aperture-to-medium coupling loss for two 30-ft antennas, one at each end of the path, with a 2° scatter angle and a 900-MHz operating frequency. The beamwidth Ω is

$$\Omega = \frac{7.3 \times 10^4}{30 \times 900}$$

$$= 2.7°, \text{ equivalent to 47 mrad}$$

The scatter angle is

$$\theta = 2°, \text{ equivalent to 35 mrad}$$

Thus the ratio $\theta/\Omega = 35/47$, or approximately 0.75. From Table 5.2 this is equivalent to a loss of 1.2 dB.

CCIR Rep. 238-3 suggests another approach to the calculation of aperture-to-medium coupling loss using solely antenna gains G_t and G_r:

$$\text{Aperture-to-medium coupling loss (dB)} = 0.07 \exp[0.055(G_t + G_r)] \tag{5.15}$$

where G_t = gain of the transmitting antenna (dB) and G_r = gain of receiving antenna (dB).

Using the previous example of two 30-ft antennas and 900-MHz operating frequency, $G_t = G_r = 36$ dB and $G_t + G_r = 72$ dB, and

$$\text{Aperture-to-medium coupling loss (dB)} = 0.07 \exp(0.055 \times 72)$$

$$= 3.67 \text{ dB}$$

The difference in value between the two methods should be noted.

5.6 TAKEOFF ANGLE (TOA)

The TOA is probably the most important factor under control of the engineer who selects a tropospheric scatter site in actual path design. The TOA is the angle between a horizontal ray extending from the radiation center of an antenna and a ray extending from the radiation center of the antenna to the radio horizon. Figure 5.7 illustrates the definition.

The TOA is computed by means of path profiling several miles out from the candidate site location. It then can be verified by means of transit siting. Path profiling is described in Section 4.3.3.

Figure 5.8 shows the effect of TOA on transmission loss. As the TOA is increased, about 12 dB of loss is added for each degree increase in TOA.

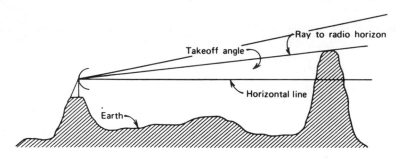

Figure 5.7 Definition of TOA.

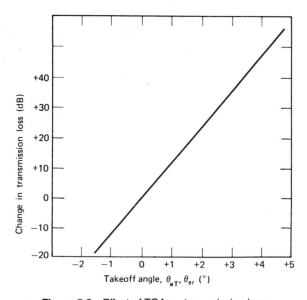

Figure 5.8 Effect of TOA on transmission loss.

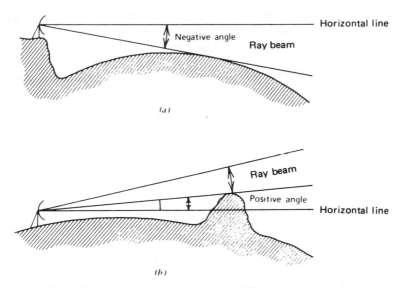

Figure 5.9 (*a*) A more desirable site regarding TOA. (*b*) A less desirable site regarding TOA (note that the beam should actually just clear the obstacle).

This loss shows up in the scatter loss term of the equation for computing the median path loss (equation 5.1). This approximation is valid at 0° in the range of +10 to −10°.

The benefit of siting a tropospheric scatter station on as high a site as possible is obvious. The idea is to minimize obstructions to the horizon in the direction of the "shot." As we shall see later, every decibel saved in median path loss may represent a savings of many thousands of dollars. Therefore the more we can minimize the TOA, the better. Negative TOAs are very desirable. Figure 5.9 illustrates this criterion.

5.7 OTHER SITING CONSIDERATIONS

5.7.1 Antenna Height

Increasing the antenna height decreases the TOA, in addition to the small advantage of getting the antenna up and over surrounding obstacles. Raising an antenna from 20 ft above the ground to 100 ft above the ground provides something on the order of less than a 3-dB improvement in median path loss at 400 MHz and about a 1-dB improvement at 900 MHz (Ref. 4).

5.7.2 Distance to Radio Horizon

The radio horizon may be considered one more obstacle that the tropospheric scatter ray beam must get over. Varying the distance to the horizon varies the TOA. If we maintain a constant TOA, the distance to the horizon can vary widely with insignificant effect on the overall transmission loss.

5.7.3 Other Considerations

If we vary the path length with constant TOA, the median path loss varies about 0.1 dB/mi. The primary effect of increasing the path length is to change the TOA, which will notably affect the total median path loss. This is graphically shown in Figure 5.8.

5.8 PATH CALCULATIONS

This section provides information to assist in determining the basic transmission parameters of a specific tropospheric scatter hop to meet a set of particular transmission objectives, usually related to an overall system plan. Such objectives may be found in the CCIR, namely, CCIR Recs. 593 (Ref. 28), 396-1 (Ref. 30), and 397-3 (Ref. 29). Another objective used by the U.S. Department of Defense is that found in Ref. 6, sec. 4.2.2.1. These performance objectives deal with noise in the voice channel, stating that it should not exceed a particular level during a particular percentage of time. It is recommended that the reader review Section 4.5.2.5. Here we showed how the carrier-to-noise ratio can be related to the signal-to-noise ratio. If the carrier-to-noise ratio can be related to the desired time frame, then, as a consequence, the signal-to-noise ratio may be related to the same time percentage.

Path calculations here are based on the same criterion and starting point as in Chapter 4, namely, the thermal noise in the far-end receiver. Thus we start with -204 dBW as the thermal noise absolute floor value for the perfect receiver at room temperature with 1 Hz of bandwidth. The receiver thermal noise threshold is then calculated by algebraically adding the receiver NF in decibels and $10 \log B_{IF}$, the IF bandwidth in hertz (see Section 1.9.6).

The next step is to calculate the EIRP of a single transmitter chain using the method shown in Section 1.14 and Section 4.5.2.6.2. Then we algebraically add the transmission loss calculated in Section 5.4 for the desired time availability and service probability. The result is the isotropic receive level at the distant receive antenna. To this value we algebraically add the antenna gain and line losses down one receive chain. (There may be as many as four in the case of quadruple diversity.) The result is the receive signal level (RSL). This can be expressed in an equation as follows:

$$RSL_{dBW} = EIRP_{dBW} + \text{transmission loss} + G_r - L_{LR} \qquad (5.16)$$

Let's consider a simple example. A certain tropo transmitter has an output of

10 kW, the line losses at each end are 1.5 dB, the antenna gain at each end is 41 dB, and the transmission loss is 221 dB. Calculate the RSL:

$$\text{EIRP} = +40 \text{ dBW} - 1.5 + 41 \text{ dB}$$
$$= 79.5 \text{ dBW}$$

Then

$$\text{RSL}_{\text{dBW}} = +79.5 \text{ dBW} - 221 \text{ dB} + 41 \text{ dB} - 1.5 \text{ dB}$$
$$= -102 \text{ dBW}$$

We have already calculated the thermal noise floor of the tropo receiver. This gives us a value for N and we will wish to derive C/N. We know that $C = \text{RSL}$, so

$$\left(\frac{C}{N}\right)_{\text{dB}} = \text{RSL}_{\text{dBW}} - N_{\text{dBW}} \qquad (5.17)$$

We will now illustrate this methodology through a worked example. The approach will be slightly different.

Example. Let us now consider the example given in Section 5.4.7 with the following information. From this example we wish to analyze some trade-offs with antenna size (aperture) and transmitter output power.

Operating frequency 4700 MHz.

FDM/FM operation with 60 VF channels. The highest modulating frequency is 300 kHz and $B_{\text{IF}} = 2.5$ MHz. Path length is 310.5 km.

Receiver noise figure is 3 dB.

The desired path time availability is 99%, and service probability 50%, with the resulting transmission loss of 232 dB.

We will disregard antenna-to-medium coupling loss.

First calculate the receiver noise threshold using equation 1.20:

$$P_n = -204 \text{ dBW} + 3 \text{ dB} + 10 \log 2.5 \times 10^6$$
$$= -137 \text{ dBW}$$

We will budget noise at 10 pWp/km in accordance with CCIR Rec. 397-3 (Ref. 29). $10 \times 310.5 = 3105$ pWp at the receive end of the link. We turn to Table 4.6 and convert pWp to (signal-to-noise ratio S/N), and the S/N in a derived voice channel is 53.2 dB (flat). (Also see "Some Notes on Noise in the Voice Channel," Section 4.5.2.5.2.)

From a given S/N we can derive a carrier-to-noise ratio C/N as follows:

$$\frac{S}{N} = \frac{C}{N} + 20\log\left(\sqrt{2}\,\frac{\Delta F_{tt}}{f_8 r c_h}\right) + 10\log\left(\frac{B_{IF}}{B_{ch}}\right) + P + W + D_{im} \quad (5.18)$$

where ΔF_{tt} = rms test tone deviation

f_{ch} = highest baseband frequency (300 kHz in this case)

B_{ch} = voice channel bandwidth (3.1 kHz in this case)

P = emphasis improvement factor (Figure 4.21, 2.8 dB)

W = psophometric weighting improvement factor (2.5 dB)

D_{im} = diversity improvement factor (quadruple diversity, 5.6 dB)

Apply equation 5.18 to calculate a value for C/N or

$$53.2 \text{ dB} = \frac{C}{N} + 20\log\left(\frac{1.41 \times 100 \times 10^3}{300 \times 10^3}\right) + 10\log\left(\frac{2500 \times 10^3}{3.1 \times 10^3}\right)$$
$$+ 2.8 + 2.5 + 5.6$$

$$53.2 = \frac{C}{N} - 6.55 + 29.06 + 2.8 + 2.5 + 5.6$$

$$53.2 = \frac{C}{N} + 33.41$$

$$\frac{C}{N} = 19.79 \text{ dB}$$

We now know that we want a carrier-to-noise ratio C/N at the front end of the tropo receiver of 19.79 dB. We also know that the thermal noise floor of the receiver is -137 dBW. Accordingly, our signal level should be 19.79 dB higher than this noise floor value or -137 dBW + 19.79 dB = -117.21 dBW. We want to maintain this RSL 99% of the time as specified above. We selected the transmission loss of 232 dB to do just this.

To achieve this RSL, allowing a total of 2 dB for line losses (1 dB at each end), we wish to select the appropriate antenna sizes for the transmit and receive antennas and the required transmitter output power. We can reduce the problem to a set of networks in series, as shown in Figure 5.10.

Figure 5.10   Network model of tropospheric scatter path analysis.

Table 5.3   Example trade-off table[a]

Trans Power Out (kW)	dBW	Antenna Aperture (ft)	Gain Y (dB)	$2Y$ (dB)	Power Level (dBW)
50	+47	6	36	72	+119
10	+40	8	39	78	+118
2	+33	12	42	84	+117
1	+30	15	44	88	+118
0.5	+27	17	45	90	+117

[a]Antenna gains calculated at 4700 MHz.

From the networks in series in the figure, a simple formula can be derived:

$$X \text{ dBW} - 1 \text{ dB} + Y \text{ dB} - 232 \text{ dB} + Y \text{ dB} - 1 \text{ dB} = -117.21 \text{ dBW}$$
$$X \text{ dBW} + 2Y \text{ dB} = +116.77 \text{ dBW}$$

The right-hand term we round off to +117 dBW. Figuratively, we must generate this level of power to meet requirements. Assume that the transmit and receive antennas are the same size, thus the $2Y$ value. We now develop a table of trade-offs of how we can reach the +117-dBW level using transmitter output power and receiver and transmitter antenna gain (Table 5.3).

The reader is advised that the antenna gains in the table above are optimistic because aperture-to-medium coupling loss was neglected. For instance, using the CCIR method (equation 5.15), we are penalized 3.67 dB for the 6-ft antennas and 5.10 dB for the 8-ft antennas, and so on. Another point that should be taken into account is that we used antenna efficiency of 55%. Many well-designed tropo antennas can achieve better than a 60% efficiency. Better efficiency can improve link performance by 1 to 2 dB. Of course the table is used as a tool to derive the most cost-effective combination of transmit power and antenna aperture to meet link requirements.

5.9   EQUIPMENT CONFIGURATIONS

5.9.1   General

As indicated in Section 5.8, tropospheric scatter equipment must be configured in such a way as to (1) meet path requirements and (2) be an economically viable installation. All tropospheric scatter installations use some form of diversity (see Section 4.5.2.8). Except for some military transportable tropospheric scatter systems, quadruple diversity is the rule in nearly every case. A typical quadruple diversity tropospheric scatter system layout is

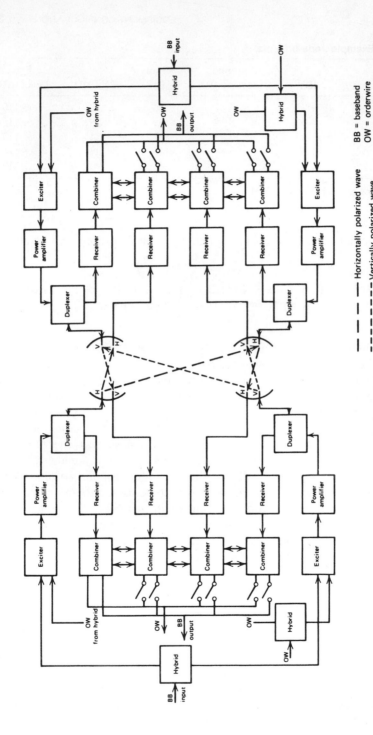

Figure 5.11 Simplified functional block diagram of a quadruple-diversity tropospheric scatter configuration.

--- --- Horizontally polarized wave
------- Vertically polarized wave

BB = baseband
OW = orderwire

shown in Figure 5.11. It is made up of identifiable sections as follows:

- Antennas, duplexer, transmission lines
- Modulator–exciters and power amplifiers
- Receivers, preselectors, and threshold extension devices
- Diversity and diversity combiners

Through proper site selection and system layout involving these four categories, realistic tropospheric scatter systems can be set up on paths up to 250 statute mi (400 km) in length.

5.9.2 Tropospheric Scatter Operational Frequency Bands

Tropospheric scatter installations commonly operate in the following frequency bands:

350–450 MHz
755–985 MHz
1700–2400 MHz
4400–5000 MHz

The reader should also consult CCIR Rec. 388 (Ref. 27) and CCIR Rep. 285-6 (Ref. 18). Also see CCIR Rep. 286 (Geneva, 1982).

5.9.3 Antenna, Transmission Lines, Duplexer, and Other Related Transmission Line Devices

Antennas The antennas used in tropospheric scatter installations are broadband high-gain parabolic reflector devices. The antennas covered here are similar in many respects to those discussed in Section 4.7, but have higher

Table 5.4 Some typical antenna gains

Reflector Diameter (ft)	Frequency (GHz)	Gain (dB)
15	0.4	23
	1.0	31
	2.0	37
	4.0	43
30	0.4	29
	1.0	37
	2.0	43
	4.0	49
60	0.4	35
	1.0	43
	2.0	49
	4.0	55

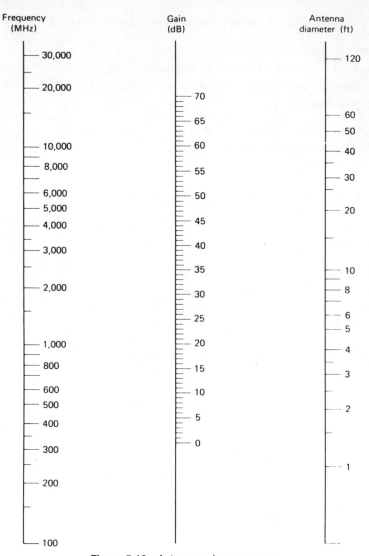

Figure 5.12 Antenna gain nomogram.

gain and consequently are larger and considerably more expensive. As we discussed in that section, the gain of this type of antenna is a function of the reflector diameter. Table 5.4 gives some typical gains for several frequency bands and several standardized reflector diameters. Figure 5.12 is a nomogram from which gain in decibels can be derived given the operating frequency and the diameter in feet of the parabolic reflector. A 55% efficiency is assumed for the antenna. It should be noted that the tendency today is to improve feed methods, especially where "decibels are so expensive," such as in the case of tropospheric scatter and earth station installations. Improved feeds illuminate the reflector more uniformly and reduce spillover, with the consequent improvement of antenna efficiency. For example, for a 30-ft reflector operating at 2 GHz, improving the efficiency from about 55% to 61% will increase the gain of the antenna about 0.5 dB.

It is desirable, but not always practical, to have the two antennas (as shown in Figure 5.11) spaced not less than 100 wavelengths apart to ensure proper space diversity operation. Antenna spillover (i.e., radiated energy in side lobes and back lobes) must be reduced to improve radiation efficiency and to minimize interference with simultaneous receiver operation and with other services.

The first side lobe should be down (attenuated) at least 23 dB and the rest of the unwanted lobes down at least 40 dB from the main lobe. Antenna alignment is extremely important because of the narrow beamwidths. These beamwidths are usually less than 2° and often less than 1° at the half-power points (see Section 4.7).

A good voltage standing wave ratio (VSWR) is also important, not only from the standpoint of improving system efficiency, but also because the resulting reflected power with a poor VSWR may damage components further back in the transmission system. Often load isolators are required to minimize the damaging effects of reflected waves. In high-power tropospheric scatter systems, these devices may even require a cooling system.

A load isolator is a ferrite device with approximately 0.5-dB insertion loss. The forward wave (the energy radiated toward the antenna) is attenuated 0.5 dB; the reflected wave (the energy reflected back from the antenna) is attenuated more than 20 dB.

Another important consideration in planning a tropospheric scatter antenna system is polarization (see Figure 5.11). For a common antenna the transmit wave should be orthogonal to the receive wave. This means that is the transmitted signal is horizontally polarized, the receive signal should be vertically polarized. The polarization is established by the feeding device, usually a feed horn. The primary reason for using opposite polarization is to improve isolation, although the correlation of fading on diversity paths may be reduced. A figure commonly encountered for isolation between polarizations on a common antenna is 30 dB. However, improved figures may be expected in the future.

Transmission Lines   In selecting and laying out transmission lines for tropospheric scatter installations, it should be kept in mind that losses must be kept to a minimum. That additional fraction of a decibel is much more costly in tropospheric scatter than in radiolink installations. The tendency, therefore, is to use waveguide on most tropospheric scatter installations because of its lower losses than coaxial cable. Waveguide is universally used above 1.7 GHz.

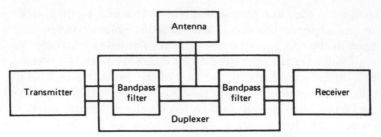

Figure 5.13 Simplified block diagram of a duplexer.

Transmission line runs should be less than 200 ft (60 m). Ideally, the attenuation of the line should be kept under 1 dB from the transmitter to the antenna feed and from the antenna feed to the receiver, respectively. To minimize reflective losses, the VSWR of the line should be 1.05:1 or better when terminated in its characteristic impedance. Figures 4.32 and 4.33 show several types of transmission lines commercially available.

The Duplexer The duplexer is a transmission line device that permits the use of a single antenna for simultaneous transmission and reception. For tropospheric scatter application a duplexer is a three-port device (see Figure 5.13) so tuned that the receiver leg appears to have an admittance approaching (ideally) zero at the transmitting frequency. At the same time the transmitter leg has an admittance approaching zero at the receiving frequency. To establish this, sufficient separation in frequency is required between the transmitted and received frequencies. Figure 5.13 is a simplified block diagram of a duplexer. The insertion loss of a duplexer in each direction should be less than 0.5 dB. Isolation between the transmitter port and the receiver port should be better than 30 dB. High-power duplexers are usually factory tuned. It should be noted that some textbooks call the duplexer a diplexer.

5.9.4 Modulator – Exciter and Power Amplifier

One type of modulation used on tropospheric scatter transmission systems is FM.* As our discussion develops, keep in mind that tropospheric scatter systems are high-gain low-noise extensions of the radiolink systems discussed in Chapter 4.

The tropospheric scatter transmitter is made up of a modulator–exciter and a power amplifier (see Figure 5.11). The power outputs are fairly well standardized at 1, 2, 10, 20, and 50 kW. For most commercial applications the 50-kW installation is not feasible from an economic point of view.

---

*New military troposcatter systems use digital modulation such as the Raytheon AN/TRC-170(V). See Section 5.13.

Installations that are 2 kW or below are usually air-cooled. Those above 2 kW are liquid-cooled, usually with a glycol–water solution using a heat exchanger. If klystron power amplifiers are used, such tubes are about 33% efficient. Thus a 10-kW klystron will require at least 20 kW of heat exchange capacity.

The transmitter frequency stability (long-term) should be $\pm 0.001\%$. Spurious emission should be down better than 80 dB below the carrier output level. Preemphasis is used as described in Section 4.5.2.2 and depends on the highest modulating frequency of the applied baseband.

The baseband configuration of the modulating signal, depending on the number of channels to be transmitted, is selected in the spectrum of 60–552 kHz (CCITT supergroups 1 and 2—see section 3.2.3.6). However, CCITT subgroup A, 12–60 kHz, is often used as well. For longer route tropospheric scatter systems, the link design engineer may tend to limit the number of voice channels, selecting a baseband configuration that lowers the highest modulating frequency to be transmitted as much as possible. This tends to increase equivalent overall system gain by reducing the bandwidth of the IF (B_{IF} equation 5.17), which is equivalent to reducing the RF bandwidth.

The modulator injects an RF pilot tone which is used for alarms at both ends as well as to control far-end combiners. 60 kHz is common in U.S. military systems. CCIR recommends 116 or 119 kHz for 24-channel systems, 304 or 331 kHz for 60-channel systems, and 607 (or 304) kHz for 120-channel systems (CCIR Rec. 401-2).

The modulator also has a service channel input. This is covered in CCIR Rec. 400-2. It recommends the use of the band 300–3400 Hz. U.S. military systems often multiplex more than one service orderwire in the band 300–12,000 Hz. Often one of these channels may be used to transmit fault and alarm information.

The power amplifier should come equipped with a low-pass filter to attenuate second harmonic output by at least 40 dB and third harmonic output by at least 50 dB.

5.9.5 The FM Receiver Group

The receiver group in tropospheric scatter installations usually consists of two or four identical receivers in dual or quadruple diversity configurations, respectively. Receiver baseband outputs are combined in maximal ratio square combiners. See Section 4.5.2.8 for a discussion of combiners and the function of the ratio-square-type combiner. A simplified functional block diagram of a typical quadruple-diversity receiving system is shown in Figure 5.14.

The receiver noise threshold may be computed as follows:

$$\text{Noise threshold (dBW)} = 10 \log kTB_{IF} + \text{NF} \qquad (5.19)$$

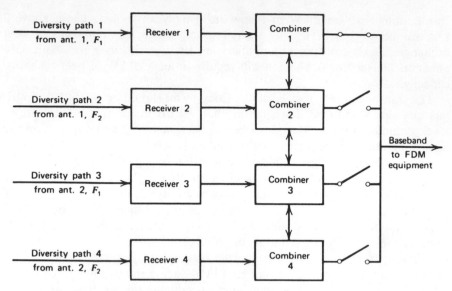

Figure 5.14   Simplified functional block diagram of a quadruple-diversity tropospheric scatter receiving system. F_1 = frequency 1; F_2 = frequency 2.

where k = Boltzmann's constant, 1.38×10^{-23} J/K
T = 290 K
B_{IF} = IF bandwidth (Hz)
NF = receiver noise figure (dB)

Typical receiver front-end NFs are given in Table 5.5. Table 5.6, which provides maximum IF bandwidths B_{IF} for several voice channel configurations, will be helpful in calculating receiver noise thresholds when receiver front-end NFs are given. For our discussion, the receiver front-end NF will be the NF for the entire receiver.

Having B_{IF} and NF, we can now calculate the noise threshold for each of the three voice channel configurations given using Tables 5.5 and 5.6. Table

Table 5.5   Typical receiver front-end NFs

Frequency Band (MHz)	Type	NF (dB)
350–450	Bipolar	1.0
775–985	Bipolar	1.0
1700–2400	GaAs FET	1.1
4400–5000	GaAs FET	1.3

Table 5.6   Maximum IF bandwidths for several voice
voice configurations

Number of Voice Channels in FDM Baseband	Maximum IF Bandwidth B_{IF}(Hz)
60	3×10^6
120	6×10^6
180	10×10^6

Source: Ref. 20.

5.7 tabulates the three bandwidths for each value of NF given in Table 5.5, giving the equivalent noise threshold. The derived formula (i.e., -204 dBW + NF + $10 \log B_{IF}$) assumes uncooled (room temperature) receivers.

From Table 5.7 it is obvious that to achieve FM improvement threshold, the receiver must have an input carrier-to-noise ratio equivalent to adding $+10$ dB algebraically to the noise threshold. For instance, if a receiver has a 3-dB noise figure, a B_{IF} of 3 MHz, its noise threshold is -136.2 dBW, and its FM improvement is -126.2 dBW.

Table 5.7   Noise and FM thresholds for several receiver noise figures
at three IF bandwidths

NF (dB)	Bandwidth (MHz)	Noise Threshold (dBW)	FM Improvement Threshold (dBW)
1.0	3	-138.2	-128.2
	6	-135.2	-125.2
	10	-133.0	-123.0
1.5	3	-137.7	-127.7
	6	-134.7	-124.7
	10	-132.5	-122.5
2.0	3	-137.2	-127.2
	6	-134.2	-124.2
	10	-132.0	-122.0
2.5	3	-136.7	-126.7
	6	-133.7	-123.7
	10	-131.5	-121.5
3.0	3	-136.2	-126.2
	6	-133.3	-123.2
	10	-131.0	-121.0
3.5	3	-135.7	-125.7
	6	-132.7	-122.7
	10	-130.5	-120.5
4.0	3	-135.2	-125.2
	6	-132.2	-122.2
	10	-130.0	-120.0

Another method of improving the equivalent gain on a path is to use threshold extension techniques. The FM improvement threshold of a receiver can be "extended" by using a more complex and costly demodulator, called a *threshold extension demodulator*. The amount of improvement that can be expected using threshold extension over conventional receivers is on the order of 7 dB.* Therefore for the above example, where the FM improvement threshold was -126.2 dBW without extension, with extension it would be -133.2 dBW.

Threshold extension works on a FM feedback principle which reduces the equivalent instantaneous deviation, thereby reducing the required bandwidth B_{IF}, which in turn effectively lowers the receiver noise threshold. A typical receiver with a threshold extension module employs a tracking filter which instantaneously tracks the deviation with a steerable bandpass filter having a 3-dB bandwidth of approximately four times the top baseband frequency. The control voltage for the filter is derived by making a phase comparison between the feedback signal and the IF input signal.

5.9.6 Diversity and Diversity Combiners

Some form of diversity is mandatory in tropospheric scatter. Most present-day operational systems employ quadruple diversity. There are several ways of obtaining some form of quadruple diversity. One of the most desirable is shown in Figures 5.11 and 5.14, where both frequency and space diversity are utilized. For the frequency diversity section, the system designer must consider the aspects of frequency separation illustrated in Figure 4.23. Space diversity is almost universally used, but the physical separation of antennas is normally in the horizontal plane with a separation distance greater than 100, and preferably 150 wavelengths.

Frequency diversity, although very desirable, often may not be permitted owing to RFI considerations. Another form of quadruple diversity, perhaps better defined as quasiquadruple diversity, involves polarization, or what some engineers call *polarization diversity*. This is actually another form of space diversity and has been found not to provide a complete additional order of diversity. However, if often will make do when the additional frequencies are not available to implement frequency diversity.

Polarization diversity is usually used in conjunction with conventional space diversity. The four space paths are achieved by transmitting signals in the horizontal plane from one antenna and in the vertical plane from a second antenna. On the receiving end two antennas are used, each antenna having dual polarized feed horns for receiving signals in both planes of polarization. The net effect is to produce four signal paths that are relatively independent.

---

*Assuming a modulation index of 3.

A discussion of diversity combiners is given in Section 4.5.2.8. There the feasibility of the maximal-ratio-square combiner is demonstrated, and consequently, it is the most commonly used combiner on tropospheric scatter communication systems.

5.10   ISOLATION

An important factor in tropospheric scatter installation design is the isolation between the emitted transmit signal and the receiver input. Normally we refer to the receiver sharing a common antenna feed with the transmitter.

A nominal receiver input level for military tropospheric scatter systems is −80 dBm (Ref. 6) for design purposes. If a transmitter has an output power of 10 kW or +70 dBm and transmission line losses are negligible, then isolation must be greater than 150 dB.

To achieve overall isolation such that the transmitted signal interferes in no way with receiver operation when the equipment is operating simultaneously, the following items aid the required isolation when there is sufficient frequency separation between transmitter and receiver:

- Polarization
- Duplexer
- Receiver preselector
- Transmit filters
- Normal isolation from receiver conversion to IF

5.11   INTERMODULATION (IM)

NPR measurements (see Section 4.5.2.5.2) are a good indication of operational IM distortion capabilities of tropospheric scatter equipment. When the NPR is measured on a back-to-back basis with 120-channel loading, we could expect a value of 55 dB. Once the same equipment is placed in operation on an active path, the NPR may drop as low as 47 dB from the near-end transmitter to the far-end receiver. The deterioration of the NPR is due to IM noise that can be traced to the intervening medium. It is just this IM distortion brought about in the medium that limits useful transmitted bandwidths in tropospheric scatter systems.

The bandwidth that a tropospheric scatter system can transmit without excessive distortion is related to the multipath delays experienced. These delays depend on the size of the scatter volume. The common volume is determined by antenna size and scattering characteristics.

5.12   MAXIMUM FEASIBLE TROPO TRANSMISSION LOSS

A maximum feasible tropo transmission loss can be calculated by establishing a model and inserting maximum feasible terminal parameters. We assume a quadruple-diversity terminal at each end of the link. The operating frequency is 2 GHz. The following are the operational parameters of the set of terminals in the model:

Antennas: 120-foot aperture. At 2 GHz, each antenna has a gain of 55 dB at 55% aperture efficiency.

Power amplifiers: 50 kW each (+47 dBW).

12 FDM/SSB voice channels with the highest modulating frequency of 60 kHz.

Receiver noise figure of 2.5 dB.

Aperture-to-medium coupling loss using CCIR method, 29 dB.

Transmission line losses at each end, 2 dB.

We now calculate the peak deviation using equation 4.13b. From Table 4.4, we find the rms deviation per VF channel is 35 kHz. Then the peak deviation D is

$$D = 4.47 \times 35\left[\log^{-1}\left(\frac{-1 + 4\log 12}{20}\right)\right]$$

$$D = 229 \text{ kHz}$$

Using equation 4.11, Carson's rule, we calculate the bandwidth of the tropo receiver IF:

$$B_{IF} = 2(229 + 60)$$
$$= 578 \text{ kHz}$$

Determine the FM improvement threshold using equation 4.10:

$$P_{FM} = -204 \text{ dBW} + 10\log 578 \times 10^3 + 2.5 \text{ dB} + 10 \text{ dB}$$
$$= -133.88 \text{ dBW}$$

Calculate the EIRP from a transmit antenna using equation 1.33:

$$\text{EIRP}_{dBW} = +47 \text{ dBW} + 55 \text{ dB} - 2 \text{ dB}$$
$$= +100 \text{ dBW}$$

We subtract from the receive antenna gain the aperture-to-medium coupling loss (29 dB) and the line losses (2 dB), which provides a net gain of 24 dB at the input to the receiver.

Now consider the following model of the tropo link, where the RSL is exactly equal to the FM improvement threshold. The transmission loss X is the unknown. Working from left to right on the model above, we set up the following equation and solve for X, the transmission loss:

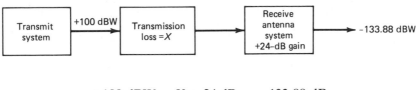

$$+100 \text{ dBW} - X + 24 \text{ dB} = -133.88 \text{ dB}$$
$$X = 257.88 \text{ dB}$$

The reader should note that the transmission loss available is very degraded due to the high value of aperture-to-medium coupling loss. We used the CCIR method (Section 5.5) to calculate aperture-to-medium coupling loss, resulting in very conservative values. If we used the first method suggested in Section 5.5, we would arrive at more liberal values. For instance, if we used the worst-case value in Table 5.2, 6 dB, the calculated transmission loss would be some 23 dB greater, or about 281 dB.

50-kW transmitters and 120-ft parabolic antennas are not feasible for commercial operation because of the immense expense of such installations. Such large installations are justifiable for military links. With the advent of satellite communication, many of these large military installations have been taken out of service.

Suppose we reduced the size of the tropo terminals in the above example to 60-ft antennas (gain 49 dB) and 10 kW transmitters (+40 dBW). Following the CCIR method, the aperture-to-medium coupling loss reduces to about 15 dB. The transmission loss is now about 253 dB.

5.13 DIGITAL TROPOSPHERIC SCATTER AND DIFFRACTION SYSTEMS

5.13.1 Introduction

Digital tropospheric scatter and diffraction links are being implemented on all new military construction and are attractive for commercial telephony as we approach the era of an all-digital network. The advantages are similar to those described in the previous chapter. However, on tropo and diffraction paths the problem of dispersion due to multipath can become acute. But the problem can be handled; in fact, it can be turned to advantage with the proper equipment.

The design of a digital tropo or diffraction link in most respects is similar to the design of its analog counterpart. Siting, path profiles, and calculation of transmission loss, including aperture-to-medium coupling loss, use identical procedures as those previously described. However, the approach to link analysis and the selection of modulation scheme differ. These issues will be discussed below.

5.13.2 Digital Link Analysis

Digital tropo/diffraction link analysis can be carried out by a method that is very similar to that described in Section 5.8:

- Calculate the EIRP
- Algebraically add the transmission loss
- Add the receiving antenna gain G_r
- Algebraically add the line losses incurred up to the input of the low-noise amplifier (LNA)

The aperture-to-medium coupling loss must be accounted for. As was previously suggested, we can add this value to the transmission loss. Another method is to subtract this loss from the receiving antenna gain G_r, but aperture-to-medium coupling loss must be accounted for.

One point of guidance: the analysis will follow a single string on a link. We mean here that the EIRP is calculated for a single transmitter and its associated antenna and then down through a single diversity branch on the receiving side as though it were operating alone.

The objective of this exercise is to calculate E_b/N_0. From equation 4.48a,

$$\frac{E_b}{N_0} = \text{RSL}_{\text{dBW}} - 10\log(\text{BR}) - (-204\,\text{dBW} + \text{NF}_{\text{dB}}) \qquad (5.20)$$

(i.e., the first two terms in equation 5.20 represent E_b and the last two terms represent N_0, where the subtraction sign implies logarithmic division). In equation 5.20 BR is the bit rate in bps and NF_{dB} is the noise figure of the receive chain or the noise figure of the LNA, the latter being sufficient in most cases.

Determine RSL (similar to equation 4.27):

$$\text{RSL}_{\text{dBW}} = \text{EIRP}_{\text{dBW}} + T_L + G_r + L_{\text{Lr}} \qquad (5.21)$$

where T_L is the transmission loss including aperture-to-medium coupling loss, G_r is the gain of the receiving antenna, and L_{Lr} is the total line losses from the antenna feed to the input of the LNA.

In tropo/diffraction receiving systems, in addition to waveguide loss, there are duplexer losses, preselector filter loss, transition losses, and possibly others.

Example. Assume a digital link with a bit rate of 1.544 Mbps operating at 5000 MHz requiring a 99% time availability and a 95% service probability. We find that the transmission loss calculated for this link for the required time availability and service probability is 243.0 dB. The aperture-to-medium coupling loss is 19.12 dB, which we add to the transmission loss, deriving a final transmission loss value of 262 dB. 30-ft antennas are used at each end of the link where the gain of each antenna is 51 dB. The receiver noise figure is 3 dB, line losses at each end are 3 dB, and the modulation is QPSK. Transmitter output is 10 kW. Determine E_b/N_0 and then bit error rate

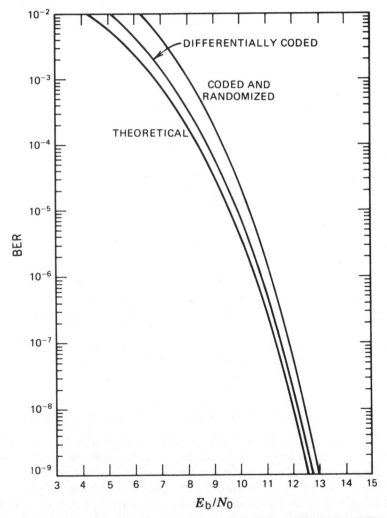

Figure 5.15 Bit error probability versus E_b/N_0 for coherent BPSK and QPSK. (From Ref. 22.)

(BER) using Figure 5.15:

First calculate EIRP using equation 1.33 and then RSL using equation 5.21:

$$EIRP_{dBW} = 10 \log 10,000 - 3 \text{ dB} + 51 \text{ dB}$$
$$= +88 \text{ dB}$$
$$RSL_{dBW} = +88 \text{ dBW} - 262 \text{ dB} + 51 \text{ dB} - 3 \text{ dB}$$
$$= -126 \text{ dBW}$$

Then calculate E_b/N_0 using equation 5.20:

$$\frac{E_b}{N_0} = 126 \text{ dBW} + 10 \log 1.544 \times 10^6 + 204 \text{ dBW} - 3 \text{ dB}$$
$$= 13.11 \text{ dB}$$

If we allow a 3-dB modulation implementation loss, we are left with a net E_b/N_0 of 10 dB. Turning to Figure 5.15 we find that the equivalent BER on the theoretical curve to by 4×10^{-6}. We would expect to have this value BER 99% of the time.

5.13.3 Dispersion

Dispersion is the principal cause of degradation of BER on digital transhorizon links. With conventional waveforms such as BPSK, MPSK, BFSK, and MFSK, dispersion may be such, on some links, that BER performance is unacceptable.

Dispersion is simply the result of some signal power from an emitted pulse that is delayed, with that power arriving later at the receiver than other power components. The received pulse appears widened or smeared or what we call dispersed. These late-arrival components spill over into the time slots of subsequent pulses. The result is intersymbol interference (ISI), which deteriorates BER.

Expected values of dispersion on transhorizon paths vary from 30 to 380 ns. The cause is multipath. The delay is a function of path length, antenna beamwidth, and the scatter angle components α_0 and β_0.

5.13.4 Some Methods of Overcoming the Effects of Dispersion

One simple method to avoid overlapping pulse energy is to time-gate the transmitted energy, which allows a resting time after each pulse. Suppose we were transmitting a megabit per second and we let the resting time be half a pulsewidth. Then we would be transmitting pulses of 500 nsec of pulse width, and there would be a 500-ns resting time after each pulse, time enough to allow the residual delayed energy to subside. The cost in this case is a 3-dB loss of emitted power.

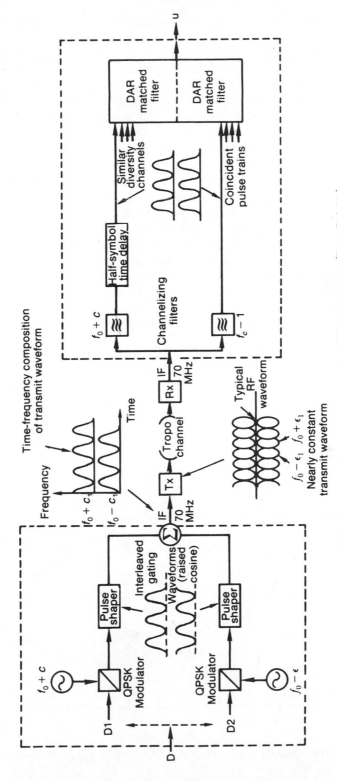

Figure 5.16 Operation of the two-frequency AN/TRC-170 modem. (From Ref. 25.)

A two-frequency approach taken to reach the same objective in the design of the Raytheon AN/TRC-170 DAR modem, which is the heart of this digital troposcatter radio terminal, is to transmit on two separate frequencies, alternatively gating each. The two-frequency pulse waveform is simply the time interleaving of two half-duty cycle pulse waveforms, each on a separate frequency. This technique offers two significant advantages over the one-frequency waveform. First, the two signals (subcarriers) are interleaved in time and are added to produce a composite transmitted signal with nearly constant amplitude, thereby nearly recovering the 3 dB of power lost due to time-gating. The operation of this technique is shown in Figure 5.16.

The second advantage is what is called intrinsic or implicit diversity. This can be seen as achieved in two ways. First, the residual energy of the "smear" can be utilized, whereas in conventional systems it is destructive (i.e., it causes intersymbol interference). Second, on lower bit rate transmission, where the bit rate is R, R is placed on each subcarrier, rather than $R/2$ for the higher bit rates. The redundancy at the lower bit rates gives an order of in-band diversity. The modulation on the AN/TRC-170 is QPSK on each subcarrier. The maximum data rate is 4.608 Mbps, which includes a digital orderwire and service channel.

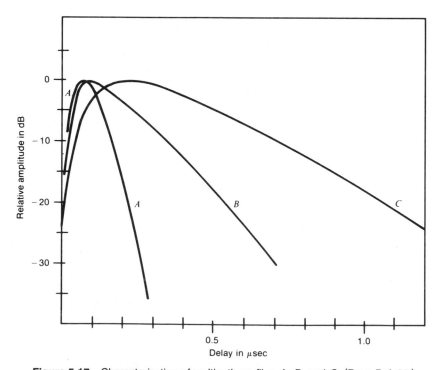

Figure 5.17   Characterization of multipath profiles A, B, and C. (From Ref. 23.)

The AN/TRC-170 operates in the 4.4–5.0-GHz band with a transmitter output power of 2 kW. The receiver noise figure is 3.1 dB. In its quadruple-diversity configuration with 9.5-ft antennas and when operating at a trunk bit rate of 1.024 Mbps, the terminal can support a path loss typically of 240 dB (BER $= 1 \times 10^{-5}$). This value is based on an implicit diversity advantage for a multipath delay spread typical of profile *B* (see Figure 5.17). On a less dispersive path based on profile *A*, the performance would be degraded by about 2.6 dB or a transmission loss of 237.4 dB. With a more dispersive path based on profile *C*, we would expect a 1.1-dB improvement over that of profile *B* (Ref. 23).

The more dispersive the path, the better the equipment operates up to about 1 μs of rms dispersion. This maximum value would be shifted upward or downward depending on the data rate.

Three multipath profiles are shown in Figure 5.17; rms values for the multipath delay spread of each profile are

Profile *A*: 65 ns

Profile *B*: 190 ns

Profile *C*: 380 ns

Table 5.8   Some typical troposcatter system parameters

Path Distance (km)	Frequency Band (MHz)	Transmitter Power (kW)	Antenna Diameter (m)[a]	Diversity	Channel Capacity	Comments
150–250	1000	1	9	4	72	
	1000	2	9	2	84	
	1000	2	14	4	108	
	1000	10	18	4	132	Parametric
	4000	5	9	4	24	amplifier
250–320	1000	10	18	2	36	
	1000	10	18	4	72–240	
	2000	10	9	4	72	
	2000	10	18	4	36	
320–420	1000	0.5	18	2	6	
	1000	10	18	2	36	
	1000	10	18	4	24–48	
	2000	10	18	4	36	
320–500	1000	10	18	4	24	Parametric amplifier
500–600	1000	10	37	4	24	Parametric amplifier
700–900	1000	50	37	4	24	
	1000	57	37	4	24[b]	

[a]Nominal diameter of parabolic reflector.
[b]Can support only 12 operational channels with a BER on data systems in excess of 1×10^{-3}, indicative of noise hits due to fades.

Table 5.9   Tropospheric scatter path loss — measured values versus calculated values using three calculation methods

World Area	Path Length (km)	Frequency (MHz)	Predicted Tropospheric Scatter Path Loss (dB)			Measured Loss
			NBS 101	CCIR	Yeh	
Northeastern United States	275	460	191.8	192.4	200.9	196.0
United Kingdom	275	3480	218.7	220.3	221.6	220.9
Japan	300	1310	204.1	206.0	207.2	211.0
Southeast Asia	595	1900	261.0	254.8	243.9	260.9
South Caribbean	314	900	184.7	186.0	186.8	189.0
Canada, New York State	465	468	218.2	218.7	223.9	220.6
South Asia	738	1840	255.0	254.5	236.8	245.7

Source: Ref. 16.

5.14   TYPICAL TROPOSPHERIC SCATTER PARAMETERS

Table 5.8 presents some of the more important path parameters for several operational tropospheric scatter paths. Note that TOAs are not given, nor is the tropospheric scatter path loss. Now consult Table 5.9, which supports the preceding section. This table compares measured tropospheric scatter path loss with calculated values using three of the accepted, tropospheric scatter path loss calculation methods.

5.15   FREQUENCY ASSIGNMENT

The problem of frequency assignment in tropospheric scatter systems is similar to that of microwave LOS systems (see Section 4.12). The problem with tropospheric scatter becomes more complex because

- Radiated power is much greater (on the order of 30–60 dB greater)
- Nearly all installations are quadruple diversity
- Receivers are more sensitive, with front-end NFs of about 3 dB versus about 10 dB for radiolink receivers

Furthermore, splatter must be controlled so as not to affect other nearby services. The splatter may be a result of side lobe radiation or from radiation on unwanted frequencies.

For CCIR references, the reader may wish to consult CCIR Rec. 388 (Ref. 27). CCIR Reps. 285.6 (Ref. 18) and 286 offer some guidance on frequency arrangement.

To reduce splatter from harmonics, Ref. 19* recommends a transmitter low-pass filter attenuating second-harmonic output at least 40 dB and third harmonics at least 50 dB. Ref. 19 also specifies that:

- The minimum separation between transmit and receive carrier frequency of the same polarization on the same antenna shall be 120 MHz
- The minimum separation between transmit and receive carrier frequency at a single station shall be 50 MHz, but in any case an integral multiple of 0.8 MHz
- To avoid interference within a single station, separation of the transmit and receive frequencies shall not be near the first IF of the receiver
- The minimum separation of transmit (receive) carrier frequencies is 5.6 MHz on systems with 36 FDM voice channels or less, 11.2 MHz for 60 voice channels, and 16.8 MHz for 120 voice channels (B_{IF} assumed as in Table 5.6).

This document further states that frequency channels shall be assigned on a hop-by-hop basis such that the median value of an unwanted signal in the receiver shall be at least 10 dB below the inherent noise of the receiver (i.e., noise threshold) when using the same or adjacent frequency channels in two relay sections.

REVIEW EXERCISES

1. List at least five applications of troposcatter/diffraction communication links.

2. Compare a typical troposcatter terminal to a typical microwave LOS terminal. There should be at least five key items compared.

3. Describe the two types of fading usually encountered on a tropo link. Give the type of time distribution associated with each.

4. Differentiate between time availability and service probability when describing the calculated performance of a tropo/diffraction link.

5. Define θd as used in the tropo transmission loss equation.

6. With microwave LOS links, fading is a function of path length. How does long-term fading vary with distance in the case of a troposcatter link?

7. What is the primary cause of aperture-to-medium coupling loss?

*This reference is given for guidance only. It is superseded by the MIL-STD-188-100 series, in which these recommendations have been omitted.

8. We are forced to increase the takeoff angle by 1.5° of the antenna on one side of a tropo link. give a quantative value of how this will affect the transmission loss.

9. Calculate the aperture-to-medium coupling loss using the CCIR method when the antennas at each end of the link display a gain of 41 dB.

10. Determine the C/N (long term) on a 2 GHz tropo link with the following parameters: transmission loss, 212.5 dB; transmitter output, 5 kW; antenna gain, 43 dB each end; line losses at each end, 1.5 dB; receiver noise figure, 1.5 dB; and IF bandwidth of the receiver, 2 MHz.

11. What is the function of a duplexer in a tropo terminal?

12. Discuss trade-offs of using the tropo frequency band of 900 MHz versus 4 GHz, considering range and varying antenna size but keeping all other parameters constant.

13. Discuss the importance of isolation on a quadruple-diversity terminal. How do we achieve the isolation? Use some dB values.

14. What is the value in dB of E_b/N_0 when the RSL is -121 dBW, the bit rate is 2.048 Mbps, and the receiving system noise figure is 2.2 dB?

15. Explain the cause of dispersion on a digital tropo path. It can bring about a serious impairment unless we design to mitigate its effects. What is the principal impairment that results from dispersion?

16. What range of dispersion in ns might be encountered on tropo links?

REFERENCES AND BIBLIOGRAPHY

1. *Reference Data for Radio Engineers*, 6th ed., Howard W. Sams, Indianapolis, IN, 1977.
2. R. L. Freeman, "Multichannel Transmission by Tropospheric Scatter," *Telecommun. J.*, ITU Geneva, June 1969.
3. P. L. Rice et al., *Transmission Loss Predictions for Tropospheric Scatter Communication Circuits*, Tech. Note 101, as revised, National Bureau of Standards, Boulder, CO, Jan. 1967.
4. K. O. Hornberg, *Sitting Criteria for Tropospheric Scatter Propagation Communication Circuits*, Memo. Rep. PM-85-15, National Bureau of Standards, Boulder, CO, Apr. 1959.
5. *General Engineering—Beyond-Horizon Radio Communications*, USAF Tech. Order TO 31Z-10-13, U.S. Department of Defense, Washington, DC, Oct. 1971.
6. MIL-STD-188-313, U.S. Department of Defense, Washington, DC, Dec. 1973.
7. A. P. Barghausen et al., *Equipment Characteristics and Their Relationship to Performance for Tropospheric Scatter Communication Circuits*, Tech. Note 103, National Bureau of Standards, Boulder, CO, Jan. 1963.

8. R. L. Marks et al., *Some Aspects of the Design for FM Line-of-Sight Microwave and Troposcatter Systems*, USAF Rome Air Development Center, NY, U.S. National Technical Information Service, AD 617-686, Springfield, VA, Apr. 1965.

9. E. F. Florman and J. J. Tory, *Required Signal-to-Noise Ratios, RF Signal Power and Bandwidth for Multichannel Radio Communication Systems*, Tech. Note 100, National Bureau of Standards, Boulder, CO, Jan. 1962.

10. A. P. Barsis et al., *Predicting the Performance of Long Distance Tropospheric Communication Circuits*, Rep. 6032, National Bureau of Standards, Boulder, CO, Dec. 1958.

11. *Forward Propagation Tropospheric Scatter Communications Systems, Handbook for Planning and Siting*, USAF Tech. Order TO 31R5-1-11, U.S. Department of Defense, Washington, DC, as revised Nov. 30, 1959.

12. E. D. Sunde, "Digital Troposcatter Transmission and Modulation Theory," *Bell Syst. Tech. J.*, vol. 43, Jan. 1964.

13. E. D. Sunde, "Intermodulation Distortion in Analog FM Tropospheric Scatter Systems," *Bell. Syst. Tech. J.*, Jan. 1964.

14. P. F. Panter, *Communication Systems Design—Line-of-Sight and Troposcatter Systems*, McGraw-Hill, New York, 1972.

15. *Naval Shore Electronics Criteria—Line-of-Sight and Tropospheric Scatter Communications Systems*, Navelex 0101.0112, U.S. Department of the Navy, Washington, DC, May 1972.

16. R. Larsen, "A Comparison of Some Troposcatter Prediction Methods," Conference Paper, IEE Conference on Tropospheric Radio Wave Propagation, Sept.–Oct. 1968.

17. "Propagation Data Required for Trans-Horizon Radio-Relay Systems," CCIR Rep. 238-3, XIIIth Plenary Assembly Kyoto, 1978.

18. "Propagation Effects on the Design and Operation of Trans-Horizon Radio-Relay Systems," CCIR Rep. 285-6, XVth Plenary Assembly, Dubrovnik, 1986.

19. *DCS Engineering Installation Manual* DCAC 330-175-1, through Change 9, U.S. Department of Defense, Washington, DC.

20. *Facility Design for Tropospheric Scatter*, MIL-HDBK-417, U.S. Department of Defense, Washington, DC, Nov. 1977.

21. L. P. Yeh, "Tropospheric Scatter Communications Systems," presented at ITU World Planning Committee Meeting, Mexico City, 1967.

22. Roger L. Freeman, *Reference Manual for Telecommunications Engineering*, Wiley, New York, 1985.

23. Roger L. Freeman, *Radio System Design for Telecommunications* (1–100 *GHz*), Wiley, New York, 1987.

24. T. E. Brand, W. J. Connor, and R. J. Sherwood, "AN/TRC-170—Troposcatter Communication System," NATO Conference on Digital Troposcatter, Brussels, Mar. 1980.

25. W. J. Connor, "AN/TRC-170(V)–A Digital Troposcatter Communication System," IEEE ICC'78 Conference Record.

26. *Propagation Influences on Digital Transmission Systems: Problems and Solution*," AGARD-CP-363, NATO, Athens, June 1984.

27. "Radio Frequency Channel Arrangements for Trans-Horizon Radio Relay Systems," CCIR Rec. 388, XVth Plenary Assembly, Dubrovnik, 1986.

28. "Noise in Real Circuits of Multi-channel Trans-Horizon FM Radio-Relay Systems of Less than 2500 km," CCIR Rec. 593, XVth Plenary Assembly, Dubrovnik, 1986.

29. "Allowable Noise Power in the Hypothetical Reference Circuit of Trans-Horizon Radio-Relay Systems for Telephony using Frequency Division Multiplex," CCIR Rec. 397-3, XVth Plenary Assembly, Dubrovnik, 1986.

30. "Hypothetical Reference Circuit for Trans-Horizon Radio-Relay Systems using Frequency Division Multiplex," CCIR Rec. 396-1, XVth Plenary Assembly, Dubrovnik, 1986.

6

SATELLITE COMMUNICATIONS

6.1 BACKGROUND AND INTRODUCTION

Satellite communication created a quantum leap forward in long-distance communication. It competed with undersea cable for transoceanic voice channel connectivity. It brought reliable, high-quality communication to countries and rural areas that previously depended on high-frequency (HF) radio and/or ground return single-wire telegraph. For instance, in Argentina there was a 500% multiplier on international traffic when the first INTEL-SAT facility was installed. Canada's TeleSat ANIK satellites brought automatic telephony and TV to its far arctic regions.

Ships at sea are provided almost instantaneous connectivity to the international public switched telephone network by means of INMARSAT (International Maritime Satellite [Organization]). INMARSAT earth stations are found in nearly all major maritime nations. The United States and the Soviet Union have a joint venture for search and rescue (SAR) alert and location by means of satellites. The U.S. Geographical Positioning System (GPS) consisting of constellations on several polar orbits provides universal coordinated time (UTC time) with several microseconds' accuracy and position within ± 10 m in three dimensions anywhere on earth to platforms equipped with the appropriate receiver.

India, Brazil, Indonesia, the Arab States, and Europe have regional or domestic satellite systems. The United States, however, is probably the leader in this area. In the United States and Canada, TV programming relay is one area of business activity that truly has mushroomed. TV relay by satellite is being widely used by

- Broadcasters
- Cable TV
- Industrial/education users
- Direct-to-home TV subscribers

The U.S. and NATO armed forces rely heavily on satellite communication for strategic, tactical, and support communications. Typical systems are the U.S. Navy's FLTSAT (fleet satellite [series]), DSCS (Defense Satellite Communication System), MILSTAR, and NATO satellites.

Telephone circuit trunking on national and international routes is another major application. Private industrial networks use satellites to provide long-haul connectivity. Bypass has become the byword. Large- and medium-sized corporations have found it cost-effective to bypass local exchange carriers (LECs) and inter-LATA carriers by using satellite transponder space allowing direct access to their own local PBX facilities.

VSAT (very small aperture terminal) satellite systems are a vastly expanding subset of bypass. Hotels, fast-food chains, chain stores, and other commercial entities that are spread far and wide across a geographical expanse connect individual low-bit-rate terminals through a large hub facility. This facility is often associated with the company's large mainframe computer. Such systems turn out to be very inexpensive alternatives to using the public switched telephone network (PSTN).

Satellite communication has grown so much and so fast that the Western Hemisphere equatorial orbit is about full. The notable advantages of terrestrial fiber optic links are tempering the further growth of satellite communications.

This chapter sets forth methods of design of satellite communication links, both analog and digital. The primary thrust is initial system design and terminal dimensioning and its rationale. The chapter also provides standard interface information for several example systems. Regarding propagation issues, this chapter deals with satellite systems operating below 10 GHz. We have arbitrarily made 10 GHz a dividing line below which excess attenuation due to rainfall and gaseous absorption can be neglected. Chapter 7 addresses the problems of propagation above 10 GHz for satellite and terrestrial links.

6.2   AN INTRODUCTION TO SATELLITE COMMUNICATION

6.2.1   Two Broad Categories of Communication Satellites

This text will deal with two broad categories of communication satellites. The first is the repeater satellite, affectionately called the "bent pipe" satellite.

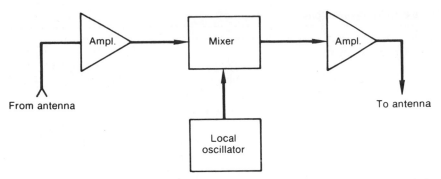

Figure 6.1 Simplified functional block diagram of the payload of a conventional translating RF repeater or bent pipe satellite.

The second is the processing satellite, which is used exclusively on digital circuits, where, as a minimum, the satellite demodulates the uplink signal to baseband and regenerates that signal. Analog circuits use exclusively "bent pipe" techniques; digital circuits may use either variety.

The bent pipe satellite is simply a frequency translating RF repeater. Figure 6.1 is a simplified functional block diagram of the payload of such a satellite.

6.2.2 Satellite Orbits

There are three types of satellite orbits:

- Polar
- Equatorial
- Inclined

The figure a satellite defines in orbit is an ellipse. Of course, a circle is a particular class of ellipse. A Molniya orbit is a highly inclined elliptical orbit.

The discussion here will dwell almost entirely on geostationary satellites. A geostationary satellite has a circular orbit. Its orbital period is one sidereal day (23 h, 56 min, 4.091 s) or nominally 24 h. Its inclination is 0°, which means that the satellite is always directly over the equator. It is geostationary. That is to say that it appears stationary over any location on earth, which is within optical view.

Geostationary satellites are conventionally located with respect to the equator (0° latitude) and a subsatellite point, which is given in longitude at the earth's surface. The satellite's range at this point, and only at this point,

Table 6.1   The geostationary satellite orbit

For the special case of a synchronous orbit—satellite in prograde circular
   orbit over the equator:

Altitude	19,322 nautical mi, 22,235 statute mi, 37,784 km
Period	23 hr, 56 min, 4.091 s (one sidereal day)
Orbit inclination	0°
Velocity	6879 statute mi/h
Coverage	42.5% of earth's surface (0° elevation)
Number of satellites	Three for global coverage with some areas of overlap (120° apart)
Subsatellite point	On the equator
Area of no coverage	Above 81° north and south latitude
Advantages	Simpler ground station tracking
	No handover problem
	Nearly constant range
	Very small Doppler shift
Disadvantages	Transmission delay
	Range loss (free space loss)
	No polar coverage

Source: Ref. 1.

is 35,784 km. Table 6.1 gives details and parameters of the geostationary satellite.

The table also outlines several of the advantages and disadvantages of this satellite. Most of these points are self-explanatory. For satellites not at geosynchronous altitude and not over the equator, there is the appearance of movement. The movement with relation to a point on earth will require some form of automatic tracking on the earth station antenna to keep it always pointed at the satellite. If a satellite system is to have full earth coverage using a constellation of geostationary satellites, a minimum of three satellites would be required to be separated by 120°. As one moves northward or southward from the equator, the elevation angle to a geostationary satellite decreases (see Section 6.2.3). Elevation angles below 5° are generally undesirable, because of fading and increase in antenna noise. This is the rationale in Table 6.1 for "area of no coverage." Handover refers to the action taken by a satellite earth station antenna when a nongeostationary (often misnamed "orbiting satellite") disappears below the horizon (or below 5° elevation angle) and the antenna slews to a companion satellite in the system that is just appearing above the opposite horizon. It should be pointed out here that geostationary satellites do have small residual relative motions. Over its subsatellite point, a geostationary satellite carries out a small apparent suborbit in the form of a figure eight because of higher space harmonics of the earth's gravitation and tidal forces from the sun and the moon. The

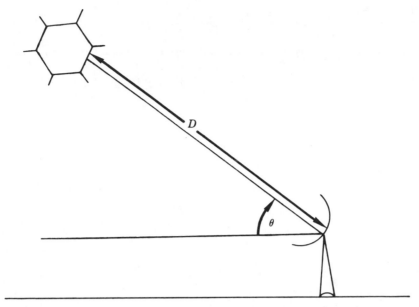

Figure 6.2 Definition of elevation angle (θ) or "look angle" and range (D) to satellite.

satellite also tends to drift off station because of the gravitational attraction of the sun and the moon as well as solar winds. Without correction the inclination plane drifts roughly 0.86° per year (Ref. 1, Section 13.4.1.9).

6.2.3 Elevation Angle Definition

The elevation angle or "look angle" of a satellite terminal antenna is the angle measured from the horizontal to the point on the center of the main beam of the antenna when the antenna is pointed directly at the satellite. This concept is shown in Figure 6.2. Given the elevation angle of a geostationary satellite, we can define the range. We will need the range, D in Figure 6.2, to calculate the free-space loss or spreading loss for the satellite radiolink.

6.2.4 Calculation of Azimuth, Elevation, and Range to a Geostationary Satellite

Figure 6.3 is a nomogram from which we can calculate from a known earth station location, azimuth, elevation angle, and range to a geostationary

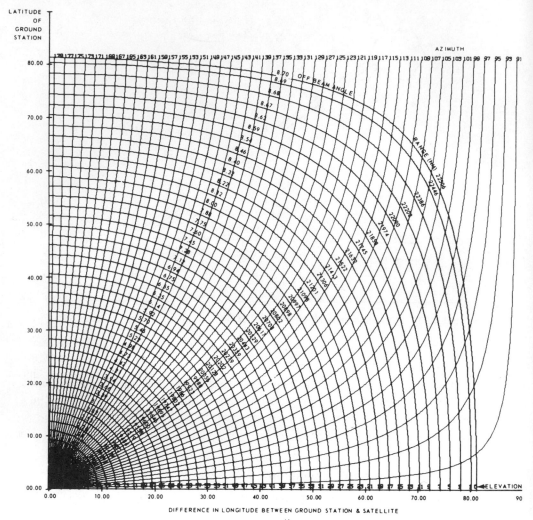

Note:
Angle off beam center is useful only if satellite
has a despun circular beam with beam center at
subsatellite point. (Intelsat III)

Satellite direction from earth station	Azimuth Transformation
East & South	None
West & South	Subtract azimuth from 360°
East & North	Subtract azimuth from 180°
West & North	Add 180° to azimuth

Example	Latitude	Longitude
Japan	37°N	141°E
Satellite	0°	176°E
Separation	176°−141° = 35°	
Azimuth	131°	
Elevation	33.5°	
Off Beam Angle	7.26°	
Range	20,680 NM	

Figure 6.3 Determination of range to a geostationary satellite, azimuth, and elevation angles. (From Ref. 1; courtesy of COMSAT, Washington, D.C..)

satellite if the subsatellite point of that satellite is known. Ref. 21 (pp. 224–225) provides a mathematical method of calculating these important parameters.

6.3 FREQUENCY BANDS AVAILABLE FOR SATELLITE COMMUNICATIONS

When making a general reference to frequency bands available for satellite communication, we speak of "frequency pairs." For instance, we refer to the 6/4 GHz band. In all cases the first number represents the uplink band or the band of frequencies available for an earth station to transmit to a satellite. The second number represents the downlink band or the band of frequencies available for a satellite to transmit to an earth station. It will be noted that the uplink frequency is always higher in frequency than the downlink. This is purposeful. The higher frequency suffers greater spreading or free-space loss than its lower frequency counterpart, and an earth station aims upward with well-controlled antenna side lobes. Obviously, an earth station has better transmitting assets than a satellite. It has unlimited prime power available and can use large aperture antennas and large power amplifiers. A satellite spews a signal to earth that must be limited in flux density so as not to interfere with terrestrial services that share the same band. A satellite does not have the transmission assets of an earth station. For example, it derives its prime power from solar cells backed up by secondary cells. One small advantage we can give the satellite is that it transmit on the lower frequency of the frequency pair with lower path loss. Much of this thinking derived, of course, when earth coverage antennas were nearly universally used.

The following lists commonly used frequency bands for satellite communication:

6/4 GHz	5925–6425 MHz	Uplink	Commercial
	3700–4200 MHz	Downlink	
8/7 GHz	7900–8400 MHz	Uplink	Military
	7250–7750 MHz	Downlink	
14/11 GHz	14.0–14.5 GHz	Uplink	Commercial
	11.7–12.2 GHz	Downward	
30/20 GHz	27.5–30.5 GHz	Uplink	Commercial
	17.7–20.2 GHz	Downlink	
30/20 GHz	30.0–31.0 GHz	Uplink	Military
	20.2–21.2 GHz	Downlink	
44/20 GHz	43.5–45.5 GHz	Uplink	Military
	20.2–21.2 GHz	Downlink	

6.4 LINK DESIGN PROCEDURES — THE LINK BUDGET

6.4.1 Introduction

To size or dimension a satellite communication terminal correctly, we will carry out a link budget analysis. The methodology is very similar to the path analyses described in Chapters 4 and 5. There are also certain legal constraints that we should be aware of.

At the outset, a certain satellite link must meet performance requirements. If the link is analog FDM/FM, the requirements will be expressed as noise in the derived voice channel; if TV, a certain value of signal-to-noise ratio S/N; if a digital link, a bit error rate. To meet these requirements, we will establish a value of carrier-to-noise ratio C/N or C/N_0 at the receiver input of the downlink. The C, of course, is the RSL measured in dBW or dBm; the N or N_0 value is based on the receiving system thermal noise threshold. Such a procedure should now be thoroughly familiar to the reader from Chapters 4 and 5.

Link budgets are carried out in a tabular manner. Figure 4.20 shows a tabular method of doing a path calculation for a line-of-sight (LOS) microwave link. A path calculation is just another name for a link budget.

The first item in a link budget table is the EIRP expressed in dBW or dBm. If we are dealing with an uplink, it is the satellite terminal EIRP; if a downlink, it is the satellite EIRP. The next item in the table is free-space loss. Several other miscellaneous losses are then included. We then calculate the IRL (isotropic receive level) by algebraically summing the column.

To the IRL value in dBm or dBW we algebraically add the G/T. Here we encounter a difference in methodology with the link budget's LOS microwave counterpart. G/T will be new to those not familiar with satellite communication. G/T is the receiving system figure of merit. It quantifies the receiving system sensitivity.

To this sum we subtract Boltzmann's constant, -228.6 dBW. The result is C/N_0, or, if you will RSL/N_0. From this value we can derive E_b/N_0 or S/N and link margin.

In the following subsections we calculate free-space loss to get some idea of the range of loss values we have to deal with. Then we show why noise is a driving factor in satellite link design by introducing limitations introduced by CCIR on satellite EIRP, resulting in very low signal levels at the satellite terminal. The reason for limiting the flux density on the earth's surface is that the satellite communication frequency bands are shared by terrestrial communication.

G/T is then introduced and example link budgets are worked.

6.4.2 Path Loss or Free-Space Loss

We introduced free-space loss in Section 4.3.1. Free-space loss formulas are repeated below for convenience. Unfortunately, we do not have a world wide (or even nationwide) standard unit of distance. We find that range to a satellite (distance) may be given in statute miles (sm), nautical miles (nm), or kilometers (km). Accordingly, we give three equations:

Range (D) unit sm:

$$L_{dB} = 36.58 + 20 \log D_{sm} + 20 \log f \text{ (MHz)} \qquad (6.1a)$$

Range (D) unit km:

$$L_{db} = 32.4 + 20 \log D_{km} + 20 \log f \text{ (MHz)} \qquad (6.1b)$$

Range (D) unit nm:

$$L_{dB} = 37.80 + 20 \log D_{nm} + 20 \log f \text{ (MHz)} \qquad (6.1c)$$

where D is the distance to the satellite in the unit indicated and f is the operating frequency in megahertz.

Often we are given the elevation angle of the satellite of interest. If this is the case, use Figure 6.3 which gives range (distance) to the satellite in nautical miles for various elevation angles.

Example. The elevation angle to a geostationary satellite is 23° and the transmitting frequency is 3840 MHz. What is the free-space loss in dB?

Turn to Figure 6.3 and we find the range (distance) to the satellite is 21,201 nm. Use this value for D in equation 6.1c. Thus

$$L_{dB} = 37.8 + 20 \log 21{,}201 + 20 \log 3840$$
$$L_{dB} = 196.01 \text{ dB}$$

6.4.3 Isotropic Receive Level — Simplified Model

Consider the example in the previous subsection where there was a downlink operating at 3840 MHz and the range to the satellite was 21,201 nm, producing a free-space loss of 196.01 dB. Assume the satellite has an EIRP of +31 dBW. If we neglect all other link losses, what is the IRL at the earth station (satellite terminal) antenna?

$$IRL \text{ (dBW)} = EIRP \text{ (dBW)} + FSL \text{ (dB)}$$
$$IRL = +31 \text{ dBW} + (-196.01) \text{ dB}$$
$$IRL = -165.01 \text{ dBW.} \qquad (6.2)$$

Some important losses have been left out. Although each may seem small in value, when totaled they will make a considerable contribution to the total link loss. Among these losses are pointing losses, polarization loss, radome loss (if radome is used), gaseous absorption loss, and excess attenuation due to rainfall.

6.4.4 Limitation of Flux Density on the Earth's Surface

The satellite communication frequency bands shown in Section 6.3 are shared with terrestrial services such as point-to-point LOS microwave. The flux density of satellite downlink signals on the earth's surface must be limited so as not to interfere with terrestrial radio services that share these same frequency bands. CCIR Rec. 358-3 recommends the following flux density limits (Ref. 2):

1. That, in frequency bands in the range 2.5 to 23 GHz shared between systems in the fixed-satellite service and line-of-sight radio-relay systems, the maximum power flux-density produced at the surface of the Earth by emissions from a satellite, including those from a reflecting satellite, for all conditions and methods of modulation, should not exceed:

   1.1 in the band 2.5 to 2.690 GHz, in any 4 kHz band:

-152	$dB(W/m^2)$ for	$\theta \leq 5°$
$-152 + 0.75(\theta - 5)$	$dB(W/m^2)$ for	$5° < \theta \leq 25°$
-137	$dB(W/m^2)$ for	$25° < \theta \leq 90°$

   1.2 in the band 3.4 to 7.750 GHz, in any 4 kHz band:

-152	$dB(W/m^2)$ for	$\theta \leq 5°$
$-152 + 0.5(\theta - 5)$	$dB(W/m^2)$ for	$5° < \theta \leq 25°$
-142	$dB(W/m^2)$ for	$25° < \theta \leq 90°$

   1.3 in the band 8.025 to 11.7 GHz, in any 4 kHz band:

-150	$dB(W.m^2)$ for	$\theta \leq 5°$
$-150 + 0.5(\theta - 5)$	$dB(W/m^2)$ for	$5° < \theta \leq 25°$
-140	$dB(W/m^2)$ for	$25° < \theta \leq 90°$

   1.4 in the band 12.2 to 12.75 GHz, in any 4 kHz band:

-148	$dB(W/m^2)$ for	$\theta \leq 5°$
$-148 + 0.5(\theta - 5)$	$dB(W/m^2)$ for	$5° < \theta \leq 25°$
-138	$dB(W/m^2)$ for	$25° < \theta \leq 90°$

   1.5 in the band 17.7 to 19.7 GHz, in any 1 MHz band:

-115	$dB(W/m^2)$ for	$\theta \leq 5°$
$-115 + 0.5(\theta - 5)$	$dB(W/m^2)$ for	$5° < \theta \leq 25°$
-105	$dB(W/m^2)$ for	$25° < \theta \leq 90°$

   Where θ is the angle of arrival of the radio-frequency wave (degrees above the horizontal);

2. That the aforementioned limits relate to the power flux-density and angles of arrival which would be obtained under free-space propagation conditions.

Note 1: Definitive limits applicable in shared frequency bands are laid down in Nos. 2561 to 2580.1 of Article 28 of the Radio Regulations. The CCIR is continuing its study of these problems, which may lead to changes in the recommended limits.

Note 2: Under Nos. 2581 to 2585 of the Radio Regulations, the power flux-density limits in the band 17.7 to 19.7 GHz shall apply provisionally to the band 31.0 to 40.5 GHz until such time as the CCIR has recommended definitive values, endorsed by a competent Administrative Conference (No. 2582.1 of the Radio Regulations).
Excerpt from CCIR Rec. 358-3, Page 28, Vol. IV/IX, Part 2, XVIth Plenary Assembly, Dubrovnik 1986.

6.4.5 Thermal Noise Aspects of Low-Noise Systems

We deal with very low signal levels in space communication systems. Downlink signal levels are in the approximate range of -154 to -188 dBW. The objective is to achieve sufficient S/N or E_b/N_0 at demodulator outputs. There are two ways of accomplishing this:

- By increasing system gain, usually with antenna gain
- Reducing system noise

In this section we will give an introductory treatment of thermal noise analytically, and later the term G/T will be discussed.

Around noise threshold, thermal noise predominates. To set the stage, we quote from Ref. 3:

The equipartition law of Boltzmann and Maxwell (and the works of Johnson and Nyquist) states that the available power per unit bandwidth of a thermal noise source is

$$p_n(f) = kT \text{ watts/Hz} \tag{6.3}$$

where k is Boltzmann's constant $(1.3806 \times 10^{-23}$ joule/K) and T is the absolute temperature of the source in kelvins.

Looking at a receiving system, all active components and all components displaying ohmic loss generate noise. In LOS radiolinks, system noise temperatures are in the range of 1500–4000 K, and the noise of the receiver front end is by far the major contributor. In the case of space communication, the receiver front end may contribute less than one-half the total system noise. Total receiving system noise temperatures range from as low as 70 up to 1000 K (for those types of systems considered in this chapter).

In Chapters 4 and 5, receiving system noise was characterized by noise figure expressed in decibels. Here, where we deal often with system noise temperatures of less than 290 K, the conventional reference of basing noise at room temperature is awkward. Therefore, noise figure is not useful at such low noise levels. Instead, it has become common to use effective noise temperature T_e (equation 6.6).

It can be shown that the available noise power at the output of a network in a band B_w is (Ref. 3)

$$p_n = g_a(f)(T + T_e)B_w \qquad (6.4)$$

where g_a = the network power gain at frequency f
  T = the noise temperature of the input source
  T_e = the effective input temperature of the network

For an antenna–receiver system, the total effective system noise temperature T_{sys}, conventionally referred to the input of the receiver, is

$$T_{sys} = T_{ant} + T_r \qquad (6.5)$$

where T_{ant} is the effective input noise temperature of the antenna subsystem and T_r is the effective input noise temperature of the receiver subsystem. The ohmic loss components from the antenna feed to the receiver input also generate noise. Such components include waveguide or other types of transmission lines, directional couplers, circulators, isolators, waveguide switches, and so forth.

It can be shown that the effective input noise temperature of an attenuator is (Ref. 3)

$$T_e = \frac{p_a}{kB_w g_a} + T_s = \frac{T(1 - g_a)}{g_a} \qquad (6.6)$$

where T_s is the effective noise temperature of the source, the lossy elements have a noise temperature T, k is Boltzmann's constant, and g_a is the gain (available loss). p_a is the noise power at the output of the network.

The loss of the attenuator l_a is the inverse of the gain or

$$l_a = \frac{1}{g_a} \qquad (6.7)$$

where l_a and g_a are numeric equivalents of the respective decibel values. Substituting into equation 6.6 gives

$$T_e = T(l_a - 1) \qquad (6.8)$$

It is accepted practice (Ref. 3) that

$$n_f = l_a + \frac{T_e}{T_0} \qquad (6.9)$$

where in Ref. 3 n_f is called the noise figure. Other texts call it the noise factor * and

$$NF_{dB} = 10 \log_{10} n_f \qquad (6.10)$$

and T_0 is standard temperature or 290 K. NF_{dB} is the conventional noise figure discussed in Section 1.12.

From equation 6.8 the noise figure (factor) is

$$n_f = 1 + \frac{(l_a - 1)T}{T_0} \qquad (6.11)$$

If the attenuator lossy elements are at standard temperature (e.g., 290 K), the noise figure equals the loss (the noise factor equals the numeric of the loss):

$$n_f = l_a \qquad (6.12a)$$

or expressed in decibels,

$$NF_{dB} = 10 \log l_a = L_{a(dB)} \qquad (6.12b)$$

For low-loss (i.e., ohmic loss) devices whose loss is less than about 0.5 dB, such as short waveguide runs, which are at standard temperature, equation 6.8 reduces to a helpful approximation.

$$T_e \approx 66.8L \qquad (6.13)$$

where L is the loss of the device in decibels.

The noise figure in decibels may be converted to effective noise temperature by

$$NF_{dB} = 10 \log_{10}\left(1 + \frac{T_e}{290}\right) \qquad (6.14)$$

Example. If a noise figure were given as 1.1 dB, what is the effective noise temperature?

$$1.1 \text{ dB} = 10 \log\left(1 + \frac{T_e}{290}\right)$$

$$0.11 = \log\left(1 + \frac{T_e}{290}\right)$$

$$1 + \frac{T_e}{290} = \log^{-1}(0.11)$$

$$1 + \frac{T_e}{290} = 1.29$$

$$T_e = 84.1 \text{ K}$$

* We like to distinguish the two. Noise figure is measured in decibels and noise factor is in decimal units (e.g., 3 dB = 2, 10 dB = 10, 13 dB = 20).

6.4.6 Calculation of C/N_0

We present two methods to carry out this calculation. The first method follows the rationale given in Section 4.6.5. C/N_0 is measured at the input of the first active stage of the receiving system. For space receiving systems this is the low-noise amplifier (LNA) or other device carrying out a similar function. Figure 6.4 is a simplified functional block diagram of such a receiving system.

C/N_0 is simply the carrier-to-noise ratio, where N_0 is the noise density in 1 Hz of bandwidth. C is the receive signal level (RSL). Restating equation 6.3,

$$N_0 = kT \tag{6.15}$$

where k is Boltzmann's constant and T is the effective noise temperature, in this case of the space receiving system. We can now state this identity:

$$\frac{C}{N_0} = \frac{C}{kT} \tag{6.16}$$

Turning to Figure 6.4, we see that if we are given the signal level impinging on the antenna, which we call the isotropic receive level (IRL), the receive signal level (RSL or C) at the input to the LNA is the IRL plus the antenna gain minus the line losses, or, stated in equation form,

$$C_{\text{dBW}} = \text{IRL}_{\text{dBW}} + G_{\text{ant}} - L_{\text{L(dB)}} \tag{6.17}$$

where L_L are the line losses in decibels. These losses will be the sum of the waveguide or other transmission line losses, antenna feed losses, and, if used, directional coupler loss, waveguide switch loss, power split loss, bandpass filter loss (if not incorporated in LNA), circulator/isolator losses, and so forth.

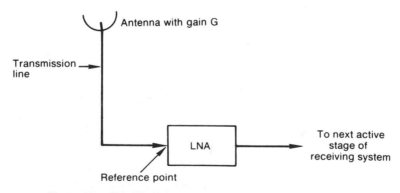

Figure 6.4 Simplified block diagram of space receiving system.

To calculate N_0, equation 6.15 can be restated as

$$N_0 = -228.6 \text{ dBW} + 10 \log T_{\text{sys}} \qquad (6.18)$$

where -228.6 dBW is the theoretical value of the noise level in dBW for a perfect receiver (noise factor of 1) at absolute zero in 1 Hz of bandwidth. T_{sys} is the receiving system effective noise temperature, often just called system noise temperature.

Example. Given a system (effective) noise temperature of 84.1 K, what is N_0?

$$N_0 = -228.6 \text{ dBW} + 10 \log 84.1$$
$$= -228.6 + 19.25$$
$$= -209.35 \text{ dBW}$$

To calculate C or RSL, consider the following example.

Example. The IRL from a satellite is -155 dBW; the earth station receiving system (space receiving system) has an antenna gain of 47 dB, an antenna feed loss of 0.1 dB, a waveguide loss of 1.5 dB, a directional coupler insertion loss of 0.2 dB, and a bandpass filter loss of 0.3 dB; the system noise temperature T_{sys} is 117 K. What is C/N_0?
    Calculate C (or RSL):

$$C = -155 \text{ dBW} + 47 \text{ dB} - 0.1 \text{ dB} - 1.5 \text{ dB} - 0.2 \text{ dB} - 0.3 \text{ dB}$$
$$C = -110.1 \text{ dBW}$$

Calculate N_0:

$$N_0 = -228.6 \text{ dBW} + 10 \log 117 \text{ K}$$
$$= -207.92 \text{ dBW}$$

Thus

$$\frac{C}{N_0} = C_{\text{dBW}} - N_{0(\text{dBW})} \qquad (6.19)$$

In this example, substituting:

$$\frac{C}{N_0} = -110.1 \text{ dBW} - (-207.92 \text{ dBW})$$
$$= 97.82 \text{ dB}$$

The second method of determining C/N_0 involves G/T, which is discussed in the next section.

6.4.7 Gain-to-Noise Temperature Ratio *G*/*T*

G/T can be called the "figure of merit" of a radio receiving system. It is most commonly used in space communication. It not only gives an experienced engineer a "feel" of a receiving system's capability to receive low-level signals effectively, it is also used quite neatly as an algebraically additive factor in space system link budget analysis.

G/T can be expressed by the following identity:

$$\frac{G}{T} = G_{dB} - 10\log T \tag{6.20}$$

where G is the receiving system antenna gain and T (better expressed as T_{sys}) is the receiving system noise temperature. Now we offer a word of caution. When calculating G/T for a particular receiving system, we must stipulate where the reference plane is. In Figure 6.4 it was called the "reference point." It is at the reference plane where the system gain is measured. In other words, we take the gross antenna gain and subtract all losses (ohmic and others) up to that plane or point. This is the net gain at that plane.

System noise is treated in the same fashion. Equation 6.5 stated

$$T_{sys} = T_{ant} + T_r$$

The antenna noise temperature T_{ant}, coming inward in the system, includes all noise contributors, including sky noise, up to the reference plane. Receiver noise T_r includes all noise contributors from the reference plane to the baseband output of the demodulator.

In most commercial space receiving systems, the reference plane is taken at the input to the LNA, as shown in Figure 6.4. In many military systems it is taken at the foot of the antenna pedestal. In one system, it was required to be taken at the feed. It can be shown that G/T will remain constant so long as we are consistent regarding the reference plane.

Calculation of the net gain G_{net} to the reference plane is straightforward. It is the gross gain of the antenna minus the sum of all losses up to the reference plane. Determining T_{sys} is somewhat more involved. We use equation 6.5. The calculation of T_{ant} is described in Section 6.4.7.1, and that of T_r in Section 6.4.7.2.

6.4.7.1 Calculation of Antenna Noise Temperature *T*ₐₙₜ   The term T_{ant}, or antenna noise, includes all noise contributions up to the reference plane. Let us assume for all further discussion in this chapter that the reference plane coincides with the input to the LNA (Figure 6.4). There are two "basic" contributors of noise: sky noise and noise from ohmic losses.

Sky noise is a catchall for all external noise sources that enter through the antenna, through its main lobe, and through its side lobes. External noise is

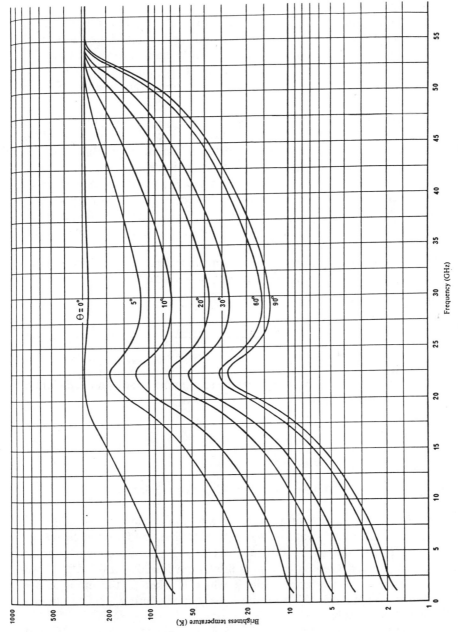

Figure 6.5 Brightness temperature (sky noise) for clear air for 7.5 g / m³ of water-vapor concentration. θ is the elevation angle. From CCIR Rep. 720-2, Figure 5, Page 185, Vol. V, XVIth Plenary Assembly, Drubrovnik 1986. (Ref. 4; courtesy of ITU – CCIR.)

largely due to extraterrestrial sources and thermal radiation from the atmosphere and the earth. Cosmic noise is extraterrestrial radiation that seems to come from all directions.

The sun is an extremely strong source of noise, and it can interrupt satellite communication when it passes behind the satellite being used and thus lies in the main lobe of an earth station's antenna receiving pattern. The moon is a much weaker source, which is relatively innocuous to satellite communication. Its radiation is due to its own temperature and reflected radiation from the sun.

The atmosphere affects external noise in two ways. It attenuates noise passing through it, and it generates noise because of the energy of its constituents. Ground radiation, which includes radiation of objects of all kinds in the vicinity of the antenna, is also thermal in nature.

For our discussion we will say that sky noise T_{sky} varies with frequency, elevation angle, and surface water-vapor concentration. Figure 6.5 gives values of sky noise for elevation angles θ of 0°, 5°, 10°, 20°, 30°, 60°, and 90° for water-vapor concentration 7.5 g/m³.* These figures do not include ground radiation contributions.

Antenna noise T_{ant} is the total noise contributed to the receiving system by the antenna up to the reference plane. It is calculated by the formula (Ref. 21)

$$T_{ant} = \frac{(l_a - 1)290 + T_{sky}}{l_a} \tag{6.21}$$

where l_a is the numeric equivalent of the system ohmic losses (in decibels) up to the reference plane. l_a may be expressed, then, as

$$l_a = \log_{10}^{-1} \frac{L_a}{10} \tag{6.22}$$

where L_a is the sum of the losses to the reference plane.

Example. Assume an earth station with an antenna at an elevation angle of 10°, clear sky, 7.5 g/m³ water-vapor concentration, and ohmic losses as follows: waveguide loss of 2 dB, feed loss of 0.1 dB, directional coupler insertion loss of 0.2 dB, and a bandpass filter insertion loss of 0.4 dB. These are the losses up to the reference plane, which is taken as the input to the LNA (Figure 6.6). What is the antenna noise temperature T_{ant}? The operating frequency is 12 GHz.

Determine the sky noise from Figure 6.5. For an elevation angle of 10° and a frequency of 12 GHz, the value is 22 K.

---

*For other water-vapor concentrations consult CCIR Rec. 720-2 (Ref. 4).

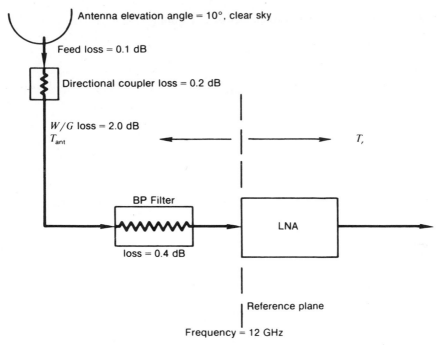

Figure 6.6 Example earth station receiving system.

Sum the ohmic losses up to the reference plane:

$$L_a = 0.1 \text{ dB} + 2 \text{ dB} + 0.2 \text{ dB} + 0.4 \text{ dB} + 2.7 \text{ dB}$$

$$l_a = \log^{-1}\left(\frac{2.7}{10}\right)$$

$$= 1.86$$

Substitute into equation 6.21:

$$T_{ant} = \frac{(1.86 - 1)290 + 22}{1.86}$$

$$= 145.9 \text{ K}$$

6.4.7.2 Calculation of Receiver Noise Temperature T_r

A receiver will probably consist of a number of stages in cascade, as shown in Figure 6.7. The effective noise temperature of the receiving system, which we will call T_r,

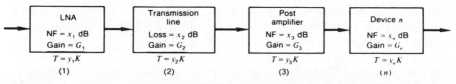

Figure 6.7 Generalized cascaded amplifiers/attenuators for noise temperature calculation.

is calculated from the traditional cascade formula

$$T_r = y_1 + \frac{y_2}{G_1} + \frac{y_3}{G_1 G_2} + \cdots + \frac{y_n}{G_1 G_2 \cdots G_{n-1}} \qquad (6.23)$$

where y is the effective noise temperature of each amplifier or device and G is the numeric equivalent of the gain (or loss) of the device.

Example. Compute T_r for the first three stages of a receiving system. The first stage is an LNA with a noise figure of 1.1 dB and a gain of 25 dB. The second stage is a lossy transmission line with 2.2-dB loss. The third and final stage is a postamplifier with a 6-dB noise figure and a gain of 30 dB.

Convert the noise figures to equivalent noise temperatures, using equation 6.14:

$$1.1 \text{ dB} = 10 \log\left(1 + \frac{T_e}{290}\right)$$

$$T_e = 83.6 \text{ K}$$

$$6.0 = 10 \log\left(1 + \frac{T_e}{290}\right)$$

$$T_e = 864.5 \text{ K}$$

Calculate the noise temperature of the lossy transmission line, using equation 6.4. First determine l_a:

$$l_a = \log^{-1}\left(\frac{L_a}{10}\right)$$

$$= 1.66$$

$$T_e = (1.66 - 1)290$$

$$= 191.3 \text{ K}$$

Calculate T_r, using equation 6.23:

$$T_r = 83.6 + \frac{191.3}{316.2} + \frac{864.5}{316.2 \times 1/1.66}$$

$$= 83.6 + 0.605 + 4.53$$

$$= 88.735 \text{ K}$$

It should be noted that in the second and third terms, we divided by the numeric equivalent of the gain, not the gain in decibels. In the third term, of course, it was not a gain, but a loss (e.g., 1/1.66 where 1.66 is the numeric equivalent of a 2.2-dB loss). It will also be found that in cascaded systems the loss of a lossy device is equivalent to its noise figure.

6.4.7.3 *Example Calculation of G/T* A satellite downlink operates at 21.5 GHz. Calculate the G/T of a terminal operating with this satellite. The

reference plane is taken at the input to the LNA. The antenna has a 3-ft aperture displaying a 44-dB gross gain. There is 2 ft of waveguide with 0.2 dB/ft of loss. There is a feed loss of 0.1 dB, a bandpass filter has 0.4-dB insertion loss, and a radome has a loss of 1.0 dB. The LNA has a noise figure of 5.0 dB and a 30-dB gain. The LNA connects directly to a downconverter/IF amplifier combination with a single sideband noise figure of 13 dB.

Calculate the net gain of the antenna to the reference plane:

$$G_{net} = 44 \text{ dB} - 1.0 \text{ dB} - 0.1 \text{ dB} - 0.4 \text{ dB} - 0.4 \text{ dB}$$
$$= 42.1 \text{ dB}$$

This will be the value for G in the G/T expression.

Determine the sky noise temperature value at the 10° elevation angle, assume clear sky with dry conditions at 21.5 GHz. Use Figure 6.5. The value is 145 K.

Calculate L_A, the sum of the losses to the reference plane. This will include, of course, the radome loss:

$$L_A = 1.0 \text{ dB} + 0.1 \text{ dB} + 0.4 \text{ dB} + 0.4 \text{ dB} = 1.9 \text{ dB}$$

Determine l_a, the numeric equivalent of L_A from the 1.9-dB value (equation 6.22):

$$L_a = \log^{-1}\left(\frac{1.9}{10}\right)$$
$$= 1.55$$

Calculate T_{ant}; use equation 6.21:

$$T_{ant} = \frac{(1.55 - 1)290 + 145}{1.55}$$
$$= 196.45 \text{ K}$$

Calculate T_r. Use equation 6.23. First convert the noise figures to equivalent noise temperatures using equation 6.14. The LNA has a 5.0-dB noise figure, and its equivalent noise temperature is 627 K; the downconverter/IF amplifier has a noise figure of 13 dB and an equivalent noise temperature of 5496 K.

$$T_r = 627 + \frac{5496}{1000}$$
$$= 632.5 \text{ K}$$

Determine T_{sys} using equation 6.5:

$$T_{sys} = 196.45 + 632.5$$
$$= 828.95 \text{ K}$$

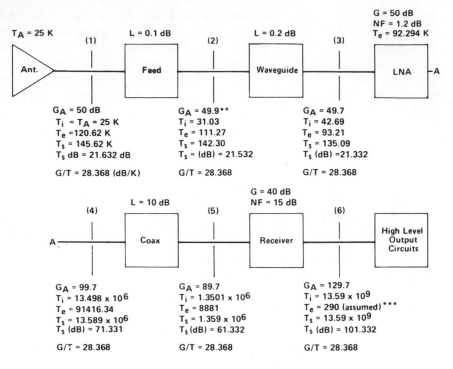

Figure 6.8 Example of noise temperature and G/T calculations for cascade of two ports. Note that G/T is independent of position in cascade. (From Ref. 5, Paper 2A, "Noise Temperature and G/T of Satellite Receiving Systems"; reprinted with permission.)

Calculate G/T using equation 6.20:

$$\frac{G}{T} = 42.1 \text{ dB} - 10\log(828.95)$$

$$= +12.91 \text{ dB/K}$$

The following discussion, taken from Ref. 5, further clarifies G/T analysis.

Figure 6.8 shows a satellite terminal receiving system and its gain/noise analysis. The notation used is that of the reference document.

The following are some observations of Figure 6.8 (using notation from the reference):

1. The value of T_S is different at every junction.
2. The value of G/T (where $T = T_S$) is the same at every junction.
3. The system noise temperature at the input to the LNA is influenced largely by the noise temperature of the components that precede the

LNA and the LNA itself. The components that follow the LNA have a negligible contribution to the system noise temperature at the LNA input junction (reference plane) if the LNA gain is sufficiently high.

The parameters that have significant influence on G/T are the following:

1. The antenna gain and the antenna noise temperature.
2. The antenna elevation angle. The lower the angle, the higher the sky noise, thus, the higher the antenna noise, and, hence, the lower the G/T for a given antenna gain.
3. Feed and waveguide insertion losses. The lower the insertion loss of these devices, the higher the G/T.
4. LNA. The lower the noise temperature of the LNA, the higher the G/T. The higher the gain of the LNA, the less the noise contribution of the stages following the LNA. For instance, in Figure 6.8, if the gain of the LNA were reduced to 40 dB, the value of T_S would increase to 144.1 K. This means that the value of G/T would be reduced by about 0.26 dB. For an LNA with a gain of only 30 dB, the G/T would then drop by an additional 1.96 dB.

6.4.8 Calculation of C/N_0 Using the Link Budget Technique

The link budget is a tabular method of calculating space communication system parameters. It is a similar approach used in Chapters 4 and 5, where it was called link analysis. In the method presented here the starting point of a link budget is the platform EIRP. The platform can be a terminal or a satellite. In an equation, it would be expressed as follows:

$$\frac{C}{N_0} = \text{EIRP} - \text{FSL}_{dB} - (\text{other losses}) + \frac{G}{T_{dB/K}} - k \qquad (6.24)$$

where FSL is the free-space loss, k is Boltzmann's constant expressed in dBW or dBm, and the "other losses" may include

- Polarization loss
- Pointing losses (terminal and satellite)
- Off-contour loss
- Gaseous absorption losses
- Excess attenuation due to rainfall (as applicable)

The off-contour loss refers to spacecraft antennas that are not earth coverage, such as spot beams, zone beams, and multiple beam antenna (MBA), and the contours are flux density contours on the earth's surface. This loss expresses in equivalent decibels the distance the terminal platform is from a contour line. Satellite pointing loss, in this case, expresses contour line error or that the contours are not exactly as specified.

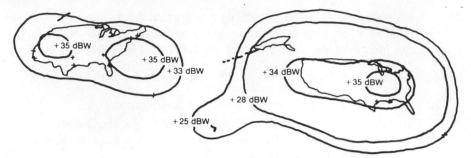

Figure 6.9 GTE SPACENET 4-GHz EIRP contours for satellite at 70°W (left) and 119°W (right). (From Ref. 1; courtesy of GTE SpaceNet.)

6.4.8.1 *Some Link Loss Guidelines*   Free-space loss is calculated using equations 6.1a–c. Care should be taken in the use of units for distance (range) and frequency.

When no other information is available, use 0.5 dB as an estimate for polarization loss and pointing losses (e.g., 0.5 dB for satellite pointing loss and 0.5 dB for terminal pointing loss).

For off-contour loss, the applicable contour map should be used, placing the prospective satellite terminal in its proposed location on the map and estimating the loss. Figure 6.9 is a typical contour map.

Atmospheric absorption losses are comparatively low for systems operating below 10 GHz. These losses vary with frequency, elevation angle, and altitude. For the 7-GHz downlink, 0.8 dB is appropriate for a 5° elevation angle, dropping to 0.5 dB for 10° and 0.25 dB for 15°; all values are for sea level. For 4 GHz, recommended values are 0.5 dB for 5° elevation angle and 0.25 dB for 10°; all values are for sea level. Atmospheric absorption losses are treated more extensively in Chapter 7.

Excess attenuation due to rainfall is rigorously treated in Chapter 7. This attenuation also varies with elevation angle and altitude. Suggested estimates are 0.5 dB at 5° elevation angle for a 4-GHz downlink band, 0.25 at 10°, and 0.15 dB at 15°, with similar values for the 6-GHz uplink. For the 7-GHz military band: 3 dB for 5°, 1.5 dB at 10°, and 0.75 dB at 15°, all values at sea level. Use similar values for the 8-GHz uplink.

6.4.8.2 *Link Budget Examples*   The link budget is used to calculate C/N_0 when other system parameters are given. It is also used when C/N_0 is given when it is desired to calculate one other parameter such as G/T or EIRP of either platform in the link.

Example 1. 4-GHz downlink, FDM/FM, 5° elevation angle. Satellite EIRP + 30 dBW. Range to satellite (geostationary) is 22,208 nautical mi or 25,573 statute mi from Figure 6.3. Terminal $G/T = +20$ dB/K. Calculate C/N_0.

Calculate free-space loss (L_{dB}) from equation (6.1c):

$$L_{db} = 36.58 + 20\log(4 \times 10^3) + 20\log(25{,}573)$$

$$= 196.78 \text{ dB}$$

EIRP of satellite	+30 dBW
Free-space loss	−196.78 dB
Satellite pointing loss	−0.5 dB
Off-contour loss	−0.5 dB
Atmospheric absorption loss	−0.5 dB
Rainfall loss	−0.5 dB
Polarization loss	−0.5 dB
Terminal pointing loss	−0.5 dB
Isotropic receive level	−169.78 dBW
Terminal G/T	+20 dB/K
Sum	−149.78 dBW
Boltzmann's constant	−(−228.6 dBW)
C/N_0	78.82 dB

Example 2. Calculate the required satellite G/T, where the uplink frequency is 6.0 GHz and the terminal EIRP is +70 dBW, and with a 5° terminal elevation angle. The required C/N_0 at the satellite is 102.16 dB (typical for an uplink video link). The free-space loss is determined as in Example 1 but with a frequency of 6.0 GHz. This loss is 200.3 dB. Call the satellite G/T value X.

Terminal EIRP	+70 dBW	
Terminal pointing loss	−0.5 dB	
Free-space loss	−200.3 dB	
Polarization loss	−0.5 dB	
Satellite pointing loss	−0.5 dB	
Atmospheric absorption loss	−0.5 dB	
Off-contour loss	−0.5 dB	
Rainfall loss	−0.5 dB	
Isotropic receive level at satellite	−133.3 dBW	
G/T of satellite	X dB/K	(initially let X = 0 dB/K)
Sum	−133.3 dBW	
Boltzmann's constant	−(−228.6 dBW)	
C/N_0 (as calculated)	95.3 dB	
C/N_0 (required)	102.16 dB	
G/T	+6.86 dB/K	(difference)

This G/T value, when substituted for X, will derive a C/N_0 of 102.16 dB. It is not advisable to design a link without some margin. Margin will compensate for link degradation as well as errors of link budget estimation. The more margin (in decibels) that is added, the more secure we are that the link will work. On the other hand, each decibel of margin costs money. Some compromise between conservatism and economic realism should be met. For this link, 4 dB might be such a compromise. If it were all allotted to satellite G/T, then the new G/T value would be $+10.86$ dB/K. Other alternatives to build in a margin would be to increase transmitter power output, thereby increasing EIRP, or to increase terminal antenna size, among other possibilities. The power of using the link budget can now be seen easily. The decibels flow through on a one-for-one basis, and it is fairly easy to carry out trade-offs.

It should be noted that when the space platform (satellite) employs an earth coverage (EC) antenna, satellite pointing loss and off-contour loss are disregarded. The beamwidth of an EC antenna is usually accepted as 17° or 18° when the satellite is in geostationary orbit.

6.4.8.3   Calculating System C/N_0   The final C/N_0 value we wish to know is that at the terminal receiver. This must include, as a minimum, the C/N_0 for the uplink and C/N_0 for the downlink. If the satellite transponder is shared (e.g., simultaneous multicarrier operation on one transponder), a value for C/N_0 for satellite intermodulation noise must also be included. The basic equation to calculate C/N_0 for the system is given as

$$\left(\frac{C}{N_0}\right)_{\text{sys}} = \frac{1}{1/(C/N_0)_u + 1/(C/N_0)_d} \qquad (6.25)$$

Example.   Consider a bent pipe satellite system where the uplink $C/N_0 = 105$ dB and the downlink $C/N_0 = 95$ dB. What is the system C/N_0?

Convert each value of C/N_0 to its numeric equivalent:

$$\log^{-1}\left(\frac{105}{10}\right) = 3.16 \times 10^{10}; \qquad \log^{-1}\left(\frac{95}{10}\right) = 0.316 \times 10^{10}$$

Invert each value and add. Invert this value and take 10 log:

$$\left(\frac{C}{N_0}\right)_{\text{sys}} = 94.6 \text{ dB/Hz}$$

Many satellite transponders simultaneously permit multiple carrier access, particularly when operated in the FDMA or SCPC regimes. FDMA (frequency division multiple access) and SCPC (single channel per carrier) systems are described subsequently in this chapter.

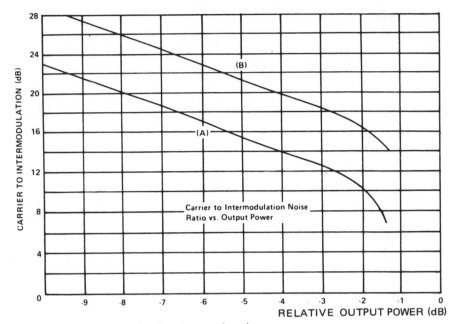

(A) Large number of uniformly spaced carriers.
(B) Two Carriers.

Figure 6.10   TWT power transfer characteristics. (From Ref. 5; courtesy of Scientific-Atlanta.)

When transponders are operated in the multiple-carrier mode, the down-link signal is rich in intermodulation (IM) products. Cumulatively, this is IM noise, which must be tightly controlled, but cannot be eliminated. Such noise must be considered when determining $(C/N_0)_{sys}$ when two or more carriers are put through the same transponder simultaneously. The following equation now applies to calculate $(C/N_0)_{sys}$:

$$\left(\frac{C}{N_0}\right)_{sys} = \frac{1}{(C/N_0)_u^{-1} + (C/N_0)_d^{-1} + (C/N_0)_i^{-1}} \qquad (6.26)$$

where the subscripts sys, u, d, and i refer to system, uplink, downlink, and intermodulation, respectively.

The principal source of IM noise (products) in a satellite transponder is the final amplifier, often called HPA (high-power amplifier). With present technology, this is usually a TWT (traveling-wave tube) amplifier, although SSAs (solid-state amplifiers) are showing increasing implementation on space platforms.

Figure 6.10 shows typical IM curves for a bent pipe type transponder using a TWT transmitter. The lower curve in the figure (curve *A*) is the IM

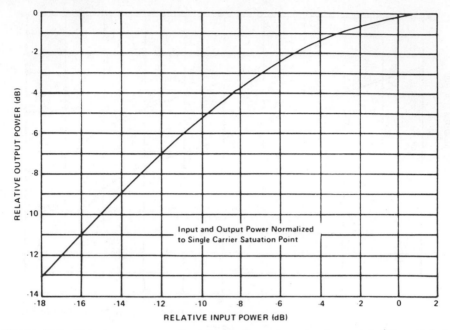

Figure 6.11 Output power normalized to single carrier saturation point. (From Ref. 5; courtesy of Scientific-Atlanta.)

performance for a large number of equally spaced carriers within a single transponder. The upper curve (curve B) shows the IM characteristics of two carriers.

In order to bring the IM products to an acceptable level and thus not degrade the system C/N_0 due to poor C/IM performance, the total power of the uplink must be "backed off" or reduced, commonly to a C/IM ratio of 20 dB or better. The result, of course, is a reduction in downlink EIRP. Figure 6.11 shows the effect of input (drive) power reduction versus output power of a TWT. To increase the C/IM ratio to 24 dB for the two-carrier case, we can see from Figure 6.10 that the input power must be backed off by 7 dB. Figure 6.11 shows that this results in a total downlink power reduction of 3 dB. It should also be noted that the resulting downlink power must be shared among the carriers actually being transmitted. For example, if two equal-level carriers share a transponder, the power in each carrier is 3 dB lower than the backed-off value.

6.4.9 Calculating S/N

6.4.9.1 In a VF Channel of an FDM/FM Configuration To calculate S/N and psophometrically weighted noise (pWp0) in a voice channel when

C/N in the IF is given, use the following formula (Ref. 5):

$$\frac{S}{N} = \frac{C}{N} + 20\log\left(\frac{\Delta F_{TT}}{f_{ch}}\right) + 10\log\left(\frac{B_{IF}}{B_{ch}}\right) + P + W \qquad (6.27)$$

where
B_{IF} = IF noise bandwidth
ΔF_{TT} = rms test tone deviation
f_{ch} = highest baseband frequency
B_{ch} = voice channel bandwidth (3.1 kHz)
P = top VF channel emphasis improvement factor
W = psophometric weighting improvement factor (2.5 dB)

Once the voice channel S/N has been calculated, the noise in the voice channel in picowatts may be determined from

$$\text{Noise} = \log^{-1}\left\{\frac{90 - (S/N)}{10}\right\} \quad (\text{pWp0}) \qquad (6.28)$$

Table 6.2 lists the transmission parameters of INTELSAT V, VA and VI, which we will use below in an example problem. These are typical bent pipe satellite system parameters for communication links.

Example. Using Table 6.2 calculate S/N and psophometrically weighted noise in an FDM/FM derived voice channel for a 972-channel system. First use equation 6.27 to calculate S/N and then equation 6.28 to calculate the noise in pWp. Use the value f_r for rms test tone deviation, f_m for the maximum baseband frequency, and B_{IF} for the allocated satellite bandwidth. The emphasis improvement may be taken from Figure 4.21. C/N is 17.8 dB in this case:

$$\frac{S}{N} = 17.8\ \text{dB} + 20\log\left(\frac{802 \times 10^3}{4028 \times 10^3}\right) + 10\log\left(\frac{36 \times 10^6}{3.1 \times 10^3}\right)$$

$$+\ 2.5\ \text{dB} + 4.5\ \text{dB}$$

$$=\ 51.43\ \text{dB}$$

$$\text{Noise} = \log^{-1}\left(\frac{90 - 51.43}{10}\right)$$

$$=\ 7194.5\ \text{pWp}$$

Table 6.2 INTELSAT V, VA, VA(IBS), and VI transmission parameters (regular FDM/FM carriers)

Carrier Capacity (Number of Channels)	Top Baseband Frequency (kHz)	Allocated Satellite BW Unit (MHz)	Occupied Bandwidth (MHz)	Deviation (Rms) for 0 dBm0 Test Tone (kHz)	Multichannel Rms Deviation (kHz)	Carrier-to-Total Noise Temperature Ratio at Operating point 8000 + 200 pW0p from RF Sources (dBW/K)	Carrier-to-Noise Ratio in Occupied Bandwidth (dB)	Ratio of Unmodulated Carrier Power to Maximum Carrier Power Density Under Full-Load Conditions
n	f_m	B_a	B_0	f_r	f_{mc}	C/T	C/N	(db/4 kHz)
12	60	1.25	1.125	109	159	−154.7	13.4	20.0
24	108	2.5	2.00	164	275	−153.0	12.7	22.3
36	156	2.5	2.25	168	307	−150.0	15.1	22.8
48	204	2.5	2.25	151	292	−146.7	18.4	22.6
60	252	2.5	2.25	136	296	−144.0	21.1	22.4
60	252	5.0	4.0	270	546	−149.0	12.7	25.3
72	300	5.0	4.5	294	616	−149.1	13.0	25.8
96	408	5.0	4.5	263	584	−145.5	16.6	25.6
132	552	5.0	4.4	223	529	−141.4	20.7	24.2[a]
								($X = 1$)
96	408	7.5	5.9	360	799	−148.2	12.7	27.0
132	552	7.5	6.75	376	891	−145.9	14.4	27.5
192	804	7.5	6.4	297	758	−140.6	19.9	25.8[a]
								($X = 1$)
132	552	10.0	7.5	430	1020	−147.1	12.7	28.0
192	804	10.0	9.0	457	1167	−144.4	14.7	28.6
252	1052	10.0	8.5	358	1009	−139.9	19.4	27.0[a]
								($X = 1$)

252	1052	15.0	12.4	577	1627	−144.1	13.6	30.0
312	1300	15.0	13.5	546	1716	−141.7	15.6	30.2
372	1548	15.0	13.5	480	1646	−138.9	18.4	30.1
432	1796	15.0	13.0	401	1479	−136.2	21.2	27.6[a]
								(X = 1)
312	1300	17.5	15.75	663	2081	−143.4	13.2	31.2
372	1548	17.5	15.75	583	1999	−140.8	15.9	31.0
432	1796	17.5	15.75	517	1919	−138.5	18.2	30.8
432	1796	20.0	18.0	616	2276	−139.9	16.1	31.5
492	2044	20.0	18.0	558	2200	−137.8	18.2	31.4
552	2292	20.0	18.0	508	2121	−136.0	20.0	30.2[a]
								(X = 1)
432	1796	25.0	20.7	729	2688	−141.4	14.1	32.2
492	2044	25.0	22.5	738	2911	−140.3	14.8	32.6
552	2292	25.0	22.5	678	2833	−138.5	16.6	32.5
612	2540	25.0	22.5	626	2755	−136.9	18.1	32.4
612	2540	36.0	32.4	983	4325	−141.0	12.5	34.3
792	3284	36.0	32.4	816	4085	−137.0	16.5	34.1
972	4028	36.0	32.4	694	3849	−133.8	19.7	32.8[a]
								(X = 1)
792	3284	36.0	36.0	930	4653	−138.3	14.8	34.7
972	4028	36.0	36.0	802	4417	−135.2	17.8	34.5

Source: Ref. 6. Courtesy of INTELSAT.

[a]This value is X dB lower than the value calculated according to the normal formula used to derive this ratio:

$$10 \log_{10}\left(f_{mc}\frac{\sqrt{2\pi}}{4}\right)$$

where X is the value in brackets in the last column and f_{mc} is the rms multichannel deviation in kHz. This factor is necessary in order to compensate for low modulation index carriers which are not considered to have a Gaussian power density distribution.

6.4.9.2 *For a Typical Video Channel*   As suggested by Ref. 5,

$$\frac{S}{N_v} = \frac{C}{N} + 10\log 3\left(\frac{\Delta f}{f_m}\right)^2 + 10\log\left(\frac{B_{IF}}{2B_v}\right) + W + CF \qquad (6.29)$$

where S/N_v = peak-to-peak luminance signal-to-noise ratio

Δf = peak composite deviation of the video

f_m = highest baseband frequency

B_v = video noise bandwidth (for NTSC systems this is 4.2 MHz)

B_{IF} = IF noise bandwidth

W = emphasis plus weighting improvement factor
(12.8 dB for NTSC North American systems)

CF = rms to peak-to-peak luminance signal conversion factor (6.0 dB)

For many satellite systems, without frequency reuse, a 500-MHz assigned bandwidth is broken down into twelve 36-MHz segments.* Each segment is then assigned to a transponder. For video transmission either a half or full transponder is assigned.

Example. A video link is relayed through a 36-MHz (full) transponder bent pipe satellite where the peak composite deviation is 11 MHz and the C/N is 14.6 dB. What is the weighted S/N? Assume NTSC standards.

Using equation 6.29,

$$\frac{S}{N} = 14.6 \text{ dB} + 10\log\left[3\left(\frac{11}{4.2}\right)^2\right] + 10\log\left(\frac{36}{8.4}\right) + 12.8 + 6$$

$$= 52.9 \text{ dB}$$

In this case the video noise bandwidth is 4.2 MHz, which is also the highest baseband frequency. European systems would use 5 MHz.

Equation 6.29 is only useful for C/N above 11 dB. Below that C/N value, impulse noise becomes apparent and the equation is not valid.

6.4.10   System Performance Parameters

Table 6.3 gives some typical performance parameters for video, aural channel, and FDM/FM for a 1200-channel configuration.

*Many new systems coming on line use wider bandwidths.

Table 6.3 **Some Typical Satellite link performance parameters**

System Parameters	Units	FDM/FM	Video
TV Video			
C/N	dB		14.6
Maximum video frequency	MHz		4.2
Overdeviation	dB		
Peak operating deviation	MHz		10.7
FM improvcment	dB		13.2
BW improvement	dB		6.3
Weighting/emphasis improvement	dB		12.8
P-rms conversion-factor	dB		6.0
Total improvement	dB		38.3
S/N (peak-to-peak/rms-luminance signal)	dB		52.9
TV Program Channel (Subcarrier)			
Peak carrier deviation	MHz		2.0
Subcarrier frequency	MHz		7.5
FM improvement	dB		11.5
BW improvement	dB		14.0
Total improvement	dB		2.5
C/N_{sc} (subcarrier)	dB		16.9
Peak subcarrier deviation	kHz		75
Maximum audio frequency	kHz		15
FM improvement	dB		18.8
BW improvement	dB		13.8
Emphasis improvement	dB		12.0
Total improvement	dB		44.6
S/N (audio)	dB		59.2
FDM/FM			
Number channels		1,200	
Test tone deviation (rms)	kHz	650	
Top baseband frequency	kHz	5,260	
FM improvement	dB	18.2	
BW improvement	dB	40.7	
Weighting improvement	dB	2.5	
Emphasis (top slot) improvement	dB	4.0	
Total improvement	dB	29.0	
TT/N (test tone to noise ratio)	dB	49.0	
Noise	pWp0	12,589	

Source: Ref. 5. Courtesy of Scientific-Atlanta.

6.5 ACCESS TECHNIQUES

6.5.1 Introduction

Access refers to the way a communication system uses a satellite transponder. There are three basic access techniques:

- FDMA (frequency division multiple access)—analog or digital operation
- TDMA (time division multiple access)—digital operation exclusively
- CDMA (code division multiple access)—digital operation exclusively

With FDMA a satellite transponder is divided into frequency band segments where each segment is assigned to a user. The number of segments can vary from one, where an entire transponder is assigned to a single user, to literally hundreds of segments, which is typical of SCPC (single channel per carrier) operation. For analog telephony operation, each segment is operated in an FDM/FM mode. In this case FDM group(s) and/or supergroup(s) are assigned for distinct distant location connectivity.

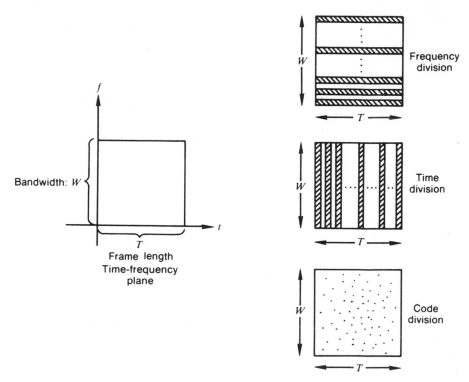

Figure 6.12 The three basic multiple access techniques depicted in a time – frequency plane.

TDMA works in the time domain. Only one user appears on the transponder at any given time. Each user is assigned a time slot to access the satellite. System timing is crucial with TDMA. It lends itself only to digital transmission, typically PCM.

CDMA is especially attractive to military users due to its antijam and low probability of intercept (LPI) properties. With CDMA, the transmitted signal is spread over part or all of the available transponder bandwidth in a time–frequency relationship by a code transformation. Typically, the modulated RF bandwidth is ten to hundreds of times greater than the information bandwidth.

CDMA had previously only been attractive to the military user because of its antijam properties. Since 1980 some interest in CDMA has been shown in the commercial sector for demand access for large populations of data circuit/network users with bursty requirements in order to improve spectral utilization. For further discussion of spread spectrum systems, refer to Ref. 24.

Figure 6.12 depicts the three basic types of satellite multiple access. The horizontal axis represents time, and the vertical axis spectral bandwidth.

6.5.2 Frequency Division Multiple Access (FDMA)

FDMA has been the primary method of multiple access for satellite communication systems for the past 25 years and will probably remain so for quite some time. In the telephone service, it is most attractive for permanent direct or high-usage (HU) routes, where during the busy hour, trunk groups to a particular, distinct location require 5 or more Erlangs (e.g., one or more FDM groups).

For this most basic form of FDMA, a single earth station transmits a carrier to a bent pipe satellite. The carrier contains an FDM configuration in its modulation envelope consisting of groups and supergroups for distinct distant locations. Each distant location receives and demodulates the carrier, but demultiplexes only those FDM groups and supergroups pertaining to it. For full duplex service, a particular earth station receives, in return, one carrier from each distant location with which it provides connectivity. So, on its downlink, it must receive, and select by filtering, carriers from each distant location, and demultiplex only that portion of each derived baseband that contains FDM channelization destined for it.

Another form of FDMA is SCPC, where each individual telephone channel independently modulates a separate radio-frequency carrier. Each carrier may be frequency modulated or digitally modulated, often PSK. SCPC is useful on low-traffic-intensity routes (e.g., less than 5 Erlangs) or for overflow from FDM/FDMA. SCPC will be discussed in Section 6.5.2.1.

FDMA is a mature technology. Its implementation is fairly straightforward. Several constraints must be considered in system design. Many of these constraints center around the use of TWT as HPA in satellite transponders.

Depending on the method of modulation employed, amplitude and phase nonlinearities must be considered to minimize IM products, intelligible crosstalk, and other interfering effects by taking into account the number and size (bandwidth) of carriers expected to access a transponder. These impairments are maintained at acceptable levels by operating the transponder TWT at the minimum input backoff necessary to ensure that performance objectives are met. However, this method of operation results in less available channel capacity when compared to a single access mode.

TWT amplifiers operate most efficiently when they are driven to an operating point at or near saturation. When two or more carriers drive the TWT near its saturation point, excessive IM products are developed. These can be reduced to acceptable levels by reducing the drive, which, in turn, reduces the amplifier efficiency. This reduction in drive power is called backoff.

In early satellite communication systems, IM products created by TWT amplitude nonlinearity were the dominant factor in limiting system operation. However, as greater power became available and narrow bandwidth transponders capable of operation with only a few carriers near saturation became practical, maximum capacity became dependent on a carefully evaluated trade-off analysis of a number of parameters, which include the following:

1. Satellite TWT impairments, including in-band IM products produced by both amplitude and phase nonlinearities, and intelligible crosstalk caused by FM–AM–PM conversion during multicarrier operation.
2. FM transmission impairments not directly associated with the satellite transponder TWT, such as adjacent carrier interference caused by frequency spectrum overlap between adjacent carriers, which gives rise to convolution and impulse noise in the baseband; dual path distortion between transponders; interference due to adjacent transponders IM earth station out-of-band emission and frequency reuse cochannel interference.
3. General constraints, including available power and allocated bandwidth; uplink power control; frequency coordination; and general vulnerability to interference.

Backoff was discussed briefly in Section 6.4.8.3. From Figures 6.10 and 6.11 we can derive a rough rule of thumb that tells us that for approximately every decibel of backoff, IM products drop 2 dB for the multicarrier case (i.e., more than three carriers on a transponder). Also, for every decibel of backoff on TWT driving power, TWT output power drops 1 dB. Of course, this causes inefficiency in the use of the TWT. As the number of carriers is increased on a transponder, the utilization of the available bandwidth becomes less efficient. If we assume 100% efficient utilization of a transponder with only 1 carrier, then with 2 carriers the efficiency drops to about 90%, with 4 to 60%, with 8 to about 50%, and with 14 carriers to about 40%.

Crosstalk is a significant impairment in FDMA systems. It can result from a sequence of two phenomena:

- An amplitude response that varies with frequency-producing amplitude modulation coherent with the original frequency modulation of an RF carrier or FM–AM transfer, and
- A coherent amplitude modulation that phase-modulates all carriers occupying a common TWT amplifier due to AM/PM conversion.

As a carrier passes through a TWT amplifier, it may produce amplitude modulation from gain–slope anomalies in the transmission path. Another carrier passing through the same TWT will vary in phase at the same rate as the AM component and thereby pick up intelligible crosstalk from any carrier sharing the transponder. Provisions should be made in specifying TWT amplifier characteristics to ensure that AM–PM conversion and gain–slope variation meet system requirements. Intelligible crosstalk should be 58 dB down or better, and with modern equipment, this goal can be met quite easily.

Out-of-band RF emission from an earth station is another issue; 500 pWp0 is commonly assigned in a communication satellite system voice channel noise budget for earth station RF out-of-band emission. The problem centers on the earth station HPA TWT. When a high-power, wideband TWT is operated at saturation, its full output power can be realized, but it can also produce severe IM RF products to the up-path of other carriers in an FDMA system. To limit such unwanted RF emission, we must again turn to the backoff technique of the RF drive to the TWT. Some systems use as much as 7-dB backoff. For example, a 12-kW HPA with 7-dB backoff is operated at about 2.4 kW. This is one good reason for overbuilding earth station TWTs and accepting the inefficiency of use.

Uplink power control is an important requirement for FDM/FM/FDMA systems. Sufficient power levels must be maintained on the uplink to meet signal-to-noise ratio requirements on the derived downlinks for bent pipe transponders. On the other hand, uplink power on each carrier accessing a particular transponder must be limited to maintain IM product generation (IM noise) within specifications. This, of course, is the backoff discussed above. Close control is required of the power level of each carrier to keep transmission impairments within the total noise budget.

One method of meeting these objectives is to study each transponder configuration on a case-by-case basis before the system is actually implemented on the satellite. Once the proper operating values have been determined, each earth station is requested to provide those values of uplink carrier levels. These values can be further refined by monitoring stations that precisely measure resulting downlink power of each carrier. The theoretical power levels are then compared to both the reported uplink and the measured downlink levels.

Energy dispersal is yet another factor required in the design of a bent pipe satellite system. Energy dispersal is used to reduce the spectral energy density of satellite signals during periods of light loading (e.g., off busy hour). The reduction of maximum energy density will also facilitate efficient use of the geostationary satellite orbit by minimizing the orbital separation needed between satellites using the same frequency band and multiple-carrier operation of broadband transponders. The objective is to maintain spectral energy density the same for conditions of light loading as for busy-hour loading. Several methods of implementing energy dispersal are described in CCIR Rep. 384-5 (Ref. 7).

One method of increasing satellite transponder capacity is by *frequency reuse*. As the term implies, an assigned frequency band is used more than once. The problem is to minimize mutual interference between carriers operating on the same frequency but accessing distinct transponders. There are two ways of avoiding interference in a channel that is used more than once:

1. By orthogonal polarization
2. By multiple exclusive spot-beam antennas

The use of opposite-hand circular or crossed-linear polarizations may be used to effect an increase in bandwidth by a factor of 2. Whether the potential increase in capacity can be realized depends on the amount of cross-polarization discrimination that can be achieved for the cochannel operation. Cross-polarization discrimination is a function of the quality of the antenna systems and the effect of the propagation medium on polarization of the transmitted signal. The amount of polarization "twisting" or distortion is a function of the elevation angle for a given earth station. The lower the angle, the more the twist.

Isolation between cochannel transponders should be greater than 25 dB. INTELSAT V specifies 27-dB minimum isolation between polarizations.

Whereas a satellite operating in a 500-MHz bandwidth might have only 12 transponders (36- or 40-MHz bandwidth each), with frequency reuse, 24 transponders can be accommodated with the same bandwidth. Thus, the capacity has been doubled.

Table 6.2 shows a typical transponder allocation for FDMA operation.

6.5.2.1 Single Channel Per Carrier (SCPC) FDMA

Many SCPC systems have been implemented and are now in operation, and many more are coming on line. There are essentially two types of SCPC systems: preassigned and demand assigned (DAMA, which stands for demand assignment multiple

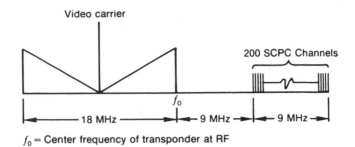

Figure 6.13   Satellite transponder frequency plan, video plus SCPC.

access). The latter requires some form of control system. Some systems occupy an entire transponder, whereas others share a transponder with video service, leaving a fairly large guard band between the video portion of the transponder passband and the SCPC portion. Such transponder sharing is illustrated in Figure 6.13.

Channel spacings on a transponder vary depending on the system. INTELSAT systems commonly use 45-kHz spacing. Others use 22.5, 30, and 36 kHz as well as 45 kHz. Modulation is FM or BPSK/QPSK. The latter lends itself well to digital systems, whether PCM or CVSD (continuous variable-slope delta modulation).

If we were to divide a 36-MHz bandwidth on a transponder into uniform 45-kHz segments, the total voice channel capacity of the transponder would be 800 VF channels. These are better termed half-channels, because, for telephony, we always measure channel capacity as full-duplex channels. Therefore 400 of the 800 channels would be used in the "go" direction, and the other 400 would be used in the reverse direction. An SCPC system is generally designed for a 40% activity factor. Hence, statistically, only 320 of the channels can be expected to be active at any instant during the busy hour. This has no effect on the bandwidth, only on the loading. Most systems use voice-activated service. In this case, carriers appear on the transponder for the activated channels only. This provides probably the worst case for IM noise for all conventional bent pipe systems during periods of full activation. A simplified drawing of an SCPC system is shown in Figure 6.14.

SCPC channels can be preassigned or DAMA. The preassigned technique is only economically feasible in situations where source/destination locations have very low traffic density during the busy hour (e.g., less than several Erlangs).

DAMA SCPC systems are efficient for source/sink locations of comparatively low traffic intensity, especially under situations with multiple location community of interest (e.g., trunks to many distinct locations but under 12 channels to any one location). This demarcation line at 12 channels, of course, is where the designer should look seriously at establishing fixed-assignment FDMA/FDM channel group allocation.

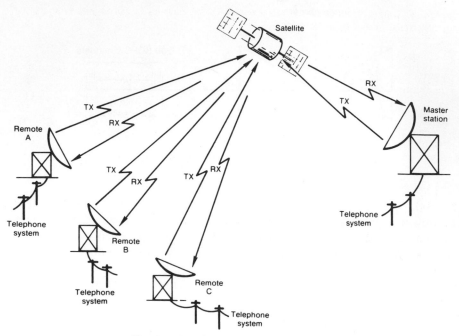

Figure 6.14 A typical FM SCPC system.

The potential improvement in communication system capacity and efficiency is the primary motivation for DAMA systems. Our primary consideration in this section is voice traffic. Call durations average several minutes. The overhead time to connect and disconnect is 1 or 2 s in an efficient channel assignment system. Some of the issues of importance in demand assignment systems are

- User requirements such as traffic intensity in Erlangs or ccs, number of destinations, and grade of service (probability of blocked calls)
- Capacity improvements due to implementation of DAMA scheme
- Assignment algorithms (centralized versus distributed control)
- Equipment cost and complexity

Consider an example to demonstrate the potential improvement due to demand assignment (see Table 6.4). An earth station is required to communicate with 40 destinations, and traffic intensity to each destination is 0.5 Erlang, with a blocking probability of $P_B = 0.01$. Assume that the call arrivals have a Poisson distribution and the call holding times have an exponential distribution. Then for a trunk group with n channels in which blocked calls are cleared, based on an Erlang B distribution (Ref. 1, Section 1-2), A, in Table 6.4, gives the traffic intensity in Erlangs. If the system is

Table 6.4   Comparison of preassigned versus DAMA channel requirements based on a grade of service of $P_B = 0.01$

Number of Destinations	Channel Requirements[a]					
	$A = 0.1$		$A = 0.5$		$A = 1.0$	
	Preassigned	DAMA	Preassigned	DAMA	Preassigned	DAMA
1	2	2	4	4	5	5
2	4	3	8	5	10	7
4	8	3	16	7	20	10
8	16	4	32	10	40	15
10	20	5	40	11	50	18
20	40	7	80	18	100	30
40	80	10	160	30	200	53

Source: Ref. 8. Reprinted with permission of IEEE Press.
[a]A = Erlang traffic intensity.

designed for preassignment, each destination will require four channels to achieve $P_B = 0.01$. For 40 destinations, 160 channels will be required for the preassigned case. In the DAMA case, the total traffic is considered, since any channel can be assigned to any destination. The total traffic load is 20 Erlangs (40×0.5) and the Erlang B formula gives a requirement of only 30 channels. The efficiency of the system with demand assignment is improved over the preassigned by a factor of 5.3 (160/30). Table 6.4 further expands on this comparison using various traffic loadings and numbers of destinations. Reference 9, Chapter 1, provides good introductory information on traffic engineering.

There are essentially three methods for controlling a DAMA system:

- Polling
- Random-access central control
- Random-access distributed control

The polling method is fairly self-explanatory. A master station (Figure 6.14) "polls" all other stations in the system sequentially. When a positive reply is received, a channel is assigned accordingly. As the number of earth stations in the system increases, the polling interval becomes longer, and the system tends to become unwieldly because of excessive postdial delay as a call attempt waits for the polling interval to run its course.

With random-access central control, the status of channels is coordinated by a central computer located at the master station. Call requests (called *call attempts* in switching) are passed to the central computer via a digital orderwire (i.e., digitally over a radio service channel), and a channel is assigned, if available. Once the call is completed and the subscriber goes on-hook, the speech path is taken down and the channel used is returned to the demand-access pool of idle channels. According to system design, there are a number of methods to handle blocked calls (all trunks busy [ATB] in telephone switching), such as queueing and second attempts.

The distributed control random-access method utilizes a processor controller at each earth station accessing the system. All earth stations in the network monitor the status of all channels via continuous updating of channel status information by means of the digital orderwire circuit. When an idle channel is seized, all users are informed of the fact, and the circuit is removed from the pool. Similar information is transmitted to all users when circuits are returned to the idle condition. One problem, of course, is the possibility of double seizure. Also the same problems arise regarding blockage (ATB) as in the central control system. Distributed control is more costly and complex, particularly in large systems with many users. It is attractive in the international environment because it eliminates the "politics" of a master station.

6.5.3   Time Division Multiple Access (TDMA)

6.5.3.1   Introduction   TDMA operates in the time domain and is applicable only to digital systems because information storage is required. With TDMA, use of a satellite transponder is on a time-sharing basis. At any given moment in time, only one earth station accesses the satellite. Individual time slots are assigned to earth stations operating with that transponder in a sequential order. Each earth station has full and exclusive use of the transponder bandwidth during its time-assigned segment. Depending on bandwidth of the transponder and type of modulation used, bit rates of 10–100 Mbps are used.

If only one carrier appears on the transponder at any time, then intercarrier IM products cannot be developed and the TWT-based HPA of the transponder may be operated at full power. This means that the TWT requires no backoff in drive and may be run at or near saturation for maximum efficiency. With multicarrier FDMA systems, backoffs of 5–10 dB are not uncommon.

TDMA utilizes the transponder bandwidth more efficiently, too. Reference 10 compares approximate channel capacities of an INTELSAT IV global beam transponder operating with a normal INTELSAT Standard A (30-m antenna) earth stations using FDMA and TDMA, respectively. Assuming 10 accesses, typical capacity using FM/FDMA is approximately 450 one-way voice channels. With TDMA, using standard 64-kbps PCM voice channels, the capacity of the same transponder is 900 voice channels. If digital speech interpolation (DSI) is now implemented on the TDMA system, the voice channel capacity is increased to approximately 1800 channels.

Still another advantage of TDMA is flexibility. Flexibility is not only a significant benefit to large systems, but is often the key to system viability in

Table 6.5   FDMA versus TDMA

Advantages	Disadvantages
FDMA	
Mature technology	IM in satellite transponder output
No network timing	Requires careful uplink power control
Easy FDM interface	Inflexible to traffic load
TDMA	
Maximum use of transponder power	Still emerging technology
No careful uplink power control	Network timing
Flexible to dynamic traffic loading	Major digital buffer considerations
Straightforward interface with digital network	Difficult to interface with FDM
More efficient transponder bandwidth usage	
Digital format compatible with	
forward error control	
Source coding	
Demand access algorithms	
Applicable to switched satellite service	

smaller systems. Nonuniform accesses pose no problem in TDMA because time-slot assignments are easy to adjust. This applies to initial network configuration, assignments, reassignments, and demand assignments. This is ideal for a long-haul system where traffic adjustments can be made dynamically as the busy hour shifts from east to west. Changes can also be made for growth or additional services.

Disadvantages of TDMA are timing requirements and complexity. Accesses may not overlap. Obviously, overlapping causes interference between sequential accesses. Guard times between accesses must be made minimal for efficient operation. The longer the guard times, the shorter the burst length and/or number of accesses. Typically, 5–15 accesses can be accommodated per transponder with guard times on the order of 50–200 ns.

As the world's telecommunication network evolves from analog to digital and as frequency congestion increases, there will be more and more demand to shift to TDMA operation in satellite communication.

Table 6.5 compares FDMA with TDMA.

6.5.3.2   TDMA Operation on a Bent Pipe Satellite

General. With a time division multiple access (TDMA) arrangement on a bent pipe satellite, each earth station accessing a transponder is assigned a time slot for its transmission, and all uplinks use the same carrier frequency on a particular transponder. We recall that a major limitation of FDMA systems is the required backoff of drive in a transponder to reduce IM products developed in the TWT final amplifier owing to simultaneous multi-carrier operation. With TDMA, on the other hand, only one carrier appears at the transponder input at any one time, and, as a result, the TWT can be run to saturation minus a small fixed backoff to reduce waveform spreading. This results in more efficient use of a transponder and permits greater system capacity. In some cases capacity can be doubled when compared to an equivalent FDMA counterpart. Another advantage of a TDMA system is that the traffic capacity of each access can be modified on a nearly instanta-neous basis. The loading of a long-haul system can be varied as the busy hour moves across it, assuming that accesses are located in different time zones. This is difficult to achieve on a conventional FDMA system.

Simplified Description of TDMA Operation. An important requirement of TDMA is that transmission bursts do not overlap. To ensure nonoverlap, bursts are separated by a guard time, which is analogous to a guardband in FDMA. The longer we make the guard time, the greater assurance we have of nonoverlap. However, this guard time reduces the efficiency of the system. Decreasing guard time improves system efficiency. The amount of guard time, of course, is a function of system timing. The better the timing system is, the shorter we can make guard times. Typical guard times for operating systems are on the order of 50–200 ns.

Figure 6.15 shows a typical TDMA frame. A frame is a complete cycle of bursts, usually with one burst per access. The burst length per access need not be of uniform duration; in fact, it is usually not. Burst length can be made a function of the traffic load of a particular access at a particular time. This nonuniformity is shown in the figure. The frame period is the time required to sequence the bursts through a frame.

The number of accesses on a transponder can vary from 3 to over 100. Obviously, the number of accesses is a function of the traffic intensity of each access, assuming a full-capacity system. It is also a function of the transpon-der bandwidth and the digital modulation employed (e.g., the packing ratio or the number of bits per hertz). For example, more bits can be packed per unit bandwidth with an 8-ary PSK signal of fixed duration than a BPSK signal of equal duration. For high-capacity systems, frame periods vary from 100 μs to over 2 ms. As an example, the INTELSAT TDMA system has a frame period of 2 ms.

Figure 6.15 also shows a typical burst format, which we call an access subframe. The first segment of the subframe is called the CR/BTR (carrier recovery/bit timing recovery). This symbol sequence is particularly necessary on a coherent PSK system, where the CR is used by the PSK demodulator in each receiver to recover local carrier and the BTR to synchronize (sync) the local clock. In the INTELSAT system CR/BTR is 176 symbols long. Other

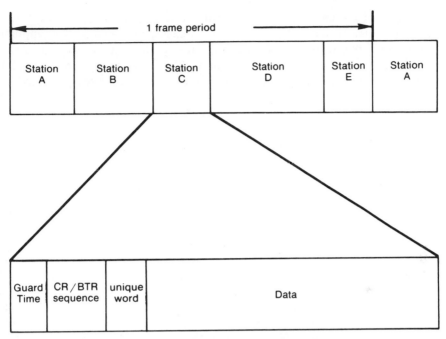

Figure 6.15 Typical TDMA frame and burst formats.

systems may use as few as 30 symbols. We must remember that CR/BTR is overhead, and thus it would be desirable to shorten its duration as much as possible.

Generally, the minimum number of bits (or symbols—in QPSK systems one symbol or baud carries two bits of information) required in a CR/BTR sequence is only roughly a function of system bit or symbol rate. Carrier recovery and bit timing recovery at the receiver must be accomplished by realizable phase-lock loops and/or filters that have a sufficiently narrow bandwidth to provide a satisfactory output signal-to-noise ratio S/N. There is a trade-off in system design between acquisition time (implying a wider bandwidth) and S/N (implying a narrower bandwidth). Ref. 10 suggests that CR/BTR bandwidths of 0.5–2% of the bit or symbol rate provide a good compromise between acquisition time and bit error rate performance resulting from a finite-loop-output S/N. Adaptive phase-lock loops that acquire in a wide-band mode and track with a narrower bandwidth can be used to reduce CR/BTR overhead.

The next bit sequence in the burst subframe (Figure 6.15) is the unique word (UW), which establishes an accurate time reference in the received burst. The primary purpose of the UW is to perform the clock alignment function. It can also be used as a transmit station identifier. Alternatively, the UW can be followed by a transmit station identifier sequence (SIC—station identification code).

The loss of either the CR/BTR or the UW is fatal to the receipt of a burst. For voice traffic, a lost burst causes clicks, and sounds like impulse noise to the listener. In the case of a data bit stream, large blocks of data can be lost due to a "skew" or slip of alignment. TDMA systems are designed for a probability of miss or false detection of 1×10^{-8} or better per burst to maintain a required threshold bit error rate of 1×10^{-4}. A major guideline we should not lose sight of is the point where supervisory signaling will be lost. This value is a BER of approximately 1×10^{-3}. The design value of 1×10^{-8} threshold will provide a mean time to miss or false detection of several hours with a frame length on the order of 1 ms (Ref. 10).

TDMA system design usually allows for some errors in the UW without loss of alignment. One approach to reduce chances of misalignment suggests a change in waveform during the UW interval. For instance, on a 8-ary PSK system, we might transmit BPSK during this interval, providing more energy per bit, ensuring an improved error performance during this important interval and thus improving the threshold performance.

Other overhead or housekeeping functions are inserted in the burst subframe between the UW and the data text. These may include voice and teleprinter orderwires, BERT (bit error rate test) and other sequences, alarms, and a control and delay channel. Preambles or burst overhead usually require between 100 and 600 bits. INTELSAT uses 288 symbols or 576 bits (with QPSK, 1 symbol is equivalent to 2 information bits).

The efficiency of a TDMA system depends largely on how well we can amortize system overhead and reduce guard times, in both length and number. Frame length affects efficiency. As the length increases, the number of overhead bits per unit time decreases. Also, as the length increases, the receiving system buffer memory size must increase, increasing the complexity and cost of terminals.

TDMA System Clocking, Timing, and Synchronization. It was previously stressed that an efficient TDMA system must have no burst overlap on the one hand, and as short a guard time as possible between bursts on the other hand. We are looking at guard times in the nanosecond regime. The satellites under discussion here are geostationary. For a particular TDMA system the range to a satellite can vary from 23,000 to 26,000 statute mi. We can express these range values in time equivalents by dividing by the velocity of propagation in free space, or 186,000 mi/s. These values are 23,000/186,000 and 26,000/186,000 or 123.469 and 139.573 ms. The time difference for a signal to reach a geostationary satellite from a very-low-elevation-angle earth station and a very-high-elevation-angle earth station is 16.104 ms (e.g., 139.573−123.469) or 16,104 μs or 16,104,335 ns. Of course, this is a worst case, but still feasible. The TDMA system must be capable of handling these orders of time differences among the accessing earth stations. How do we do it and meet the guidelines set out previously? We must also keep in mind that geostationary satellites actually are in motion in a suborbit, causing an additional time difference and doppler shift, both varying dynamically with time.

There are two generic methods used to handle the problem: "open loop" and "closed loop." Open-loop methods are characterized by the property that an earth station's transmitted burst is not received by that station. We mean here that an earth station does not monitor the downlink of its own signal for sync and timing purposes. By not using its delayed receiving signal for timing, the loop is not closed; hence, it is open.

Closed loop covers those synchronization techniques in which the transmitted signals are returned through the bent pipe transponder repeater to the transmitting station. This permits nearly perfect synchronization and high-precision ranging. The term *closed loop* derives from the looping back through the satellite of the transmitted signal, permitting the transmitting TDMA station to compare the time of the transmitted-burst leading edge to that of the same burst after passing through the satellite repeater. The TDMA transmitter is then controlled by the result, an early or late arrival relative to the transmitting station's time base. (*Note*: These definitions of open loop and closed loop should be taken in context and not confused with open-loop and closed-loop satellite tracking.)

One open-loop method uses no active form of synchronization. It is possible to achieve accuracies from 5 μs to 1 ms (Ref. 11) through what can be termed "coarse sync." The system is based on very stable free-running clocks, and an approximation is made of the orbit parameters where burst positioning can be done to better than 200-μs accuracy. The method was used on some early TDMA trial systems and on some military systems and will probably be employed on satellite-based data networks, especially with long frames.

One of the most common methods used to synchronize a family of TDMA accesses is by a reference burst. A reference burst is a special preamble only, and its purpose is to mark the start of frame with a burst code word. The station transmitting the reference burst is called a reference station. The reference bursts are received by each member of the family of TDMA accesses, and all transmissions of the family are locked to the time base of the reference station. This, of course, is a form of open-loop operation. Generally, a reference burst is inserted at the beginning of frame. Since reference bursts pass through the bent pipe repeater and usually occupy the same bandwidth as traffic bursts, they provide each station in the family with information on doppler shift, time-delay variations due to satellite motion, and channel characteristics. However, the reference burst technique has some drawbacks. It can only serve one repeater and a specific pair of uplinks and downlinks. There are difficulties with this technique in transferring such results accurately to other repeaters, other beams, or stations in different locations. The reference bursts also add to system overhead by using a bandwidth/time product not strictly devoted to the transfer of useful, revenue-bearing data/information. However, we must accept that some satellite capacity must be devoted to achieve synchronization.

6.6 PROCESSING SATELLITES

Processing satellites, as distinguished from "bent pipe" satellites, operate in the *digital* mode, and as a minimum, demodulate the uplink signal to baseband for regeneration. As a maximum, at least as we envision today, they operate as digital switches in the sky. In this section we will discuss satellite systems that demodulate and decode in the transponder and then we'll present some ideas on switching schemes suggested in the NASA 30/20-GHz system. This will be followed by a section on coding gain and a section on link analysis for processing satellites.

6.6.1 Primitive Satellite Processing

The most primitive form of satellite processing is the implementation of on-board regenerative repeaters. This only requires that the uplink signal be demodulated and passed through a hard limiter or a decision circuit. The implementation of regenerative repeaters accrues the following advantages:

- Isolation of the uplink and downlink by on-board regeneration prevents the accumulation of thermal noise and interference. Cochannel interference is a predominant factor of signal degradation because of the measures taken to augment communication capacity by such means as frequency reuse.
- Isolating the uplink and downlink makes the optimization of each link possible. For example, the modulation format of the downlink need not be the same as that for the uplink.
- Regeneration at the satellite makes it possible to implement various kinds of signal processing on board the satellite. This can add to the communication capacity of the satellite and provide a more versatile set of conveniences for the user network.

Ref. 12 points out that it can be shown on a PSK system that a regenerative repeater on board can save 6 dB on the uplink budget and 3 dB on the downlink budget over its bent pipe counterpart, assuming the same BER on both systems.

Applying this technique to a digital TDMA system requires carrier recovery and bit timing recovery. Although the carrier frequencies and clocking are quite close among all bursts, coherency of the carrier and the clocking recovery may not be anticipated between bursts. To regenerate baseband

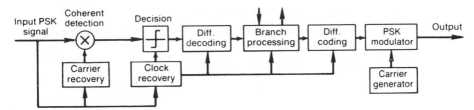

Figure 6.16   Configuration of ideal regenerative repeater.

signals on TDMA systems effectively, carrier and bit timing recovery are done in the preamble of each burst and must be done very rapidly to maintain a high communication efficiency. There are two methods that can be implemented to resolve the correct phase of the recovered carrier. One uses reference code words in the transmitted bit stream at regular intervals. The other solution is to use differential encoding on the transmitted bit stream. Although this latter method is simpler, it does degrade BER considering equal C/N_0 of each approach.

An ideal regenerative repeater for a satellite transponder is shown in Figure 6.16 for PSK operation. It will carry out the following functions:

- Carrier generation
- Carrier recovery
- Clock recovery
- Coherent detection
- Differential decoding
- Differential encoding
- Modulation
- Signal processing
- Symbol/bit decision

The addition of forward error correction (FEC) coding/decoding on the uplink and on the downlink carries on-board processing one step further. FEC coding and decoding are discussed subsequently. In a fading environment such as one might expect with satellite communication systems operating above 10 GHz during periods of heavy rainfall, an interleaver would be added after the coder and a deinterleaver before the decoder. Fading causes burst errors, and conventional FEC schemes handle random errors. Interleavers break up a digital bit stream by shuffling coded symbols such that symbols in error due to the burst appear to the decoder as random errors. Of course, interleaving intervals or spans should be significantly longer than the fade period expected.

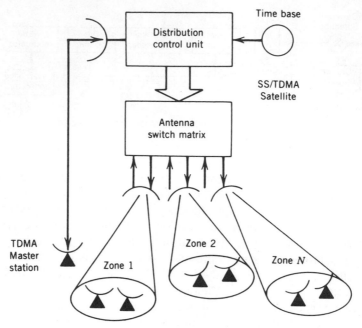

Figure 6.17 SS/TDMA concept.

6.6.2 Switched-Satellite TDMA (SS/TDMA)

We now carry satellite processing one step further by employing antenna beam switching in conjunction with TDMA. This technique provides bulk trunk routing increasing satellite capacity by frequency reuse. Figure 6.17 depicts this concept. TDMA signals from a geographical zone are cyclically interconnected to other beams or zones so that a set of transponders appears to have beam-hopping capability. A sync window or reference window is usually required to synchronize the TDMA signals from earth terminals to the on-board switch sequence.

Sync window is a generic method to allow earth stations to synchronize to a switching sequence being followed in the satellite. A switching satellite, as described here, consists of a number of transmitters cross-connected to receivers by a high-speed time-division switch matrix. The switch matrix connections are changed throughout the TDMA frame to produce the required interconnections of earth terminals. A special connection at the beginning of the frame is the sync window, during which signals from each spot-beam zone are looped back to their originating spot-beam zone, thus forming the timing reference for all zones. This establishes closed-loop synchronization.

.Figure 6.18 shows a scheme for locking and tracking a sync window in a satellite switching sequence. A burst of two tones, F_1 and F_2, is transmitted

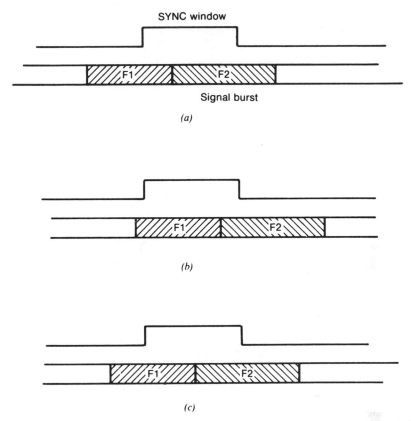

Figure 6.18 Sync window for SS/TDMA and use of bursts of two frequencies. (*a*) Signal burst too early; (*b*) signal burst too late; (*c*) synchronized to sync window. (From Ref. 11; reprinted with permission.)

by a single access station. Only the portion that passes through the sync window is received back at the access station.

The basic concept is to measure and compare the received subbursts F_1 and F_2 as shown in the figure. Although a very narrow bandwidth and full RF power are used, digital averaging over many frames still is required. The difference is used to control F_1/F_2 burst to a resolution of one symbol, and the process is continually repeated in closed loop. The sync bursts to the TDMA network are slaved, and the network is therefore synchronized to the sync window and the switching sequence on the satellite.

With SS/TDMA the network connectivity and the traffic volume between zones can be adapted to changing needs by reprogramming the control processor of the antennas. Also, of course, the narrow beams (e.g., higher-gain antennas) increase the uplink C/N and the downlink EIRP for a given transponder HPA power output. However, the applicability of the system

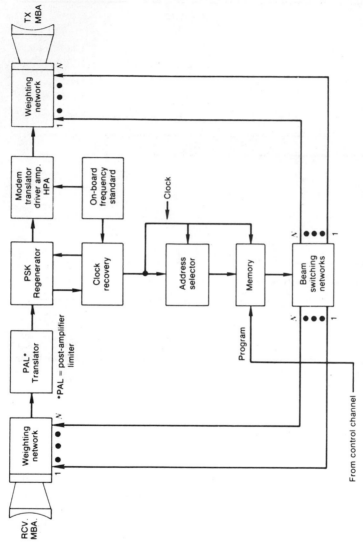

Figure 6.19 Block diagram, beam-switching processor.

must be carefully analyzed. Since there is a limit to the speed at which the antenna beam can be switched as the data symbol rate increases, guard times become an increasing fraction of the message frame, resulting in a loss of efficiency. This can be offset somewhat by the use of longer frames, but this requires increased buffering and increases the transmission time.

The scheme becomes very inefficient if a large proportion of the traffic is to be broadcast to many zones. SS/TDMA is not applicable to channelized satellites where multiple transponders share common transmit and receive antennas and the signals in each transponder cannot be synchronized for transmission between common terminal zones.

Figure 6.19 is a block diagram of a beam-switching processor in which narrow scanning beams are implemented by the use of a processor-controlled multiple-beam antenna (MBA) for both uplinks and downlinks. The output of the address selector is modified by the memory to provide control signals to the beam-switching network, which, in turn, controls the antenna-weighting networks. This process steers the uplink and downlink antennas to the selected zones for a time interval that matches the time interval of a data

Table 6.6   Comparison of switch matrix architectures

Description	Advantages	Disadvantages
Fan-out/fan-in	Broadcast mode capability	High-input VSWR High insertion loss Redundancy difficult to implement
Single pole/multiple throw	Low insertion loss	Poor reliability
Rearrangeable switch	Low insertion loss	Poor reliability Random interruptions Control algorithm complicated
Coupler crossbar	Planar construction (minimum size, weight, and volume) Broadcast mode capability Redundancy easy to implement Good input/output VSWR[a] Signal output level independent of the path Enhanced reliability	Difficult feedthroughs Difficult broadbanding Isolation hard to maintain

Source: Ref. 13.

[a]VSWR = voltage standing wave ratio.

frame to be exchanged between the selected zones. Initial synchronization of the system is obtained from a synchronizing signal transmitted to the processor from the TDMA master station. The memory can be updated at any time by signals from the master terminal via a control channel. Often, in practice, two memories are provided so that one can be updated while the other is on-line, thus preserving traffic continuity.

6.6.3 IF Switching

When the principal requirement for switching is for traffic routing rather then just for antenna selection, one method is to convert all uplink signals to a common IF, followed by a switching matrix and upconverting translators.

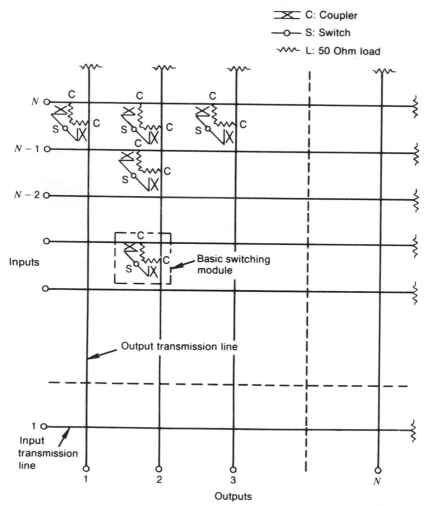

Figure 6.20 Coupler crossbar switching matrix for satellite IF switching. (From Ref. 13.)

This provides greater switching capability and flexibility. It allows for a wider choice of switching components in the design, since many more device types show good performance at the lower IF frequencies than at the higher uplink and downlink RF frequencies. Table 6.6 lists four switch architectures and their advantages and disadvantages. Of these, Ref. 13 states that the coupler crossbar offers the best performance for a large switching matrix (e.g., on the order of 20×20—20 inlets, 20 outlets). Figure 6.20 is a block diagram of such a crossbar matrix. In this case the crosspoints could be PIN diodes or dual-gate field effect transistors (FETs).

6.6.4 Intersatellite Links

Carrying the on-board processing concept still further, we now consider intersatellite links that can greatly extend the coverage area of a network. An intersatellite link (ISL) is a full-duplex link between two satellites, and other similar links can be added, providing intersatellite connectivity among satellites in a large constellation. A number of military satellite constellations (e.g., MILSTAR) have cross-linking capability. The 58–62-GHz band has been assigned for this purpose by the ITU. Laser cross-links are actively under consideration by the U.S. Department of Defense.

An ISL capability can have a significant impact on system design. The following three system design characteristics are most directly affected:

1. *Connectivity.* The ISL can be used to provide connectivity among users served by regional satellites, still retaining the required level of interconnectivity within a community of users. It is assumed that each satellite in the system will serve a region with a high intraregion community of interest. The community of interest among regions will be lower. Thus each satellite will handle the high-traffic-intensity intraregion, and cross-links will serve as tandem relays for the lower intensity traffic between regions.

2. *Capacity.* As described previously, uplinks and downlinks will have a high fill factor, and with cross-links implemented, low-traffic-intensity uplinks and downlinks will not be required to serve other regions with low interregional community of interest.

3. *Coverage.* Many satellite systems are designed for worldwide coverage where users of one satellite footprint require connectivity to users not in view of that satellite. The use of cross-links eliminates the need of earth relay at a dual-antenna earth station. The cross-link saves money and reduces propagation delay. Generally, an ISL is more economic than adding the additional uplink and downlink to the next satellite.

The cross-link concept incurs other advantages:

- Not affected by climatic conditions (e.g., the link does not pass through the atmosphere, assuming geostationary satellites)
- Low antenna noise temperature
- Low probability of earth-based intercept

- For military systems using the 60-GHz band, there is a low probability of ground-based jamming owing to satellite antenna discrimination and the high atmospheric absorption in the 60-GHz band

An intersatellite link (ISL) or cross-link system consists of four subsystems: receiver, transmitter, antenna or lens subsystem, and acquisition and tracking subsystem. The subsystems perform the following functions (Ref. 13):

1. *Receiver*. The receiver, in the generic case, detects, demodulates, and decodes the received signal. It interfaces the uplink/downlink system by cross-strapping at baseband. In some implementations this may also require format/protocol conversion. In a primitive cross-link system there may be no decoding and minimal format conversion.

2. *Transmitter*. In the general case, the transmitter selects the traffic for cross-linking. This may be accomplished by detecting a unique header word for routing. The signal for cross-linking is then encoded, which, in turn, modulates a carrier (or a laser); the signal is then up-converted and amplified. In a primitive system, there may not be any coding step.

3. *Antenna/Lens Subsystem*. For conventional RF transmission a suitable dish or lens antenna is required to radiate the transmitted signal and receive the incoming signal. For an optical system, a suitable lens or mirror, or combination, would be required to direct the laser signal.

4. *Tracking and Acquisition*. The two satellites involved, whether in a geostationary or nongeostationary orbit, require an antenna system that acquires and tracks to an appropriate accuracy, usually specified to some fraction of a beamwidth. Such techniques as raster scan and monopulse can be used for tracking. Initial pointing may be ground controlled from the TT&C* or master station(s). It should be noted that it is not necessary to orient the entire spacecraft to these accuracies, but only the ISL antenna reflector or feed.

6.7 CODING AND CODING GAIN

6.7.1 Introduction

Error rate is a principal design factor for digital transmission systems. PCM speech systems can tolerate a BER as degraded as 10^{-2} for intelligibility, and CVSD (continuous variable-slope delta modulation) remains intelligible with BERs of nearly 10^{-1}. For conventional speech telephony, the gating error rate value is determined by supervisory signaling. This value of BER should be better than 1×10^{-3}. CCITT gives a value of 1×10^{-4} for telegraph/telex traffic end-to-end and 1×10^{-6} for computer data. Designers of transmission

---

*TT&C = telemetry, tracking, and command (the satellite control subsystem).

systems for computer networks will differ with this latter value, and BER values of 10^{-7} end-to-end or better may be the performance values required for the transmission of computer data.

To design a transmission system to meet a specified error performance, we must first consider the cause of errors. Let us assume that intersymbol interference is negligible. Then errors derive from insufficient signal-to-noise ratio, which results in bit mutilation by thermal noise peaks (additive white Gaussian noise, or AWGN). Error rate is a function of signal-to-noise ratio or E_b/N_0. Obviously, we can achieve a desired BER on a link by specifying a very high E_b/N_0. On many satellite and tropospheric scatter links, this may not be economically feasible.

Errors derive from insufficient E_b/N_0 because of inferior design, equipment deterioration, or fading. On unfaded links or during unfaded conditions, these errors are random in nature, and, during fading, errors are predominantly bursty in nature. The length of the burst can be related to the fade duration.

There are several tools available to the design engineer to achieve a desired link error rate. Obviously, the first is to specify sufficient E_b/N_0 for the waveform selected, adding a margin for link deterioration due to equipment aging. We can use a similar approach for fading as described in Chapter 4.

Another approach is to specify a lower E_b/N_0 for a BER say in the range of 1×10^{-4} and implement automatic repeat request (ARQ). This is usually done on an end-to-end or a section-by-section basis when multilayer protocols are implemented. With ARQ, data messages are built up at the originating end on a block or packet basis. Each block or packet has appended a "block check count" or parity tail. This tail is generated at the originating end of the link by a processor that determines the parity characteristics of the message of uses a cyclic redundancy check (CRC), and the tail is the remainder of that check, often 16 bits in length. At the receiving end a similar processing technique is used, and the locally derived remainder is compared to the remainder received from the distant end. If they are the same, the message is said to be error free, and if not, the block or packet is in error. (See Chapter 12.)

There are several ARQ implementations. One is called stop-and-wait ARQ. In this case, if the block is error free, the receiver transmits an ACK (acknowledgment) to the transmitting end, which, in turn, sends the next block or packet. If the block is in error, the receiver sends a NACK (negative acknowledgment) to the transmitter, which now repeats the block just sent. It continues to repeat it until the ACK signal is received.

A second type of ARQ is called "continuous ARQ." At the transmit end, in this case, sending of blocks is continuous. There is also accounting information or sequential block or packet numbering in each block header. Similar parity checking is carried out as before. When the receive end encounters a block in error, it identifies the errored block to the transmitter, which intersperses the repeated blocks (with proper identifier) with its regular, continuous transmissions. Obviously, continuous ARQ is a more complex implementation than stop-and-wait ARQ. Considerably more processing and buffer memory are required at each end. It also follows that the more circuit delay, the more memory is required.

On satellite circuits carrying data, delay is an important design consideration. The delay from earth to a geostationary satellite is 125 ms or somewhat more. On a stop-and-wait ARQ system, considering the satellite link only, the period of "wait" is 4×125 ms, or 0.5 s. This is wasted time and inefficient. Continuous or go-back-N ARQ does not waste this time, but is more expensive to implement.

Another approach is to use error-correction coding. It offers a number of advantages but has at least one major disadvantage. It is up to the system designer to consider system overbuild, ARQ implementations, and error correction in implementing the optimum error-control technique.

6.7.2 Basic Forward Error Correction (FEC)

Forward error correction (FEC) is a method of error control that employs the adding of systematic redundancy at the transmit end of a link such that errors caused by the medium can be corrected at the receiver by means of a decoding algorithm.

Figure 6.21 shows a digital communication system with FEC. The binary data source generates information bits at R_s bits per second. These information bits are encoded for FEC at a code rate R. The output of the encoder is a binary sequence at R_c symbols per second. This output is related to the bit rate by the following expression:

$$R_c = \frac{R_s}{R} \tag{6.30}$$

Code rate R is the ratio of the number of information bits to the number of encoded symbols for binary transmission. For example, if the information bit rate were 2400 bps and code rate were $\frac{1}{2}$, then the symbol rate (R_c) would be 4800 symbols per second.

The encoder output sequence is then modulated and transmitted over the transmission medium or channel. Demodulation is then performed at the receive end. The output of the demodulator is R_c symbols per second, which is fed to the decoder. The decoder output to the data sink is the recovered 2400 bits per second (R_s).

Figure 6.21   Simplified diagram of the FEC technique.

The major advantages of an FEC system are

- No feedback channel required as with the ARQ system
- Constant information throughput (e.g., no stop-and-wait gaps)
- Decoding delay is generally small and constant
- Significant coding gain for an AWGN channel

There are two basic disadvantages to an FEC system. To effect FEC, for a fixed information bit rate, the bandwidth must be increased, because, by definition, the symbol rate on the transmission channel is greater than the information bit rate. There is also the cost of the added complexity of the coder and decoder.

6.7.3 Coding Gain

For a given information bit rate and modulation waveform (i.e., QPSK, 8-ary FSK, etc.), the required E_b/N_0 for a specified BER with FEC is less than the E_b/N_0 required without FEC. The coding gain is the difference in E_b/N_0 values.

The use of a selected FEC method in satellite communication can effect major savings simply by adding processors at each end of a link. There is no reason why FEC cannot be used on other digital systems where from 1 to 6 dB of additional gain is required under unfaded conditions. FEC with interleaving can show an even greater improvement under fading conditions.

Consider a satellite downlink where, by coding, the satellite EIRP can be reduced by half (e.g., 3 dB) without affecting performance. This could allow the use of a satellite transmitter with half the output power that would be required without FEC implemented. Transponder transmitters are a major weight factor in satellites. Reducing the output power by half could possibly reduce the transmitter weight by 75% (including its power supply). Battery weight may be reduced by perhaps 50% (batteries are used to power the transponder during satellite eclipse), with the concurrent reduction of solar cells. It is not only the savings in the direct cost of these items, but also the savings in lifting the weight of the satellite to place it in orbit, whether by space shuttle or rocket booster.

With unfaded conditions, assuming random errors, coding gains from 1 to 6 dB or more can be realized. The amount of gain achievable under these conditions is a function of the modulation type (waveform), the code employed, the coding rate (equation 6.30), the constraint length, the type of decoder, and the demodulation approach.

An FEC system can use one of two broad classes of codes: block and convolutional. These are briefly described below.

6.7.4   FEC Codes

6.7.4.1   *Block Codes*

With block-coding techniques each group of K consecutive information bits is encoded into a group of N symbols for transmission over the channel. Normally, the K information bits are located at the beginning of the N-symbol block code, and the last $N - K$ symbols correspond to the parity check bits formed by taking the modulo-2 sum of certain sets of K information bits. Block codes containing this property are referred to as systematic codes. The encoded symbols for the $(K + 1)$th information bit and beyond are completely independent of the symbols generated for the first K information bits and, hence, cannot be used to help decode the first group of K information bits at the far-end decoder. This

Table 6.7   Primitive BCH codes of lengths 255 and less

N	K	t	N	K	t	N	K	t
7	4	1	127	64	10	255	87	26
				57	11		79	
15	11	1		50	13		71	29
	7	2		43	14		63	30
				36	15		55	31
31	26	1		29	21		47	42
	21	2		22	23		45	43
	16	3		15	27		37	45
	11	5		8	31		29	47
	6	7	255	247	1		21	55
63	57	1		239	2		13	59
	51	2		231	3		9	63
	45	3		223	4			
	39	4		215	5			
	36	5		207	6			
	30	6		199	7			
	24	7		191	8			
	18	10		187	9			
	16	11		179	10			
	10	13		171	11			
	7	15		163	12			
				155	13			
127	120	1		147	14			
	113	2		139	15			
	106	3		131	18			
	99	4		123	19			
	92	5		115	21			
	85	6		107	22			
	78	7		99	23			
	71	9		99	25			

Source: Ref. 13.

essentially says that blocks are independent entities, and one block has no enhancement capability on another.

Because N symbols are used to represent K bits, the code rate R of such a block code is K/N bits per symbol, or

$$R = \frac{K}{N} \tag{6.31}$$

Notice that equation 6.31 is just a restatement of equation 6.30, but with different notation.

Block codes are often described with the notation such as $(7, 4)$, meaning $N = 7$ and $K = 4$. Here the information bits are stored in $K = 4$ storage devices and the device is made to shift $N = 7$ times. The first K symbols of the block output are the information symbols, and the last $N - K$ symbols are a set of check symbols that form the whole N-symbol word. A block code may also be identified with the notation (N, K, t), where t corresponds to the number of errors in a block of N symbols that the code will correct.

6.7.4.2 Bose–Chaudhuri–Hocquenghem (BCH) Codes
Binary BCH codes are a large class of block codes with a wide range of code parameters. The so-called primitive BCH codes, which are the most common, have code word lengths of the form $2^m - 1$, $m \geq 3$, where m describes the "degree" of the generating polynomial (e.g., the highest exponent value of the polynomial). For BCH codes, there is no simple expression relating to the N, K, and t parameters. Table 6.7 gives these parameters for all binary BCH codes of lengths 255 and less. In general, for any m and t, there is a BCH code of length $2^m - 1$ that corrects any combination of t errors and requires no more than m/t parity check symbols.

BCH codes are cyclic and are characterized by a generating polynomial. A selected group of primitive generating polynomials is given in Table 6.8. The encoding of a BCH code can be performed with a feedback shift register of length K or $N - K$ (Ref. 13).

6.7.4.3 Golay Code
The Golay $(23, 12)$ code is a linear cyclic binary block code that is capable of correcting up to three errors in a block of 23 binary symbols. There are two possible generator polynomials for the Golay code as follows:

$$g_1(x) = 1 + x^2 + x^4 + x^5 + x^6 + x^{10} + x^{11} \tag{6.32}$$

$$g_2(x) = 1 + x + x^5 + x^6 + x^7 + x^9 + x^{11} \tag{6.33}$$

The encoding of the Golay code can be performed with an 11-stage feedback shift register with feedback connections determined by the coefficients of either $g_1(x)$ or $g_2(x)$. There are efficient decoding techniques available that

Table 6.8   Selected primitive binary polynomials

m	$p(x)$
3	$1 + x + x^3$
4	$1 + x + x^4$
5	$1 + x^2 + x^5$
6	$1 + x + x^6$
7	$1 + x^3 + x^7$
8	$1 + x^2 + x^3 + x^4 + x^8$
9	$1 + x^4 + x^9$
10	$1 + x^3 + x^{10}$
11	$1 + x^2 + x^{11}$
12	$1 + x + x^4 + x^6 + x^{12}$
13	$1 + x + x^3 + x^4 + x^{13}$
14	$1 + x + x^6 + x^{10} + x^{14}$
15	$1 + x + x^{15}$

Source: Ref. 13.

are implemented in hardware. It should be noted that often an overall parity check bit is appended to each codeword.

6.7.4.4 *Convolutional Codes*   Viterbi (Ref. 14) defines a convolutional encoder as

> a linear finite state machine consisting of a K-stage shift register and n linear algebraic function generators. The input data which are usually, but not necessarily always, binary, are shifted along the register b bits at a time.

Figure 6.22 is an example of a convolutional encoder. If there were a five-stage shift register where the input data were shifted along 1 bit at a time and we had three modulo-2 adders (e.g., $n = 3$), using the Viterbi notation, the code would be described as a 5, 3, 1 convolutional code (e.g., $K = 5$, $n = 3$, $b = 1$).

In Figure 6.22 information bits are shifted to the right 1 bit at a time ($b = 1$) through the K-stage shift register as new information bits enter from the left. Bits out of the last stage are discarded. The bits are shifted one position each T s, where $1/T$ is the information rate in bits per second. The modulo-2 adders are used to form the output coded symbols, each of which is a binary function of a particular subset of the information bits in the shift register. The output coded symbols can be seen to depend on a sequence of K information bits, and hence we define the constraint length K. The constraint length K is defined as the total number of binary register stages in the encoder.

If we feed in 1 bit at a time to the encoder (e.g., $b = 1$), each coded symbol carries an average of $1/n$ information bits, and the code is said to

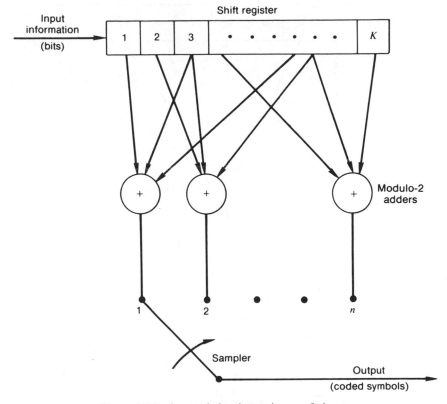

Figure 6.22   A convolutional encoder, $n = 3$, $b = 1$.

have a rate R of $1/n$. For the more generalized case where $b \neq 1$, the rate R is expressed as

$$R = \frac{b}{n} \tag{6.34}$$

where b = number of bits shifted into the register at a time. (See equation 6.31, and the similarity.)

In Figure 6.22, when the first modulo-2 adder is replaced by a direct connection to the first stage of the shift register, the first symbol becomes a replica of the information bit. Such an encoder is called a systematic convolutional encoder, as shown in Figure 6.23.

Let us examine a rate $\frac{1}{2}$ constraint length $K = 3$ encoder shown in Figure 6.24. This figure indicates the outputs for a particular binary input sequence assuming the state (i.e., the previous 2 data bits in the register) were zero. Modulo-2 addition (e.g., $0 \oplus 0 = 0, 0 \oplus 1 = 1, 1 \oplus 0 = 1$, and $1 \oplus 1 = 0$) is used. With the input and output sequences defined from right to left, the first

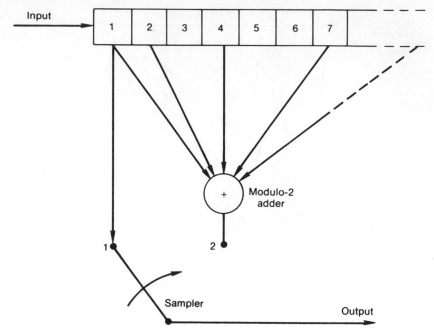

Figure 6.23 A systematic convolutional encoder.

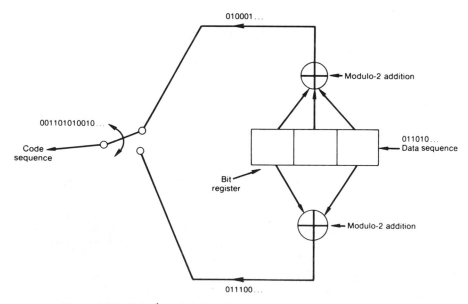

Figure 6.24 Rate $\frac{1}{2}$ convolutional encoder with constraint length $K = 3$.

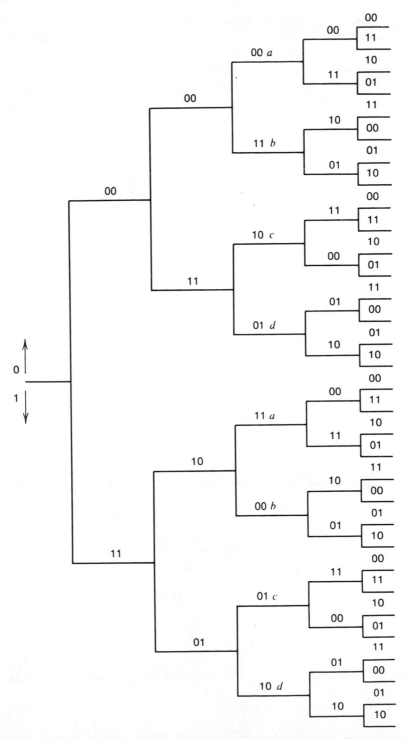

Figure 6.25 Tree representation for the encoder shown in Figure 6.24. (From Ref. 15.)

3 input bits—0, 1 and 1—generate the code outputs 00, 11, and 01, respectively. The outputs are then multiplexed (commutated) into a single code sequence. For this rate $\frac{1}{2}$ case, of course, the output code sequence has twice the symbol rate as the incoming data sequence.

A convolutional code is often shown by means of a tree diagram, as shown in Figure 6.25. At each branch (node) of the tree the input information bit determines which direction (i.e., which branch) will be taken, following the convention "up" for zero and "down" for one. Restated, if the first input bit is a zero, the code symbols are those shown on the first upper branch, while if it is a one, then the output symbols are those shown on the first lower branch. Similarly, if the second input bit is a zero, we trace the tree diagram to the next upper branch, while if it is a one, we trace the diagram downward. In this manner all 32 possible outputs for the first five inputs may be traced.

From the tree diagram in Figure 6.25 it also becomes clear that after the first three branches the structure becomes repetitive. In fact, we readily recognize that beyond the third branch the code symbols on branches emanating from the two nodes labeled "*a*" are identical, and soon, for all the similarly labeled pairs of nodes. The reason for this is obvious from examination of the encoder. As the fourth bit enters the coder at the right, the first data bit falls off on the left and no longer influences the output code symbols. Consequently, the data sequences $100xy\ldots$ and $000xy\ldots$ generate the same code symbols after the third branch and, as is shown in the tree diagram, both nodes labeled "*a*" can be joined together.

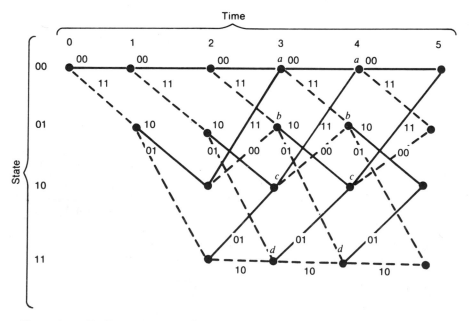

Figure 6.26   Trellis representation for the encoder shown in Figure 6.24. (From Ref. 15.)

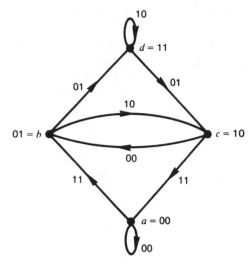

Figure 6.27 State diagram representation for the encoder shown in Figure 6.24. (From Ref. 15.)

This leads to redrawing the tree diagram as shown in Figure 6.26. This is called a trellis diagram, since a trellis is a treelike structure with remerging branches. A convention is adopted here where the code branches produced by a zero input bit are shown as solid lines and code branches produced by a one input bit are shown as dashed lines.

The completely repetitive structure of the trellis diagram suggests a further reduction in the representation of the code to the state diagram in Figure 6.27. The "states" of the state diagram are labeled according to the nodes of the trellis diagram. However, since the states correspond merely to the last 2 input bits to the coder, we may use these bits to denote the nodes or states of this diagram.

It can be observed that the state diagram can be drawn directly observing the finite-state machine properties of the encoder and, particularly, the fact that a four-state directed graph can be used to represent uniquely the input–output relation of the finite-state machine, since the nodes represent the previous 2 bits, while the present bit is indicated by the transition branch. For example, if the encoder (synonomous with finite-state machine) contains the sequence 011, this is represented in the diagram by the transition from state $b = 01$ to state $d = 11$ and the corresponding branch indicates the code symbol outputs 01.

This example will be used when we describe the Viterbi decoder.

6.7.4.5 Convolutional Decoding

Decoding algorithms for block and convolutional codes are quite different. Because a block code has a formal structure, advantage can be taken of the known structural properties of the words or the algebraic nature of the constraints among the symbols used to represent an information sequence. One class of powerful block codes with well-defined decoding algorithms is the BCH codes (Section 6.7.4.2).

The decoding of convolutional codes can be carried out by a number of different techniques. Among these are the simpler threshold and feedback decoders and those more complex decoders with improved performance (coding gain) such as the Viterbi decoder and sequential decoder. These techniques depend on the ability to home in on the correct sequence by designing efficient search procedures that discard unlikely sequences quickly. The sequential decoder differs from most other types of decoders in that when it finds itself on the wrong path in the tree, it has the ability to search back and forth, changing previously decoded information bits until it finds the correct tree path. The frequency with which the decoder has to search back and the depth of the backward searches are dependent on the value of the channel BER.

An important property of the sequential decoder is that, if the constraint length is large enough, the probability that the decoder will make an error approaches zero (i.e., a BER better than 1×10^{-9}). One cause of error is overflow, being defined as the situation in which the decoder is unable to perform the necessary number of computations in the performance of the tree search. If we define a computation as a complete examination of a path through the decoding tree, a decoder has a limit on the number of computations that it can make per unit time. The number of searches and computations is a function of the number of errors arriving at the decoder input, and the number of computations that must be made to decode 1 information bit is a random variable. An important parameter for a decoder is the average number of computations per decoded information bit. As long as the probability of bit error is not too high, the chances of decoder overflow will be low, and satisfactory performance will result.

For the previous discussion it has been assumed that the output of a demodulator has been a hard decision. By "hard" decision we mean a firm, irrevocable decision. If these were soft decisions instead of hard decisions, additional improvement in error performance (or coding gain) on the order of several decibels could be obtained. By a "soft" decision we mean that the output of a demodulator is quantized into four or eight levels (e.g., 2- or 3-bit quantization respectively), and then certain decoding algorithms can use this additional information to improve the output BER. Sequential and Viterbi decoding algorithms can use this soft decision information effectively, giving them an advantage over algebraic decoding techniques, which are not designed to handle the additional information provided by the soft decision (Ref. 15).

The soft decision level of quantization is indicated conventionally by the letter Q, which indicates the number of bits in the quantized decision sample. If $Q = 1$, we are dealing with a hard decision demodulator output; $Q = 2$ indicates a quantization level of 4; $Q = 3$ a quantization level of 8; and so forth.

6.7.4.6 Viterbi Decoding The Viterbi decoder is one of the more common decoders on links using convolutional codes. The Viterbi decoding algorithm is a path maximum-likelihood algorithm that takes advantage of the remerging path structure (see Figure 6.26) of convolutional codes. By path maximum-likelihood decoding, we mean that of all the possible paths through the trellis, a Viterbi decoder chooses the path, or one of the paths, most likely in the probabilistic sense to have been transmitted. A brief description of the operation of a Viterbi decoder using a demodulator giving hard decisions is now given.

For this description our model will be a binary symmetric channel (i.e., BPSK, BFSK). Errors that transform a channel code symbol 0 to 1 or 1 to 0 are assumed to occur independently from symbols with a probability of p. If all input (message) sequences are equally likely, the decoder that minimizes the overall path error probability for any code, block, or convolutional is one that examines the error-corrupted received sequence, which we may call $y_1, y_2, \ldots, y_j, \ldots$, and chooses the data sequence corresponding to that sequence that was transmitted or $x_1, x_2, \ldots, x_j \ldots$ which is closest to the received sequence as measured by the Hamming distance. The Hamming distance can be defined as the transmitted sequence that differs from the received sequence by the minimum number of symbols.

Consider the tree diagram (typically Figure 6.25). The preceding statement tells us that the path to be selected in the tree is the one whose code sequence differs in the minimum number of symbols from the received sequence. In the derived trellis diagram (Figure 6.26) it was shown that the transmitted code branches remerge continually. Thus the choice of possible

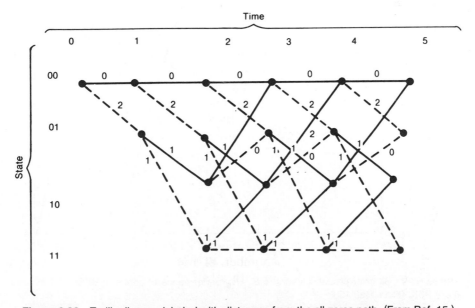

Figure 6.28 Trellis diagram labeled with distances from the all-zeros path. (From Ref. 15.)

paths can be limited in the trellis diagram. It is also unnecessary to consider the entire received sequence at any one time to decide upon the most likely transmitted sequence or minimum distance. In particular, immediately after the third branch (Figure 6.26) we may determine which of the two paths leading to node or state *a* is more likely to have been sent. For example, if 010001 is received, it is clear that this is a Hamming distance 2 from 000000, while it is a distance 3 from 111011. As a consequence, we may exclude the

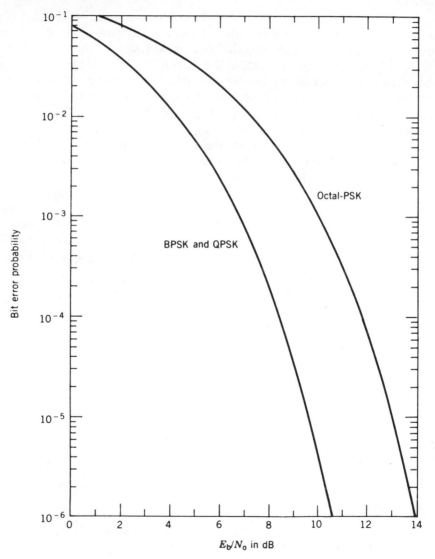

Figure 6.29 Bit error probability versus E_b/N_0 performance of coherent BPSK, QPSK, and 8-ary PSK. (From Ref. 15.)

lower path into node *a*. For no matter what the subsequent received symbols will be, they will affect the Hamming distances only over subsequent branches after these two paths have remerged and, consequently, in exactly the same way. The same can be said for pairs of paths merging at the other three nodes after the third branch. Often, in the literature, the minimum distance path of the two paths merging at a given node is called the "survivor." Only

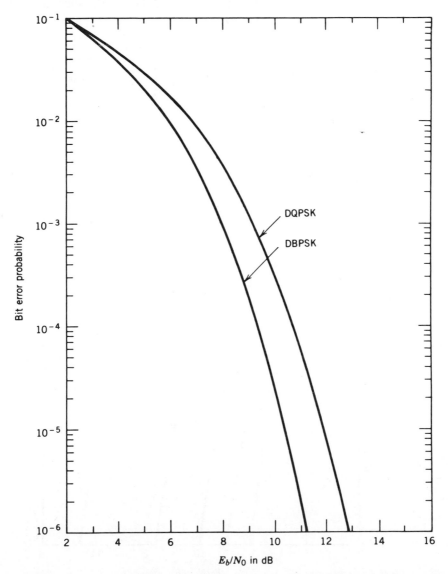

Figure 6.30 Bit error probability versus E_b/N_0 performance of DBPSK and DQPSK. (From Ref. 15.)

two things have to be remembered: the minimum distance path from the received sequence (or survivor) at each node and the value of that minimum distance. This is necessary because at the next node level we must compare two branches merging at each node level, which were survivors at the previous level for different nodes. This can be seen in Figure 6.26 where the comparison at node a after the fourth branch is among the survivors of the comparison of nodes a and c after the third branch. For example, if received sequence over the first four branches is 01000111, the survivor at the third node level for node a is 000000 with distance 2 and at node c it is

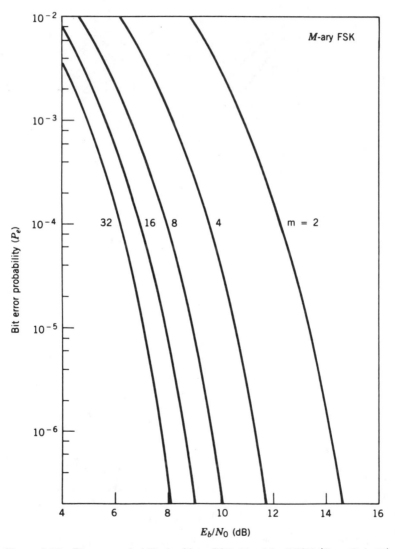

Figure 6.31 Bit error probability for M-ary FSK; $M = 2$ for BFSK. (From Ref. 13.)

110101, also with distance 2. In going from the third node level to the fourth, the received sequence agrees precisely with the survivor from c but has distance 2 from the survivor from a. Hence the survivor at node a of the fourth level is the data sequence 1100 that produced the code sequence 11010111, which is at minimum distance 2 from the received sequence.

In this way we may proceed through the received sequence and at each step preserve one surviving path and its distance from the received sequence,

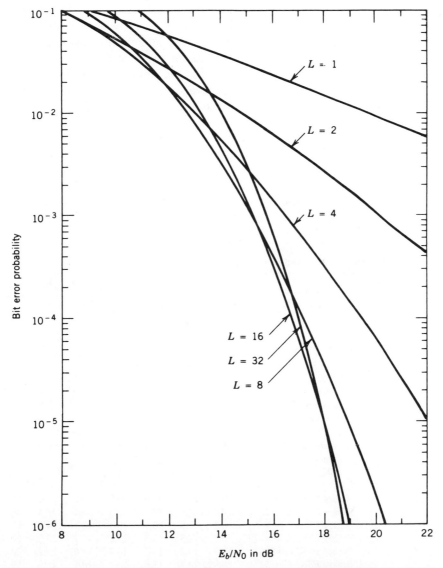

Figure 6.32 Bit error probability versus $\bar{E}_b / N_0$ performance of binary FSK on a Rayleigh fading channel for several orders of diversity. $L =$ order of diversity. (From Ref. 15.)

which is more generally called a "metric." The only difficulty that may arise is the possibility that, in a given comparison between merging paths, the distances or metrics are identical. In this case we may simply flip a coin, as is done for block code words at equal distances from the received sequence. For even if both equally valid contenders were preserved, further received symbols would affect both metrics in exactly the same way and thus not further influence our choice.

Another approach to the description of the algorithm can be obtained from the state diagram representation given in Figure 6.27. Suppose we sought that path around the directed state diagram arriving at node a after the kth transition, whose code symbols are at a minimum distance from the received sequence. But, clearly, this minimum distance path to node a at time k can only be one of two candidates: the minimum distance path to node a at time $k - 1$ and the minimum distance path to node c at time $k - 1$. The comparison is performed by adding the new distance accumulated in the kth transition by each of these paths to their minimum distances (metrics) at time $k - 1$.

Therefore it appears that the state diagram also represents a system diagram for this decoder. With each node or state, we associate a storage register that remembers the minimum distance path into the state after each transition as well as a metric register that remembers its (minimum) distance from the received sequence. Furthermore, comparisons are made at each

Table 6.9 Summary of uncoded system performance

Channel	Modulation/ Demodulation	E_b/N_0 (dB) Required for Given BER						
		10^{-1}	10^{-2}	10^{-3}	10^{-4}	10^{-5}	10^{-6}	10^{-7}
Additive white Gaussian noise (AWGN)	BPSK and QPSK	−0.8	4.3	6.8	8.4	9.6	10.5	11.3
	Octal-PSK	1.0	7.3	10.0	11.7	13.0	13.9	14.7
	DBPSK	2.1	5.9	7.9	9.3	10.3	11.2	11.9
	DQPSK	2.1	6.8	9.2	10.8	12.0	12.9	13.6
	Noncoherently demodulated binary FSK	5.1	8.9	10.9	12.3	13.4	14.2	14.9
	Noncoherently demodulated 8-ary FSK	2.0	5.2	7.0	8.2	9.1	9.9	10.5
Independent Rayleigh fading	Binary FSK, $L = 1$	9.0	19.9	30.0	40.0	50.0	60.0	70.0
	Binary FSK, $L = 2$	7.9	14.8	20.2	25.3	30.4	35.4	40.4
	Binary FSK, $L = 4$	8.1	13.0	16.5	19.4	22.1	24.8	27.3
	Binary FSK, $L = 8$	8.7	12.8	15.3	17.2	18.9	20.5	22.0
	Binary FSK, $L = 16$	9.7	13.2	15.3	16.7	18.0	19.1	20.0
	Binary FSK, $L = 32$	10.9	14.1	15.8	17.1	18.1	18.9	19.7

Source: Ref. 15.

step between the two paths that lead into each node. Consequently, four comparators must also be provided.

We will expand somewhat on the distance properties of convolutional codes following the example given in Figure 6.24. It should be noted that as with linear block codes, there is no loss in generality in computing the distance from the all-zeros code word to all other code words, for this set of

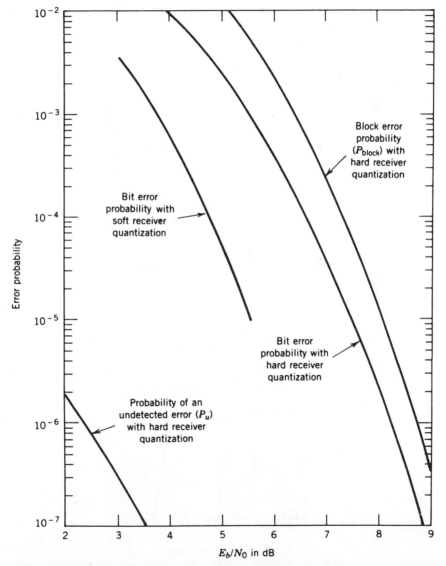

Figure 6.33 Block, bit, and undetected error probabilities versus E_b / N_0 for BPSK and QPSK modulation using extended Golay coding. AWGN channel assumed. (From Ref. 15.)

distances is the same as the set of distances from any specific code word to all the others.

For this purpose we may again use either the trellis diagram, or the state diagram. First of all, we redraw the trellis diagram in Figure 6.26 labeling the branches according to their distances from the all-zeros path. Now consider all the paths that merge with the all-zeros path for the first time at some arbitrary node j. From the redrawn trellis diagram (Figure 6.28), it can be seen that of these paths there will be just one path at distance 5 from the

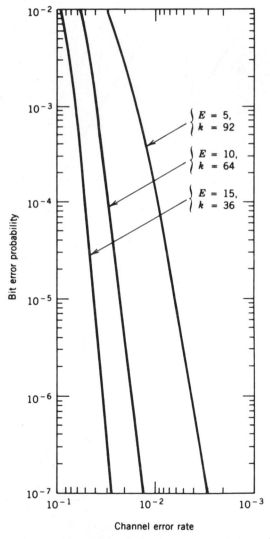

Figure 6.34 Decoded bit error probability versus channel (uncoded) error rate performance of several BCH codes with block length 127. (From Ref. 15.)

all-zeros path, and this diverged from it three branches back. Similarly, there are two at distance 6 from it, one that diverged four branches back and the other that diverged five branches back, and so fourth. It should be noted that the input bits for the distance 5 path are 00...01000 and thus differ in only 1 input bit from the all-zero path. The minimum distance, sometimes called the "minimum free distance," among all paths is therefore seen to be 5. This implies that any pair of channel errors can be corrected, for two errors will cause the received sequence to be at a distance 2 from the transmitted (correct) sequence, but it will be at least at distance 3 from any other possible code sequence. In this way the distances of all paths from an all-zeros (or any arbitrary) path can be determined from the trellis diagram.

6.7.5 Channel Performance of Uncoded and Coded Systems

For uncoded systems a number of modulation implementations are reviewed in the presence of additive white Gaussian noise (AWGN) and with Rayleigh fading. The AWGN performance of BPSK, QPSK, and 8-ary PSK is shown in Figure 6.29. AWGN is typified by thermal noise or wideband white noise jamming. The demodulator for this system requires a coherent phase reference.

Another similar implementation is differentially coherent phase shift keying. This is a method of obtaining a phase reference by using the previously

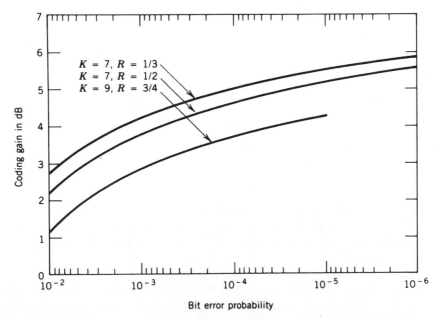

Figure 6.35 Coding gain for several convolutional codes with BPSK modulation, AWGN, and 3-bit receiver quantization. (From Ref. 15.)

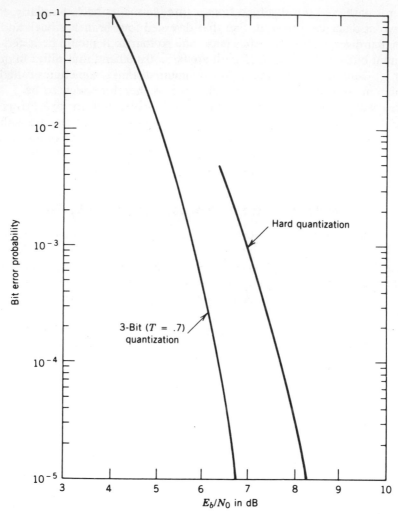

Figure 6.36 Bit error probability versus E_b / N_0 performance of a $K = 7$, $R = \frac{1}{2}$ convolutional coding system with DBPSK modulation and an AWGN channel. (From Ref. 15.)

received channel symbol. The demodulator makes its decision based on the change in phase from the previous to the present received channel symbol. Figure 6.30 gives the performance of DBPSK* and DQPSK* with values of BER versus E_b/N_0.

Figure 6.31 gives performance for M-ary FSK.

Independent Rayleigh fading can be assumed during periods of heavy rainfall on satellite links operating above about 10 GHz (see Chapter 7). Such

---

*D stands for differential.

fading can severely degrade error rate performance. The performance with this type of channel can be greatly improved by providing some type of diversity. Here we mean providing several independent transmissions for each information symbol. In this case we will restrict the meaning to some form of time diversity that can be achieved by repeating each information symbol several times and using interleaving/deinterleaving for the channel symbols. Figure 6.32 gives binary bit error probability for several orders of diversity (L = order of diversity; $L = 1$, no diversity) for the mean bit energy-to-noise ratio $\overline{E}_b/N_0$. This figure shows that for a particular error rate there is an optimum amount of diversity. The modulation is binary FSK.

Table 6.9 recaps error performance versus E_b/N_0 for the several modulation types considered. The reader should keep in mind that the values for E_b/N_0 are theoretical. A certain modulation implementation loss should be added for each case to derive practical values. The modulation implementation loss value in each case is equipment driven. (See Table 4.10.)

6.7.5.1 Theoretical Performance for Coded Systems

Figure 6.33 gives the theoretical performance for extended Golay (block) coding with

Table 6.10 Summary of the E_b / N_0 requirements and coding gains of $K = 7$, $R = \frac{1}{2}$ Viterbi-Decoded Convolutional Coding Systems with several modulation types at a BER of 10^{-5}

Modulation	Number of Bits of Receiver Quantization per Binary Channel Symbol (See Table 6.11 Note)	E_b/N_0 (dB) Required for $P_b = 10^{-5}$	Coding Gain (dB)
Coherent			
BPSK or QPSK	3	4.4	5.2
BPSK or QPSK	2	4.8	4.8
BPSK or QPSK	1	6.5	3.1
Octal-PSK[a]	1	9.3	3.7
DBPSK[a]	3	6.7	3.6
DBPSK[a]	1	8.2	2.1
Differentially[a] coherent QPSK	1	9.0	3.0
Noncoherently demodulated binary FSK	1	11.2	2.1

Source: Ref. 15.

[a]Interleaving/deinterleaving assumed.

Table 6.11 Summary of E_b/N_0 requirements of several Coded Communication Systems for a BER of 10^{-5} with BPSK modulation[a]

Coding Type		Number of Bits of Receiver Quantization	Coding[b] Gain (dB)
$K = 7,\ R = \frac{1}{2}$	Viterbi-decoded convolutional	1	3.1
$K = 7,\ R = \frac{1}{2}$	Viterbi-decoded convolutional	3	5.2
$K = 7,\ R = \frac{1}{3}$	Viterbi-decoded convolutional	1	3.6
$K = 7,\ R = \frac{1}{3}$	Viterbi-decoded convolutional	3	5.5
$K = 9,\ R = \frac{3}{4}$	Viterbi-decoded convolutional	1	2.4
$K = 9,\ R = \frac{3}{4}$	Viterbi-decoded convolutional	3	4.3
$K = 24, R = \frac{1}{2}$	Sequential-decoded convolutional 20 kbps,[c] 1000-bit blocks	1	4.2
$K = 24, R = \frac{1}{2}$	Sequential-decoded convolutional 20 kbps,[c] 1000-bit blocks	3	6.2
$K = 10, L = 11, R = \frac{1}{2}$	Feedback-decoded convolutional	1	2.1
$K = 8,\ \ L = 8, R = \frac{2}{3}$	Feedback-decoded convolutional	1	1.8
$K = 8,\ \ L = 9, R = \frac{3}{4}$	Feedback-decoded convolutional	1	2.0
$K = 3,\ \ L = 3, R = \frac{3}{4}$	Feedback-decoded convolutional	1	1.1
(24, 12) Golay		3	4.0
(24, 12) Golay		1	2.1
(127, 92) BCH		1	3.3
(127, 64) BCH		1	3.5
(127, 36) BCH		1	2.3
(7, 4) Hamming		1	0.6
(15, 11) Hamming		1	1.3
(31, 26) Hamming		1	1.6

Source: Ref. 15.

Key:

[a]K Constraint length of a convolutional code defined as the number of binary register stages in the encoder for such a code. With the Viterbi decoding algorithm, increasing the constraint length increases the coding gain but also the implementation complexity of the system. To a much lesser extent the same is also true with sequential and feedback decoding algorithms.

L Look-ahead length of a feedback-decoded convolutional coding system defined as the number of received symbols, expressed in terms of the corresponding number of encoder input bits, that are used to decode an information bit. Increasing the look-ahead length increases the coding gain but also the decoder implementation complexity.

(n, k) denotes a block code (Golay, BCH, or Hamming here) with n decoder output bits for each block of k encoder input bits.

Receiver
quantization Describes the degree of quantization of the demodulator outputs. Without coding and biphase ($0°$ or $180°$) modulation, the demodulator output (or intermediate output, if the quantizer is considered as part of the demodulator) is quantized to 1 bit (i.e., the sign if provided). With coding, a decoding decision is based on several demodulator outputs, and the performance can be improved if in addition to the sign the demodulator provides some magnitude information.

[b]9.6 dB required for uncoded system.

[c]The same system at a data rate of 100 kbps has 0.5 dB less coding gain.

BPSK/QPSK modulation. Figure 6.34 provides uncoded channel error rate versus BER for several BCH codes with block length of 127. k represents the largest number of information bits per block and E the number of channel errors that each code is capable of correcting.

Figure 6.35 gives the coding gain for several convolutional codes where the demodulator provides 3-bit quantizing. The modulation is BPSK. Figure 6.36 shows the performance of a rate $\frac{1}{2}$ convolutional code with $K = 7$; the modulation is DBPSK. T is a quantization parameter, usually some value of the standard deviation of the unquantized demodulator outputs. In this case $T = 0.5$ or 0.5 times the standard deviation.

Table 6.10 is a summary of E_b/N_0 requirements and coding gains of $K = 7$, rate $\frac{1}{2}$ Viterbi-decoded convolutional coding systems with several modulation types at a BER of 10^{-5}. Table 6.11 summarizes a number of FEC schemes with their respective coding gains.

6.8 LINK BUDGETS FOR DIGITAL LINKS ON PROCESSING SATELLITES

The link budget is a tool used to dimension or size a satellite communication system. In Section 6.4.8 link budgets for analog systems were described. There an uplink with its associated downlink was considered jointly (equation 6.25) to calculate system C/N. With digital systems utilizing a processing satellite, uplinks and downlinks are treated separately, calculating the required E_b/N_0 on each for a specified BER. Otherwise the approach is quite similar as that described in Section 6.4.8.

The uplink and downlink are just two more links of a larger system unless the user is collocated at the satellite terminal on each end. For the larger system, link BERs are usually specified in such a way that the user BER meets a need requirement.

Example 1. A specific uplink at 6 GHz working into a processing satellite is to have a BER of 1×10^{-5}. The modulation is QPSK and the data rate is 10 Mbps. The terminal EIRP is $+65$ dBW. The free-space loss to the satellite is 199.2 dB. What satellite G/T will be required without coding? What G/T will be required when FEC is employed? The satellite will use an earth coverage antenna. See Table 6.12.

Select the required E_b/N_0 first for no coding and then with convolutional coding rate $\frac{1}{2}$ and $K = 7$, Viterbi-decoded. Use a modulation implementation loss of 2.0 dB in both cases. From Figure 6.29, lower left curve, the E_b/N_0 is 9.6 dB for QPSK (uncoded), and from Table 6.10 for the coded system use 6.5 dB. Allow 4 dB of margin. Initially set the satellite G/T at 0 dB/K.

Table 6.12 Example 1, uplink power budget

Terminal EIRP	$+65$ dBW	
Terminal pointing loss	0.5 dB	
Free-space loss	199.2 dB	
Satellite pointing loss	0.0 dB	(earth coverage)
Polarization loss	0.5 dB	
Atmospheric losses	0.5 dB	
Rainfall (excess attenuation)	0.25 dB	(10° elevation angle)
Isotropic receive level	-135.95 dBW	
G/T	0.0 dB	
Sum	-135.95 dBW	
Boltzmann's constant	$-(-228.6$ dBW$)$	
C/N_0	92.65 dB	
$-10\log$(bit rate)	-70.00 dB	
E_b/N_0	$+22.65$ dB	
Required E_b/N_0	-9.6 dB	
Implementation loss	-2.0 dB	
Margin	11.05 dB	
Allowable margin	-4 dB	
Excess margin	7.05 dB	

The satellite G/T can be -7.05 dB for the uncoded system.
For the coded system with a 3.1-dB coding gain, the satellite G/T can be degraded to -10.15 dB $[-7.05$ dB $+(-3.1$ dB$)]$.

Example 2. A satellite has a $+30$ dBW EIRP downlink at 7.3 GHz. The desired BER is 1×10^{-6}; the modulation is coherent BPSK; FEC is implemented with convolutional encoding and 3-bit receiver quantization; the bit rate is 45 Mbps; and the free-space loss is 202.0 dB. What is the terminal G/T assuming a 5-dB margin? See Table 6.13. Assume that the satellite uses a spot beam, and a rate $\frac{1}{2}$, $K = 7$ convolutional code.

The required G/T is the sum of $26.18 + 4.9 + 2.0 + 5.0$ dB or $+38.08$ dB/K. If that value is now substituted for the initial G/T of 0 dB/K, the last entry in the table or "sum" would drop to 0. The value for the required E_b/N_0 was derived first from Figure 6.29, lower left curve, using BER $= 1 \times 10^{-6}$, thence to Figure 6.35 where we found that the coding gain for $K = 7$, $R = \frac{1}{2}$ is 5.6 dB, and we then subtracted. The rainfall loss value was, in a way, arbitrary. As we will show in Chapter 7, 3 dB in this band, at a 10° elevation angle, would provide this performance 99.9% of the time for central North America. However, with interleaving using a sufficient interleaving interval (about 4 or 5 s) and the coding employed, we can decrease the excess attenuation due to rainfall to zero. The value would be accommodated in the

Table 6.13   Example 2, downlink power budget

Satellite EIRP	+30 dBW
Satellite pointing loss	0.5 dB
Footprint error	0.25 dB
Off-footprint center loss	1.0 dB
Terminal pointing loss	0.5 dB
Polarization loss	0.5 dB
Atmospheric losses	0.5 dB
Free-space loss	202.0 dB
Rainfall loss	3.0 dB
Isotropic receive level	−178.25 dBW
Terminal G/T	0.0 dB/K
Sum	−178.25 dBW
Boltzmann's constant	−(−228.6) dBW
C/N_0	50.35 dB
10 log(bit rate)	−76.53 dB
Difference	−26.18 dB
Required	−4.9 dB
Modulation implementation loss	−2.0 dB
Margin	−5.0 dB
Sum	38.08 dB

coding gain. The coding gain used was for AWGN conditions only. It will also provide a fading improvement. Heavy rain manifests itself in fading, which, in the worst case, can be considered a Rayleigh distribution. The incidental losses (i.e., polarization, pointing errors) are good estimates and probably can be improved upon with a real system where firm values can be used.

6.9   VSAT (VERY SMALL APERTURE TERMINAL)

6.9.1   Introduction

A very small aperture terminal (VSAT) is a satellite terminal with a small antenna aperture, from 0.6 to 2.4 m (2 to about 8 ft) in diameter. The rationale for a system of VSATs is low cost.

VSATs have such modest performance that each VSAT in a network must access a large earth terminal called a *hub*. VSATs usually operate in a star network topology with the hub at the center. A simplified conceptual drawing of a VSAT star network is shown in Figure 6.37. VSAT systems operate in both the 6/4-GHz and the 14/12-GHz bands. However, the 14/12-GHz band is preferred because higher power satellite EIRPs are permitted.

The hub in a VSAT system compensates, in a manner of speaking, for the modest performance of distant-end VSATs under its control. Whereas typical

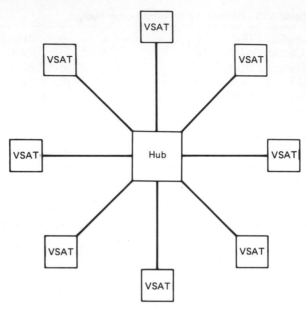

Figure 6.37   VSAT star network topology.

VSATs have comparatively low G/Ts and EIRPs, the hub has much more robust performance. Tables 6.14A–D compares typical performance of hubs and associated VSATs.

6.9.2   VSAT System Applications

We stated above that the basic rationale for implementing a VSAT system is that it is more economic for a given situation than utilizing some other form

Table 6.14A   Some typical VSAT G/T values

Antenna Aperture		Antenna Gains		G/T Values	
(m)	(ft)	G_4 (dB)	G_{12} (dB)	G/T_4	G/T_{12}
1.2	3.94	31.85	41.39	+9.0 dB/K	+15.5 dB/K
1.5	4.92	33.78	43.32	+11.0 dB/K	+17.5 dB/K
1.8	5.90	35.36	44.90	+12.56 dB/K	+19.1 dB/K
2.0	6.56	36.28	45.82	+13.5 dB/K	+20.0 dB/K
2.4	7.87	37.86	47.40	+15.1 dB/K	+21.6 dB/K

[a]*Assumptions*: Transmission line losses: 1 dB
        Sky noise: 15 K at 4 GHz; 180 K at 12 GHz
        Receiver noise temperature: 80 K at 4 GHz; 180 K at 12 GHz

Table 6.14B   VSAT hub G/T values[a]

Antenna Aperture		Antenna Gains		G/T Values	
(m)	(ft)	G_4 (dB)	G_{12}	G/T_4	G/T_{12}
6	19.7	45.82	55.37	+21.1 dB/K	+29.3 dB/K
7	23.0	47.17	56.71	+22.5 dB/K	+30.7 dB/K
8	26.25	48.32	57.86	+23.6 dB/K	+31.9 dB/K
9	29.52	49.34	58.9	+24.6 dB/K	+32.9 dB/K
10	32.81	50.26	59.8	+25.6 dB/K	+33.8 dB/K

[a]*Assumptions*: Transmission line losses: 2 dB
  Sky noise: 15 K at 4 GHz; 30 K at 12 GHz
  Receiver noise temperature 75 K at 4 GHz; 120 K at 12 GHz

Table 6.14C   VSAT EIRP values[a]

Antenna Aperture		Antenna Gains (dB)		EIRP$_6$ (dBW)	EIRP$_{14}$ (dBW)
(m)	(ft)	G_6	G_{14}		
1.2	3.94	35.37	42.73	+37	+44
1.5	4.92	37.3	44.66	+39	+47
1.8	5.9	38.88	46.23	+41	+48
2.0	6.56	39.8	47.15	+42	+49
2.4	7.37	41.38	48.74	+43	+51

[a]*Assumptions*: Transmission line losses: 1 dB
  Transmitter power output: 2 W (+3 dBW)

Table 6.14D   VSAT hub EIRP values[a,b]

Antenna Aperture		Antenna Gains (dB)		EIRP$_6$ (dBW)	EIRP$_{14}$ (dBW)
(m)	(ft)	G_6	G_{14}		
6	19.7	49.35	56.71	+74.35	+81.71
7	23.0	50.69	58.1	+75.69	+83.1
8	26.23	51.84	59.2	+76.84	+84.2
9	29.52	52.86	60.2	+77.86	+85.2
10	32.81	53.78	61.1	+78.78	+86.1

[a]*Assumptions*: Transmission line losses: 2 dB
  Transmitter power output: 500 W (+27 dBW)
  Antenna aperture efficiency: 60%
[b]If a VSAT hub terminal were to have 1-kW transmitter power output, then add 3 dB to the appropriate EIRP value.

of transmission. Continuing with the rationale, we wish to make two additional points:

- A VSAT system may be more cost-effective for data link connectivity than conventional connectivity through the PSTN (public switched telephone network). A VSAT system provides true *bypass* (it bypasses the telephone company or national PTT administration).
- With the proper design, a VSAT system can demonstrate superior performance, measured in error-free seconds and availability, than its PSTN counterpart.

The primary application of a VSAT system is for a large family of outstations that must send short, bursty messages to a central location and may also require a reply from the central location. Credit card verification is one good example of such an application. Another is reservations: hotel, airline, car rental, and so on. Some users are even finding it attractive for batch file transfer. Typical user groups encompass

- Fast-food chains
- Hotel chains
- Banks
- Brokerage firms
- Automobile dealerships
- Retail store chains
- Electronic mail
- Oil/gas/water pipeline telemetry and control
- Point-to-multipoint corporate and institutional networks.

Some VSAT systems are one-way (simplex): hub-to-VSAT for data, facsimile, or TV.

A VSAT network can accommodate from hundreds to thousands of remote users. Nominal bit rates, depending on system requirements and design, vary from about 600 bps to 64 kbps or more from VSAT to hub (inbound). In the opposite direction (outbound), 56 kbps, 64 kbps, or higher rates may be used. A TDM (time division multiplex) format is commonly encountered for the outbound regime.

6.9.3   Access Techniques

In our discussion of access techniques, we will only consider the full-duplex or quasi-full-duplex operation. By "quasi" we mean that a hub response to a VSAT may be delayed due to queueing or other delays. The reader will note the use of the terms *inbound* and *outbound*. When we say "outbound" we mean in the direction from the hub to one or more VSATs, and by "inbound" we mean in the direction from a VSAT to the hub. Commonly, the outbound link is a continuous TDM serial bit stream received by all active VSAT users. VSATs may also be separated into subnetworks, each with its own TDM outbound link. The TDM bit stream is made up of a series of data messages or packets, each with a header. The header carries, as a minimum, the address sequence of bits indicating the destination(s) of the message or packet. Each VSAT extracts only those messages or packets destined for it. Even when the hub has no outbound traffic, it may continue to send a serial bit sequence that is a predetermined idle sequence (i.e., no traffic). On many VSAT networks the outbound TDM bit stream is also a timing source for each VSAT.

Inbound satellite transponder access can be handled in a number of different ways. Many such access methods are some derivative of TDMA. (See Section 6.5.3). One of the simplest approaches is TDMA using random access (RA), which is often called *Aloha*. The name is derived from original work on a radio random access scheme at the University of Hawaii. This access scheme is simply one of contention. A VSAT terminal transmits a short message or packet at random to the hub. If another VSAT transmits at the same time or during the period of transmission of the first VSAT, both transmissions will be corrupted. Neither will receive an acknowledgment from the VSAT, so they both know that they have to try again. We call these collisions. Sometimes when a collision has been detected, each station backs off a random, predetermined time. This is one way to try to eliminate a second collision by the same two stations sending the same two messages. Collisions cause transmission delay and reduced throughput.

Another form of Aloha is called *slotted Aloha*. In this case, users can transmit only in discrete time slots. With such a scheme, two (or more) users can only collide with each other if they start transmitting exactly at the same time. One disadvantage of slotted Aloha is the wasted periods of time when a message or packet does not use up the total slot time allowed. Such a system requires a synchronized time reference, which can be considered a disadvantage.

R-Aloha or Aloha with capacity reservation is still another variation of Aloha. It is useful when there are only a few high-traffic-intensity users and other low-intensity (sporadic) users. The high-traffic-intensity users have reserved slot space(s) and the remainder of the slots are for the low-intensity users. This latter group operates on a contention basis. There are many variations of this reservation system. One efficient derivative is the dynamic reservation scheme where a user can request reserved slot space for sequential multiple messages/packets or long files. The remaining slots use random access (contention).

Another, quite different access scheme serves users that have a greater frequency of usage and longer messages to send. It is similar in many respects to the DAMA/SCPC techniques described in Section 6.5.2.1. With this

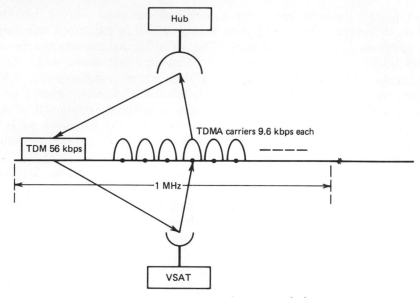

Figure 6.38   VSAT system operation with a 1-MHz allocation on a satellite transponder.

scheme, inbound data rates are from 1.2 to 9.6 kbps. Outbound data rates are 56/64 kbps on one or more TDM serial bit streams. A VSAT, in this scheme, requests service from the hub, and the hub assigns a vacant full-time channel from a pool of free channels. When the VSAT has completed its transaction(s), that channel is returned to the pool of available channels and is ready for assignment to another VSAT user.

6.9.4   VSAT Transponder Operation and Link Budget

A total VSAT system may occupy no more than 1 MHz of transponder frequency space. Other, larger VSAT systems may require more frequency allocation on a satellite transponder. A typical 1-MHz assignment may be apportioned as shown in Figure 6.38.

The TDMA inbound carriers to the hub (see the figure) can be configured for one user per carrier, many users per carrier using one of the several TDMA techniques described above or a DAMA discipline. There are many subnetworking possibilities as well. For example, each TDMA carrier can be associated with a family of users. Forward error correction (FEC) is commonly employed and the use of automatic repeat ARQ (see Chapter 12) is almost universal.

Link availabilities exceed 99% and some exceed 99.9%. Excess attenuation due to rainfall impacts availability on the 14/12-GHz operation. Other factors are interference, equipment reliability, sun transit, and packet/message collision on random access schemes.

Many VSAT systems are modeled with message lengths on the order of 800 to 2048 bits (Ref. 17). With messages of this length, transmit time from the VSAT is well under 1 s. Of course, message length and message rate per user are very operationally driven.

Modulation is commonly some form of PSK, either BPSK or QPSK. BPSK is more robust and is generally favored over QPSK due to power density limits imposed by official regulatory bodies to protect satellite and terrestrial systems from excessive interference.

We now review a typical K_u band VSAT outbound power budget for 56-kbps operation. The satellite has a carrier EIRP of $+17.5$ dBW. The modulation is BPSK using convolutional coding with rate $= 1/2$, soft-decision Viterbi decoder. The E_b/N_0 required for 1×10^{-7} operation is 7.0 dB, including 0.5 dB of modulation implementation loss.

Link Budget—Hub-to-VSAT

Downlink

Satellite EIRP	$+17.5$ dBW
Free-space loss (12 GHz)	-205.76 dB
Other link losses	-2.0 dB
Isotropic receive level	-190.26 dBW
VSAT G/T	$+21.6$ dB/K (2.4-m antenna aperture)
Sum	-168.66 dBW
Minus Boltzmann's constant	$-(-228.6$ dBW$)$
$C/N_{0(d)} =$	59.94 dB
Typical $C/I_0 =$	67.9 dB for satellite interference

Uplink

Hub EIRP	$+81.7$ dBW
Free-space loss (14 GHz)	-207.1 dB
Other losses	-2.0 dB
Isotropic receive level	-127.4 dBW
Satellite G/T	0.0 dB/K
Sum	-127.4 dBW
Minus Boltzmann's constant	$-(-228.6$ dBW$)$
$C/N_0 =$	101.2 dB

Table 6.15 INTELSAT V, VA, VA(IBS), and VI satellite parameter summary

Beam Connection Uplink/Downlink	Satellite	Bandwidth (MHz)	EIRP (dBW)	Saturation Flux Density (dBW/m² ± 2 dB)			G/T (dB/K)	Comments (Table Notes in Parentheses)
				Extra-High	High	Low		
Global/global	V(F1–F4)	36	23.5	−82.5	−75.0	−67.5	−18.6	All channels except 7–8
Global/global	V and VA(F5–F15)	36	23.5	−85.1	−77.6	−70.1	−16.0	All channels except 7–8
Global/global	VI(F1–F5)	36	26.5	−82.1	−77.6	−70.1	−14.0	All channels except 1'–2' (1)
Global/global	V(F1–F4)	72	26.5	−79.5	−72.0	−64.5	−18.6	Channel 7–8
Global/global	V and VA(F5–F15)	72	26.5	−82.1	−74.6	−67.1	−16.0	Channel 7–8
Pseudoglobal	VI(F1–F5)	72	31.0	−78.8	−74.3	−63.8	−12.4	Channel 1'–2' (15)
Hemi/hemi	V(F1–F4)	72	29.0	−79.5	−72.0	−64.5	−11.6	All channels except 9
Hemi/hemi	V(F5–F9)	72	29.0	−82.1	−74.6	−67.1	−9.0	All channels except 9
Hemi/hemi	VA(F10–F15)	72	29.0	−84.1	−76.6	−67.1	−9.0	All channels except 9 (2)
Hemi/hemi	VI(F1–F5)	72	31.0	−82.1	−77.6	−67.1	−9.2	All channels except 9 (3)
Hemi/hemi	V(F1–F4)	36	26.0	−82.5	−75.0	−67.5	−11.6	Channel 9
Hemi/hemi	V and VA(F5–F15)	36	26.0	−85.1	−77.6	−70.1	−9.0	Channel 9
Hemi/hemi	VI(F1–F5)	36	28.0	−82.1	−77.6	−70.1	−9.2	Channel 9 (3)
Zone/zone	V(F1–F4)	72	29.0	−79.5	−72.0	−64.5	−8.6	All channels
Zone/zone	V(F5–F9)	72	29.0	−82.1	−74.6	−67.1	−6.0	All channels
Zone/zone	VA(F10–F15)	72	29.0	−84.1	−76.6	−67.1	−6.0	All channels except 9
Zone/zone	VI(F1–F5)	72	31.0	−82.1	−77.6	−67.1	−2.0/− 7.0	All channels except 9 (4–6)
Zone/zone	VA(F10–F15)	36	26.0	−85.1	−77.6	−70.1	−6.0	Channel 9
Zone/zone	VI(F1–F5)	36	28.0	−82.1	−77.6	−70.1	−2.0/− 7.0	Channel 9 (4–6)
Global/spot (4 GHz)	VA(F10–15)	36	32.5	−85.1	−77.6	−70.1	−16.0	All channels except 7–8
Global/spot (4 GHz)	VA(F10–F15)	72	35.5	−82.1	−74.6	−67.1	−16.0	Channel 7–8
Circ. w. spot/circ. w. spot	V and VA(F1–F12)	72	44.4	N/A	−80.3	−75.3	3.3	(7)
Ellip. W. spot/ellip. W. spot	VA(F13–F15)	72	43.6	−84.0	−77.0	−72.0	0.0	Inner, 11 GHz, high (7, 11, 12)
Ellip. W. spot/ellip. W. spot	VA(F13–15)	72	39.0	−79.6	−72.6	−67.6	−4.4	Outer, 11 GHz, high (7, 11, 12)
Ellip. W. spot/ellip. W. spot	VA(F13–15)	72	41.1	−84.0	−77.0	−72.0	0.0	Inner, 11 GHz, normal (7, 11)
Ellip. W. spot/ellip. W. spot	VA(F13–F15)	72	36.5	−79.6	−72.6	−67.6	−4.4	Outer, 11 GHz, normal (7, 11)
Ellip. W. spot/ellip. W. spot	VA(F13–F15)	72	43.0	−84.0	−77.0	−72.0	0.0	Inner, 12 GHz, high (12, 13)
Ellip. W. spot/ellip. W. spot	VA(F13–F15)	72	39.0	−79.6	−72.6	−67.6	−4.4	Outer, 12 GHz, high (12, 13)
Ellip. W. spot/ellip. spot	VA(F13–F15)	72	40.5	−84.0	−77.0	−72.0	0.0	Inner, 12 GHz, normal (13)
Ellip. W. spot/ellip. W. spot	VA(F13–F15)	72	36.0	−79.6	−72.6	−67.6	−4.4	Outer, 12 GHz, normal (13)
Ellip. W. spot/ellip. W. spot	VI(F1–F5)	72	44.7	−84.0	−78.0	−73.0	1.7	Inner (8, 10)

Ellip. W. spot/ellip. W. spot	VI(F1–F5)	72	41.7	–83.3	–77.3	–72.3	–1.3	Outer (8–10)
Ellip. E. spot/ellip. E. spot	V and VA(F1–F12)	72	41.1	N/A	–77.0	–72.0	0.0	(7)
Ellip. E. spot/ellip. E. spot	VA(F13–F15)	72	41.1	–84.0	–77.0	–72.0	0.0	11 GHz (7, 11)
Ellip. E. spot/ellip. E. spot	VA(F13–F15)	72	38.5	–84.0	–77.0	–72.0	0.0	12 GHz (11, 14)
Ellip. E. spot/ellip. E. spot	VI(F1–F5)	72	44.7	–84.0	–78.0	–73.0	1.0	(8, 10)

Source: Appendix A to Document IESS-410 (Ref. 19, pp. 2–4). Courtesy of INTELSAT.

Notes:

1. The EIRP for channel 9 is 3 dB less (i.e., 23.5 dBW).

2. INTELSAT VA(F13–F15) saturation flux density values for channels 3–4 and 7–8 are 0.8 dB less sensitive (e.g., –76.6 becomes –75.8 dBW/m^2) than the other channels in all gain steps.

3. The G/T values for all channels including channel 9 decreases to –9.5 dB/K in the low-gain mode.

4. The G/T for Zone 1(NW) and Zone 3(NE) is –2 dB/K, whereas for Zone 2(SW) and Zone 4(SE) it is –7 dB/K.

5. In the low-gain mode the G/T of Zone 1(NW) and Zone 3(NE) decreases to –3 dB/K, while that of Zone 2(SW) and Zone 4(SE) decreases to –7.5 dB/K.

6. INTELSAT VI SSPAs operate with the Zone 1 and Zone 3 beams and each have an additional 2.5-dB gain step that can be switched "in" and "out" by ground command. When this gain step is switched "out," then all flux density values decrease by 2.5 dB (e.g., –77.6 becomes –80.1).

7. The available bandwidth for channel 7–12 is 241 MHz.

8. The EIRP for channel 9–12 is 3 dB higher.

9. All values of RF flux density are ±2.0 dB except for the west spot outer beam of INTELSAT V(F1–F5), which is ±4.25 dB.

10. The available bandwidth for channels 9–12 is 150 MHz.

11. The saturation flux density of channels 7–12 is 0.8 dB less sensitive (e.g., –72.0 becomes –71.2 dBW/m^2) than the indicated values for all gain steps.

12. "High" power refers to the mode of operation wherein the primary and redundant *K*-band west spot TWTAs are operated in parallel. It is possible to select this mode of operation independently for each of the *K*-band west spot channels. It is also possible to operate these paralleled TWTAs with the *K*-band east spot beam when the *K*-band payload is configured in the inverted business service mode. In the inverted mode the high-power capability is available in the east beam.

13. The satellite transmit frequencies are 11.70–11.95 GHz. It is also possible to operate at 12.50–12.75 GHz using the inverted business mode (see note 12). For the satellite EIRPs available at this frequency band, refer to Appendix B of IESS-401.

14. The satellite transmit frequencies are 12.50–12.75 GHz. It is also possible to operate at 11.70–11.95 GHz using the inverted business mode (see note 12). For the satellite EIRPs available at this frequency band, refer to Appendix B IESS-401.

15. On INTELSAT VI it is possible to electrically connect the west and east hemispheric beam transponders in channels 1′–2′ to form one transponder. With this configuration a carrier transmitted in the uplink of either the east or west hemispheric beams will be simultaneously transmitted, on the downlink, on both the east and west hemispheric beams. This mode of operation of channels 1′–2′ is called pseudoglobal. The receive polarization is LHCP and transmit polarization is RHCP. The flux density values are ±2.2 dB. The G/T for the low-gain mode decreases to –12.7 dB/K.

16. SSPA = Solid State Power Amplifier.

We calculate $C/N_{0(\text{sys})}$:

$$C/N_{0(\text{sys})} = \frac{1}{1/[C/N_{0(u)}] + 1/[C/N_{0(d)}] + 1/[C/N_{0(I)}]} = 59.29 \text{ dB}$$

$C/N_0(\text{sys})$ $\quad = 59.29 \text{ dB}$
$-10 \log(\text{BR*})$ $\quad = 47.48 \text{ dB} \quad \text{BR*} = 56,000 \text{ bps}$
E_b/N_0 $\quad\quad\quad = 11.81 \text{ db}$
E_b/N_0 required $= 7.0 \text{ dB}$
Margin $\quad\quad\quad = 4.81 \text{ dB}$

K_u band (14/12 GHz) is above the 10-GHz dividing line we have set where excess attenuation due to rainfall becomes a factor in the link budget. The 4.78-dB margin we obtained above provides some measure to improve link availability during periods of heavy rainfall. If we would also add interleaving with the proper interleaving interval to bridge rainfall fades, even greater protection would be available from excess attenuation owing to rainfall. The calculation of excess attenuation due to rainfall is discussed in Chapter 7.

We intended to show and it is nearly universally true that VSAT systems are downlink limited. What we mean by this is the "weak link in the chain" is the downlink. The natural tendency would be to increase the size of the antenna to achieve greater margin. Then the apparent economic savings of the *VSAT* are lost.

6.10   INTELSAT INTERFACES AND STANDARDS

6.10.1   Introduction

INTELSAT, or International Telecommunication Satellite (Consortium), is an international organization with member nations including nearly all countries associated with the "West" and the Third World. Some eastern bloc nations have also recently joined INTELSAT. Originally, INTELSAT provided transoceanic services, which it still does. Now it is also making significant inroads in domestic and regional systems and some national network connectivities. Roman numerals identify satellite models, each with improvements over its predecessors. The series INTELSAT II, III, and IV/IVA have dropped or are dropping out of service for the larger capacity and more robust INTELSAT V and VI. INTELSAT VII is expected to be in operation by 1994.

Table 6.15 gives basic space segment parameters for present and planned INTELSAT satellites. Satellite transponder performance is a function of the

*Bit rate.

Table 6.16 INTELSAT VII Summary

Beam/Coverage	Bandwidth (MHz)	EIRP (dBW)	Saturation Flux Density (dBW/m²)		G/T (dB/K)	Band
			Low	High		
Hemi 1	77, 72(3), 38	+33	−87	−73	−8.5	C
Hemi 2	77, 72(3), 38	+33(32.1)[1]	−87	−73	−6	C
Zone 1	77, 72(3), 38	+33	−87	−73	−5.5	C
Zone 2	77, 72(3), 38	+33	−87	−73	−4	C
Zones 1 and 1A		+33	−87	−73	−8.5	C
Zones 2 and 2A		+33	−87	−73	−7.0	C
C band spot A and B (channel 9)	36	+35	−87	−73	−4.5	C
C-band spot A and B (channels 10 and 11)	36	+26	−87	−73	−4.5	C
C-band spot A and B (channel 12)	36	+29	−87	−73	−4.5	C
Global A (channel 9)	36	+29	−87	−73	−11.5	C
Global C (channel 9)	36	+27	−87	−73	−11.5	C
Global A and B (channels 10 and 11)	36	+26	−87	−73	−11.5	C
Global A and B (channel 12)	36	+29	−87	−73	−11.5	C
Spot 1, inner	77, 72, 112	+48/+46	−90	−76	+5.5	Ku
Spot 1, outer	77, 72, 112	+41.5/+44	−87	−73	+2	Ku
Spot 2, inner	77, 72, 112	+44.5/+46.5	−90	−76	+2.5	Ku
Spot 2, outer	77, 72, 112	+41.5/+43	−87	−73	−1	Ku
Spot 2 + 2A, inner	77, 72, 112	+44.5	−90	−76	+0.5	Ku
Spot 2 + 2A, outer	77, 72, 112	+41.5	−87	−73	−3	Ku
Spot 3, inner	77, 72, 112	+48	−90	−76	−5.5	Ku
Spot 3, outer	77, 72, 112	+46	−87	−73	0.8	Ku

Source: (From Refs. 22 and 23. Courtesy of INTELSAT.)

antenna type connected to the transponder. It is also a function of the transponder bandwidth. The type of satellite is identified in column two: V, VA, and VI. The "F" followed by a number identifies a particular satellite. The table also gives G/T, saturation flux density, and G/T for transponder/satellite series/antenna configuration. Table 6.16 gives an overview of some of the basic characteristics of the communication payload of INTELSAT VII for flights F1–F5.

6.10.2 INTELSAT Earth Station Standards (IESS)

INTELSAT has issued six categories of earth station standards which are broken down by function as follows:

IESS-101 Introduction and Approved Document List

IESS-200 series: Antenna and RF Equipment Characteristics

IESS-300 series: Modulation and Access Characteristics

IESS-400 series: "Supplementary" (deals with such subjects as IM products, engineering orderwires, and satellite characteristics)

IESS-501 Digital Circuit Multiplication Equipment Specification 32 Kbps ADPCM with DSI

IESS-600 series: Generic Earth Station Standards

Highlights of some of these standards are covered below.

6.10.2.1 *Standard A — Antenna and Wideband RF Performance Characteristics*

Frequency band: 6/4 GHz

$G/T \geq 35.0 + 20 \log f/4 \ dB/K$

where f is the receive frequency in gigahertz. The reader should note that there is an extension to the 6/4-GHz band as follows:

Previous: 3.700–4.200-GHz downlink
           5.925–6.425-GHz uplink

Extended: 3.625–4.200-GHz downlink
           5.850–6.425-GHz uplink

Bandwidth: The transmitter shall be able to transmit one or more RF carriers anywhere in the specified 6-GHz band; the receiver shall be able to receive one or more RF carriers anywhere in the specified 4-GHz band.

Table 6.17 gives earth station polarization requirements in the 6/4-GHz bands for INTELSAT V, VA, and VI satellites.

The applicable standard is IESS-201 (Ref. 20).

Table 6.17   Earth station polarization requirements to operate with INTELSAT V, VA, VA(IBS), and VI satellites (6/4 GHz)[a]

Coverage	INTELSAT V		INTELSAT VA/VA(IBS)		INTELSAT VI	
	Earth Station Transmit	Earth Station Receive	Earth Station Transmit	Earth Station Receive	Earth Station Transmit	Earth Station Receive
1. Global A	LHCP	RHCP	LHCP	RHCP	LHCP	RHCP
2. Global B			RHCP	LHCP	RHCP	LHCP
3. West hemisphere	LHCP	RHCP	LHCP	RHCP	LHCP	RHCP
4. East hemisphere	LHCP	RHCP	LHCP	RHCP	LHCP	RHCP
5. NW Zone (Z1)[b]	RHCP	LHCP	RHCP	LHCP	RHCP	LHCP
6. NE Zone (Z3)[b]	RHCP	LHCP	RHCP	LHCP	RHCP	LHCP
7. SW Zone (Z2)[b]					RHCP	LHCP
8. SE Zone (Z4)[b]					RHCP	LHCP

Source: IESS-201, Rev. 1 (Ref. 20). Courtesy of INTELSAT.

[a]Key: LHCP = left-hand circularly polarized.
  RHCP = right-hand circularly polarized.
[b]Z1, Z2, Z3, Z4 nomenclature applies to INTELSAT VI only.

6.10.2.2   Standard B — Antenna and Wideband RF Performance Characteristics

Frequency band: 6/4 GHz
$$G/T \geq 31.7 + 20\log(f/4) \quad dB/K$$

where f is the receive frequency in gigahertz.
  For transmitting and receiving bandwidths, refer to Section 6.10.2.1.
  The polarization requirements are shown in Table 6.17.
  The applicable standard is IESS-202 (Ref. 20).

6.10.2.3   Standard C — Antenna and Wideband RF Performance Characteristics

Frequency band: 14/12 and 14/11 GHz
$$G/T \geq 37.0 + 20\log(f/11.2) \quad dB/K$$

where f is the frequency in gigahertz. The G/T value from the above equation applies only to clear sky conditions. For degraded weather conditions (i.e., heavy rain storms; see Chapter 7), the following G/T requirement should be met:

$$G/T \geq 37.0 + 2\log\left(\frac{f}{11.2}\right) + X \, dB \quad (dB/K)$$

Table 6.18 Reference downlink degradation margin for G/T determination[a]

Satellite Orbital Location	Earth Station Location in Satellite Beam	Reference Downlink Degradation Margin (dB)	Percentage of Year Margin Can Be Exceeded (%)
325.5°–341.5°	West spot	13	0.03
and 174°–180°E	East spot	11	0.01
307.0°–310.0°E	West spot	13	0.02
	East spot	11	0.02

Source: IESS-203. Courtesy of INTELSAT.

[a]The margins shown in this table are those available under clear-sky conditions.

where X is the excess (attenuation) of the downlink degradation predicted by local rain statistics over the reference downlink degradation margin given in Table 6.18 for the same time percentage. Downlink degradation is defined as the sum of the precipitation attenuation in decibels and the increase in receiving system noise temperature in decibels for the given time percentage.

Minimum earth station transmitting and receiving bandwidths are given in Table 6.19.

The applicable standard is IESS-203 (Ref. 20).

Table 6.19 Minimum bandwidth requirements for standard C earth stations

Satellite	Earth Station Frequency ITU Region	Earth Station Frequency Transmit Band (GHz)	Receive Band (GHz)	Available Transponders
V, VA, and VI	All	14.00–14.50	10.95–11.20 and 11.45–11.70	1–2, 3–4,[a] 5–6, 7–12
VA(IBS)[b]	2[c]	14.00–14.25[d]	11.70–11.95	1–2, 3–4, 5–6, 7–8[d]
	1 and 3[c]	14.00–14.25[d]	12.50–12.75	1–2, 3–4, 5–6, 7–8[d]

Source: IESS-203 (Ref. 20). Courtesy of INTELSAT.

[a]Transponder 3–4 is only available in INTELSAT VI.
[b]If INTELSAT VA(IBS) satellites are configured to operate at 11 GHz (as opposed to 12 GHz), the earth station receive frequency band will be the same as for INTELSAT V, VA, and VI satellites.
[c]In INTELSAT VA(IBS), the receiver band segments of 11.70–11.95 GHz and 12.50–12.75 GHz are interchangeable between the east and west spot beams, so that this spacecraft series can be operated in any ocean region.
[d]Standard C earth stations intending to operate in the spot-to-hemi multibeam mode available in transponder 7–8 on INTELSAT VA(IBS) should have this transmit bandwidth extended to 14.35 GHz.

6.10.2.4 *Standard D — Antenna and Wideband RF Performance Characteristics*

Frequency band: 6/4 GHz

There are two versions of the Standard D earth station: D-1 and D-2.

$$G/T \geq 22.7 + 20\log(f/4) \quad dB/K \qquad D\text{-}1$$
$$G/T \geq 31.7 + 20\log(f/4) \quad dB/K \qquad D\text{-}2$$

where f is the receiving frequency in GHz.
Polarization requirements:

Polarization A: uplink LHCP; downlink, RHCP
Polarization B: uplink, RHCP; downlink, LHCP

where LHCP = left-hand circularly polarized and RHCP = right-hand circularly polarized. Simultaneous operation in both polarization senses is not normally required.
The applicable standard is IESS-204 (Ref. 20).

6.10.2.5 *Standard E — Antenna and Wideband RF Performance Characteristics*

Frequency bands: 14/12 and 14/11 GHz

There are three versions of the Standard E earth station: E-1, E-2, and E-3.

$$G/T \geq 25.0 \text{ (but } < 29.0) + 20\log(f/11) \quad dB/K \qquad E\text{-}1$$
$$G/T \geq 29.0 \text{ (but } < 34.0) + 20\log(f/11) \quad dB/K \qquad E\text{-}2$$
$$G/T \geq 34.0 + 20\log(f/11) \quad dB/K \qquad E\text{-}3$$

where f is the frequency in gigahertz. The G/T values shown above are for clear sky conditions. For degraded conditions (i.e., heavy rainfall), the following G/T values are used:

$$G/T \geq 25.0 + 20\log(f/11) + X \quad dB/K \qquad E\text{-}1$$
$$G/T \geq 29.0 + 20\log(f/11) + X \quad dB/K \qquad E\text{-}2$$
$$G/T \geq 34.0 + 20\log(f/11) + X \quad dB/K \qquad E\text{-}3$$

where X is the excess of the downlink degradation predicted by local rain statistics over the reference downlink degradation margin shown in Table 6.18.
The minimum bandwidth requirements for Standard E earth stations are given in Table 6.20; the minimum tracking requirements are given in Table 6.21.

Table 6.20   Minimum bandwidth requirements for standard E earth stations

Satellite	ITU Region	Earth Station Transmit Frequency Band (GHz)	Earth Station Receive Frequency Band (GHz)	Available Transponders
V, VA, and VI	All	14.00–14.25	10.95–11.2	1–2, 3–4,[a] 5–6, 7–8[b]
VA(IBS)[c]	2[d]	14.00–14.25[e]	11.70–11.95	1–2, 3–4, 5–6, 7–8[e]
	1 and 3[d]	14.00–14.25[e]	12.50–12.75	1–2, 3–4, 5–6, 7–8[e]

Source: IESS-205 (Ref. 20). Courtesy of INTELSAT.

[a]Transponder 3–4 is only available in INTELSAT VI.

[b]Earth station owners should consider in their design the possibility of extending their usable bandwidth to 14.35 GHz in the transmit band and to 11.55 GHz in the receive band.

[c]If INTELSAT VA(IBS) satellites are configured to operate at 11 GHz (as opposed to 12 GHz), the earth station receive frequency band will be the same as for INTELSAT V, VA, and VI satellites.

[d]In INTELSAT VA(IBS), the receive band segments of 11.70–11.95 GHz and 12.50–12.75 GHz are interchangeable between the east and west spot beams, so that this spacecraft series can be operated in any ocean region.

[e]Standard E earth stations intending to operate in the spot-to-hemi multibeam mode available in transponder 7–8 on INTELSAT VA(IBS) should have this transmit bandwidth extended to 14.35 GHz.

Table 6.21   Minimum tracking requirements for standard E earth stations

Earth Station	INTELSAT V, VA, and VA(IBS) ($\pm 0.1°$ N/S and $\pm 0.1°$ E/W)	INTELSAT VI (See Note 3) ($\pm 0.02°$ N/S and $\pm 0.06°$ E/W)
E-1	Manual, E/W. only (weekly peaking)	Fixed antenna (see Note 2)
E-2	Manual, E/W and N/S (peaking every 3 to 4 hours)	Fixed antenna (see Note 2)
E-3	Auto track (see Note 1)	Manual, E/W only (weekly peaking)

Source: IESS-205 (Ref. 20). Courtesy of INTELSAT.

Notes:

1. Program tracking, which would operate from pointing information supplied by INTELSAT, is recommended due to the uncertainty of step track operation in a K-band environment, which may be subject to severe fading and scintillations.
2. "Fixed" antenna mounts will still require the capability to be steered from one satellite position to another, as dictated by operational requirements (typically once or twice every two to three years). These antennas should also be capable of being steered at least over a range of $\pm 5°$ from beam center for the purpose of verifying that the antenna pointing is correctly set toward the satellite, and for providing a means of verifying the side lobe characteristics in this range.
3. Antenna tracking requirements are based on provisional INTELSAT VI stationkeeping tolerances. These tolerances will be reviewed after operational experience is gained with this new series of spacecraft.

Polarization is linear and the transmit beam is orthogonal to the receive beam.

The applicable standard is IESS-205 (Ref. 20).

6.10.2.6 Standard F—Antenna and Wideband Performance Characteristics (for Business Services)

Frequency band: 6/4 GHz

There are three versions of the Standard F earth station: F-1, F-2, and F-3.

$$G/T \geq 22.7 + 20\log(f/4) \quad \text{dB/K} \qquad \text{F-1}$$
$$G/T \geq 27.0 + 20\log(f/4) \quad \text{dB/K} \qquad \text{F-2}$$
$$G/T \geq 29.0 + 20\log(f/4) \quad \text{dB/K} \qquad \text{F-3}$$

where f is the frequency in gigahertz.

Polarization: An F-type earth station shall be capable of LHCP (left-hand circular polarization) and RHCP (right-hand circular polarization) on the transmit and receive beams. However, this is not required simultaneously on the uplink nor on the downlink. Polarization requirements are shown in Table 6.17.

The minimum bandwidth requirements for the F-type earth station are shown in Table 6.22; the minimum tracking requirements are shown in Table 6.23.

The applicable standard is IESS-206 (Ref. 20).

Table 6.22 Minimum bandwidth requirements for standard F earth stations

Earth Station Mode	Band Band Segment	Frequency Band Bandwidth (MHz)	Available (MHz)	Transponders
Transmit (uplink)	1 or	5925–6256	331	1–2[a], 3–4, 5–6, 7–8,
	2	6094–6425	331	5–6, 7–8, 9, 10, 11, 12
Receive (downlink)	1 or	3700–4031	331	1–2[a], 3–4, 5–6, 7–8,
	2	3869–4200	331	5–6, 7–8, 9, 10, 11, 12

Source: IESS-206 (Ref. 20). Courtesy of INTELSAT.

[a]Earth station owners should consider in their design the possibility of extending their usable bandwidth to include transponders (1′–2′) of INTELSAT VI to 3.625 GHz for receive and 5.850 GHz for transmit.

Table 6.23   Minimum tracking requirements for standard F earth stations

Earth Station Standard	INTELSAT IVA	INTELSAT V, VA, and VB	INTELSAT VI (See Note 2)
F-1	"Fixed" antenna	"Fixed" antenna	"Fixed" antenna
F-2	Manual E/W only (weekly peaking)	Manual E/W only (weekly peaking)	"Fixed" antenna
F-3	Autotrack	Autotrack	Manual E/W only (weekly peaking)

Source: IESS-206 (Ref. 20). Courtesy of INTELSAT.

Notes:

1. "Fixed" antenna mounts will still require the capability to be steered from one satellite position to another, as dictated by operational requirements (typically once or twice every 2 to 3 years). These antennas should also be capable of being steered at least over a range of $\pm5°$ from beam center for the purpose of verifying that the antenna pointing is correctly set toward the satellite, and for providing a means of verifying the side-lobe characteristics in the range.
2. Antenna tracking requirements are based on provisional INTELSAT VI stationkeeping tolerances. These tolerances will be reviewed after operational experience is gained with this new series of spacecraft.
3. The minimum tracking requirements of the table above are subject to the earth station meeting the axial ratio requirements.

6.10.2.7   Frequency Division Multiplex / Frequency Modulation (FDM / FM) Telephony Carriers   Telephone channel quality:

- Total noise in any FDM/FM telephone channel shall not exceed 10,000 pW0p. Of this value, 1000 pW0p is allocated to the terminal segment and the allocation is budgeted as follows:

  Earth station transmitter excluding multicarrier intermodulation noise and group delay: 250 pW0p

  Noise due to total system group delay after any necessary equalization: 500 pW0p

  Other earth station receiver noise: 250 pW0p.

Following CCIR recommendations, an allowance of 1000 pW0p is reserved for possible interference from terrestrial systems sharing the same frequency bands.

Of the 10,000-pW0p total noise in a derived voice channel, 8000 pW0p is allocated to uplink and downlink thermal noise, transponder intermodulation, earth station out-of-band emission, cochannel interference within the operating satellite, and interference from adjacent satellite networks.

- 14/11-GHz links shall operate above the demodulator (FM improvement) threshold for more than 99.96% of a year.

- Regular and high-density carriers. Regular FDM/FM carriers are defined as carriers whose representative EIRP values are included in the mandatory characteristics of specification IESS-301. High-density carriers require more EIRP but less RF bandwidth per channel than regular carriers. The assignment of each high-density carrier is subject to agreement by the owner of the transmitting earth station and depends on the availability of necessary additional EIRP in the satellite transponder.
- Carrier frequency assignments. Maximum use will be made of the multidestination carrier concept. The receiving system of an earth station should be so designed that carriers in the same transponder whose center frequencies are only 1.25 MHz apart can be received.

For frequency selection purposes it is recommended that earth stations using synthesizers for frequency generation equip some transmit and receive chains with the capability of transmitting and receiving RF carriers whose frequency is a multiple of 125 kHz.

- Effective isotropically radiated power (EIRP). INTELSAT specifies EIRPs at a 10° elevation angle and at satellite beam edge. For elevation angles other than 10° and earth station locations other than at satellite beam edge, correction factors K_1 and K_2 are used. Consult INTELSAT IESS-402 for the derivation of K_1 and K_2 for specific earth station installations.

Adverse weather effects on EIRP: For 6-GHz operation during severe local weather conditions, EIRP levels can drop to 2 dB below nominal value. Recognize, of course, that this will result in a degraded channel performance at cooperating receiving stations.

At 14 GHz, it is mandatory that means be provided to prevent the power flux density at the satellite from falling more than M dB below the nominal clear sky value for more than $K\%$ of the time. M varies between 5 and 7 dB, depending on satellite longitude, and K varies between 0.01 and 0.04%. Consult Table 4 in IESS-301 for appropriate values. One method recommended to control uplink EIRP automatically to maintain the required flux density values at the satellite transponder is to provide an automatic power control circuit based on the RSL of the 11-GHz satellite beacon signals.

- EIRP values, 6/4-GHz links. Typical values are given in Table 6.24. The maximum EIRP value for FDM/FM waveforms on the 6/4-GHz links is +90.2 dBW when using high-density carriers, each with a 972-voice-channel FDM configuration when using hemispherical and zone beams on the satellite.
- Baseband composition of carriers (see Table 6.2). Based on the CCITT modulation plan, all carriers use CCITT subgroup A (12–60 kHz) plus

Table 6.24   Standard A maximum EIRP (dBW) for direct 6/4 GHz links; global beam carriers

Receive Earth Station	Previous A $(G/T \geq 40.7 \, \text{dB/K})$			Revised A $(G/T \geq 35.0 \, \text{dB/K})$	
Transponder Gain	Low	High		High	
S/C Category	(1)	(2)	(3)	(2)	(3)
Carrier Size (MHz/Ch)					
1.25/12	69.4	69.1	66.8	71.7	69.0
2.5/24	71.1	70.7	68.4	73.3	70.6
2.5/36	74.1	73.7	71.4	76.3	73.6
2.5/48	77.4	77.0	74.7	79.6	76.9
2.5/60	80.1	79.7	77.4	82.2	79.6
5.0/60	74.2	74.7	72.4	77.3	74.6
5.0/72	75.0	74.6	72.3	77.2	74.5
5.0/96	78.6	78.2	75.9	80.8	78.1
5.0/132	82.7	80.0	80.0	86.5[1.5]	82.2
7.5/96	75.9	75.5	73.2	78.1	75.4
7.5/132	78.2	77.8	75.5	80.4	77.7
7.5/192	83.5	82.3	80.1	87.3[0.7]	84.8
10.0/132	77.0	76.6	74.3	79.2	76.5
10.0/192	79.7	79.3	77.0	81.8	79.2
10.0/252	84.2	84.5	80.8	*	85.5
15.0/252	79.2	79.6	77.3	82.1	79.5
15.0/312	81.6	81.2	79.7	86.2	81.9
15.0/372	82.2	85.5	81.8	*	86.5
17.5/312	79.3	80.3	78.0	84.3	80.2
17.5/372	82.0	82.1	79.9	87.1	84.6
17.5/432	84.4	85.9	82.2	*	86.9
20.0/432	83.0	84.5	80.8	*	85.5
20.0/492	84.2	86.6	84.5	*	87.6
20.0/552	85.4	*	86.3	*	*
25.0/432	81.5	81.5	80.0	86.5	82.2
25.0/492	82.6	82.6	82.4	87.6	85.1
25.0/552	83.6	85.9	82.2	*	87.1
25.0/612	84.7	87.5	85.4	*	*
36.0/612(4)	81.7	81.9	79.7	86.9	92.6
36.0/792(4)	84.7	84.7	85.3	*	*
36.0/972(4)	87.9	*	*	*	*

Table 6.24 *(Continued)*

Receive Earth Station	Previous A $(G/T \geq 40.7\,\text{dB/K})$			Revised A $(G/T \geq 35.0\,\text{dB/K})$	
Transponder Gain	Low	High		High	
S/C Category	(1)	(2)	(3)	(2)	(3)
Carrier Size (MHz/Ch)					
36.0/792(5)	84.5	86.1	82.4	*	87.1
36.0/972(5)	86.5	*	87.1	*	*

Notes:

(1) INTELSAT V, VA, VA(IBS), and VI.

(2) INTELSAT V, VA, and VA(IBS).

(3) INTELSAT VI (F1–F5).

(4) 36 MHz carrier with 32.4 MHz occupied bandwidth.

(5) 36 MHz carrier with 36.0 MHz occupied bandwidth.

(6) * Denotes carriers not designed for this link.

The numbers between brackets [ ] indicate the amount by which relevant CCIR criteria are exceeded based on the minimum earth station antenna parameters shown below. Earth station antennas exceeding the minimum performance criteria, to the extent that the excess margin no longer applies, can transmit these carriers.

Earth station Standard	CCIR Criteria (dBW/4 kHz)	Sidelobe Envelope (dBi)	Main Beam Gain (dBi)	Δ Off-Axis Relative to Baseline (dB)
Revised A (> 1988)	$32 - 25\log\theta$	$29 - 25\log\theta$	57.8	Baseline
Retrofitted Rev A	$35 - 25\log\theta$	$32 - 25\log\theta$	57.8	0.0
Revised A (< 1989)	$35 - 25\log\theta$	$29 - 25\log\theta$	57.8	3.0
Previous A	$35 - 25\log\theta$	$32 - 25\log\theta$	63.0	5.2
Retrofitted Rev C	$29 - 25\log\theta$	$32 - 25\log\theta$	62.3	Baseline
Revised C (< 1989)	$29 - 25\log\theta$	$32 - 25\log\theta$	62.3	0.0
Previous C	$29 - 25\log\theta$	$32 - 25\log\theta$	65.0	2.7
Revised C (> 1988)	$29 - 25\log\theta$	$29 - 25\log\theta$	62.3	3.0

where:

- Retrofitted Rev A and C = Previous Standard A and C antennas that have been retrofitted to become revised Standard A and C earth stations. Earth station applicants with antennas which do not meet the main beam value of 57.8 dBi will be required to show how they meet the CCIR criteria for the carrier size they transmit.
- CCIR Rec 524-1 criteria at C-Band are:

    $35 - 25\log\theta$, dBW/4 kHz (before 1 January 1989 = < 1989)

    $32 - 25\log\theta$, dBW/4 kHz (after 31 December 1988 = > 1988)

- CCIR proposed (under study) criteria at K-Band is $39 - 25\log\theta$, dBW/40 kHz ($29 - 25\log\theta$, dBW/4 kHz). If this criteria is adopted, all categories of Standard C will meet the off-axis limit.
- Δ Off-Axis Relative to Baseline (dB) = The values shown in this column can be subtracted from the baseline value in [ ]. For example, if a baseline value in [ ] is 2.2 dB for a Revised A (> 1988), then a Previous A would meet the off-axis criteria with a margin of 3.0 dB.
- θ = is the angle in degrees between the main beam axis and the direction considered.

the requisite contiguous CCITT 12-channel groups and supergroups as required up to a maximum of 1332 FDM voice channels on one RF carrier (which fills a 36-MHz transponder).

The applicable standard is IESS-301 (Ref. 20).

6.10.2.8 The INTELSAT SPADE System

The *s*ingle channel per carrier *P*CM multiple-*a*ccess *d*emand assignment *e*quipment (SPADE) is a good example of a DAMA system. SPADE channels can all be preassigned. SPADE was developed by INTELSAT and is in use on INTELSAT IVA and V satellites.

An INTELSAT IVA/V transponder has its nominal 36-MHz bandwidth divided into 45-kHz segments for SPADE operation, giving the transponder an 800-channel (one-way) capacity (e.g., 800×45 kHz = 36 MHz) or 400 full-duplex circuits. Thus SPADE is really a subset of the FDMA technique discussed earlier. Channel encoding is PCM (64 kbps per channel), and four-phase CPSK (continuous-phase shift keying) is used for modulating each SPADE carrier.

Figure 6.39 illustrates the SPADE channelization plan. It will be noted that a SPADE network normally uses common channel signaling, which has a 160-kHz bandwidth located 18.045 MHz below the pilot frequency shown in the figure. The pilot is a constant carrier and is provided by the common signaling channel (CSC) reference station. It is used by all SPADE receivers in the network to control satellite and earth station down-converter local oscillator frequency drift. Channels 1 and 2 and channels 1′ and 2′ are left

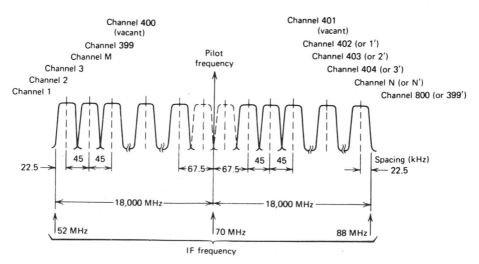

Figure 6.39   SPADE frequency plan at IF. (Based on IESS-304, Rev. 1, Ref. 20; courtesy of INTELSAT.)

vacant to allow a wider bandwidth for the CSC. The first possible carrier is designated channel 3 and is located 112.5 kHz above the lower end of the band. The 399th channel is located (center frequency) 67.5 kHz below the pilot. The first possible channel above the pilot is designated channel 3′ and is located 157.6 kHz above the pilot. Channel 399′ is located 22.5 kHz inside the upper edge of the allocated band. Any two like-numbered channels (primed and unprimed) are 18.045 MHz apart and are used to form a duplex circuit. Therefore there are really only 794 VF channels or 397 useful full-duplex telephone circuits available in the system (INTELSAT V).

The 6-GHz maximum required global-beam EIRP for SPADE carriers at 10° elevation angle and satellite antenna beam edge are

Spacecraft Series	EIRP (dBW)
INTELSAT IVA	+ 62.0
INTELSAT V, VA, VB, and VI	+ 60.5

Figure 6.40 is a functional block diagram of a SPADE terminal. Its operation is described below. Note that on the left-hand side of the figure is the interface equipment with the four-wire telephone circuits deriving from an international switching center and on the right-hand side are the 70-MHz inputs/outputs of the earth station up-converter/down-converter, respectively.

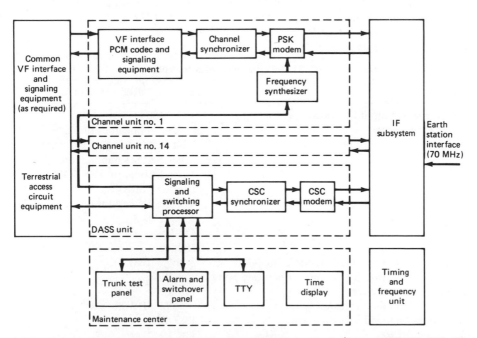

Figure 6.40 Functional block diagram of a SPADE terminal. (From IESS-304 Ref. 20; courtesy of INTELSAT.)

Table 6.25   Technical characteristics of SPADE

Channel encoding	PCM, 7-bit; A-law companding
Modulation	4-phase PSK (coherent)
Bit rate	64 kbps
Bandwidth per channel	38 kHz
Channel spacing	45 kHz
Stability requirement	± 2 kHz (with AFC[a])
Bit error rate at threshold	10^{-4}
TDMA common signaling channel (CSC)	
Bit rate	128 kbps
Modulation	2-phase PSK
Frame length	50 ms
Burst length	1 ms
Number of accesses	48
Bit error rate	10^{-7}

[a]Automatic frequency control.

An incoming voice channel has its signaling information removed and is passed to the PCM codec (coder–decoder), where the voice channel is converted into a digital bit stream in PCM format. The format consists of a 7-level code and A-law companding. The output of the transmit side of the codec is a bit stream of 64,000 bps that modulates a four-phase PSK modem (coherent) in a 38-kHz noise bandwidth (the 45 kHz mentioned earlier

Table 6.26   Summary of SCPC/QPSK characteristics for voice

Parameter	Requirement
Audio channel input bandwidth	300–3400 Hz
Transmission rate	64 kbit/s (includes preamble)
Encoding	7-bit PCM $A = 87.6$ companding law 8-kHz sampling rate
Modulation	4-phase coherent PSK (QPSK)
Ambiguity resolution	Unique words
Carrier control	Voice activated for voice channels
Channel spacing	45 kHz
Channel bandwidth	45 kHz
IF noise bandwidth	38 kHz
C/T per channel at nominal operating point	-167.3 dBW/K
C/N in IF bandwidth at nominal operating point	15.5 dB
Nominal BER at operating point	1×10^{-6}
C/T per channel at threshold	-169.3 dBW/K
C/N in IF bandwidth at threshold	13.5 dB
Threshold BER	1×10^{-4}

Source: IESS-303, Rev. 2, Ref. 20. Courtesy of INTELSAT.

includes guard bands). The output may be on any one of the 800 frequencies in the "pool." Frequency control is by means of a synthesizer.

For the receive side of a full-duplex voice channel, the above process is reversed. The 70-MHz output of the down-converter is fed to the SPADE terminal input (receive) of the PSK modem. The receive section of the modem, the demodulator, provides an output that is a digital bit stream. It is fed to the decoder section of the codec where the digital-to-analog processing takes place. The resulting analog voice channel is fed to the terrestrial interface unit and thence to the international exchange after proper signaling information has been restored. The basic technical characteristics of a SPADE voice channel are reviewed in Table 6.25.

How does the SPADE terminal know what frequency slot is vacant at a particular interval of time? This information is provided by the demand assignment signaling and switching (DASS) unit. The terrestrial interface equipment routes signaling information to the DASS unit. The DASS unit uses a separate information channel for signaling and provides constant status updating information of the busy–idle condition of channels in the pool, so that busy frequencies become unavailable for assignment of new calls. DASS provides signaling interface on one common RF channel for all voice channels transmitted. This common signaling channel is actually a TDMA broadcast channel. It receives status updating information as well. Such a separate channel is required at all SPADE installations and operates on a TDMA basis, allowing up to 49 stations plus a reference station to

Table 6.27   Summary of SCPC / QPSK characteristics for data

Parameter	Requirement
Data rate: $\frac{3}{4}$	48 or 50 kbit/s
Data rate $\frac{7}{8}$	56 kbit/s
Clock recovery	Clock timing must be recovered from the received data stream
Threshold C/N:	
48 kbit/s: BW[a] = 38 kHz	13.5 dB
56 kbit/s: BW[a] = 38 kHz	13.5 dB
50 kbit/s: BW[a] = 38 kHz	13.6 dB
Threshold BER before	
$\frac{3}{4}$ or $\frac{7}{8}$ decoding	1×10^{-4}
C/N at nominal operating point	15.5 dB
Nominal BER at operating point without coding or scrambling	1×10^{-6}
Nominal BER at operating point with coding	1×10^{-9} (without scrambling) 3×10^{-9} (with scrambling)

Source: IESS-303, Rev. 3, Ref. 20.

[a]BW = bandwidth.

Table 6.28   Maximum 6-GHz global beam EIRP requirements (dBW)

(Low Gain) (4)

Carrier information rate (kbit/s)	Receive Earth Station					
	Previous A			B		
	(1)	(2)	(3)	(1)	(2)	(3)
				[−3.3]	[−0.8]	
64	58.6	56.0	52.9	65.2	62.7	59.4
192	63.4	60.8	57.7	70.0	67.5	64.1
384	66.4	63.8	60.7	73.0	70.5	67.2
1544	72.5	69.9	66.7	79.1	76.5	73.2
2048	73.7	71.1	67.9	80.3	77.7	74.4
6312	78.6	76.0	72.8	85.2	82.6	79.3
8448	79.9	77.3	74.1	86.4	83.9	80.6
32064	85.7	83.1	79.9	92.3	89.7	86.4
34368	86.0	83.4	80.2	92.6	90.0	86.7
44736	87.1	84.5	81.3	98.7	91.1	87.8
X (for other info. rates)	10.6	8.0	4.8	17.2	14.6	11.3

(High Gain) (4)

Carrier Information Rate (kbit/s)	Previous A			Revised A			B			F-3			F-2		
	(1)	(2)	(3)	(1)	(2)	(3)	(1)	(2)	(3)	(1)	(2)	(3)	(1)	(2)	(3)
										[−0.6]			[−2.5]		
64	55.4	52.8	50.2	58.3	55.7	52.7	60.2	57.6	54.4	62.5	60.0	56.5	64.4	61.9	58.7
192	60.1	57.5	55.0	63.0	60.4	57.5	64.9	62.3	59.1	67.3	64.7	61.3	69.2	66.6	63.5
384	63.2	60.6	58.0	66.1	63.5	60.5	68.0	65.4	62.2	70.3	67.7	64.3	72.2	69.6	66.2
1544	69.2	66.6	64.1	72.1	69.5	66.6	74.0	71.4	68.2	76.3	73.8	70.3	78.2	75.7	72.6
2048	70.4	67.8	65.3	73.3	70.7	67.8	75.2	72.6	69.4	77.6	75.0	71.5	79.5	76.9	73.8
6312	75.3	72.7	70.2	78.2	75.6	72.7	80.1	77.5	74.3	82.5	79.9	76.4	84.4	81.8	78.7
8448	76.6	74.0	71.5	79.5	76.9	73.9	81.4	78.8	75.6	83.7	81.2	77.7	85.6	83.1	80.0
32064	82.4	79.8	77.3	85.3	82.7	79.8	87.2	84.6	81.4	+	+	+	+	+	+
34368	82.7	80.1	77.6	85.6	83.0	80.1	87.5	84.9	81.7	+	+	+	+	+	+
44736	83.8	81.2	78.7	86.7	84.1	81.2	88.6	86.0	82.8	+	+	+	+	+	+
X (for other info. rates)	7.3	4.7	2.2	10.2	7.6	4.7	12.1	9.5	6.3	14.4	11.9	8.4	16.4	13.8	10.7

Source: IESS-308, Rev. 4, Ref. 20. Courtesy of INTELSAT.

Notes:
(1) INTELSAT V (F1–F4).
(2) INTELSAT V (F5–F9), VA (F10–F12), VA(IBS) (F13–F15), and VI (F1–F5) channel 9.
(3) INTELSAT VI (F1–F5) channels 10–12.
(4) See general notes following Table 5, in IESS-308.
(5) + denotes carriers not intended for this connection.
[ ] = Off-axis EIRP density indicator, dB (see note 8 following Table 5 in Ref. 20). Applies to all values in the column below the number.

access the channel in 50 successive 1-ms bursts per frame at a bit rate of 128 kbps. The channel modem is different from the voice channel modem. It uses biphase CPSK (see Chapter 12 for a discussion of modems). The BER is on the order of 1×10^{-7}. The CSC channel operates at an RF level 7 dB higher than the ordinary SPADE voice channel to maintain the lower error rate at the higher bit rate.

6.10.2.9 SCPC/QPSK and SCPC/PCM/QPSK System
This is a digital transmission system similar in many respects to SPADE, but without the capability of demand assignment. Individual RF carriers are allocated to either a digitized voice channel or a data channel on a preassigned basis. The modulation is PCM/QPSK/FDMA for voice operation and QPSK/FDMA for high-speed data (e.g., above 4800 bps).

Table 6.29A   Maximum 14/11 or 14/12 GHz direct links (west to east) EIRP requirements, dBW
(72 and 77 MHz Transponders) (4, 5)

Carrier Information Rate (kbit/s)	Receive Earth Stations											
	Previous C			Revised C			E-3			E-2		
	(1)	(2)*	(3)*	(1)	(2)*	(3)*	(1)	(2)*	(3)*	(1)	(2)*	(3)*
64	45.7	47.1	46.2	48.3	49.0	47.9	51.2	51.1	49.7	56.0	54.3	53.8
192	50.4	51.9	51.0	53.0	53.8	52.7	56.0	55.8	54.5	60.7	59.1	58.6
384	53.4	54.9	54.0	56.0	56.8	55.7	59.0	58.9	57.5	63.7	62.1	61.6
1544	59.5	61.0	60.0	62.1	62.8	61.7	65.0	64.9	63.6	69.8	68.1	67.6
2048	60.7	62.2	61.3	63.3	64.0	63.0	66.3	66.1	64.8	71.0	69.4	68.8
6312	65.6	67.1	66.2	68.2	68.9	67.8	71.2	71.0	69.7	75.9	74.3	73.7
8448	66.9	68.3	67.4	69.5	70.2	69.1	72.4	72.3	70.9	77.2	75.5	75.0
32064	72.7	74.2	73.3	75.3	76.0	74.9	78.3	78.1	76.8	83.0	81.4	80.8
34368	73.0	74.5	73.6	75.6	76.3	75.2	78.6	78.4	77.1	83.3	81.7	81.1
44736	74.1	75.6	74.7	76.7	77.4	76.3	79.7	79.5	78.2	84.4	82.8	82.2
X (for other info. rates)	−2.4	−0.9	−1.8	0.2	0.9	−0.2	3.2	3.0	1.7	7.9	6.3	5.7

Notes:
(1) INTELSAT V (F1–F9), and VA (F10–F12).
(2) INTELSAT VA(IBS), (F13–F15).
(3) INTELSAT VI (F1–F5).
(4) All EIRP values assume the use of the low gain step except for those associated with INTELSAT VA(IBS) which assume the use of the high gain step.

*The EIRP values for INTELSAT VA(IBS) (F13–F15) and VI (F1–F5) have been calculated for uplink earth stations located on the inner contour of the satellite receive antenna beam. The EIRP for uplink stations located between the inner and outer contour of the satellite receive antenna beam can be up to: 5 dB higher for VA(IBS) (F13–F15); and 3 dB higher for VI (F1–F5). The actual adjustment will be supplied by INTELSAT when the location of the earth station is known.

The SCPC/QPSK and SPADE systems use compatible frequency assignment schemes. The SCPC/QPSK terminal channels have four functional configurations:

1. Voice and voice band data at or below 4800 bps using the conventional-SPADE-type voice channel unit
2. Voice or voice band data above 4800 bps using (120, 112) FEC in the voice channel unit
3. Digital data at 48 or 50 kbps using rate $\frac{3}{4}$ convolutional coding
4. Digital data at 56 kbps using rate $\frac{7}{8}$ convolutional coding

Table 6.26 summarizes characteristics for voice operation and Table 6.27 summarizes those for data operation. The applicable standard is IESS-303.

6.10.2.10 Intermediate Data Rate (IDR) Digital Carrier System

Intermediate data rate (IDR) digital carriers utilize coherent QPSK modulation operating at information rates ranging from 64 kbps to 44.736 Mbps. The information rate is defined as the bit rate entering a channel unit prior to the

Table 6.29B   Maximum 14.11 or 14/12 GHz direct links (east to west) EIRP requirements, dBW (72 and 77 MHz Transponders) (4, 5)

Carrier Information Rate (kbit/s)	Receive Earth Station											
	Previous			Revised C			E-3			E-2		
	(1)	(2)	(3)	(1)	(2)	(3)	(1)	(2)	(3)	(1)	(2)	(3)
64	47.6	48.0	47.7	49.6	50.7	51.2	52.8	54.0	54.6	57.4	59.0	59.4
192	52.4	52.8	52.5	54.4	55.4	56.0	57.6	58.7	59.4	62.2	63.7	64.1
384	55.4	55.8	55.5	57.4	58.5	59.0	60.6	61.8	62.4	65.2	66.7	67.1
1544	61.4	61.8	61.6	63.5	64.5	65.0	66.7	67.8	68.5	71.3	72.8	73.2
2048	62.6	63.1	62.8	64.7	65.7	66.2	67.9	69.0	69.7	72.5	74.0	74.4
6312	67.5	67.9	67.7	69.9	70.6	71.1	72.8	73.9	74.6	77.4	78.9	79.3
8448	68.8	69.2	69.0	70.8	71.9	72.4	74.0	75.2	75.8	78.6	80.2	80.6
32064	74.6	75.0	74.8	76.7	77.7	78.2	79.9	81.0	81.7	84.5	+	+
34368	74.9	75.3	75.1	77.0	78.0	78.5	80.2	81.3	82.0	84.8	+	+
44736	76.0	76.6	76.2	78.1	79.1	79.6	81.3	82.4	83.1	85.9	+	+
X (for other info. rates)	−0.5	−0.1	−0.3	1.6	2.6	3.1	4.8	5.9	6.6	9.4	10.9	11.3

Notes:
(1) INTELSAT V (F1–F9), and VA (F10–F12).
(2) INTELSAT VA(IBS) (F13 and F15).
(3) INTELSAT VI (F1–F5).
(4) All EIRP values assume the use of the low gain step except for those associated with INTELSAT VA(IBS) which assume the use of the high gain step.
(5) + denotes carriers not intended for this connection.

Table 6.30   QPSK Characteristics and transmission parameters for IDR carriers

Parameter	Requirement
1. Information rate (IR)	64 kbit/s to 44.736 Mbit/s
2. Overhead data rate for carriers with IR $\geq$ 1.544 Mbit/s	96 kbit/s
3. Forward error correction encoding	Rate 3/4 convolutional encoding/Viterbi decoding
4. Energy dispersal (scrambling)	As per CCITT Rec. V. 35
5. Modulation	4-Phase Coherent PSK
6. Ambiguity resolution	Combination of differential encoding (180°) and FEC (90°)
7. Clock recovery	Clock timing must be recovered from the received data stream
8. Minimum carrier bandwidth (allocated)	0.7 R Hz or [0.933 (IR + Overhead)]
9. Noise bandwidth (and occupied bandwidth)	0.6 R Hz or [0.8 (IR + Overhead)]
10. E_b/N_0 at BER(Rate 3/4 FEC)	10^{-3} 10^{-7} 10^{-8}
a. Modems back-to-back	5.3 dB 8.3 dB 8.8 dB
b. Through satellite channel	5.7 dB 8.7 dB 9.2 dB
11. C/T at nominal operating point	$-219.9 + 10\log_{10}(\text{IR} + \text{OH})$, dBW/K
12. C/N in noise bandwidth at nominal operating point	9.7
13. Nominal bit-error-rate at operating point	1×10^{-7}
14. C/T at threshold	$-222.9 + 10\log_{10}(\text{IR} + \text{OH})$, dBW/K
15. C/N in noise bandwidth at threshold	6.7 dB
16. Threshold bit-error-rate	1×10^{-3}

Source: IESS-308, Rev. 4, Ref. 20. Courtesy of INTELSAT.

Notes:
(1) IR is the information rate in bits per second.
(2) R is the transmission rate in bits per second and equals (IR + OH) times 4/3 for carriers employing Rate 3/4 FEC.
(3) The allocated bandwidth will be equal to 0.7 times the transmission rate, rounded up to the next highest odd integer multiple of 22.5 kHz increment (for information rates less than or equal to 10 Mbit/s) or 125 kHz increment (for information rates greater than 10 Mbit/s).
(4) Rate 3/4 FEC is mandatory for all IDR carriers.
(5) OH = overhead.

Table 6.31 Transmission parameters for INTELSAT recommended IDR carriers (with rate 3/4 FEC)

Information Rate (bit/s)	Overhead Rate (kbit/s)	Data Rate, (IR + OH) (bit/s)	Transmission Rate (bit/s)	Occupied Bandwidth (Hz)	Allocated Bandwidth (Hz)	C/T (dBW/K)	C/N$_0$ (dB – Hz)	C/N (dB)
64 k	0	64 k	85.33 k	51.2 k	67.5 k	−171.8	56.8	9.7
192 k	0	192 k	256.00 k	153.6 k	202.5 k	−167.1	61.5	9.7
384 k	0	384 k	512.00 k	307.2 k	382.5 k	−164.1	64.5	9.7
1.544 M	96	1.640 M	2.187 M	1.31 M	1552.5 k	−157.8	70.8	9.7
2.048 M	96	2.144 M	2.859 M	1.72 M	2002.5 k	−156.6	72.0	9.7
6.312 M	96	6.408 M	8.544 M	5.13 M	6007.5 k	−151.8	76.8	9.7
8.448 M	96	8.544 M	11.392 M	6.84 M	7987.5 k	−150.6	78.0	9.7
32.064 M	96	32.160 M	42.880 M	25.73 M	29125.0 k	−144.8	83.8	9.7
34.368 M	96	34.464 M	45.952 M	27.57 M	32250.0 k	−144.1	84.1	9.7
44.736 M	96	44.332 M	59.776 M	35.87 M	41875.0 k	−148.4	84.8	9.7

Source: IESS-308, Rev. 4, Ref. 20. Courtesy of INTELSAT.

Notes:

(1) The above table illustrates parameters for recommended carrier sizes. However, any other information rate between 64 kbit/s and 44.736 Mbit/s can be used.

(2) C/T, C/N$_0$ and C/N values have been calculated for a 10^{-7} BER and assume the use of Rate 3/4 FEC.

(3) For carrier information rates of 10 Mbit/s and below, carrier frequency spacings will be odd multiples of 22.5 kHz. For greater rates, they will be on multiples of 125 kHz.

(4) Rate 3/4 FEC is manadatory for all IDR carriers.

application of any forward error correction (FEC) coding. All IDR carriers are required to employ rate $\frac{3}{4}$ FEC convolutional coding.

Any information rate from 64 kbps to 44.736 Mbps inclusive may be transmitted. INTELSAT, however, recommends data rates based on CCITT: 64, 192, 384, 1544, 2048, 6312, and 8448 kbps, and 32.064, 34.368, and 44.736 Mbps.

The occupied satellite bandwidth unit for IDR carriers is approximately equal to 0.6 times the transmission rate. The transmission rate is defined as the coded symbol rate. To provide guard bands between adjacent carriers the nominal satellite bandwidth unit is equal to 0.7 times the transmission rate. The actual carrier spacing may be larger, and is determined by INTELSAT based upon the particular transponder frequency plan.

IDR carriers have been designed to provide a service in accordance with CCIR Recs. 522-2, 614, and 579-1. To achieve these requirements, the system is designed to provide a nominal BER of 10^{-7} under clear sky conditions. Under degraded sky conditions, a worst-case BER of 10^{-3} for all but 0.04% of the year is provided. IDR carriers are for operation on the following INTELSAT standard earth stations: A, B, C, E-3, E-2, F-3, and F-2.

Table 6.28 shows typical EIRP requirements for direct 6-GHz links and Table 6.29 shows typical EIRP requirements for direct 14-GHz links. Table 6.30 shows QPSK characteristics and transmission parameters for IDR carriers and Table 6.31 gives other important transmission parameters.

6.11   SATELLITE COMMUNICATIONS, TERMINAL SEGMENT

6.11.1   Functional Operation of a "Standard" Earth Station

6.11.1.1   The Communication Subsystem   Figure 6.41 is a simplified functional block diagram of an earth station showing the communication subsystem only. We shall use this figure to trace a signal through the equipment chain from antenna to baseband. Figure 6.42 is a more detailed functional block diagram of a typical earth station. By "standard" we can assume typically INTELSAT A, B, or C service or regional/national domestic satellite service.

The operation of an earth station communication subsystem in the FDMA/FM mode really varies little from that of an LOS radiolink system. The variances are essentially these:

- Use of low-noise front ends on receiving systems, in some cases cryogenically cooled for large earth stations (e.g., INTELSAT A) and GaAs FETS for smaller terminals
- An HPA with capability of from 200- to 8000-W output
- Larger high-efficiency antennas and feeds
- Careful design to achieve as low a noise as possible

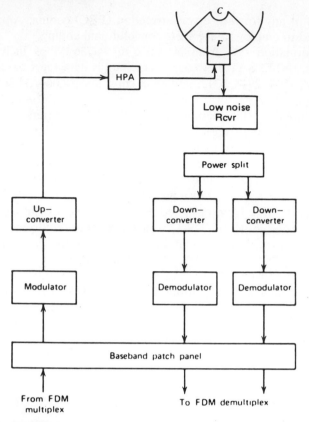

Figure 6.41 Simplified functional block diagram of an earth station communication subsystem. F = feed, HPA = high-power amplifier, C = Cassegrain subreflector.

- Use of a signal-processing technique that allows nearly constant transmitter loading (e.g., spreading waveform) (FM systems)
- Use of threshold extension demodulators in some cases (FM systems only)
- Use of forward error correction (FEC) on many digital systems, and above 10 GHz, possibly with interleaving to mitigate rainfall fading

Now let us trace a signal through the communication subsystem typical of Figure 6.41. On the transmit side the FDM baseband is fed from the mulitplex equipment through the baseband patch facility to the modulator. A spreading waveform is added to the very low end of the baseband to achieve constant loading. The baseband signal is then shaped with a preemphasis network (see Section 4.5.2.2). The baseband so shaped frequency-modulates a carrier, and the resultant is then up-converted to a 70-MHz IF. Patching facilities usually are available at the IF to loop back through the receiver

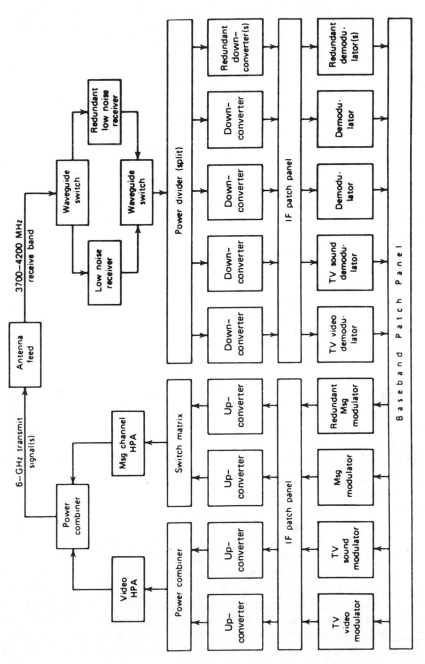

Figure 6.42 More detailed functional block diagram of a communication subsystem, typical of an INTELSAT Standard A earth station.

subsystem or through a test receiver for local testing or troubleshooting. The 70-MHz IF is then fed to an up-converter, which translates the IF to the output frequency (6 or 14 GHz). The signal is then amplified by the HPA, filtered by a low-pass filter, directed to the feed, and radiated by the antenna.

For reception, the signal derives from the feed and is fed to a low-noise receiver. In the case of an INTELSAT Standard A earth station, the low-noise receiver looks at the entire 500-MHz band (i.e., in the case of 4-GHz operation, the band from 3700 to 4200 MHz), amplifying this broadband signal 20–40 dB. When there are long waveguide runs from the antenna to the equipment building, the signal is amplified further by a low-level TWT or solid-state amplifier (SSA) called a driver. The low-noise receiver is placed as close as possible to the feed to reduce ohmic noise contributions to the system. On previous INTELSAT Standard A earth stations (Standard A prior to 1986), the low-noise amplifier (LNA) was commonly cryogenically cooled, usually with liquid helium, achieving a physical temperature of about 12–17 K. The equivalent noise temperature of the receiver, in this case, is on the order of 17–20 K. Smaller earth station facilities as well as new INTELSAT Standard earth stations use GaAs FET LNAs with a noise temperature typically from 70 to 90 K at 4 GHz. The decision regarding which type of LNA to use is driven by the G/T requirements.

The comparatively high-level broadband receive signal is then fed to a power split. There is one output from the power split for every down-converter–demodulator chain. In addition, there is often a test receiver available as well as one or several redundant receivers in case of failure of an operational receiver chain. It should be kept in mind that every time the broadband incoming signal is split into two equal-level paths, there is a 3-dB loss due to the split, plus an insertion loss of the splitter. A splitter with eight outputs will incur a loss of something in the order of 10 dB.

A down-converter is required for each receive carrier, and there will be at least one receive carrier from each distant end. Each down-converter is tuned to its appropriate carrier and converts the signal to the 70-MHz IF. In some instances dual conversion is used.

The 70-MHz IF is then fed to the demodulator on each receive chain. The resulting demodulated signal, the baseband, is reshaped in the deemphasis network (see Section 4.5.2.2) and spreading waveform signal is removed. The resulting baseband output is then fed to the baseband patch facilities and thence to the demultiplex equipment. In many large facilities, threshold extension demodulators are used in lieu of conventional demodulators to achieve a *station margin* (link margin). Threshold extension is a method used to improve FM performance by effectively lowering the FM improvement threshold. With a threshold extension demodulator where the signal modulation index is 3 or better, the amount of improvement that can be expected over conventional FM demodulators is on the order of 7 dB. For example, if the FM improvement threshold for a receiver without extension were − 126.2 dBW, then with extension it would be − 133.2 dBW.

Threshold extension works on an FM feedback principle that reduces the equivalent instantaneous deviation, thereby reducing the required IF bandwidth B_{IF}, which in turn effectively lowers the receiver noise threshold. A typical receiver with a threshold extension module may employ a tracking filter that instantaneously tracks the deviation with a steerable bandpass filter having a 3-dB bandwidth of approximately four times the top baseband frequency. The control voltage for the filter is derived by making a phase comparison between the feedback signal and the IF input signal.

It should be noted that the term often used to describe service is "message" service, which connotes common telephony service. When we see the terms "message up-converter" or "message demodulator," they refer to message service, which is actually telephony service. This is in opposition to the equipment used to carry video picture, TV sound, or program channel traffic.

6.11.1.2 The Antenna Subsystem

The antenna subsystem is one of the most important component parts of an earth station, since it provides the means of radiating signals to the satellite and/or collecting signals from the satellite. The antenna not only must provide the gain necessary to allow proper transmission and reception, but also must have the radiation characteristics that discriminate against unwanted signals and minimize interference into other satellite or terrestrial systems.

Earth stations most commonly use parabolic dish reflector antennas or derivatives thereof. Dish diameters range from 1 to 30 m.

The sizing of the antenna and its design are driven more by the earth station required G/T than the EIRP. The gain is basically determined by the aperture (e.g., diameter) of the dish, but improved efficiency can also add to the gain on the order of 0.5–2 dB. For this reason the Cassegrain feed technique is almost always used on larger earth terminal installations. Smaller military terminals also resort to the use of Cassegrain. In some cases efficiencies as high as 70% and more have been reported. The Cassegrain feed working into a parabolic dish configuration (Figure 6.43) permits the feed to look into cool space as far as the spillover from the subreflector is concerned.

Antennas of generally less than 30-ft diameter often use a front-mounted feed horn assembly in the interest of economy. The most common type is the prime focus feed antenna (Figure 6.44). This is a more lossy arrangement, but, since the overall requirements are more modest, it is an acceptable one.

In order to keep interference levels on both the up- and downlinks to acceptable levels, antenna side-lobe envelope limits (in dB) of $32 - 25 \log \theta$ relative to the main beam maximum level have been internationally adopted; θ is the angular distance in degrees from the main beam lobe maximum.

Figure 6.43 shows the functional operation of a Cassegrain-fed antenna. Such an antenna consists of a parabolic main (prime) reflector and a hyperbolic subreflector. Here, of course, we refer to truncated parabolic and

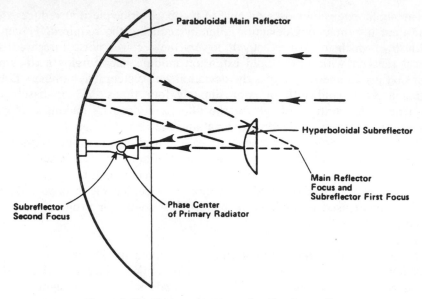

Figure 6.43 Cassegrain antenna functional operation.

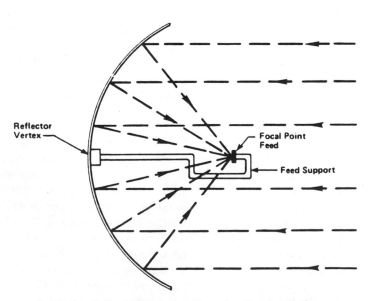

Figure 6.44 Prime focus antenna, functional operation.

hyperbolic surfaces. The subreflector is positioned between the focal point and the vertex of the prime reflector. The feed system is situated at the focus of the subreflector, which also determines the focal length of the system. Spherical waves emanating from the feed are reflected by the subreflector. The wave then appears to be emanating from the virtual focus. These waves are then, in turn, reflected by the primary reflector into a plane wave traveling in the direction of the axis of symmetry. The size of the aperture (diameter) of the prime reflector determines the gain.

The gain of a uniformly illuminated parabolic dish antenna, which has a diameter of D ft operating at a frequency of F MHz, can be expressed by

$$G_{dB} = 20 \log F_{MHz} + 20 \log D_{ft} + 10 \log \eta - 49.92 \text{ dB} \qquad (6.35)$$

where η is the aperture efficiency expressed as a percentage (a decimal). The value of η is usually in the range of 0.50 to 0.70.

It can be shown that a uniform field distribution over the reflector gives the highest gain of all constant phase distributions. The aperture efficiency can be shown to be a function of

- Phase error loss
- Illumination loss due to a nonuniform amplitude distribution over the aperture
- Spillover loss
- Cross-polarization loss
- Blockage loss due to the feed, struts, and subreflector
- Random errors over the surface of the reflector (e.g., surface tolerance)

Nearly all very high gain reflector antennas are of the Cassegrain type. Within the main beam the antenna behaves essentially like a long focal-length front-fed parabolic reflector. Slight shaping of the two reflector surfaces can lead to substantial gain enhancement. Such a design also leads to more uniform illumination of the main reflector and less spillover. Typically, efficiencies of Cassegrain-type antennas are from 65 to 70%, which is at least 10% above most front-fed designs. It also permits the LNA to be placed close to the feed, if desired. The ability of the antenna to achieve these characteristics rests largely with the feed horn design. The feed horn radiation pattern has to provide uniform aperture illumination and proper tapering at the edges of the aperture.

The simplest type of feed for the antenna is a waveguide, which can be either open ended or terminated with a horn. Both rectangular and circular waveguide have been used, with the circular considered superior, since it produces a more uniform illumination pattern over the aperture and provides better cross-polarization loss characteristics. The illumination pattern should taper to a value of -10 dB at an angle corresponding to the edge of the reflector (subreflector). This results in an asymmetric feed radiation pattern, causing a loss in efficiency, increased cross-polarization losses, and moderate reflector spillover. Cross-polarization loss can be reduced by careful selection of waveguide radius, while improvement in efficiency can be achieved by using a corrugated horn. The positive aspect to 10-dB taper is to reduce side lobe levels.

Smaller earth stations with less stringent G/T requirements often resort to using the less expensive prime-focus-fed parabolic reflector antenna. The functional operation of this antenna is shown in Figure 6.44. For intermediate aperture sizes, this type of antenna has excellent side lobe performance in all angular regions except the spillover region around the edge of the reflector. Even in this area a side lobe suppression can be achieved that will satisfy FCC/CCIR pattern requirements. The aperture efficiency for apertures greater than about 100 wavelengths is around 60%. Hence it represents a good compromise choice between side lobes and gain. For aperture sizes less than about 40 wavelengths, the blockage of the feed and feed support structure raises side lobes with respect to the peak of the main beam such that it becomes exceedingly difficult to meet the FCC/CCIR side lobe specification. However, the CCIR specification can be met, since it contains a modifier that is dependent on the aperture size.

Polarization. By use of suitable geometry in the design of an antenna feedhorn assembly, it is possible to transmit a plane electric wavefront in which the E and H fields have a well-defined orientation. For linear polarization of the wavefront, the convention of vertical or horizontal electric (E) field is adopted. The generation of linearly polarized signals is based on the ability of a length of square section waveguide to propagate a field in the $TE_{1,0}$ and $TE_{0,1}$ modes, which are orthogonally oriented. By exciting a short length of square waveguide in one mode by the transmitted signal and extracting signals in the orthogonal mode for the receiver, a means is provided for cross-polarizing the transmitted and received signals. The square waveguide is flared to form the antenna feed horn. Similarly, by exciting orthogonal modes in a section of circular waveguide, a left- and right-hand rotation is imparted to the wavefront, providing two orthogonal circular polarizations.

The waveguide assembly used to obtain this dual polarization is known as a diplexer or an "orthomode junction." In addition to generating the polarization effect, the diplexer has the advantage of providing isolation between the transmitter and the receiver ports that could be in the range of 50 dB if the antenna presented a true broadband match. In practice, some 25–30 dB or more of isolation can be expected.

For linear polarization, discrimination between received horizontal and vertical fields can be as high as 50 dB, but diurnal effects coupled with precipitation can reduce this value to 30 dB. To maintain optimum discrimination, large antennas are equipped with a feed-rotating device drive by a polarization-sensing servo loop. Circularly polarized fields do not provide much more than 30-dB discrimination, but they tend to be more stable and do not require polarity tracking. This makes circular polarization more suitable for systems that include mobile terminals. One of the important parameters of antenna performance is how well cross-polarization is preserved across the operating spectrum.

Polarization discrimination can be used to obtain frequency reuse. Alternate transponder channel spectra are allowed to overlap symmetrically and are provided with alternate polarization. The channelization schemes of INTELSAT V and VI are typical. The number of transponders in the satellite is effectively doubled by this process. Polarization discrimination can also provide a certain degree of interface isolation between satellite networks if the nearest satellite (in terms of angular orbit separation) uses orthogonal uplink and downlink polarizations (Ref. 13).

Antenna Pointing and Tracking. Satellites orbiting the earth are in motion. They can be in geostationary orbits and inclined orbits. Those in geostationary orbits appear to be stationary with respect to a point on earth. Those that are in inclined orbits are in motion with respect to a point on earth. All satellite terminals working with this latter class of satellite require a tracking capability.

Even though we have said that geostationary satellites appear stationary relative to a point on earth, they do tend to drift in small suborbits (figure eights). However, even with improved satellite stationkeeping, the narrow beamwidths encountered with large earth station antennas, such as the INTELSAT Standard A 30-m (100-ft),* require precise pointing and subsequent tracking by the earth station antenna to maximize the signal to the satellite and from the satellite. The basic modes of operation to provide these capabilities are

- Manual pointing
- Programmed tracking (open-loop tracking)
- Automatic tracking (closed-loop tracking)

Pointing deals with "aiming" the antenna initially on the satellite. Tracking keeps it that way. Programmed tracking (open-loop tracking) may assume both duties. With programmed tracking, the antenna is continuously pointed by interpolation between values of a precomputed time-indexed ephemeris. With adequate information as to the actual satellite position and true satellite terminal position, pointing resolutions are on the order of 0.03–0.05°.

Manual pointing may be effective for initial satellite acquisition or "capture" for later active tracking (closed-loop tracking). It is also effective for

*Standard A prior to 1986.

wider beamwidths antennas, where the beamwidth is sufficiently wide to accommodate the entire geostationary satellite suborbit. Midsized installations may require a periodic trim up, and some smaller installations need never be trimmed up, assuming, of course, that the satellite in question is keeping good stationkeeping.

We will now discuss three types of active or closed-loop tracking: monopulse, step-track, and conscan.

MONOPULSE TRACKING.   Monopulse is the earliest form of satellite tracking, and today it is probably still the most accurate. In monopulse tracking, multiple antenna feed elements are used to obtain multiple received signals. The relative signal levels received by the various feed elements are compared to provide azimuth and elevation-angle-pointing error signals. The error signals are used to control the servo system, which operates the antenna drive motors.

Monopulse has taken its name from radar technology, and it derives from the fact that all directional information is obtained from a single radar pulse. Beam switching or mechanical scanning is not necessary for its operation. In a three-channel monopulse system, RF signals are received by four antenna elements, usually horn feeds, located symmetrically around the boresight axis as an integral part of the antenna feed system. The multiple RF receive signals derived from these horns are combined in a beam-forming network (hybrid comparators) to produce sum and difference signals simultaneously in orthogonal planes. Figure 6.45 is a functional block diagram of the front end of a typical monopulse tracking system. The sum of the four-element radiation pattern is characterized by a single beam whose maximum lies on the antenna axis. The difference patterns are characterized by a null on the antenna axis with the lobes on opposite side in antiphase. Because the difference radiation pattern is in phase with respect to the sum pattern on one side of the antenna axis and out of phase on the other side, bearing angle sense information can be derived. By applying the sum signal to one input of a detector and the difference signal to the other input, an error voltage is produced that is proportional to the angle off-center and whose polarity is determined by the direction off-center.

In large earth stations such as INTELSAT Standard A, the sum and the two difference RF signals are kept separate. They are down-converted and applied to a three-channel tracking receiver to generate azimuth and elevation error signals for the antenna servo drive system. In other satellite terminals employing the monopulse technique, the azimuth and elevation difference signals are commutated onto a single channel, often by ferrite switches controlled by a digital scan generator. The commutated output of the ferrite switches is added to the communications (sum) channel by a directional coupler. Since the difference and sum signals are phase coherent, this has the effect of amplitude modulating the sum signal. This latter

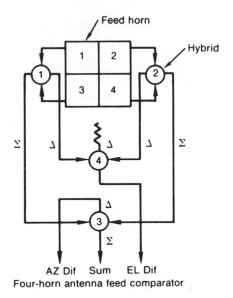

Four-horn antenna feed comparator

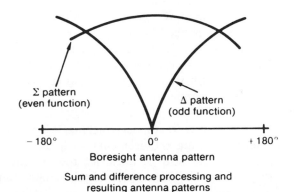

Boresight antenna pattern

Sum and difference processing and
resulting antenna patterns

Figure 6.45 Functional operation of three-channel monopulse tracking. (From Ref. 21.)

implementation of monopulse reduces the needed equipment from a three-channel tracking receiver to a single-channel receiver.

The satellite signal used for tracking is usually the satellite beacon channel. Thus, in this single-channel monopulse tracking design, the modulated sum channel is amplified by an LNA, down-converted, and demodulated in the beacon receiver. The signals are next decommutated and phase detected to obtain error voltages that are used to drive antenna so that the azimuth and elevation difference signals are minimized. Figure 6.46 is a functional block diagram of a typical satellite terminal monopulse tracking receiver subsystem.

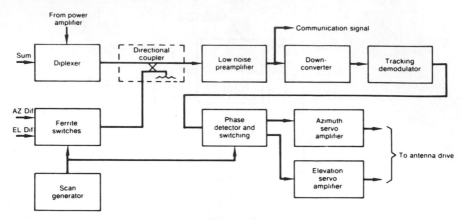

Figure 6.46 Functional block diagram of a satellite terminal monopulse tracking receiver subsystem.

The complexity and cost of monopulse feeds arise from the fact that they must be packaged into a small volume and, at the same time, provide low mutual coupling among the units without obstructing the illumination characteristics of the communications feed horn.

With monopulse tracking the beam scanning can be performed at almost any arbitrarily high rate, therefore providing the potential for high tracking rates. On the other hand, with step-track the tracking rate is limited by the dynamics of offsetting the antenna.

STEP-TRACK.   Despite the limitation of step-tracking, it is cost-effective where there are low dynamic tracking requirements such as with geostationary antennas of medium aperture size. It does not require the complex feed arrangement of the monopulse system and requires only simple, low-cost electronics. The only input signal required from the satellite terminal is the automatic gain control (AGC) voltage or other dc signal proportional to the received RF signal level, such as a signal level indication from a communications demodulator or from a beacon receiver. The output of the step-track processor algorithm can be as simple as a periodic step function for each antenna axis or as complex as a pseudorandom sinusoid. This type of output applied to the antenna servo drive subsystem will result in smooth, continuous antenna motion. Figure 6.47 is a simplified functional block diagram of a step-track system.

Step-tracking is a considerably lower cost tracking system when compared to its monopulse counterpart. It is also less accurate. It lends itself to midsized satellite terminal installations operating with geostationary satellites and to tracking some inclined satellites with relatively slow relative orbital motion.

In the step-track technique the antenna is periodically moved a small

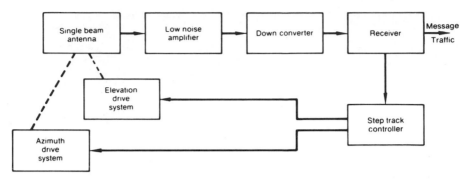

Figure 6.47 Functional block diagram of a step-track system.

amount along each axis, and the level of the received signal is compared to its previous level. A microprocessor, or part of the terminal control processor, provides processing to convert these level comparisons into input signals for the servo system, which will drive the antenna in directions that maximize the received signal level. In contrast to the monopulse technique, which seeks the null of the antenna difference pattern, step-tracking seeks the signal peak. Locating a beam maximum can never be as accurate as finding a sharp null as with the monopulse tracking technique.

CONICAL SCAN TRACKING (CONSCAN). Conical scan tracking is a refinement of the old antenna lobing technique used in World War II radars. Some of these original tracking radars used an array of radiating elements that could be switched in phase to provide two beam positions for the lobing operation. The radar operator observed on his display the same target side by side, which were the returns of the two beam positions. When the target was on axis, the two pulses were of even amplitude, and when moved off axis, the two pulses became unequal. To track a remote target, all the operator did was to maintain a balance between the two pulses by steering the antenna correctly.

This lobing technique was refined to a continuous rotation of the beam around the target, which is the basis of conscan tracking. Angle-error-detection circuitry is provided to generate error voltage outputs proportional to the tracking error and with a phase or polarity to indicate the direction of the error. The error signal actuates a servo system to drive the antenna in the proper direction to null the error to zero.

One method to accomplish this continuous beam scanning is by mechanically moving the antenna feed, since the antenna beam will move off axis as the feed is moved off the focal point. The feed is typically moved in a circular path around the focal point, causing a corresponding movement of the antenna beam in a circular path around the satellite to be tracked.

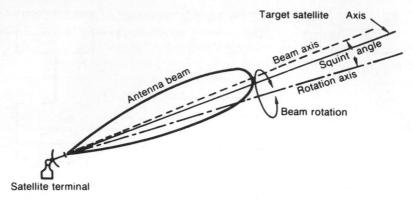

Figure 6.48   Conscan tracking operation.

The feed scan motion may be either by a rotation or a nutation. A rotating feed turns as it moves with a circular motion, causing the polarization to rotate. A nutating feed does not rotate with the plane of polarization during the scan; it has an oscillatory movement of the axis of a rotating body or wobble. This is sometimes accomplished in the subreflector. The rotation or wobble modulates the received signal. The percentage of modulation is proportional to the angle tracking error, and the phase of the envelope function relative to the beam-scanning position contains direction information. In other words, this modulation is compared in phase with quadrature reference signals generated by the nutating mechanism to obtain error direction, and the amplitude of the modulation is proportional to the magnitude of the error. Figure 6.48 shows the conscan tracking technique and Figure 6.49 is a simplified functional block diagram of a conscan subsystem.

Conscan tracking may be used to track geosynchronous or polar orbit satellites with high or low target dynamics. Its primary advantages are low

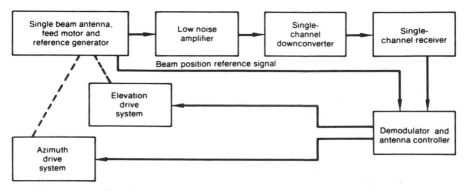

Figure 6.49   Simplified functional block diagram of a conical scanning subsystem.

cost, that it requires only one RF channel, and that it has a single-beam feed. Its principal disadvantages are that it requires four pulses or short-duration continuous signal to obtain tracking information and is subject to mechanical reliability problems due to continuous feed rotation, and to tracking loss due to propeller and other low-rate modulation sources.

PROGRAM TRACKING (OPEN-LOOP TRACKING).   Program tracking is a processor-based tracking system where the processor calculates the required azimuth and elevation angles as a function of time. Such a processor is called an ephemeris processor. Ephemeris (plural: ephemerides) refers to a tabulation of satellite locations referenced to a time scale. The ephemeris processor uses an algorithm that calculates the relative direction of the satellite with respect to the terminal on a continuous and real-time basis. It contains in memory the forecast satellite location with respect to time, which requires periodic updating every 30 or 60 days. For fixed terminal sites, site location latitude and longitude are programmed into the processor at installation; for mobile terminals the processor requires continuous positional updates, often provided by an inertial navigation system.

Since program tracking is an open-loop process, it is subject to several sources of error that are automatically corrected in autotrack (closed-loop) systems. Some of these error sources are atmospheric refractions; structural deformation due to wind, ice, and gravity loads on the antenna; misalignment of mechanical axes; errors in axis-angle measuring devices; and errors in input data relating to terminal position, satellite ephemeris, and absolute time.

In practice, a method of closed-loop tracking is usually included as well, since the accuracy required in the program tracking system is greatly relaxed if the last few tenths of a decibel in tracking accuracy are not required. In such cases program tracking is primarily used as an aid to initial satellite acquisition. However, for a geostationary satellite, acquisition causes little difficulty. In most cases a "lookup table," such as shown in Figure 6.3, suffices. Program tracking, therefore, usually has application for terminals that have to rapidly acquire nongeostationary satellites or rapidly slew from one satellite to another, as in the case of many military applications. Military ephemeris processors may have in memory ephemeris data for up to 20 satellites.

REVIEW EXERCISES

1. Give at least six major applications of satellite communication.

2. What transmission technique/medium is now tempering the growth of satellite communication?

3. What are the two broad generic categories of communication satellites? (Hints: Consider function. One is fully mature and the second is just beginning to emerge.)

4. Draw a simplified block diagram of a "bent pipe" communication satellite (transponder).

5. The text listed three types of satellite orbits. Which is almost universally used by Western nations? Give at least two advantages of using this type of orbit; give one disadvantage.

6. Give one reason why satellite down links are always in a lower frequency band than their companion uplinks.

7. Calculate the EIRP in dBW of an earth terminal with an antenna with a gain of 40 dB, transmitter output of 200 W, and transmission line losses of 2.3 dB.

8. What is the free-space loss on a 12-GHz downlink for a geostationary satellite where the earth terminal uses a 23° elevation angle?

9. What is the isotropic receive level (IRL) at an earth terminal antenna where the free-space loss to the associated satellite is 196.4 dB and other link losses are 2.6 dB? The satellite EIRP is +34 dBW.

10. Why must the flux density of a satellite signal at the earth's surface be limited?

11. What is the receiving system noise temperature (T_{sys}) if the antenna noise temperature is 300 K and the receiver noise temperature is 100 K?

12. Why is receiving system noise temperature so important in satellite communication, when it was a very secondary issue in the design of LOS microwave systems?

13. The noise figure of a low-noise amplifier (LNA) is 1.2 dB. What is its equivalent noise temperature in K?

14. List at least three contributors to receiving system noise temperature.

15. Calculate N_0 of a receiving system with a noise temperature of 100 K.

16. Determine C/N_0 at an earth terminal of a satellite downlink where the RSL is -139 dBW and the system noise temperature is 400 K.

17. State G/T as a mathematical identity (equation). Where would we measure G in dB?

18. Calculate the antenna noise temperature (T_a) for an earth terminal receiving system where the sum of the ohmic losses from the antenna feed to the input of the LNA is 1.7 dB. The sky noise contribution is 15 K.

19. Determine G/T for a satellite downlink receiving system where the antenna gain is 42 dB, ohmic losses from the antenna feed to the input to the LNA are 2.1 dB, sky noise is 12 K, and the LNA noise figure is 1.5 dB and has 30 dB of gain.

20. Consider problem 19. Why, especially in this case, can we generally neglect noise contributions from receiver stages that follow the LNA? Suppose the LNA only had a 10-dB gain. Explain why, in this case, where latter stages may make some considerable noise contribution to the system noise temperature, we now must take these contributions into consideration.

21. Name at least four parameters of an earth terminal that will affect G/T.

22. Besides free-space loss, name at least four other loss contributors to a link budget.

23. Calculate the required geostationary satellite G/T where the uplink frequency is 14 GHz, the terminal EIRP is + 76 dBW, and the terminal antenna elevation angle 10°. The required C/N_0 at the satellite is 90 dB. Allow 6 dB for excess attenuation due to rainfall (margin), terminal pointing loss of 0.2 dB, satellite pointing loss of 0.5 dB, polarization loss of 0.5 dB, and atmospheric absorption loss of 0.5 dB. (Use Figure 6.3 in the text to compute range to satellite.)

24. Determine system C/N_0 where the uplink C/N_0 is 90 dB and the downlink C/N_0 is 85 dB.

25. What is the major drawback to the use of TWT-based high-power amplifiers in a satellite using multicarrier operation?

26. List the three basic access techniques of a satellite transponder (SCPC does not count).

27. Why is uplink power control necessary for multicarrier FDMA operation?

28. Give at least two important major applications of an SCPC/DAMA system.

29. List three methods of control for a DAMA system.

30. Compare TDMA with FDMA. List at least three advantages and three disadvantages of each.

31. Why must TDMA operate exclusively in a digital mode?

32. Discuss TDMA guard time between bursts. Why is it desirable to shorten the guard time as much as possible and how does this impact total system cost?

33. What is the function of the CR/BTR in TDMA frame overhead?

34. Differentiate open-loop and closed-loop timing on TDMA systems.

35. What is the most primitive form of a processing satellite. What advantages accrue for the system? What functions are required that are not required in a bent pipe satellite?

36. List at least four advantages of a processing satellite.

37. What are the two transmission media that are candidates for satellite crosslinks?

38. Define error correction coding, often called FEC. List three advantages of FEC.

39. What is the two generic types of FEC coding? Name some advantages and disadvantages of each.

40. Define code rate. Give a simple formula for code rate.

41. Coding gain can offer one advantage and one disadvantage. What are they?

42. FEC is essentially designed to correct random errors. How can it be used to correct burst errors such as fading caused by rainfall or multipath?

43. Block codes are often defined with the notation (N, K, t). What parameter is given by each letter in this notation?

44. With convolutional codes, there are three common types of decoders. List them in ascending order of available coding gain and complexity.

45. A rate $\frac{3}{4}$ convolutional code is used on a 300-bps information rate bit stream. It is fed to a coder. What is the symbol rate out of the coder?

46. Coding gain is a function of _____. List at least four factors.

47. Assume a processing satellite. Calculate the terminal G/T under the following conditions: digital downlink at 14 GHz; 10-Mbps information data rate; BER $= 1 \times 10^{-6}$; convolutional coding $R = \frac{1}{2}$; $K = 7$; soft-decision sequential decoder where $Q = 3$; elevation angle 19°; total margin 10 dB, including rainfall margin. Modulation is QPSK with a modulation implementation loss of 1.5 dB. Other miscellaneous link losses are 2.5 dB. Radome loss is 1.0 dB. The satellite EIRP is $+44$ dBW.

48. What is/are the purpose(s) of a hub in a VSAT network?

49. Give six typical applications of VSATs.

50. What are typical bit rates on VSAT networks?

51. List and discuss three inbound VSAT time division access schemes. Compare these with SCPC/DAMA access schemes.

52. Why is the 14/12-GHz band preferred for VSAT operation, as opposed to the 6/4-GHz band? Discuss the pros and cons.

53. Discuss advantages of VSAT networks over other means of connectivity. We expect to see in the answer terms such as availability, cost, survivability (e.g., point failure), and customer satisfaction.

54. The INTELSAT SPADE earth terminal requires an EIRP of +60.5 dBW. Prepare a table with transmit powers of 10 W, 50 W, 100 W, 200 W, 500 W, and 1 kW, allowing 2-dB line losses. Assume antenna efficiencies of 65%. Show the corresponding antenna gains to meet the EIRP requirement. Allow a 2-dB uplink margin for deteriorated (non-clear sky) conditions.

55. We wish to have 1.544-Mbps connectivity between New York and Hamburg, Germany. We will lease the necessary space segment from INTELSAT. Using the text, assume INTELSAT VI at about 40° west longitude. Configure these earth station facilities using an INTELSAT standard earth station. What INTELSAT (standard) earth station and list features that INTELSAT offers. Compare these with what a customer may require.

56. Show at least six ways a satellite earth terminal differs from its LOS microwave counterpart.

57. Calculate the gain of a parabolic dish antenna operating at 14 GHz with a 7-m aperture and an aperture efficiency of 65%.

58. Why is a Cassegrain-type feed favored over the more conventional prime focus feed for satellite earth terminals?

59. Give the three common methods of pointing/tracking for a satellite earth terminal. Discuss advantages and disadvantages of each. (Hint: Remember cost!)

60. There are three methods of active tracking used on satellite communication terminals. List the three in declining order of complexity and briefly describe how each operates.

REFERENCES AND BIBLIOGRAPHY

1. Roger L. Freeman, *Reference Manual for Telecommunications Engineering*, Wiley, New York, 1985.
2. "Maximum Permissible Values of Power Flux-Density at the Surface of the Earth Produced by Satellites in the Fixed-Satellite Service Using the Same Frequency Bands above 1 GHz as Line-of-Sight Radio-Relay Systems," CCIR Rec. 358-3, XVIth Plenary Assembly, Dubrovnik, 1986.
3. *Transmission Systems for Communications*, 5th ed., Bell Telephone Laboratories, Holmdel, NJ, 1982.

4. "Radio Emission from Natural Sources in the Frequency Range above about 50 MHz," CCIR Rep. 720-2, XVIth Plenary Assembly, Dubrovnik, 1986, vol. V.

5. "Satellite Communication Symposium," Scientific Atlanta, Atlanta, GA, 1982.

6. "Performance Characteristics of Frequency Division Multiplex/Frequency Modulation (FDM/FM) Telephony Carriers (6/4 GHz and 14/11 GHz Frequency Bands)," IESS-301, Rev. 1, INTELSAT, Washington, D.C., 1986.

7. "Energy Dispersal in the Fixed Satellite Service," CCIR Rep. 384-5, XVIth Plenary Assembly, Dubrovnik, 1986, vol. IV.

8. H. L. Van Trees, Section 3.6.5 in "Demand Assignment," *Satellite Communications*, IEEE Press, New York, 1979.

9. Roger L. Freeman, *Telecommunication System Engineering*, 2nd ed., Wiley, New York, 1989.

10. H. L. Van Trees, Section 3.6.3.1 in "Time Division Multiple Access, *Satellite Communications*, IEEE Press, New York, 1979.

11. V. K. Bhargava et al., *Digital Communications by Satellite*, Wiley, New York, 1981.

12. K. Koga, T. Muratani, and A. Ogawa, "On-board Regenerative Repeaters Applied to Digital Satellite Communications," *Proceedings of the IEEE*, Mar. 1977.

13. N. R. Richards, *Satellite Communications Reference Data Handbook*, Computer Science Corp., Falls Church, VA., 1988, under DCA contract DCA-100-81-C-0044.

14. A. J. Viterbi, "Convolutional Codes and Their Performance in Communication Systems," *IEEE Trans. Commun.*, Vol. 19, Oct. 1971.

15. J. P. Odenwalder, *Error Control Coding Handbook*, Linkabit Corp., San Diego, CA, July 1976, under USAF contract F44620-76-C-0056.

16. X. T. Vyong, F. S. Zimmerman, and T. M. Shimabukuro, "Performance Analysis of Ku-band VSAT Networks," *IEEE Commun. Mag.*, vol. 26, no. 5, May 1988.

17. D. Raychaudhuri and K. Joseph, "Channel Access Protocols for Ku-band VSAT Networks: A Comparative Evaluation," *IEEE Commun. Mag.*, vol. 26, no. 5, May 1988.

18. D. Chakraborty, "VSAT Communication Networks: An Overview," *IEEE Commun. Mag.*, vol. 26, no. 5, May 1988.

19. INTELSAT Earth Station Standards, "INTELSAT Space Segment Leased Transponder Definition and Associated Operating Conditions (INTELSAT V, VA, VA(IBS) and VI Satellites)," INTELSAT, Washington, D.C., Mar. 15, 1988.

20. INTELSAT Earth Station Standards: Standard A, IESS-201; Standard B, IESS-202; Standard C, IESS-203; Standard D, IESS-204; Standard E, IESS-205; Standard F, IESS-206, and appropriate 300 series IESS. INTELSAT, Washington, D.C., (various dates).

21. Roger L. Freeman, *Radio System Design for Telecommunications (1–100 GHz)*, Wiley, New York, 1987.

22. INTELSAT Earth Station Standards, "INTELSAT VII Satellite Characteristics," IESS-409, Washington, D.C., June 1989.

23. P. T. Thompson and R. Silk, "INTELSAT VII: Another Step in the Development of Global Communications," *J. Br. Interplanet. Soc.*, vol. 43, no. 9, Aug. 1990.

24. R. C. Dixon, *Spread Spectrum Systems*, Wiley, New York, 1984.

7

RADIO SYSTEM DESIGN
ABOVE 10 GHz

7.1 THE PROBLEM — AN INTRODUCTION

There is an ever-increasing demand for radio-frequency spectrum in the industrialized nations of the world. This is due to the information transfer explosion in our society, resulting in a rapid increase in telecommunication connectivity, and the links satisfying that connectivity are required to have ever-greater capacity.

The most desirable spectrum to satisfy these needs is the band between 1 and 10 GHz. It is called the "noise window," where galactic and man-made noise are minimum. Atmospheric absorption may generally be neglected in this region.

Congestion in the 1–10 GHz region has forced us to look above 10 GHz for operational frequencies. By careful engineering we have found that frequencies above 10 GHz can give equivalent performance or nearly equivalent performance to those below 10 GHz.

We have arbitrarily selected 10 GHz as a demarcation line. Generally, below 10 GHz, in radiolink design, we can neglect excess attenuation due to rainfall and atmospheric absorption. For frequencies above 10 GHz, excess attenuation due to rainfall and atmospheric absorption can have an overriding importance in radiolink design. In fact, certain frequency bands display so much gaseous absorption that they are unusable for many applications.

The principal thrust of this chapter is to describe techniques for band selection and link design for line-of-sight (LOS) microwave and earth–space–earth links for frequencies above 10 GHz. The chapter also deals with low-elevation-angle space links, and how to deal with the effects resulting from elevation angles under 5° or so.

7.2 THE GENERAL PROPAGATION PROBLEM ABOVE 10 GHz

Propagation of radio waves through the atmosphere above 10 GHz involves not only free-space loss but several other important factors. As expressed in Ref. 1, these are

1. The gaseous contribution of the homogeneous atmosphere due to resonant and nonresonant polarization mechanisms
2. The contribution of inhomogeneities in the atmosphere
3. The particulate contributions due to rain, fog, mist, and haze (dust, smoke, and salt particles in the air)

Under item (1) we are dealing with the propagation of a wave through the atmosphere under the influence of several molecular resonances, such as water vapor (H_2O) at 22 and 183 GHz, oxygen with lines around 60 GHz, and a single oxygen line at 119 GHz. These points with their relative attenuation are shown in Figure 7.1.

Other gases display resonant lines as well, such as N_2O, SO_2, O_3, NO_2, and NH_3, but because of their low density in the atmosphere, they have negligible effect on propagation.

The major offender is precipitation attenuation (under items 2 and 3). It can exceed that of all other sources of attenuation in the atmosphere above 18 GHz. Rainfall and its effect on propagation are covered at length in this chapter.

We will first treat total loss due to absorption and scattering. It will be remembered that when an incident electromagnetic wave passes over an object that has dielectric properties different from the surrounding medium, some energy is absorbed and some is scattered. That that is absorbed heats the absorbing material; that that is scattered is quasi-isotropic and relates to the wavelength of the incident wave. The smaller the scatterer, the more isotropic it is in direction with respect to the wavelength of the incident energy.

We can develop a formula from equation 4.1 to calculate total transmission loss for a given link:

$$\text{Attenuation (dB)} = 92.45 + 20 \log f_{\text{GHz}} + 20 \log D_{\text{km}} + a + b + c + d + e \tag{7.1}$$

where f is in gigahertz and D is in kilometers. Also,

a = excess attenuation (dB) due to water vapor
b = excess attenuation (dB) due to mist and fog
c = excess attenuation (dB) due to oxygen (O_2)
d = sum of the absorption losses (dB) due to other gases
e = excess attenuation (dB) due to rainfall

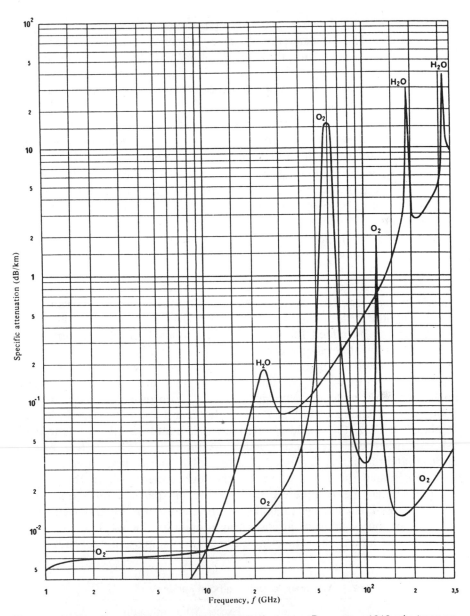

Figure 7.1 Specific attenuation due to atmospheric gases. Pressure = 1013 mb; temperature = 15°C; water vapor = 7.5 g / m³. From CCIR Rep. 719-2, Figure 3, Page 172, Vol. V, XVIth Plenary Assembly, Dubrovnik 1986 [Ref. 2, p. 172, Figure 3]. Courtesy of ITU – CCIR.

Notes and comments on equation 7.1:

1. a varies with relative humidity, temperature, atmospheric pressure, and altitude. The transmission engineer assumes that the water vapor content is linear with these parameters and that the atmosphere is homogeneous (actually horizontally homogeneous but vertically stratified). There is a water vapor absorption band about 22 GHz caused by molecular resonance.

2. c and d are assumed to vary linearly with atmospheric density, thus directly with atmospheric pressure, and are a function of altitude (e.g., it is assumed that the atmosphere is homogeneous).

3. b and e vary with the density of the rainfall cell or cloud and the size of the rainfall drops or water particles such as fog or mist. In this case the atmosphere is most certainly not homogeneous. (Droplets less than 0.01 cm in diameter are considered mist/fog, and more than 0.01 cm, rain.) Ordinary fog produces about 0.1-dB/km excess attenuation at 35 GHz, rising to 0.6 dB/km at 75 GHz.

In equation 7.1 terms b and d can often be neglected; terms a and c are usually lumped together and called "atmospheric attenuation." If we were to install a 10-km LOS link at 22 GHz, in calculating transmission loss, 1.6 dB would have to be added for what is called atmospheric attenuation but is predominantly water vapor absorption, as shown in Figure 7.1.

It will be noted in Figure 7.1 that there are frequency bands with relatively high levels of attenuation per unit distance; some are rather narrow bands and some fairly wide. For example, the O_2 absorption band covers from about 58 to 62 GHz and with skirts down to 50 and up to 70 GHz. At its peak at about 60 GHz, the sea level attenuation is about 15 dB/km. One could ask, "of what use are these bands?" Actually, the 58–62-GHz band is appropriately assigned for satellite cross-links. These links operate out in space far above the limits of the earth's atmosphere, where the terms a through e may be completely neglected. It is particularly attractive on military cross-links having an inherent protection from earth-based enemy jammers by that significant atmospheric attenuation factor. It is also useful for very-short-haul military links such as a ship-to-ship secure communication

Table 7.1 Windows for point-to-point service

Band (GHz)	Excess Attenuation due to Atmospheric Absorption (dB/km)
20 <	0.08 <
28–42	0.13
75–95	0.4
125–140	1.8

system. Again, it is the atmospheric attenuation that offers some additional security for signal intercept (low probability of intercept, LPI) and against jamming.

On the other hand, Figure 7.1 shows a number of bands that are relatively open. These openings are often called windows. Three such windows are suggested for point-to-point service in Table 7.1.

7.3 RAINFALL LOSS

7.3.1 Basic Rainfall Considerations

Of the factors a through e in equation 7.1, factor e, excess attenuation due to rainfall, is the principal one affecting path loss. For instance, even at 22 GHz, the water vapor line, excess attenuation due to atmospheric gases accumulates at only 0.165 dB/km, and for a 10-km path only 1.65 dB must be added to free-space loss to compensate for water vapor loss. This is negligible when compared to free-space loss itself, such as 119.3 dB for the first kilometer at 22 GHz, accumulating thence roughly 6 dB each time the path length is doubled (i.e., add 6 dB for 2 km, 12 dB for 4 km, etc.). Accordingly, a 10-km path would have a free-space loss of 139.3 dB plus 1.65 dB added for excess attenuation due to water vapor (22 GHz), or a total of 140.95 dB.

Excess attenuation due to rainfall is another matter. It has been common practice to express rainfall loss as a function of precipitation rate. Such a rate depends on the liquid water content and the fall velocity of the drops. The velocity, in turn, depends on raindrop size. Therefore our interest in rainfall boils down to drop size and drop-size distribution for point rainfall rates. All this information is designed to lead the transmission engineer to fix an excess attenuation due to rainfall on a particular path as a function of time and time distribution. This is a method similar to that used in Chapter 4 for overbuilding a link to accommodate fading.

An earlier approach dealt with rain on a basis of rainfall given in millimeters per hour. Often this was done with rain gauges, using collected rain averaging over a day or even periods of days. For path design above 10 GHz such statistics are not sufficient where we may require path availability better than 99.9% and do not wish to resort to overconservative design procedures (e.g., assign excessive link margins).

As we mentioned, there is a fallacy in using annual rainfall rates as a basis for calculation of excess attenuation due to rainfall. For instance, several weeks of light drizzle will affect the overall long-term path availability much less than several good downpours that are short lived (i.e., 20-min duration). It is simply this downpour activity for which we need statistics. Such downpours are cellular in nature. How big are the cells? What is the rainfall rate in the cell? What are the size of the drops and their distribution?

Hogg (Ref. 3) suggests the use of high-speed rain gauges with outputs readily available for computer analysis. These gauges can provide minute-by-minute analysis of the rate of fall, something lacking with conventional types of gauges. Of course, it would be desirable to have several years' statistics for a specific path to provide the necessary information on fading caused by rainfall that will govern system parameters such as LOS repeater spacing, antenna size, and diversity.

Some such information is now available and is indicative of a great variation in short-term rainfall rates from one geographical location to another. As an example, in one period of measurement it was found that Miami, FL, has maximum rain rates about 20 times greater than those of heavy showers occurring in Oregon, the region of heaviest rain in the United States. In Miami a point rainfall rate may exceed 700 mm/h. The effect of 700 mm/h on 70- and 48-GHz paths can be extrapolated from Figure 7.2. In the figure the rainfall rate in millimeters per hour extends to 100, which at 100 mm/h provides an excess attenuation of from 25 to 30 dB/km.

When identical systems were compared (Ref. 3) at 30 GHz with repeater spacings of 1 km and equal desired signals (e.g., producing a 30-dB signal-to-noise ratio), 140 min of total time below the desired level was obtained at Miami, FL; 13 min at Coweeta, NC; 4 min at Island Beach, NJ; 0.5 min at Bedford, UK; and less than 0.5 min at Corvallis, OR. Such outages, of course, can be reduced by increasing transmitter output power, improving receiver noise factor (NF), increasing antenna size, implementing a diversity scheme, etc.

One valid approach to lengthen repeater sections (space between repeaters) and still maintain performance objectives is to use path diversity.

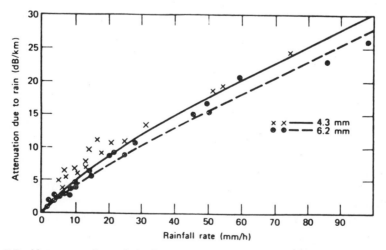

Figure 7.2 Measurements made by Bell Telephone Laboratories of excess attenuation due to rainfall at wavelengths of 6.2 and 4.2 mm (48 and 70 GHz) compared with calculated values. (From Ref. 3, copyright © 1968 by American Telegraph and Telephone Company.)

This is the most effective form of diversity for downpour rainfall fading. Path diversity is the simultaneous transmission of the same information on paths separated by at least 10 km, the idea being that rain cells affecting one path will have a low probability of affecting the other at the same time. A switch would select the better path of the two. Careful phase equalization between the two paths would be required, particularly for the transmission of high-bit-rate information.

7.3.2 Calculation of Excess Path Attenuation due to Rainfall for LOS Paths

When designing radio links (or satellite links), a major problem in link engineering is to determine the excess path attenuation due to rainfall. The adjective "excess" is used to denote path attenuation in *excess* of free-space loss (i.e., the terms a, b, c, d, and e in equation 7.1 are in excess of free-space attenuation [loss]).

Before treating the methodology of calculation of excess rain attenuation, we will review some general link engineering information dealing with rain. When discussing rainfall here, all measurements are in millimeters per hour of rain and are point rainfall measurements. From our previous discussion we know that heavy downpour rain is the most seriously damaging to radio propagation above 10 GHz. Such rain is cellular in nature and has limited coverage. We must address the question as to whether the entire hop is in the storm for the whole period of the storm. Light rainfall (e.g. less than 2 mm/h), on the other hand, is usually widespread in character, and the path average is the same as the local value. Heavier rain occurs in convective storm cells which are typically 2–6 km across and are often embedded in larger regions measured in tens of kilometers (Ref. 4). Thus for short hops (2–6 km) the path-averaged rainfall rate will be the same as the local rate, but for longer paths it will be reduced by the ratio of the path length on which it is raining to the total path length.

This concept is further expanded upon by CCIR Rep. 563-1 (Ref. 5), where rain cell size is related to rainfall rate. This is shown in Figure 7.3. This concept of rain cell size is very important, whether engineering a LOS link or a satellite link, in particular when the satellite link has a low elevation angle. CCIR Rep. 338-3 (Ref. 6) is quoted in part below:

> Measurements in the United Kingdom over a period of two years... at 11, 20 and 37 GHz on links of 4–22 km in length show that the attenuation due to rain and multipath, which is exceeded for 0.01% of the time and less, increased rapidly with path length up to 10 km, but further increase up to 22 km produced a small additional effect.

The use and application of specific rain cell size was taken from CCIR 1978. We will call this method the "liberal method" because when reviewing CCIR

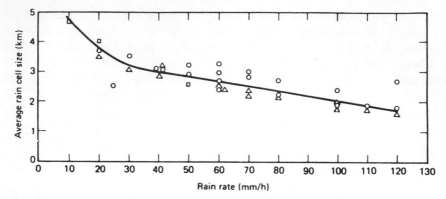

Figure 7.3 Average rain cell size as a function of rain rate. (From CCIR Rep. 563-1 [Ref. 5], courtesy of ITU – CCIR.)

Vol. V of 1982, CCIR Rep. 563-2 (Ref. 9), we can see that the CCIR has turned more cautious and conservative. Our Figure 7.3 does not appear in the report. Let us partially quote from this report:

> For attenuation predictions the situation is generally more complex (than that of interference scattering by precipitation). Volume cells are known to cluster frequently within small mesoscale areas.... Terrestrial links exceeding 10 km may therefore traverse more than one volume cell within a mesoscale cluster. In addition, since the attenuating influence of the lower intensity rainfall surrounding the cell must be taken into account, any model used to calculate attenuation must take these larger rain regions into account. The linear extent of these regions increases with decreasing rain intensity and may be as large as several tens of kilometers.

One of the most accepted methods of dealing with excess path attenuation A (dB) due to rainfall is an empirical procedure based on the approximate relation between A and the rain rate R:

$$A = aR^b \qquad (7.2)$$

where a and b are functions of frequency f and rain temperature T. Allowing a rain temperature of 20°C and allowing for the Laws and Parsons drop-size distribution, Table 7.2 gives the regression coefficients for estimating specific attenuations from equation 7.2.

We note that horizontally polarized waves suffer greater attenuation than vertically polarized waves because large rain drops are generally shaped as oblate spheroids and are aligned with a vertical rotation axis. Hence we use the subscript notation h and v for horizontal and vertical polarizations in Table 7.2 for the values a and b. A is the specific attenuation in dB/km. We obtain a and b from Table 7.2. Next we must obtain a value for the rain rate.

Table 7.2 Regression coefficients for estimating specific attenuations in equation 7.2[a]

Frequency (GHz)	a_h	a_v	b_h	b_v
1	0.0000387	0.0000352	0.912	0.880
2	0.000154	0.000138	0.963	0.923
4	0.000650	0.000591	1.121	1.075
6	0.00175	0.00155	1.308	1.265
7	0.00301	0.00265	1.332	1.312
8	0.00454	0.00395	1.327	1.310
10	0.0101	0.00887	1.276	1.264
12	0.0188	0.0168	1.217	1.200
15	0.0367	0.0335	1.154	1.128
20	0.0751	0.0691	1.099	1.065
25	0.124	0.113	1.061	1.030
30	0.187	0.167	1.021	1.000
35	0.263	0.233	0.979	0.963
40	0.350	0.310	0.939	0.929
45	0.442	0.393	0.903	0.897
50	0.536	0.479	0.873	0.868
60	0.707	0.642	0.826	0.824
70	0.851	0.784	0.793	0.793
80	0.975	0.906	0.769	0.769
90	1.06	0.999	0.753	0.754
100	1.12	1.06	0.743	0.744
120	1.18	1.13	0.731	0.732
150	1.31	1.27	0.710	0.711
200	1.45	1.42	0.689	0.690
300	1.36	1.35	0.688	0.689
400	1.32	1.31	0.683	0.684

Source: (Ref. 7). CCIR Rep. 721-2 Courtesy of ITU–CCIR.

[a]Raindrop size distribution (Laws and Parsons, 1943); terminal velocity of raindrops (Gunn and Kinzer, 1949); index of refraction of water at 20°C (Ray 1972). Values of *a* and *b* for spheroidal drops (Fedi, 1979; Maggiori, 1981) based on regression for the range 1 to 150 mm/h. [*Note*: These references are from Ref. 7.]

This is obtained from local data sources, and we need a value of rain intensity exceeded 0.01% of the time with an integration time of 1 min. If this information cannot be obtained from local sources, an estimate can be obtained by identifying the region of interest from the maps appearing in Figures 7.4–7.6, then selecting the appropriate rainfall intensity for the specified time percentage from Table 7.3, which gives the 14 regions in the maps. This gives a value for *R* in equation 7.2. Knowing the frequency and polarization, we calculate *A* in dB/km from values of *a* and *b* from Table 7.2 using equation 7.2. Figure 7.7 and/or the nomogram in Figure 7.8 may also be used to calculate *A*.

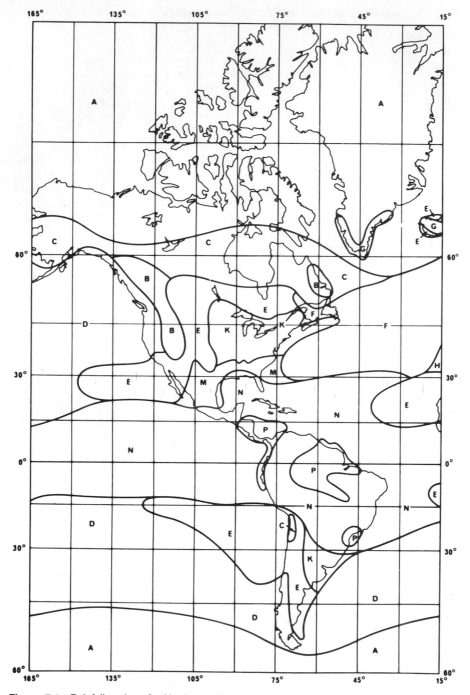

Figure 7.4   Rainfall regions for North and South America. From CCIR Rep. 563-3, Figure 12, Page 125, Vol. V, XVIth Plenary Assembly, Dubrovnik 1986.

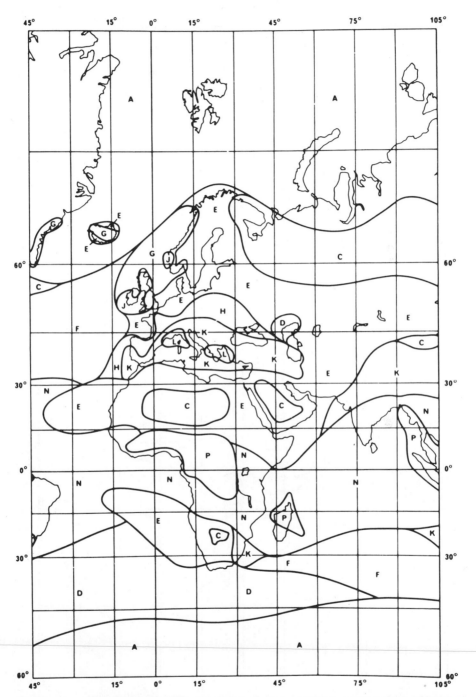

Figure 7.5 Rainfall regions for Europe and Africa. From CCIR Rep. 563-3, Figure 13, Page 126, Vol. V, XVIth Plenary Assembly, Dubrovnik 1986.

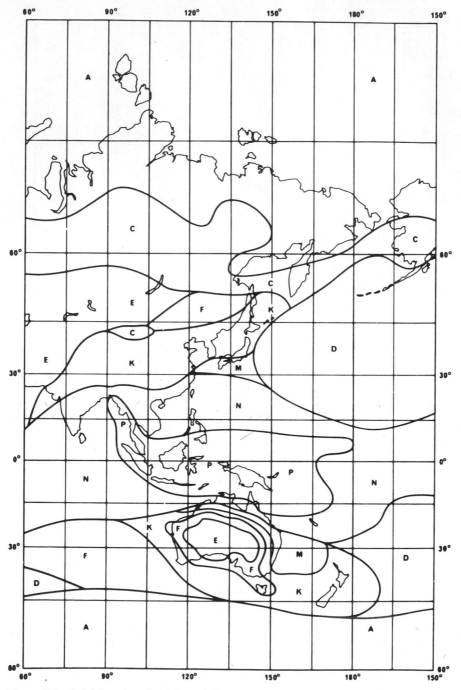

Figure 7.6 Rainfall regions for Asia and Oceana. From CCIR Rep. 563-3, Figure 14, Page 127, Vol. V, XVIth Plenary Assembly, Dubrovnik 1986.

Table 7.3 Rain climatic zones — rainfall intensity exceeded (mm / h) (reference to Figures 7.4 to 7.7)

Percentage of time (%)	A	B	C	D	E	F	G	H	J	K	L	M	N	P
1.0	< 0.5	1	2	3	1	2	3	2	8	2	2	4	5	12
0.3	1	2	3	5	3	4	7	4	13	6	7	11	15	34
0.1	2	3	5	8	6	8	12	10	20	12	15	22	35	65
0.03	5	6	9	13	12	15	20	18	28	23	33	40	65	105
0.01	8	12	15	19	22	28	30	32	35	42	60	63	95	145
0.003	14	21	26	29	41	54	45	55	45	70	105	95	140	200
0.001	22	32	42	42	70	78	65	83	55	100	150	120	180	250

Source: From CCIR Rep. 563-2, Table I, Page 132, Vol. V, XVIth Plenary Assembly, Dubrovnik 1986.

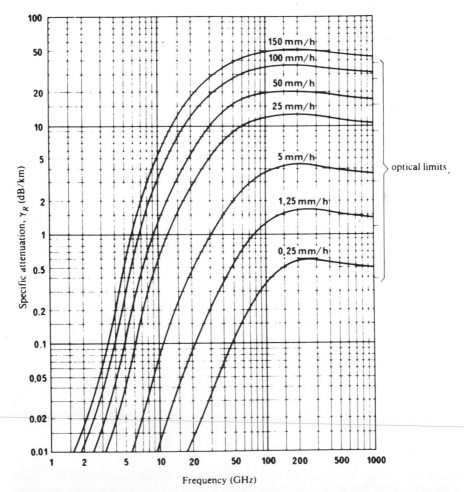

Figure 7.7 Specific attenuation *A* due to rain. Raindrop-size distribution (Laws and Parsons, 1943); terminal velocity of raindrops (Gunn and Kinzer, 1949); index of refraction of water at 20°C (Ray, 1972); spherical drops. From CCIR Rep. 721-1, Figure 1, Page 200, Vol. V, XVIth Plenary Assembly, Dubrovnik 1986.

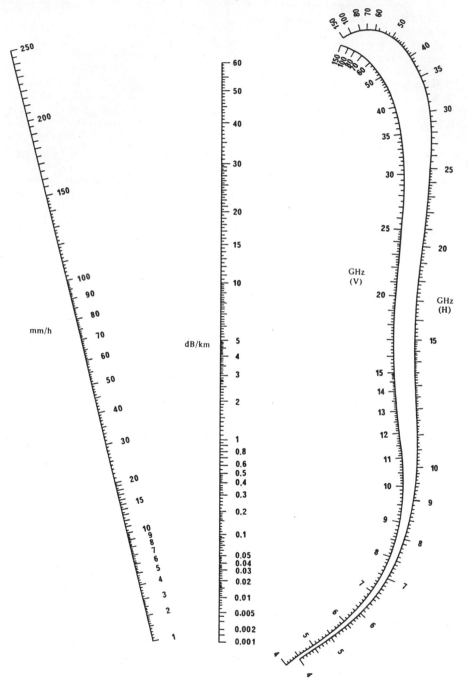

Figure 7.8 Specific attenuation due to rain. (H) = horizontal polarization; (V) = vertical polarization. From CCIR Rep. 721-2, Figure 2, Page 201, Vol. V, XVIth Plenary Assembly, Dubrovnik 1986.

We now turn again to the problem of effective path length or L_{eff} (Ref. 8). This is obtained by multiplying the actual path length L by a reduction factor r. A first estimate to calculate r is given as

$$r = \frac{1}{1 + 0.045L} \tag{7.3}$$

The attenuation exceeded for 0.01% of the time is found from

$$A_{0.01} = aR_{0.01}^b \tag{7.4}$$

$$A_{\text{eff}} = A \times L \times r \tag{7.5}$$

where A is the value calculated in equation 7.4. Attenuation exceeded for other percentages P can be found by the following power law:

$$A_{\text{p}} = A_{0.01}(0.12P^{-(0.546 + 0.0743 \log P)}) \tag{7.6}$$

We may substitute the following values for the complex right-hand factor: 0.12, 0.39, 1, and 2.14 for 1%, 0.1%, 0.01%, and 0.0001%, respectively.

Example 1. Consider a path in the FRG (what was formerly West Germany) 10 km long operating at 30 GHz. Use equation 7.2. Obtain values of a and b and assume vertical polarization. Use Table 7.2 and $a = 0.167$ and $b = 1.000$. Assume a time availability of 99.99% and thus the rainfall intensity exceeded is 0.01%, and from Table 7.3, from climate region E (Figure 7.5), $R = 22$ mm/h.

Calculate $A_{0.01}$ using equation 7.4:

$$A_{\text{dB}} = 0.167(22)^{1.000}$$
$$A = 3.674 \text{ dB/km}$$

Determine the effective distance. First calculate the reduction factor r using equation 7.3:

$$r = \frac{1}{1 + 0.045L}$$
$$r = 0.69$$

Determine the excess attenuation due to rainfall A_{eff} using equation 7.5:

$$A_{\text{eff}} = A \times 10 \times r$$
$$= 3.674 \times 10 \times 0.69$$
$$= 25.35 \text{ dB}$$

Example 2. Calculate the excess attenuation due to rainfall for a 15-km path operating at 18 GHz in the Florida panhandle of the United States. Assume horizontal polarization and a time availability of 99.9% (an exceedance of 0.1%).

Turn to Figure 7.4 and the climatic region is N, thence to Table 7.3 and we obtain a rainfall rate of 95 mm/h for an exceedance of 0.01%. Obtain the values of a and b from Table 7.2 and interpolate between the values of 15 and 20 GHz. $a = 0.05974$ and $b = 1.121$. Use equation 7.4 to calculate A:

$$A = 0.05974(95)^{1.121}$$
$$A_{0.01} = 9.85 \text{ dB/km}$$

Calculate the reduction factor r, where $L = 15$; use equation 7.3:

$$r = \frac{1}{1 + 0.045 \times 15}$$
$$r = 0.6$$

Use equation 7.6 to calculate $A_{0.1}$:

$$A_{0.1} = 9.85 \times 0.12 \times 0.1^{-0.39}$$
$$A_{0.1} = 2.9 \text{ dB/km}$$

Determine A_{eff} for the 15-km path; use equation 7.5:

$$A_{\text{eff}} = 2.9 \times 0.6 \times 15$$
$$A_{\text{eff}(0.1)} = 26.1 \text{ dB}$$

Consequently, we would have to add 26.1 dB to the free-space loss for a path with 99.9% availability. For the worst-case situation, to this total value, we would also have to add the fade margin for the same path availability. Conventional space and frequency diversity would mitigate against multipath fading. The use of vertical polarization would reduce the excess attenuation due to rainfall. Path diversity, with a path separation of at least 10 km, would be a major mitigating factor. The value of 10-km separation has been taken from CCIR Vol. V (1986); however, it can be shown that separations of 2 km or more will provide some diversity improvement for rainfall. A more in-depth discussion is presented on path diversity in Section 7.3.4.

7.3.3 Calculation of Excess Attenuation due to Rainfall for Satellite Paths

7.3.3.1 *Introduction* Rainfall attenuation on satellite paths is a function of frequency and elevation angle. The calculation of excess attenuation due to rainfall for uplinks and downlinks is somewhat similar to the exercise described in Section 7.3.2. The principal difference is that the path is elevated (e.g., a function of elevation angle). The preferred model is described in this section is basically based on work done by R. K. Crane (Refs. 10 and 16), embellished by Feldman and characterized by Kaul (Refs. 11 and 14). The reader is also encouraged to review CCIR Rep. 564-3, Section 6 (Ref. 12).

7.3.3.2 Scattering Raindrops both attenuate and scatter microwave energy along an earth–space path. From the basic Rayleigh scattering criteria (the dimensions of the scatterer are much smaller than the wavelength) and the fact that the median raindrop diameter is approximately 1.5 mm, one would expect that Rayleigh scattering theory should be applied in the frequency range above 10 GHz. However, Rayleigh scattering also requires that the imaginary component of the refractive index be small, which is not the case for water drops. Because of this effect and the wide distribution of raindrop diameters, the Rayleigh scattering theory seems to apply only up to 3 GHz. Above 3 GHz, Mie scattering applies and is the primary technique utilized for calculations of excess attenuation due to rainfall. Mie scattering accounts for the deficiencies of Rayleigh scattering and has proven to be the most accurate technique (Ref. 11).

7.3.3.3 Drop-Size Distribution Several investigators have studied the distribution of raindrop sizes as a function of rain rate and type of storm activity. The three most commonly used distributions are

- Laws and Parsons (LP) (Section 7.3.2)
- Marshall–Palmer (MP)
- Joss-thunderstorm (J-T) and drizzle (J-D)

The LP distribution is generally more favored for design purposes because it has been widely tested by comparison to measurements for both widespread (lower rain rates) and convective rain (higher rain rates), and at frequencies above 10 GHz, the LP values give higher values for excess attenuation due to rainfall than the J-T values. In addition, it has been observed that the raindrop temperature is most accurately modeled by the 0°C data rather than the 20°C data, since for most high-elevation-angle satellite links, the raindrops are cooler at high altitudes and warm as they fall to earth.

7.3.3.4 The Global Model (Crane) The global model (Crane) (Ref. 10) uses a specific rain model based entirely on meteorological observations, not attenuation measurements. The rain model, combined with the attenuation estimation, was tested by comparison with attenuation measurements. This procedure was used to circumvent the requirement for attenuation observa-

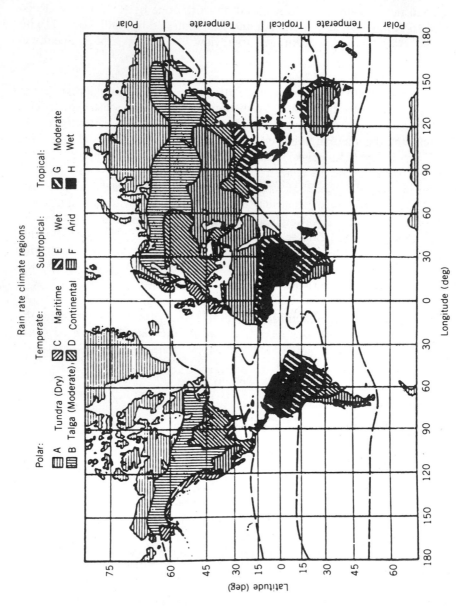

Figure 7.9 The global rain rate regions for the continental areas. (From Ref. 11.)

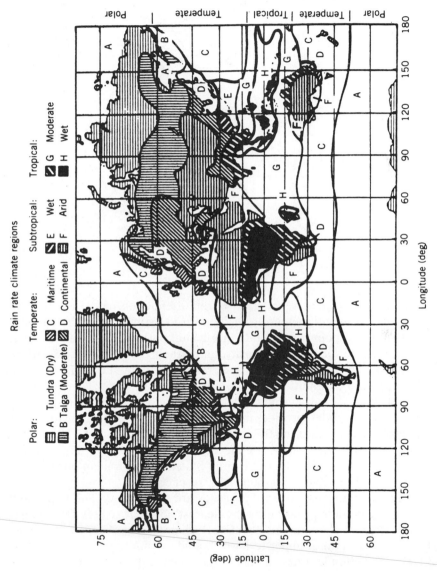

Figure 7.10 Global rain rate climate regions for the ocean areas. (From Ref. 11.)

511

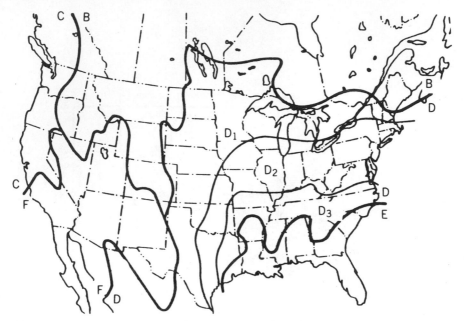

Figure 7.11 Rain rate climate regions for the contiguous United States showing the subdivision of region D. (From Ref. 11.)

tions over a span of many years. The overall attenuation model is based on the use of independent, meteorologically derived estimates for the cumulative distributions of point rainfall rate, horizontal-path-averaged rainfall rate, the vertical distribution of rain intensity, and a theoretically derived relationship between specific attenuation and rain rate obtained using median observed drop-size distributions at a number of rain rates.

The first step is to determine the instantaneous point rain rate R_p distribution. The model provides median distribution estimates for eight rainfall regions, A–H, covering the entire globe. Figures 7.9 and 7.10 give the geographic rain climate regions for the continental and ocean areas of the earth. The continental (contiguous) United States and European portions are further expanded in Figures 7.11 and 7.12, respectively.

The climate regions shown in the figures are very broad. The upper and lower rain rate bounds provided by the nearest adjacent region have a ratio of 3.5 at 0.01% of the year for the CCIR climate region D, for example, which produces a ratio of upper-to-lower bound attenuation values of 4.3 dB at 12 GHz. The attendant uncertainty in the estimated attenuation value can be reduced by using actual rain rate distributions for an area of interest if these long-term statistics are available. Region D for the United States has been further broken down into regions D_1, D_2, and D_3 for convenience.

Once the region of interest has been identified, R_p values may then be obtained from the rain rate distribution curves in Figure 7.13. Figure 7.13*a*

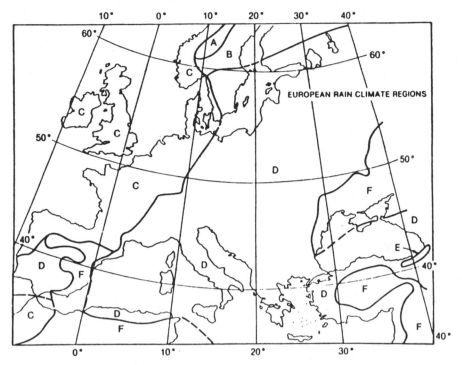

Figure 7.12 Rain rate climate regions for Europe. (From Ref. 11.)

gives the curves for the eight global climate regions designated A–H for 1-min averaged surface rain rate as a function of the percentage of a year that the rain rate is exceeded. For the region-D subdivision distributions, Figure 7.13*b* should be used. Numerical values for R_p are provided in Table 7.4.

7.3.3.5 *Discussion of the Model*

In Section 7.3.2 we used a path-averaged rainfall rate using an effective path averaging factor *r*. This was a valid approach for a LOS microwave path, but will not apply for the estimation of attenuation on a slant path to a satellite. Here account must be taken of the variation of specific attenuation with height. As we are aware, atmospheric temperature decreases with height, and above some height, the precipitation particles will all be ice particles. Ice and snow do not produce significant attenuation. Only regions in the atmosphere with liquid water precipitation particles are of interest in the estimation of attenuation. The size and number of raindrops per unit volume may vary with height. Weather radar measurements have shown that the reflectivity of a rain volume may vary with height, but, on the average, the reflectivity is roughly constant up to the 0°C isotherm and decreases above that height. The rain rate can be considered to be constant to the height of the 0°C isotherm at low rain rates, and this

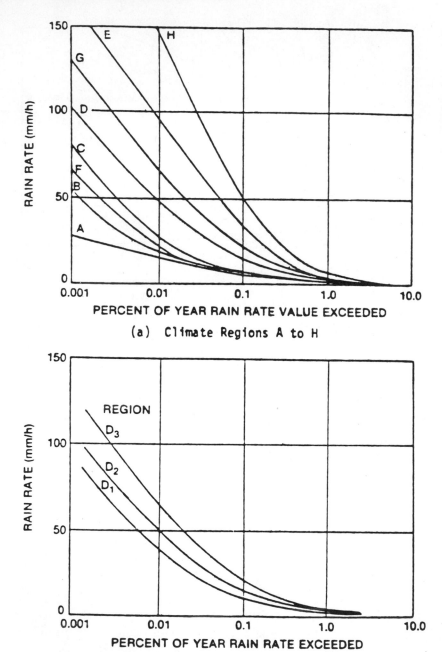

(a) Climate Regions A to H

(b) Climate Region D divided into three subregions
(D$_2$ = D above)

Figure 7.13 Point rain rate distributions as a function of percentage of year exceeded. (From Ref. 11.)

Table 7.4   Point rain rate distribution values (R_p, mm / h) versus percent of year rain rate is exceeded

Percentage of Year	Rain Climate Region										Minutes Per Year	Hours Per Year
	A	B	C	D_1	D_2	D_3	E	F	G	H		
0.001	28	54	80	90	102	127	164	66	129	251	5.3	0.09
0.002	24	40	62	72	86	107	144	51	109	220	10.5	0.18
0.005	19	26	41	50	64	81	117	34	85	178	26	0.44
0.01	15	19	28	37	49	63	98	23	67	147	53	0.88
0.02	12	14	18	27	35	48	77	14	51	115	105	1.75
0.05	8	9.5	11	16	22	31	52	8.0	33	77	263	4.38
0.1	6.5	6.8	72	11	15	22	35	5.5	22	51	526	8.77
0.2	4.0	4.8	4.8	7.5	9.5	14	21	3.8	14	31	1,052	17.5
0.5	2.5	2.7	2.8	4.0	5.2	7.0	8.5	2.4	7.0	13	2,630	43.8
1.0	1.7	1.8	1.9	2.2	3.0	4.0	4.0	1.7	3.7	6.4	5,260	87.66
2.0	1.1	1.2	1.2	1.3	1.8	2.5	2.0	1.1	1.6	2.8	10,520	175.3

Source: (Ref. 11.)

height is used to define the upper boundary of the region of attenuation. However, a high correlation between the 0°C height and the height to which liquid rain drops exist in the atmosphere should not be expected for the higher rain rates because large liquid raindrops are carried aloft above the 0°C height in the strong updraft cores of intense rain cells (Ref. 11). We must estimate the rain layer height appropriate to the path in question before proceeding with the attenuation calculation, since even the 0°C isotherm height depends on latitude and the general rain conditions.

With the global model the average height of the 0°C isotherm for days with rain was taken to correspond to the height expected for 1% of the year. The highest height observed with rain was taken to correspond to the value expected 0.001% of the year, the average summer height of the -5°C isotherm. The resultant curves of the latitude dependences of the heights to be expected for surface point rain rates exceeded 1% of the year and 0.001% of the year are shown in Figure 7.14. The model has a seasonal rms uncertainty for the 0°C isotherm height of 500 m or roughly 13% of the average estimated height. This value of 13% is used to estimate the expected uncertainties associated with Figure 7.14.

The correspondence between the 0°C isotherm height values and the excessive precipitation events showed a tendency toward a linear relationship between R_p and the 0°C isotherm height H_0 for high values of R_p. Since, for high rain rates, the rain rate distribution function displays a nearly linear relationship between R_p and $\log(P)$, where P is the probability of occurrence, the interpolation model used for the estimation of H_0 for P between 0.001% and 1% is assumed to have the form

$$H_0 = a + b \log(P) \tag{7.7}$$

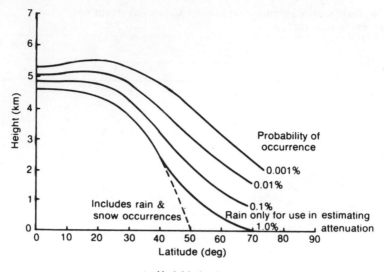

(a) **Variable Isotherm**

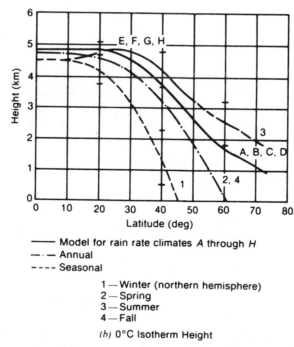

────── Model for rain rate climates *A* through *H*
─ · ─ Annual
─ ─ ─ Seasonal

1 — Winter (northern hemisphere)
2 — Spring
3 — Summer
4 — Fall

(b) 0°C Isotherm Height

Figure 7.14 Effective heights for computing path lengths through rain events. (From Ref. 11.)

The relationship was used to provide the intermediate values displayed in Figure 7.14a. In Figure 7.14b the 0°C isotherms are shown for various latitudes and seasons.

7.3.3.6 The Attenuation Model

To estimate the excess attenuation due to rainfall on a satellite link, we first determine the vertical distance between the height (or altitude) of the earth terminal and the 0°C isotherm height or $H_0 - H_g$, where H_g is the earth terminal height for the percentage of the year (or R_p) value of interest. The path horizontal projection distance D can then be calculated by

$$D = \begin{cases} \dfrac{H_0 - H_g}{\tan \theta} & \theta \geq 10° \quad (7.8a) \\[2mm] E\psi \ (\psi \text{ in radians}) & \theta < 10° \quad (7.8b) \end{cases}$$

where H_0 = height of 0°C isotherm

H_g = height of ground terminal

θ = path elevation angle

E = effective earth radius (8500 km)

ψ = path central angle

and

$$\psi = \sin^{-1}\left\{ \frac{\cos \theta}{H_0 + E}\left[(H_g + E)^2 \sin^2\theta + 2E(H_0 - H_g) \right.\right.$$
$$\left.\left. + H_0^2 - H_g^2 - (H_g + E)\sin \theta \right]^{1/2}\right\} \qquad (7.9)$$

The specific attenuation may then be calculated for an ensemble of raindrops if their size and shape densities are known. Experience has shown that adequate results may be obtained using the LP number density model for the attenuation calculations based on Crane (Ref. 16), and a power-law relationship is fit to calculated values to express the dependence of specific attenuation on rain rate based on the work of Olsen et al. (Ref. 17). The parameters a and b (Section 7.3) in the power-law relationship

$$A = aR_p^b \qquad (7.10)$$

where A, specific attenuation (dB/km), and R_p, point rain rate (mm/h) (also see equation 7.2), are both functions of operating frequency. The appropriate values of a and b may be taken from Table 7.5. For consistency, it is recommended that the reader use this table rather than Table 7.2 for satellite uplink and downlink calculations.

Table 7.5   Parameters *a* and *b* for computing specific attenuation
$A = aR^b$, 0°C, distribution

Frequency f (GHz)	Multiplier $a(f)$	Exponent $b(f)$
1	0.00015	0.95
4	0.00080	1.17
5	0.00138	1.24
6	0.00250	1.28
7.5	0.00482	1.25
10	0.0125	1.18
12.5	0.0228	1.145
15	0.0357	1.12
17.5	0.0524	1.105
20	0.0699	1.10
25	0.113	1.09
30	0.170	1.075
35	0.242	1.04
40	0.325	0.99
50	0.485	0.90
60	0.650	0.84
70	0.780	0.79
80	0.875	0.753
90	0.935	0.730
100	0.965	0.715

Source: Ref. 16.

7.3.3.7   Calculation of Excess Attenuation due to Rainfall by the Variable Isotherm Height Technique

The variable isotherm height technique is based upon the fact that the effective height of the attenuating medium changes depending on the type of rainfall event. It also takes into consideration that various types of rainfall events selectively influence various percentages of time throughout the rainfall cycle. Therefore, a relation exists between the effective isotherm height and the percentage of time that the rain event occurs. This relation has been shown earlier in Figure 7.14*a*. The total attenuation is obtained by integrating the specific attenuation along the path. The resulting equation that is used for estimating the slant path attenuation (A) is

$$A = \frac{aR_p^b}{\cos \theta} \left[\frac{e^{UZb} - 1}{Ub} - \frac{X^b e^{YZb}}{Yb} + \frac{X^b e^{YDb}}{Yb} \right] \qquad \theta \geq 10° \qquad (7.11)$$

where U, X, Y, and Z are empirical constants that depend on the point rain rate. These constants are

$$U = \frac{1}{Z}\left[\ln(Xe^{YZ})\right] \tag{7.12}$$

$$X = 2.3R_{\mathrm{p}}^{-0.17} \tag{7.13}$$

$$Y = 0.026 - 0.03\ln R_{\mathrm{p}} \tag{7.14}$$

$$Z = 3.8 - 0.6\ln R_{\mathrm{p}} \tag{7.15}$$

For lower elevation angles ($\theta < 10°$),

$$A = \frac{L}{D}aR_{\mathrm{p}}^{b}\left[\frac{e^{UZb} - 1}{Ub} - \frac{X^{b}e^{YZb}}{Yb} + \frac{X^{b}e^{YDb}}{Yb}\right] \tag{7.16}$$

where

$$L = \left[(E + H_{\mathrm{g}})^{2} + (E + H_{0})^{2} - 2(E + H_{\mathrm{g}})(E + H_{0})\cos\psi\right]^{1/2} \tag{7.17}$$

and ψ = path central angle defined above.

The following steps apply the variable isotherm height rain attenuation model to a general earth–space path:

Step 1. At the satellite earth terminal's geographic latitude and longitude, obtain the appropriate climate region from the eight regions (A–H). Use Figure 7.9, 7.10, 7.11, or 7.12. However, if long-term rain rate statistics are available for the location of the earth terminal, they should be used instead of the model distribution functions.

Step 2. Select the probabilities of occurrence (P) covering the range of interest in terms of the percentage of time rain rate is exceeded (e.g., 0.01%, 0.1%, or 1%). This, of course, will be based on the specified path time availability. If the path availability is specified as 99.9%, P would then equal 0.1%, or 100 − 99.9%.

Step 3. Obtain the terminal point rain rate R_{p} (mm/h) using Table 7.4, or long-term measured values, if available, of rain rate versus the percentage of the year rain rate is exceeded at the climate region and probabilities of occurrence (step 2).

Step 4. For a satellite link through the entire atmosphere, obtain the rain layer height from the height of the 0°C isotherm (melting layer) H_{0} at the path latitude from Figure 7.14a. The heights will vary correspondingly with the probabilities of occurrence (step 2). To interpolate, plot $H_{0}(P)$ versus log P and use a straight line to relate H_{0} to P.

Step 5. Obtain the horizontal path projection D of the oblique path through the rain volume:

$$D = \frac{H_0 + H_g}{\tan \theta} \quad \theta \geq 10° \tag{7.18}$$

$$H_0 = H_0(P) = \text{height (km) of isotherm for probability } P \tag{7.19}$$

$$H_g = \text{height of ground terminal (km)}$$

$$\theta = \text{path elevation angle}$$

Step 6. Test $D \leq 22.5$ km; if true, proceed to the next step. If $D \geq 22.5$ km, the path is assumed to have the same attenuation value as for a 22.5-km path, but the probability of occurrence is adjusted by the ratio of 22.5 km to the path length:

$$\text{New probability of occurrence, } P' = P\left(\frac{D}{22.5 \text{ km}}\right) \tag{7.20}$$

where D = path length projected on surface (> 22.5 km).

Step 7. Obtain the parameters $a(f)$ and $b(f)$, relating the specific attenuation to the rain rate from Table 7.5.

Step 8. Compute the total attenuation due to rain using R_p, a, b, θ, and D:

$$A = \frac{aR_p^b}{\cos \theta}\left[\frac{e^{UZb} - 1}{Ub} - \frac{X^b e^{YZb}}{Yb} + \frac{X^b e^{YDb}}{Yb}\right] \quad \theta \geq 10° \tag{7.21}$$

where A = total path attenuation due to rain (dB)

a, b = parameters relating the specific attenuation to rain rate (from step 7), $\alpha = aR_p^b$ = specific attenuation

R_p = point rain rate (step 3)

θ = elevation angle of path

D = horizontal path distance (step 5) $Z \leq D \leq 22.5$ km

or alternatively, if $D < Z$, then

$$A = \frac{aR_p^b}{\cos \theta}\left[\frac{e^{UbD} - 1}{Ub}\right] \tag{7.22}$$

or if $D = 0$, $\theta = 90°$,

$$A = (H - H_g)(aR_p^b) \tag{7.23}$$

Example. Calculate the excess attenuation due to rainfall for a downlink operating at 21 GHz where the elevation angle is 10° and the desired path time availability is 99.9%. The earth terminal is in southeastern Minnesota (45N, 90W).

Select the appropriate climate region. Use Figure 7.11 and the climate region is D_1. Turn to Table 7.4 and select the value for R_p for an exceedance of 0.1% (e.g., $100.0 - 99.9$), and this value is $R_p = 11$ mm/h. Obtain the rain layer height from the height of the 0°C isotherm (melting layer) H_0 at the path latitude (45°N). Use Figure 7.14a and $H_0 = 2.8$ km. Assume $H_g = 0.2$ km.

Calculate D (step 5); use equation 7.18:

$$
\begin{aligned}
D &= \frac{2.8 - 0.2}{\tan 10°} \\
&= \frac{2.6}{0.176} \\
&= 14.77 \text{ km}
\end{aligned}
$$

D is less than 22.5 km, thus proceed to step 7. Obtain the parameters $a(f)$ and $b(f)$; use Table 7.5 and interpolate values for 21 GHz: $a = 0.0785$ and $b = 1.098$.

Turn to equations 7.12–7.15. (Remember that ln implies the natural logarithm, that is the logarithm to the base e.) Start with equation 7.15 and proceed backward:

$$
\begin{aligned}
Z &= 3.8 - 0.6\ln(11) \\
Z &= 2.36 \\
Y &= 0.026 - 0.03\ln(11) \\
Y &= -0.046 \\
X &= 2.3(11)^{-0.17} \\
X &= 1.53 \\
U &= \frac{1}{2.36}\left[\ln 1.53 e^{-0.10856}\right] \\
U &= \frac{1}{2.36}\left[\ln 1.53 \times 0.897\right] \\
U &= \frac{1}{2.36}\left[\ln 1.37\right] \\
U &= 0.134
\end{aligned}
$$

Apply these calculated values, the values of a and b, and $\theta = 10°$ to determine A, the total excess attenuation due to rainfall for the path in question. Use equation 7.21; θ is equal to $10°$:

$$A = \frac{0.0785(11)^{1.098}}{\cos 10°}\left[\frac{e^{0.134 \times 2.36 \times 1.098} - 1}{0.134 \times 1.098} - \frac{1.53^{1.098}e^{-0.046 \times 2.36 \times 1.098}}{-0.046 \times 1.098} + \frac{1.53^{1.098}e^{-0.046 \times 14.77 \times 1.098}}{-0.046 \times 1.098}\right]$$

It is often easier to calculate by pieces and simplify factors as follows: $\cos 10° = 0.985$. The factor in front of the brackets is 1.11. The first term inside the brackets is $0.415/0.147 = 2.82$. The second term is $(-1.6 \times 0.888)/-0.051 = +27.86$. The third term is $(1.6 \times 0.474)/-0.051 = -14.87$. Simplify:

$$A = 1.11(2.82 + 27.86 - 14.87)$$
$$= 17.55 \text{ dB}$$

This value would be entered into the link budget for the downlink.

7.3.3.8 CCIR Method CCIR Rep. 564-3 (Ref. 12) presents a less cumbersome method to calculate excess attenuation due to rainfall. The method uses the effective path length concept, similar to that described in Section 7.3.2.

The CCIR method is used to determine slant path excess attenuation due to rainfall at a given location. The following parameters are used:

$R_{0.01}$: point rainfall rate for the location for 0.01% of an average year (mm/h)
h_s: height above mean sea level of the earth station (km)
θ: elevation angle
φ: latitude of the earth station (deg)
f: frequency (GHz)

We then calculate:

h_R: rain height (km) from latitude of φ
L_s: slant path length (km)
L_G: the horizontal projection of the slant path L_s (km)
$r_{0.01}$: reduction factor for 0.01% of the time
γ_r: specific attenuation (dB/km)
$A_{0.01}$: attenuation exceeded 0.01% of an average year

The CCIR method consists of seven steps to predict the attenuation exceeded 0.01% of the time and an eighth step for other time percentages. This method, taken from CCIR Rep. 564-3 (Ref. 12), is presented here (also see

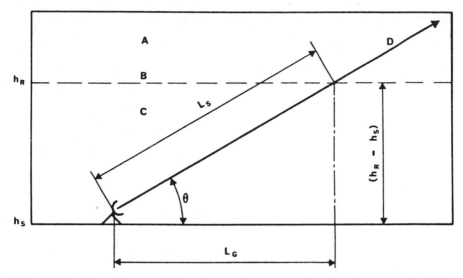

Figure 7.15 Schematic presentation of an earth – space path giving the parameters to be input into the attenuation prediction process. *A*, Frozen precipitation; *B*, rain height; *C*, liquid precipitation; *D*, earth – space path. (From CCIR Rep. 564-3 [Ref. 12, p. 395, Figure 1, Vol. V, XVIth Plenary Assembly, Dubrovnik 1986].)

Figure 7.15):

Step 1. The rain height h_R is calculated from the latitude of the station φ (assuming $h_R \simeq h_{FR}$, following §4.4.3 in Rep. 563):

$$h_R(\text{km}) = \begin{cases} 4.0 & 0 < \varphi < 36° \\ 4.0 - 0.075(\varphi - 36) & \varphi \geq 36° \end{cases} \qquad (7.24)$$

A considerable deviation from the above value may be expected if the important rainy season is very different from the summer season.

Step 2. For $\theta \geq 5°$ the slant path length L_s below the rain height is obtained from

$$L_s = \frac{(h_R - h_s)}{\sin \theta} \quad (\text{km}) \qquad (7.25)$$

For $\theta < 5°$ a more accurate formula should be used:

$$L_s = \frac{2(h_R - h_s)}{\left(\sin^2 \theta + 2(h_R - h_s)/R_e\right)^{1/2} + \sin \theta} \quad (\text{km}) \qquad (7.26)$$

Step 3. The horizontal projection L_G of the slant path length is found from

$$L_G = L_s \cos \theta \quad \text{km} \qquad (7.27)$$

Step 4. The reduction factor $r_{0.01}$, for 0.01% of the time can be calculated from

$$r_{0.01} = \frac{1}{1 + 0.045L_G} \tag{7.28}$$

Step 5. Obtain the rain intensity $R_{0.01}$ exceeded for 0.01% of an average year (with an integration time of 1 min). If this information cannot be obtained from local data sources, an estimate can be obtained from the maps with rainfall contours in Figures 7.4–7.6 and Table 7.3.

Step 6. Obtain the specific attenuation γ_R, using the frequency-dependent coefficients given in Table 7.2 and the rainfall rate $R_{0.01}$ determined from step 5, by using

$$\gamma_R = a(R_{0.01})^b \tag{7.29}$$

Step 7. The attenuation exceeded for 0.01% of an average year may then be obtained from

$$A_{0.01} = \gamma_R L_s r_{0.01} \quad \text{(dB)} \tag{7.30}$$

Step 8. The attenuation to be exceeded for other percentages of an average year, in the range 0.001% to 1.0%, may be estimated from the attenuation to be exceeded for 0.01% for an average year by using

$$\frac{A_p}{A_{0.01}} = 0.12p^{-(0.546 + 0.043 \log p)} \tag{7.31}$$

This interpolation formula has been determined to give factors of 0.12, 0.38, 1, and 2.14 for 1%, 0.1%, 0.01%, and 0.001%, respectively.

Example (CCIR Method). A satellite downlink operates at 21 GHz. The satellite earth station is located in southeastern Minnesota (45N, 90W, and 10° elevation angle). The desired time availability is 99.9%. Determine the excess attenuation due to rainfall.

Step 1. Calculate the rain height h_R using equation 7.24. $h_R = 4$ km.

Step 2. Determine the slant path length L_s below the rain height. Use equation 7.25. The height of the earth station above mean sea level h_s is 0.2 km.

$$L_s = \frac{3.8}{\sin(10°)}$$
$$L_s = 22 \text{ km}$$

Step 3. Calculate L_G, the horizontal projection of the path length, using equation 7.27.

$$L_G = 22 \times \cos(10°)$$
$$L_G = 21.6 \text{ km}$$

Step 4. Determine the reduction factor $r_{0.01}$ for 0.01% of the time. Use equation 7.28.

$$r_{0.01} = \frac{1}{1 + 0.045 \times 21.6}$$
$$r_{0.01} = 0.507$$

Step 5. Obtain the rain intensity $R_{0.01}$ exceeded for 0.01% of an average year. From Figure 7.4 we find that the rain region is K. From Table 7.3 we find that

$$R_{0.01} = 42 \text{ mm/h}$$

Step 6. Obtain the specific attenuation γ_R using the frequency-dependent coefficients given in Table 7.2 and $R_{0.01} = 42$ mm/h. Interpolating, $a_v = 0.07788$ and $b_v = 1.058$. Use equation 7.29 and substitute values.

$$\gamma_R = 0.07788(42)^{1.058}$$
$$\gamma_R = 4.06 \text{ dB/km}$$

Step 7. The attenuation exceeded for 0.01% of an average year is obtained from equation 7.30.

$$A_{0.01} = 4.06 \times 22 \times 0.507 = 45.28 \text{ dB}$$

Step 8. Calculate the excess attenuation due to rainfall for the 99.9% value or an exceedance of 0.1% of the time. Use value 0.38 and multiply the 45.28 dB value from step 7. Thus $A_{0.1} = 17.2$ dB. Compare this value with the 17.55 dB from the example used in Section 7.3.3.7.

7.3.4 Utilization of Path Diversity to Achieve Performance Objectives

Excess attenuation due to rainfall often degrades satellite uplinks and downlinks operating above 10 GHz so seriously that the requirements of optimum economic design and reliable performance cannot be achieved simultaneously. Path diversity can overcome this problem at some reasonable cost compromise. Path diversity advantage is based on the hypothesis that rain cells and, in particular, the intense rain cells that cause the most severe fading are rather limited in spatial extent. Furthermore, these rain cells do not occur immediately adjacent to one another. Therefore the probability of simultaneous fading on two paths to spatially separated earth stations would be less than that associated with either individual path. The hypothesis has been borne out experimentally (Ref. 18).

Let us define two commonly used terms: diversity gain and diversity advantage. Diversity gain is defined (in this context) as the difference between the rain attenuation exceeded on a single path and that exceeded jointly on separated paths for a given percentage of time. Diversity advantage

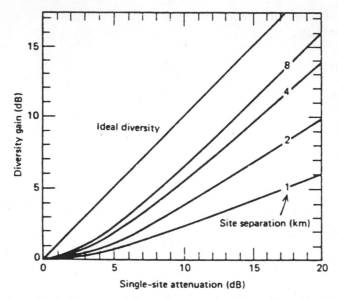

Figure 7.16 Diversity gain for various site separations. (From Ref. 14.)

is defined (in this context) as the ratio of the percentage of time exceeded on a single path to that exceeded jointly on separated paths for a given rain attenuation level.

Diversity gain may be interpreted as the reduction in the required system margin at a particular percentage of time afforded by the use of path diversity. Alternatively, diversity advantage may be interpreted as the factor by which the fade time is improved at a particular attenuation level due to the use of path diversity.

The principal factor to achieve path diversity to compensate for excess attenuation due to rainfall is separation distance. The diversity gain increases rapidly as the separation distance d is increased over a small separation distance, up to about 10 km. Thereafter the gain increases more slowly until a maximum value is reached, usually between about 10 and 30 km. This is shown in Figure 7.16.

The uplink/downlink frequencies seem to have little effect on diversity gain up to about 30 GHz. (Ref. 11). This same reference suggests that for link frequencies above 30 GHz attenuation on both paths simultaneously can be sufficient to create an outage. Hence extrapolation beyond 30 GHz is not recommended, at least with the values given in Figure 7.16.

7.4 EXCESS ATTENUATION DUE TO ATMOSPHERIC GASES ON SATELLITE LINKS

The zenith one-way attenuations for a moderately humid atmosphere (e.g., 7.5-g/m^3 surface water vapor density) at various starting heights above sea level are given in Figure 7.17 and in Table 7.6. These curves were computed by Crane and Blood (Ref. 10) for temperate latitudes assuming the U.S. standard atmosphere, July, 45°N latitude. The range of values shown in Figure 7.17 refers to the peaks and valleys of the fine absorption lines. The range of values for starting heights above 16 km is even greater.

Figure 7.17 also shows the standard deviation of the clear-air zenith attenuation as a function of frequency. The standard deviation was calculated

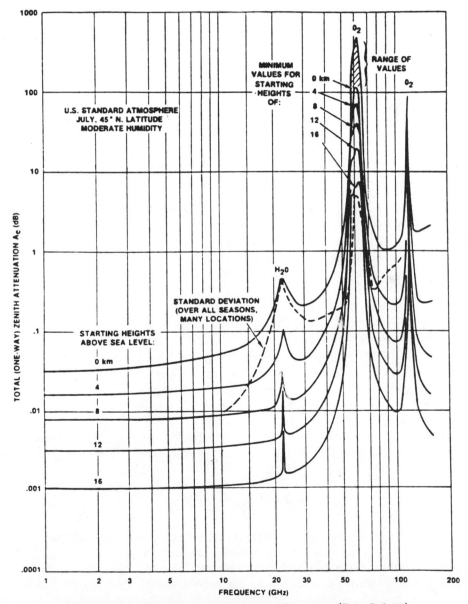

Figure 7.17 Total zenith attenuation versus frequency. (From Ref. 11.)

Table 7.6 Typical one-way clear air total zenith attenuation values (7.5 g / m³ H₂O, July, 45°N latitude, 21°C)

Frequency (GHz)	Altitude				
	0	0.5	1.0	2.0	4.0
10	0.055	0.05	0.043	0.035	0.02
15	0.08	0.07	0.063	0.045	0.023
20	0.30	0.25	0.19	0.12	0.05
30	0.22	0.18	0.16	0.10	0.045
40	0.40	0.37	0.31	0.25	0.135
80	1.1	0.90	0.77	0.55	0.30
90	1.1	0.92	0.75	0.50	0.22
100	1.55	1.25	0.95	0.62	0.25

Source: Ref. 11.

from 220 measured atmosphere profiles spanning all seasons and geographical locations (by Crane, Ref. 15). The zenith attenuation is a function of frequency, earth terminal altitude above sea level, and water vapor content. Compensating for earth terminal altitudes can be done by interpolating between the curves in Figure 7.17.

The water vapor content is the most variable component of the atmosphere. Corrections should be made to the values derived from Figure 7.17 and Table 7.6 in regions that notably vary from the 7.5-g/m³ value given. Such regions would be arid or humid, jungle or desert. This correction to the total zenith attenuation is a function of the water vapor density at the surface p_0 as follows:

$$\Delta A_{c1} = b_p \left(p_0 - 7.5 \text{ g/m}^3 \right) \tag{7.32}$$

where ΔA_{c1} is the additive correction to the zenith clear air attenuation that accounts for the difference between the actual surface water vapor density and 7.5 g/m³. The coefficient b_p is frequency dependent and is given in Figure 7.18. To convert from the more familiar relative humidity or partial pressure of water vapor, refer to Section 7.4.2.

The surface temperature T_0 also affects the total attenuation because it affects the density of both the wet and dry components of the gaseous attenuation. This relation is (Ref. 10)

$$\Delta A_{c2} = C_T (21° - T_0) \tag{7.33}$$

where ΔA_{c2} is an additive correction to the zenith clear-air attenuation. Figure 7.18 gives the frequency-dependent values for C_T.

The satellite earth terminal elevation angle has a major impact on the gaseous attenuation value for a link. For elevation angles greater than about

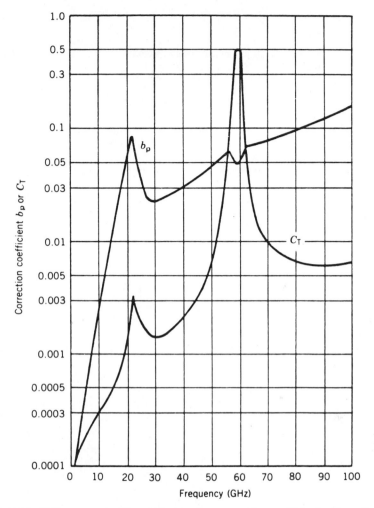

Figure 7.18   Water vapor density and temperature correction coefficients. (From Ref. 11.)

5°, the zenith clear air attenuation value A_c is multiplied by the cosecant of the elevation angle θ. The total attenuation for an elevation angle θ is given by

$$A_c = A'_c \csc \theta \qquad (7.34)$$

7.4.1   Example Calculation of Clear-Air Attenuation – Hypothetical Location

For a satellite downlink, we are given the following information: frequency, 20 GHz; altitude of earth station, 600 m; relative humidity (RH), 50%;

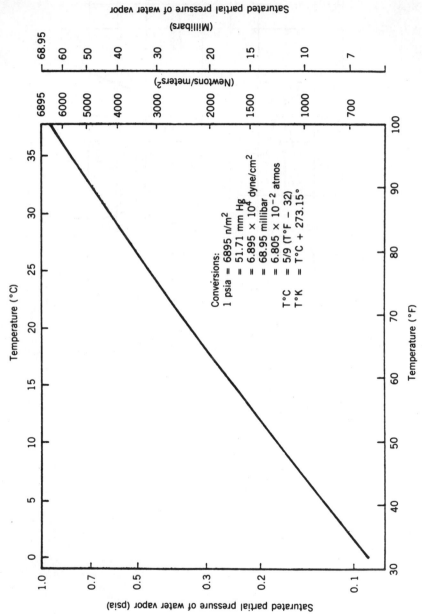

Figure 7.19 The saturated partial pressure of water vapor versus temperature. (From Ref. 11.)

temperature (surface, T_0), 70°F (21.1°C); and elevation angle, 25°. Calculate clear-air attenuation.

Obtain total zenith attenuation A'_c from Table 7.6, and interpolate value for altitude: $A'_c = 0.24$ dB.

Find the water vapor density p_0. From Figure 7.19, the saturated partial pressure of water vapor at 70°F is $e_s = 2300$ N/m². Apply formula 7.35 (Section 7.4.2) and

$$p_0 = \frac{(0.5)2300}{(0.461)(294.1)}$$

$$p_0 = \frac{1150}{135.6}$$

$$p_0 = 8.48 \text{ g/m}^3$$

Calculate the water vapor correction factor ΔA_{c1}. From Figure 7.18, for a frequency of 20 GHz, correction coefficient $b_p = 0.05$. Then use equation 7.32:

$$\Delta A_{c1} = (0.05)(8.48 - 7.5) = 0.05 \text{ dB}$$

Compute the temperature C_T using Figure 7.18. At 20 GHz $C_T = 0.0015$. As can be seen, this value can be neglected in this case.

Calculate the clear-air zenith attenuation corrected A'_c:

$$A'_c = 0.24 \text{ dB} + 0.05 \text{ dB} + 0 \text{ dB}$$

$$= 0.29 \text{ dB}$$

Compute the clear-air slant attenuation using equation 7.34:

$$A_c = 0.29 \csc 25°$$

$$A_c = 0.29 \times 2.366$$

$$= 0.69 \text{ dB}$$

This value would then be used in the link budget for this hypothetical link.

7.4.2 Conversion of Relative Humidity to Water Vapor Density

The surface water vapor density p_0 (g/m³) at a given surface temperature T_0 may be calculated from the ideal gas law:

$$p_0 = \frac{(\text{RH})e_s}{0.461 \text{ J} \cdot \text{g}^{-1}/\text{K}^{-1}(T_0 + 273)} \tag{7.35}$$

where RH = relative humidity and e_s (N/m²) = the saturated partial pressure of water vapor that corresponds to the surface temperature T_0 (°C). See Figure 7.19. The relative humidity corresponding to 7.5 g/m³ at 20°C (68°F) is RH = 0.42 or 42% (Ref. 11).

7.5 ATTENUATION DUE TO CLOUDS AND FOG

Water droplets that constitute clouds and fog are generally less than 0.01 cm in diameter (Ref. 11). This allows a Rayleigh approximation to calculate the attenuation due to clouds and fog for frequencies up to 100 GHz. The specific attenuation a_c is, unlike the case of rain, independent of drop-size distribution. It is a function of liquid water content p_1 and can be expressed by

$$a_c = K_c p_1 \quad (\text{dB/km}) \tag{7.36}$$

where p_1 is normally expressed in g/m³. K_c is the attenuation constant which is a function of frequency and temperature and is given in Figure 7.20.

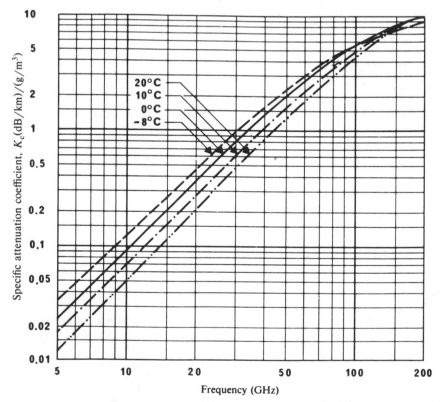

Figure 7.20 Attenuation coefficient K_c due to water vapor droplets. (From CCIR Rep. 721-2 [Ref. 7]; courtesy of ITU – CCIR.)

The curves in Figure 7.20 assume pure water droplets. The values for salt-water droplets, corresponding to ocean fogs and mists, are higher by approximately 25% at 20°C and 5% at 0°C (Ref. 20).

The liquid water content of clouds varies widely. Stratiform or layered clouds display ranges of 0.05 to 0.25 g/m^3 (Ref. 11). Stratocumulus, which is the most dense of this cloud type, has shown maximum values from 0.3 to 1.3 g/m^3 (Ref. 11). Cumulus clouds, especially the large cumulonimbus and cumulus congestus that accompany thunderstorms, have the highest values of liquid content. Fair-weather cumulus clouds generally have a liquid water content of less than 1 g/m^3. Ref. 19 reported values exceeding 5 g/m^3 in cumulus congestus and estimates an average value of 2 g/m^3 for cumulus congestus and 2.5 g/m^3 for cumulonimbus clouds.

Care must be exercised in estimating excess attenuation due to clouds when designing uplinks and downlinks. First, clouds are not homogeneous masses of air containing uniformly distributed droplets of water. Actually, the liquid water content can vary widely with location in a single cloud. Even sharp differences have been observed in localized regions on the order of 100 m across. There is a fairly rapid variation with time as well, owing to the complex patterns of air movement taking place within cumulus clouds.

Typical path lengths through cumulus congestus clouds roughly fall between 2 and 8 km. Using equation 7.36 and the value given for water vapor density and the attenuation coefficient K_c from Figure 7.20, an added path loss at 35 GHz from 4 to 16 dB would derive. Fortunately, for the system designer, the calculation grossly overestimates the actual attenuation that has been observed through this type of cloud structure. Table 7.7 provides values that seem more dependable. In the 35- and 95-GHz bands, cloud attenuation, in most cases, is 40% or less of the gaseous attenuation values. One should not lose sight of the fact, in these calculations, of the great variability

Table 7.7   Atmospheric attenuation in the vertical direction for 95 and 150 GHz, Slough, UK, October 1975 – May 1976

Frequency	95 GHz	150 GHz
Attenuation (dB) in clear air; water vapor content at ground-level 4–11 g/m^3	0.7–1	1–3
Additional attenuation (dB) due to clouds:		
Stratocumulus	0.5–1	0.5–1
Small, fine weather cumulus	0.5	0.5
Large cumulus	1.5	2
Cumulonimbus	2–7	3–8
Nimbostratus (rain cloud)	2–4	5–7

Source: CCIR Rep. 721-2 [Ref. 7, p. 209, Table 2, vol. V, XVIth Plenary Assembly, Dubrovnik 1986].

in the size and state of development of the clouds observed. Data from Table 7.7 may be roughly scaled in frequency, using the frequency dependence of the attenuation coefficient from Figure 7.20.

Fog results from the condensation of atmospheric water vapor into water droplets that remain suspended in air. The water vapor content of fog varies from less than 0.4 up to as much as 1 g/m^3.

The attenuation due to fog in dB/km can be estimated using the curves in Figure 7.20. The 10°C curve is recommended for summer, and the 0°C curve should be used for the other seasons. Typical liquid water content values for fog vary from 0.1 to 0.2 g/m^3. Assuming a temperature of 10°C, the specific attenuation would be about 0.08–0.16 dB/km at 35 GHz and 0.45–0.9 dB/km for 95 GHz. In a typical fog layer 50 m thick, a path at a 30° elevation angle would have only 100-m extension through fog, producing less than 0.1-dB excess attenuation at 95 GHz. In most cases, the result is that fog attenuation is negligible for satellite links.

7.6 CALCULATION OF SKY NOISE TEMPERATURE AS A FUNCTION OF ATTENUATION

The effective sky noise (see Section 6.4.7.1 and Figure 6.5) due to the troposphere is primarily dependent on the attenuation at the frequency of observation. Reference 23 shows the derivation of an empirical equation relating specific attenuation A to sky noise temperature:

$$T_s = T_m[1 - 10^{(-A/10)}] \qquad (7.37)$$

where T_s is the sky noise, T_m the mean absorption temperature of the attenuating medium (e.g., gaseous, clouds, rainfall), and A the specific attenuation that has been calculated in the previous subsections. Temperatures are in kelvins. The value

$$T_m = 1.12(\text{surface temperature in K}) - 50 \text{ K} \qquad (7.38)$$

has been empirically determined by Ref. 23.

Some typical values taken in Rosman, NC (Ref. 11) are given in Table 7.8 for rainfall. Also, CCIR Rep. 564-3 (Ref. 12), Section 3, should be consulted.

Example. From Table 7.8, with a total rain attenuation of 11 dB, what is the sky noise at 20 GHz? Assume $T_m = 275$ K.

Use equation 7.37:

$$T_s = 275(1 - 10^{-11/10})$$
$$T_s = 253.16 \text{ K}$$

Table 7.8   Cumulative statistics of sky temperature due to rain for Rosman, NC, at 20 GHz ($T_m = 275$ K)

Percentage of Year	Point Rain Rate Values (mm/h)	Average Rain Rate (mm/h)	Total Rain Attenuation[a] (dB)	Sky Noise Temperature[b] (K)
0.001	102	89	47	275
0.002	86	77	40	275
0.005	64	60	30	275
0.01	49	47	23	274
0.02	35	35	16	269
0.05	22	24	11	252
0.1	15	17	7	224
0.2	9.5	11.3	4.6	180
0.5	5.2	6.7	2.6	123
1.0	3.0	4.2	1.5	82
2.0	1.8	2.7	0.93	53

Source: Ref. 11.

[a]At 20 GHz the specific attenuation $A = 0.06 R_{av}^{1.12}$ dB/km and for Rosman, NC, the effective path length is 5.1 km to ATS-6.

[b]For a ground temperature of 17°C = 63°F, the T_m is 275 K.

7.7   THE SUN AS A NOISE GENERATOR

The sun is a white-noise jammer of an earth terminal when the sun is aligned with the downlink terminal beam. This alignment occurs, for a geostationary satellite, twice a year near the equinoxes, and in the period of the equinox will occur for a short period each day. The sun's radio signal is of sufficient level to nearly saturate the terminal's receiving system, wiping out service for that period. Figure 7.21 gives the power flux density of the sun as a function of frequency. Above about 20 GHz the sun's signal remains practically constant at -188 dBW/Hz$^{-1} \cdot$ m^{-2} for "quiet sun" conditions.

Reception of the sun's signal or any other solar noise source can be viewed as an equivalent increase in a terminal's antenna noise temperature by an amount T_s. T_s is a function of terminal antenna beamwidth compared to the apparent diameter of the sun (e.g., 0.48°C), and how close the sun approaches the antenna boresight. The following formula, taken from Ref. 24, provides an estimate of T_s when the sun or any other extraterrestrial noise source is aligned in the antenna beam:

$$T_s = \frac{1 - \exp\left[-(D/1.2\theta)^2\right]}{f^2 D^2}\left(\log^{-1}\frac{S + 250}{10}\right) \qquad (7.39)$$

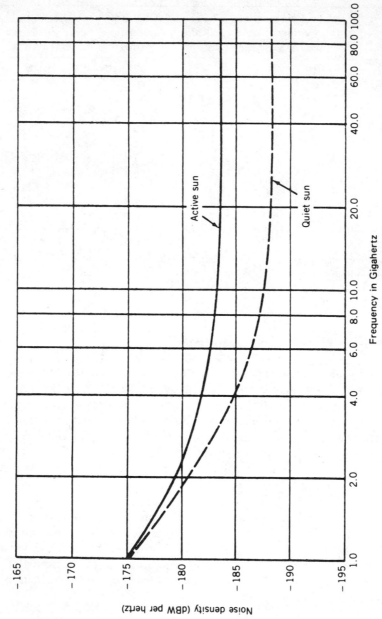

Figure 7.21 Values of noise from quiet and active sun. Sun fills entire antenna beam. (From Ref. 11.)

where D = apparent diameter of the sun or $0.48°$

f = frequency (GHz)

S = power flux density, $\mathrm{dBW\text{-}Hz^{-1} \cdot m^{-2}}$

θ = half-power beamwidth of the terminal antenna (deg)

Example. An earth station operating with a 20-GHz downlink has a 2-m antenna (beamwidth of $0.5°$). What is the maximum increase in antenna noise temperature that would be caused by a quiet sun transit?

Use formula 7.39:

$$T_s = 8146 \text{ K}$$

7.8 PROPAGATION EFFECTS WITH A LOW ELEVATION ANGLE

As the elevation angle of an earth terminal is lowered, the ray beam penetrates an ever-increasing amount of atmosphere. Below about $10°$, fading on the downlink signal must be considered. Fading or signal fluctuations apply only to the ground terminal downlink because its antenna is in close proximity to a turbulent medium. The companion uplink satellite path will suffer uplink fluctuation gain degradation only due to scattering of energy out of the path (Ref. 11). Because of the large distance traversed by the uplink signal since leaving the troposphere, the signal arrives at the satellite as a plane wave and with only a small amount of angle-of-arrival effects.

Phase variations must also be expected for the low-elevation-angle condition. Phase variations arise due to the variable delay as the wave passes through a medium with variable refractivity. Phase scintillation can also occur.

7.9 DEPOLARIZATION ON SATELLITE LINKS

Depolarization is an effect wherein the polarization of a satellite link's waves is altered by the troposphere. Some texts refer to depolarization as cross-polarization. For the case of a linearly polarized wave passing through a medium, components of the opposite polarization will be developed. In the case of a circularly polarized wave, there will be a tendency to develop into an elliptical wave. This is particularly important on frequency reuse systems where the depolarization effectively reduces the polarization isolation and can tend to increase crosstalk on the signal.

Depolarization on a satellite link can be caused by rain, ice, snow, and multipath and refractive effects.

REVIEW EXERCISES

1. Give at least two causes of excess attenuation on satellite paths that must be taken into account for satellite systems operating above 10 GHz.

2. Identify the two frequency bands between 10 and 100 GHz where excess attenuation due to atmospheric gases is high, one of which is excessive.

3. List some uses one might make of these high-attenuation bands.

4. Argue why cumulative annual rainfall rates may not be used for calculation of excess attenuation due to rainfall and why we must use point rainfall rates.

5. Name at least four ways a system design engineer can build a link margin.

6. In early attempts to build in sufficient margin on satellite and LOS links operation above 10 GHz, it was found that the required margins were excessively large because excess attenuation per kilometer was integrated along the entire path (the entire path in the atmosphere for satellite links). Describe how statistics on rain cell size assisted to better estimate excess attenuation due to rainfall.

7. Calculate the specific attenuation per kilometer for a path operating at 30 GHz on a LOS basis with a time availability for the path of 99.9%. Neglect path length considerations, of course. The path is located in northeastern United States. Carry out the calculation for both horizontal and vertical polarizations.

8. Determine the excess attenuation due to rainfall for a LOS path operating at 50 GHz in central Australia. The path length is 20 km and the desired time availability 99.99%. Assume vertical polarization.

9. Calculate the excess attenuation due to rainfall for each polarization for an LOS path 25 km long with an operating frequency of 18 GHz; the desired path availability (propagation reliability) is 99.99%. The path is located in Massachusetts.

10. Name five ways to build a rainfall margin for the path in question 9.

11. Calculate the excess attenuation due to rainfall for a satellite path with a 20° elevation angle for a 21-GHz downlink. The earth station is located in southern Minnesota and the desired time availability for the link is 99%.

12. An earth station is to be located near Bonn, FRG, and will operate at 14 GHz. The desired uplink time availability is 99.95% and the subsatellite point is 10°W. Assume an elevation angle of 20°. What is the excess attenuation due to rainfall?

13. An earth station is to be installed in Diego Garcia, an island in the Indian Ocean, with an uplink at 44 GHz. The elevation angle is 15° and the desired path (time) availability is 99%. What value of excess attenuation due to rainfall should be used in the link budget?

14. There is an uplink at 30 GHz and the required excess attenuation due to rainfall is 15 dB. Path diversity is planned. Show how the value of excess attenuation due to rainfall for a single site can be reduced for site separations of 1, 2, 4, and 8 km.

15. For an earth station, calculate the excess attenuation due to atmospheric gases for a site near sea level. The site is planned for 30/20-GHz operation. The elevation angle is 15°. The relative humidity is 60% and the surface temperature is 27°C (70°F).

16. Determine the sky noise contribution for the attenuation of gases calculated in question 15. Calculate the sky noise temperature due to the excess attenuation due to rainfall from question 13.

REFERENCES AND BIBLIOGRAPHY

1. H. J. Liebe, *Atmospheric Propagation Properties in the 10 to 75 GHz Region: A Survey and Recommendations,"* ESSA Tech. *Report ERL 130-ITS, Boulder, CO,* 1969.

2. "Attenuation by Atmospheric Gases," CCIR Rep. 719-2, XVIth Plenary Assembly, Dubrovnik, 1986, vol. V.

3. D. C. Hogg, "Millimeter Wave Propagation through the Atmosphere," *Science,* Mar. 1968.

4. R. K. Crane, "Prediction of the Effects of Rain on Satellite Communication Systems," *Proc. IEEE,* vol. 65, 456–474, 1977.

5. "Radiometeorological Data," CCIR Rep. 563-1, XIVth Plenary Assembly, Kyoto, 1978, vol. V.

6. "Propagation Data and Prediction Methods Required for Line-of-Sight Radio Relay Systems," CCIR Rep. 338-3, vol. V, XVth Plenary Assembly, Geneva, 1982.

7. "Attenuation by Hydrometers, in Particular, Precipitation and Other Atmospheric Particles," CCIR Rep. 721-2, XVIth Plenary Assembly, Dubrovnik, 1986, vol. V.

8. "Propagation Data and Prediction Methods Required for Line-of-Sight Radio Relay Systems," CCIR Rep. 338-5, XVIth Plenary Assembly, Dubrovnik, 1986.

9. "Radiometeorological Data," CCIR Rep. 563-2, vol. V, XVth Plenary Assembly, Geneva, 1982.

10. R. K. Crane and D. W. Blood, *Handbook for the Estimation of Microwave Propagation Effects—Link Calculations for Earth–Space Paths,* Environmental Research and Technology Report no. 1, DOC no. P-7376-TRI, U.S. Department of Defense, Washington, DC, 1979.

11. R. Kaul, R. Wallace, and G. Kinal, *A Propagation Effects Handbook for Satellite Systems Design: A Summary of Propagation Impairments on* 10–100 *GHz Satellite Links, with Techniques for System Design,* NTIS N80-25520, ORI Inc., Silver Spring, MD, 1980.

12. "Propagation Data and Prediction Methods Required for Earth–Space Telecommunication Systems," CCIR Rep. 564-3, vol. V, XVIth Plenary Assembly, Dubrovnik, 1986.

13. A. H. Ort and E. M. Rasmusson, "Atmospheric Circulation Statistics," NOAA Professional Paper no. 5, U.S. Department of Commerce, Washington, DC, 1971.

14. Roger L. Freeman, *Reference Manual for Telecommunications Engineering*, Wiley, New York, 1985.

15. R. K. Crane, *An Algorithm to Retrieve Water Vapor Information from Satellite Measurements*, NEPRF Tech. Report 7-76, Project no. 1423, Environmental Research and Technology, Inc., Concord, MA, 1976.

16. R. K. Crane, *Microwave Scattering Parameters for New England Rain*, MIT Lincoln Laboratory Tech. Report 426, AD 647798, 1966.

17. R. L. Olsen, D. V. Rogers, and B. R. Hodge, "The aR^b Relation in Calculation of Rain Attenuation," *IEEE Trans. Antennas Propag.*, AP-26, 1978.

18. D. B. Hodge, "The Characteristics of Millimeter Wavelength Satellite-to-Ground Space Diversity Links," IEE Conf. no. 98, London, Apr. 1978.

19. Roger L. Freeman, *Radio System Design for Telecommunications* (1–100 *GHz),* Wiley, New York, 1987.

20. K. L. Koester and L. H. Kosowsky, "Millimeter Wave Propagation in Ocean Fogs and Mists," *Proceedings of IEEE Antenna Propagation Symposium*, AP-26, 1978.

21. B. J. Mason, *The Physics of Clouds*, Clarendon Press, Oxford, UK, 1971.

22. H. K. Weickmann and H. J. Kaumpe, "Physical Properties of Cumulus Clouds," *J. Meteorol.*, vol. 10, 1953.

23. K. H. Wulfsberg, "Apparent Sky Temperatures at Millimeter-Wave Frequencies," Physical Science Research paper no. 38, Air Force Cambridge Research Laboratory, no. 64-590, 1964.

24. J. W. M. Baars, "The Measurement of Large Antennas with Cosmic Radio Sources," *IEEE Trans. Antennas Propag.*, AP-21, no. 4, 1973.

25. S. Perlman et al., "Concerning Optimum Frequencies for Space Vehicle Communications," *IRE Trans. Mil. Electron.*, Mil-4, nos. 2–3, 1960.

8

HIGH-FREQUENCY RADIO

8.1 GENERAL

Radio frequency (RF) transmission between 3 and 30 MHz by ITU convention is called high frequency (HF) or shortwave. Many in the industry extend the HF band downward to just above the standard broadcast band from 2.0 to 30 MHz. This text holds with the ITU convention.

HF communication is a unique method of radio transmission because of its peculiar characteristics of propagation. This propagation phenomenon is such that many radio amateurs at certain times carry out satisfactory communication better than halfway around the world with 1–2 W of radiated power.

8.2 APPLICATIONS OF HF RADIO COMMUNICATION

HF is probably the most economic means of low information rate transmission over long distances (e.g., > 200 mi). It might be argued that meteor burst communication is yet more economic in some circumstances. Performance makes the difference. As we will learn in Chapter 9, meteor burst transmission links have the disadvantage of waiting time between short data packets. HF does not.

Traditionally, since the 1930s, HF has been the mainstay of ship–shore–ship communication. It still is. Satellite communication offered by INMARSAT (International Marine Satellite [organization]) certainly provides a more reliable service, but HF continues to hold sway. We will not argue rationale one way or the other.

Ship–shore HF communication service includes:

- CW or continuous wave, which traditionally connotes keying an RF carrier on and off, forming the dots and dashes of the international Morse code.
- Selective calling teleprinter service (CCIR Rec. 493-3 [Ref. 1]) using frequency shift keying (FSK) with a wide frequency shift (see Chapter 12).
- Simplex teleprinter service with narrow and wide shift FSK as used by many of the world's navies such as the U.S. Navy's Fox broadcast, merchant marine broadcast (MERCAST), and Hydro broadcasts (Refs. 31 and 38).
- Single-sideband (SSB) voice telephony. This is often used for access to a national public switched telephone network.

HF is also used on ground–air–ground circuits for secondary, backup, and primary communication on some transoceanic paths.

HF is widely employed for propaganda broadcasts such as the U.S. Voice of America, Radio Free Europe, and the USSR Voice of Moscow, to only mention a few. Many of these HF installations are very large, with effective isotropically radiated power (EIRPs) in excess of a megawatt.

HF also can provide very inexpensive point-to-point radio teleprinter service: simplex, half-duplex, full-duplex; four-channel time division multiplex (TDM), and 16-channel frequency division multiplex (FDM) using voice frequency carrier techniques (see Chapter 12).

Weather maps to ships and other entities are broadcast on HF by facsimile transmission using narrow-band frequency modulation. It is used for the transmission of time signals such as WWV on 5, 10, 15, and 20 MHz. HF also remains one of the principal means of military communication, both tactical and strategic, for voice, record traffic, data, and facsimile.

HF has many disadvantages. Some of these are

- Low information rate. The maximum bandwidth by radio regulation is four 3-kHz independent sideband voice channels in a quasi-frequency division configuration with low data rates (e.g., about 2400 bps per 3-kHz voice channel).
- Degraded link time availability when compared to satellite, fiber optic, coaxial cable, wire pair, troposcatter, and line-of-sight (LOS) microwave communication. HF link time availabilities vary from 80% up to better than 95% for some new spread spectrum wide-band adaptive systems.
- Impairments include dispersion in both the time and frequency domain. Fading is endemic on skywave links. Atmospheric, galactic, and man-made interference noise are among the primary causes of low availability besides the basic propagation phenomenon itself.

8.3 COMPOSITION OF TYPICAL HF TERMINALS

8.3.1 Basic Equipment

An HF installation for two-way communication consists of one or more transmitters and one or more receivers. The most common type of modulation/waveform is single-sideband suppressed carrier (SSBSC). The operation of this equipment may be half- or full-duplex. We define half-duplex as the operation of a link in one direction at a time. In this case, the near-end transmitter transmits and the far-end receiver receives; then the far-end transmitter transmits and the near-end receiver receives. There are several advantages with this type of operation. A common antenna may be shared by a transmitter and receiver. Under many circumstances, both ends of the link use the same frequency. A simplified diagram of half-duplex operation with a shared antenna is shown below.

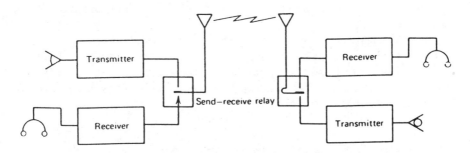

Full-duplex is when there is simultaneous two-way operation. At the same time the near end is transmitting to the far end, it is receiving from the far end. Usually, a different antenna is used for transmission than for reception. Likewise, the transmit and receive frequencies must be different. This is so to prevent the near-end transmitter from interfering with its own receive frequency. Typical full-duplex operation is shown below.

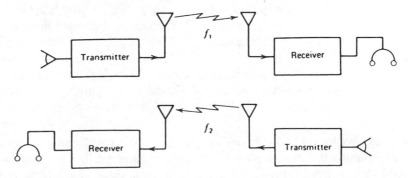

Being interfered with by one's own transmitter is called *cosite interference.* Care must be taken to assure sufficient isolation. There are many measures to be taken to mitigate cosite interference. Among those that should be considered are frequency separation, receiver selectivity, use of separate antennas and their sufficient separation, shielding, filtering transmit output, power amplifier linearity and grounding, bonding, and shielding. The worst environments for cosite interference and other forms of electromagnetic interference (EMI) include airborne and shipboard situations, particularly on military platforms.

When there is multiple transmitter and receiver operation required from a common geographical point, we may have to resort to two- or three-site operation. With this approach, isolation is achieved by physical separation of transmitters from receivers, generally by 2 km or more. However, such separations may even reach 20–30 km. At more complex installations, a third site may serve as an operational center, well isolated from the other two sites. In addition, we may purposely search for a "quiet" site to locate the receiver installation; "quiet" in this context means quiet regarding man-made noise. Both the transmitter site and receiver site may require many acres of cleared land for antennas.

The sites in question are interconnected by microwave LOS links and/or by coaxial cable or fiber optic cable. Where survivability and reliability are very important, such as at key military installations, at least two distinct transmission media may be required. Generally, one is LOS microwave and the other some type of cable.

8.3.2   Basic Single-Sideband (SSB) Operation

Figures 8.1*a* and *b* are simplified functional block diagrams of a typical HF SSB transmitter and receiver, respectively.

Transmitter Operation   A nominal 3-kHz voice channel amplitude modulates a 100-kHz* stable intermediate-frequency (IF) carrier; one sideband is filtered and the carrier is suppressed. The resultant sideband couples the RF spectrum of $100 + 3$ kHz or $100 - 3$ kHz, depending on whether the upper or lower sideband is to be transmitted. This signal is then up-converted in a mixer to the desired output frequency, F_o. Thus the local oscillator output to the mixer is $F_o + 100$ kHz or $F_o - 100$ kHz. Note that frequency inversion takes place when lower sidebands are selected. The output of the mixer is fed to a linear power amplifier and then radiated by an antenna. Transmitter power outputs can vary from 10 W to 100 kW or more.

Receiver Operation   The incoming RF signal from the distant end, consisting of a suppressed carrier plus a sideband, is received by the antenna and filtered by a bandpass filter, often called a preselector. The signal is then amplified and mixed with a stable oscillator to produce an IF (assume again

*Other common IF frequencies are 455 and 1750 kHz.

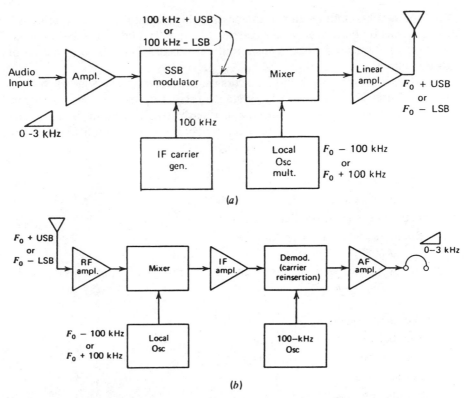

Figure 8.1 Simplified functional block diagram of (a) a typical SSB transmitter (100-kHz IF) and (b) a typical SSB receiver (100-kHz IF). USB = upper sideband; LSB = lower sideband; F_0 = operating frequency; IF = intermediate frequency.

that the IF is 100 kHz*). Several IF amplifiers increase the signal level. Demodulation takes place by reinserting the carrier at IF. In this case, it would be from a stable 100-kHz source, and detection is usually via a product detector. The output of the receiver is the nominal 3-kHz voice channel.

8.3.3 SSB System Considerations

One of the most important considerations in the development of an SSB signal at the near-end transmitter and its demodulation at the far-end receiver is accurate and stable frequency generation for use in frequency sources.

Generally, SSB circuits can maintain tolerable intelligibility when the transmitter and companion far-end receiver have an operating frequency difference no greater than 50 Hz. Narrow shift FSK operation such as voice frequency carrier telegraph (VFTG) will not tolerate frequency differences greater than 2 Hz.

*Other common IF frequencies are 455 and 1750 kHz.

Prior to 1960, end-to-end frequency synchronization for high-quality SSB circuits was by means of a pilot carrier. At the far end, the receiver would lock on the semisuppressed transmitted carrier and slave its local oscillators to this circuit.

Today, HF transmitters and receivers use synthesizers as frequency sources. A synthesizer is a tunable, highly stable oscillator. It gives one or several simultaneous sinusoidal RF outputs on *discrete* frequencies in the HF range. In most cases it provides the frequency supply for all RF carrier needs in SSB applications. For example, it will supply

- Transmitter IF carrier
- IF carrier reinsertion supply
- Transmitter local oscillator supply
- Receiver local oscillator supplies

The following are some of the demands we must place upon an HF synthesizer:

- Frequency stability
- Number of frequency increments
- Spectral purity of RF outputs
- Frequency accuracy
- Supplementary outputs
- Capability of being slaved to a frequency standard

Synthesizer RF output stability should be better than 1.5×10^{-8}. Frequency increments may be in 1-kHz steps for less expensive equipment and ranges down to every 10 Hz for higher quality operational equipment.

8.3.4 Linear Power Amplifiers

The power amplifier in an SSB transmitter raises the power of a low-level signal with minimum possible added distortion. That is, the envelope of the signal output must be as nearly as possible an exact replica of the signal input. Therefore, by definition, a linear power amplifier is required. Such power amplifiers used for HF communication display output powers in the range of 10 W to 100 kW or more. More commonly, we would expect to find this range narrowed to 100 to 10,000 W. Power output of the transmitter is one input used in link calculations and link prediction computer programs described later in the chapter.

The trend in HF power amplifiers today, even for high-power applications, is the use of solid-state amplifier modules. A single module may be used for the lower power applications and groups of modules with combiners for higher power use. One big advantage of using solid state is that it eliminates the high voltage required for vacuum tube operation. Another benefit is improved reliability and graceful degradation. Here we mean that solid-state devices, especially when used as modular building blocks, tend to degrade rather than suffer complete failure. This latter is more the case for vacuum tubes.

8.3.4.1 *Intermodulation Distortion* Nonlinearity in an HF transmitter results in intermodulation (IM) distortion when two or more signals appear in the waveform to be transmitted. Intermodulation noise or intermodulation products are discussed in Chapter 1.

IM distortion may be measured in two different ways:

- Two-tone test
- Tests using white noise loading

The two-tone test is carried out by applying two tones simultaneously at the audio input of the SSB transmitter. A 3 : 5 frequency ratio between the two tones is desirable so that the IM products can be identified easily. For a 3-kHz input audio channel, a 3 : 5 frequency ratio could be tones of 1500 and 2500 Hz.

The test tones are applied at equal amplitude, and their gains are increased to drive the transmitter to full power output. Exciter or transmitter output is sampled and observed on a spectrum analyzer. The amplitudes of the undesired products (see Section 1.9.1) and the carrier products are measured in terms of decibels below either of the equal-amplitude test tones as they appear in the exciter or transmitter output. The decibel difference is the signal-to-distortion ratio (S/D). This should be 40 dB or better. As one might expect, the highest level product is the third-order product. As discussed in Section 1.9.2, this product is two times the frequency of one tone minus the frequency of the second tone. For example, if the two test tones are 1500 and 2500 Hz, then

$$2 \times 1500 - 2500 = 500 \, \text{Hz} \quad \text{or} \quad 2 \times 2500 - 1500 = 3500 \, \text{Hz}$$

and, consequently, the third-order products are 500 and 3500 Hz. The presence of IM products numerically lower than 40 dB indicates maladjustment or deterioration of one or several transmitter stages, or overdrive.

The white noise test for IM distortion more nearly simulates operating conditions of a complex signal such as voice. The approach here is similar to that used to determine noise power ratio (described in Chapter 4). The 3-kHz audio channel is loaded with uniform-amplitude white noise and a slot is cleared, usually the width of a VFTG channel (e.g., 170 Hz). The signal-to-distortion ratio is the ratio of the level of the white noise signal outside the slot to the level of the distortion products in the slot.

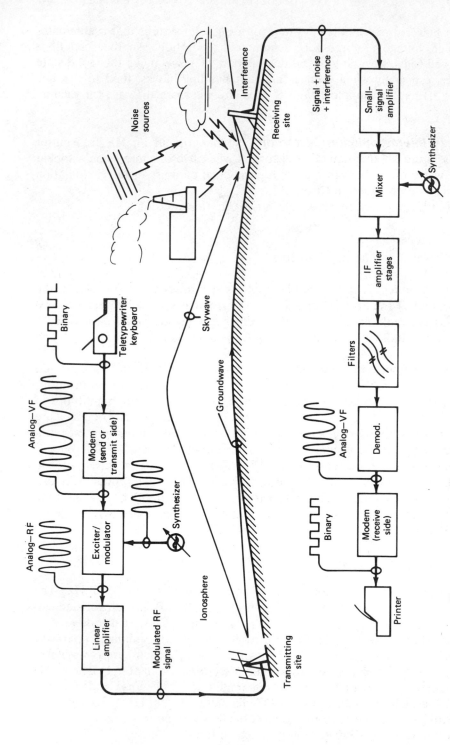

Figure 8.2 A typical HF radio link providing teleprinter service. Courtesy of Radio General Company.

8.3.5 HF Configuration Notes

In large HF communication facilities, space diversity reception is the rule. A rule of thumb for space diversity is that the antennas must be separated by a distance greater than 6λ. Each antenna terminates in its own receiver. The outputs of the receivers are combined either at the receiver site or at the operational center.

One receive antenna can serve many receivers by means of a multicoupler. A multicoupler, in this case, is a broadband amplifier with many output ports to which we can couple receivers.

A major impairment at HF receiver sites is man-made noise. For this reason, we select receiver sites that are comparatively quiet or quiet rural, as described in CCIR Rep. 258.

8.3.6 HF Link Operation

Figure 8.2 illustrates the operation of an HF link carrying teleprinter traffic. The upper side of the drawing shows the transmit side of the link. One-hop skywave operation is portrayed. The lower part of the drawing shows the receive side. The receive site is at the right. Its several noise impairments are shown, such as cosmic noise, man-made noise, atmospheric noise propagated into the site from long distances, local thunderstorm noise, and interference from other users.

In the figure the basic elements of a transmitter and of a receiver are shown as described in Sections 8.3.1–8.3.4. Antennas are described in Section 8.15.

The drawing shows a modem that converts the binary bit stream from a teleprinter keyboard to a signal compatible with the transmitter audio input. The modem develops an audio tone that is frequency shifted (FSK). A higher tone frequency represents a mark (binary 1) and the lower frequency tone represents a space (binary 0). This is further discussed in Chapter 12.

Another method is to offset the transmit synthesizer frequency by one and the other tone frequency. However, the first method is by far the most common. Several other tone formats are discussed in Section 8.14.

A companion modem is used on the receive side which converts the receiver audio tone output to a binary serial bit stream compatible with the teleprinter. The modem need not be a separate entity, but may be incorporated in the transmit exciter and receiver after the demodulator.

8.4 BASIC HF PROPAGATION

8.4.1 Introduction

An HF wave emitted from an antenna is characterized by a groundwave and a skywave component. The groundwave follows the surface of the earth and can provide useful communication over saltwater up to about 650 mi (1000 km) and over land from some 25 (40 km) to 100 mi (160 km) or more, depending on RF power, antenna type and height off the ground, atmospheric noise, man-made noise, and ground conductivity. Generally, the lowest part of the HF band is the most desirable for groundwave communication. However, as we progress lower in frequency, depending on geographical location, atmospheric noise becomes a serious impairment. Well-designed groundwave links, with their shorter range, will achieve a better time availability than skywave links.

Skywave links are used for long circuits from about 100 mi (160 km) to over 8000 mi (12,800 km). We have had successful connectivity by continuous wave (CW) from a ship in the Ross Sea (Antarctica) to WCC (Chatham, MA) in the early morning hours (local time) for over 8 consecutive weeks of the Austral summer, and SSB voice communication from Frobisher Bay (Canada NWT) to Little America (Antarctica) in 1958 for periods of several hours a day for about $1\frac{1}{2}$ weeks. In both cases, modest antennas were used with transmitter outputs on the order of several hundred watts.

8.4.2 Skywave Transmission

The skywave transmission phenomenon of HF depends on ionospheric refraction. Transmitted radio waves hitting the ionosphere are bent or refracted. When they are bent sufficiently, the waves are returned to earth at a distant location. Often at the distant location they are reflected back to the sky again, only to be returned to earth still again, even further from the transmitter.

The ionosphere is the key to HF skywave communication. Look at the ionosphere as a layered region of ionized gas above the earth. The amount of refraction varies with the degree of ionization. The degree of ionization is primarily a function of the sun's ultraviolet (UV) radiation. Depending on the intensity of the UV radiation, more than one ionized layer may form (see Figure 8.3). The existence of more than one ionized layer in the atmosphere is explained by the existence of different UV frequencies in the sun's radiation. The lower frequencies produce the upper ionospheric layers, expending all their energy at high altitude. The higher frequency UV waves penetrate the atmosphere more deeply before producing appreciable ionization. Ionization of the atmosphere may also be caused by particle radiation from sunspots, cosmic rays, and meteor activity.

For all practical purposes four layers of the ionosphere have been identified and labeled as follows:

D Region or D-layer. Not always present, but when it does exist, it is a daytime phenomenon. It is the lowest of the four layers. When it exists, it occupies an area between 50 and 90 km above the earth. The D region is usually highly absorptive due to its high collision frequency.

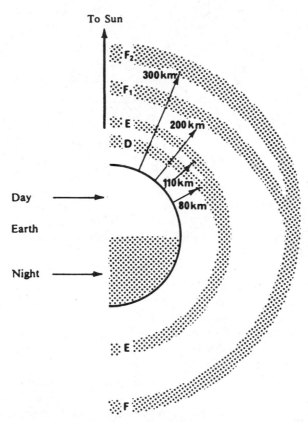

Figure 8.3 Ionized layers of the atmosphere as a function of nominal height above the earth's surface.

E Region or E-layer. A daylight phenomenon, existing between 90 and 140 km above the earth. It depends directly on the sun's UV radiation and hence it is most dense directly under the sun. The layer all but disappears shortly after sunset. Layer density varies with seasons owing to variations in the sun's zenith angle with seasons.

F1 Layer. A daylight phenomenon existing between 140 and 250 km above the earth. Its behavior is similar to that of the E layer in that it tends to follow the sun (i.e., most dense under the sun). At sunset the F1 layer rises, merging with the next higher layer, the F2 layer.

F2 Layer. This layer exists day and night between 150 and 250 km (night) and 250 and 300 km above the earth (day). During the daytime in winter, it extends from 150 to 300 km above the earth. Variations in height are due to solar heat. It is believed that the F2 layer is also strongly influenced by the earth's magnetic field. The earth is divided into three magnetic zones representing different degrees of magnetic

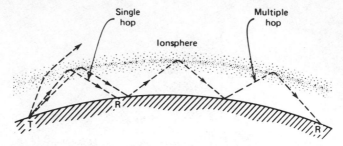

Figure 8.4 Single and multihop HF skywave transmission, T = transmitter; R = receiver.

intensity called east, west, and intermediate. Monthly F2 propagation predictions are made for each zone.* The north and south auroral zones are also important for F2 propagation, particularly during high sunspot activity.

Consider these layers as mirrors or partial mirrors, depending on the amount of ionization present. Thus transmitted waves striking an ionospheric layer, particularly the F layer, may be refracted directly back to earth and received after their first hop, or they may be reflected from the earth back to the ionosphere again and repeat the process several times before reaching the distant receiver. The latter phenomenon is called multihop transmission. Single and multihop transmission are illustrated diagrammatically in Figure 8.4.

To obtain some idea of the estimated least possible number of F-layer hops as related to path length, the following may be used as a guide:

Number of Hops	Path Length (km)
1	< 4000
2	4000–7000
3	7000–12,000

(see Section 8.10.2.4).

HF propagation above about 8 MHz encounters what is called a *skip zone*. This is an "area of silence" or a zone of no reception extending from the outer limit of groundwave communication to the inner limit of skywave communication (first hop). The skip zone is shown graphically in Figure 8.5.

The region of coverage from an HF transmitter can be extended through the "skip zone" by a subset of skywave transmission called "near-vertical incidence" (NVI) or "quasi-vertical incidence" (QVI). The objective is to launch a wave nearly directly upward from the antenna. Therefore special

*Monthly propagation forecasts are made by the Central Radio Propagation Laboratory (CRPL), U.S. National Bureau of Standards.

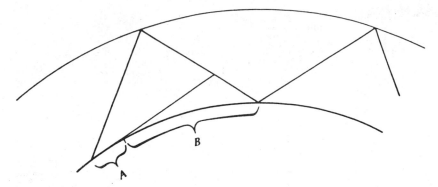

Figure 8.5   Skip zone. A = limit of groundwave communication; B = skip zone.

antennas that have high takeoff angles are required (see Section 8.15). By radiating the HF RF energy nearly vertically, we can achieve "reflection" from the F layer almost overhead. NVI propagation provides HF connectivity from 20 mi or less to over 500 mi. The best performance for this type of operation is to use frequencies from 3 to 6 MHz. Receive signals using this mode of propagation will fade rapidly (1 Hz), producing time dispersion on the order of 100 μs and with some Doppler spread. With proper signal processing at the receiver, these impairments can be mitigated. Using some of the new wide-band HF techniques that are just now emerging can resolve the multipath and much of the dispersion.

8.5   CHOICE OF OPTIMUM OPERATING FREQUENCY

One of the most important elements for the successful operation of an HF link using skywave transmission is the selection of an operating frequency that will assure a link with 100% time availability, day, night, and year-round. Seldom, if ever, can this be achieved.

Optimum HF propagation between points X and Y anywhere on the earth varies with

1. Location, in particular, latitude
2. Season
3. Diurnal variations (time of day)
4. Cyclical variations (relating to the sunspot number)
5. Abnormal (disturbed) propagation conditions

Location   The intensity of ionizing radiation that strikes the atmosphere varies with latitude. The intensity is greatest in equatorial regions, where the sun is more directly overhead than in the higher latitudes.

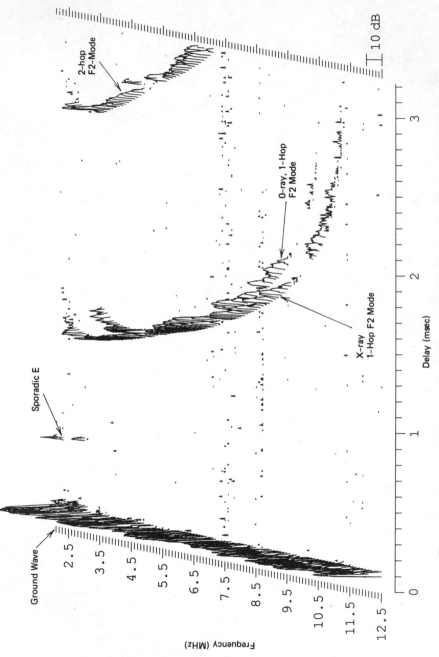

Figure 8.6 An ionogram of a one-hop path, temperate zone, evening. Note that the O-ray (trace) and X-ray (trace) are clearly visible. Ionogram courtesy of Dr. L. S. Wagner, Naval Research Laboratories, Washington, D.C. (Ref. 39).

We find that the critical frequencies for E and F1 regions vary directly with the sun's elevation, being highest in equatorial regions and decreasing as a function of the increase in latitude. The *critical frequency* is the highest frequency, using vertical incidence (transmitting a wave directly overhead, at a 90° elevation angle), at a certain time and location, at which RF energy is reflected back to earth.

The F2 layer also varies with latitude. Here the issue is more complex. It is postulated that the variance in ionization is caused by other sources, such as X-rays, cosmic rays, and the earth's magnetic field. However, F2-layer critical frequency does not have a strong variance as a function of latitude but is more associated with longitude. Critical frequencies related to the F2 layer are generally higher in the Far East than in Africa, Europe, and the Western Hemisphere.

HF transmission engineers consider regions (location) in a more general sense, again related to latitude. The most attractive region which displays relatively low values of dispersion and Doppler spread is the temperate region, especially overwater paths in that region. The second region covers HF paths that are transequatorial. These paths encounter "*spread F*" propagation, which causes relatively high time dispersion because the F-layer over the equator is more diffuse, resulting in much greater multipath effects.

The third difficult region covers those HF paths that cross the auroral oval. There are two auroral ovals, one in the Southern Hemisphere and one in the Northern Hemisphere. Each is centered on its respective magnetic pole. Transauroral paths are difficult paths in terms of meeting performance requirements and are even more difficult during magnetic storms and other solar flare disturbances.

The most difficult of all HF paths are those that cross the polar cap where Doppler spread has been measured in excess of 10 Hz and time dispersion over 1–2 ms (Ref. 37 and 40).

Figure 8.6 shows an ionogram of a benign one-hop path in the evening. It shows the groundwave return appearing vertically on the far left. The one-hop F2 return(s) consisting of an O-ray and an X-ray. O stands for "ordinary" and the X for "extraordinary." How these rays are generated are discussed below. The ionogram also shows a vestige of multimode transmission appearing as a two-hop mode return on the far right side of the figure. Figure 8.7 shows a typical auroral oval, in this case, north of the equator. It should be kept in mind that the oval is indeed centered on the magnetic pole but that its boundary can shift; it is not fixed (Ref. 2).

Seasonal Factors Our earth orbits the sun with an orbital period of 1 yr. It is this orbit that brings about our seasons: spring, summer, autumn, and winter. The sun is a major controlling element on the behavior of the ionosphere. For instance, E-layer ionization of sufficient magnitude to support skywave propagation depends almost entirely on the sun's elevation in the sky; it is stronger in summer than in winter.

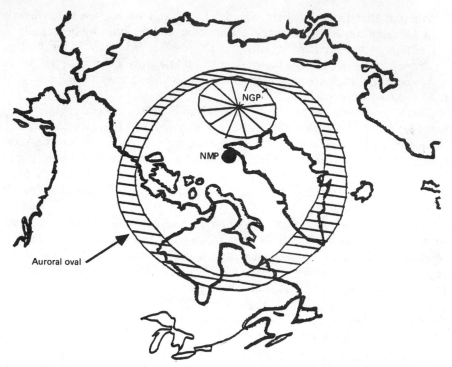

Figure 8.7 The auroral oval. NMP = northern magnetic pole. NGP = northern geographical pole.

The F1 layer, in general, exists only during daylight hours. During the winter its critical frequency varies in a similar manner as the E layer, being dependent on the sun's elevation. Ionization, as one might imagine, is greater in the summer months than in the winter months, when there is less sunlight. Often, in the winter months the F1 layer merges with the F2 layer and cannot be specifically identified except in equatorial regions.

The F2 layer is the reverse; daytime ionization is very intense and critical frequencies are higher than its F1 counterpart. Because of the extended periods of darkness in winter months, the ionosphere has more time to lose its electrical charge (recombination), resulting in nighttime critical frequencies dipping to very low values.

During the summer with its extended daylight hours, the F2 layer heats and expands, resulting in notably lower ionization density than in winter. This makes summer daytime F2 critical frequencies lower than winter values. The general assumption is that the F1 and F2 layers combine at night. This certainly is true in the winter months. In summer months, because of the longer daylight hours, this recombination does not occur to the extent that it does in winter. This results in nighttime F2 critical frequencies that are significantly higher in summer than in winter. Also, the difference between day and night critical frequencies is much smaller in summer than in winter.

Diurnal Variations (Time of Day)   We break the 24-h day into three generalized time periods:

1. Day
2. Transitions
3. Night

The transition periods occurs twice: once around sunrise and again around sunset.

Again we are dealing with the intensity and duration of sunlight through the daytime hours. It is primarily the UV radiation from the sun that ionizes the atmosphere. During the daylight hours this radiation reaches a maximum intensity; during the hours of darkness, there is little UV radiation impinging the atmosphere and the E and F regions decrease to a relatively weak single layer through which propagate clouds with greater electron densities, which gives rise to sporadic E (E_s) propagation modes.

The F2 region critical frequency rises steeply at sunrise, reaching its maximum after the sun has reached its zenith. It starts to drop off sharply when the sun is starting to set. The transition periods, around sunrise and again around sunset, are when we encounter rapidly changing critical frequencies.

Cyclical Variations   The results of sunspots lead to the one phenomenon that affects atmospheric behavior more than any other. Sunspot activity is cyclical and is therefore referred to as the solar cycle. We characterize solar activity (i.e., number, intensity, and duration of sunspots) by the sunspot number. Sunspots appear on the sun's surface and are tremendous eruptions of whirling electrified gases. These gases are cooled to temperatures below those of the sun's surface, resulting in darkened areas appearing to us as spots on the sun. The whirling gases are accompanied by extremely intense magnetic fields. At times there are sudden flare-ups on the sun that some call *magnetic storms*. Magnetic storms will be discussed below.

Sunspots last from several days to several months. The sun has a rotational period of 27 days. Larger sunspots can remain visible for several rotations of the sun, giving rise to propagation anomalies with 27-day cycles.

Galileo was the first to report sunspots in modern recorded history. Some sources credit the Chinese, long before the birth of Christ, for being the first to observe sunspots.

The Zurich observatory began reporting sunspot data on a regular basis in 1749. Here Rudolf Wolf devised a standard method of measure of relative sunspot activity. This is called the Wolf sunspot number, or the Zurich sunspot number or just the sunspot number. The sunspot number is more of an index of solar activity than a measure of the number of sunspots observed.

The official sunspot number is derived from daily recording of sunspot numbers at a solar observatory in Locarno, Switzerland. The raw sunspot data is processed and disseminated by the Sunspot Index Data Center in Brussels. There has been a continuity of these measurements since 1749 when they were begun in Zurich. The sunspot cycle is roughly 11 years. A cycle is the time in years between contiguous solar activity minimums. A minimum sunspot number has an average value of 5 and an average maximum value of 109 (Ref. 3). Solar cycles are numbered, with cycle No. 1 starting in 1749; cycle 21 began in June 1976 and cycle 22 in September 1986, reaching a maximum value of about 190 in January 1990. Sunspot cycles are not of uniform duration. Some are as short as 9 years and others as long as 14 years (Ref. 3). Figure 8.8 shows observed and predicted sunspot numbers up through 1991 and extrapolated until 1998.

Sunspots have a direct bearing on UV radiation intensity of the sun which, as we mentioned above, is the principal cause of ionization of the upper atmosphere, which we call the ionosphere. It has a direct relation to the critical frequency and consequently to the *maximum usable frequency* (MUF). During periods of low sunspot numbers, the daytime MUF may not reach 17 MHz; during periods of sunspot maximum, the MUF has been known to far exceed 50 MHz. In sunspot cycle 19, state police radios operating in the 50-MHz region in Massachusetts and Rhode Island were made virtually

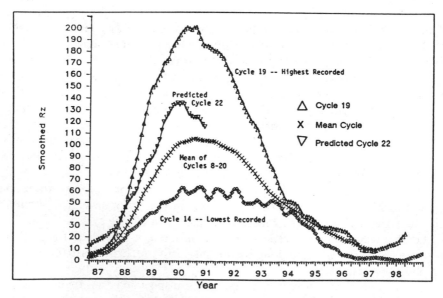

Figure 8.8 Observed and predicted sunspot numbers from McNish–Lincoln analysis. R_z is relative sunspot number (smoothed). (From Ref. 4.)

useless by interfering signals from the West Coast, some 3000 mi away. This is one reason most dispatching systems, especially in public safety, are now using 150 MHz and some low ultrahigh frequencies (UHF). Thus they avoid the cyclical problem of those periods when sunspot numbers are high.

Ionospheric Disturbances We have classified four factors dealing with skywave HF transmission. If we neglect interference, there is some predictability for the design of an HF path. The computer has helped us along the way with such excellent programs as IONCAP and PROPHET (Refs. 5–7). They take into account sunspot number, latitude/longitude, time of day, season, and system characteristics. However, the transmission engineer feels nearly helpless when faced with the unexpected, the ionospheric disturbance.

There are three types of ionospheric disturbances we have classified:

1. Ionospheric storm (magnetic storm)
2. Sudden ionospheric disturbance (SID)
3. Polar blackout

The author was in the Arctic during cycle 19 peak (1958), aboard ship. Whenever there was a good display of aurora borealis, the HF band would black out and would remain that way from 12 to 72 h. All one would hear is a high level of thermal noise. Curiously, it seemed that signals in the low medium-frequency (MF) and low-frequency (LF) bands were enhanced in these periods of HF blackout.

There is fair correlation with the number and intensity of SIDs, ionospheric storms, and blackouts with the sunspot cycle. The higher the sunspot number, the greater the probability of occurrence and the greater the intensity. Again, we believe the cause to be solar flare activity, a sudden change in the magnetic field around sunspots (Ref. 3). These produce high levels of X-rays, UV radiation, and cosmic noise. It takes about 8 min for the effects of this radiation to degrade the ionosphere. This is an SID. The result is heavy HF skywave signal attenuation.

Slower-traveling charged particles arrive some 18–36 h after the sudden flare-up on the sun. The result of these particles hitting the ionosphere is often blackout, especially in the higher latitudes. In other areas we notice a remarkable lowering of the MUF.

WWV transmitting from Boulder, CO, on 5, 10, 15, and 20 MHz is normally used for time and frequency calibration. At 18 min after each hour, WWV of the U.S. National Bureau of Standards (NBS) transmits timely solar and geomagnetic activity with updates every 6 h. WWV transmits the K index and the A index. The K index is a single digit from 0 through 9 and is a measure of current geomagnetic activity; 0 is the lowest activity factor and 9 the highest. The A index is a measure of solar flux activity.

8.5.1 Frequency Management

8.5.1.1 Definitions In the context of this chapter, frequency management is the art/technology of selecting an optimum frequency for HF communication at a certain time of day between any two points on the earth. One important factor is omitted in the process, that is, the interference that may be present in the path, affecting one or both ends of the path.

We will be using the terms *MUF*, *LUF*, and *FOT* (*OWF*). The MUF and LUF are the upper and lower limiting frequencies for skywave communication between points *X* and *Y*. The MUF is the maximum usable frequency on an oblique incidence path; the LUF is the lowest usable frequency.

The concept of MUF is most important for the HF link design engineer for skywave links. We will find that the MUF is related to the critical frequency by the secant law. The optimum working frequency (OWF; original derivation from a French term), sometimes called FOT (from the French, *fréquence optimum de travail*), is often taken as $0.85 \times$ MUF. Little is mentioned in the literature about the LUF, since it is so system sensitive.

The MUF is a function of the sunspot number, time of day, latitude, day of the year, and so on, which are things completely out of our control. The LUF, on the other hand, is somewhat under our control. If we hold a transmitting station EIRP constant, as the operating frequency decreases, the available power at the distant receiver normally decreases owing to increased ionospheric absorption. Furthermore, the noise power increases so that the signal-to-noise ratio deteriorates and the circuit reliability decreases. The minimum frequency below which the reliability is unacceptable is called the lowest useful frequency (LUF). The LUF depends on transmitter power; antenna gain at the desired takeoff angle (TOA); factors that determine transmission loss over the path, such as frequency, season, sunspot number, geographical location, and so forth; and the external noise level. One of the primary factors is ionospheric absorption and hence, since this varies with the solar zenith angle, the LUF peaks at about noon. Consequently, in selecting a frequency, it is necessary to ascertain whether the LUF exceeds this frequency.

When applying computer prediction programs such as IONCAP, under certain situations we will find the LUF exceeds the MUF. This tells us, given the input parameters to the program, that the link is unworkable. We may be able to shift the LUF downward in frequency by relaxing link reliability (time availability), reducing signal-to-noise ratio requirements, increasing EIRP, and increasing receive antenna directional gain.

One factor in the transmission loss formula for HF is D-layer absorption. It varies as $1/f^2$. For this reason we want to operate at the highest possible frequency to minimize D-layer absorption. Suppose we operated at the MUF. It is a boundary limit and would be unstable with heavy fading. We choose a frequency as close as possible to the MUF yet stay out of boundary conditions. The operating frequency goes under two names: FOT or OWF, both defined earlier. The OWF is usually 0.85 the value of the MUF for F2 operation. Our objective is to keep our transmitter frequency as close to the OWF as possible. This is done to minimize atmospheric absorption but yet not too close to the MUF to reduce ordinary–extraordinary ray fading. As the hours of the day pass, the OWF will move upward or downward and through the cycle of our 24-h day. It could move one or even two octaves from maximum to minimum frequency. How do we know when and where to move to in frequency?

8.5.1.2 Methods   There are six general methods in use today to select the best frequency. We might call these methods to carry out *frequency management*. The six methods are

1. By experience
2. Use of CRPL (Central Radio Propagation Laboratory) predictions
3. Carrying out one's own predictions by one of several computer programs available
4. Use of ionospheric sounders
5. Use of distant broadcast facilities
6. Self- and embedded sounding

Experience. Many old-time operators still rely on experience. First they listen to their receivers, then they judge if an operating frequency change is required. It is the receive side of a link that commands a distant transmitter. An operator may well feel that "yesterday I had to QSY* at 7 P.M. to 13.7 MHz, so today I will do the same." Listening to his/her receiver will confirm or deny this belief. When he/she hears his/her present operating frequency (from the distant end) start to take deep fades and/or the signal level starts to drop, it is time to start searching for a new frequency. The operator checks other assigned frequencies on a spare receiver and listens for other identifiable signals near these frequencies to determine conditions to select a better frequency. If conditions are found to be better on a new frequency, the transmitter operator is ordered to change frequency (QSY at the distant end). This, in essence, is the experience method. We will address this issue further on from a somewhat different perspective.

CRPL Predictions. Here, of course, we are dealing with the use of predictions issued by the U.S. Institute of Telecommunication Sciences, located in Boulder, CO. The predictions are published monthly, three months in advance of their effective dates (Ref. 8).

---

*A "Q" signal is an internationally recognized three-letter operational signal used among operators. A typical Q signal is QSY, which means "change your frequency to _____." Q signals were used almost exclusively on CW (Morse) circuits and their use has extended to teleprinter and even voice circuits.

Consider an HF circuit designed for 95% time availability* (propagation reliability) and assume a minimum signal-to-noise ratio of 12 dB Hz. The median receive signal level (RSL) must be increased on the order of 14 dB to overcome slow variations of skywave field intensity and atmospheric noise, and 11 dB to overcome fast variations of skywave field intensity. Therefore a rough order of magnitude value of signal-to-median-noise ratio with sufficient margin for 90% of the days is 37 dB for M-ary-frequency shift keying (MFSK) data/telegraph transmission for a bit error rate (BER) of 1×10^{-4}. See CCIR Rec. 339-6 (Ref. 9).

Predictions by PC. Quite accurate propagation predictions for an HF path can be carried out on a personal computer such as the IBM/IBM clone PC-AT. A very widely accepted program is IONCAP, which stands for Ionospheric Communication Analysis and Prediction Program. It is written in Fortran (ANSI) and is divided into seven largely independent subsections (Refs. 5 and 6):

1. Input subroutines
2. Path geometry subroutines
3. Antenna pattern subroutines
4. Ionospheric parameter subroutines
5. Maximum usable frequency (MUF) subroutines
6. System performance subroutines
7. Output subroutines

Table 8.1 is a listing of 30 available output methods. The IONCAP computer program performs four basic analysis tasks. These tasks are summarized below. Note that E_s indicates highest observed frequency of the ordinary component of sporadic E and HPF means highest probable frequency.

1. *Ionospheric Parameters.* The ionosphere is predicted using parameters that describe four ionospheric regions: E, F1, F2, and E_s. For each sample area, the location, time of day, and all ionospheric parameters are derived. These may be used to find an electron density profile, which may be integrated to construct a predicted ionogram. These options are specified by methods 1 and 2 in Table 8.1.

2. *Antenna Patterns.* The user may precalculate the antenna gain pattern needed for the system performance predictions. These options are specified by methods 13–15 in Table 8.1. If the pattern is precalculated, then the antenna gain is computed for all frequencies (1–30 MHz) and elevation

---

*We define "time availability" as the percentage of time a certain performance objective (e.g., BER) is met.

Table 8.1   IONCAP — available output methods

Method	Description of Method
1	Ionospheric parameters
2	Ionograms
3	MUF–FOT lines (nomogram)
4	MUF–FOT graph
5	HPF–MUF–FOT graph
6	MUF–FOT–E_s graph
7	FOT–MUF table (full ionosphere)
8	MUF–FOT graph
9	HPF–MUF–FOT graph
10	MUF–FOT–ANG graph
11	MUF–FOT–E_s graph
12	MUF by magnetic indices, K (not implemented)
13	Transmitter antenna pattern
14	Receiver antenna pattern
15	Both transmitter and receiver antenna patterns
16	System performance (S.P.)
17	Condensed system performance, reliability
18	Condensed system performance, service probability
19	Propagation path geometry
20	Complete system performance (C.S.P.)
21	Forced long-path model (C.S.P.)
22	Forced short-path model (C.S.P.)
23	User-selected output lines (set by TOPLINES and BOTLINES)
24	MUF–REL table
25	All modes table
26	MUF–LUF–FOT table (nomogram)
27	FOT–LUF graph
28	MUF–FOT–LUF graph
29	MUF–LUF graph
30	Create binary file of variables in "COMMON/MUFS/" (allows the user to save MUFs–LUFs for printing by a separate user written program)

Source: Ref. 6.

angles. If the pattern is not precalculated, then the gain value is determined for a particular frequency and elevation angle (takeoff angle) as needed.

3. *Maximum Usable Frequency (MUF)*. The maximum frequency at which a skywave mode exists can be predicted. The 10% (FOT), 50% (MUF), and 90% (HPF [highest probable frequency]) levels are calculated for each of the four ionospheric regions predicted. These numbers are a description of the state of the ionosphere between two locations on the earth and not a statement of the actual performance of any operational communication circuit. These options are specified by methods 3–12 in Table 8.1.

4. *Systems Performance*. A comprehensive prediction of radio system performance parameters (up to 22) is provided. Emphasis is on the statistical performance over a month's time. A search to find the lowest usable frequency (LUF) is provided. These options are specified by methods 16–29 in Table 8.1.

Table 8.2 shows a typical run for method 16. The path is between Santa Barbara, CA, and Marlborough, MA.

There are several other programs that can be run on a PC. One is Minimuf, which predicts only the MUF in midlatitudes (Ref. 10). PROPHET and Advanced PROPHET are two HF prediction codes for military application (Refs. 7 and 11). The accuracy is not as good as IONCAP, but they are a useful tool for propagation prediction and for the specific military applications for which they are designed. They provide area coverage easily, which IONCAP does not, and also allow interaction for various details of electronic warfare.

Ionospheric Sounders. Ionospheric oblique sounders give real-time data on the MUF, transmission modes propagating (F1, F2, one-, two- or three-hop), and the LUF between two points that are sounder-equipped. One location has a sounder transmitter installed that sweeps the entire HF band using an FM/CW signal. Transmitter power output usually is in the range of 1–10 W. A sounder receiver and display is operated at the distant end. The receiver is time synchronized with the distant transmitter. It displays a time history of the sweep of the band. The x-axis of the display is the HF frequency band 2 (3) to 30 MHz and the y-axis is time delay measured in milliseconds. Here received power is displayed as a function of frequency and time delay. Of course, the shortest delay is that power reflected off the E layer, the next shortest the F layer (F1 and F2 in daytime, in that order), then multihop power from the F layers. A multihop signal is delayed even more. Figure 8.9 shows a conceptual diagram of oblique incidence sounder operation.

One system designed and manufactured by BR Communications is the AN/TRQ-35. This unit includes a low-power transmitter and, at the far end of the circuit, a receiver with a spectrum monitor that compiles channel occupancy statistics for 9333 channels spaced 3 kHz from 2–30 MHz. BR Communications calls their spectrum monitor a "signal-to-noise ratio" analyzer that determines the background noise level—usually atmospheric noise —and then compiles statistics on the percentage of time that level is exceeded in each channel. The monitor has memory and can display any channel usage up to 30 min (Ref. 12).

A typical oblique ionospheric display (ionogram) is shown in Figure 8.10.

Another type of sounder is the backscatter or vertical incidence sounder, where the receiver and transmitter are collocated. In this case, the HF spectrum is swept as before, but the RF energy is launched vertically for refraction by the ionosphere directly overhead. Backscatter sounding pro-

Table 8.2   Samples of an IONCAP run, method 16

```
                    METHOD 16  IONCAP PC.20  PAGE 2

           OCT 1990              SSN = 192.
SANTA BARBARA TO MARLBOROUGH            AZIMUTHS      N. MI.       KM
34.50 N  119.00 W - 42.50 N  71.50 W   63.19 273.90 2252.5    4171.2
                        MINIMUM ANGLE 1.0  DEGREES
ITS- 1 ANTENNA PACKAGE
XMTR   2.0  TO  30.0  CONST. GAIN H  .00 L  .00 A  .0  OFF AZ   .0
RCVR   2.0  TO  30.0  CONST. GAIN H  .00 L  .00 A  .0  OFF AZ   .0
POWER =    1.000 KW 3 MHZ NOISE = -150.0 DBW REQ. REL = .80   REQ. SNR = 48.0
MULTIPATH POWER TOLERANCE = 6.0 DB MULTIPATH DELAY TOLERANCE = .500 MS
  UT   MUF


 5.0 18.9  2.0   3.4   4.2   6.3   8.5 10.7 12.8 15.0 17.2 19.4 21.5 FREQ
          1F2   4 E  1F2  2ES  2F2  2F2  2F2  2F2  2F2  2F2  1F2  1F2 MODE
          3.8   8.7   3.2   1.3 11.1 11.4 12.2 14.0 19.0 19.0  3.5  3.5 ANGLE
         14.8  14.3  14.7  14.1 14.8 14.8 14.9 15.1 15.7 15.7 14.8 14.8 DELAY
         513.  103.  487.  110. 302. 309. 326. 363. 475. 475. 500. 500. V HITE

          .50 1.00 1.00 1.00 1.00 1.00 1.00  .99  .93  .74  .41  .10 F DAYS

         151.  158.  153.  157. 143. 142. 142. 145. 177. 235. 153. 173. LOSS
          16.  -15.    0.    2.  12.  15.  16.  14. -16. -73.  10.  -9. DBU
         -116  -128  -117  -117 -110 -111 -111 -114 -147 -204 -123 -143 S DBW
         -171  -136  -143  -145 -151 -158 -163 -166 -168 -170 -172 -173 N DBW
          55.    9.   25.   28.  40.  47.  51.  51.  22. -34.  49.  30. SNR
          10.   48.   28.   24.  12.   5.   2.   8.  43.  99.  16.  34. RPWRG
          .64   .00   .00   .01  .16  .45  .68  .59  .11  .00  .52  .20 REL
          .00   .00   .00   .01  .03  .00  .00  .00  .00  .00  .00  .00 MPROB


11.0 13.7  2.0   3.5   5.1   6.6   8.2   8.6   9.7 11.2 12.8 14.3 15.9 FREQ
          1F2   3 E  2ES  2F2  2F2  1F2  1F2  2F2  1F2  1F2  1F2  1F2 MODE
          4.9   4.1   1.3  16.7 15.0  4.9   3.7 17.9  1.2  2.5  4.9  4.9 ANGLE
         14.9  14.1  14.1  15.4 15.2 14.9 14.8 15.6 14.5 14.7 14.9 14.9 DELAY
         557.   89.  110.  422. 385. 559. 510. 450. 407. 457. 557. 557. V HITE

          .50 1.00  .91 1.00 1.00 1.00 1.00  .98  .90  .68  .35  .10 F DAYS

         148.  172.  168.  147. 144. 142. 141. 150. 139. 142. 153. 169. LOSS
          12.  -28.   -6.    7.  10.  17.  18.   8.  19.  17.   8.  -8. DBU
         -117  -140  -123  -114 -113 -107 -107 -118 -109 -112 -122 -139 S DBW
         -167  -144  -148  -151 -154 -157 -158 -161 -164 -166 -168 -169 N DBW
          49.    4.   25.   36.  41.  49.  50.  42.  55.  54.  45.  30. SNR
          15.   54.   31.   19.  15.   6.   5.  22.   2.   8.  19.  35. RPWRG
          .53   .00   .01   .09  .18  .56  .61  .32  .74  .64  .43  .20 REL
          .00   .00   .00   .02  .00  .00  .00  .00  .00  .00  .00  .00 MPROB


17.0 42.3  2.0   4.6   7.1   9.7 12.2 14.8 17.4 23.0 28.7 34.3 40.0 FREQ
          1F2   3 E  2ES  2ES  2ES  2ES  4F2  2F2  2F2  2F2  2F2  1F2 MODE
          3.9   3.3   1.3   1.3  1.3  1.3 23.8  9.4  9.2 11.2 18.3  8.4 ANGLE
         14.8  14.1  14.1  14.1 14.1 14.1 15.8 14.6 14.6 14.8 15.6 15.4 DELAY
         517.   79.  110.  110. 110. 110. 261. 267. 264. 304. 459. 715. V HITE
```

Table 8.2 *(Continued)*

.50	1.00	1.00	1.00	1.00	.92	1.00	1.00	.99	.95	.83	.61	F DAYS
152.	626.	348.	253.	213.	195.	186.	159.	152.	151.	188.	160.	LOSS
22.	****	****	-96.	-46.	-26.	-26.	5.	13.	16.	-21.	12.	DBU
-117	-355	-317	-220	-172	-154	-156	-126	-120	-119	-158	-127	S DBW
-181	-145	-155	-160	-163	-165	-167	-169	-174	-176	-179	-180	N DBW
63.	****	****	-60.	-9.	11.	11.	42.	53.	56.	21.	54.	SNR
1.	262.	221.	118.	67.	48.	50.	17.	8.	9.	44.	11.	RPWRG
.78	.00	.00	.00	.00	.00	.00	.25	.62	.66	.10	.61	REL
.00	.00	.00	.00	.00	.00	.00	.00	.00	.00	.00	.00	MPROB

24.0	38.2	2.0	4.0	6.1	10.3	14.6	18.8	23.0	27.3	31.5	35.8	40.0	FREQ
1F2	3 E	2ES	4F2	2F2	2F2	2F2	2F2	2F2	2F2	1F2	1F2	MODE
3.7	4.1	1.3	24.4	9.3	9.3	9.8	11.0	14.1	18.9	9.2	3.7	ANGLE
14.8	14.1	14.1	15.9	14.6	14.6	14.7	14.8	15.1	15.7	15.5	14.8	DELAY
510.	89.	110.	269.	266.	265.	276.	300.	366.	473.	749.	510.	V HITE

.50	1.00	.89	1.00	1.00	1.00	1.00	.99	.97	.87	.66	.31	F DAYS
157.	227.	177.	175.	151.	148.	148.	149.	155.	210.	169.	164.	LOSS
11.	-84.	-27.	-22.	8.	13.	15.	15.	11.	-43.	2.	5.	DBU
-127	-196	-146	-145	-119	-116	-117	-119	-123	-179	-136	-134	S DBW
-180	-144	-149	-153	-161	-167	-171	-174	-176	-178	-179	-180	N DBW
53.	-53.	3.	9.	42.	50.	54.	55.	52.	-2.	43.	46.	SNR
12.	111.	55.	47.	13.	6.	3.	3.	13.	67.	22.	18.	RPWRG
.60	.00	.00	.00	.22	.58	.72	.71	.57	.01	.38	.46	REL
.00	.00	.00	.00	.00	.00	.00	.00	.00	.00	.00	.00	MPROB

Source: Radio General Company, Stow, MA.

Notes: Header information largely supplied by user.

Line 1: month, year, and sunspot number.

Line 2: Label as supplied by user and headings for next line.

Line 3: Transmitter location, receiver location, the azimuth of the transmitter to the receiver in degrees east of north; path distance in nm and km.

Line 4: minimum radiation angle in degrees.

Line 5: Antenna subroutine used (use IONCAP antenna subroutine).

Line 6: Transmitter antenna data.

Line 6A: Receiver antenna data.

Line 7: System line which has: transmitter power in kW, man-made noise level at 3 MHz in dBW, required reliability, and required S/N in dB.

System performance lines which are repeated for each hour.

Line 0: FREQ: Time and frequency line as associated with each column. The first four lines always refer to the most reliable mode (MRM). The system performance parameter usually comes from the sum of all six modes.

Line 1, MODE: E is E-layer, F1 is F1 layer, F2 is F2 layer, E_s is E_s layer; N is a one-hop E_s with n F1 or F2 hops (MRM).

Line 2, ANGLE: The radiation angle in degrees (MRM).

Line 3, DELAY: Time delay in ms (MRM).

Line 4, V HITE: Virtual height in km (MRM).

Line 5, F DAYS: The probability that the operating frequency will exceed the predicted MUF.

Line 6, LOSS: Median system loss in dB for the most reliable mode (MRM).

Line 7, DBU: Median field strength expected at the receiver location in dB above 1 $\mu V/m$.

Line 8, S DBW: Median signal power expected at the receiver input terminals in dB above a watt.

Line 9, N DBW: Median noise power expected at the receiver in dB above a watt.

Line 10, SNR: Median signal-to-noise ratio in dB.

Line 11, RPWRG: Required combination of transmitter power and antenna gains needed to achieve the required reliability.

Line 12, REL: Reliability. The probability that the SNR exceeds the required SNR. Note this applies to all days of the month and includes the effect of all mode types: E, F1, F2, E_s, and over-the-MUF modes.

Line 13, MPROB: The probability of an additional mode within the multipath tolerances (short paths only).

vides data on the critical frequency, from which we can derive the MUF for any one-hop path. This is done with the following formula, which is based on Snell's law:

$$F_0 = F_n \sec \phi \qquad (8.1)$$

where F_n = the maximum frequency (critical frequency) that will be reflected back at vertical incidence

F_0 = the maximum usable frequency (MUF) at oblique angle ϕ

ϕ = the angle of incidence (i.e., the angle between the direction of propagation and a line perpendicular to the earth)

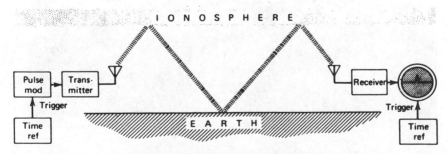

Figure 8.9 Synchronized oblique sounding.

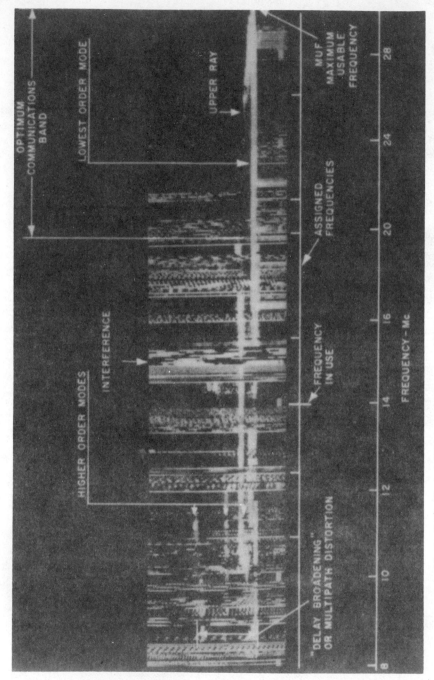

Figure 8.10 Typical HF ionogram from an oblique sounding system. (Courtesy of Granger Associates, Menlo Park, CA.)

Backscatter techniques can be used on any one-hop path where one point is where the backscatter sounder is located. On-the-other hand, oblique incidence sounders provide information on the single, equipped path, which may be one-hop or multiple hops.

Another approach that can be used on some occasions is to know the locations of other operating oblique sounder transmitters near the desired location for HF connectivity. "Near" can mean several hundred miles either side of north–south direction and about 150 mi either side in the east–west direction of the desired distant end. Naturally, synchronization and interface data on these other sounders are necessary if we wish to make use of their signals.

Use of Distant Broadcast Facilities. We can obtain rough-order-of-magnitude propagation data derived from making comparative signal-level measurements on distant HF broadcast facilities "near" the desired distant point with which we wish to establish HF connectivity. Many facilities broadcast simultaneously on multiple HF bands, so a comparison can be made from one band to another on signal level, fade rate, and fade depth to determine the optimum band for operation. Many HF receivers are equipped with S-meters* from which we can derive some crude comparative level measurements.

For example, if we wish to communicate with an area in or near the U.S. state of Colorado, we can use WWV, which has excellent time and frequency dissemination service transmitting on 5, 10, 15, and 20 MHz simultaneously 24 h a day.

International broadcast such as the BBC and the Voice of America are other candidates. With present state-of-the-art synthesized HF receivers, we can tune directly to them. If one can read CW (copy international Morse code), there are hundreds of marine coastal stations that will have continuous transmissions on 4, 6, 8, 12, 17, and 22 MHz (the marine bands), depending on which bands each facility considers to be "open" (useful) for the service that facility provides. These facilities will drop off automatic operation and into manual CW operation when they are "in traffic" (operating with a distant ship station). We may call this method "poor man's sounding."

Embedded Sounding, Early Versions. These earlier devices, still very widely used, are based on the idea that a facility or facility grouping has only a limited number of assigned operating frequencies. These devices transmit low RF energy, usually using FSK at a low bit rate, to sample each frequency to derive link quality assessment (LQA) data and, in some instances, may also use the same device for selective calling. LQA involves the exchange of link quality information on a one-on-one basis or networkwide. SELSCAN, a Rockwell-Collins trademark, is a good example of such a system. It measures

---

*An S-meter gives a comparative measurement of received signal level.

signal-to-noise ratio and multipath delay distortion on up to 30 stored preset channels for use by automatic frequency selection algorithms. SELSCAN also has an ALE (automatic link establishment) feature—(Ref. 13).

It is suggested that the reader consult MIL-STD-188-141A (Ref. 14) for some excellent methods of ALE and LQA. We are really dealing here with an OSI layer-2 data link access protocol (see Chapter 12). SELSCAN, however, has been designed as an adjunct to SSB voice operation on HF for setting up a link on an optimum frequency. Harris (United States) and Tadiran (Israel), among others, have similar ALE/LQA devices.

More Advanced Embedded Sounding. These are self-sounding systems designed for HF data/telegraph circuits that are half- or full-duplex. We will consider such circuits, for this argument, to have as near as possible to 100% duty cycle (i.e., they are active nearly 100% of the time). Such systems dedicate pure overhead, perhaps 5 or 10%, for self-sounding. Self-sounding, in this context, means that no separate sounding equipment is used and that a cooperative distant end is required.

Two cooperating facilities tied by an HF link operate on a fully synchronous basis and have their clocks, and hence time of day (TOD), slaved to a common system time or to universal coordinated time (UTC). At intervals, every 5 min, for example, both facilities take time out from exchanging traffic and search, first above their operating frequency and then below. Such systems work well with wideband HF, which operates over a full 1 MHz or 500 kHz with very low level PN spread phase shift keying (PSK) signals. Here then each station moves up 1 MHz, for instance, then down 1 MHz in synchronism, each exchanging LQA data with the other regarding new frequencies.

LQA, for instance, can be a 3-bit group contained in the header information. A 1 1 1 exchange from each would indicate to move to the higher frequency, and a 1 1 0 to the lower frequency. Thus both stations are constantly homing in on the optimum working frequency (OWF). During transition periods, the search is carried out more frequently during calm periods, such as from say 10 A.M. to 3 P.M. At midpath (local time) the search may only be required only every 10 min. Such systems often exchange only short message blocks and LQA data is updated in every header. As the binary number starts to drop, an algorithm kicks off a retimer to shorten the period between searches. If the first two digits hold in at binary 1 for long periods (e.g., 15 min), the timer is extended. The LQA three-digit word can be controlled by BER measurement or signal-to-noise ratio monitor (S/N). The problem with the use of S/N only is that it may not be indicative of multipath where BER or automatic repeat request (ARQ) negative acknowledgments (NACKs) (see Chapter 12 for ARQ) will be indicative of BER, which often can be traced to intersymbol interference, a typical effect of multipath.

8.6 PROPAGATION MODES

There are three basic modes of HF propagation:

1. Groundwave
2. Skywave (oblique incidence)
3. Near-vertical incidence (NVI). This is a distinct subset of the skywave mode (2)

8.6.1 Basic Groundwave Propagation

The spacewave (not skywave) intensity decreases with the inverse of the distance, the groundwave with the inverse of the distance squared. Therefore at long distances and nonzero elevation angles, the intensity of the spacewave exceeds that of the groundwave. The groundwave is diffracted somewhat to follow the curvature of the earth. The diffraction increases as frequency decreases. Diffraction is also influenced by the imperfect conductivity of the ground. Energy is absorbed by currents induced in the earth so that energy flow takes place from the wave downward. The loss of energy dissipated in the earth leads to attenuation dependent on conductivity and dielectric constant. With horizontal polarization the wave attenuation is greater than with vertical polarization due to the different behavior of Fresnel reflection coefficients for both polarizations.

To summarize, groundwave is an excellent form of HF propagation where we can, during daylight hours, achieve 99% or better time availability. Groundwave propagation decreases with increasing frequency and with decreasing ground conductivity. As we go down in frequency, atmospheric noise starts to limit performance in daytime, and skywave interference at night is a basic limiter of groundwave performance on whichever the lower frequencies we wish to use, providing we are not operating above the MUF.

8.6.2 Skywave Propagation

A wave that has been reflected from the ionosphere is commonly called a skywave. The reflection can take place at the E region and the F1 and/or F2 regions. In some circumstances RF energy can be reflected back from any two or all three regions at once. Figure 8.3 shows the three regions or layers of interest as well as the D layer, which is absorptive. Figure 8.4 illustrates skywave communication.

HF skywave communications can be by one-hop, two-hop, or three-hop, depending on path length and ionospheric conditions. Figure 8.11 shows eight possible skywave modes of propagation. On somewhat longer paths ($\approx > 1000$ km) we can receive RF energy from two or more modes at once, giving rise to multipath reception which causes signal dispersion. Dispersion

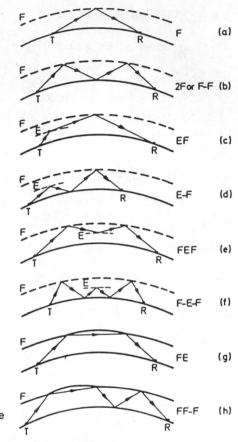

Figure 8.11 Examples of possible skywave multihop/multimode paths.

results in intersymbol interference (ISI) on digital circuits. Multimode reception is shown a bit later in Figure 8.13.

In general, we can say that multipath propagation can arise from

- Multihop, especially when transmit and receive antennas have low gain and low takeoff angles
- Low and high angle paths (the low ray and the high ray)
- Multilayer propagation
- Ordinary (O) and extraordinary (X) rays from one or more paths

These effects, more often than not, are existent in combination. An example of the high ray and low ray as they might be seen on an ionogram is shown in Figure 8.12. It also shows the MUF.

Typical one-hop ranges are 2000, 3400, and 4000 km for E-, F1-, and F2-layer reflections, respectively. These limits depend on the layer height of

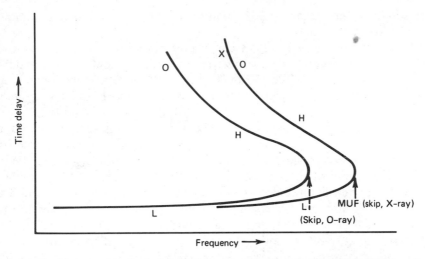

Figure 8.12 Ionogram idealized sketch. Oblique propagation along magnetic field for a fixed distance (point-to-point), F region. High ray (H) and low ray (L) and splitting into X-ray and O-ray are shown.

maximum electron density for rays launched at grazing incidence. The range values take into account the poor performance of HF antennas at low elevation angles.

Distances beyond the values given above can be achieved by utilizing consecutive reflections between the ionosphere and the earth's surface (see Figure 8.4 and Figure 8.11b–h). For each ground reflection the signal must pass through the absorptive D layer twice, adding significantly to signal attenuation. The ground reflection itself is absorptive. It should also be noted that the elevation angle increases as a function of the hop number, which results in the lowering of the path MUF.

The skywave propagation based on F-layer modes is identified by a three-character notation such as 1F1, 1F2, 2F2, and 3F2. The first digit is the hop number and the second two characters identify the dominant mode (i.e., F1 or F2 reflection). Accordingly, 2F2 means two hops where the dominant mode is F2 reflection.

An HF receiving installation will commonly receive multiple modes simultaneously, typically 1F2 and 2F2 and at greater distances 2F2 and 3F2. The strongest mode on a long path is usually the lowest order F2 mode unless the antenna discriminates against this. It can be appreciated that higher order F2 modes suffer greater attenuation due to absorption by D-layer passage and ground reflection. The result is a lower level signal than the lower order F2 propagating modes. In other words, a 3F2 mode is considerably more attenuated at a certain location than a 2F2 mode if it can be received at the same location (assuming isotropic antennas). It has traveled through the D layer two more times than its 2F2 counterpart and been absorbed one more time by ground reflection.

E-layer propagation is rarely of importance beyond one hop. Reflections from the F1 layer occur only under restricted conditions, and the 1F1 mode is less common than the 1E and 1F2 modes. The 1F1 mode is more common at high latitudes at ranges of 2000–2500 km. Multiple-hop F1 modes are very rare (Ref. 15).

Long-distance HF paths ($\approx > 4000$ km) involve multiple hops and a changing ionosphere as a signal traverses a path. This is especially true on east–west paths, where much of the time some portion of the path will be in transition, part in daylight, part in darkness. At the equator, a 4000-km path extends across nearly three time zones, further north or south, four and five time zones. Mixed-mode operation is a common feature of transequatorial paths. Figure 8.11 illustrates single-mode paths (*a* and *b*) and mixed-mode paths (*d–f*). Figure 8.11*c* shows a path with asymmetry. This occurs when a wave frequency exceeds the E-layer MUF only slightly, so that the wave does not penetrate the layer along a rectilinear path but will be bent downward, resulting in the asymmetry.

8.6.2.1 *Ray or Wave Splitting*   Because of the presence of the geomagnetic field, the ionosphere is a doubly refracting medium. Magnetoionic theory shows that a ray entering the ionosphere is split into two separate waves owing to the influence of the earth's magnetic field. One ray of the split rays is called the ordinary wave (O-wave) and the other the extraordinary wave (X-wave). The O-wave is refracted less than its X-wave counterpart and this is reflected at a greater altitude; the O-wave will have a lower critical frequency, hence a lower MUF. Both waves experience different amounts of refraction and thus travel independently along different ray paths displaced in time at the receiver, typically from 1 to 10 μs. The O-wave suffers less absorption and is usually the stronger and therefore the more important. Figure 8.12 is a conceptual sketch of O- and X-wave propagation as well as the high ray and the low ray. Figure 8.13 is a simplified diagram of the arrival at a receiver of the X-ray and O-ray (or wave).

So even if we have a one-mode path, say somewhat under 1000 km, the received signal will still be impaired by multipath arising from the O-ray and X-ray; and if we shift down in frequency somewhat from the MUF, there is further multipath impairment of the high ray and the low ray, which are both split into O and X.

Simple multipath fading is shown in Figure 8.14, which is a snapshot sketch of 1 MHz of HF spectrum showing the effects of multipath on an emitted signal. At certain frequencies, the multipath components are in phase and the signal level constructs and builds to a higher level than predicted by calculation. At other frequencies the multipath energy components are out of phase and the signal amplitude resultant shows the out-of-phase destruction. Where multipath components are just 180° out of

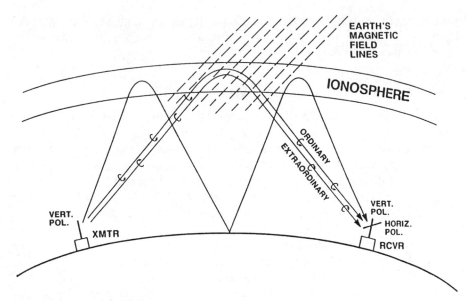

Figure 8.13 The skywave channel, showing the formation of the X- and O-waves, and simultaneous reception of one hop and two hops at receiver.

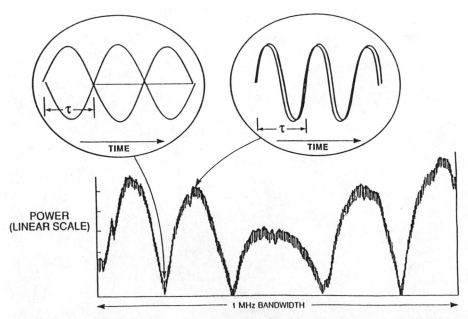

Figure 8.14 A snapshot sketch of a 1-MHz bandwidth, showing the frequency dependence of multipath effects on a transmitted wideband HF (WBHF) signal.

phase, there is full destruction and no signal present at that snapshot moment in time at that frequency. There are five such nulls shown in the snapshot picture (Figure 8.14).

8.6.3 Near-Vertical Incidence (NVI) Propagation

NVI propagation is used for short range HF communication. It can fill the so-called *skip zone* or *zone of silence*; here groundwave propagation is no longer effective, to the point where one-hop skywave, using oblique incidence propagation, may be used. NVI utilizes the same skywave principles of propagation discussed above. The key factor in NVI operation is the antenna. For effective HF communication using the NVI mode, the antenna must radiate its main beam energy at a very high angle (TOA), near vertical, if you will.

NVI circuits suffer the same impairments as oblique skywave circuits, but in the case of NVI the fading is more severe, particularly polarization fading. One rough rule of thumb is that if an equivalent oblique path experiences say 5 μs of dispersion, an NVI path will experience about 10 times as much, or 50 μs. NVI paths in a temperate zone have been known to suffer as much as 100 μs of dispersion. In equatorial areas such paths may experience even greater dispersion. We can also have dispersion from a second hop where the principal signal power is being derived from a 1-hop dominant mode. Figure 8.15 shows diagrammatically the operation of NVI propagation. The letter A

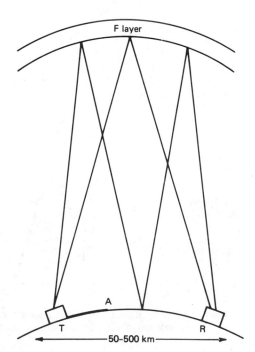

Figure 8.15 A near-vertical incidence (NVI) path showing one and two hops from the F layer. The letter A shows the maximum effective range of groundwave.

in the figure shows the extent of useful communication by means of a groundwave component if we were to transmit with a low-elevation-angle antenna with vertical polarization, such as a whip.

We will generally favor the lower frequencies for NVI operation, from 2 to 7 MHz. Higher-elevation-angle circuits tend to lower MUFs to start with. However, atmospheric noise limitations may drive us up in frequency as it does with groundwave operation. We will discuss atmospheric noise in Section 8.7. Atmospheric and man-made noise levels decrease with increasing operating frequencies.

8.6.4 Reciprocal Reception

Many HF system engineers assume reciprocity on an HF path. What is meant that if there is a good path from point X to point Y, the path is equally good from point Y to point X. This serves as the basis of one-way sounding and we do not quarrel with the assumption. There are several points that should be considered. The interference levels (including atmospheric noise) may be different at X than at Y. Also, propagation loss may differ owing to the influence of the earth's magnetic field. These two points may be particularly true on long paths. Using proper link quality assessment (LQA) procedures in both directions can alleviate such asymmetric situations by each far-end receiver optimizing its companion near-end transmitter operation both in frequency management and RF output power.

8.7 HF COMMUNICATION IMPAIRMENTS

8.7.1 Introduction

There are a number of important impairments on an HF channel that affect received signal quality. Different signals are affected in different ways and in severity. Digital data transmission can be severely affected by dispersion, both in the time domain and in the frequency domain. Analog transmission is much less affected, as is CW. FSK is much more tolerant than coherent PSK. Signal-to-noise ratio requirements are a function of signal type. Table 8.3 reviews some typical HF waveforms (signal modulation types), required bandwidths, and signal-to-noise ratios.

Much of the thrust in this section is on digital transmission. At the end of the section we will briefly discuss medium-related impairments to analog transmission.

8.7.2 Fading

HF skywave (and NVI) signals suffer from fading, a principal impairment on HF. It results from the characteristics of the ionosphere. The amplitude and

Table 8.3 Emission types, bandwidths, and required signal-to-noise ratios

Class of Emission	Pre-detection Bandwidth of Receiver (Hz)	Post-detection Bandwidth of Receiver (Hz)	Grade of Service	Audio Signal-to-Noise Ratio (1) (dB)	RF Signal-to-Noise Density Ratio (2) (3) (dB)		
					Stable Condition	Fading Condition (4) (5) Non-Diversity	Dual Diversity
A1A telegraphy 8 bauds	3000	1500	Aural reception (6)	−4	31	38	
A1B telegraphy 50 bauds, printer	250	250	Commercial grade (7)	16	40		58
A1B telegraphy 120 bauds, undulator	600	600		10	38		49
A2A telegraphy 8 bauds	3000	1500	Aural reception (6) (19)	−4	35	38	
A2B telegraphy 24 bauds	3000	1500	Commercial grade (7) (19)	11	50	56	
F1B telegraphy 50 bauds, printer $2D$ = 200–400 Hz	1500	100	$P_C = 0.01$ $P_C = 0.001$ (8) $P_C = 0.0001$		45 51 (9) 56	53 63 (9) 74	45 52 (9) 59
F1B telegraphy 100 bauds, printer $2D$ = 170 Hz, ARQ	300	300	(10)		43	52	
F7B telegraphy 200 bauds, printer $2D$ = ..., ARQ			(10)				
F1B telegraphy MFSK 33-tone ITA2 10 characters/s	400	400	$P_C = 0.01$ $P_C = 0.001$ (8) $P_C = 0.0001$		23 24 26	37 45 (25) 52	29 34 39
F1B telegraphy MFSK 12-tone ITA5 10 characters/s	300	300	$P_C = 0.01$ $P_C = 0.001$ (8) $P_C = 0.0001$		26 27 29	42 49 (25) 56	32 36 42
F1B telegraphy MFSK 6-tone ITA2 10 characters/s	180	180	$P_C = 0.01$ $P_C = 0.001$ (8) $P_C = 0.0001$		25 26 28	41 48 (25) 55	31 35 41
F7B telegraphy							
R3C phototelegraphy 60 rpm	3000	3000			50	59	
F3C phototelegraphy 60 rpm	1100	3000	Marginally commercial (22) Good commercial (22)	15 20	50 55	58 65	

Table 8.3   *(Continued)*

Class of Emission	Pre-detection Bandwidth of Receiver (Hz)	Post-detection Bandwidth of Receiver (Hz)	Grade of Service	Audio Signal-to-Noise Ratio (1) (dB)	RF Signal-to-Noise Density Ratio (2) (3) (dB)		
					Stable Condition	Fading Condition (4) Non-Diversity	(5) Dual Diversity
A3E telephony double-sideband	6000	3000	Just usable (11) Marginally (12) commercial Good commercial (13)	6 15 }(18) 33	50 59 67 (14)	51 64 }(20) 75 (14)	48 60 }(15) 70 (14) }(20)
H3E telephony single-sideband full carrier	3000	3000	Just usable (11) Marginally (12) commercial Good commercial (13)	6 15 }(18) 33	53 62 }(23) 70 (14)	54 67 }(20) 78 (14)	51 63 }(15) 73 (14) }(20)
R3E telephony single-sideband reduced carrier	3000	3000	Just usable (11) Marginally (12) commercial Good commercial (13)	6 15 }(18) 33	48 57 }(24) 65 (14)	49 62 }(20) 73 (14)	46 58 }(15) 68 (14) }(20)
J3E telephony single-sideband suppressed carrier	3000	3000	Just usable (11) Marginally (12) commercial Good commercial (13)	6 15 }(18) 33	47 56 64 (14)	48 61 }(20) 72 (14)	45 57 }(15) 67 (14) }(20)
B8E telephony independent-sideband 2 channels	6000	3000 per channel	Just usable (11) Marginally (12) commercial Good commercial (13)	6 15 }(18) 33	49 58 66 (14)	50 63 }(20) 74 (14)	47 59 }(15) 69 (14) }(20)
B8E telephony independent-sideband 4 channels	12000	3000 per channel	Just usable (11) Marginally (12) commercial Good commercial (13)	6 15 }(18) 33	50 59 67 (14)	51 64 }(20) 75 (14)	48 60 }(15) 70 (14) }(20)
J7B multichannel VF telegraphy 16 channels 75 bauds each	3000	110 per channel	$P_C = 0.01$ $P_C = 0.001$ }(8) $P_C = 0.0001$		59 65 }(21) 69	67 77 }(21) 87	59 66 }(21) 72

Source: CCIR Rep. 339-6 (Ref. 9, Table 1, p. 22) Vol. III, XVIth Plenary Assembly, Dubrovnik 1986.

Note:

(1) Noise bandwidth equal to postdetection bandwidth of receiver. For an independent-sideband telephony, noise bandwidth equal to the postdetection bandwidth of one channel.

(2) The figures in this column represent the ratio of signal peak envelope power to the average noise power in a 1-Hz bandwidth except for double-sideband A3E emission where the figures represent the ratio of the carrier power to the average noise power in a 1-Hz bandwidth.

(3) The values of the radio-frequency signal-to-noise density ratio for telephony listed in this column apply when conventional terminals are used. They can be reduced considerably (by amounts as yet undetermined) when terminals of the type using linked compressor–expanders (Lincompex) are used. A speech-to-noise (rms voltage) ratio of 7 dB measured at audio-frequency in a 3-kHz band has been found to correspond to just marginally commercial quality at the output of the system, taking into account the compandor improvement.

(4) The values in these columns represent the median values of the fading signal power necessary to yield an equivalent grade of service, and do not include the intensity fluctuation factor (allowance for day-to-day fluctuation) which may be obtained from Report 252-2 + Supplement (published separately) in conjunction with Report 322 (published separately). In the absence of information from these reports, a value of 14 dB may be added as the intensity fluctuation factor to the values in these columns to arrive at provisional values for the total required signal-to-noise density ratios which may be used as a guide to estimate required monthly-median values of hourly-median field strength. This value of 14 dB has been obtained as follows:

The intensity fluctuation factor for the signal, against steady noise, is 10 dB, estimated to give protection for 90% of the days. The fluctuations in intensity of atmospheric noise are also taken to be 10 dB for 90% of the days. Assuming that there is no correlation between the fluctuations in intensity of the noise and those of the signal, a good estimate of the combined signal and noise intensity fluctuation factor is

$$\sqrt{10^2 + 10^2} = 14 \text{ dB}$$

(5) In calculating the radio-frequency signal-to-noise density ratios for rapid short-period fading, a log-normal amplitude distribution of the received fading signal has been used (using 7 dB for the ratio of median level to level exceeded for 10% or 90% of the time) except for high-speed automatic telegraphy services, where the protection has been calculated on the assumption of a Rayleigh distribution. The following notes refer to protection against rapid or short-period fading.

(6) For protection 90% of the time.

(7) For A1B telegraphy, 50 baud printer: for protection 99.99% of the time. For A2B telegraphy, 24 bauds: for protection 98% of the time.

(8) The symbol P_C stands for the probability of character error.

(9) Atmospheric noise ($V_d = 6$ dB) is assumed (see Report 322).

(10) Based on 90% traffic efficiency.

(11) For 90% sentence intelligibility.

(12) When connected to the public service network: based on 80% protection.

(13) When connected to the public service network: based on 90% protection.

(14) Assuming 10-dB improvement due to the use of noise reducers.

(15) Diversity improvement based on a wide-spaced (several kilometers) diversity.

(16) Transmitter loading of 80% of the rated peak envelope power of the transmitter by the multichannel telegraph signal is assumed.

(17) Required signal-to-noise density ratio based on performance of telegraphy channels.

(18) For telephony, the figures in this column represent the ratio of the audio-frequency signal, as measured on a standard VU-meter, to the rms noise, for a bandwidth of 3 kHz. (The corresponding peak signal power, i.e., when the transmitter is 100% tone-modulated, is assumed to be 6 dB higher.)

(19) Total sideband power, combined with keyed carrier, is assumed to give partial (two-element) diversity effect. An allowance of 4 dB is made for 90% protection (8 bauds), and 6 dB for 98% protection (24 bauds).

(20) Used if Lincompex terminals will reduce these figures by an amount yet to be determined.

(21) For fewer channels these figures will be different. The relationship between the number of channels and the required signal-to-noise ratio has yet to be determined.

(22) Quality judged in accordance with article 23.1 of ITU publication "Use of the Standardized Test Chart for Facsimile Transmissions."

phase of skywave signals fluctuate with reference to time, space, and frequency. These effects collectively are described as fading and have a decisive influence on the performance of HF radio communication systems.

We consider four types of fading here (Ref. 16):

1. *Interference Fading*. This is the most common type of fading encountered on HF circuits. It is caused by mixing of two or more signal components propagating along different paths. This is multipath fading which may arise from multiple-mode and multiple-layer propagated rays, high- and low-angle modes, ground- and skywaves. This latter phenomenon is usually encountered during transition periods and at night.

2. *Polarization Fading Due to Faraday Rotation*. This is brought about by the earth's magnetic field and the split into the O-ray and the X-ray, which become two elliptically polarized components. Both components can interfere to yield an elliptically polarized resultant wave. The major axis of the resulting ellipse will have continuous changes in direction due to changes in the electron density encountered along its propagation paths. HF antennas are ordinarily linearly polarized. However, when an elliptically polarized wave with a constantly changing electric field vector shifts from perpendicular to the receiving HF antenna to parallel to the HF receiving antenna, the input signal voltage to the receiver will vary from maximum (perpendicular case) to zero (parallel case). Consequently, the receiver input voltage will vary according to the spatial rotation of the ellipse of polarization. Fading periods vary from a fraction of a second to seconds.

3. *Focusing and Defocusing Due to Atmospheric Irregularities*. These deformed layers can focus or defocus a signal wave if they encounter deformities that are concave or convex, respectively. The motion of these structures can cause fades with periods up to some minutes.

4. *Absorption Fading Is Caused by Solar Flare Activity*. This type of fading particularly affects the lower frequencies, and fades may last from minutes to more than an hour.

Of interest to the communication engineer are fading depth, duration, and frequency (fade rate). For short-term fading, the fading depth is the difference in decibels between the signal levels exceeded for 10% and 90% of the time. Measurements have confirmed that we may expect about 14 dB for a Rayleigh distribution short-term fading, which is the most common form of such fading. This value is valid for paths 1500–6000 km long and does not vary much with the time of day or season.

For long-term variations in signal level (i.e., variations of hourly median signal values of a month), the log-normal distribution provides a best fit. A good value is an 8-dB margin for HF paths below 60° geomagnetic latitude and 11 dB for paths above 60° and especially over the polar cap region (Ref. 16). This suggests that a 22-dB fade margin should provide better than a 99%

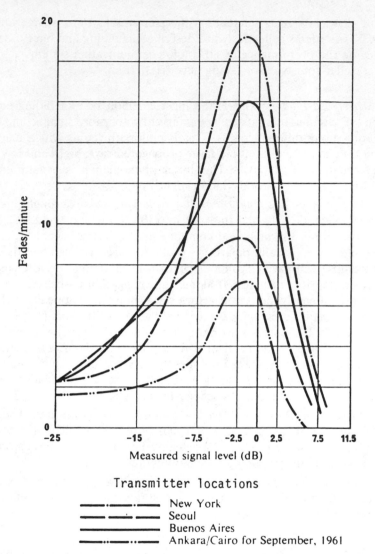

Figure 8.16 Number of fades per minute as a function of the signal level for various circuits terminating in Frankfurt, Germany. (From CCIR Rep. 197-4 [Ref. 18, Figure 3, p. 261]; Vol. III, XVIth Plenary Assembly, Dubrovnik, 1986.)

time availability (for fading, not frequency management) assuming a 3-σ point and for temperate zone paths. This decibel value is referenced to the median signal level.

Some fade rate values on typical HF circuits are provided in Figure 8.16. Figure 8.17 gives a sampling of typical fade durations on a medium-to-long path.

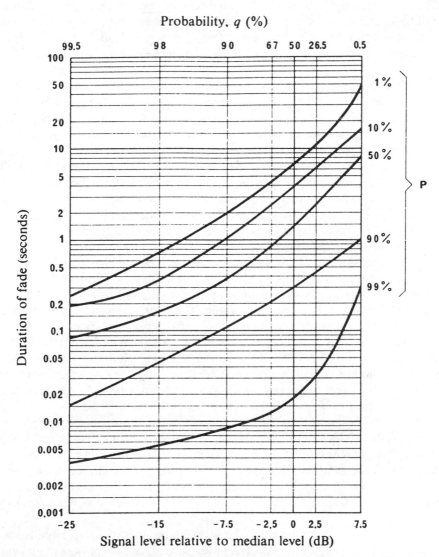

Figure 8.17 Duration of fades as a function of the level of the test signal. Circuit: New York–Frankfurt, Germany; September 14, 1961; 1100 h Central European Time; frequency 13.79 MHz. The figures on the right-hand side of the curves represent the percentage p of the number of fades for which a given duration of fade is exceeded. The measured values of signal levels are shown, together with the probability q that these levels will be exceeded. (From CCIR Rep. 197-4 [Ref. 18, Figure 4, p. 262]; Vol. III, XVIth Plenary Assembly, Dubrovnik 1986.)

8.7.3 The Effects of Impairments at the HF Receiver

8.7.3.1 General In Section 8.7.2 we discussed fade rate and depth. This is amplitude fading. It is accompanied by associated group path delay and phase path delay. Doppler shift arises from ionospheric movement as well as movement if one or both ends of the path are mobile (i.e., in motion).

8.7.3.2 Time Dispersion On skywave paths the primary cause of time dispersion is multipath propagation, which derives from differences in transit time between different propagation paths, as discussed in Section 8.4. The multipath spread causes amplitude and phase variations in the signal spectrum owing to interference of the multipath wave components. When these fluctuations are correlated within the signal bandwidth (e.g., 3 kHz) and all the spectral components behave more or less in the same manner, we then call this *flat fading*. When these fluctuations have little correlation, the fading is called *frequency selective fading*. Time dispersion is characterized by a delay power spectrum and is measured as multipath delay spread in microseconds or milliseconds.

Time dispersion is an especially serious and destructive impairment to digital communication signals on HF. One rule of thumb that is useful is that if the delay spread exceeds half the time width (period) of a signal element (baud), the error rate becomes intolerable. This is one rationale for extending the width of a signal element (e.g., by lowering the *baud* rate), to combat time dispersion. For instance, if a serial bit stream is transmitted at 100 bauds, a baud period is 0.01 s (10 ms). In this case, the circuit will remain operational, although with a degraded BER, if the time dispersion remains under 5 ms (i.e., half the period of a signal element or bit). At 200 baud the half-baud period value drops to 2.5 ms and at 50 baud it is 10 ms.

Multipath has been shown to be a function of the operating frequency relative to the MUF. Multipath delay tends to approach zero as the operating frequency approaches the MUF value. Turn to Figure 8.12 and we see that as we approach the MUF moving upward in frequency, we will receive only one signal power component. (*Note:* There are really two MUFs, one for the O and one for the X-ray trace. The X is usually below the O in signal strength.)

8.7.3.3 Frequency Dispersion Experience on operating HF circuits has shown that frequency dispersion (Doppler spread) nearly always is present when there is time dispersion. But the converse does not necessarily hold true. The Doppler shift and spread are due to a drifting ionosphere. As the signal encounters elemental surfaces of the ionosphere, each with a different velocity vector, the result is a Doppler spread. Such Doppler shifts and spreads can have disastrous effects, particularly on narrow-band FSK systems which can tolerate no more than $\pm$ 2-Hz total frequency departure. If the 2-Hz value is exceeded, the BER will approach 5×10^{-1}. Typical values of Doppler spread on midlatitude paths are 0.1–0.2 Hz.

Table 8.4   A voice frequency carrier telegraph (VFCT) modulation plan for HF

Channel Designation	Mark Frequency (Hz)	Center Frequency (Hz)	Space Frequency (Hz)
1	382.5	425	467.5
2	552.5	595	637.5
3	722.5	765	807.5
4	892.5	935	977.5
5	1062.5	1105	1147.5
6	1232.5	1275	1317.5
7	1402.5	1445	1487.5
8	1572.5	1615	1657.5
9	1742.5	1785	1827.5
10	1912.5	1955	1997.5
11	2082.5	2125	2167.5
12	2252.5	2295	2337.5
13	2422.5	2465	2507.5
14	2592.5	2635	2677.5
15	2762.5	2805	2847.5
16[a]	2932.5	2975	3017.5
17[b]	3012.5	3145	3187.5
18[b]	3272.5	3315	3357.5

Diversity

Pair 1(9), 2(10), 3(11), 4(12), 5(13), 6(14), 7(15), and 8(16)
Note: Connect loop to lower numbered channel of each pair

Source: MIL-STD-188-342 (Ref. 20).
[a] Marginal over HF (nominal 3-kHz) channels.
[b] Not usable over HF channels.

8.8   MITIGATION OF PROPAGATION-RELATED IMPAIRMENTS

The general approach microwave engineers use to overcome propagation impairments is to increase link margin (i.e., S/N). Generally, this approach has drawbacks with HF propagation because many of the impairments may be difficult to overcome with increasing power. Other measures also merit consideration.

One measure commonly used at HF installations is to employ diversity. The basis of diversity is to take advantage of signals that are not correlated. There are three basic types of diversity: space, time, and frequency. Other types of diversity are variants of the three basic types. Several types of diversity are discussed below.

Space Diversity. The same signals are received by at least two antennas separated in space. Separation should be > 6 wavelengths.

Polarization Diversity. Signals are received on antennas with different polarizations.

Frequency Diversity. Information is transmitted simultaneously on different frequencies. Some refer to this as in-band diversity, where subcarrier tones are separated by from 300 to 7000 Hz. CCIR recommends at least 400-Hz separation between redundant FSK tones in CCIR Rec. 106-1 (Ref. 19). MIL-STD-188-342 (Ref. 20) suggests a voice frequency (VF) carrier modulation plan shown in Table 8.4. The tone pairing plan is given at the bottom of the table. In this case redundant data is transmitted on pairs of FSK tones separated by 1360 Hz.

Time Diversity. The signal is transmitted several times. Some techniques are available that can dramatically improve error performance using time diversity. These techniques not only take advantage of the decorrelation of fading with time, but also the simple redundancy.

Channel coding with interleaving is only now being exploited on HF. We discussed some typical channel coding and interleaving techniques in Section 6.7.4. This coupled with automatic repeat request (ARQ) schemes using short message blocks can bring error performance on HF links into manageable bounds.

One method of mitigation of the dispersive effects of multipath propagation by pulse width extension was discussed in Section 8.7.3.2.

8.9   HF IMPAIRMENTS — NOISE IN THE RECEIVING SYSTEM

8.9.1   Introduction

In previous chapters on radio systems, the primary source of noise was thermal noise generated in the receiver front end. Except under certain special circumstances, this is not the case for HF receiving systems. External noise is by far dominant for HF receivers. In declining importance, we categorize this noise as follows:

1. Interference from other emitters
2. Atmospheric noise
3. Man-made and galactic noise
4. Receiver thermal noise

8.9.2   Interference

The HF band has tens of thousands of users who are assigned operating frequencies by national authorities such as the FCC in the United States and by the International Frequency Registration Board (IFRB), a subsidiary

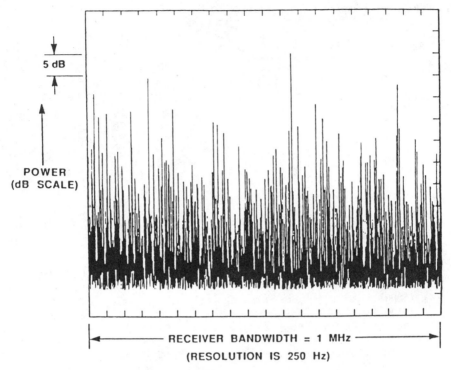

Figure 8.18 Spectrum analyzer snapshot of European HF interference spectrum.

organization of the ITU. The emitters operate in frequency bands in accordance with the service they perform and the region of the world they are in. The ITU has divided the world into three regions. We must also take into consideration noise from ionospheric sounder transmitters, harmonics, and spurious emissions from licensed emitters. Of primary importance in establishing an HF link is a "clear" frequency. Clear means interference-free, and the next consideration is how interference-free. On several tests of HF data modems it was found that interference even 15 or more dB below desired signals corrupted BERs. If we were to try to use a 3-kHz channel, we could well find an emitter 5000 km away can place a signal 30, 40, or more dB over our desired signal. With nominal 1-MHz wideband HF systems, it is estimated that interference from in-band emitters may contribute an integrated noise level over 30 dB above atmospheric noise (Ref. 21). A typical spectrum analyzer snapshot of a European interference spectrum is shown in Figure 8.18.

One can look at the HF spectrum between the LUF and MUF as a picket fence in which the pickets are the interferers. Unfortunately, by the very nature of HF propagation, these "pickets" randomly appear and disappear, their location changes, and their amplitudes increase and decrease not only

Table 8.5   Percentage congestion values at defined threshold[a]

Frequency band (kHz)	Service (dBm)	Day −117	−107	−97	−87	−77
5950–6200	broadcast	63	41	23	9	2
8195–8500	maritime mobile	24	11	5	1	1
10150–10600	fixed mobile	70	33	17	10	4
12050–12230	fixed	82	50	24	11	5
13800–14000	fixed/mobile	60	30	17	6	2
15000–15100	aero mobile	57	31	10	4	1
15100–15600	broadcast	96	80	55	33	16
17410–17550	fixed	44	21	8	5	1
		Night				
5950–6200	broadcast	100	100	93	75	51
8195–8500	maritime mobile	100	92	52	29	13
10150–10600	fixed/mobile	100	91	49	24	11
12050–12230	fixed	100	88	60	35	16
13800–14000	fixed/mobile	72	15	6	2	1
15000–15100	aeronautical mobile	97	63	32	7	5
15100–15600	broadcast	96	78	52	33	16
17410–17550	fixed	53	32	17	7	2

[a] − 107 dBm is equivalent here to a received field strength of 2 μV/m.

owing to fading but also to constant changes in the ionosphere as the MUF changes with time, sporadic E propagation, transition F-layer changes, and so on.

Ref. 22 gives congestion values in Manchester, UK, in July 1982 for the period between about noon and midnight local time. The HF band was scanned by stepping a receiver in 1-kHz increments and with a bandwidth of 1 kHz. It shows that the broadcast bands are the most congested and have the highest interference levels. In this study receive levels starting at − 117 dBm were recorded in decades up to − 77 dBm. Several examples of their measurements are given in Table 8.5, which shows the percentage of congestion values at defined thresholds.

There are several techniques to overcome or mitigate interference. The most desirable, of course, is to operate on a clear frequency with no interference at the distant-end receiver at the optimum working frequency (OWF). This is not often fully achievable.

In nearly all cases a facility is assigned a group of frequencies, usually with one of the following bandwidths, dependent on the service: 1, 3, 6, 9, or 12 kHz. Also see Table 8.3. Broadcast installations have other bandwidths. Broadcast is not a concern of this text. Distant-end receivers usually "command" near-end transmitters. The receiver site checks a new frequency for propagation and occupancy. More modern systems use advanced oblique sounders such as the AN/TRQ-35 or SELSCAN, Autolink, or MESA (Rockwell, Harris, and Tadiran trademarks, respectively), which also provide occupancy data. Thus modern systems have the propagation and occupancy check automated.

We can see from Table 8.5 that there is a fair probability that an optimum propagating frequency may be occupied by other users. There are several means to overcome this problem:

1. Find a clear frequency with suboptimum propagation.
2. Increase transmit power to achieve the desired signal-to-noise ratio at the distant end.
3. Use directional antennas where the interferer is in a side lobe and hence attenuated compared to the desired signal.
4. Use antenna nulling. This is a form of electronic beam steering that creates a null in the direction of the interferer.
5. Use sharp front-end preselection on receiver(s).

Some military circuits use FEC coding rich in redundancy. Rather than using code rates such as $\frac{7}{8}$, $\frac{3}{4}$, $\frac{2}{3}$, or $\frac{1}{2}$, we resort to code rates of $\frac{1}{8}$, $\frac{1}{16}$, $\frac{1}{32}$, or $\frac{1}{64}$. This highly redundant bit stream phase shifts an RF signal that is frequency hopped (spread spectrum) over a fairly wide frequency band in the vicinity of the OWF (for the desired connectivity). Typical spreads are 100 kHz, 500 kHz, up to 2 MHz. In theory, some of the hopped energy must get through the holes in the picket fence. Such coding multiplies the symbol rate of transmission. For example, if we wish to maintain 75-bps information rate, at rate $\frac{1}{16}$ we will have to transmit 16 × 75, or 1200 symbols a second. If we do not increase our RF power output, the energy per symbol is reduced by 16, or 6% of the energy if we had no coding. To equate energy per bit in this situation, if we use a transmitter with a 100-W output at 75 bps, we then would require a 1.6-kW transmitter output to maintain the same energy per symbol (bit) as when rate $\frac{1}{16}$ was employed.

On the other hand, the emerging wideband-HF (WBHF) systems excise the high-level interferers with only a loss of several tenths of a decibel, up to 2 dB of equivalent receive power. For instance, with 30% excision, only about 1.5 dB of receive power is lost. A 1-MHz PN spread system using a chip rate of 512 kchips in effect has 512,000 pieces of redundant RF-energy-carrying information spread across 1 MHz of HF spectrum. With so much redundancy, certainly we can effectively assure some portion of the desired transmit power to successfully get through the picket fence. Such systems have very low transmitted spectral energy density and are compatible with conventional narrow-band HF systems. Processing gains are on the order of 60–66 dB when information bit rates are on the order of 75–150 bps. We should not equate processing gain to power gain. In fact there is no power gain. Processing gain describes how well it will work against a jammer.

8.9.3 Atmospheric Noise

Atmospheric noise is a result of numerous thunderstorms occurring at various points on the earth, but concentrated mainly in tropical regions. These electrical disturbances are transmitted long distances via the ionosphere in the same manner as HF skywaves. Because the resulting field intensities of the noise decrease with the distance traveled, the level of atmospheric noise encountered from this source becomes progressively smaller in the higher latitudes of the temperate zones and in the polar regions. As we are aware, skywave propagation varies with time of day and season. Hence the intensity of atmospheric noise varies with both location and time.

Since the major portion of atmospheric noise is traced to thunderstorm activity, the atmospheric noise level at a particular location is due to contributions from both local and from distant sources. During a local thunderstorm, the average noise level is about 10 dB higher than the average noise of the same period in the absence of local thunderstorm activity (Ref. 23). From this it can be seen that the atmospheric noise level is related directly to weather conditions. The position of the equatorial weather front greatly affects atmospheric noise at all locations. This front varies in position from day to day and its general location seasonally moves north and south with the sun.

The degree of activity varies from time to time and from place to place, being much greater over land than over sea. The main areas of thunderstorm activity lie in equatorial regions, notably the East Indies, equatorial Africa, equatorial South America, and Central America. Thunderstorms are present about 50% of the days at locations in these equatorial belts and this activity is the principal source of long-distance atmospheric noise. It has been estimated that there are about 2000 thunderstorms in progress at each instant throughout the world (Ref. 23).

Thunderstorm activity is located over tropical land masses during local summer season and is more active over land, usually between 1200 and 1700 local time. Thunderstorm activity over the sea generally occurs at night and can last for more than a day.

Atmospheric noise from local sources shows discrete crashes similar to impulse noise, while long-distance atmospheric noise consists of rapid and irregular fluctuations with a frequency of 10 or 20 kHz per second and a damped wave train of oscillations. The amplitude of lightning disturbances varies approximately inversely with the frequency squared and is propagated in all directions both for ground- and skywaves.

8.9.3.1 Calculating Atmospheric Noise Using CCIR REP. 322

CCIR Rep. 322 is the most widely accepted data source and methodology on atmospheric noise. It has been derived from years of noise-monitoring data taken from monitoring stations around the world. The data is presented graphically as atmospheric noise contours at 1 MHz and then may be scaled graphically or by formula to the desired frequency.

CCIR divides a 24-h day into six time blocks of 4 h each starting at 12 midnight. For seasonal variations in atmospheric noise, there are charts for the four seasons. Thus there are 24 charts (6 × 4), each of which is followed by a frequency scaling chart accompanied by a noise variability chart. A sample of these is given in Figure 8.19. Figure 8.20 is for frequency scaling and Figure 8.21 for noise variability and they accompany Figure 8.19.

It is noted in Figure 8.19 that the noise contours given represent *median* noise levels and are measured in decibels above kT_0b at 1 MHz; their notation is F_{am}. We will derive a formula for F_{am}.

We can express the antenna noise factor f_a, which expresses the noise power received from sources *external* to the antenna:

$$f_a = \frac{P_n}{kT_0b} = \frac{T_a}{T_0} \tag{8.2}$$

where P_n = the noise power available from an equivalent loss-free antenna (W)

k = Boltzmann's constant = -228.6 dBW/Hz

b = the effective noise bandwidth (Hz)

T_0 = 288 K (we have used 290 K in previous chapters)

T_a = the effective antenna temperature in the presence of external noise

$10 \log kT_0 = -204$ dBW

$F_a = 10 \log f_a$

Both f_a and T_a are independent of bandwidth because the available noise power from all sources may be assumed to be proportional to bandwidth, as is the reference power level.

The antenna noise factor F_a, in decibels, is for a short vertical antenna over a perfectly conducting ground plane. This parameter is related rms noise field strength along the antenna by

$$E_n = F_a - 65.5 - 20 \log F_{MHz} \tag{8.3}$$

where E_n is the rms field strength for a 1-kHz bandwidth (dB[μv/m]) and F_a is the noise factor for the frequency f in question (dB). F_{MHz} is the frequency.

The value of field strength for any bandwidth B_{Hz} other than 1 kHz can be derived by adding $(10 \log b - 30)$ to E_n. For instance, to derive E_n in 1 Hz of bandwidth, we subtract 30 dB. CCIR Rep. 322 cautions that E_n is the vertical component of the noise field at the antenna.

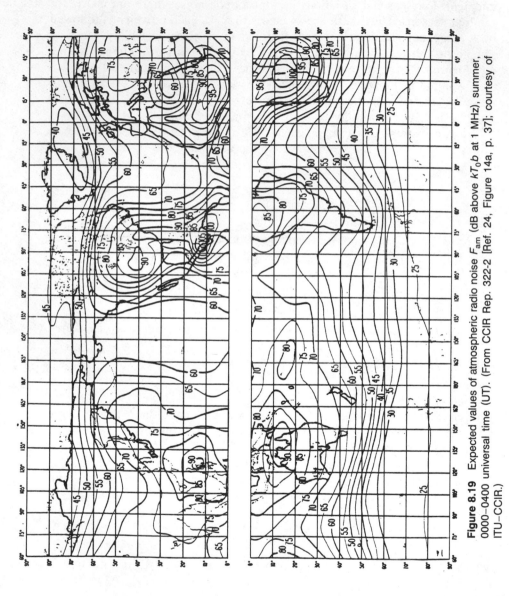

Figure 8.19 Expected values of atmospheric radio noise F_{am} (dB above kT_0b at 1 MHz), summer, 0000–0400 universal time (UT). (From CCIR Rep. 322-2 [Ref. 24, Figure 14a, p. 37]; courtesy of ITU–CCIR.)

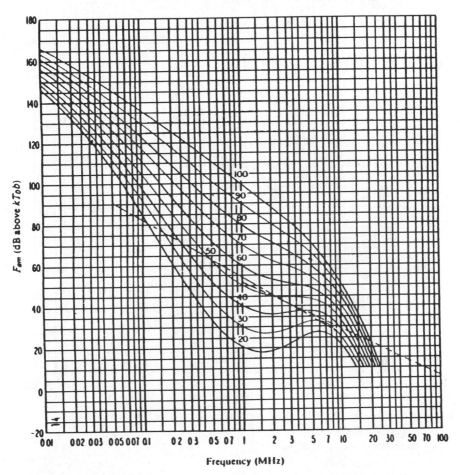

Figure 8.20 Variation of radio noise with frequency (summer, 0000–0400 h).

_____ Expected values of atmospheric noise

_·_·_·_·_ Expected values of man-made noise at a quiet receiving location

_ _ _ _ _ _ Expected values of galactic noise

(From CCIR Rep. 322-2 [Ref. 24, Figure 14b, p. 38]; courtesy of ITU–CCIR.)

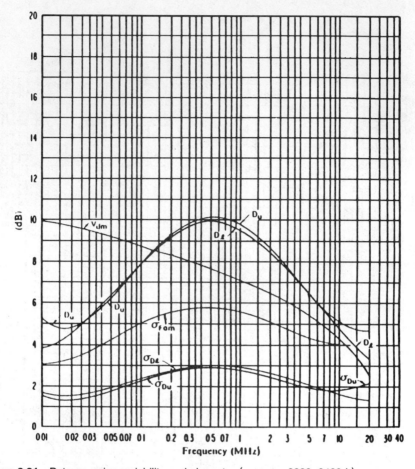

Figure 8.21 Data on noise variability and character (summer, 0000–0400 h).

$\sigma_{F_{am}}$: Standard deviation of values of F_{am}

D_u: ratio of upper decile to median value, F_{am}

σ_{D_u}: standard deviation of values of D_u

D_l: ratio of median value, F_{am}, to lower decile

σ_{D_l}: standard deviation of value of D_l

V_{dm}: expected value of median deviation of average voltage. The values shown are for a bandwidth of 200 Hz.

From CCIR Rep. 322-2 [Ref. 24, Figure 14c, p. 38].

In predicting the expected noise levels, the systematic trends with time of day, season, frequency, and geographical location are taken into account explicitly. There are other variations that must be taken into account statistically. The value of F_a for a given hour of the day varies from day to day because of random changes in thunderstorm activity and propagation conditions. The median of the hourly values within a time block (the time block median) is designated F_{am}. Variations of the hourly values within the time block can be represented by values exceeded for 10% and 90% of the hours, expressed as deviations D_u and D_l from the time block median plotted on normal probability graph paper (level in dB), the amplitude distribution of the deviations, D, above the median can be represented with reasonable accuracy by a straight line through the median and upper decile values, and a corresponding line through the median and the lower decile values can be used to represent values below the median. Extrapolation beyond D_u and D_l, however, yields only very approximate values of noise.

The value of F_a is average noise power. For digital circuits, it is useful to have knowledge of the amplitude probability distribution (APD) of the noise. This will show the percentage of time for which any level is exceeded, which usually is the noise envelope described. The APD is dependent on the short-term characteristics of the noise, and, as a result, cannot be deduced from the hourly values of F_a alone. The reader should consult CCIR Rep. 322 if more detailed analysis is desired in this area.

From equation 8.2 we can state the noise threshold (noise power) of a receiver (P_n):

$$P_n = F_{am} + X\sigma + 10 \log B_{Hz} \qquad (8.4)$$

If $X = 1$ (one standard deviation), 68% (of the time)

$X = 2$ (two standard deviations), 95% (of the time)

$X = 3$ (three standard deviations), 99% (of the time)

Example. Calculate the receiver noise threshold for 95% time availability where the receiver is connected to a lossless quarter-wavelength whip antenna equipped with a good ground plane. The receiver is tuned to 10 MHz with a 1-kHz bandwidth. The receiving facility is located in Massachusetts, and we are interested in the summer time block of 0000–0400 (A.M.). From Figure 8.19 we see that the F_{am} noise contour is 85 dB at 1 MHz. We turn to Figure 8.20 and F_{am} (10 MHz) scales down to 48 dB. Figure 8.21 gives $\sigma_{F_{am}}$ as 4 dB and we wish two standard deviations (95%), which is 8 dB for the variability.

$$P_n = 48 + 8 + 10 \log 1000 - 204 \text{ dBw}$$

$$P_n = -118 \text{ dBW in 1-kHz bandwidth}$$

We have shown much of the methodology given in CCIR Rep. 322, enough for an initial system design. However, several comments are in order. F_{am} decreases rapidly with increasing frequency. Besides atmospheric noise sources, other noise sources affecting total receiver noise power must start to be considered as frequency increases. These are

f_c, the noise factor of the antenna (function of its ohmic loss)
f_t, the noise factor of the transmission line (function of its ohmic loss)
f_r, noise factor of the receiver

The operating noise factor f, then, equals

$$f_a - 1 + f_c f_t f_r \quad \text{(numerics)} \qquad (8.5)$$

At low frequencies (e.g., < 10 MHz) atmospheric noise predominates and will determine the value of f. Because f_a decreases with increasing frequency, transmission line noise and receiver thermal noise become more important and f_c tends to approach unity. The values of f_t and f_r can be determined from calculations involving design features of the transmission line and receiver or by direct measurement.

The effective noise factor of the antenna, insofar as it is determined by atmospheric noise, may be influenced in several ways. If the noise sources were distributed isotropically, the noise factor would be independent of directional properties. In practice, however, the azimuthal direction of the beam may coincide with the direction of an area where thunderstorms are prevalent, and the noise factor will be measured accordingly compared with an omnidirectional antenna. On the other hand, the converse may be true. The directivity in the vertical plane may be such as to differentiate in favor or against reception of noise from a strong source.

Unfortunately, CCIR Rep. 322 does not take into account the sunspot number, and as mentioned previously, horizontal polarization.

8.9.4 Man-Made Noise

Above about 10 MHz we will often find that man-made noise is predominant. Man-made noise can be generated by many sources, such as electrical machinery, automobile ignitions, all types of electronic processors/computers, high-power electric transmission lines, and certain types of lighting. As a result, man-made noise is a function of industrialization and habitation density.

The most recognized reference on man-made noise is CCIR Rep. 258 (Ref. 25). Figure 8.22 gives median values of man-made noise, expressed in decibels, above −204 dBW/Hz. The figure shows five curves:

A. Business
B. Residential
C. Rural
D. Quiet rural
E. Galactic (extending upward in frequency from 10 MHz)

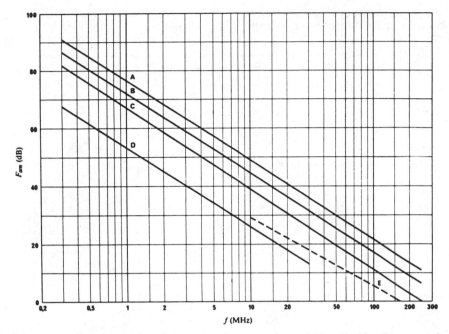

Figure 8.22 Median values of man-made noise power for a short vertical lossless grounded monopole antenna. Environmental category: A, business; B, residential; C, rural; D, quiet rural; E, galactic. (From CCIR Rep. 258-4 [Ref. 25, Figure 1, p. 208]; courtesy of ITU–CCIR.)

Business areas are defined as any area used predominantly for any type of business, such as stores, offices, industrial parks, large shopping centers, main streets and highways lined with various business enterprises, and so on. Residential areas are those predominantly used for single- or multiple-family dwellings of at least two single-family units per acre (five per hectare) and with no large or busy highways. Rural areas are defined as areas in which dwellings number no more than one per 5 acres (2 hectares) and where there

Table 8.6 Values of constants c and d (equation 8.6)

Environmental Category	c	d
Business (curve A)	76.8	27.7
Interstate highways	73.0	27.7
Residential (curve B)	72.5	27.7
Parks and university campuses	69.3	27.7
Rural (curve C)	67.2	27.7
Quiet rural (curve D)	53.6	28.6
Galactic noise (curve E)	52.0	23.0

Source: CCIR Rep. 258-4 (Ref. 25, Table I, p. 209). Courtesy of ITU–CCIR.

Table 8.7   Representative values of selected measured noise parameters for business, residential, and rural environmental categories[a]

Frequency (MHz)	Business				Residential				Rural			
	F_{am} (dB[kT_0])	D_u (dB)	D_l (dB)	σ_{NL} (dB)	F_{am} (dB[kT_0])	D_u (dB)	D_l (dB)	σ_{NL} (dB)	F_{am} (dB[kT_0])	D_u (dB)	D_l (dB)	σ_{NL} (dB)
0.25	93.5	8.1	6.1	6.1	89.2	9.3	5.0	3.5	83.9	10.6	2.8	3.9
0.50	85.1	12.6	8.0	8.2	80.8	12.3	4.9	4.3	75.5	12.5	4.0	4.4
1.00	76.8	9.8	4.0	2.3	72.5	10.0	4.4	2.5	67.2	9.2	6.6	7.1
2.50	65.8	11.9	9.5	9.1	61.5	10.1	6.2	8.1	56.2	10.1	5.1	8.0
5.00	57.4	11.0	6.2	6.1	53.1	10.0	5.7	5.5	47.8	5.9	7.5	7.7
10.00	49.1	10.9	4.2	4.2	44.8	8.4	5.0	2.9	39.5	9.0	4.0	4.0
20.00	40.8	10.5	7.6	4.9	36.5	10.6	6.5	4.7	31.2	7.8	5.5	4.5
48.00	30.2	13.1	8.1	7.1	25.9	12.3	7.1	4.0	20.6	5.3	1.8	3.2
102.00	21.2	11.9	5.7	8.8	16.9	12.5	4.8	2.7	11.6	10.5	3.1	3.8
250.00	10.4	6.7	3.2	3.8	6.1	6.9	1.8	2.9	0.8	3.5	0.8	2.3

Source: CCIR Rep. 258-4 (Ref. 25, Table II, p. 210). Courtesy of ITU–CCIR.
[a]*Key*:

F_{am}: median value.
D_u, D_l: upper, lower decile deviations from the median value within an hour at a given location.
   σ_{nl}: standard deviation of location variability.

are no intense noise sources. Minimum man-made noise may be found in "quiet" rural areas. It is in these areas where galactic noise predominates above about 10 MHz. This is the dashed line (E) shown in Figure 8.22.

We can also calculate the median man-made noise from the following expression:

$$F_{am} = c - d \log f \tag{8.6}$$

where f is the operating frequency in megahertz. The values of the constants c and d may be taken from Table 8.6. Table 8.7 provides some selected values of F_{am} and deviations of the median value. CCIR Rep. 258 advises that equation 8.6 may give erroneous values for quiet rural (D) and galactic noise (E) environments.

Example. A receiver operates at 15 MHz with a 1-kHz bandwidth. The receiver is located in a rural environment and uses a lossless whip antenna one-quarter wavelength long with a ground plane. Calculate the noise power threshold of a receiver under these circumstances. Assume a lossless transmission line.

Use equation 8.4 and replace $X\sigma_{am}$ with the expression $(D_u + \sigma_{nl})$, whose values we take from Table 8.7.

$$P_n = 34 \text{ dB} + (8.2 \text{ dB} + 4.2 \text{ dB}) + 10 \log 1000 - 204 \text{ dBW}$$
$$= -127.6 \text{ dBW}$$

8.9.5   Receiver Thermal Noise

Only under very special circumstances does receiver thermal noise become a consideration under normal HF operation. If we consider that atmospheric, man-made, and cosmic noise have a nonuniform distribution around an antenna (i.e., it tends to be directional), then the antenna gain at a certain frequency will generally favor signal and discriminate against noise.

Some HF antennas are designed with a very low gain; actually, they have a directional loss. The gain is expressed in negative dBi units. If we couple this antenna to a long or otherwise lossy transmission line, the noise at the receiver will have a significant thermal noise component. Seldom, however, is receiver-generated thermal noise a significant contributor to total HF receiving system noise. External noise is the dominant noise in determining the noise floor in the case of HF when calculating signal-to-noise ratio or E_b/N_0. In the design of HF receivers, thermal noise is not an overriding issue, and we find noise figures for these receivers are on the order of 12–16 dB.

8.10   NOTES ON HF LINK TRANSMISSION LOSS CALCULATIONS

8.10.1   Introduction

To predict the performance of an HF link or to size (dimension) link terminal equipments to meet some performance objectives, a first step is to calculate the link transmission loss. The procedure is similar to the transmission loss calculations for line-of-sight (LOS) microwave or troposcatter links. In other words, we must determine the signal attenuation in decibels between the transmitting antenna and its companion far-end receiving antenna. For an LOS link our concern was essentially free-space loss. The calculation for an HF link is considerably more involved.

8.10.2   Transmission Loss Components

The unfaded net transmission loss (L_{TL}) of an HF link can be expressed by

$$L_{TL} = L_{FSL} + L_D + L_B + L_M \quad \text{(dB)} \qquad (8.7)$$

where L_D = the D-layer absorption losses
$\quad L_B$ = the ground reflection losses
$\quad L_M$ = miscellaneous losses
$\quad L_{FSL}$ = the free-space loss expressed by the familiar formula

$$L_{FSL} = 32.45 + 20 \log d_{km} + 20 \log F_{MHz} \quad \text{(dB)} \qquad (8.8)$$

where d is the distance in kilometers and F the frequency in megahertz.

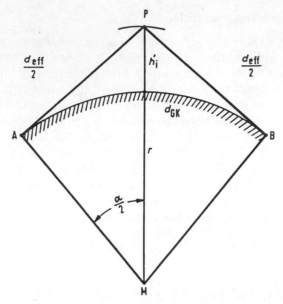

Figure 8.23 Geometric representation of a one-hop HF path where d_{eff} is the distance APB.

8.10.2.1 Free-Space Loss When we calculate L_{FSL} (equation 8.8), d_{eff} is the total distance a signal travels from transmitter A to its far-end receiver B. For short near-vertical incidence (NVI) paths, d_{eff} will be notably greater than the great circle distance from A to B. As the distance from A to B increases, and we start to use "normal" skywave modes, the great circle and the effective distance (d_{eff}) between A and B start to converge. Thus for short paths (e.g., < 1000-km great circle distance), we will substitute d_{eff} in equation 8.8 in place of d. For paths greater than 1000 km, we can just use the great circle distance between transmitter A and distant receiver B (Ref. 26).

A geometric representation of a one-hop HF path is shown in Figure 8.23 where we see d_{eff} is the path APB and P is the reflection point h_i' above the earth's surface at midpoint. d_{eff} can be calculated:

$$d_{eff} = 2\sqrt{8.115 \times 10^7 + 12{,}740h_i' + h_i'^2 - \cos\frac{\alpha}{2} \times (8.115 \times 10^7 + 12{,}740h_i')}$$

$$(8.9)$$

Equation 8.9 assumes r_i, the radius of the earth in Figure 8.23, to be 6370 km, α is the great circle arc from A to B, and h_i' the virtual reflection height. We can calculate α from the great circle equation:

$$\cos\alpha = \sin A \sin B + \cos A \cos B \cos \Delta L \qquad (8.10)$$

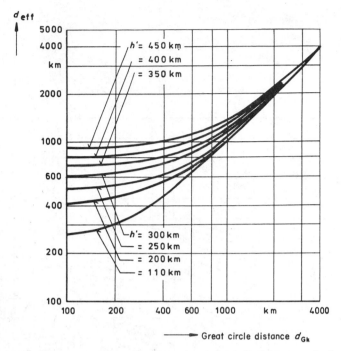

Figure 8.24 Effective distance d_{eff} when given the great circle distance and reflection height h'. (From Ref. 26.)

where α = angle of the great circle arc (see Figure 8.23)
      A = latitude of station A
      B = latitude of station B
      ΔL = difference in longitude between stations A and B

d_{eff} can also be derived from Figure 8.24 where h' is the reflection point height above the earth's surface. The height of the reflection point h', can be derived from Figures 8.28a and b.

8.10.2.2 D-layer Absorption Losses
D-layer absorption is a daytime phenomenon. The D layer disappears at night. D-layer absorption varies with the zenith angle of the sun, the sunspot number, the season, and the operating frequency. In fact, it varies as the inverse of the square of the operating frequency. This is one reason we are driven to use higher frequencies for skywave links, to reduce D-layer absorption.

To calculate D-layer absorption on a particular skywave path, we first compute the absorption index I:

$$I = (1 + 0.0037R)(\cos 0.881\,\chi)^{1.3} \qquad (8.11)$$

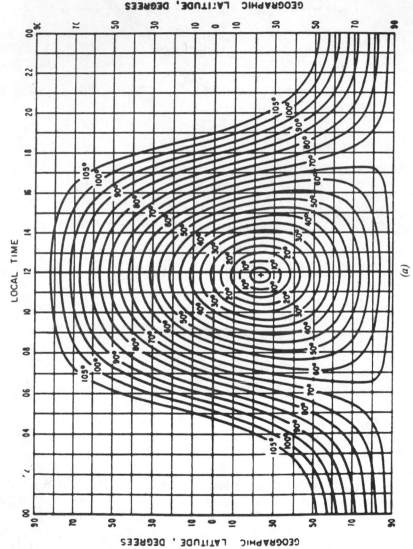

Figure 8.25   Solar zenith angle for (a) December and for (b) June. (From Ref. 26.)

(a)

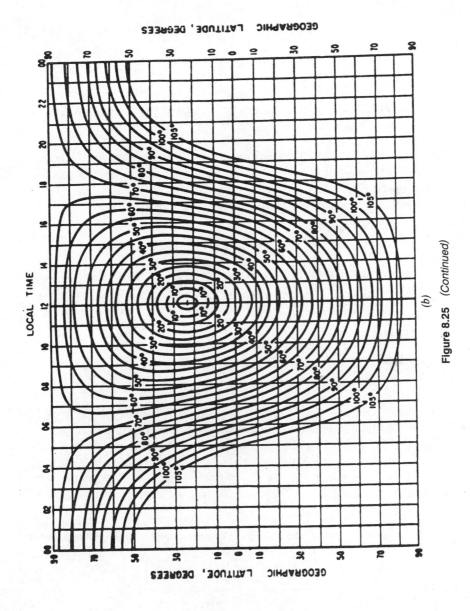

Figure 8.25 *(Continued)*

where R is the sunspot number and χ the solar zenith angle of the sun. If χ is greater than 100°, it is nighttime and we can neglect D-layer absorption.

From Ref. 37, we calculate the solar zenith angle with the following formula:

$$\cos \chi = \sin \phi \sin \varepsilon + \cos \phi \cos \varepsilon \cos h \qquad (8.12)$$

where ϕ = geographical latitude

ε = solar declination

h = the local hour angle of the sun measured westward from apparent noon, which is mean noon corrected for the equation of time and the standard time used at the location of interest

Tables of hourly values of $\cos \chi$ from sunrise to sunset for the 15th day of each month for most of the ionosphere vertical incidence sounding stations are given in the URSI Ionosphere Manual (Ref. 26, p. 19).

From Ref. 27, the hour angle of the sun, h, is calculated:

$$h = (\text{LST} - \text{right ascension})\left(\frac{15.0}{\pi}\right) \qquad (8.13)$$

where the local sidereal time, LST, is

$$\text{LST} = \{[(\text{DN} \times 24 + \text{GMT})(1.002737909) - (\text{DN} \times 24)] - \phi\} + S \qquad (8.14)$$

where ϕ = longitude of the point (in this case) (rad)

DN = the number of the day in the year

GMT = Greenwich mean time in decimal hours

If the local sidereal time is less than 0, then 24 h is added to it. If it is more than 24, then 24 h is subtracted from it. S is a correction factor.

Figures 8.25a and b can also be used to determine the solar zenith angle (Ref. 26). We now can calculate I, the absorption index, using equation 8.11. This value is then corrected for the winter anomaly, if required. These correction factors are given in Table 8.8. To use the nomogram in Figure 8.26

Table 8.8 Correction factors of the absorption index to account for the winter anomaly

Month		
Northern Hemisphere	Southern Hemisphere	Factor
November	May	1.2
December	June	1.5
January	July	1.5
February	August	1.2

Source: Ref. 26.

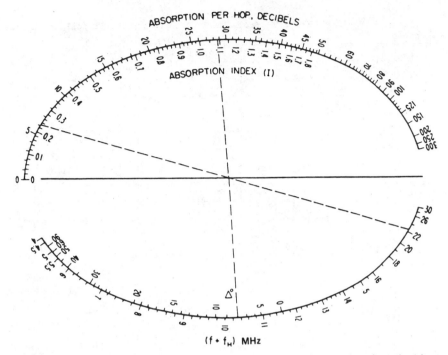

Figure 8.26 Nomogram for obtaining the ionospheric absorption per hop from the absorption index I, the effective wave frequency $f + f_H$, and the angle of elevation Δ. Example: (1) Enter with the absorption index I ($1 = 1.09$) and the elevation angle Δ ($= 8°$). (2) Mark the reference point of intersection with the center line. (3) With reference point and effective frequency $f + f_H$ ($= 22.4$), draw a straight line as far as the curve marked absorption per hop ($= 6$ dB). (From Ref. 26.)

effectively, we need to know the gyrofrequency for the region of interest. We can derive this value from Figure 8.27.

Example. A certain HF path operates during a sunspot number 100, the operating frequency is 18 MHz, and the elevation or takeoff angle (TOA) is 10°. Calculate the D-layer absorption value for 12 noon local time (at midpath) for the month of January at 40° north latitude. First we must derive the solar zenith angle from Figure 8.25a, 70°. Now we can substitute numbers in equation 8.11:

$$I = (1 + 0.0037 \times 100)(\cos 0.881 \times 70°)^{1.3}$$
$$I = 0.52$$

Multiply this value by the factor 1.5 taken from Table 8.8 to correct for the winter anomaly. I (corrected) is 0.78. Derive the gyrofrequency (F_H) for the latitude from Figure 8.27; it is 1.4 MHz. The value $F + F_H$ is 18 + 1.4 MHz,

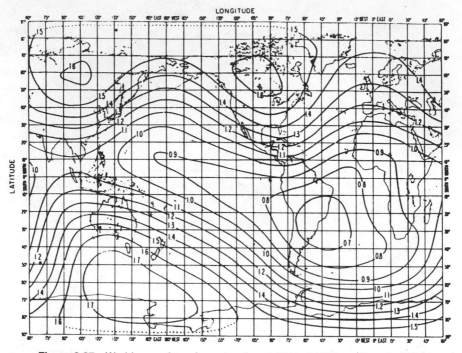

Figure 8.27   World map of gyrofrequency for a height of 100 km. (From Ref. 26.)

or 19.4 MHz. This value we use in Figure 8.26 to derive the D-layer absorption. We do this by first entering the takeoff angle (elevation angle) of the antenna, which is 10°. Now draw a line vertically to the corrected absorption index (I). We now have a reference point at the intersection of the solid horizontal lines. Through this reference point draw another line (right to left) from the frequency value of 19.4 MHz; it intersects "Absorption per Hop, Decibels" at 6 dB. We have a one-hop path, so the absorption value is 6 dB. If it were a 2-hop path with these parameters, it would be approximately 6×2, or 12 dB for D-layer absorption. Figures 8.28*a* and *b* can be used to determine the F2 layer height h'.

8.10.2.3 Ground Reflection Losses and Miscellaneous Skywave Propagation Losses   The following guidelines may be used to calculate ground reflection losses (Ref. 28):

> One-hop = 0 dB
> Two-hop = 2 dB
> Three-hop = 4 dB (etc.)

For miscellaneous skywave propagation losses, if no other information is available, use the value 7.3 dB. We will use this value for L_M in equation 8.7.

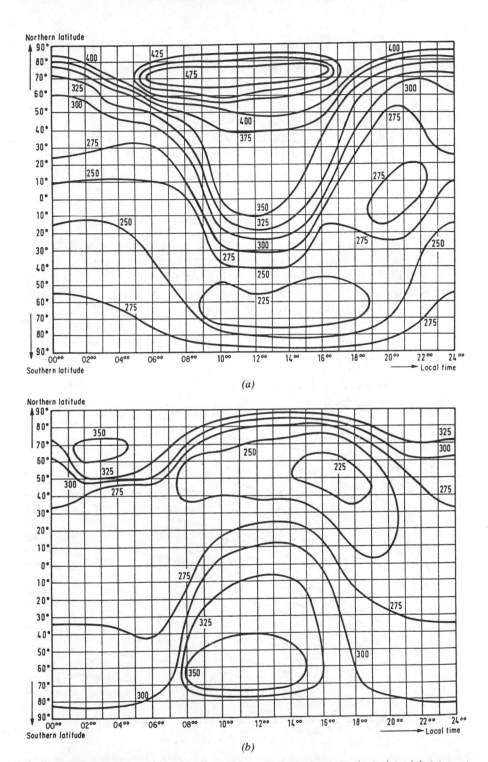

Figure 8.28 Approximate values of the virtual layer altitude of h' F_2 (in km) for (a) July and (b) January. (From Ref. 29.)

8.10.2.4   *Guidelines to Determine Dominant Hop Mode*   We only consider the lower order E- and F-layer modes (Ref. 28):

  For path lengths up to 2000 km: 1E, 1F2, and 2F2
  For path lengths between 2000 and 4000 km: 2E, 1F2, and 2F2
  For path lengths between 4000 and 7000 km: 2F2 and 3F2
  For path lengths between 7000 and 9000 km: 3F2

8.10.3   A Simplified Example of Transmission Loss Calculation

Consider a 1500-km path with 1F2 dominant mode in the temperate zone (40°N) operating over land in June with a sunspot number of 100 ($R12$) at 12 noon local time (midpath). This is an application of equation 8.7. Use Figure 8.29 to determine the radiation angle when given the great circle distance and virtual (F2) layer height. We can derive the layer height from Figures 8.28*a* and *b*. In this case, of course, we use Figure 8.28*a* for a value of 375 km. The elevation angle is $\approx 17°$. We now calculate the free-space loss using

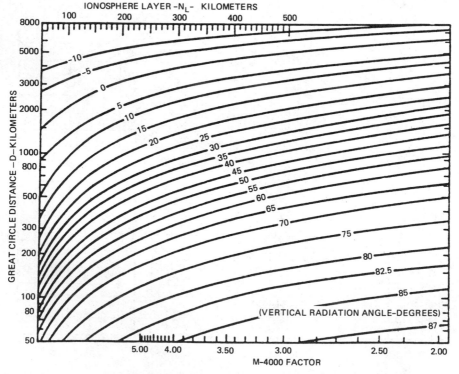

Figure 8.29   Nomogram to derive radiation angle as a function of great circle distance and ionospheric layer height. (From Ref. 26.)

equation 8.8:

$$\text{FSL}_{dB} = 32.45 + 20\log F_{MHz} + 20\log d_{km}$$

The optimum working frequency (OWF) is determined to be 17 MHz. We now have the distance and frequency, so we can proceed to calculate FSL, which is 120.57 dB.

We now want to calculate the D-layer absorption loss, L_D in equation 8.7. Use Figure 8.25b to calculate the solar zenith angle. This is 15°. Now we use formula 8.11 to calculate the absorption index (I):

$$I = (1 + 0.0037 \times 100)(\cos 0.881 \times 15)^{1.3}$$
$$I = 1.37(0.97)^{1.3}$$
$$I = 1.45$$

Because the month of interest is June, we can neglect the winter anomaly.

The next step is to derive the gyrofrequency using Figure 8.27. The value is 1.5 MHz, so $F + F_H = 18.5$ MHz. We now derive the D-layer absorption from the nomogram, Figure 8.26. It is 8 dB. This is a one-hop path, so ground reflections are zero. L_M is 7.3 dB (see Section 8.10.2.3). Therefore the total transmission loss (unfaded) from equation 8.7 is

$$L_{TL} = 120.57 \text{ dB} + 8 \text{ dB} + 0 \text{ dB} + 7.3 \text{ dB}$$
$$L_{TL} = 135.87 \text{ dB}$$

8.10.4 Groundwave Transmission Loss

8.10.4.1 Introduction Groundwave or surface wave transmission by HF is particularly effective over "short" ranges. The range or distance for effective transmission decreases with increasing frequency and decreases with decreasing ground conductivity. Seawater is highly conductive and we find that ranges from 500 to 800 mi (800 to 1200 km) can be achieved during daytime. Skywave interference at night, man-made and atmospheric noise, can reduce the range. The link design engineer must find an optimum frequency trading off atmospheric noise with frequency and range. We must also note that vertical polarization is more effective than horizontal polarization for groundwave paths.

Overland groundwave path transmission range is a function of ground conductivity and roughness. The performance prediction of such paths is a rather imperfect art.

8.10.4.2 Calculation of Groundwave Transmission Loss We use a very simple method to calculate groundwave transmission loss. We first turn to CCIR Rec. 368-5 (Ref. 30) and use the appropriate curves. Three such

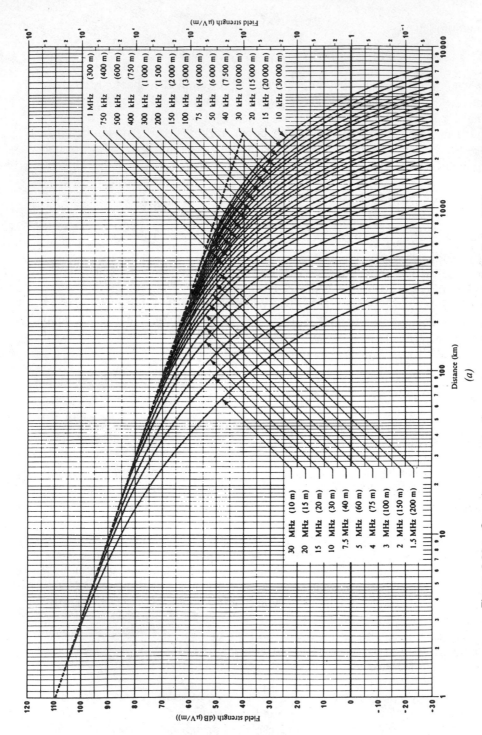

Figure 8.30a Groundwave propagation curves; seawater, average salinity, 20°C, $\sigma = 5$ S/m, $\varepsilon = 70$. ---, inverse distance curve. (From CCIR Rec. 368-5, [Ref. 30]; Volume V, Figure 1, page 40, XVIth Plenary Assembly, Dubrovnik, 1986.)

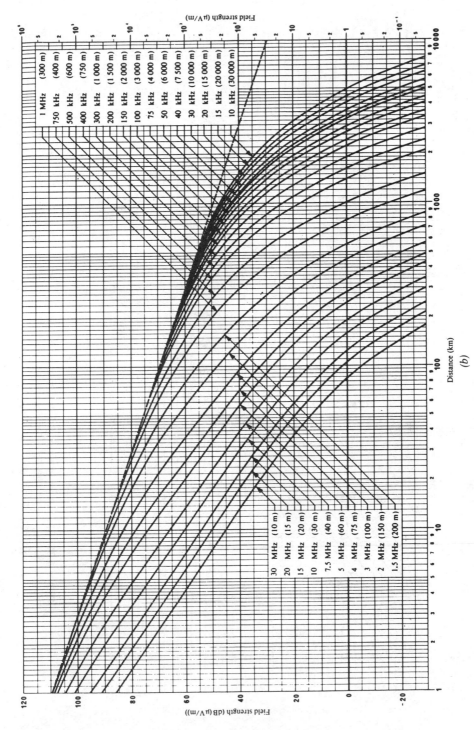

Figure 8.30b Groundwave propagation curves; land, $\sigma = 3 \times 10^{-2}$ S/m, $\varepsilon = 30$. ----, inverse distance curve. (From CCIR Rec. 368-5, [Ref. 30]; Figure 2, page 41, Volume V, XVIth Plenary Assembly, Dubrovnik, 1986.)

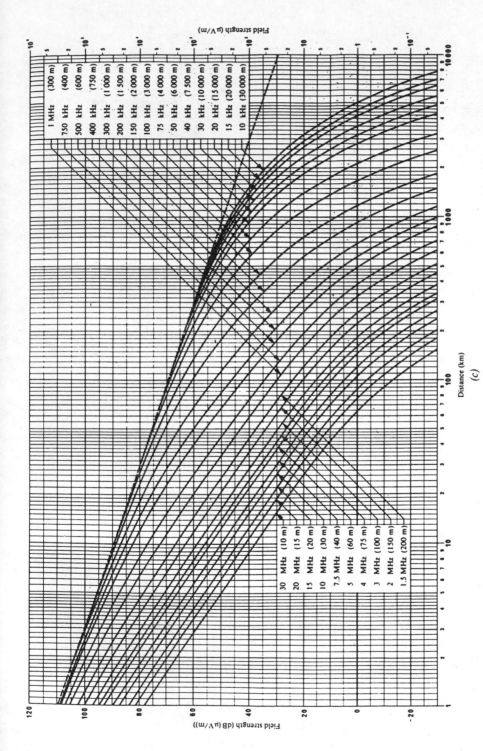

Figure 8.30c Groundwave propagation curves; medium dry ground, $\sigma = 10^{-3}$ S/m, $\varepsilon = 15$. -----, inverse distance curve. (From CCIR Rec. 368-5, [Ref. 30]; Figure 5, page 45, Volume V, XVIth Plenary Assembly, Dubrovnik, 1986.)

Table 8.9   Guidelines in the selection of ground conductivity and dielectric constant for various soil types

Type of Terrain[a]	Conductivity σ (S/m^{-1})	Dielectric Constant ε
Seawater of average salt content	4	80
Fresh water (20°C)	3×10^{-3}	80
Moist ⎫	10^{-2}	30
Medium ⎪ soil	10^{-3}	15
Dry ⎬	10^{-4}	4
Very dry ⎭	10^{-5}	4
	3×10^{-5}	4

Source: Ref. 26.

[a]These types of terrain apply to the following areas:

Seawater:	practically all oceans
Fresh water:	inland waters, such as large lakes, wide rivers, river estuaries, and so forth.
Moist soil:	marsh and fenland, regions with high groundwater level, floodplains, and so on.
Medium soil:	agricultural areas, wooded land, typical of countries in the temperate zones
Dry soil:	dry, sandy regions such as coastal areas, steppes, and also arctic regions
Very dry soil:	deserts, industrial regions, large towns, and high mountains

curves are given in Figures 8.30*a*, *b*, and *c*. Figure 8.30*a* is for a path over seawater; the *b* and *c* figures are for overland paths with different ground conductivities. Table 8.9 will help in the selection process for the appropriate curve (ground conductivity).

We then select the operational frequency and distance, from which we will derive a field strength *E*, in decibels above a microvolt. To be consistent with other chapters dealing with radiolinks, we will want to calculate the transmission loss. Note 4 to CCIR Rec. 368-5 provides the following equation to derive transmission loss given the value for *E*:

$$L_b = 137.2 + 20 \log F_{\text{MHz}} - E \qquad (8.15)$$

We illustrate this procedure by example. Suppose we have a 10-km link and have selected 3 MHz for our operational frequency over "normal" land (Figure 8.30*b*). We first determine the value *E* from the figure. It is 82.5 dB (μV/m). We apply this value of *E* to the formula:

$$L_b = 137.2 + 20 \log 3 - 82.5 \text{ dB} (\mu\text{V/m})$$
$$= 137.2 + 9.5 - 82.5$$
$$= 64.24 \text{ dB}$$

Hence the transmission loss over *smooth* earth with *homogeneous* ground conductivity and permittivity is 64.24 dB. Because the intervening terrain is not smooth, we add a terrain roughness degradation factor of 6 dB, for a total transmission loss of 70.24 dB.

Some notes of caution: If we were to measure transmission loss on this circuit, we would probably find a much greater loss as measured at the input port of the far-end receiver. The following lists some variables we may encounter on our path and in our equipment which may help explain such disparities:

1. The CCIR Rec. 368-5 curves are for vertical polarization. A horizontally polarized wave will have a notably higher transmission loss, on the order of 50 dB higher* (Ref. 26, p. 269).
2. With the exception of overwater paths, we will deal with ground conductivities/permittivities that can vary by more than an order of magnitude as we traverse the path.
3. The CCIR Rec. 368 curves are for a short vertical monopole at earth's surface. The earth is assumed to be flat and *perfectly conducting* in the immediate area of the antenna but not along the path.

Thus the curves are idealized. Unless an excellent ground screen is installed under the antenna at both ends of the path, a degradation factor should be added for less than perfect ground. Degradation factors must also be added for antenna matching and coupling losses.

8.11 LINK ANALYSIS FOR EQUIPMENT DIMENSIONING

8.11.1 Introduction

In this section we will determine such key equipment/system design parameters as transmitter output power, type of modulation, impact of modulation selection on signal-to-noise ratio, bandwidth, fading and interference, use of FEC coding/interleaving, and type of diversity and diversity improvement. We assume in all cases that we are using the optimum working frequency (OWF) and have calculated transmission loss for that frequency, time, season, and sunspot number. We also assume that we have calculated the appropriate atmospheric noise level for time, season, and frequency.

We will use Table 8.3 as a guide for required bandwidths and S/N requirements for some of the more common types of modulation used for HF point-to-point operation. Section 8.12 will deal with more sophisticated modulation schemes and how they can improve performance.

---

*In theory—in practice, the gap between vertical and horizontal polarization may be much smaller.

8.11.2 Methodology

The signal-to-noise ratio at the distant receiver can be expressed by

$$\frac{S}{N} = S_{dBm} - N_{dBm} \tag{8.16}$$

Substitute RSL_{dBm} for S_{dBm}, where RSL is the receive signal level. This assumes, of course, that $S/N = C/N$ (carrier-to-noise ratio). To permit this equality, some value k is added to our transmission loss. For lack of another term, we may call this modulation implementation loss, where we sum up all the inefficiencies in the modulation–demodulation process (including coding and interleaving) to come up with a value for k. Unless we have other values, we'll use 2 dB for the value of k.

In equation 8.16 N is the noise floor of the receiver, which is probably dominated by atmospheric noise and/or man-made noise rather than thermal noise. Refer to Sections 8.9.3 and 8.9.4.

RSL is calculated in the conventional manner:

$$RSL_{dBm} = EIRP_{dBm} - L_{TL} + G_{rec} \tag{8.17}$$

where L_{TL} = total transmission loss in decibels, including k dB of modulation implementation loss

EIRP = effective isotropically radiated power from the transmit antenna

G_{rec} = net antenna gain at the receiver

Remember that the antenna gains used in the calculation of EIRP and the receiver antenna gain must be that gain at the proper azimuth and the elevation of the takeoff angle (TOA) (or radiation angle) (Figure 8.29).

Try the following example (refer to Section 8.10.3): There is a 1500-km path with 1F2 as the predominant mode, in the temperate zone (40°N), operating over land in June with a sunspot number of 100 ($R12$), and the time of day is noon. The receiver has a 15-dB noise figure and a 3-kHz bandwidth and operates in a rural setting. The OWF is 17 MHz and the TOA is about 17°. FSL is 120.57 dB, D-layer absorption is 9 dB, miscellaneous losses are 7.3 dB, and the modulation implementation loss is 2 dB; thus L_{TL} is 138.87 dB.

The atmospheric noise at 17 MHz is on a 10-dB contour for summer, 12 noon (Figures 8.19 and 8.20) with a variability of 8 dB for 95% of the time. Man-made noise has a value F_{am} of 42 dB with a variability of 8 dB. The man-made noise component of the noise floor is by far the dominant and it is

$$p_{mm} = 42 + 8 + 10\log 3000 - 174 \text{ dBm}$$
$$= -89.23 \text{ dBm}$$

We will use this value for N_{dBm} in equation 8.16.

Let's assume that the transmitter has 1-kW RF output power and the net antenna gain at the angles of interest is 10 dB (including transmission line losses). The net receiver antenna gain is also assumed to be 10 dB. The EIRP is then

$$EIRP_{dBm} = +60 \text{ dBm} + 10 \text{ dB}$$
$$= +70 \text{ dBm}$$
$$L_{TL} = 138.87 \text{ dB}$$
$$RSL = +70 \text{ dBm} - 138.87 \text{ dB} + 10 \text{ dB}$$
$$RSL = -58.87 \text{ dBm}$$
$$\frac{S}{N} = -58.87 - (-138.87 \text{ dB})$$
$$= 80 \text{ dB}/3\text{-kHz}$$

Suppose the link we are analyzing was going to use single-sideband (SSB) operation. Now turning to Table 8.3, we find for J3E (SSB) operation, stable condition, an S/N of 64 dB is required in *1 Hz* of bandwidth. Our S/N value is for 3000-Hz bandwidth. The Table 8.3 (CCIR) value becomes 29.2 dB for a 3000-Hz bandwidth. We arrive at this value by subtracting $10 \log 3000$ from the CCIR value in 1-Hz bandwidth. Our margin is 80 dB − 29.22 dB, or 50.78 dB (unfaded condition).

Still on the J3E line in Table 8.3 we now look to the column that indicates good commercial service with fading and now we require 72 dB S/N in a 1-Hz bandwidth, or 37.2 dB in a 3-kHz bandwidth. Our new margin is 42.8 dB.

We can trade off some of the margin for transmit power or antenna gain. For the case of transmit power, we can reduce it by 20 dB and still have a 22.8-dB margin. The link now operates with 10 W of power. We also can reduce the antenna gain at each end by 5 dB (total of 10 dB for both ends) and the margin would now be 12.8 dB.

The assumption is made, of course, that the HF facility has a frequency assignment near 17 MHz and that it is an "interference-free channel." This latter statement, in practice, is highly problematical.

8.12   SOME ADVANCED MODULATION AND CODING SCHEMES

8.12.1   Two Approaches

There are two approaches to the transmission of binary message traffic or data on a conventional HF link: serial tone transmission and parallel tone transmission. Today it is still probably more common to encounter links using binary frequency shift keying (FSK) on a single channel or multiple channels in the 3-kHz passband with voice frequency carrier telegraph (VFCT). (See

Table 8.4.) On single-channel systems, when transmitting 150 bps or less, the center tone frequency is at 1275 Hz with a frequency shift of ± 42.5 Hz (Ref. 33). For 600-bps operation the center frequency is at 1500 Hz with a frequency shift of ± 200 Hz; for 1200 bps, the center tone frequency is at 1700 Hz shifted ± 400 Hz.

We will briefly describe two parallel tone systems and two serial tone systems. The first parallel tone system transmits 16 simultaneous tones and each tone is differentially phase shifted. The second method uses 39 tones. In either case these tones are contained in the nominal 3-kHz passband envelope. Of the two serial tone techniques, one uses 8-ary PSK and the other 8-ary FSK modulation.

8.12.2 Parallel Tone Operation

The first technique referenced here (MIL-STD-188-110 [Ref. 31]) operates from 75 to 2400 bps. The modulator accepts serial binary data (see Chapter

Table 8.10   Data tone frequencies and bit locations for HF data modems

Tone Frequency (Hz)	Function	Even and Odd Bit Locations of Serial Binary Bit Stream, Encoded and Phase Modulated on Each Data Tone Employing:					
		Quadrature-Phase Modulation		Biphase Modulation			
		2400 bps	1200 bps	600 bps	300 bps	150 bps	75 bps
605	Continuous Doppler Tone		←—————————————	In-Band Diversity		—————————————→	
825[a]	Synchronization Slot						
935	Data tone 1	1st and 2nd	1st and 2nd	1st	1st	1st	1st
1045	Data tone 2	3rd and 4th	3rd and 4th	2nd	2nd	2nd	1st
1155	Data tone 3	5th and 6th	5th and 6th	3rd	3rd	1st	1st
1265	Data tone 4	7th and 8th	7th and 8th	4th	4th	2nd	1st
1375	Data tone 5	9th and 10th	9th and 10th	5th	1st	1st	1st
1485	Data tone 6	11th and 12th	11th and 12th	6th	2nd	2nd	1st
1595	Data tone 7	13th and 14th	13th and 14th	7th	3rd	1st	1st
1705	Data tone 8	15th and 16th	15th and 16th	8th	4th	2nd	1st
1815	Data tone 9	17th and 18th	1st and 2nd	1st	1st	1st	1st
1925	Data tone 10	19th and 20th	3rd and 4th	2nd	2nd	2nd	1st
2035	Data tone 11	21st and 22nd	5th and 6th	3rd	3rd	1st	1st
2145	Data tone 12	23rd and 24th	7th and 8th	4th	4th	2nd	1st
2255	Data tone 13	25th and 26th	9th and 10th	5th	1st	1st	1st
2365	Data tone 14	27th and 28th	11th and 12th	6th	2nd	2nd	1st
2475	Data tone 15	29th and 30th	13th and 14th	7th	3rd	1st	1st
2585	Data tone 16	31st and 32nd	15th and 16th	8th	4th	2nd	1st

Source: MIL-STD-188-110 (Ref. 31).

[a]No tone is transmitted at this frequency.

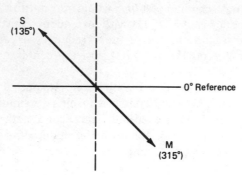

A. For data signaling rates of 75, 150, 300, or 600 bps.

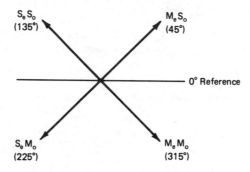

B. For data signaling rates of 1200 or 2400 bps.

Notes:
1. M = logic sense of mark; S = logic sence of space.

2. The subscripts refer to the even (e) or odd (o) bit locations of the serial binary bit stream. (see table 8.10)

Figure 8.31 Phase modulation vectors for an HF data modem. (From MIL-STD-188-110 [Ref. 31].)

12) and converts the bit stream into 16 parallel data streams. The signal element interval on each data bit stream is 13.33 ms and its modulation rate is 75 baud. The modulator provides a separate tone combination for initial synchronization and, if required, a separate tone for Doppler correction. Tone frequencies and bit locations are shown in Table 8.10. For data signaling rates of 75, 150, 300, and 600 bps at the modulator input, each data tone signal element is biphase modulated, as shown in the upper part of Figure 8.31. Each bit of the serial binary input signal is encoded, depending on the mark or space logic sense of the bit, into a phase change of the data tone signal element as listed in Table 8.11. For data signaling rates of 1200

Table 8.11   Modulation characteristics for the 16-parallel-tone HF modem

Input Data Signaling Rate (bps)	Degree of In-Band Diversity Combining	Type of Modulation	Logic Sense of Dibits or Bits in Serial Binary Bit Stream Depending on:		Phase of Data Tone Signal Element Relative to Phase of Preceding Signal Element (deg)
			Even Bit Locations	Odd Bit Locations	
2400	N/A	Four-phase	Mark	Space	+45
			Space	Space	+135
			Space	Mark	+225
1200	2		Mark	Mark	+315
600	2	Two-phase	Mark[a]		+315
300	4				
150	8				
75	16		Space[a]		+135

Source: MIL-STD-188-110 (Ref. 31).
[a]Regardless of even or odd bit locations.

and 2400 bps at the modulator input, each data tone signal is four-phase modulated (QPSK), as shown in the lower part of Figure 8.31. In this case, each dibit of the serial binary input signal is encoded, depending on the mark or space logic sense and the even or odd bit location of each bit, into a phase change of the data tone signal element as listed in Table 8.11. The phase change of a data tone signal element is relative to the phase of the immediately preceding signal element. This is called differential phase shift keying. In-band diversity (see Section 8.8) is provided for all but the 2400-bps data rate.

There are two arguments in favor of the parallel tone approach. The first is that the transmitted signal element is long (13.33 ms) for all user data rates. Consequently, in theory, a system using this technique can withstand up to 6 ms of multipath dispersion (see Section 8.7.3.2). The second argument is that equalization across the 3-kHz audio passband becomes a moot point because the band is broken down into 110-Hz segments and each segment carries a very low modulation rate, so that delay and amplitude equalization are unnecessary.

The second technique uses 39 parallel tones each with quadrature differential phase-shifted modulation. A 40th unmodulated tone is added for Doppler correction. In-band diversity is available for all data rates below 1200 bps. Forward error correction (FEC) coding employing a shortened Reed–Solomon (15, 11) block code with appropriate interleaving is incorporated in the modem. A means is provided for synchronization of the signal element and interleaved block timing.

Table 8.12   Probability of error versus signal-to-noise ratio[a]

Signal-to-Noise Ratio (dB in 3-kHz Bandwidth)	Probability of Bit Error	
	2400 bps	1200 bps
5	8.6 E-2	6.4 E-2
10	3.5 E-2	4.4 E-3
15	1.0 E-2	3.4 E-4
20	1.0 E-3	9.0 E-6
30	1.8 E-4	2.7 E-6
Signal-to-Noise Ratio (dB in 3-kHz Bandwidth)	Probability of Bit Error	
	300 bps	75 bps
0	1.8 E-2	4.4 E-4
2	6.4 E-3	5.0 E-5
4	1.0 E-3	1.0 E-6
6	5.0 E-5	1.0 E-6
8	1.5 E-6	1.0 E-6

Source: MIL-STD-188-110 (Ref. 31).

[a]Two independent, equal, average-power Rayleigh fading paths, with 2-Hz fading bandwidth and 2-ms multipath spread.

Each of the 39 tones is assigned a 52.25-Hz channel. The lowest frequency data tone is at 675.00 Hz and the highest is at 2812.50 Hz. The Doppler correction tone is at 393.75 Hz. The frequency accuracy of a data tone must be maintained within ± 0.05 Hz. For operation below 1200 bps both time and in-band frequency diversity are used.

Table 8.12 gives bit error probabilities versus signal-to-noise ratio for the 39-tone modem based on an HF baseband simulator, assuming two independent, equal, average-power Rayleigh fading paths with a 2-Hz fading bandwidth and 2-ms multipath dispersion (Ref. 31).

8.12.3   Serial Tone Approaches

The first technique we describe is covered in MIL-STD-188-110 (Ref. 31) and employs *M*-ary phase-shift keyed (PSK) on a single carrier frequency (or tone). The modem accepts serial binary data at 75×2^n up to 2400 bps and converts this bit stream into an 8-ary PSK modulated output signal. The serial bit stream, before modulation, is coded with a convolutional code with a constraint length of 7 ($K = 7$). Table 8.13 gives the coding rates for the various input data rates. An interleaver is used with storage of 0.0-, 0.6-, and 4.8-s block storage (interleaving interval). The carrier tone frequency is 1800 Hz ± 1 Hz. Figure 8.32 show the modulation state diagram for *M*-ary PSK as employed by this modem.

Table 8.13   Error-correction coding, fixed-frequency operation

Data Rate (bps)	Effective Code Rate	Method for Achieving the Code Rate
2400	$\frac{1}{2}$	Rate $\frac{1}{2}$ code
1200	$\frac{1}{2}$	Rate $\frac{1}{2}$ code
600	$\frac{1}{2}$	Rate $\frac{1}{2}$ code
300	$\frac{1}{4}$	Rate $\frac{1}{2}$ code repeated 2 times
150	$\frac{1}{8}$	Rate $\frac{1}{2}$ code repeated 4 times
75	$\frac{1}{2}$	Rate $\frac{1}{2}$ code

Source: MIL-STD-188-110 (Ref. 31).

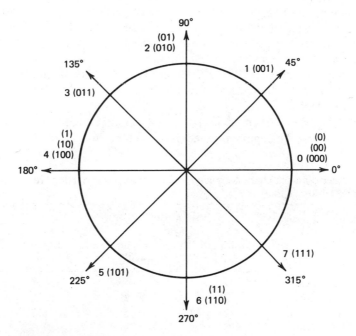

Legend:
0°... 315°= Phase (degrees)
0 ... 7 = Tribit numbers
(000) ... (111) = Three–bit channel symbols
(00) ... (11) = Two–bit channel symbols
(0) ... (1) = One–bit channel symbols

Figure 8.32   State constellation diagram. (From MIL-STD-188-110 [Ref. 31].)

The waveform (signal structure) of the modem has four functionally distinct sequential transmission phases. The time phases are

1. Synchronization preamble phase
2. Data phase
3. End-of-message (EOM) phase
4. Coder and interleaver flush phase

The length of the preamble depends on the interleaver setting. For the 0.0 setting, it is 0.0-s duration; for the 0.6 setting, 0.6-s duration; and for the 4.8-s setting, 4.8-s duration. This known preamble sequence allows the distant-end receive modem to achieve time and frequency synchronization.

During the data transmission phase, the desired data message is interspersed with known data sequences to train the far-end modem for channel equalization. We are aware, of course, that on an HF medium channel group delay and amplitude distortion are constantly changing with time. For instance, at 2400 bps 16 symbols of a known sequence (called known data) are followed by 32 symbols of user data (called unknown data). After the last unknown data bit has been transmitted, a special 32-bit sequence is sent to the coder, which performs the EOM function. It informs the distant receiving modem of end of message. The EOM sequence consists of the hexadecimal number 4B65A5B2. The final transmission phase is used to flush (reset) the far-end FEC decoder.

Table 8.14 gives bit error probabilities versus signal-to-noise ratio for various bit rates. This performance is based on the Waterson simulator

Table 8.14 Performance characteristics for serial tone modem using Waterson simulator (CCIR Rep. 549-2)

User Bit Rate	Channel Paths	Multipath (ms)	Fading[a] Bandwidth (Hz)	SNR[b] (dB)	Coded Bit Error Rate
2400	1 fixed			10	1.0 E-5
2400	2 fading	2	1	18	1.0 E-5
2400	2 fading	2	5	> 30	1.0 E-3
2400	2 fading	5	1	> 30	1.0 E-5
1200	2 fading	2	1	11	1.0 E-5
600	2 fading	2	1	7	1.0 E-5
300	2 fading	5	5	7	1.0 E-5
150	2 fading	5	5	5	1.0 E-5
75	2 fading	5	5	2	1.0 E-5

Source: MIL-STD-188-110 (Ref. 31).

[a] Per CCIR Rep. 549-2.

[b] 3-kHz bandwidth.

described in CCIR Rep. 549-2 (Ref. 32). Here the modeled multipath spread values and fading (2σ) bandwidth values derive from two independent but equal average-power Rayleigh paths (Ref. 31).

The second serial tone approach is taken from Annex A of MIL-STD-188-141A (Ref. 14). The primary purpose of the standard is to automate a low-data-rate HF system, whether a single link, a star network with polling, or a large grid network. A fully detailed open system interconnection (OSI) layer 2 (data link layer; see Chapter 12) is outlined. Two important functions are incorporated: automatic link establishment (ALE) and a method of self-sounding, including a method of exchanging link quality assessment (LQA) data among network members.

The modulation is 8-ary FSK and the eight tones are as follows: 750 Hz (000), 1000 Hz (001), 1250 Hz (011), 1500 Hz (010), 1750 Hz (110), 2000 Hz (111), 2250 Hz (101), and 2500 Hz (100). It will be appreciated that just one tone is transmitted at a time. The tone transitions are phase continuous with a baud rate of 125 baud, which is a transmission rate of 375 coded symbols per second (1 baud = 3 symbols).

The system uses block coding FEC with the Golay (24, 12, 3) rate $\frac{1}{2}$ code. In other words, 1 data bit is represented by two coded symbols. In the data text mode (DTM), automatic message display (AMD) mode, and the basic ALE mode, an auxiliary coding is also employed: redundant $\times 32$ with $\frac{2}{3}$ majority voting (with 49 transmitted symbols).

The uncoded data rate (user data rate) is 61.22 bps in the DTM, the AMD mode, and the ALE modes. In the data block mode (DBM), the uncoded data rate is 187.5 bps (375/2). The throughput maximum data rate is 53.57 bps in the DTM, AMD, and basic ALE modes. The source coding is a subset of ASCII (see Chapter 12).

The LQA is built into the protocol header. LQA information includes BER, SINAD (signal + noise + distortion to noise + distortion ratio), and multipath value (MP). The LQA field consists of 24 bits, including 11 overhead bits, 3 multipath bits, 5 SINAD bits, and 5 BER bits. The system incorporates a sounding probe for frequency optimization/management.

Table 8.15 Probability of linking

Probability of Linking (%)	Signal-to-Noise Ratio (dB)		
	Gaussian Noise Channel	CCIR Good Channel	CCIR Poor Channel
≥ 25	−2.5	+0.5	+1.0
≥ 50	−1.5	+2.5	+3.0
≥ 85	−0.5	+5.5	+6.0
≥ 95	0.0	+8.5	+11.0

Source: MIL-STD-188-141A (Ref. 14).

Table 8.15 gives performance data regarding probability of linking (ALE) using the standard HF simulator described in CCIR Rep. 549 (Ref. 32).

8.13 LINCOMPEX

Lincompex is an acronym for "link compression and expansion." It is a technique that provides a uniquely controlled companding (compression–expansion) function on single-sideband (SSB) voice systems. Performance improvement on links using Lincompex is on the order of 7 dB.

On systems using Lincompex, speech is compressed to a comparatively constant amplitude and the compressor control current is utilized to frequency modulate an oscillator in a separate control channel carried in a slot just above the voice channel. The speech channel, which contains virtually all the frequency information of the speech signal, and the control channel, which contains the speech amplitude information, are combined for transmission of the nominal HF 3-kHz channel. As each speech syllable is individually compressed, the transmitter is more effectively loaded than in current SSB practice. On reception both the speech and the control signals are amplified to constant level, the demodulated control signal being used to determine the expander gain and thus restore the original amplitude variations to the speech signal. Because the output level at the receiving end depends solely on the frequency of the control signal, which is itself directly related to the input level at the transmitting end, the overall system gain and loss can be maintained at a constant value. Operation with a slight loss (two-wire to two-wire) eliminates the need for singing suppressors, although echo suppressors will still be needed on long delay circuits.

The Lincompex speech channel is contained in the band 250–2700 Hz; the control channel is a tone of 2900 Hz, which is frequency modulated. It maximum deviation is ± 60 Hz. Applicable CCIR documentation is CCIR Rec. 455-1 and Rep. 354-5, Refs. 35 and 36, respectively.

8.14 BASIC REQUIREMENTS FOR HF EQUIPMENT

8.14.1 Introduction

Interoperability, operability, and compatibility set the most demanding requirements on HF systems and equipment. Interoperability is the ability, in this context, of a near-end transmitter to operate with a far-end receiver. In other words, traffic will be passed between the two in an optimum manner. Operability refers to all facets of efficient operation of the equipment. Compatibility refers to how well the equipment functions with collocated radio equipment and with other HF services sharing the medium.

Most of the requirements covered below were selected from U.S. military standards, especially MIL-STD-188-141A (Ref. 14) rather than recognized international standards. The reason for this is the U.S. military standards, in this case, place much more demanding and severe requirements on HF equipment.

8.14.2 Frequency Accuracy

The frequency accuracy of HF transmitters and receivers is ± 10 Hz, measured during a period of not less than 30 days.

8.14.3 Phase Stability

The probability that the phase difference will exceed 5° over any two successive 10-ms periods shall be less than 1%. Measurements should be performed over a sufficient number of adjacent periods to establish the specified probability with a confidence of at least 95%.

8.14.4 Phase Noise

The synthesizer and mixer phase noise should not exceed the limits shown in Figure 8.33 under continuous carrier single-tone output conditions.

8.14.5 Bandwidth and Channel Response

Bandwidths for the most common types of emission encountered on HF are given in Table 8.3, which was taken from CCIR. Figures 8.34 and 8.35 give guidance on channel response for the nominal 3-, 6-, and 12-kHz SSB/ISB (ISB = independent sideband) waveforms. These figures are taken from MIL-STD-188-141A (Ref. 14). Group delay should not vary by more than 0.5 ms over the passband 300–3050 Hz.

8.4.6 Transmitter Characteristics

8.14.6.1 In-Band Noise Broadband noise in a 1-Hz bandwidth with the selected sideband should be at least 85 dBc below the level of the rated peak envelope power of the HF transmitter.

8.14.6.2 IM Distortion The IM products produced by any two equal-level audio test signals should be at least 30 dB below each reference tone when the transmitter is operated at PEP (peak envelope power). The frequencies

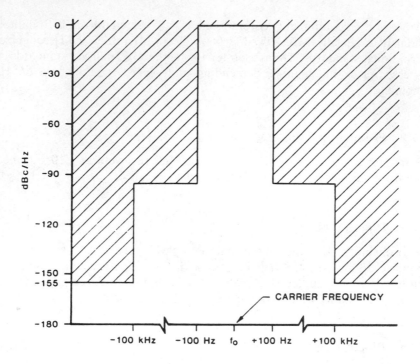

NOTE:

dBc = DECIBELS REFERENCED TO A FULL-RATED PEP CARRIER OUTPUT.

Figure 8.33 Phase noise limit mask for fixed-site and transportable long-haul radio transmitters with temperature-controlled frequency-determining elements. (From MIL-STD-188-141A [Ref. 14].)

of the audio test tones should not be harmonically or subharmonically related and should have a minimum separation of 300 Hz. (Ref. 14.)

8.14.6.3 Spectral Purity

Broadband Emissions. When a transmitter is driven with a single tone to the rated PEP, the power spectral density of the transmitter broadband emission should not exceed the level shown in Table 8.16. Discrete spurs are excluded from the measurement. The measurement bandwidth should be within 1 Hz.

Discrete Spurious Emissions. For HF transmitters, when driven by a single tone to produce an RF output of 25% rated PEP, all discrete frequency

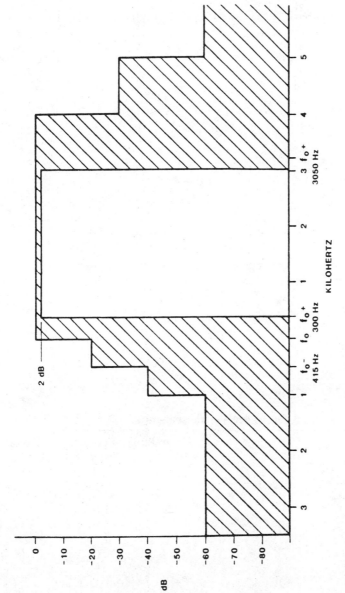

Figure 8.34 Overall channel response for single- or dual-channel equipment (From MIL-STD-188-141A [Ref. 14].)

NOTES:

1. CHANNEL RESPONSE SHALL BE WITHIN SHADED PORTION OF CURVE

2. f_o FOR A SINGLE CHANNEL IS THE CARRIER FREQUENCY.

3. f_o FOR 2 CHANNEL ISB IS THE CENTER FREQUENCY.

627

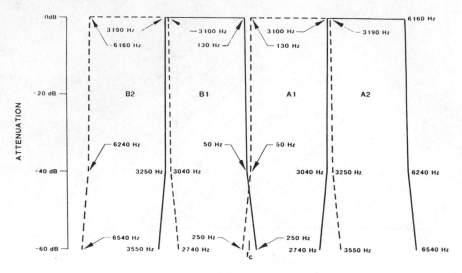

NOTES:

1. THE VIRTUAL SUBCARRIER FOR THE A2 AND B2 INVERTED CHANNELS SHALL BE $f_c \pm$ 6290 Hz.

2. FREQUENCIES SHOWN ARE AT THE FILTER dB (BREAK POINT) LEVELS NOTED.

Figure 8.35 Overall channel characteristics (four-channel equipment). f_c = center frequency. (From MIL-STD-188-141A [Ref. 14].)

Table 8.16 Out-of-band spectral density limits of transmitters

Frequency[a] (Hz)	Attenuation Below In-Band Power Density (dB)
$f_m = f_c \pm (0.5B + 500)$	40 (DO: 43)
$f_m = f_c \pm 1.0B$	45 (DO: 48)
$f_m = f_c \pm 2.5B$	60 (DO: 80)
$(f_c + 4.0B) \leq f_m \leq 1.05\,f_c$	
$0.95\,f_c \leq f_m \leq (f_c - 4.0B)$	70 (DO: 80)
$f_m \leq 0.95\,f_c$	
$f_m \geq 1.05\,f_c$	90 (DO: 120)

Source: MIL-STD-188-141A (Ref. 14).

f_m = frequency of measurement (Hz)
f_c = center frequency of bandwidth (Hz)
B = bandwidth (Hz)
DO = design objective

spurious emissions should be suppressed as follows:

- Between the carrier frequency and $4B$ (where B = bandwidth), at least 40 dBc
- Between $4B$ and $\pm 5\%$ of f_c removed from the carrier frequency, at least 60 dBc
- Beyond $\pm 5\%$ removed from the carrier frequency, at least 80 dBc. (Ref. 14)

8.14.6.4 *Carrier Suppression*   The carrier suppression on SSB/ISB systems should be at least 50 dBc below the output level of a single tone modulating the transmitter to its rated PEP. (Ref. 14)

8.14.7   Receiver Characteristics

8.14.7.1 *Image Rejection*   The rejection of image signals should be at least 80 dB, with a design objective of 100 dB.

8.14.7.2 *Intermediate-Frequency Rejection*   Signals at the intermediate frequency (frequencies) should be rejected by at least 80 dB, with 100 dB as a design objective.

8.14.7.3 *Adjacent-Channel Rejection*   The HF receiver should reject any signal in the undesired sideband and adjacent channel in accordance with Figure 8.34.

8.14.7.4 *Other Single-Frequency External Spurious Responses*   Receiver rejection of spurious frequencies, other than IF and image, should be at least 65 dB for frequencies $+2.5\%$ to $+30\%$ and from -2.5% to -30% of the center frequency, and at least 80 dB for frequencies $\pm 30\%$ of the center frequency.

8.14.7.5 *Desensitization Dynamic Range*   The following requirement applies to receivers operating in the SSB mode with an intermediate-frequency (IF) passband setting providing at least 2750 Hz (i.e., 300–3050 Hz) of bandwidth at the 2-dB points. With the receiver tuning centered on a sinusoidal input test signal and with the test signal level adjusted to produce an output SINAD* of 10 dB, a single interfering sinusoidal signal, offset from

---

*SINAD = signal-to-noise-and-distortion ratio.

the test signal by an amount equal to ±5% of the carrier frequency, is injected into the receiver input. The output SINAD should not be degraded by more than 1 dB as follows:

- For radios whose frequency-determining elements are temperature controlled, the interfering signal is equal to or less than 100 dB above the test signal
- For radios whose frequency-determining elements are not temperature controlled, the interfering signal is equal to or less than 90 dB above the test signal

8.14.7.6 Receiver Sensitivity   The sensitivity of the receiver over the operating frequency range, in the SSB mode of operation (3-kHz bandwidth), should be such that a −111-dBm (design objective, −121 dBm) unmodulated signal at the antenna terminal, adjusted for a 1000-Hz audio output, produces an audio output with a SINAD of at least 10 dB over the operating frequency range.

8.14.7.7 Receiver Out-of-Band Intermodulation Distortion   Second- and higher-order responses shall require a two-tone signal amplitude with each tone at least 80 dB greater than the required for a single-tone input to produce an output SINAD of 10 dB. This requirement is applicable for equal-amplitude input signals with the closest signal spaced 30 kHz or more from the operating frequency.

8.14.7.8 Third-Order Intercept Point   Using test signals within the IF passband, the worst-case third-order intercept point should not be less than +10 dBm.

8.14.7.9 Automatic Gain Control (AGC)   The steady-state output level of the receiver, for a single tone, should not vary by more than 3 dB over an RF input range from −103 dBm to +13 dBm. In the nondata mode, the AGC attack time should not exceed 30 ms, in the data mode, 10 ms. The AGC release time for nondata modes should be between 800 and 1200 ms for SSB voice and interrupted continuous wave (ICW) operation. This is the period from the RF signal downward transition until the audio output is simply receiver noise being amplified in the absence of any RF input signal.

8.14.7.10 Receiver Linearity   The following applies with the receiver operating at maximum sensitivity, and with a reference input signal that produces a SINAD of 10 dB at the receiver output. The output SINAD should increase monotonically and linearly within $+10\%$ for a linear increase in input signal level until the SINAD is equal to at least 40 dB (with a design objective of 60 dB). The requirement applies over the operating frequency range of the receiver. (Ref. 14.)

8.15 HF ANTENNAS

8.15.1 Introduction

The HF antenna installation can impact link performance more than any other system element of an HF system. It is also often the least understood and appreciated.

The antenna subsystem can be a "force-multiplier" if you will. A 10-dB net gain antenna system can make a 1-kW RF power output behave like 10 kW. It can attenuate interfering signals entering side lobes. With an arrayed antenna system using advanced interference nulling techniques, interference rejection can be even more effective.

The selection of a particular type of HF antenna is application driven. We consider three generalized applications: (1) point-to-point, (2) near-vertical-incidence (NVI), (3) multipoint and subsets of skywave and groundwave. Table 8.17 reviews these applications and some antenna types appropriate to the application.

Important parameters in the selection of antennas are

- Radiation patterns: azimuthal and vertical (elevation)
- Directive gain (receiver)
- Power gain (tuner efficiency, transmission line loss—transmitter)
- Polarization
- Impedance
- Ground effects
- Instantaneous bandwidth

Other important factors are size and cost.

In previous chapters we have used the isotropic radiator as the reference antenna, where gain is expressed in dBi (dB related to an isotropic). The isotropic antenna has a gain of 1 or 0 dB and radiates uniformly in all directions in free space.

HF engineers often use other antennas as reference antennas. Table 8.18 compares some of these with the isotropic.

The material in Section 8.15 is primarily based on Ref. 33.

Table 8.17 Antenna applications

Application	Antenna Type	Notes
Point-to-point skywave one-hop	Horizontal LP[a] Terminated long wire Dipole and doublet	
Point-to-point skywave, multihop	Vertical LP[a] Rhombic Sloping vee	
Point-to-point groundwave	Whip ⎫ Tower ⎭	Vertical polarization generally preferred
NVI	Doublet ⎫ Frame ⎭	Reasonable gain at high TOAs[b]
Multipoint skywave, one-hop	Conical monopole Discone Cage antenna	⎫ Broadband
Multipoint skywave, multihop	LP[a] rosette Array of rhombics	omnidirectional (azimuth)
Multipoint groundwave	Whip with counterpoise Tower	⎭

[a]Log periodic.
[b]Takeoff angles.

8.15.2 Antenna Parameters

8.15.2.1 Radiation Patterns Radiation patterns are usually provided by the antenna manufacturer for a specific model antenna. These are graphical plots, one for the horizontal plane or azimuthal and one for the vertical plane (elevation). Figures 8.36a and b show typical plots. The vertical plane should have some or total correspondence with the calculated elevation angle(s) or TOAs (takeoff angles).

One quantity determined uniquely by the radiation pattern is the antenna *directivity* that an antenna can attain. It is defined as the ratio of the

Table 8.18 Reference antennas

Antenna Type	Gain (dBi)	Notes
Isotropic	0	In free space
Half-wave dipole	2.15	In free space
Full-wave dipole	3.8	In free space
Short vertical	4.8	In free space
Vertical	5.2	Quarter wave on perfectly conducting flat ground; lossless tuning device

Source: DCAC 330-175-1 (Ref. 33).

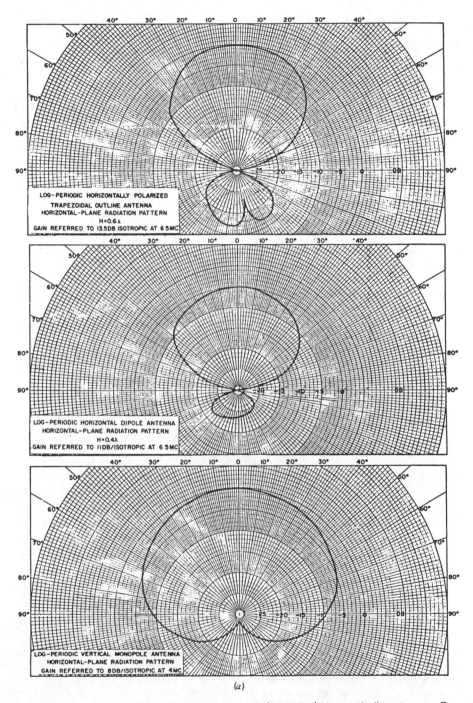

LOG–PERIODIC HORIZONTALLY POLARIZED
TRAPEZOIDAL OUTLINE ANTENNA
HORIZONTAL–PLANE RADIATION PATTERN
H=0.6λ
GAIN REFERRED TO 13.5DB ISOTROPIC AT 6.5MC

LOG–PERIODIC HORIZONTAL DIPOLE ANTENNA
HORIZONTAL–PLANE RADIATION PATTERN
H=0.4λ
GAIN REFERRED TO 11DB/ISOTROPIC AT 6.5MC

LOG–PERIODIC VERTICAL MONOPOLE ANTENNA
HORIZONTAL–PLANE RADIATION PATTERN
GAIN REFERRED TO 8DB/ISOTROPIC AT 4MC

(a)

Figure 8.36a Radiation patterns, horizontal plane (azimuthal), log periodic antennas. For directive values (dB): upper curve, add 13.5 dB; middle curve, add 11 dB; and lower curve, add 8 dB to readings. (From DCAC 330-175-1 [Ref. 33].)

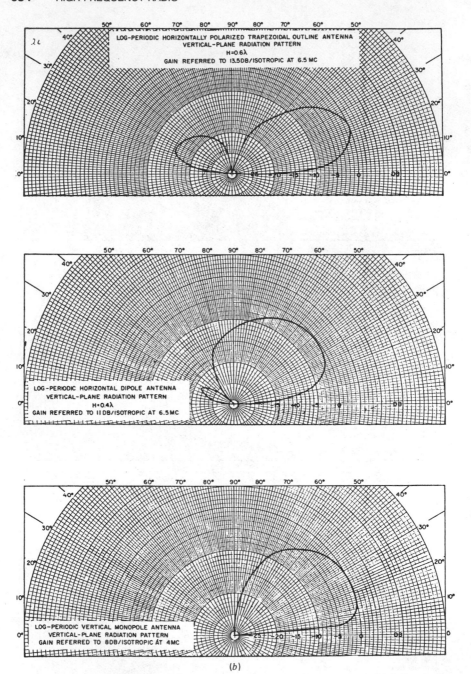

(b)

Figure 8.36b Radiation patterns, vertical plane (elevation or TOA), log periodic antennas. For directive gain values (dB): upper curve, add 13.5 dB; middle curve, add 11 dB; and lower curve, add 8 dB to the readings. (From DCAC 330-175-1 [Ref. 33].)

maximum radiated power density to the average radiated power density. Whether a gain is in fact attained depends on losses with the antenna system.

8.15.2.2 *Polarization* Polarization of the radiation from an antenna is produced by virtue of the fact that the current flow direction in an antenna is a vector quantity to which spatial orientation of the electric and magnetic fields are related. Single linear antennas in free space produce linear polarized waves in the far field, with the electric vector in a plane parallel to and passing through the axis of the radiating element. Antennas containing radiating elements with different spatial orientations and time phases produce elliptically polarized waves, with the ellipticity being a function of direction.

A linear antenna in free space used for receiving responds most strongly to another antenna with the same polarization and not at all to one polarized at $90°$ to the first antenna. Right-hand circularly (RHC) polarized antennas produce maximum receiving response when receiving transmission from a right-hand polarized antenna, and none from a left-hand circularly (LHC) polarized antenna.

Several other effects are also important. A circularly polarized antenna receiving a signal from a linear polarized antenna delivers 3 dB less power for a given transmitting power than would be received from a circular transmitted wave of the correct rotation sense. Conversely, of course, the same received signal loss applies to a linear receiving antenna and a circularly polarized transmitting antenna.

Vertically polarized waves radiated by a vertical antenna over imperfect ground are elliptically polarized because of the presence of a small electric field along the ground in space quadrature and a different time phase from the major electric component.

For HF skywave propagation, the received signal is usually elliptically polarized even when the transmitting antenna produces linear polarization. This condition arises from the effect of the earth's magnetic field in the ionized layers which splits the incident linearly polarized signal into the ordinary and extraordinary wave. The two waves travel at different velocities and experience different polarization rotations. The resulting received signal is elliptically polarized.

8.15.2.3 *Impedance* The impedance of an antenna depends on the radiation resistance, the reactive storage field, antenna conductor losses, and coupled impedance effects from nearby conductors. For simple antennas such as dipoles, integration of the power pattern over a spherical surface will yield a quantity called the radiation resistance which appears as a multiplying factor for the antenna current in the integrated power pattern. The radiation resistance obtained from the antenna impedance can be measured by an RF bridge placed at the current reference terminals of the antenna.

The antenna impedance as measured with a bridge will contain the radiation resistance, a resistance proportional to antenna conductor losses and, if the antenna is near other conducting objects, a component representing an interaction between the antenna and the nearby conductors. This latter component may be either positive or negative resistance and will usually have a reactive component.

The reactive storage field of an antenna depends to a considerable extent on the antenna diameter and physical shape. The storage field represents energy that flows into the antenna and then is returned to the generator during each cycle.

Antennas that have small diameters in terms of wavelength have large storage fields, whereas thicker antennas have lesser storage energy. Since the storage field represents circulating energy, which is analogous to that flowing in a physical reactance, the storage field contributes a reactive component to the antenna impedance.

Dipole antennas and other balanced standing wave antennas have reactive behavior similar to that of an open-circuited transmission line with a length equal to half the antenna length. For example, the reactive component of the impedance of a dipole is capacitive for lengths shorter than half-wavelength, is zero at about half-wavelength, and is inductive for lengths between one-half and one wavelength. The antenna reactance continues to alternate cyclically in a half-wavelength period for longer lengths. For convenience in matching the standing wavelength antenna, it is usual at high frequencies to use antennas at a frequency where resonance occurs, that is, where the reactance is zero.

Traveling-wave antennas, such as rhombics and terminated vees, have wide-impedance bandwidths. In fact, for these antennas, impedance bandwidths usually exceed radiation pattern bandwidths. Log-periodic (LP) antennas achieve wide-impedance bandwidth by selectively choosing only elements that are nearly resonant at a specific frequency.

Nominal impedance values of some common HF antennas are as follows (Ref. 33):

Horizontal rhombic and terminated vee	600 Ω
Horizontal LP	50/300 Ω
Vertical (dipole and monopole)	50 Ω
Yagi and half-wave dipole	50 Ω
Conical monopole, discone, and inverted discone	50 Ω
Vertical tower	50 Ω

8.15.2.4 Gain and Bandwidth   The gain of an antenna is defined as the ratio of the maximum power density radiated by the antenna to the maximum power density radiated by a reference antenna (see Table 8.18). The directivity of an antenna, which is sometimes confused with antenna gain, is the ratio of the maximum power density radiated by the antenna to the *average power*

Table 8.19 Antenna comparison

Type	Power Gain (dB)	Usable Rad Angle (deg)	Bandwidth Ratio	Horizontal Beamwidth	Side-lobe Suppression
Horizontal rhombic	8–23	4–35	≥ 2 : 1	6–26°	> 6 dB
Terminated Vee	4–10	4–35	≥ 2 : 1	8–36°	> 6 dB
Horizontal LP	10–17	5–45	≥ 8 : 1	55–75°	> 12 dB
Vert. LP (dipole)	6–10	3–25	≥ 8 : 1	90–140°	> 12 dB
Horiz. half wave dipole	5–7	5–80	≥ 5%	80–180°/lobe	N/A
Discone	2–5	4–40	≥ 4 : 1	N/A	N/A
Conical monopole	−2–+ 2	3–45	≥ 4 : 1	N/A	N/A
Inverted monopole	1–5	5–45	≥ 4 : 1	N/A	N/A
Vertical tower	−5–+ 2	3–30	≥ 4 : 1	N/A	N/A

Source: DCAC 330-175-1 (Ref. 33).

Notes

1. Typical power gains are gains over good earth for vertical polarization and poor earth for horizontal polarization.

2. Usable radiation angles are typical over good earth for vertical polarization and poor earth for horizontal polarization.

3. Normal bandwidth is the ratio of the two frequencies within which the specified voltage standing wave ratio (VSWR) will not be exceeded or within which the desired pattern will not suffer more than 3-dB degradation.

radiated by the antenna. The distinction between the two terms arises from the fact that directivity will exceed antenna gain. Since all antennas have some losses, the directivity will exceed antenna gain. The directivity of an antenna can be obtained from the antenna radiation patterns alone, without consideration of antenna circuit losses. Because directivity is a ratio, absolute power values are not required in determination of directivity, and convenient normalizing factors can be applied to radiation power values.

Rhombic and vee antennas, which dissipate portions of the antenna input power in terminators, will have lower gain values than directivity values by about the termination losses. Cophased dipole arrays have low conductor losses, and directivity, as a result, only slightly exceeds gain values for such antennas. LPs and Yagis can have 1-dB or more difference between directivity and gain. Table 8.19 compares power gain, usable radiation angles, nominal bandwidth, horizontal beamwidth, and side-lobe suppression.

8.15.2.5 Ground Effects

The free-space radiation pattern efficiency and impedance of an antenna are modified when the antenna is placed near ground. The impedance change is small for antennas placed at least one wavelength above ground, but the change becomes increasingly greater as that height is reduced. Since the ground appears as a lossy dielectric medium for HF, location of the antenna near ground may increase the losses a considerable extent unless a ground plane (a wire mesh screen or ground radials) is used to reduce ground resistance. Vertical monopole antennas, which are often fed at the ground surface, require a system of ground radials extending from the antenna to a sufficient distance to provide a low-resistance ground return path for the ground currents produced by the induction fields. For short vertical antennas, the radial length should be approximately $\lambda/2\pi$ long; for longer antennas the length should be approximately the antenna height. In addition, near the antenna base a wire mesh ground screen is recommended to reduce I^2R losses. For medium- or long-distance HF links, horizontal antennas should be mounted greater than one-half wavelength above ground and usually do not require a ground screen or ground plane.

In addition to losses, the presence of ground causes a change in antenna impedance. The change is brought about by the interaction between the antenna fields and fields of the ground currents.

Ground-reflected energy combines with the direct radiation of an antenna to modify the significantly vertical radiation pattern of the antenna. The magnitude of the image antenna current is equal to the magnitude of the real antenna current multiplied by the ground-reflection amplitude coefficient. An additional phase difference between the direct and the ground-reflected field at a distant point results from the difference of the two path lengths.

8.15.2.6 *Bandwidth*

Antenna bandwidth is specified as the frequency band over which the voltage standing wave ratio (VSWR) criteria are met and the radiation pattern provides the required performance. The frequency band is, to some extent, determined by the application, since greater deterioration of antenna characteristics can be tolerated for receiving use than for transmitting use. Antenna bandwidth is usually limited by the change of either or both the radiation shape and impedance with change of frequency.

Simple antennas, such as dipole, when in free space, have a figure-eight pattern broadside to the antenna. For dipoles, the pattern maxima are oriented normal to the dipole with lengths up to approximately $1\frac{1}{4}$ wavelengths. With greater lengths, the pattern maxima may not be oriented in this direction.

Linear conductors with traveling wave distributions have a continuous shift of the distribution in the direction of radiation pattern maximum, and the maximum approaches the axis of the conductor with increasing frequency. For this type of antenna and all other linear antennas, the pattern maximum can never fall exactly along the conductor axis because of the inherent null of the conductor in the direction of current flow, except when located near the earth.

When simple antennas are arrayed with other like antennas, the pattern changes increase with frequency because the changing electrical spacing between the arrayed elements also becomes a factor in the determination of the radiation pattern. In an array with discrete elements, an additional complication is produced by grating lobes. These lobes can occur whenever the interelement spacing exceeds a half-wavelength. If the element spacing equals one wavelength, the grating lobes can have intensities equal to the main radiation lobe. These lobes arise because the wide interelement spacing allows all element radiations to combine in phase in more than one direction.

Another factor in pattern change with frequency is produced by antenna energy which is reflected from the ground and combined with the direct radiated energy. Since the electrical path difference between the direct and the ground-reflected energy is proportional to the frequency, the elevation pattern of an antenna above ground is a function of frequency.

In a communication circuit, two degrees of severity in pattern change with frequency can occur. In one the pattern shape changes, but the maximum is in the required radiation direction. In the second the maximum is deflected from the desired direction with frequency. The first can be tolerated at the expense of increased circuit interference and noise. The second can cause complete circuit outages with frequency change.

Considerations of pattern change with frequency usually limit the use of broadband monopoles or dipoles to no more than two octaves. This usable frequency range is reduced to no more than one octave when these elements are arrayed. Rhombic antennas have a tolerable pattern shift, if used within a frequency range of less than one octave. LP antennas escape the directivity limitations of antennas using conductors of fixed physical length by selectively changing the active conductor length with frequency. However, unless the LP antenna is specially designed for use over ground, the antenna will have a pattern shift caused by ground reflection.

Input impedance change of an antenna may limit the coverage of an antenna to a smaller frequency range than would be expected from the antenna radiation pattern. This is particularly true of transmitting antennas when the antenna supplies the load for the transmission line and transmitter, and any inability to load the transmitter. Input impedance requirements for receiving antennas are not so severe because the receiver is the termination for the line and the antenna VSWR causes only a reduction in received power transfer.

8.15.3 Several Typical HF Antennas

8.15.3.1 Rhombic Antennas   A typical rhombic antenna is illustrated in Figure 8.37. As shown there, one apex is connected to an open-wire (balanced) transmission line which usually has a characteristic impedance of 600 Ω. In order to make a rhombic antenna nonresonant and unidirectional, the opposite apex is terminated with a resistance slightly larger in value than Z_0. The radiation efficiency of a typical rhombic is about 67%. Critical physical parameters are the height above ground (H), the tilt angle, which is $\frac{1}{2}$ the interior obtuse angle at the side pole, and the leg length L. A rhombic antenna works most efficiently with radiation angles (TOAs) between 4° and

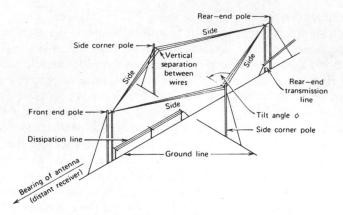

Figure 8.37 Rhombic antenna (transmitting).

35°. One rhombic requires from 5 to 15 acres of land. Power gains are on the order of 8–17 dB and are a function of frequency.

8.15.3.2 The Log-Periodic Antenna Family

Several typical log-periodic (LP) antennas are shown in Figure 8.38. An LP may be horizontally or vertically polarized. The variation of electrical properties of an antenna is a function of the variation of geometric properties of shape and dimension expressed in units of wavelength. When the radiating elements of an antenna are arranged such that their physical dimensions form a geometric progression, the electrical properties of the antenna repeat themselves at intervals determined by a scaling factor. From such a dimensional progression there results an antenna that is substantially frequency independent over many intervals or periods. Antennas so arranged exhibit a small periodic variation in impedance and radiation patterns with the natural logarithm of frequency, and are termed log-periodic antennas. We discuss three such types: the dipole array, the trapezoidal array, and the vertical monopole array.

The LP dipole antenna (Figure 8.38, middle) consists of parallel and coplanar half-wave elements in which the element length and spacing are logarithmically periodic, proceeding from the small apex out to the longest elements. This antenna is fed with a balanced two-wire line entering the apex and running through the center of the structure, transposed between adjacent elements so that all adjacent elements are fed 180° out of phase. Feeding the antenna with a coaxial cable transmission line requires a balun to transpose from the balanced transmission line of the antenna to the unbalanced transmission line of the coaxial cable.

The LP monopole configuration is similar to the LP dipole arrangement just discussed except that the plane containing the radiating elements is vertical (lower part of Figure 8.38), and the longest element is roughly

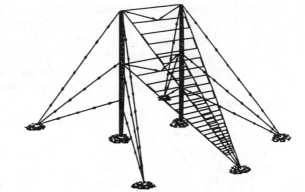

TRAPEZOIDAL OUTLINE LOG-PERIODIC ANTENNA

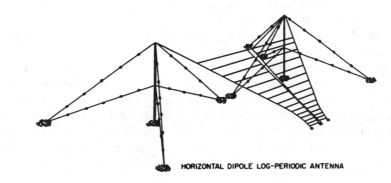

HORIZONTAL DIPOLE LOG-PERIODIC ANTENNA

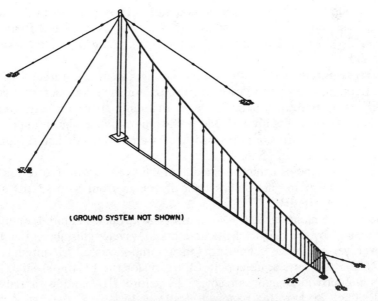

(GROUND SYSTEM NOT SHOWN)

LOG-PERIODIC VERTICAL MONOPOLE ANTENNA

Figure 8.38   Log-periodic (LP) antennas.

one-quarter wavelength at the lowest cutoff frequency of operation. Geometrically, the monopole arrangement is one-half of the dipole system, but in the monopole arrangements a ground system provides the *image* equivalent of the other half-dipoles. A single vertical LP requires only one tower and is less difficult to install. In order to reduce the earth current losses, however, an artificial ground system is normally required for almost all vertical monopole-type systems.

Another log-periodic configuration commonly used is the horizontal trapezoidal outline (top of Figure 8.38). In this configuration the radiating system consists of two planes of parallel elements, with the elements in each plane arranged in a repeating trapezoid. This configuration is similar to the horizontal dipole LP. For horizontal polarization the long dimension of the element is oriented horizontally, but the two planes containing the elements are arranged such that they join near ground level and extend outward and upward in two bays, one under the other.

The characteristic impedance of an LP varies between 100 and 300 Ω, balanced. The radiation efficiency of an LP approaches 100% except for the vertical LP, which may be limited because of high ground losses.

The performance of an LP antenna is very dependent on the proper choice of physical parameters for each application. The most fundamental physical parameter, upon which all others depend, is the geometric scaling factor τ, sometimes called the "design ratio." As is usually the case for antennas operating under the influence of ground reflections, the height of the array H, in wavelengths above ground, determines the vertical radiation angle (TOA).

The ground system for a vertical monopole LP requires about 3–5 acres of land in the immediate vicinity of the antenna. A typical LP monopole covering 2–30 MHz requires a tower 140 ft high, and the antenna length is about 300 ft.

Space requirements for the dipole configuration are rather high when the antenna must operate down to 2 MHz. An antenna operating down to 2.5 MHz may require 2 acres of land, including space for guy wires, and tower heights are generally 100–140 ft. A trapezoidal configuration operating in the range 4–40 MHz requires two towers approximately 230 ft high and a total land area of about 5 acres.

The LP monopoles exhibit a maximum gain ranging from 6 to 8 dB. The LP dipole maximum gain values typically range from 8 to 13 dB and the trapezoid up to 16 dB.

The vertical plane directivity of almost any ground-mounted antenna will be influenced by the height of the antenna in wavelengths above the ground. However, vertical plane beamwidths for ground-mounted LP antennas range from 40° for the vertical monopole LP to 25° for the LP trapezoidal.

LPs properly designed can cover the entire HF band with reasonable VSWRs (e.g., 2.0:1) with a bandwidth ratio of 15:1!

8.15.3.3 Dipole and Doublet Antennas Figure 8.39 shows three examples of dipole and doublet antennas. Here we refer to a generic group of antennas that are half-wavelength resonant or the electrical length of the antenna is half-wavelength. It is center fed with a midrange Z_0 of 73 Ω. Measured impedance over earth with midrange conductivity is in the range of 50–90 Ω depending on height above ground. These antennas are often fed with unbalanced coaxial cable, usually 50 Ω. A variation is the *folded dipole*, where a two-conductor version (Figure 8.39, middle) has an impedance of 300 Ω and a three-conductor version (Figure 8.39, bottom), 600 Ω. These latter types of antennas often use a balanced 300- or 600-Ω transmission line. An antenna of this type must be at least quarter-wavelength above ground. The bandwidth of the antenna is about 5%. Varying the antenna height above ground can vary the elevation angle of maximum radiation. At about quarter-wavelength above ground, we can expect radiation angles up to 90°, and at half-wavelength, 30°.

8.15.3.4 Broadband Vertical Antennas Figure 8.40 shows a family of broadband vertical antennas, including the discone, conical monopole, and inverted discone. Each presents to an unbalanced transmission line a nominal impedance of about 50 Ω with a maximum VSWR of 3:1 for a nominal bandwidth ratio of 3:1. Each antenna is a vertically polarized omnidirectional radiator. α is the conic flare and there are the conic base diameters C_{max} and C_{min} and disk-to-cone spacing s.

These antennas require a ground system made up of radial wire that should be no less than one-quarter wavelength long and preferably $\frac{4}{10}$ wavelength long. Such an antenna operating as low as 2 MHz may require up to 3 acres of land. The inverted discone requires as many as eight supporting poles and is up to 80 ft in height. The discone has gains in the range of +2 to +5 dBi and the conical monopole from −2 to +2 dBi. Bandwidth ratios are from 3:1 to 5:1.

8.15.3.5 Yagi Antennas A Yagi antenna is shown in Figure 8.41. It is tower mounted and the antenna can be considered an array of dipoles where only one is driven and the other elements are parasites. Generally, the input impedance at resonance can range between 10 and 60 Ω, balanced. The significant physical properties are the height above ground H, number of elements n, length of elements l_n, element spacing d, and radius of the elements a. Element length is approximately half-wavelength. The required length of the array L depends on the desired gain and directivity. Typical array lengths range from 0.3 wavelength for a three-element array to 3 wavelengths for multielement arrays. For HF applications, the maximum practical array length is about 2 wavelengths. Since the radiation angle is a function of the height above ground H, required array heights vary between 0.25 and 2.5 wavelengths. Dipole elements are normally made of tubing and

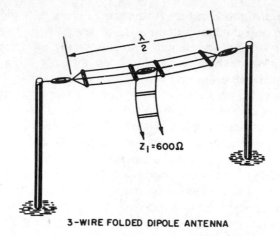

3-WIRE FOLDED DIPOLE ANTENNA

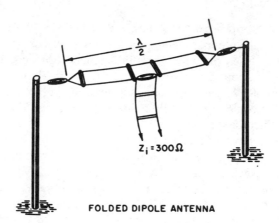

FOLDED DIPOLE ANTENNA

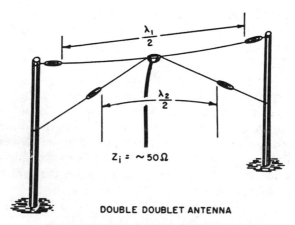

DOUBLE DOUBLET ANTENNA

Figure 8.39 Dipole and doublet antennas.

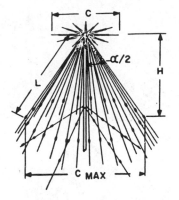

(A). HF DISCONE ANTENNA

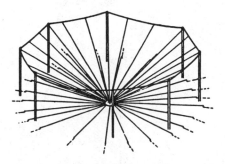

(B). INVERTED DISCONE ANTENNA

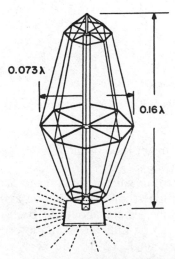

(C). CONICAL MONOPOLE ANTENNA

Figure 8.40 Broadband vertical antennas.

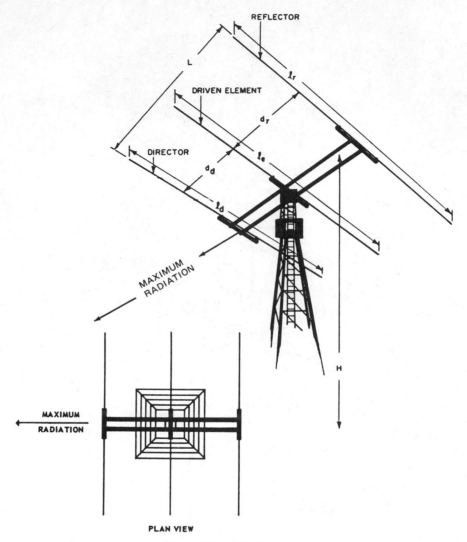

Figure 8.41 A Yagi antenna.

the smaller the length-to-diameter ratio of the tubing, the better the gain and bandwidth characteristics of the array.

A typical Yagi is optimized for gain or for front-to-back ratio, with power gains from 8 to 18 dB. Side-lobe suppression can be as great as 20 dB, with the exception of the back lobe, where about the best achievable is 12 dB. Low takeoff angles (TOA) can be difficult to attain, especially at lower HF frequencies. For a radiation angle of 5°, the array height must be 2.75 wavelengths above ground level.

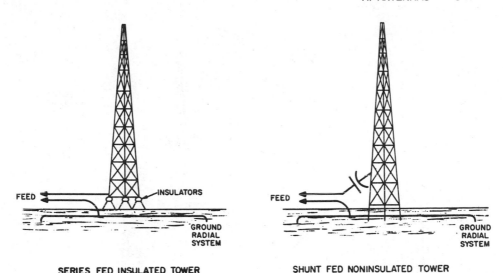

Figure 8.42 Vertical tower antennas.

The most limiting factor of a Yagi antenna is bandwidth, which is on the order of 3% (0.03:1).

8.15.3.6 *Vertical Tower Antennas* These antennas radiate a vertically polarized groundwave over a wide range of azimuthal angles. It is important to mount the antenna on high-conductivity ground to attain maximum field strength along the horizontal plane. Ground screens are commonly used to improve ground conductivity directly beneath the antenna. Figure 8.42 presents two types of typical tower antennas.

The vertical tower is commonly mounted on an insulator and is series fed between the base and the ground system via an appropriate impedance-matching device. The impedance at the base is approximately 36 Ω for a thin vertical wire quarter-wavelength high. The input impedance is lower for the small length-to-diameter ratios typical of tower structures. Radiation resistance decreases rapidly with decreasing radiator height. A low-resistance ground system is essential.

Power gain at low radiation angles is dependent on tower height, as is radiation efficiency. For a vertical antenna operating at the low end of the HF band, the required height could exceed 200 ft. A grounded vertical antenna produces maximum field intensity in the horizontal plane at $\frac{5}{8}$ wavelength; at $\frac{3}{4}$ wavelength high-angle lobes predominate.

8.15.3.7 *The Horizontal Vee Antenna* A typical horizontal vee antenna is shown in Figure 8.43. The apex of the vee is connected to a balanced

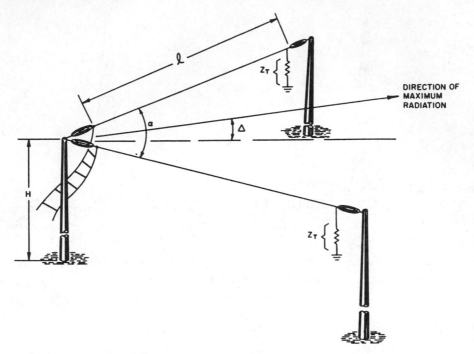

Figure 8.43 Nonresonant horizontal vee antenna.

transmission line between where Z_0 is about 500 Ω. The radiation efficiency is on the order of 35–50%. Resistive terminators of about 500 Ω are required at the end of each leg tied to ground. Significant physical parameters are height above ground H, apex angle α, leg length L, and vertical radiation angle Δ. The supporting tower height is on the order of 75 ft and the antenna requires from 3 to 7 acres. Power gains are on the order of 5–9 dB and are a function of frequency.

8.16 REFERENCE FIELDS — THEORETICAL REFERENCES

The following material is extracted from Ref. 33.

In most propagation problems, the methods of solution are based on a transmitting antenna that will produce some standard value of field intensity at a standard distance. This standard field may be one of several values produced by one of several types of reference antenna. Three of the most commonly used reference antennas are the omnidirectional radiator (isotropic source) in free space, the omnidirectional radiator over perfectly conducting earth, and the short lossless vertical antenna over perfect earth. Each field is expressed in reference to 1-kW power input at a standard distance of 1 mi or 1 km.

The most often used reference antenna is the omnidirectional radiator, a theoretical antenna that radiates equally well in all directions. The field intensity produced by this antenna may be found by equating the radiated power to the surface integral of a uniform field over a spherical surface surrounding the radiator:

$$E^2 = \frac{P_t \eta}{4\pi d^2} \text{ mV/m} \tag{8.18}$$

where P_t = the transmit power (W)
η = the impedance of free space
d = the distance (km)

Thus if the transmit power P_t is 1 kW and the distance is 1 km, the field intensity is found by

$$E = \left(\frac{1000\eta}{4\pi} \right)^{1/2} \text{ mV/m} \tag{8.19}$$

and since $\eta = 120\pi$ (impedance of free space),

$$E = [1000(30)]^{1/2}$$
$$E = 173.2 \text{ mV/m at 1 km.}$$

If this omnidirectional antenna is placed on the surface of perfectly conducting earth, we obtain

$$E = 173.2\sqrt{2}$$
$$E = 245.0 \text{ mV/m at 1 km}$$

as the field intensity, since all the power is radiated in a hemisphere and, therefore, the power per unit area is doubled.

The field of a short vertical antenna over perfect earth may be found by

$$E_\Delta = \frac{60\pi I}{d} \left(\frac{l}{\lambda} \right) \cos \Delta (1 + e^{-j4\pi h/\lambda \sin \Delta}) \tag{8.20}$$

where l/λ = length in wavelengths
Δ = vertical radiation angle
I = input current
h/λ = effective height in wavelengths
d = the distance in kilometers

But since the height of the differential element for a ground-based antenna is zero, the field intensity in the ground plane over perfect earth is

$$E_\Delta = \frac{120\pi I}{d}\left(\frac{l}{\lambda}\right). \tag{8.21}$$

The radiation resistance R_r is

$$R_r = 160\pi^2\left(\frac{l}{\lambda}\right)^2 \tag{8.22}$$

and since $I = \sqrt{P_t/R_r}$ then

$$I = \frac{\sqrt{P_t}}{\sqrt{160\pi^2(l/\lambda)^2}} - \frac{\sqrt{P_t}}{4\sqrt{10\pi l/\lambda}} \tag{8.23}$$

and the field intensity becomes

$$E_\Delta = \frac{120\pi}{d}\left(\frac{l}{\lambda}\right)\frac{\sqrt{P_t}\cdot\lambda}{4\sqrt{10}\,\pi l} \tag{8.24}$$

and

$$E_\Delta = \frac{30}{d}\sqrt{\frac{P_t}{10}} \tag{8.25}$$

where d is in kilometers

P is in watts

E_Δ is in mV/m

Solving for a power input of 1 kW at a distance of 1 km in the ground plane $(\Delta = 0°)$,

$$E_\Delta = 30\sqrt{\frac{1000}{10}} = 300 \text{ mV/m}.$$

8.17 CONVERSION OF RADIO FREQUENCY (RF) FIELD STRENGTH TO POWER

The following material is extracted from Ref. 34.

Many radio engineers are accustomed to working in the power domain (e.g., dBm, dBW). We may wish to know the receive signal level (RSL) at the input to the first active stage of an HF receiver. In the power domain, the characteristic impedance is not a consideration by definition.

HF engineers traditionally work with field strength usually expressed in microvolts per meter (μV/m). When we convert μV/m to dBm, characteristic impedance becomes important. We remember the familiar formula

$$\text{Power}_{(W)} = \frac{E^2}{R} = I^2 R \tag{8.26}$$

where E is expressed in volts and I in amperes, and we can consider R to be the characteristic impedance.

Carrying this one step further,

$$\text{Power}_{(W)} = \frac{[E(\text{V/m})]^2 [\text{effective antenna area } (\text{m}^2)]}{(\text{impedance of free space})} \tag{8.27}$$

The impedance of free space is 120π or 377 Ω.

The effective antenna area is

$$A_{(\text{m}^2)} = \frac{G\lambda^2}{4\pi} \tag{8.28}$$

where A = effective antenna area (m^2)
G = antenna gain (numeric, *not* dB)
λ = wavelength (m)

If $G = 1$ (0 dBi), then

$$P_{(W)} = \frac{E^2 \lambda^2}{377 \times 4\pi} \tag{8.29}$$

We rewrite equation 8.29 expressed in frequency rather than wavelength:

$$P_{(W)} = \frac{E^2 (c^2)}{4737.5(f^2)} \tag{8.30}$$

where c = velocity of propagation in free space, or 3×10^8 m/s. Convert to more useful units: express P in milliwatts, E in microvolts per meter, and f in megahertz. Then

$$P_{(\text{mW})} = \frac{1.89972 \times 10^{-8}(E)^2}{f^2_{(\text{MHz})}} \tag{8.31}$$

Here E is expressed in microvolts per meter. We now derive

$$P_{(\text{dBm})} = -77 + 20\log(E) - 20\log(f) \tag{8.32}$$

where E is the field strength in microvolts per meter and f is expressed in megahertz. To convert back to field strength

$$E = 10^{[P + 77 + 20 \log(f)]/20}$$

(8.33)

REVIEW EXERCISES

1. Give three advantages and three disadvantages of HF as a communication medium.

2. Identify at least five applications of HF for telecommunication that are used very widely.

3. Give at least three of the variables that affect range of HF groundwave communication links.

4. In the text, four ionospheric layers are described that affect HF propagation. What are they? One of these layers disappears at night. It affects an HF link in several ways. What are these effects and how can they be mitigated?

5. Describe two completely different ways we can communicate short distances by HF, say out to 100–200 km.

6. Classify by region on the earth's surface where it is most desirable and least desirable to communicate by HF.

7. Describe how season affects HF propagation.

8. Considering transition periods, why are north–south paths better behaved than east–west paths?

9. How does sunspot activity affect HF propagation?

10. Describe the average length of a sunspot cycle and variation in sunspot number (average) over the period.

11. List the five methods given in the text to carry out frequency management.

12. Using Table 8.2, at 1100 universal time (UT), what frequency would be recommended?

13. How can we derive the maximum usable frequency (MUF) when given the critical frequency?

14. How can one get comparative signal strength data for a particular connectivity without the use of an ionospheric sounder or by the "cut and try" method?

15. Describe how an embedded ionospheric sounding system operates. Some trademark sounding devices give other important data. Explain what

some of the other data is and how it can be used to aid path performance.

16. Describe an advanced self-sounding system. What does link quality
assessment (LQA) mean and how can it be used?

17. What is the weakness in the approach of basing LQA entirely on
received signal strength?

18. Identify the three basic HF transmission modes. Distinguish each from
the other and give their application.

19. What detracts from groundwave operation at night?

20. What three effects give rise to multipath propagation?

21. What is the effect of D-layer absorption on multihop propagation?

22. Justify why we would be better off choosing the lowest order F2 mode on
multimode reception.

23. What is the meaning of 3F2?

24. Explain the basic impairments due to propagation on transequatorial
skywave paths, especially if they have only small latitude changes.

25. Explain the cause of the formation of the O-ray and X-ray. Which arrives
first at the distant receiver?

26. Suppose we operate well below the MUF on a 1F2 path. How many
modes could we receive? (Assume 1F2 only.)

27. What sort of antenna would we require for optimum near-vertical incidence (NVI) operation?

28. What frequency band would we favor (approximately) for NVI operation?

29. Discuss the *reciprocity* of HF operation.

30. What fade margin is recommended by CCIR on an HF skywave circuit
for 99% time availability?

31. How much dispersion will a 50-baud FSK signal tolerate (i.e., binary
FSK, or BFSK)?

32. Give at least three mitigation techniques that we might use to combat the
effects of HF multipath.

33. Identify four types of diversity we may consider using for HF operation.
Describe each in one or two sentences.

34. Why is receiver noise figure a secondary concern on HF systems?

35. There is a fair probability that the optimum operational frequency may be occupied by at least one other user. Name six methods to overcome this problem.

36. What is the cause of atmospheric noise and how does it propagate? How does it vary with frequency?

37. Calculate the receiver noise threshold at 8 MHz, summer at 7 P.M. local time in Massachusetts for a receiver with a 3-kHz bandwidth, valid for 95% of the time.

38. Above 20 MHz, what types of external noise may be predominant?

39. Man-made noise is a function of what two items?

40. What are the components (contributors) of HF link transmission loss?

41. When we calculate the free-space loss for a skywave path 100 km long, what is the most significant range component? For a 1500-km path?

42. How does D-region absorption vary with frequency? It also varies as a function of what? Give at least five items.

43. For a three-hop path, what is the estimated value of ground-reflection losses?

44. Using the text, calculate D-layer absorption for the following path: 1F2 mode, sunspot number 10, operating frequency 14.5 MHz, 12-noon midpath at 45° north latitude with an elevation angle of 9°.

45. Determine the transmission loss for a groundwave path over the ocean 100 km long at 4 MHz.

46. Calculate the optimum groundwave frequency over medium dry land in Colombia at 1400 local time assuming BFSK service, a signal-to-noise ratio (S/N) of 15 dB, a receiver noise figure of 15 dB, 75-bps operation, 1-kW EIRP, and a receiving antenna of 0 dBi, which includes line losses to receiver front end, for a time availability of 95%.

47. On long groundwave paths at night, leaving aside atmospheric noise, what must we be particularly watchful for? What are some compensating methods for this impairment?

48. Give two basic arguments in favor of parallel tone transmission (versus serial tone transmission) for 2400-bps operation. Give at least one argument against the parallel tone approach.

49. The text describes a serial tone system for synchronous transmission of digital traffic. Why is "known data" interspersed with "unknown data?"

50. What is LQA and what is its purpose? Describe how one might use an LQA sequence for a self-sounding system. Argue both sides: Using LQA,

is a (PN) spread wideband system easier or more difficult than a conventional narrow-band system for self-sounding?

51. Explain in six sentences or less the operation of Lincompex. Where is it applied and what approximate improvement does it provide (dB)?

52. Why is the control of out-of-band emission so important? Give two related answers.

53. Give two reasons why a high-gain, low side-lobe antenna, can, if you will, be a force-multiplier on a point-to-point HF link.

54. Name at least five of the six important antenna parameters given in the text.

55. Differentiate between gain and directivity of an antenna.

56. Select an optimum antenna for: (1) a groundwave path, multipoint; (2) an NVI path; (3) a short one-hop skywave path (e.g., 1200 km); (4) a very long (> 5000 km) path; and (5) an HF star network operating half-duplex over one- and two-hop paths.

57. Identify the reference value of field strength for the 1-km reference field with a 0° takeoff angle (groundwave).

58. Give the equivalent values in dBm of the following field strengths: 300 mV/m, 300 μV/m, 55 μV/m, 10 μV/m. -114 dBm equals how many microvolts per meter?

REFERENCES AND BIBLIOGRAPHY

1. "Digital Selective Calling System for Use in the Maritime Mobile Service," CCIR Rec. 493-3, XVIth Plenary Assembly, Dubrovnik, 1986, vol. VIII-2.
2. L. S. Wagner, J. S. Goldstein, and W. D. Meyers, *Wideband Probing of the Trans-auroral HF Channel*, *Solar Minimum*, Naval Research Laboratories, Washington, DC, 1987.
3. George Jacobs and Theodore Cohen, *The Shortwave Propagation Handbook*, CQ Publishing, Hicksville, NY, 1979.
4. Solar Geophysical Data, Prompt Reports, National Geophysical Data Center, Boulder, CO, March 1990.
5. John L. Lloyd et al., *Estimating the Performance of Telecommunication Systems Using the Ionospheric Transmission Channel—Techniques for Analyzing the Ionospheric Effects upon HF Systems*, Institute of Telecommunication Sciences, NTIA, Boulder, CO, 1983.
6. Larry Teters et al., *Estimating the Performance of Telecommunication Systems Using the Ionospheric Transmission Channel—Ionospheric Analysis and Prediction Program User's Manual*, Institute of Telecommunication Sciences, NTIA, Boulder, CO, 1983.

7. PROPHET Software Description Document, IWG Corp., San Diego, CA, 1984.

8. *CRPL Predictions*, NBS Circular 462, National Bureau of Standards, Washington, DC, 1948.

9. "Bandwidths, Signal-to-Noise Ratios and Fading Allowances in Complete Systems," CCIR Rec. 339-6, XVIth Plenary Assembly, Dubrovnik, 1986, vol. III.

10. P. H. Levine, R. B. Rose, and J. N. Martin, *Minimuf-3—A Simplified HF MUF Prediction Algorithm*, IEE Conference on Antennas and Propagation, Pub. no. 78, 1978.

11. *Operator's Manual for the USCG Advanced PROPHET System*, Naval Ocean Systems Command, San Diego, CA, 1987.

12. *Real-time Frequency Management for Military HF Communications*, BR Communications Tech. Note no. 2, BR Communications, Sunnyvale, CA, June 1980.

13. Product Sheet (SELSCAN registered trademark), Collins 309L-2 & 4 and 514A-12 SELSCAN Adaptive Communications Processor, Rockwell International, Cedar Rapids, IA, 1981.

14. "Interoperability and Performance Standards for MF and HF Radio Equipment," MIL-STD-188-141A (Fed. Std. 1045), U.S. Department of Defense, Washington, DC, Sept. 1988.

15. Klaus-Juergen Hortenbach, *HF Groundwave and Skywave Propagation*, AGARD-R-744, Cologne, FRG, Oct. 1986.

16. "Ionospheric Propagation Characteristics Pertinent to Terrestrial Radio Communication Systems Design (Fading)," CCIR Rep. 266-6, XVIth Plenary Assembly, Dubrovnik, 1986, vol. VI.

17. *Radio Regulations*, ITU, Geneva 1982, revised 1988.

18. "Factors Affecting the Quality of Performance of Complete Systems in the Fixed Service," CCIR Rep. 197-4, XVIth Plenary Assembly, Dubrovnik, 1986, vol. III.

19. "Voice Frequency Telegraphy over Radio Circuits," CCIR Rec. 106-1, XVIth Plenary Assembly, Dubrovnik, 1986, vol. III.

20. "Equipment Technical Design Standards for Voice Frequency Carrier Telegraph (FSK)," MIL-STD-188-342, U.S. Department of Defense, Washington, DC, Feb. 1972.

21. B. D. Perry and L. G. Abraham, *Wideband HF Interference and Noise Model Based on Measured Data*, Rep. M-88-7, MITRE Corp., Bedford, MA, 1988.

22. P. J. Laycock et al., "A Model for HF Spectral Occupancy," *Fourth International Conference on HF Systems and Technology*, IEE, London, 1988.

23. *Electrical Communication System Engineering*, *Radio*, Department of the Army, Washington, DC, Aug. 1956.

24. "Characteristics and Applications of Atmospheric Radio Noise," CCIR Rep. 322-2 (issued separately from other CCIR documents), ITU, Geneva, 1983.

25. "Man-made Radio Noise," CCIR Rep. 258-4, XVIth Plenary Assembly, Dubrovnik, 1986, vol. VI.

26. K. Davies, *Ionospheric Radio Propagation*, NBS Monograph 80, U.S. Department of Commerce, National Bureau of Standards, Boulder, CO, 1965.

27. "Technical Description of the Communication Assessment Program (CAP)," Rev. 3.0, Defense Communications Agency, Arlington Hall Station, VA, Jan. 24, 1986.

28. "Simple HF Propagation Prediction Method for MUF and Field Strength," CCIR Rep. 894-1, XVIth Plenary Assembly, Dubrovnik, 1986, vol. V.

29. Technical Report No. 9, U.S. Army Signal Corps, Radio Propagation Agency, Ft. Monmouth, NJ, 1956.

30. "Groundwave Propagation Curves for Frequencies between 10 kHz and 30 MHz," CCIR Rec. 368-5, XVIth Plenary Assembly, Dubrovnik, 1986, vol. V.

31. "Equipment Technical Design Standards for Common Long-haul/Tactical Data Modems," MIL-STD-188-110, through notice 2, U.S. Department Defense, Washington, DC, Nov. 1988.

32. "HF Ionospheric Channel Simulators," CCIR Rep. 549-2, XVIth Plenary Assembly, Dubrovnik, 1986, vol. III.

33. "MF/HF Communications Antennas," DCAC 330-175-1, Addendum no. 1 to DCS Engineering—Installation Standards Manual, Defense Communications Agency, Washington, DC, May 1966.

34. *Technical Issues #89-1* (Dave Adamy), Association of Old Crows, Alexandria, VA, 1989.

35. "Improved Transmission for RF Radio Telephone Circuits," CCIR Rec. 455-1, vol. III, XVIth Plenary Assembly, Dubrovnik, 1986.

36. "Improved Transmission Systems for Use over HF Radiotelephone Circuits," CCIR Rep. 354-5, vol. III, XVIth Plenary Assembly, Dubrovnik, 1986.

37. Gerhard Braun, *Planning and Engineering of Shortwave Links*, Siemens—Heyden, London, 1982.

38. "Voice Frequency Telegraphy on Radio Circuits," CCITT Rec. R.39, Fascicle VII.1, IXth Plenary Assembly, Melbourne, 1988.

39. Private communication. Dr. L. Wagner, Naval Research Laboratories, Washington, DC 1/5/91.

40. Roy F. Basler et al., "Ionospheric Distortion of HF Signals," Final Rept., Contract DNA-008-85-C-0155, SRI International, Menlo Park, CA 1987.

9

METEOR BURST
COMMUNICATION

9.1 INTRODUCTION

Meteor burst communication (MBC) can provide inexpensive very low data rate connectivity for links up to about 1000 sm (1600 km) long. Very low data rates in this context are in the throughput range of tens to hundreds of bits per second.

MBC utilizes the phenomenon of scattering of a radio signal from the ionization trails caused by meteors entering the atmosphere. The trails are of short duration from tens of milliseconds to several seconds. A meteor trail must have some form of common geometry between one end of a link and the other. Thus a particular link can use this trail, common to both ends, for a very short period of time and then the users will have to wait for another trail entering the atmosphere with similar common geometry characteristics. On a particular link a transmitter bursts data when a common trail is discovered, waits for another trail, and bursts data again. The time between useful trails is called *waiting time*.

The useful radio frequency range for meteor burst operation is between about 20 and 120 MHz. The lower frequencies are ideal and provide the best performance. As we have seen in the previous chapter, receivers are externally noise limited up to about 40–50 MHz. The types of noise that concern us are man-made and galactic noise. The intensity of both noise types varies inversely with frequency. It is the presence of this noise that drives MBC system designers to use frequencies in the range of 40–50 MHz.

MBC transmitters have output powers ranging from 100 W to 5 kW or more. Antennas for fixed-frequency operation are usually Yagis; and horizontally polarized log periodics (LPs) if we wish to cover a frequency range of more than 5 MHz. Figure 9.1 shows the concept of MBC communication.

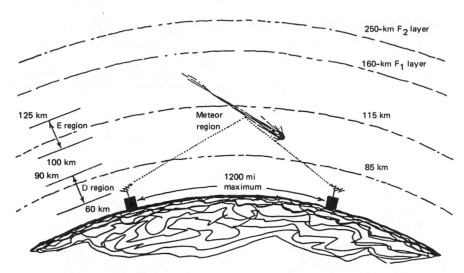

Figure 9.1 The concept of operation of a meteor burst communication link.

The implementation of MBC communication systems is attractive, especially from the standpoint of economy. The low data rate and the waiting times are disadvantages. One common application is remote sensing of meteorological conditions and/or seismic conditions. One large MBC system is installed in the Rocky Mountains in the United States to provide data on snowfall and accumulated snow. It is called the SnoTel system. MBC networks can also serve as orderwires for larger networks, particularly for the military.

9.2 METEOR TRAILS

9.2.1 General

Billions of meteors enter the earth's atmosphere every day. One source (Ref. 1) states that each day the earth sweeps up some 10^{12} objects that, upon entering the atmosphere, produce sufficient ionization to be potentially useful for reflecting/scattering radio signals.

Meteors enter the atmosphere and cause trails at altitudes of 70–140 km. The trails are long and thin, generating heat which causes the ionization. They sometimes emit visible light. The forward scatter of radio waves from these trails can support communication. The trails quickly dissipate by diffusion into the background ionization of the earth's atmosphere.

Meteor trails are classified into two categories, underdense and overdense, depending on the line density of free electrons. The dividing line is 2×10^{14}

Table 9.1   Estimate of properties of sporadic meteors

Notes		Mass (g)	Radius (cm)	Number Swept Up by Earth per Day	Electron Line Density (Electrons/Meter)
Particles that survive passage through atmosphere		10^4	8	10	
	Overdense visual	10^3	4	10^2	
		10^2	2	10^3	
		10	0.8	10^4	10^{18}
		1	0.4	10^5	10^{17}
Particles totally disintegrated in upper atmosphere		10^{-1}	0.2	10^6	10^{16}
		10^{-2}	0.08	10^7	10^{15}
	Underdense nonvisual	1^{-3}	0.04	10^8	10^{14}
		10^{-4}	0.02	10^9	10^{13}
		10^{-5}	0.008	10^{10}	10^{12}
		10^{-6}	0.004	10^{11}	10^{11}
		10^{-7}	0.002	10^{12}	10^{10}
Particles that can't be detected by radio means		10^{-8} to 10^{-13}	0.004 to 0.0002	Total about 10^{20}	Practically none

Source: Ref. 2. Courtesy of Meteor Communications Corporation. Reprinted with permission.

electrons per meter. Trails with a line density less than the value are *underdense*; those with a line density greater than 2×10^{14} electrons per meter are termed *overdense*. The dividing line of about 2×10^{14} electrons per meter corresponds to the ionization produced by a meteor whose weight is about 1×10^{-3} g. When averaged over 24 h, the number of meteors is almost inversely proportional to weight. As a result, we would expect that the number of underdense trails would far exceed the number of overdense trails. However, the signals reflected from underdense trails fall off roughly in proportion to the square of the weight, whereas signals from overdense trails increase only a little with weight. In practice, though, we find perhaps only 70% of received MBC signals are from underdense trails. Even so, the mainstay of an MBC system is the underdense trail.

Another interesting and useful fact about sporadic meteors is their mass distribution. This distribution is such that the total masses of each size of particle are approximately equal (i.e., there are 10 times as many particles of mass that is 10^{-4} g as there are particles of mass 10^{-3} g). Table 9.1 lists the approximate relationship between mass, size, electronic density, and number (Ref. 2).

9.2.2 Distribution of Meteors

At certain times of the year meteors occur as showers and may be prolific over durations of hours. Typical of these are the Quadrantids (early January), Arietids (May, June), Perseids (July, August) and Geminids (December) (Ref. 3). However, for MBC planning purposes, we use the type of meteors discussed in Section 9.2.1, which CCIR calls "sporadic meteors."

Meteor intensity (i.e., the number of usable meteor trails per hour) varies with latitude, becoming less in number but more uniform diurnally for the high latitudes. With the midlatitude case, there is roughly a sinusoidal diurnal variation of incidence, with the maximum intensity about 0600 local time and the minimum some 12 hours later, or 1800 local time. The ratio of maximum to minimum is about 4 : 1. There is a seasonal variation of similar magnitude with a minimum in February and a maximum in July. Considerable day-to-day variability exists in the incidence of sporadic and shower meteors (Ref. 4).

These variabilities are explained by the earth's rotation and its orbit around the sun. The diurnal variability is a consequence of the earth's motion in its orbit around the sun. All meteor particles are in some form of orbit about the sun. At about 0600 local time an MBC site in question is on the forward side of the earth and at that time it sweeps up slower moving particles as well as a random number of particles colliding with the earth. At 1800 local time the same MBC site is on the back side of the earth, where the slower moving particles cannot catch up with the earth. Thus 1800 is the least productive time of the day. There is always the value of random meteor counts, with the earth's velocity modulating the mean, increasing the morning count and decreasing the evening count. Sometimes at midday we see a day-to-day variation of 5 : 1, and this is due to useful returns from sporadic E (layer) rather than just meteor intensity variation.

The season maximum in the July/August period is due to the fact that the earth's orbit takes it through a region of more dense solar orbit material.

9.2.3 Underdense Trails

MBC links basically depend on reflections from underdense trails. An underdense trail does not actually *reflect* energy; instead, radio waves excite individual electrons as they pass through the trail. These excited electrons act as small dipoles, reradiating the signal at an angle equal, but opposite, to the incident angle of the trail.

Signals received from an underdense trail rise to a peak value in a few hundred microseconds, then tend to decay exponentially (Figure 9.2). Decay times from a few milliseconds to a few seconds are typical. The decay in signal strength results from the destructive phase interference caused by radial expansion or diffusion of the trail's electrons.

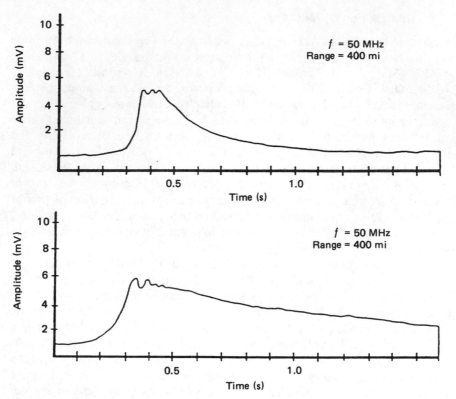

Figure 9.2 Typical underdense trails / returns where received signal intensity versus time is plotted. (From Ref. 2; courtesy of Meteor Communications Corporation.) Reprinted with permission.

In Section 9.5.6 of this chapter we will show that the received power at an MBC terminal is proportional $1/f^3$, which limits the maximum useful frequency to below 80 MHz. Commonly used frequencies are in the range of 30–50 MHz.

From equation 9.8 in Section 9.5.6 an amplitude–range relationship can also be found. An approximate normalized plot (Ref. 2) of range versus amplitude is shown in Figure 9.3. Ref. 2 also shows that range and frequency also affect time duration. Such a plot is shown in Figure 9.4 for four different frequencies.

9.2.4 Overdense Trails

We defined overdense meteor trails as those with electron line densities greater than 2×10^{14} electrons/meteor. In this case the line density is so great that signal penetration is impossible and reflection occurs rather than reradiation. Donich (Ref. 2) reports that there are no distinctive patterns for

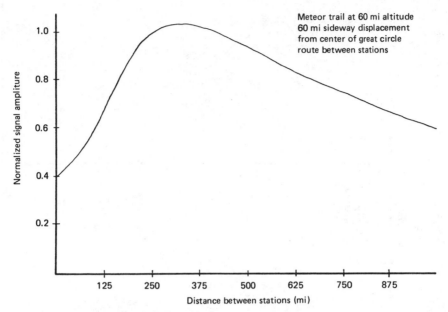

Meteor trail at 60 mi altitude
60 mi sideway displacement
from center of great circle
route between stations

Figure 9.3 Normalized underdense reflected signal power. (From Ref. 2; courtesy of Meteor Communications Corporation.) Reprinted with permission.

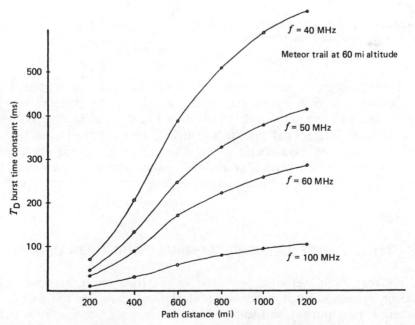

f = 40 MHz

Meteor trail at 60 mi altitude

f = 50 MHz

f = 60 MHz

f = 100 MHz

Figure 9.4 Burst time constant versus range. $T_D = \lambda^2 \sec^2 \phi / 32\pi^2 D$. (From Ref. 2; courtesy of Meteor Communications Corporation.) Reprinted with permission.

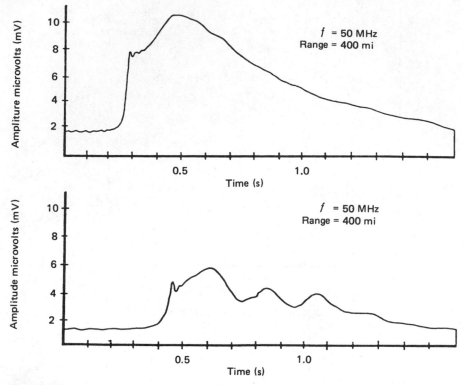

Figure 9.5 Overdense meteor reflections. (From Ref. 2; courtesy of Meteor Communications Corporation.) Reprinted with permission.

overdense trails except that they may reach a higher amplitude of signal level and last longer. With long-lasting trails we can expect fading of received signals. Some of the fading may be attributed to destructive interference of reflections off different parts of the trail and the breakup of a trail during late periods because of ionospheric winds. The period of fade is over several hundred milliseconds, permitting an MBC system to operate between nulls. Figure 9.5 shows plots of received power level versus time for overdense trails.

9.3 TYPICAL METEOR BURST TERMINALS AND THEIR OPERATION

A meteor burst terminal consists of a transmitter, receiver, and modem/processor. Such a terminal is shown in Figure 9.6. For half-duplex operation, one antenna will suffice. However, a fast-operating T/R (transmit/receive) switch is required. We will appreciate why the switch should be *fast* in a moment.

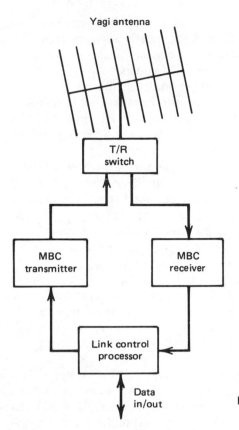

Yagi antenna

T/R switch

MBC transmitter

MBC receiver

Link control processor

Data in/out

Figure 9.6 A typical meteor burst terminal (half-duplex).

The most efficient modulation is phase shift keying, either binary (BPSK) or quaternary (QPSK). Fixed-frequency (assigned-frequency) facilities commonly use Yagi antennas optimized for the frequencies of interest. If the transmit and receive frequency are separated by more than 3 or 4 MHz, a second, optimized Yagi may be desirable, because, as described in Chapter 8, a Yagi is a narrow-band antenna. The transmitter and receiver in Figure 9.6 are conventional. The receiver noise figure may be in the range of 2 or 3 dB. In nearly all situations, external noise is dominant. In the quiet rural condition, we would expect galactic noise to predominate; in other locations, man-made noise.

The link control processor contains the software and firmware necessary to make an MBC link work. It detects the presence of the probe from the distant end, identifies that it is its companion probe, and then releases the message. Several methods of MBC link operation are described below.

There are several ways a meteor burst system can operate. One of the most common is the "master–remote" technique. Here a master station, usually more robust than the remote, sends a continuous probe in the direction of the remote. A probe is a transmitted signal on frequency F_1, usually with identification information. The remote is continuously monitoring frequency F_1, and when it hears the probe, properly authenticated with the identification information, it knows that a common meteor trail exists between the two and transmits a burst of data at frequency F_2.

Another method is quasi-half-duplex, where we transmit and receive on the same frequency. In this case a probe is sent in bursts, with sufficient resting time between bursts to accommodate the propagation time from the distant end. When the remote terminal hears a probe burst, it replies with its message, and the poll bursts from the master station cease.

Of course, we can expand this concept to a polling regime where a master station polls remotes, one at a time, from a large family of remotes. Such an operation is carried out on the SnoTel system mentioned above. One master has 600 remotes under its jurisdiction. One can carry this concept to full-duplex operation. In this case the transmitter and receiver each have separate antennas separated and isolated as much as possible. Cosite interference can be a major problem with this type of operation.

Another method of operation is broadcast. In this case a master station broadcasts traffic to "silent" remotes. Such broadcasts take advantage of the statistics of an MBC channel. A short message is continuously repeated over a comparatively long period of time. It is expected that the recipient will receive various pieces of the message and will have to reconstruct the message when the last piece is received.

"Message piecing" is a common technique used on MBC systems. If a message has any length at all—let's say a burst rate of 8 kbps and a burst duration of 100 ms or $1/10$ s, thus a maximum message length of 800 bits (this includes all overhead)—then message piecing must occur if a message is longer than 800 bits. This is the same concept as packet transmission. Each message piece is a "packet" and these packets must be reassembled at the receive location to reconstruct the originator's complete message.

9.4   SYSTEM DESIGN PARAMETERS

9.4.1   Introduction

MBC link performance is defined as the "waiting time" required to transfer a message with a specified reliability. The principal parameters affecting performance are operating frequency, data burst rate (bps), transmitter power, antenna gain, and receiver sensitivity threshold.

9.4.2   Operating Frequency

Meteor trails will reradiate or reflect very high frequency (VHF) radio signals in the 20–200-MHz frequency range. However, since the reflected signal amplitude is proportional to $1/f^3$ and its time duration to $1/f^2$, the message waiting time increases sharply as frequency is increased. Frequencies in the 20–50-MHz range are most practical for minimum waiting time. The lower limit exists, as we mentioned previously, due to external noise conditions. The noise types that external noise consists of in this region are man-made and galactic. In fact, in many instances, the lower limit of 20 MHz must be increased to 30 or 40 MHz in typical urban noise environments.

9.4.3   Data Rate

Here we refer to the burst rate or burst data rate. An MBC terminal transmits data in high-rate bursts, ideally throughout the duration of the usable portion of the meteor trail event, from 0.2 to 2 s, typically. Data burst rates generally are in the range of 2–16 kbps. The upper limit may be constrained by legal bandwidth considerations dictated by national regulatory authorities such as the FCC. Each data burst must contain overhead information, and the amount of overhead will restrict useful data throughput. Typical modulation is coherent PSK, either BPSK or QPSK.

9.4.4   Transmit Power

The higher the transmit power, the shorter the waiting time. Many MBC terminals with moderate performance features operate with RF power output in the range of 150–200 W. Larger facilities operate at 0.5, 1, 5, or even 10 kW. Naturally, there is a trade-off between performance and economy.

9.4.5   Antenna Gain

As we are aware, antenna gain and beamwidth are inversely proportional. We want our MBC antennas to encompass a large portion of the sky to take advantage of as many meteor trails as possible. As we increase antenna gain, we decrease beamwidth and decrease the "slice of the sky." The trade-off between amount of sky encompassed by an antenna ray beam and antenna gain seems to be in the region of +13 dBi gain. Donich (Ref. 2) gives +16 dBi for short links (400–600 mi) and +21 to +24 dBi for long-range links (600–1200 mi).

9.4.6   Receiver Threshold

Receiver threshold can be defined as the receive signal level (RSL) required to achieve a certain bit error rate (BER). It is a function of the type of modulation used, bandwidth, and the receiver noise, which is usually externally limited. Of course, the lower the receiver threshold, the lower the waiting time. Often a receiver is noise limited by man-made noise. Methodology for calculating threshold using man-made noise as the overriding noise contributor is given in Section 8.8.4. Donich (Ref. 2) shows that an MBC system operating at 40 MHz using coherent BPSK at 2 kbps would have a noise threshold at -121 dBm for a BER of 1×10^{-3}.

9.5 PREDICTION OF MBC LINK PERFORMANCE

9.5.1 Introduction

The starting point in the methodology we present for predicting link performance is the calculation of receiver noise threshold and a threshold for the required BER.

We next provide a basic set of MBC relationships, taken from CCIR Rep. 251-4, used to calculate MBC link transmission loss. Several generalized shortcuts are also presented. Methods of calculating other prediction parameters, such as meteor rate, burst duration time, and waiting time probability, are also described.

9.5.2 Receiver Threshold

We use a similar method to calculate MBC receiver noise threshold as a high-frequency (HF) receiver. The receiver is externally noise limited; receiver thermal noise is of secondary importance. In most cases the type of noise will be man-made. Only in a quiet rural environment will we find a receiver galactic noise limited. Therefore we turn to the methodology given in CCIR Rep. 258, which we discuss in Section 8.9.4. Use equation 8.4 modified.

$$P_n = F_{am} + (D_u + \sigma_{a1}) + 10 \log B_{Hz} - 204 \text{ dBW} \qquad (9.1)$$

Values for D_u and σ_{a1} are taken from Table 8.7.

Example. An MBC receiver operates in a residential environment at 48 MHz and its bandwidth is 8 kHz. Find the noise threshold of the receiver. Neglect transmission line losses. From Table 8.7 we find $F_{am} = 25.9$ dB, $D_u = 12.3$ dB, and $\sigma_{a1} = 4$ dB. Hence,

$$P_n = 25.9 + 12.3 + 4 + 10 \log 8000 - 204 \text{ dBW}$$

$$P_n = -148.6 \text{ dBW or } -118 \text{ dBm}$$

If we were to assume BPSK modulation and a 2-dB modulation implementation loss, we would require an E_b/N_0 of $9 + 2 = 11$ dB for a BER of

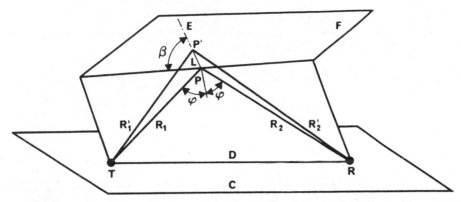

Figure 9.7 Ray geometry for a meteor-burst propagation path. C, earth's surface; D, plane of propagation; E, trail; F, tangent plane; β, angle between the trail axis and the plane of propagation; T, transmitter; R, receiver. For the terms L, P, P', R_1, R_2, R_1', R_2', and φ, see Section 9.5.4. From CCIR Rep. 251-4, Figure 1, Page 197, Vol. VI, XVIth Plenary Assembly, Dubrovnik 1986.

1×10^{-4}. (See Figure 6.29). From the example, we can calculate N_0 by subtracting $10 \log 8000$ from the P_n value or -157 dBm. Then the threshold for a BER of 1×10^{-4} must be 11 dB above the noise threshold in 1 Hz of bandwidth or -157 dBm $+ 11$ dB $= -146$ dBm, or the RSL threshold value for a BER $= 1 \times 10^{-4}$ is -146 dBm $+ 39$ dB or -107 dBm.

9.5.3 Positions of Regions of Optimum Scatter

The scattering of straight meteor ionization trails is strongly aspect sensitive. To be effective, it is necessary for the trails approximately to satisfy a specular reflection condition. This requires the ionized trail to be tangential to a prolate spheriod whose foci are at the transmitter and receiver terminals (Figure 9.7). The fraction of incident meteor trails that are expected to have usable orientations is about 5% in the area of the sky that is most effective (Ref. 5). Figure 9.8 shows the estimated percentages of useful trails for a terminal separation of 1000 km. This figure aptly shows that the optimum scattering regions ("hot spots") are situated about 100 km to either side of the great circle, independent of path length.

This feature, together with the fact that the trails lie mainly in the height range of 85–110 km, serve to establish the two hot-spot regions toward which both antennas should be directed. The two hot spots vary in relative importance according to time of day and path orientation (Ref. 5). Generally, antennas used in practice have beams broad enough to cover both hot spots. Thus the performance is not optimized, but on the other hand the need for beam swinging does not arise.

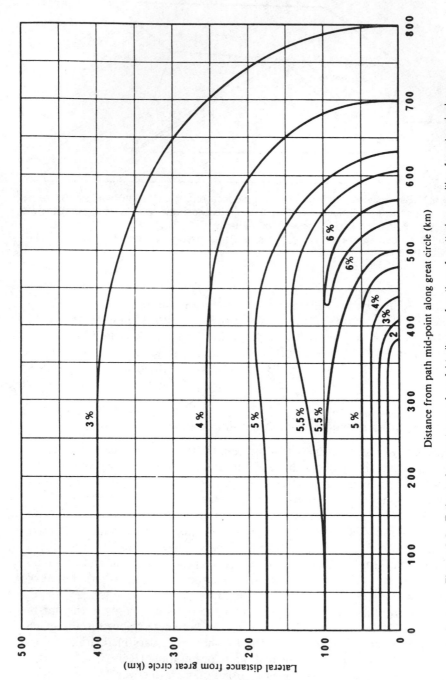

Figure 9.8 Estimated percentages of useful trails as a function of scattering position for a terminal separation of 1000 km. From CCIR Rep. 251-4, Figure 2, Page 197, Vol. VI, XVIth Plenary Assembly, Dubrovnik 1986.

Distance from path mid-point along great circle (km)

Lateral distance from great circle (km)

9.5.4 Effective Length, Average Height, and Radius of Meteor Trails

This section is adapted from CCIR Rep. 251-4 (Ref. 4):

Consider the ray geometry for a meteor burst propagation path as shown in Figure 9.7 between transmitter T and receiver R. P represents the tangent point and P' a point further along the trail such that $(R_1' + R_2')$ exceeds $(R_1 + R_2)$ by half a wavelength. Thus PP' (of length L) lies within the principal Fresnel zone and the total length of the trail within this zone is $2L$. Provided R_1 and R_2 are much greater than L, it follows that

$$L = \left[\frac{\lambda R_1 R_2}{(R_1 + R_2)(1 - \sin^2 \varphi \cos^2 \beta)} \right]^{1/2} \tag{9.2}$$

where φ = angle of incidence

β = angle between the trail axis and the plane of propagation

λ = wavelength

In order to evaluate the scattering cross section of the trail it is usual to assume that ambipolar diffusion causes the radial density of electrons to have a Gaussian distribution and that the volume density is reduced while the line density remains constant. These assumptions lead to an equation for the volume density N_v in electrons per cubic meter as a function of radius r and time from the instant of formation t, which is

$$N_v(r, t) = \frac{q}{\pi(4Dt + r_0^2)} \exp\left[\frac{-r^2}{(4Dt + r_0^2)} \right] \tag{9.3}$$

where q = electron line density per meter

D = ambipolar diffusion coefficient in m^2/s

r_0 = initial radius of trail in meters

Both D and r_0 are marked functions of height. From experimental results the following empirical formula for evaluating the average height of trails, which is a function of frequency, can be derived:

$$h = -17 \log f + 124 \tag{9.4}$$

where h = average trail height (km) and

f = wave frequency (MHz). The average trail height is a function of other system parameters in addition to frequency. However, equation 9.4 is a good approximation.

Various empirical relationships have been derived between the initial trail radius and the meteor height. An average expression is

$$\log r_0 = 0.035h - 3.45 \tag{9.5}$$

9.5.5 Ambipolar Diffusion Constant

This section is adapted from CCIR Rep. 251-4 (Ref. 4):

A good estimate of the ambipolar diffusion constant is provided by the expression

$$\log D = 0.067h - 5.6 \tag{9.6}$$

Based on the results of Greenhow and Hall the ratio of the ambipolar diffusion constant D to the velocity of the meteor V (required in the evaluation of received power) can be approximated by

$$\frac{D}{V} = \left[0.0015h + 0.035 + 0.0013(h - 90)^2\right]10^{-3} \tag{9.7}$$

where V = velocity of the meteor (m/s).

9.5.6 Received Power

Note that the next three subsections are adapted from CCIR Rep. 251-4 (Ref. 4).

9.5.6.1 Underdense Trails

$$p_R(t) = \frac{p_T g_T g_R \lambda^2 \sigma a_1 a_2(t) a_2(t_0) a_3}{64\pi^3 R_1^2 R_2^2} \tag{9.8}$$

where λ = wavelength (m)

σ = echoing area of the trail (m^2)

a_1 = loss factor due to finite initial trail radius

$a_2(t)$ = loss factor due to trail diffusion

a_3 = loss factor due to ionospheric absorption

t = time measured from the instant of complete formation of the first Fresnel zone(s)

t_0 = half the time taken for the meteor to traverse the first Fresnel zone

p_T = transmitter power (W)

$p_R(t)$ = power available from the receiving antenna (W)

g_T = transmit antenna gain relative to an isotropic antenna in free space

g_R = receive antenna gain relative to an isotropic antenna in free space

R_1 and R_2 = see Figure 9.7.

(Lossless transmitting and receiving antennas are assumed).

The echoing area σ is given as

$$\sigma = 4\pi r_e^2 q^2 L^2 \sin^2 \alpha \qquad (9.9)$$

where r_e = effective radius of the electron = 2.8×10^{-15} m and
$\quad\alpha$ = angle between the incident electric vector at the trail and the direction of the receiver from that point.

Since L^2 is directly proportional to λ, the echoing area σ is also proportional to λ and hence the received power for underdense trails varies as λ^3. Horizontal polarization normally is used at both terminals. The $\sin^2\alpha$ term in equation 9.9 is then nearly unity for trails at the two hot spots.

The loss factor a_1 is given by

$$a_1 = \exp\left(-\frac{8\pi^2 r_0^2}{\lambda^2 \sec^2\varphi}\right) \qquad (9.10)$$

It represents losses arising from interference between the reradiation from the electrons wherever the thickness of the trail at formation is comparable with the wavelength.

The factor $a_2(t)$ allows for the increase in radius of the trail by ambipolar diffusion. It may be expressed

$$a_2(t) = \exp\left(-\frac{32\pi^2 Dt}{\lambda^2 \sec^2\varphi}\right) \qquad (9.11)$$

for angle φ, see Figure 9.7.

The increase in radius due to ambipolar diffusion can be appreciable even for as short a period as is required for the formation of the trail. The overall effect with regard to the reflected power is equal to that which would arise if the whole trail within the first Fresnel zone had expanded to the same extent as at its midpoint. Since this portion of trail is of length $2L$ the midpoint radius is that arising after a time lapse of L/V s. Calling the time lapse t_0 gives, for trails near the path midpoint ($R_1 \approx R_2 \approx R$):

- For trails at right angles to the plane of propagation ($\beta = 90°$):

$$t_0 \simeq \left(\frac{\lambda R}{2V^2}\right)^{1/2} \qquad (9.12)$$

- For trails in the plane of propagation ($\beta = 0$):

$$t_0 \simeq \left(\frac{\lambda R}{2}\right)^{1/2} \times \frac{\sec\varphi}{V} \qquad (9.13)$$

Substituting t_0 from equation (9.12) into equation (9.11) gives for the $\beta = 90°$ case

$$a_2(t_0) = \exp\left[-\frac{32\pi^2}{\lambda^{3/2}}\left(\frac{D}{V}\right)\left(\frac{R}{2}\right)^{1/2}\frac{1}{\sec^2\varphi}\right] \qquad (9.14)$$

For $\beta = 0°$ the exponent in this expression is $\sec^2\varphi$ times greater.

$a_2(t)$ is the only time-dependent term and gives the decay time of the reflected signal power. Defining a time constant T_{un} for the received power to decay by a factor e^2 (i.e., 8.7 dB) leads to

$$T_{un} = \frac{\lambda^2 \sec^2\varphi}{16\pi^2 D} \qquad (9.15)$$

With reflection at grazing incidence, $\sec^2\varphi$ will be large and hence so is the echo-time constant. The echo-time constant is also increased by the use of lower frequencies.

9.5.6.2 Overdense Trails The formula for the received power in the case of overdense meteor trails is usually based on the assumption of reflection from a metallic cylinder whose surface coincides with the region for which the dielectric constant is zero. The effect of refraction in the underdense portion of the trail is usually ignored. As in the underdense case, the received power varies as λ^3 and again the echo duration varies as λ^2. However, the maximum received power is now proportional to $q^{1/2}$, in contrast to q^2 for underdense trails. Thus the increase in received signal power with ionization intensity is more modest.

9.5.6.3 Typical Values of Basic Transmission Loss Since any practical meteor burst communication system will rely mainly on underdense trails, the overdense formulas are of less importance. Satisfactory performance esti-mates can be made using formulas for the underdense case with assumed values of q in the range of 10^{13}–10^{14} electrons per meter according to the prevailing system parameters.

Basic transmission loss curves derived from equation (9.8) with $q = 10^{14}$ electrons per meter are given in Figure 9.9. As the angle β can take any value between 0° and 90° only these two extreme cases are shown. The advantage of lower propagation loss at the lower frequencies is clearly seen. Average meteor heights given from equation 9.4 have been used in deriving the curves. It should be noted that the prediction of system performance depends critically on the heights assumed.

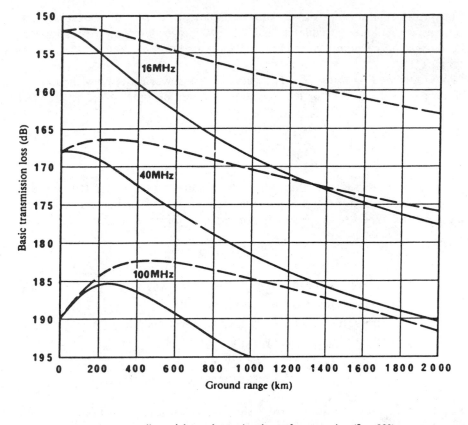

trails at right angles to the plane of propagation (β = 90°)

— — trails in the plane of propagation (β = 0°)

Figure 9.9 Basic transmission loss for underdense trails derived from equation 9.8 with electron density 1 × 10^{14} electrons / m and horizontal polarization. From CCIR Rep. 251-4, Figure 3, Page 201, Vol. VI, XVIth Plenary Assembly, Dubrovnik 1986.

9.5.7 Meteor Rate

The meteor rate or meteor burst per unit time (M_c) is related to system parameters through the following expression (Ref. 6):

$$M_c = \left(P_T \times G_T \times G_R / F_c^3 \times T_R \right)^{1/2} \qquad (9.16)$$

where P_T = the transmitter power (W)

T_R = the receiver threshold (W)

G_T = the transmit antenna gain relative to an isotropic

F_c = the carrier frequency (MHz)

G_r = the receiver antenna gain relative to an isotropic.

From this relationship and a known meteor rate M_T of a test system, operating at a known frequency and a known power level, an expression can be derived to calculate the meteor rate M_c at different values and parameters.

We now let $P = P_T G_T G_R / T_R$, where P is a power ratio. Now the meteor rate ratio of the desired system to the test system becomes

$$\frac{M_c}{M_T} = \left(\frac{P_c}{P_T}\right)^{1/2} \left(\frac{f_T}{f_c}\right)^{3/2} \tag{9.17}$$

This method advises that the higher the system power factor (PF), where $PF = 10 \log P$, and/or the lower the frequency, the higher the observed meteor rate. When power factor is expressed in terms of decibels,

$$M_c/M_T = 10^{(PF_c - PF_T)} + 10^{1.5 \log(f_T/f_c)} \tag{9.18}$$

where PF_c = power factor of the unknown system (dB)

$\quad\quad PF_T$ = power factor of the test system (dB)

$\quad\quad f_c$ = operating frequency of the unknown system

$\quad\quad f_T$ = operating frequency of the test system

$\quad\quad M_c$ = meteor rate of the desired system

$\quad\quad M_T$ = meteor rate of the test system (see Table 9.2)

Values of M obtained from a test system operated for over 1 yr are given in Table 9.2. The test system PF was 180 dB and its operating frequency f_T was 47 MHz. The values given in Table 9.2 show the diurnal and seasonal variations. Plots M_c/M_T versus the power factor of the desired system are given in Figure 9.10 for various operating frequencies. Thus, if the operating frequency and the power factor are known, the meteor rate M_c can be obtained. Conversely, if a desired M_c and operating frequency are given, the required power factor can be defined.

9.5.8 Burst Time Duration

Signals from undersense trails rise to an initial peak value in a few hundred microseconds, then decay exponentially in amplitude. Decay times from a few milliseconds to a few seconds are typical. This time variation must be taken into account in order to predict message waiting times; accordingly, the next step is to determine the average burst decay times (time above a threshold) for the specific MBC system in question. Eshleman's (Ref. 7) analytic expression for this varying signal strength is given as

$$V(t) = V_D e^{-(t/T_D)} \tag{9.19}$$

Table 9.2   Test system data[a]

	Meteor Bursts[b]/h	
	Daily Average[c]	Daily Minimum
January	50	19
February	50	19
March	50	19
April	50	19
May	50	19
June	55	20
July	65	22
August	70	25
September	70	25
October	70	25
November	65	25
December	70	25
Yearly	60	23

Source: Ref. 6. Courtesy of Meteor Communications Corporation. Reprinted with permission.
[a]System power factor $(PF)_T = 180$ dB; frequency $f_T = 47$ MHz.
[b]A burst is defined as the recognition of the coded synchronization signal from the master station.
[c]The daily average is defined by the 12 hours per day the average will be exceeded and the 12 hours per day the performance will be less than average.

where V_D is the peak value of the signal strength and T_D the average time constant in seconds.

T_D is related to both frequency and range between the two MBC stations, as given by

$$T_D = \lambda^2 \sec^2 \phi / 32^2 D \qquad (9.20)$$

where D = the diffusion coefficient, which is $8M^2/s$
$\phi = \frac{1}{2}$ the forward scattering angle as a function of range
λ = the wavelength

The value of D given above was obtained from test data given in Ref. 6. It should be noted that D exhibits a diurnal variation, with a maximum in the afternoon and a minimum in the morning. The value of $8M^2/s$ is a daily average and is the value used for all T_D calculations in this section. Plots of T_D versus range with frequency as a parameter are given in Figure 9.11. Thus, once a specific operating frequency is selected and an operating range established, the average burst time can be defined (Ref. 6).

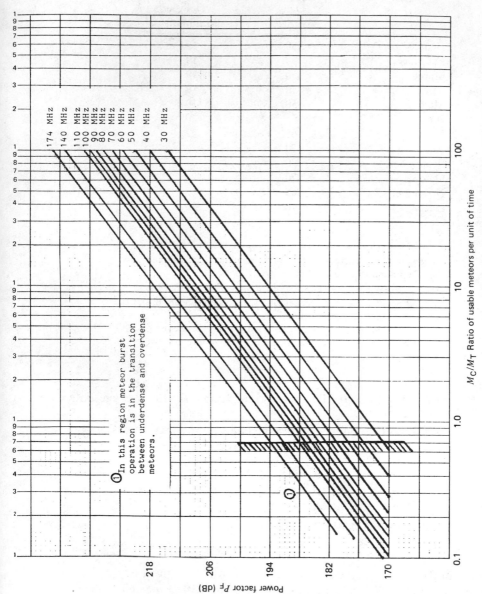

Figure 9.10 Power factor–meteor rate. M_T established from empirical data; P_F for test data = 180 dB; frequency for test data =47 MHz. In this region marked meteor burst operation is in the transition between underdense and overdense meteors. Courtesy of Meteor Communications Corp. Reprinted with permission.

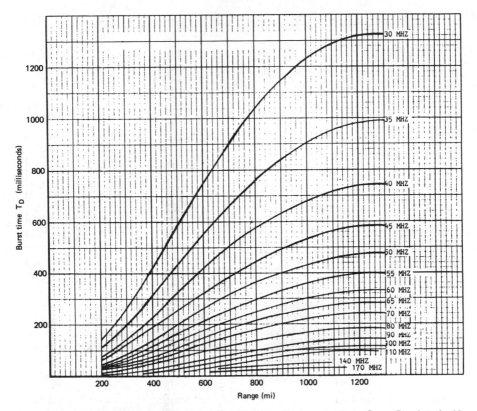

Figure 9.11 Burst time constant. Courtesy of Meteor Communications Corp. Reprinted with permission.

9.5.9 Burst Rate Correction Factor

The number of bursts per hour obtained from Figure 9.10 will require a correction factor where we can calculate the number of bursts that are sufficiently long to transfer a complete message. Of course, we will have to stipulate a certain message length. The distribution of burst durations is an exponential function and can be expressed as

$$M = M_1 e^{-(t/T_D)} \tag{9.21}$$

where M = the number of meteors exceeding the specified threshold for t seconds

$\quad M_1$ = the total number of meteors exceeding the specified threshold

$\quad T_D$ = the burst time constant

$\quad t$ = the time to transfer the complete message

A normalized plot of M/M_1 is shown in Figure 9.12. Therefore M/M_1 becomes a scaling factor for M_c derived from Figure 9.10 as a function of

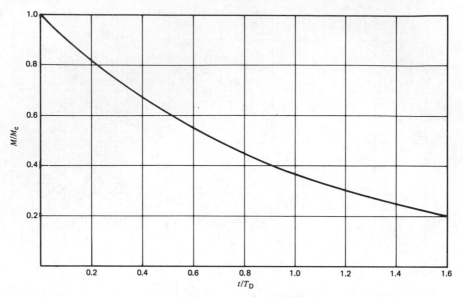

Figure 9.12 Burst duration distribution. $M / M_c = e^{-t/T_D}$. M = total number of bursts / h exceeding a specified threshold for t s; M_c = total number of bursts / h exceeding a specified threshold (the threshold is defined as the recognition of the synchronization code from the transmitting or probing station); T_D = burst time constant; t = time (s). Courtesy of Meteor Communications Corp. Reprinted with permission.

message transaction time. The value of M_c is reduced to remove bursts that have insufficient time duration to complete a message transmission by setting the value of t in equation 9.21.

9.5.10 Waiting Time Probability

Underdense meteor-burst occurrences are random in nature and follow a Poisson distribution as a function of time. The fundamental Poisson equation is

$$P = 1 - e^{-Mt} \tag{9.22}$$

where P is the probability of a meteor occurrence in time t. M is the meteor density or number of bursts per hour and t is the time in hours. Equation 9.22 provides the probability relationship to derive message waiting time. If time t is given in minutes, the equation 9.22 becomes

$$P = 1 - e^{-Mt/60} \tag{9.23}$$

In the preceding paragraphs, the meteor burst communication performance prediction approach resulted in a value M. Using the Poisson distribution given above, a family of curves can be generated for a broad set of values for M. Figure 9.13–9.15 show these relationships, with primary interest, of course, where the probability is greater than 0.9 (Ref. 6).

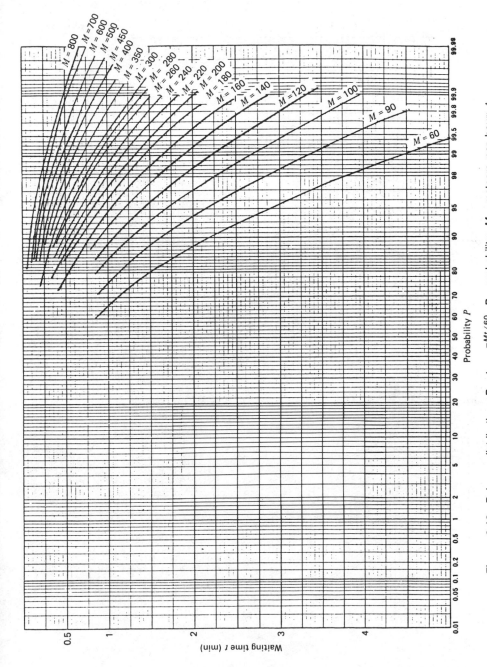

Figure 9.13 Poisson distribution. $P = 1 - e^{-Mt/60}$. P = probability; M = meteors per hour; t = waiting time. Courtesy of Meteor Communication Corp. Reprinted with permission.

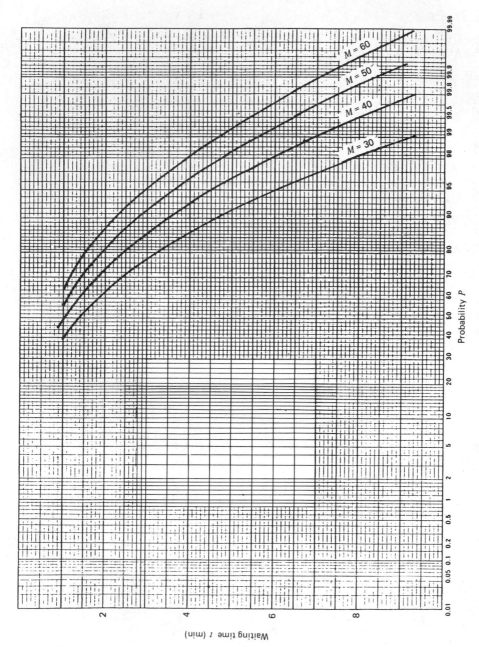

Figure 9.14 Poisson distribution. $P = 1 - e^{-Mt/60}$. P = probability; M = meteors per hour; t = waiting time. Courtesy of Meteor Communication Corp. Reprinted with permission.

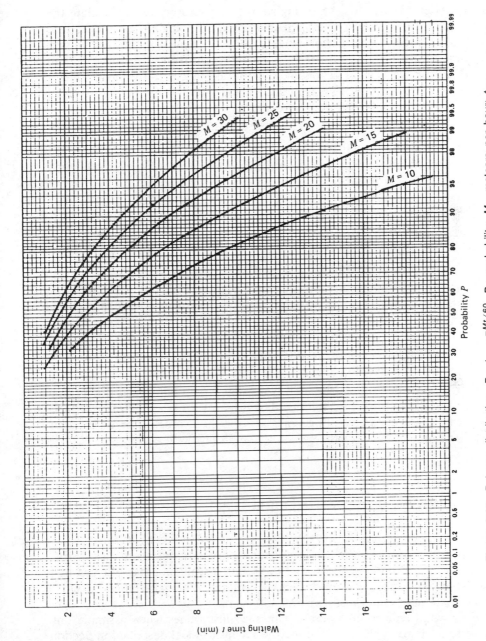

Figure 9.15 Poisson distribution. $P = 1 - e^{-Mt/60}$. P = probability; M = meteors per hour; t = waiting time. Courtesy of Meteor Communication Corp. Reprinted with permission.

9.6 DESIGN / PERFORMANCE PREDICTION PROCEDURE

Donich (Ref. 6) offers the following step-by-step procedure to calculate the performance of a meteor burst link operating in a single-burst mode. One can either predict the performance given the power factor parameters (Section 9.5.7), or in reverse, specify a desired performance level and work to the desired power factor to obtain that performance. If the power factor parameters are defined, the procedure below may be followed to predict performance within those given parameters:

1. Calculate the minimum signal levels for a BER of 1×10^{-3} by the methodology shown in Section 9.5.2 given the data rate, modulation type, receiver noise figure, and ambient external noise.

2. Determine the average M for the defined power factor parameters, including line losses, related to long-term test data. Use Section 9.5.7, Table 9.2, and Figure 9.10.

3. Calculate the average burst time constant, given the operating frequency and range. Use Section 9.5.8 and Figure 9.11.

4. Determine the message transfer time required to complete the message reception, including propagation time and message overhead requirements.

5. Remove from the value of M, obtained from step 2, the number of bursts of insufficient duration, using the message transfer time, obtained from step 4, and correct M. Use Figure 9.12.

6. Use the value of M, obtained from step 5, and determine the waiting times for required reliabilities. Use Figures 9.13–9.15. Ref. 2 reports that these derived waiting times are for the average time of the year and time of day. For the worst case, multiply by three.

9.7 NOTES ON MBC TRANSMISSION LOSS

Some insight is given in Ref. 8 regarding transmission loss at 40 MHz. Table 9.3 shows link distance; the two components of MBC transmission loss, MBC scatter loss and free-space loss; and the total transmission loss. We see that the values range around Donich's 180-dB reference model (180 dB) (Ref. 6) and the 40-MHz values of Figure 9.9.

How will a typical link operate with such loss values when a simple link budget technique is applied? We use the receiver model described in Section 9.5.2, where the RSL threshold is -107 dBm, and there is a 8-kHz bandwidth, BPSK modulation, and a BER of 1×10^{-4}. Antenna gains at each

Table 9.3   **MBC transmission loss at 40 MHz**[a]

Distance (km)	MBC Scatter Loss (dB)	Free-Space Loss (dB)	Total Loss (dB)
300	52	114.03	166
500	53	118.46	171.5
1000	57	124.49	181.5
1500	58.5	128.01	186.5
2000	61.5	130.5	192

[a]*Notes*: The free-space loss column uses great circle distance. For links under 300 km, R_1 and R_2 should be used (the distance up and down from the reflection height). The MBC scatter loss is taken from Ref. 8.

end are 13 dB and the transmitter output power is 1 kW ($+60$ dBm). Line losses are neglected.

EIRP	$+73$ dBm
MBC transmission loss	-180 dB
Receiver antenna gain	$+13$ dB
RSL	-94 dBm
Threshold	-107 dBm
Margin	13 dB

If we assume the parameters given, once the return from the trail falls below -107 dBm, the link performance is unacceptable. We also see that with this system the maximum transmission loss is $180 + 13$ dB, or 193 dB.

Forward error correction (FEC) coding as well as its attendant coding gain has been offered as one means to extend a trail's useful life. FEC requires symbol redundancy and thus more bandwidth. Of course, the more we open up a receiver's bandwidth, the more thermal noise, degrading operation. Therefore coding gain derived from FEC (in decibels) must be greater than the required noise bandwidth increase (in decibels) due to redundancy. (See Ref. 9).

Ince (Ref. 8) also aptly points out that MBC link maximum range is often more constrained by antenna gain degradation due to low takeoff angles than MBC transmission loss per se. For a 1200-km link, the takeoff angle is 5°, but for a 2000-km link the optimum takeoff angle is less than 1°. If we wish to implement such long links, we must use very elevated antennas to avoid ground reflections which cause the gain degradation. The optimum range of MBC systems is really in the range of 400–800 km.

MBC links will often experience scatter from the E layer and such scatter is very sporadic. With E scatter, continuous connectivity can last from minutes to hours.

Another phenomenon that MBC links can experience is multipath effects (fading/dispersion) during late trail conditions. This is likely to occur on more intense trails, such as some returns from overdense trails. Such trails still give useful returns even when solar winds start to break up the trail. Multipath derives from returns from different trail segments after breakup, occurring, of course, toward the end of a trail's useful life. Coding with appropriate interleaving is one method of mitigating these multipath effects.

D-layer absorption often is neglected in link budget analyses of meteor burst links. We discuss D-layer absorption and its calculation in Chapter 8. Another source is CCIR Rep. 252-2 (Ref. 10). D-layer absorption, a daytime-only phenomenon, may exceed 3 dB at 40 MHz at midlatitudes during comparatively high sunspot number periods. It is about half this value at 60 MHz.

Yet another path loss is due to Faraday rotation. The D region and the earth's magnetic field cause a linearly polarized VHF wave to be rotated both before and after meteor trail reflection. These rotations result in an overall end-to-end loss due to polarization mismatching between an incident wave and a linearly polarized receiving antenna. An excellent paper dealing with Faraday rotation effects on meteor burst communication links was published by Cannon in 1985 (Ref. 11).

As in the case of D-layer absorption, polarization rotation loss is also affected by path length (secant of the takeoff angle), the sun's zenith angle, and the sunspot number. Like D-layer absorption, Faraday rotation losses also disappear at night. Faraday rotation losses decrease rapidly as frequency increases. Cannon (Ref. 11) shows Faraday rotation losses varying from 1 dB to over 15 dB at 40 MHz (very path dependent) and dropping to a maximum of 1.6 dB at 60 MHz on the same paths.

9.8 MBC CIRCUIT OPTIMIZATION

Obviously, to help optimize MBC link operation, we want to take as much advantage of trail duration as possible. Among the details we should observe is turnaround time. Here we mean the time from receiving a valid probe to the time transmission begins. At the probe transmitter there may be a short period of receiver desensitization after transmitting. This can be minimized in the receiver design and by having a separate transmit and receive antenna. T/R (transmit/receive) switches must be fast operating. Message overhead must be as short as possible.

Adaptive trail operation has also been suggested (Refs. 12, 13 and 17). Underdense meteor trails are idealized with an exponential decay. One method measures the signal-to-noise ratio (S/N) of the probe to provide a measure of trail intensity. Burst rate is adjusted for the initial intensity. More intense trails can support higher burst rates. Another system, operating in the full-duplex mode, makes periodic measures of probe intensity (which operates throughout the message transmission), periodically adjusting the burst rate accordingly. Some of the economy of MBC systems is being given up for better throughput performance. However some argue that the performance improvement is marginal.

9.9 METEOR BURST NETWORKS

Since MBC links are limited to 1000 miles per hop, longer range communications have been achieved using several master stations "chained" together to form a network. Message piecing software has been expanded to provide routing and relay functions. Message piecing is similar to the packet relay function described in Chapter 12. For example, seven MBC master stations have been chained together to provide a network stretching from Tampa, FL (USA) to Anchorage, AK. (USAF NORAD-SAC network—Ref. 18).

REVIEW EXERCISES

1. Describe the basis of meteor burst communication (MBC) link operation.

2. What kind of average data throughput in bits per second can we expect from an MBC link?

3. "Waiting time" describes exactly what?

4. Describe the operating radio frequency relationship with MBC path loss. Why, then, do we go higher in frequency on most operational links when returns are more intense the lower we go in frequency?

5. For fixed-frequency operation, what is the most commonly used antenna? What is one type of antenna we might use to cover a broad band of frequencies for MBC link operation? Remember polarization.

6. What is the principal advantage of selecting MBC over other means of communication? Name at least two major disadvantages.

7. Identify the two categories of meteor trails. How are they described (i.e., in what units?)?

8. Give the seasonal and daily variation in performance of an MBC link regarding time of day and months of the year. How do we explain these variations?

9. What is the range of useful time duration of an underdense meteor trail?

10. What transmission impairment can we expect from trails with comparatively long time durations?

11. What type of modulation is commonly used on MBC links?

12. Describe one of the most commonly used techniques of MBC link operation, typically from a remote sensor. (Hint: Consider how we know the presence of a trail and then transmit traffic.)

13. MBC links often carry very short, often "canned" messages that can be accommodated by one common trail. How are longer messages handled?

14. In urban and suburban environments, what type of external noise is predominant? In quiet, rural environments?

15. What are the two major constraints on MBC bandwidth?

16. What is the ideal gain of an MBC link antenna? Why not more gain? Include the concept of "hot spots" in the discussion.

17. Calculate the BER threshold of an MBC receiver with a 10-kHz bandwidth, BER = 1×10^{-3}, and BPSK modulation, operating in a quiet, rural environment. Use a modulation implementation loss of 1 dB.

18. What is the range of altitude of effective meteor burst trails?

19. What is the "two hot spot" theory? How can we practically accommodate both?

20. For an MBC link operating at 40 MHz, determine the average height of useful meteor burst trails.

21. Meteor rate is a function of what parameters? It is directly proportional to three parameters and inversely proportional to two. Give all five parameters.

22. Meteor rate arrival can be defined by what (mathematical) type of distribution?

23. One simple way of viewing MBC link transmission loss is that it is made up of two loss components. What are they? Roughly what range of decibel loss values would we expect?

24. Discuss the use of forward error correction (FEC) to extend the useful life of an meteor trail. Offer some trade-offs.

25. What is a cause of multipath on an MBC link? When is multipath most likely to occur?

26. An MBC link can take advantage of other radio transmission phenomena. Name three (only one is covered in the text).

27. Name at least two additional link losses we can expect.

28. Give at least four ways to increase MBC link throughput.

29. Why would we wish to raise antennas either on towers or high ground (or both) for long MBC links?

30. What complications are seen in the design of a full-duplex MBC link besides economic ones? What advantages?

31. How does Faraday rotation loss vary with frequency? Time of day?

REFERENCES AND BIBLIOGRAPHY

1. D. W. Brown and W. P. Williams, "The Performance of Meteor-Burst Communications at Different Frequencies," *Aspects of Electromagnetic Wave Scattering in Radio Communications*, AGARD Conf. Proceedings 244, 24-1 to 24-26, Brussels, 1978.

2. Thomas G. Donich, *Theoretical and Design Aspects for a Meteor Burst Communications System*, Meteor Communications Corp., Kent, WA, 1986.

3. D. W. R. McKinley, *Meteor Science and Engineering*, McGraw-Hill, New York, 1961.

4. "Communication by Meteor-Burst Propagation," CCIR Rep. 251-4, XVIth Plenary Assembly, Dubrovnik, 1986, vol. VI.

5. V. R. Eshleman and L. L. Manning, "Meteors in the Ionosphere," *Proc. IRE*, vol. 49, Feb. 1959.

6. Thomas G. Donich, *MCBS Design/Performance Prediction Method*, Meteor Communications Corp., Kent, WA, 1986.

7. V. R. Eshleman, *Meteors and Radio Propagation*, Stanford University Rept. no. 44, contract N60NR-25132, Feb. 1955.

8. E. Nejat Ince, "Communications through EM-wave Scattering," *IEEE Commun. Mag.*, May 1982.

9. Scott L. Miller and Laurence B. Milstein, "Performance of a Coded Meteor Burst System," IEEE MILCOM'89, Boston, MA, Oct. 1989.

10. "CCIR Interim Report for Estimating Sky-wave Field Strength and Transmission Loss at Frequencies between the Approximate Limits of 2 and 30 MHz," CCIR Rep. 252-2, 1970 (out of print).

11. P. S. Cannon, *Polarization Rotation in Meteor Burst Communication Systems*, Royal Aircraft Establishment Tech. Report 85082 (TR85082), London, Sept. 1985.

12. W. B. Birkemeier et al., "Feasibility of High Speed Communications on the Meteor Scatter Channel," University of Wisconsin, Madison, 1983.

13. *Efficient Communications Using the Meteor Burst Channel*, STS Telecom, Port Washington, NY, 1987 (NSF ISI-8660079).

14. G. R. Sugar, "Radio Propagation by Reflection from Meteor Trails," *Proc. IRE*, Feb. 1964.

15. Michael R. Owen, "VHF Meteor Scatter—An Astronomical Perspective," *QST*, June 1986.

16. David W. Brown, "A Physical Meteor-Burst Propagation Model and Some Significant Results for Communication System Design," *IEEE Journal on Selected Areas in Communications*, vol. SAC-3, no. 5, Sept. 1985.

17. Dale K. Smith and Thomas G. Dovich, *Variable Data Rate Applications in Meteor Burst Communications*, Meteor Communications Corp., Kent, WA, 1989.

18. Dale K. Smith and Richard J. Fulthorp, *Transport, Network and Link Layer Considerations in Medium and Large Meteor Burst Communications Networks*, Meteor Communications Corp., Kent, WA, 1989.

10

COAXIAL CABLE SYSTEMS

10.1 INTRODUCTION

A coaxial cable is simply a transmission line consisting of an unbalanced pair made up of an inner conductor surrounded by a grounded outer conductor, which is held in a concentric configuration by a dielectric. The dielectric can be of many different types, such as solid "poly" (polyethylene or polyvinyl chloride), foam, Spirafil, air, or gas. In the case of air/gas dielectric, the center conductor is kept in place by spacers or disks.

Systems have been designed to use coaxial cable as a transmission medium with a capability of transmitting a frequency division multiplex (FDM) configuration ranging from 120 to 13,200 voice channels. Community antenna television (CATV) systems use single cables for transmitted bandwidths on the order of 300–500 MHz.

FDM was developed originally as a means to increase the voice channel capacity of wire systems. At a later date the same techniques were applied to radio. Then for a time, the 20 years after World War II, radio systems became the primary means for transmitting long-haul toll telephone traffic. Then for some 15 years coaxial cable made a strong comeback in this area. Now fiber optic cable is the medium of choice for terrestrial and undersea transmission systems. In most cases fiber is more cost-effective and efficient, particularly for digital transmission.

One advantage of coaxial cable systems is reduced noise accumulation when compared to radiolinks. For point-to-point multichannel telephony the FDM line frequency (see Chapter 3) configuration can be applied directly to the cable without further modulation steps as required in radiolinks, thus substantially reducing system noise.

In most cases radiolinks will prove more economical then coaxial cable. Nevertheless, owing to the congestion of centimetric radio wave systems (radiolinks) (see Chapter 4), coaxial cable is a viable alternative. Coaxial cable should be considered in lieu of radiolinks using the following general guidelines:

- In areas of heavy microwave (including radiolink) RFI
- On high-density routes where it may be more economical than radiolinks (Think here of a system that will require 5000 circuits at the end of 10 years), and
- On long national or international backbone routes where the system designer is concerned with noise accumulation.

Coaxial cable systems may be attractive for the transmission of TV or other video applications. Some activity has been noted in the joint use of TV and FDM telephone channels on the same conductor. Another advantage in some circumstances is that system maintenance costs may prove to be less than for equal-capacity radiolinks.

One deterrent to the implementation of coaxial cable systems, as with any cable installation, is the problem of getting the right-of-way for installation, and its subsequent maintenance (gaining access), especially in urban areas. Another consideration is the possibility of damage to the cable once it is installed. Construction crews may unintentionally dig up or cut the cable.

10.2 BASIC CONSTRUCTION DESIGNS

Each coaxial line is called a *tube*. A pair of these tubes is required for full-duplex long-haul application. One exception is the CCITT small-bore coaxial cable system where 120 voice channels, both "go" and "return," are accommodated in one tube. For long-haul systems more than one tube is included in a sheath. In the same sheath filler pairs or quads are included, sometimes placed in the interstices, depending on the size and lay-up of the cable. The pairs and quads are used for orderwire and control purposes as well as for local communication. Some typical cable lay-ups are shown in Figure 10.1. Coaxial cable is usually placed at a depth of 90–120 cm, depending on frost penetration, along the right-of-way. Tractor-drawn trenchers or plows are used to open the ditch where the cable is placed, using fully automated procedures.

Cable repeaters are spaced uniformly along the route. Secondary or "dependent" repeaters are often buried. Primary power feeding or "main" repeaters are installed in surface housing. Cable lengths are factory cut so that the splice occurs right at repeater locations.

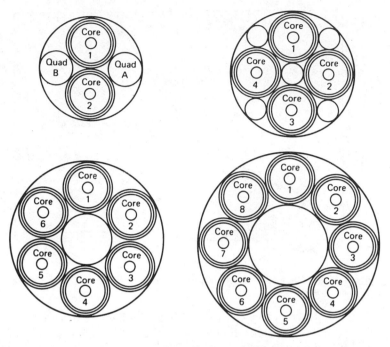

Figure 10.1 Some basic coaxial lay-ups.

10.3 CABLE CHARACTERISTICS

For long-haul transmission, standard cable sizes are as follows:

(inches)	(mm)
0.047/0.174	1.2/4.4 (small diameter)
0.104/0.375	2.6/9.5

The fractions express the outside diameter of the inner conductor over the inside diameter of the outer conductor. For instance, for the large-bore cable the outside diameter of the inner conductor is 0.104 in. and the inside diameter of the outer conductor is 0.375 in. This is shown in Figure 10.2. As can be seen from equation 10.1, which relates to Figure 10.2, the ratio of the diameters of the inner and outer conductors has an important bearing on attenuation. If we can achieve a ratio of $b/a = 3.6$, a minimum attenuation per unit length will result.

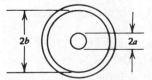

Figure 10.2 Basic electrical characteristics of coaxial cable.

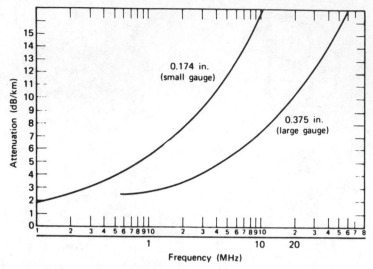

Figure 10.3 Attenuation–frequency response per kilometer of coaxial cable.

For an air dielectric cable pair, $\varepsilon = 1.0$, the outside diameter of the inner conductor $= 2a$, the inside diameter of the outer conductor $= 2b$, and $f =$ frequency. The attenuation constant α is then

$$\alpha = 2.12 \times 10^{-5} \frac{\sqrt{f}\left(\dfrac{1}{a} + \dfrac{1}{b}\right)}{\log b/a} \quad \text{(dB/mi)} \tag{10.1}$$

where $a =$ radius of inner conductor and $b =$ radius of outer conductor.

The characteristic impedance is

$$Z_0 = \left(\frac{138}{\sqrt{\varepsilon}}\right)\log\left(\frac{b}{a}\right) = 138\log\left(\frac{b}{a}\right) \quad \text{(in air)} \tag{10.2}$$

where ε is the dielectric constant of the insulating material.

The characteristic impedance of coaxial cable is $Z_0 = 138\log(b/a)$ for an air dielectric. If $b/a = 3.6$, then $Z_0 = 77\ \Omega$. Using dielectric other than air reduces the characteristic impedance. If we use the disks mentioned above to support the center conductor, the impedance lowers to $75\ \Omega$.

Figure 10.3 is a curve giving the attenuation per unit length in decibels versus frequency for the two most common types of coaxial cable discussed in this chapter. Attenuation increases rapidly as a function of frequency and is a function of the square root of frequency as shown in Figure 10.3. The transmission system engineer is basically interested in how much bandwidth is available to transmit an FDM line frequency configuration (Chapter 3). For instance, the 0.375-in. cable has an attenuation of about 5.8 dB/mi at 2.5 MHz and the 0.174-in. cable, 12.8 dB/mi. At 5 MHz the 0.174-in. cable has about 19 dB/mi and the 0.375-in. cable, 10 dB/mi. Attenuation is specified for the highest frequency of interest.

Coaxial cable can transmit signals down to dc, but in practice, frequencies below 60 kHz are not used because of difficulties of equalization and shielding. Some engineers lift the lower limit to 312 kHz. The high-frequency (HF) limit of the system is a function of the type and spacing of repeaters as well as cable dimensions and the dielectric constant of the insulating material. It will be appreciated from Figure 10.3 that the gain frequency characteristics of the cable follow a root frequency law, and equalization and preemphasis should be designed accordingly.

10.4 SYSTEM DESIGN

Figure 10.4 is a simplified application diagram of a coaxial cable system in long-haul point-to-point multichannel telephone service. To summarize system operation, an FDM line frequency (Chapter 3) is applied to the coaxial cable system via a line terminal unit. Dependent repeaters are spaced uniformly along the length of the cable system. These repeaters are fed power from the cable itself. In the ITT design (Ref. 1) the dependent repeater has a plug-in automatic level control unit. In temperate zones, where cable laying is to a sufficient depth and where diurnal and seasonal temperature variations are within the "normal" (a seasonal swing of $\pm 10°C$), a plug-in level control (regulating) unit is incorporated in every fourth dependent amplifier (see Figure 10.5). We use the word "dependent" for the dependent repeater for two reasons. It depends on a terminal or main repeater for power and it provides to the terminal or main repeater fault information.

Let us examine Figures 10.4 and 10.5 at length. Assume that we are dealing with a nominal 12-MHz system on a 0.375-in. (9.5-mm) cable. Up to 2700 voice channels can be transmitted. To accomplish this, two tubes are required, one in each direction. Most lay-ups, as shown in Figure 10.1, have more than two tubes. Consider Figure 10.4 from left to right. Voice channels in a four-wire configuration connect with the multiplex equipment in both the "go" and the "return" directions. The output of the multiplex equipment is the line frequency (baseband) to be fed to the cable. Various line frequency configurations are shown in Figures 3.11, 3.12, 3.19, and 3.20. The line signal is fed to the terminal repeater, which performs the following functions:

- Combines the line control pilots with the multiplex line frequency
- Provides preemphasis to the transmitted signal, distorting the output signal such that the higher frequencies get more gain than the lower frequencies, as shown in Figure 10.3
- Equalizes the incoming wideband signal
- Feeds power to dependent repeaters

The output of the terminal repeater is a preemphasized signal with required pilots along with power feed. In the ITT design this is a dc voltage up to 650 V with a stabilized current of 110 mA. A main (terminal) repeater feeds, in

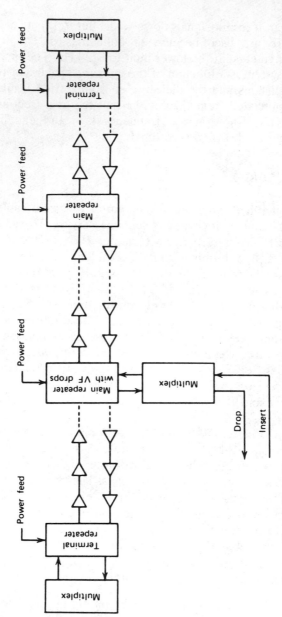

Figure 10.4 Simplified application diagram of a long-haul coaxial cable system for multichannel telephony.

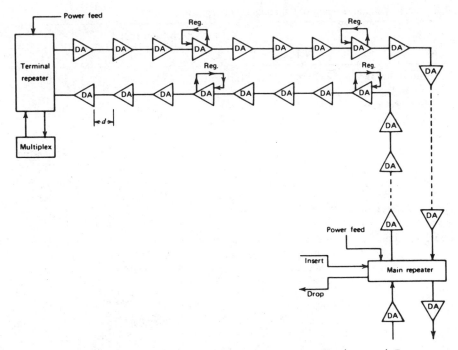

Figure 10.5 Detail of application diagram. DA = dependent amplifier (repeater); Reg. = regulation circuitry; d = distance between repeaters.

this design, up to 15 dependent repeaters in each direction. Thus a maximum of 30 dependent repeaters appear in a chain for every main or terminal repeater. Other functions of a main repeater are to equalize the wideband signal and to provide access for drop and insert of telephone channels by means of throughgroup filters.

Figure 10.5 is a blowup of a section of Figure 10.4, showing each fourth repeater with its automatic level regulation circuitry. The distance d between DA repeaters is 4.5 km or 2.8 mi for a nominal 12-MHz system (0.375-in. cable). Amplifiers have gain adjustments of ± 6 dB, equivalent to varying repeater spacings of ± 570 m (1870 ft).

As can be seen from the above, the design of coaxial cable systems for both long-haul multichannel telephone service as well as CATV systems has become, to a degree, a "cookbook" design. Basically, system design involves

- Repeater spacing as a function of cable type and bandwidth
- Regulation of signal level
- Temperature effects on regulation
- Equalization
- Cable impedance irregularities
- Fault location or the so-called supervision
- Power feed

Table 10.1 Characteristics of L coaxial cable systems[a]

Item	L System Identifier			
	L1	L3	L4	L5
Maximum design line length	4000 mi	1000 mi	4000 mi	4000 mi
Number of 4-kHz FDM VF channels	600	1860	3600	10,800[b]
TV NTSC	Yes	Yes plus 600 VF	No	Not stated
Line frequency	60–2788 kHz	312–8284 kHz	564–17,548 kHz	1590–68,780 kHz
Nominal repeater spacing	8 mi	4 mi	2 mi	1 mi
Power feed points	160 mi or every 20 repeaters	160 mi or every 42 repeaters	160 mi or every 80 repeaters	75 mi or every 75 repeaters

[a]*Notes*:
1. Cable type of all L systems, 0.375 in.
2. Number of VF channels expressed per pair of tubes, one tube "go" and one tube "return."
[b]L5E = 13,200 VF channels.

Table 10.2 Characteristics of CCITT specified coaxial cable systems (large-diameter cable)[a]

Item	Nominal Top Modulation Frequency, MHz				
	2.6	4	6	12	60
CCITT Rec.	G.337A	G.338	G.337B	G.332	G.333
Repeater type	Tube	Tube	Tube	Transistor	Transistor
Video capability	No	Yes	Yes	Yes	Not stated
Video + FDM capability	No	No	No	Yes	Not stated
Nominal repeater spacing	6 mi/ 9 km	6 mi/ 9 km	6 mi/ 9 km	3 mi/ 4.5 km	1 mi/ 1.55 km
Main line reg.[b] pilot	2604 kHz	4092 kHz	See CCITT Rec. J.72	12,435 kHz	12,435/4287 kHz
Auxiliary reg.[b] pilot(s)		308, 60 kHz	See Rec. J.72	4287, 308 kHz	61,160, 40,920 and 22,372 kHz

[a]*Note*: Cable type for all systems, 0.104/0.375 in. = 2.6/9.5 mm.
[b]Reg. = regulation.

Other factors are, of course, the right-of-way for the cable route with access for maintenance and the laying of the cable. With these factors in mind, Tables 10.1 and 10.2 review the basic parameters of the Bell System approach (L-carriers) and the CCITT approach, respectively.

For the 0.375-in. coaxial cable systems practical noise accumulation is less than 1 pWp/km, whereas radiolinks allocate 3 pWp/km. These are good guideline numbers to remember for gross system considerations. Noise in coaxial cable systems derives from the active devices in the line (e.g., the repeaters and terminal equipment, both line conditioning and multiplex). Noise design of these devices is a trade-off between thermal and IM noise. IM noise is the principal limiting parameter forcing the designer to install more repeaters per unit length with less gain per repeater.

Refer to Chapter 3 for CCITT recommended FDM line frequency configurations, in particular Figures 3.19 and 3.20, valid for 12-MHz systems. CCITT pilot frequencies and system levels are covered in Section 10.7.

10.5 REPEATER DESIGN — AN ECONOMIC TRADE-OFF FROM OPTIMUM

10.5.1 General

Consider a coaxial cable system 100 km long using 0.375-in. cable capable of transmitting up to 2700 VF channels in an FDM/SSB configuration (12 MHz). At 12 MHz cable attenuation per kilometer is approximately 8.3 dB (from Figure 10.3). The total loss at 12 MHz for the 100-km cable section is $8.3 \times 100 = 830$ dB. Therefore one approach the system design engineer might take would be to install a 830-dB amplifier at the front end of the 100-km section. This approach is rejected out of hand. Another approach would be to install a 415-dB amplifier at the front end and another at the 50-km point. Suppose that the signal level was -15 dBm composite at the originating end. Hence -15 dBm $+ 415$ dB $= +400$ dBm or $+370$ dBW. Remember that $+60$ dBW is equivalent to 1 MW; otherwise we would have an amplifier with an output of 10^{37} W or 10^{31} MW. Still another approach is to have 10 amplifiers with 83-dB gain, each spaced at 10-km intervals. Another would be 20 amplifiers or $830/20 = 41.5$ dB each; or 30 amplifiers at $830/30 = 27.67$ dB, each spaced at 3.33-km intervals. As we shall see later, the latter approach begins to reach an optimum from a noise standpoint, keeping in mind that the upper limit for noise accumulation is 3 pWp/km. The gain most usually encountered in coaxial cable amplifiers is 30–35 dB.

If we remain with the 3-pWp/km criterion, in nearly all cases radiolinks (Chapter 4) will be installed because of their economic advantage. Assuming 10 full-duplex RF channels per radio system at 1800 VF channels per RF channel, the radiolink can transmit 18,000 full-duplex channels, and do it probably more cheaply on an installed cost basis. On the other hand, if we can show noise accumulation less on coaxial cable systems, these systems will prove in at some number of channels less than 18,000 if the reduced cumulative noise is included as an economic factor. There are other considerations, such as maintenance and reliability, but let us discuss noise further.

Suppose that we design our coaxial cable systems for no more than 1 pWp/km. Most long-haul coaxial cable systems being installed today meet this figure. However, we will use the CCITT figure of 3 pWp/km in some of the examples that follow.

As discussed in Chapter 1, noise for this discussion consists of two major components, namely,

- Thermal noise
- IM noise

Coaxial cable amplifier design, to reach a goal of 1 pWp/km of noise accumulation, must walk a "tightrope" between thermal and IM noise. It is also very sensitive to overload with its consequent impact on IM noise.

The purpose of the abbreviated and highly simplified discussion in this section is to give the transmission system engineer some appreciation of coaxial repeater design. For a deeper analysis, the reader should refer to (Ref. 2, Chaps. 12–16; Ref. 3, Chaps. 3–7, and Ref. 15, Chap. 12).

10.5.2 Thermal Noise

From Section 1.9.6 the thermal noise threshold P_n may be calculated for an active two-port device as follows:

$$P_n = -174 \text{ dBm/Hz} + \text{NF} + 10 \log B_w \qquad (10.3)$$

where B_w = bandwidth (Hz) and NF_{dB} = noise figure of the amplifier.

Restating equation (10.3) for a voice channel with a nominal B_w of 3000 Hz, we can then have

$$P_n = -139 \text{ dBm} + \text{NF} \quad (\text{dBm/3 kHz}) \qquad (10.4)$$

Assume a coaxial cable system with identical repeaters, each with gain G_r, spaced at equal intervals along a uniform cable section. Here G_r exactly equals the loss of the intervening cable between repeaters. The noise output of the first repeater is $P_n + G_r$ (in dBm). For N repeaters in cascade, the total noise (thermal) output of the Nth repeater is

$$P_n + G_r + 10 \log N \quad (\text{dBm}) \qquad (10.5)$$

An important assumption all along is that the input–output impedance of the repeaters just equals the cable impedance Z_0.

10.5.3 Overload and Margin

The exercise of this section is to develop an expression for system noise and discuss methods of reducing it. In Section 10.5.2 we developed a term for the thermal noise for a string of cascaded amplifiers ($P_n + G_r + 10 \log N$). The next step is to establish 0 dBm as a reference, or more realistically -2.5 dBmp, because we are dealing with a voice channel nominally 3 kHz wide and we want it weighted psophometrically (see Section 1.9.6). Now we can establish a formula for a total thermal noise level as measured at the end of a coaxial cable system with N amplifiers in cascade:

$$P_t = P_n + G_r + 10 \log N - 2.5 \text{ dBmp} \qquad (10.6)$$

As before we assume that all the amplifiers are identical and spaced at equal intervals and that the gain of each is G_r, which is exactly equal to the loss of the intervening cable between each amplifier.

Examining equation (10.6) we see that the operating level is high. The next step is to establish an operating level that should never be exceeded and call it L. A margin to that level must also be established to take into account instability of the amplifiers caused by aging effects, poor maintenance, temperature variations, misalignment, and so forth. The margin to the maximum operating level point is M_g. All units are in decibels. A more realistic equation can now be written for total thermal noise including a suitable margin:

$$P_t = P_n + G_r + 10 \log N - 2.5 \text{ dBmp} + L + M_g \qquad (10.7)$$

These levels are shown graphically in Figure 10.6.

A number of interesting relationships can be developed if we consider a hypothetical example. CCITT permits 3-pWp/km noise accumulation

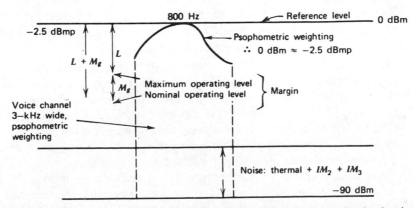

Figure 10.6 Graphic representation of reference level, signal levels, and noise levels in a coaxial cable system. (*Note*: Levels are not drawn to scale.)

(CCITT Rec. G.222). Allow 2 pWp of that figure to be attributed to thermal noise. If we were to build a system 100 km long, we could then accumulate 200 pWp of the thermal noise. Now set 200 pWp equal to P_t in equation (10.7). First convert 200 pWp to dBmp (-67 dBmp). Thus

$$-67 \text{ dBmp} = P_n + G_r + 10\log N - 2.5 \text{ dBmp} + L + M_g$$
$$P_r + G_r + 10\log N + L + M_g = -64.5 \text{ dBmp}$$

Let us assign some numbers to the equation which are somewhat reasonable. To the 100-km system install 20 repeaters at equal intervals. Cable loss is 5 dB/km, or 500 dB total loss, at the highest operating frequency. Therefore repeater gain G_r is 25 dB, with $N = 20$. Let

$$L + M = 15 \text{ dB}$$

From Section 10.5.2,

$$P_n = -139 \text{ dBm} + NF$$

Thus

$$-139 + NF + 25 + 10\log 20 + 15 = -64.5 \text{ dBmp}$$
$$NF = 21 \text{ dB or less}$$

This is a NF that is fairly easy to meet.

Let us examine this exercise a little more closely and see whether we cannot derive some important relationships that can offer the system and amplifier design engineer some useful guidance:

1. By doubling the length of the system, system noise increases 3 dB, or by doubling the number of amplifiers, G_r being held constant, system noise doubles (i.e., $10\log 2N$).

2. By making the terms L and M_g smaller, or in other words, increasing the maximum operating level, reducing the margin, system thermal noise improves on a decibel-for-decibel basis.

3. Of course, by reducing NF, system noise may also be reduced. But suppose that NF turned out to be very small in the calculations, a figure that could not be met or would imply excessive expense. Then we would have to turn to other terms in the equation, such as reducing terms G_r, L, and M_g. However, there is little room to maneuver with the latter two, 15 dB in the example. That leaves us with G_r. Of course, reducing G_r is at the cost of increasing the number of amplifiers (or increasing the size of the cable to reduce attenuation, etc.). As we reduce G_r, the term $10\log N$ increases because we are increasing the number of amplifiers N. The trade-off between the term $10\log N$ and G_r occurs where G_r is between 8 and 9 dB.

Another interesting relationship is that of the attenuation of the cable. It will be noted that the loss in the cable is roughly inversely proportional to the cable diameter. As an example, let us assume that the loss of a cable section between repeaters is 40 dB. By increasing the cable diameter 25%, the loss of the cable section becomes $40/(1 + 0.25) = 32$ dB. In our example above, by increasing the cable diameter, repeater gain may be decreased with the consequent improvement in system noise (thermal). (*Note*: The above examples are given as exercises and may not necessarily be practicable owing to economic constraints.)

10.5.4 IM Noise

The second type of noise to be considered in coaxial cable system and repeater design is IM noise. IM noise on a multichannel FDM system may be approximated by a Gaussian distribution (see Section 3.2.4.4) and consists of second-, third-, and higher-order IM products. Included in these products, in the wide-band systems we cover here, are second and third harmonics. IM products (e.g., IM noise) are a function of the nonlinearity of active devices* (see Section 1.9.6).

To follow our argument on IM noise in coaxial cable repeater design, the reader is asked to accept the following (Ref. 2). If a simple sinusoid wave is introduced at the input port of a cable amplifier, the output of the amplifier could be expressed by an equation with three terms, the first of which is linear, representing the desired amplification. The second and third terms are quadratic and cubic, representing the nonlinear behavior of the amplifier (i.e., second- and third-order products). On the basis of this power series, for each 1-dB change of fundamental input to the amplifier, the second harmonic changes 2 dB, and the third harmonic 3 dB. Furthermore, for two waves A and B a second-order sum $(A + B)$ or difference $(A - B)$ is equivalent to the second harmonic of A at the output plus 6 dB. Likewise, the sum of $A + B + C$ would be equivalent to the level of the third harmonic at the output plus 15.6 dB of one of the waves. We consider that all inputs are of equal level. The situation for $2A + B$ would be equivalent to $3A + 9.6$ dB, and so forth. These last three power series may be more clearly expressed when set down as follows, where P_H is the harmonic IM power:

$$P_{H(A \pm B)} = P_{H(2A)} + 6 \text{ dB} \tag{10.8a}$$

$$P_{H(A \pm B \pm C)} = P_{H(3A)} + 15.6 \text{ dB} \tag{10.8b}$$

$$P_{H(2A \pm B)} = P_{H(3A)} + 9.6 \text{ dB} \tag{10.8c}$$

For this discussion let IM_2 and IM_3 express the nonlinearity of a repeater;

---

*IM products may also be produced in passive devices, but to simplify the argument in this chapter, we have chosen to define IM products as those derived in active devices.

they are, respectively, the power of the second and third harmonics corresponding to a 0 dBm fundamental (-2.5 dBmp). Adjusted to the maximum operating level L (see Figure 10.6), the second harmonic power P_{2A} is

$$P_{2A} = IM_2 + 10\log N + L \quad \text{(dBmp)} \tag{10.9}$$

L is assumed to be positive as in our argument in the preceding section.

Now decrease the applied signal level to a repeater by L dB, assuming that the power of the fundamental of a wave at the 0 TLP was L dBmp. It follows that, by decreasing the applied power L dB so that the fundamental of a wave is now 0 dBmp at the 0 TLP, the magnitude of the fundamental is decreased L dB at the input of the first amplifier.

For every decrease of 1 dB in the fundamental, the second harmonic decreases by 2 dB. Therefore the new power of the second harmonic amplitude will be decreased by $2L$ dB, or

$$P_{2A} = IM_2 + 10\log N + L - 2L \quad \text{(for the system)}$$
$$= IM_2 + 10\log N - L \quad \text{(dBmp)} \tag{10.10}$$

which is the second harmonic noise power level.

Let us consider the $A + B$ product. For fundamentals of equal magnitudes, such a product in a single repeater will be 6 dB higher than a $2A$ product. It also varies by 2 dB per 1-dB variation in both fundamentals and adds in power addition as a function of the number of repeaters. Consequently,

$$P_{(A \pm B)} = IM_3 + 10\log N - L + 6 \text{ dBmp} \tag{10.11}$$

Similarly for the $3A$ condition (e.g., third harmonic) of a wave fundamental A,

$$P_{3A} = IM_3 + 10\log N - 2L \quad \text{(dBmp)} \tag{10.12}$$

For the $A + B - C$ condition,

$$P_{(A+B-C)} = IM_3 + 20\log N - 2L + 15.6 \text{ dBmp} \tag{10.13}$$

Again we find $2L$ because third-order products vary 3 dB for every 1-dB change in the fundamental. We use $20\log$ rather than $10\log$, assuming that the products add in phase (i.e., voltagewise) versus the number of repeaters. The $10\log$ term represented power addition. For the $2A - B$ condition,

$$P_{2A-B} = IM_3 - 2L + 20\log N + 9.6 \text{ dBmp} \tag{10.14}$$

10.5.5 Total Noise and Its Allocation

In summary there are three noise components to be considered:

- Thermal noise
- Second-order IM noise
- Third-order IM noise

Let us consider them two at a time. If a system is thermal and second-order IM noise limited, minimum noise is achieved, allowing an equal contribution. For the 3-pWp/km case we would assign 1.5 pW to each component.

For thermal and third-order IM-noise-limited systems, twice the contribution is assigned to thermal noise as to third-order IM noise. Again for the 3-pWp case 2 pWp is assigned to thermal noise and 1 pWp to third-order IM noise.

Expressed in decibels with P_t equal to total noise, the following table expresses these relationships in another manner:

System	Noise Assigned to Thermal	to IM
Thermal and second-order limited	$P_t - 3$	$P_t - 3$
Thermal and third-order limited	$P_t - 1.8$	$P_t - 4.8$
Overload limited	P_t	

The parameter L is established such that these apportionments can be met by adjusting repeater spacing and repeater design. As an example in practice, Figure 10.7 shows noise allocation of the North American L4 system (Ref. 5).

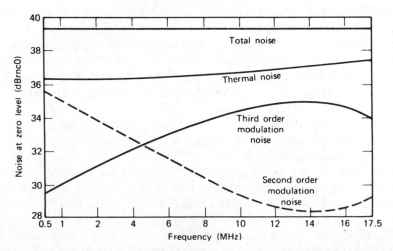

Figure 10.7 Allocation of noise in a practical system. Accumulated noise over 4000 mi of the North American L4 system. (Copyright © 1969 by American Telephone and Telegraph Company.)

10.6 EQUALIZATION

10.6.1 Introduction

Consider the result of transmitting a signal down a 12-MHz coaxial cable system with the amplitude–frequency response shown in Figure 10.3. The noise per voice channel would vary from an extremely low level for the channels assigned to the very lowest frequency segments of the line frequency (baseband) to extremely high levels for those channels that were assigned to the spectrum near 12 MHz. For long systems there would be every reason to believe that these higher frequencies would be unusable if nothing was done to correct the cable to make the amplitude response more uniform as a function of frequency. Ideally, we would wish it to be linear.

Equalization of a cable deals with the means used to assure that the signal-to-noise ratio (S/N) in each FDM telephone channel is essentially the same no matter what its assignment in the spectrum (see Chapter 3). In the following discussion we consider both fixed and adjustable equalizers.

10.6.2 Fixed Equalizers

The coaxial cable transmission system design engineer has three types of cable equalization available that fall into the category of fixed equalizers:

- Basic equalizers
- Line build-out (LBO) networks
- Design deviation equalizers

The basic equalizer is incorporated in every cable repeater. It is designed to compensate for the variation of the loss frequency characteristic of uniform cable sections. This is done by simply making the fixed gain proportional to the square root of frequency, matching the loss for the nominal length. For the North American L4 system the gain characteristic is shown in Figure 10.8. For the case of 12-MHz cable, the section nominal length would be 4.5 km (CCITT Rec. G.332).

The key word in the preceding paragraph is *uniform*. Unfortunately, some cable sections are not uniform in length. It is not economically feasible to build tailor-made repeaters for each nonuniform section. This is what LBO networks are used for. Such devices are another class of fixed equalizer for specific variations of nominal repeater spacing. One way of handling such variations is to have available LBO equalizers for 5, 10, 15%, and so on, of the nominal distance.

The third type of equalizer compensates for design deviations of the nominal characteristics which are standard for dependent repeaters and the actual loss characteristic of the cable system in which the repeaters are to be installed. Such variations are systematic so that the third level of equaliza-

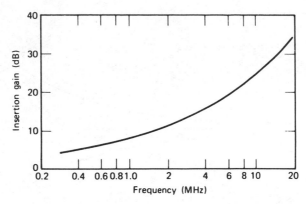

Figure 10.8 Repeater gain characteristic for North American L4 system.

tion, the design deviation equalizer, is installed one for each 10, 15, or 20 repeaters to compensate for gross design deviations over that group of repeaters.

10.6.3 Variable Equalizers

Figure 10.9*a* shows the change of loss of cable as a function of temperature variations, and Figure 10.9*b* shows approximate earth temperature variations with time. Adjustable equalizers are basically concerned with gain frequency variations with time. Besides temperature, variations due to the aging of components may also be a problem. However, this is much less true with solid-state equipment. The cable loss per kilometer shown in Figure 10.3 is the loss at a mean temperature. The $\frac{3}{8}$-in cable used in the L-system application has a mean variation of $\pm 20°$F from the nominal ($+55°$F) buried at a 4-ft depth in the United States. At 20 MHz the loss due to temperature effects is about ± 0.38 dB/mi, and it is about ± 0.67 dB/mi at 60 MHz (Ref. 2). This loss can be estimated at $0.11\%/°$F.

The primary purpose of automatic regulation is to compensate for the gain variation due to temperature changes. Such automatic regulation usually is controlled by a pilot tone at the highest cable frequency. For instance, in the ITT cable design for 12 MHz (Ref. 1),

> The pilot controlled system will always apply exact compensation . . . at the pilot frequency of 12,435 kHz, an error may occur at other frequencies. On a single amplifier this error is very small but will add systematically along the route.

Such an error is usually corrected by manually adjustable equalizers.

In the North American L4 system (Ref. 5) the regulation is controlled by a 11,648-kHz line pilot. The gain frequency characteristic is varied to compensate for temperature-associated changes in the loss of a regulated section.

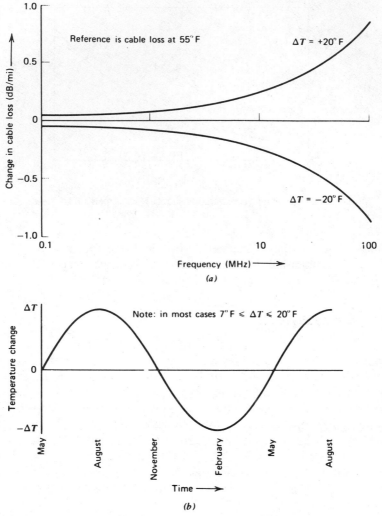

Figure 10.9 (a) Change in loss of 1 mi of $\frac{3}{8}$-in coaxial cable for $\pm 20°$F change in temperature. (b) Approximate earth temperature variations with time at 4-ft depth. (Copyright © 1970 by Bell Telephone Laboratories. Ref. 2.)

Another regulator is controlled by a thermistor buried in the ground near the repeater to monitor ground temperature. This latter regulator provides about half the temperature compensation necessary.

The L4 system also uses additional repeaters called equalizing repeaters, which are spaced up to 54 mi apart. The repeater includes six networks for adjusting the gain frequency characteristics to mop up collective random deviations in the 54-mi section. The equalization of the repeater is done remotely from manned stations while the system is operational.

10.7 LEVEL AND PILOT TONES

Intrasystem levels are fixed by cable and repeater system design. These are the L and M_g established in Section 10.5.

Modern 12-MHz systems display an overload point of $+24$ dBm or more.* Remember that

$$\text{Overload point} = \text{equivalent peak power level}$$
$$+ \text{relative sending level} + \text{margin}$$

The margin is M_g, as in Section 10.5, and L may be related to relative sending level.

M_g can be reduced, depending on how well system regulation is maintained. System pilots, among other functions (covered in Section 10.8), provide a means for automatic gain control (AGC) of some or all cable repeaters so as to compensate (partially or entirely; see Section 10.6) for transmission loss deviations due to temperature effects on the system and aging of active components (e.g., repeaters).

Typically pilot levels are -10 dBm0. The level is a compromise, bearing in mind system loading, to minimize the pilots' contribution to IM noise in a system that is multichannel in the frequency domain (FDM). Another factor tending to force the system designer to increase the pilot level is the signal-to-noise ratio of the pilot tone required to actuate level-regulating circuitry effectively. Pilot level adjustment at the injection point usually requires a settability better than 0.1 dB. Internal pilot stability should display a stability improved over the desired cable system level stability. If the system level stability is to be ± 1 dB, then the internal pilot stability should be better than ± 0.1 dB.

The number of system pilots assigned and their frequencies depend on bandwidth and the specific system design. Commonly, 12,435 kHz is used for regulation and 13.5 MHz for supervisory in the L4 system. In the same system an auxiliary pilot is offered at 308 kHz and, as an option, a frequency-comparison pilot at 300 kHz.

The only continuous in-band pilot in the L4 system is located at 11,648 kHz. Supervisory pilot tones are transmitted in the band of 18.50–18.56 MHz. An L multiplex synchronizing pilot is located at 512 kHz.

10.8 SUPERVISORY

The term *supervisory* in coaxial cable system terminology refers to a method of remotely monitoring the repeater condition at some manned location. In the case of the L4 system, 16 pilot tones are brought up on command, giving the status of 16 separate buried repeaters.

*CCITT Rec. G.223 calls for at least a $+20$-dBm overload point.

Table 10.3 Supervisory system parameters (ITT)

Response oscillator frequency	13.5 MHz
Response oscillator level	−20 dBm0
Response pulse duration	10 μs
Response pulse delay	250 μs
Interrogation pulse repetition rate	6/s
Interrogation pulse amplitude	~ 0.5 V
Interrogation pulse rise time	~ − 10 μs

The ITT method uses a common oscillator frequency (13.5 MHz) and relies on time separation to establish the identity of each repeater being monitored. An interrogation signal is injected at the terminal repeater or other manned station. At the first dependent repeater the signal is filtered off and, after a delay, regenerated and passed on to the next repeater. Simultaneously, on the receipt of the regenerated pulse, a switch is closed, connecting a local oscillator signal to the output of the repeater for a short time interval. This local oscillator pulse is transmitted back to the terminal or other manned station. The delay added at each repeater is added to the natural delay of the intervening cable. This added delay allows for a longer return pulse from each repeater, thereby simplifying circuitry. This same interrogating pulse, delayed, regenerated, and then passed on to the next repeater, carries on down the line of dependent repeaters, causing returning "tone bursts" originating from successive amplifiers along the cable route.

The tone burst response pulses are rectified and fed to a counter at the manned station. The resulting count is compared with the expected count, and an alarm is indicated if there is a discrepancy. The faulty amplifier is identified automatically. Table 10.3 gives basic operating parameters of the system. Such a system can be used for coaxial cable system segments up to 280 km in length.

10.9 POWERING THE SYSTEM

Power feeding of buried repeaters in the ITT system permits the operation of 15 dependent repeaters from each end of a feed point (12-MHz cable). Thus up to 30 dependent repeaters can be supplied power between power feed points. A power feed unit at the power feed point (see Figure 10.4) provides up to 650-V dc voltage between center conductor and ground, using 110-mA stabilized dc. Power feed points may be as far apart as 140 km (87 mi) on large-diameter cable.

10.10 60-MHz COAXIAL CABLE SYSTEMS

Wide-band coaxial cable systems have been implemented to satisfy the ever-increasing demand for long-haul toll-quality telephone channels. Such

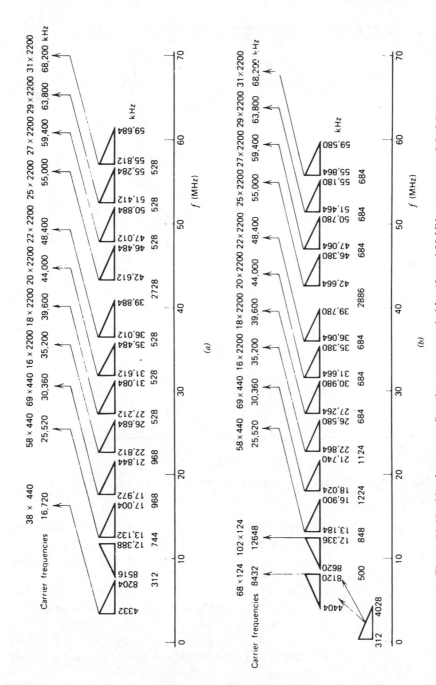

Figure 10.10 Line frequency allocation recommended for 40- and 60-MHz systems on 2.6/9.5-mm coaxial cable pairs. (a) Using plan 1. (b) Using plan 2. (From CCITT Rec. G.333; courtesy of ITU–CCITT.)

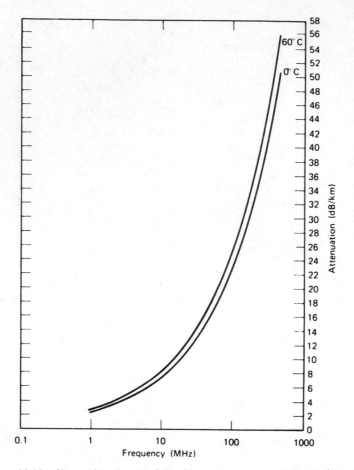

Figure 10.11   Attenuation characteristic of large-diameter coaxial cable (0.375 in).

systems are designed to carry 10,800* FDM nominal 4-kHz VF channels. The line frequency configuration for such a system, as recommended in CCITT Rec. G.333, is shown in Figure 10.10. To meet long-haul noise objectives the large-diameter cable is recommended (e.g., 2.6/9.5 mm).

When expanding a coaxial cable system, a desirable objective is to use the same repeater locations as with the old cable and add additional repeaters at intervening locations. For instance, if we have 4.5-km spacing for a 12-MHz system and our design shows that we need three times the number of repeaters for an equal-length 60-MHz system, then repeater spacing should be at 1.5-km intervals.

The ITT 12-MHz system uses 4.65-km spacing. Therefore its 60-MHz system will use 1.55-km (0.95-mi) spacing with a mean cable temperature of

---

*The AT&T L5E system is designed for 13,200 VF channels.

10°C. The attenuation characteristic of the large-gauge cable is shown in Figure 10.11. This is an extension of Figure 10.3.

Repeater gain for the ITT system is nominally 28.5 dB at 60 MHz and can be varied ±1.5 dB. LBO networks allow still greater tolerance. The overload point, following CCITT Rec. 223, is taken at +20 dBm with a transmit level of −18 dBm.

The system pilot frequency is 61.160 MHz for regulation. A second pilot frequency of 4.287 MHz corrects the level of the lower frequency range. Pilot regulation repeaters are installed at from 7 to 10 nonregulated repeaters, with deviation equalization at every 24th repeater. All repeaters have temperature control (controlled by the buried ambient).

Power feeding is planned at every 100 km (63 mi). Hence 64 repeaters will be fed remotely using constant dc feed over the conductors. Each repeater will tap off about 15 V, requiring 2 W. Thirty-two repeaters at 2 × 15 V each will require 960 V. An additional 120-V dc is required for pilot-regulated repeaters plus one repeater with deviation equalization. Added to this is the 50-V IR drop on the cable. The total feed voltage adds to 1226-V dc. Fault location is similar to that for the 12-MHz ITT system. (For the ITT 60-MHz system, see Ref. 14.)

10.11   THE L5 COAXIAL CABLE TRANSMISSION SYSTEM

A good example of a 60-MHz coaxial cable transmission system that is presently operational, carrying traffic, is the L5 system operating on a transcontinental route in North America (see Table 10.1). In its present lay-up it consists of 22 tubes, of which 20 are on line and 2 are spare. Each tube has the capacity to transmit 10,800 VF channels* in one direction. For full-duplex operation, two tubes are required for 10,800 VF channels, or the total system capacity is $[(22 − 2)/2] × 10,800$ or 108,000 VF channels.

The system is designed for a 40-dBrnC0 (8000-pWp) noise objective in the worst VF channel at the end of a 4000-mi (6400-km) system. This is a noise accumulation of 8000/6400 or 1.25 pWp/km. The system design is such that it is second-order modulation and thermal noise limited. Repeater overload is at the +24-dBm point.

Compared to the L4 system, L5 provides three times the voice channel capacity at three times the L4 spectrum, and twice the repeaters with a small deterioration in noise accumulation.

The modulation plan is an extension of that shown in Chapter 3, Figure 3.22. The basis of the plan is the development of the "jumbo group" (JG), made up of six master groups (Bell System FDM hierarchy). Keep in mind that the basic mastergroup consists of 600 voice channels (in this case) or 10 standard supergroups and occupies the band of 564–3084 kHz. The basic

*13,200 VF channels in the case of L5E.

jumbo group occupies the band of 564–17,548 kHz with a level control pilot at 5888 kHz. The three jumbo groups are assigned the following line frequencies:

JG 1   3,124–20,108 kHz

JG 2   22,068–39,056 kHz

JG 3   43,572–60,556 kHz

Equalizing pilots are at 2976, 20,992, and 66,048 kHz as transmitted to the line. There is a temperature pilot at 42,880 kHz.

The basic jumbo group frequency generator is built around an oscillator that has an output of 5.12 MHz. This oscillator has a drift rate of less than 1 part in 10^{10} per day after aging and a short-term stability of better than 1 part in 10^{8} per millisecond. Excessive frequency offset is indicated by an alarm.

Automatic protection of the 10 operating systems is afforded by the line protection switching system (LPSS) on a 1:10 basis. The maximum length of switching span is 150 mi. Power feeds are at 150-mi intervals, feeding power in both directions. Thus a power span is 75 mi long, or 75 repeaters. Power is 910 mA on each cable, ± 1150 V operating against ground.

The basic repeater is a fixed-gain amplifier, spaced at 1-mi intervals. Typically, every fifth repeater is a regulating repeater, and this regulation is primarily for temperature compensation.

10.12   COAXIAL CABLE OR RADIOLINK — THE DECISION

10.12.1   General

One major decision that the transmission system engineer often faces is whether to install on a particular point-to-point circuit a radiolink or coaxial cable.

What are the factors that will determine the choice? Most obviously, they fall into two categories, technical and economic. Table 10.4 compares the two media from a technical viewpoint. These comparisons can serve as a fundamental guide for making a technical recommendation in the selection of a facility. System mixes may also be of interest. (Refer to Chapter 4 for a discussion of radiolink engineering.)

The discussion that follows is an expansion of some of the points covered in Section 10.1. Table 10.4 summarizes the factors listed below.

10.12.2   Land Acquisition as a Limitation to Coaxial Cable Systems

Acquisition of land detracts more from the attractiveness of the use of coaxial cable than any other consideration; it adds equally to the attractive-

Table 10.4   Comparison of coaxial cable versus radio link

Item	Cable	Radio
Land acquisition	Requires land easements or right-of-way along entire route and recurring maintenance access later	Repeater site acquisition every 30–50 km with building, tower, access road at each site
Insert and drop	Insert and drop at any repeater. Should be kept to minimum. Land buys, building required at each insert location	Insert and drop at more widely spaced repeater sites
Fading	None aside from temperature variations	Important engineering parameter
Noise accumulation	Less, 1 pWp/km	More, 3 pWp/km
Radio frequency interference (RFI)	None	A major consideration
Limitation on number of carriers or basebands transmitted	None	Strict, band-limited plus RFI ambient limitations
Repeater spacing	1.5, 4.5, 9 km	30–50 km
Comparative cost of repeaters	Considerably lower	Considerably higher
Power considerations	High voltage dc in milli-ampere range	48 V dc static no-break at each site in ampere range
Cost versus traffic load	Full load proves more economical than radiolink	Less load proves more economical than cable
Multiplex	FDM-CCITT	FDM-CCITT
Maintenance and engineering	Lower level, lower cost	Higher level, higher cost
Terrain	Important consideration in cable laying	Can jump over, even take advantage of difficult terrain

ness of selecting radiolinks (LOS microwave). With a radiolink system large land areas are jumped and the system engineer is not concerned with what goes on in between. One danger that many engineers tend to overlook is that of the chance building of a structure in the path of the radio beam after installation on the routes has been completed.

Cable, on the other hand, must physically traverse the land area that intervenes. Access is necessary after the cable is laid, particularly at repeater locations. This may not be as difficult as it first appears. One method is to follow parallel to public highways, keeping the cable lay on public land. Otherwise, with a good public relations campaign, easement or rights-of-way often are not hard to get.

This leads to another point. The radiolink relay sites are fenced. Cable lays are marked, but the chances of damage by the farmer's plow or construction activity are fairly high with the cable alternative.

10.12.3 Fading

Radiolinks are susceptible to fading. Fades of 40 dB on long hops are not unknown. Overbuilding a radiolink system tends to keep the effect of fades on system noise within specified limits.

On coaxial cable systems signal level variation is mainly a function of temperature variation. Level variations are well maintained by regulators controlled by pilot tones and, in some cases, auxiliary regulators controlled by ground ambient.

10.12.4 Noise Accumulation

Noise accumulation has been discussed in Section 10.5. Either system will serve for long-haul backbone routes and meet the minimum specific noise criteria established by CCITT/CCIR. However, in practice the engineering and installation of a radio system may require more thought and care to meet those noise requirements. Modern coaxial cable systems have a design target of 1 pWp/km of noise accumulation. If good design techniques are applied and IF repeaters used, radio systems can meet the 3-pWp criterion. Besides fading, radiolinks, by definition, have more modulation steps and consequently are noisier.

10.12.5 Group Delay — Attenuation Distortion

Group delay is less of a problem with radiolinks. Figure 10.3 shows the amplitude response of a cable section before amplitude equalization. The cable plus amplifiers as well as amplitude equalizers add to the group delay problem.

It should be noted that for video transmission on cable an additional modulation step is required to translate the video to the higher frequencies and invert the band (see Section 13.8). While on radiolinks, video can be transmitted directly without additional translation or inversion besides RF modulation.

10.12.6 Radio Frequency Interference (RFI)

There is no question as to the attractiveness of broadband buried coaxial cable systems over equivalent radiolink systems when the area to be traversed by the transmission medium is one of dense RFI. Usually, these areas are built-up metropolitan areas with high industrial/commercial activity. Unfortunately, as a transmission route enters a dense RFI area, land values increase disproportionately, as do construction costs for cable laying. Yet the trade-off is there.

10.12.7   Maximum VF Channel Capacity

In heavily populated areas of highly developed nations frequency assignments are becoming severely limited or unavailable. Although some of the burden on assignment will be removed as the tendency toward usage of the millimeter region of the spectrum is increased, coaxial cable remains the most attractive of the two for high-density FDM configurations.

If it is assumed that there are no RFI of frequency assignment problems, a radiolink can accommodate up to eight carriers in each direction (CCIR Rec. 384-1) with 2700 VF channels per carrier. Thus the maximum capacity of such a system is $8 \times 2700 = 21,600$ VF channels.

Assume a 12-MHz coaxial cable system with 22 tubes, 20 operative, that is, 10 "go" and 10 "return." Each coaxial tube has a capacity of 2700 channels. Therefore the maximum capacity is $10 \times 2700 = 27,000$ VF channels.

It should be noted that the radio system with a full 2700 VF channels may suffer from some multipath problems. Coaxial cable systems have no similar degradation problems. However, cable impedance must be controlled carefully when cable sections are spliced. Such splices usually are carried out at repeater locations.

Consider now 60-MHz cable systems with 20 active tubes, 10 "go" and 10 "return." Assume a 10,000-channel capacity per tube; hence $10,000 \times 10 = 100,000$ VF channels, or the equivalent to five full radiolink systems.

10.12.8   Repeater Spacing

As discussed in Chapter 4, a high average for radiolink repeater spacing is 50 km (30 mi), depending on drop and insert requirements as well as an economic trade-off between tower height and hop distance. For coaxial cable systems, repeater separation depends on the highest frequency to be transmitted, ranging from 9.0 km for 4-MHz systems to 1.5 km for 60-MHz systems. A radiolink repeater is much more complex than a cable repeater.

Coaxial cable repeaters are much cheaper than radiolink repeaters, considering tower, land, and access roads for radiolinks. However, much of this advantage for coaxial cables is offset because radiolinks require many fewer repeaters. It also should be kept in mind that a radiolink system is more adaptable to difficult terrain.

10.12.9   Power Considerations

The 12-MHz ITT coaxial cable system can have power feed points separated by as much as 140 km (87 mi) using 650-V dc at 150 mA. In a 140-km section of a radiolink route at least four power feed points would be required, one at each repeater site. About 2 A is required for each transmitter–receiver combination using standard 48-V dc battery, usually with static no-break power. Power also will be required for tower lights and perhaps for climatizing equipment enclosures.

10.12.10 Engineering and Maintenance

Cable systems are of "cookbook" design. Radiolink systems require a greater engineering effort prior to and during installation. Likewise, the level of maintenance of radio systems is higher than that for cable.

10.12.11 Multiplex Modulation Plans

Interworking or tandem working of radio and coaxial cable systems is made easier because both broadband media use the same standard CCITT or L-system modulation plans (see Chapter 3).

REVIEW EXERCISES

1. Describe the electrical/physical construction of a coaxial cable.

2. The text implies coaxial cable is a medium for the transmission of analog signals. What is the voice channel capacity of one cable on the largest system described?

3. What is replacing/supplementing coaxial cable as the transmission medium of choice for terrestrial transmission?

4. Where and under what conditions would we use coaxial cable (or fiber optic cable) rather than line-of-sight (LOS) microwave radiolinks?

5. What are two major deterrents/disadvantages of using coaxial cable and fiber optic cable for transmission?

6. What is the characteristic impedance of coaxial cable commonly used for long-distance transmission (with air dielectric and supported by disks)?

7. How would one characterize attenuation–frequency characteristic of coaxial cable (i.e., what type of mathematical expression?)?

8. How do dependent repeaters derive their power for operation?

9. Name at least three functions of a terminal repeater.

10. What is the repeater spacing for the highest FDM capacity coaxial cable system?

11. Why are pilot tones used on coaxial cable systems?

12. Coaxial cable system design involves seven basic functions/parameters. Name at least six.

13. For analog FDM systems on coaxial cable, what is the noise design objective for the transmission medium (and its amplifiers) on a per-kilometer basis?

14. What is the range of gain in dB most likely to be encountered on long-haul coaxial cable system amplifiers?

15. What two types of noise are encountered on coaxial cable systems?

16. If the noise figure of a coaxial cable amplifier were 20 dB, repeater gain 25 dB, and there were 20 identical repeaters in tandem, what would the noise be in a 3-kHz voice channel at the output of the 20th repeater? What would the noise be if we doubled the length of the system at the output of the 40th repeater? The second part should be answered without resorting to calculation.

17. How does cable diameter affect loss at a particular frequency?

18. We said that by increasing system operating level, we could reduce the thermal noise contribution to noise in the derived voice channel. Comment on the effect on IM noise.

19. How do third-order products vary with a 1-dB change in level of the fundamental (frequencies)? How do second harmonics vary?

20. For budgetary purposes, how much noise (percentage or fraction) is allocated to thermal noise and how much to IM noise?

21. What is the primary purpose of the *basic* equalizer?

22. In the design of an equalizer, how is gain related to frequency?

23. Basic equalizers are designed for a *uniform* length of coaxial cable. How is nonuniform length compensated for?

24. How does one compensate for nonuniform design (not length)?

25. What is the range of overload points in dBm? (Hint: AT&T versus CCITT.)

26. Why is system level so important on long-haul coaxial cable systems carrying an FDM configuration?

27. What is the meaning of *supervisory* in the context of coaxial cable systems?

28. Describe one method of finding the location of a faulty distant dependent repeater.

29. How is repeater gain specified?

30. Discuss at least eight factors to be considered when selecting coaxial cable or LOS microwave radiolink as the transmission medium given a certain set of circumstances.

REFERENCES AND BIBLIOGRAPHY

1. P. J. Howard, M. F. Alarcon, and S. Tronsli, "12-Megahertz Line Equipment," *Electr. Commun.*, vol. 48, no. 1/2, 1973.
2. *Transmission Systems for Communications*, 4th ed., Bell Telephone Laboratories, American Telephone and Telegraph Co., New York, 1971.
3. W. A. Rheinfelder, *CATV System Engineering*, TAB Books, Blue Ridge Summit, PA, 1970.
4. P. Norman and P. J. Howard, "Coaxial Cable System for 2700 Circuits," *Electr. Commun.*, vol. 42, no. 4, 1967.
5. "The L-4 Coaxial System," *Bell Syst. Tech. J.*, vol. 48, Apr. 1969.
6. CCITT Orange Books, Geneva, 1976, vol. III, G recommendations, particularly the G.200 and G.300 series (see Appendix A).
7. J. A. Lawlor, *Coaxial Cable Communication Systems*, *Management Overview*, Tech. Memorandum, ITT, New York, Feb. 1972.
8. *Lenkurt Demodulator*, Lenkurt Electric Corp., San Carlos, CA, June 1967, May and June 1970, and May 1971.
9. *Data Handbook for Radio Engineers*, 6th ed., Howard W. Sams, Indianapolis, IN, 1977.
10. F. J. Herr, "The L5 Coaxial System Transmission System Analysis," *IEEE Trans. Commun.*, Feb. 1974.
11. F. C. Kelcourse and T. A. Tarbox, "Design of Repeatered Lines for Long-Haul Coaxial Systems," *IEEE Trans. Commun.*, Feb. 1974.
12. E. H. Angell and M. M. Luniewicz, "Low Noise Ultralinear Line Repeaters for the L5 Coaxial System," *IEEE Trans. Commun.*, Feb. 1974.
13. Y.-S. Cho et al., "Static and Dynamic Equalization of the L5 Repeatered Line," *IEEE Trans. Commun.*, Feb. 1974.
14. L. Becker, "60-Megahertz Line Equipment," *Electr. Commun.*, vol. 48, no. 1/2, 1973.
15. *Telecommunications Transmission Engineering*, 2nd ed., vol. 2, American Telephone and Telegraph Co., New York, 1977.
16. Roger L. Freeman, *Telecommunication System Engineering*, 2nd ed., Wiley, New York, 1989.

11

FIBER OPTIC
COMMUNICATION LINKS

11.1 OVERVIEW

Fiber optic transmission has grown from a nascent technology to the transmission medium of choice for trunk traffic over a 15-year period. In our technology of transmission the commodity is bandwidth. Compared to other media, fiber optics can be considered to have an infinite bandwidth. Refs. 4 and 14 describe bandwidths in excess of 1 THz. The useful band for radio is 100 GHz (i.e., 3 kHz–100 GHz); for coaxial cable, about 500 MHz; and wire pair, some 10 MHz. Considering these values of competing media, fiber does indeed have an infinite bandwidth.

Fiber lends itself particularly well to digital transmission, which we will emphasize in this chapter. For instance, coaxial cable transmission requires far more repeaters than its glass fiber cable counterpart with a ratio of 10 or 20 to 1 or better. British Telecom has successfully transmitted 20 Gbps 100 km without repeaters (Ref. 20). Operational trunks are carrying 560 Mbps, and some trunks operate with 1.7- and 2.4-Gbps transmission rates.

A simplified functional block diagram of a typical fiber optic communication link is shown in Figure 11.1. The optical source (transmitter) may be a light emitting diode (LED) or injection laser diode (ILD) coupling light to a fiber strand consisting of a glass core covered with a cladding. Two or more such strands, are bundled into a cable, each with its own light source. Each strand is connected at the far end to an optical detector or receiver. Such detectors may be a PIN diode or avalanche photodiode (APD).

Fiber optic links may be as short as several feet or provide intercontinental connectivity. Fiber optic cable is a transmission medium used on LANs (local area networks), MANs (metropolitan area networks), and WANs (wide area

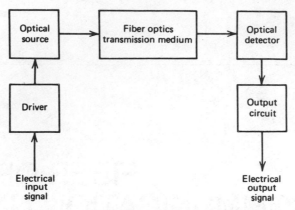

Figure 11.1   Typical fiber optic communication link.

networks). The cable television industry uses fiber on trunk circuits bringing multichannel TV to distribution points.

Fiber optic cable is much smaller and lighter than its metallic counterparts. Rather than a pair or tube, it uses a strand of glass, slightly larger than a human hair. Fiber is immune to electromagnetic interference (EMI) and the light signal is impossible to detect unless the glass strand is penetrated. Thus it is secure. It does not require equalization. When compared to coaxial cable, it has a flat response over the small band of interest.

In all previous chapters of this book we have used frequency (Hz) to describe where a radio frequency emission is located in the electromagnetic spectrum or to describe bandwidth. When working with light transmission, such as fiber optics, we use wavelength (m). The reason for this, it is said, is that fiber and other types of light transmission were developed by physicists.

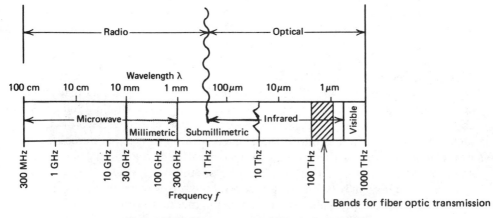

Figure 11.2   Frequency spectrum above 300 MHz.

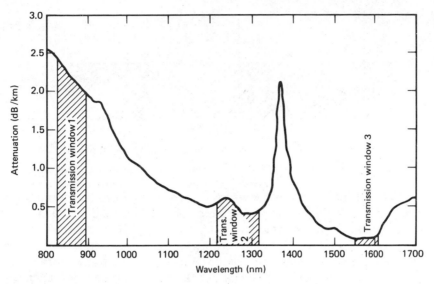

Figure 11.3 Optical fiber attenuation versus wavelength.

The region in which fiber optic transmission systems operate is in the infrared band, which is shown in Figure 11.2.

Figure 11.3 shows the loss per kilometer achievable across the near-infrared from 800- to 1700-nm (nanometer) band and three fiber optic transmission windows. The comparatively low loss per kilometer in the wavelength bands of these windows is the reason why we choose these regions of the spectrum to operate fiber optic links. The windows are in the following wavelength bands:

810–850 nm   Nominal wavelength: 820 nm
1220–1340 nm Nominal wavelength: 1330 nm
1540–1610 nm Nominal wavelength: 1550 nm

It will be noted in Figure 11.3 that these windows occur at absorption minimums. The derived wavelength units we use are nanometers (nm), or 10^{-9} m, and micrometers or microns (μm), 10^{-6} m.

11.2 INTRODUCTION TO OPTICAL FIBER AS A TRANSMISSION MEDIUM

The practical propagation of light through an optical fiber might best be explained using ray theory and Snell's law. Simply stated, we can say that when light passes from a medium of higher refractive index into a medium of

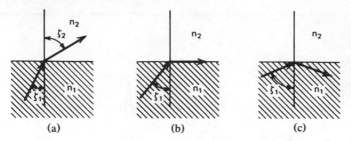

Figure 11.4 Ray paths for several angles of incidence, $n_1 > n_2$.

lesser refractive index, the refracted ray is bent away from the normal. For instance, a ray traveling in water and passing into an air region is bent away from the normal to the interface between the two regions. As the angle of incidence becomes more oblique, the refracted ray is bent more until finally the refracted ray emerges at an angle of 90° with respect to the normal and just grazes along the surface. Figure 11.4 shows various incidence angles. Figure 11.4b illustrates what is called the critical angle, where the refracted ray just grazes along the surface. Figure 11.4c is an example of total reflection. This is when the angle of incidence exceeds the critical angle. A glass fiber, when utilized as a medium for the transmission of light, requires total internal reflection.

For effective transmission of light a glass fiber is made up of a glass fiber core that is covered with a jacket called *cladding*. If we let the core have a refractive index of n_1 and the cladding a refractive index of n_2, the structure (core with cladding) will act as a light waveguide when $n_1 > n_2$.

Another property of the fiber for a given wavelength λ is the normalized frequency V, then

$$V = \frac{2\pi a}{\lambda} \sqrt{n_1^2 - n_2^2} \approx n_1 \sqrt{2\Delta} \qquad (11.1)$$

where a = the core radius, n_2 for unclad fiber = 1; and $\Delta = (n_1 - n_2)/n_1$.

The term $\sqrt{n_1^2 - n_2^2}$ in equation 11.1 is called the numerical aperture (NA). In essence the numerical aperture is used to describe the light-gathering ability of a fiber. In fact, the amount of optical power accepted by a fiber varies as the square of its numerical aperture. It is also interesting to note that the numerical aperture is independent of any physical dimension of the fiber.

As shown in Figure 11.1, there are three basic elements in an optical fiber transmission system: the source, the fiber link, and the optical detector. Regarding the fiber link, there are two basic design parameters that limit the length of a link without repeaters, or limit the distance between repeaters. These most important parameters are loss, usually expressed in dB/km, and dispersion, which is often expressed as an equivalent bandwidth-distance product in MHz/km. A link may be power limited (loss limited) or it may be dispersion limited.

Dispersion, manifesting itself with intersymbol interference at the far end, is brought about by two factors. One is material dispersion and the other is modal dispersion. Material dispersion is caused by the fact that the refractive index of the material changes with frequency. If the fiber waveguide supports several modes, we have a modal dispersion. Since the different modes have different phase and group velocities, energy in the respective modes arrives at the detector at different times. Consider that most optical sources excite many modes, and if these modes propagate down the fiber waveguide, delay distortion (dispersion) will result. The degree of distortion depends on the amount of energy in the various modes at the detector input.

One way of limiting the number of propagating modes in the fiber is in the design and construction of the fiber waveguide itself. Return again to equation 11.1. The modes propagated can be limited by increasing the radius a and keeping the ratio n_1/n_2 as small as practical, often 1.01 or less.

We can approximate the number of modes N that a fiber can support by applying formula 11.1. If $V = 2.405$, only one mode will propagate (HE_{11}). If V is greater than 2.405, more than one mode will propagate, and when a reasonably large number of modes propagate,

$$N = (\tfrac{1}{2})V^2 \qquad (11.2)$$

Dispersion is discussed in more detail later in the chapter.

11.3 TYPES OF FIBER

There are three categories of fiber as distinguished by their modal and physical properties:

- Single mode (monomode)
- Step index (multimode)
- Graded index (multimode)

Single-mode fiber is designed such that only one mode is propagated. To do this, $V < 2.405$. Such a fiber exhibits little modal dispersion. Typically, we might encounter a fiber with indices of refraction of $n_1 = 1.48$ and $n_2 = 1.46$. If the optical source wavelength is 820 nm, for single-mode operation the maximum core diameter would be 2.6 μm, a very small diameter indeed.

Step-index fiber is characterized by an abrupt change in refractive index, and graded-index fiber is characterized by a continuous and smooth change in refractive index (i.e., from n_1 to n_2). Figure 11.5 shows the fiber construc-

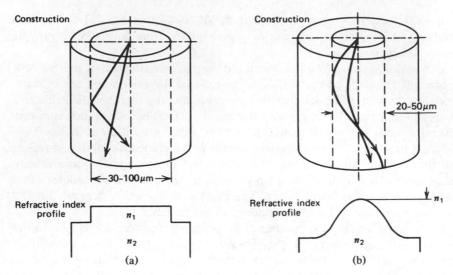

Figure 11.5 Construction and refractive index properties for (a) step-index fiber and (b) graded-index fiber.

tion and refractive index profile for step-index fiber (Figure 11.5a) and graded-index fiber (Figure 11.5b).

Step-index multimode fiber is more economical than graded-index fiber. For step-index fiber the multimode bandwidth distance product, the measure of dispersion discussed above, is on the order of 10–100 MHz/km. With repeater spacings on the order of 10 km, only a few megahertz of bandwidth is possible.

Graded-index fiber is more expensive than step-index fiber, but it is one alternative for improved distance–bandwidth products. When a laser diode source is used, values of from 400 to 1000 MHz/km are possible. If an LED source is used with its much broader emission spectrum, distance–bandwidth products with graded-index fiber can be achieved up to 300 MHz/km or better. Material dispersion in this case is what principally limits the usable bandwidth.

There are two additional criteria for optical fiber that are important in system design. These are minimum bending radius and fiber strength.

Radiation losses at fiber waveguide bends are usually quite small and may be neglected in system design unless the bending radius is smaller than that specified by the manufacturer. Minimum bending radii vary from about 2 to 10 cm, depending on the cable characteristics, or, as a rule of thumb, around 10 times the cable diameter.

Fiber cable strength is also specified by the manufacturer. For example, one manufacturer for a specific cable type specifies a maximum pulling tension of 1780 newtons (N) (400 lb) at 20°C, a maximum permissible

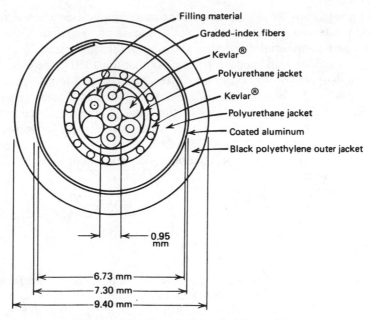

Filling material
Graded-index fibers
Kevlar®
Polyurethane jacket
Kevlar®
Polyurethane jacket
Coated aluminum
Black polyethylene outer jacket

0.95 mm

6.73 mm
7.30 mm
9.40 mm

Figure 11.6   Direct-burial optical fiber cable.

compression load of 655-N/cm (375-lb/in.) flat plate, and a maximum permissible impact force of 280 N · cm (160 lb/in.).

Figure 11.6 shows a typical five-fiber cable for direct burial.

11.3.1   Notes on Monomode Transmission

Ninety percent of the fiber produced as of this writing is single mode or monomode (Ref. 14). The same reference states that single-mode fiber with an injection laser diode (ILD) source is on the order of 5% more expensive than multimode fiber with a light emitting diode (LED) source.

Using 1550-nm wavelength, the scattering limit (attenuation) is 0.15 dB/km. The theoretical product is 1 THz/km or more but limited by the spectral width of the light source and second-order material and waveguide dispersion effects (Ref. 4).

Commonly, single-mode fiber is of step-index design. There is a dispersion null near 1300 nm, but minimum fiber attenuation is experienced in the 1550-nm window. It is therefore convenient in these fibers to shift the wavelength minimum dispersion region to 1550 nm. This is accomplished by adjusting the refractive index profile. Zero dispersion occurs when waveguide dispersion cancels material dispersion (Ref. 4). For our discussion, we neglect dispersion as an impairment on single-mode fiber.

It is on very high bit rate systems using monomode fiber where chromatic dispersion becomes an issue. Chromatic dispersion consists of two terms discussed above, material dispersion and waveguide dispersion.

Monomode fiber employed for transmission in the 1330/1550-nm windows has a core diameter on the order of 10 μm, and the cladding thickness is 10 times this value. So monomode (single-mode) fiber has a total diameter on the order of 110–120 μm. Because of this small diameter, we can expect a numerical aperture on the order of 0.11, whereas the larger monomode fiber has numerical apertures between 0.20 and 0.3.

Because of its high rate of production, monomode fiber is more economic than multimode fiber.

11.4 SPLICES AND CONNECTORS

Optical fiber cable is commonly available in 1-km sections. There are two methods of connecting these sections in tandem and connecting the fiber to the source at one end and to the optical detector at the other. These are by splicing or using special connectors. The objective in either case is to transfer as much light as possible through the coupling. Good splices generally couple more light than connectors.

A good splice can have an insertion loss as low as 0.2 dB, whereas connectors, depending on the type and on how well they are installed, can have insertion losses as low as 0.3 dB and some as high as 1 dB or more.

An optical fiber splice requires highly accurate alignment and an excellent end finish to the fibers. There are two causes of loss at a splice:

- Lateral displacement of fiber axes
- Angular misalignment

A good fusion splice has an average insertion loss of 0.09 dB with a standard deviation of 0.12 dB. Ninety-four percent of all splices have insertion losses below 0.3 dB per splice (Ref. 4). With experienced technicians and a single-mode fusion splicer, 10 to 12 min of time per splice should be allocated.

Mechanical splicing involves installation of connectors at the cable factory. Use of connectorized cable can save time in field installation and the expense of a fusion splice and special technician training. Insertion losses of such connector-type splices, such as silicon array chip splicing, for a single-mode ribbon cable shows a mean insertion loss of 0.38 dB with a standard deviation of 0.31 dB.

11.5 LIGHT SOURCES

A light source, perhaps more properly called a photon source, has the fundamental function in a fiber optic communication system to convert electrical energy (current) into optical energy (light) efficiently, in a manner that permits the light output to be launched into the optical fiber effectively. The light signal so generated must also accurately track the input electrical signal so that noise and distortion are minimized.

The two most widely used light sources for fiber optic communication systems are the etched-well surface LED and the ILD. LEDs and ILDs are fabricated from the same basic semiconductor compounds and have similar heterojunction structures. They do differ considerably in their performance characteristics. LEDs are less efficient than ILDs, but are cheaper. The spatial intensity distribution of an LED is Lambertian (cosine), whereas a laser diode exhibits a relatively high degree of waveguiding and hence, for a given acceptance angle, can couple more power into a fiber than the LED. In other words, the LED has a comparatively broad output spectrum and the ILD has a narrow spectrum, on the order of 1 or 2 nm wide.

With present technology the LED is capable of launching about 100 μW of optical power into the core of a fiber with a numerical aperture of 0.2 or greater. A laser diode with the same input power can couple up to 7 mW into the same cable. However, ILDs usually are operated with outputs of 1 or 2 mW (0 or +3 dBm) at the longer wavelengths. The coupling efficiency of an LED is on the order of 2%, whereas the coupling efficiency of an ILD is better than 50%, with one manufacturer reportedly achieving about 70%.

Methods of coupling a source into an optical fiber vary as do coupling efficiencies. To avoid ambiguous specifications on source output powers, such powers should be stated as out of the "pigtail." A pigtail is a short piece of optical fiber coupled to the source at the factory and, as such, is an integral part of the source. Of course, the pigtail should be the same type of fiber as that specified for the link.

Component lifetimes for LEDs are on the order of 200,000 h, with up to a million hours reported in the literature. Many manufacturers guarantee an ILD for 50,000 h, and up to 100,000 h of lifetime. The life expectancy of an ILD is reduced when it is overdriven to derive more coupled output power (i.e., more than 5-7 mW).

The ILD is a temperature-dependent device. Its threshold current increases nonlinearly with temperature. Rather than attempt to control the device's temperature, a negative feedback circuit is used whereby a portion of the emitted light is sampled, detected, and fed back to control the drive current. Such circuits are similar to the familiar automatic gain control (AGC) circuits used in radio receivers.

Laser diodes used in earlier 1330-nm light wave systems emitted light in several wavelength modes. These modes are called by some *side lobes*.

Remember that the 1330-nm window is the nominal dispersion null wavelength. Using lasers of this type in the 1500-nm low-loss window would result in indistinguishable pulses at the receiver due to chromatic dispersion. Reducing the bandwidth (spectral width) of the emitted light in effect will reduce chromatic dispersion.

One approach to reduce bandwidth is to use a distributed feedback (DFB) laser which suppresses the side lobes, making the laser output very nearly single mode. A pulse of optical power from a single-mode DFB laser has limited pulse broadening at 1550 nm. However, even though the side lobes are suppressed, the DFB laser is not monochromatic. The single central mode of a modulated DFB laser has a spectral width that is usually less than 0.1 nm caused by laser chirp. Modulation of injection current of a DFB laser produces variations in both intensity and wavelength at its output. It is the wavelength variation that is called *laser chirp*. Laser chirp results in pulse broadening at the receiver, and the link length will be limited at some bit rate–distance product (similar to bandwidth distance product), such as 500 MHz/km. DFB lasers can have typically less than 0.1-nm central mode spectral width and side lobes down 30 dB. With these spectral characteristics, transmission penalties caused by a 15–20-ps/km × nm dispersion coefficient at 1550 nm should be negligible until the bit rate–distance product is well above about 60 Gbps/km (Ref. 19).

11.6 LIGHT DETECTORS

The most commonly used detectors (receivers) for fiber optic communication systems are photodiodes, either PIN or avalanche photodiode (APD). The terminology *PIN* derives from the semiconductor construction of the device where an intrinsic (I) material is used between the p–n junction of the diode.

A photodiode can be considered a photon counter. The photon energy E is a function of frequency and is given by

$$E = h\nu \tag{11.3}$$

where h = Planck's constant (W/s^2) and ν-frequency (Hz). E is measured in watt-seconds or kilowatt-hours.

The receiver power in the optical domain can be measured by counting, in quantum steps, the number of photons received by a detector per second. The power in watts may be derived by multiplying this count by the photon energy, as given in equation 11.3.

The efficiency of the optical-to-electrical power conversion is defined by a photodiode's *quantum efficiency* η, which is a measure of average number of electrons released by each incident photon. A highly efficient photodiode would have a quantum efficiency near 1, and decreasing from 1 indicates progressively poorer efficiencies. The quantum efficiency, in general, varies with wavelength and temperature.

For the fiber optic communication system engineer, *responsivity* is a most important parameter when dealing with photodiode detectors. Responsivity is expressed in amperes per watt or volts per watt and is sometimes called sensitivity. Responsivity is the ratio of the rms value of the output current or voltage of a photodetector to the rms value of the incident optical power.

In other words, responsivity is a measure of the amount of electrical power we can expect at the output of a photodiode, given a certain incident light power signal input. For a photodiode the responsivity R is related to the wavelength λ of the light flux and to the quantum efficiency η, the fraction of the incident photons that produce a hole–electron pair. Thus

$$R = \frac{\eta\lambda}{1234}(A/W) \qquad (11.4)$$

with λ measured in nanometers.

Responsivity can also be related to electron charge Q by the following:

$$R = \frac{\eta Q}{h\nu} \qquad (11.5)$$

where $h\nu$ = photon energy (equation 11.3) and Q = electron charge, 1.6×10^{-19} coulombs (C).

Figure 11.7 plots typical responsivities for four photodetectors. The two upper curves are for semiconductor photodiodes shown with material codes

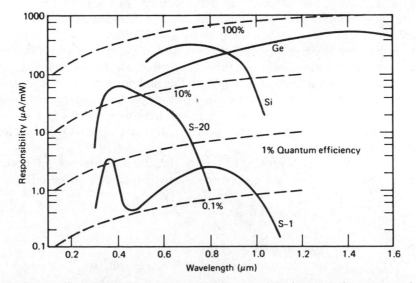

Figure 11.7 Typical responsivities plotted for four photodetectors. (From Ref. 11.)

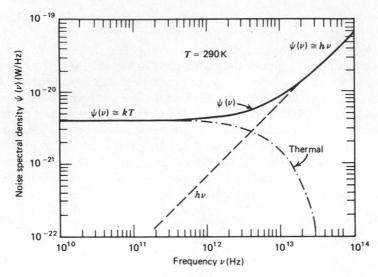

Figure 11.8   Noise in the frequency domain.

Ge and Si. Curves S-1 and S-20 are for the two photodiode materials AgOCs and Na_2KsbCs, respectively. The curves are plotted with quantum efficiency η as a parameter. The dashed lines are for comparative purposes, where η is assumed to be constant with wavelength.

As in all transmission systems, noise is a most important consideration. Just as in the other systems treated in this text, the noise analysis of a fiber optic system is centered on the receiver. We must treat noise for these types of systems in both the optical and the electrical domain. In the optical domain, as shown in Figure 11.8, quantum noise is dominant. Quantum noise manifests itself as shot noise on the primary photocurrent from the detector. There is also shot noise on the dark current, and in the case of an APD there is excess noise from avalanche multiplication. Because the optical detector converts light energy into electrical energy, thermal noise must be treated on the electrical side of the detector. In most cases this will be the noise generated in the preamplifier that follows the light detector. In fact in PIN detectors, because their gain is very nearly 1, thermal noise is the principal contributor.

The drawing below (Figure 11.9) is a simplified model of an optical receiver. The optical detector is the two intertwining circles at the left.

$$\text{Thermal noise} = \frac{4kT_{\text{eff}}B}{R_{\text{eq}}} = i_{\text{NA}}^2 \qquad (11.6)$$

where R_{eq} = equivalent resistance of driver amplifier
T_{eff} = effective noise temperature of load resistor (K)
B = bandwidth (Hz)
k = Boltzmann's constant

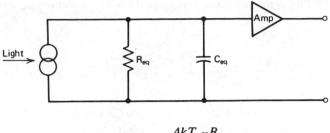

$$\text{Thermal noise} = \frac{4kT_{\text{eff}}B}{R_{\text{eq}}} = i_{\text{NA}}^2$$

Figure 11.9 An optical receiver model.

Noise is related, in this case, to the mean-square value of the current of the load resistor i_{NA}^2.

Shot noise can also be related to the mean-square value of current and consists of two parts, one from the fluctuations in the signal, i_{NS}^2,

$$i_{\text{NS}}^2 = 2q\left(2P_{\text{opt}}\frac{\eta q}{h\nu}\right)B \tag{11.7}$$

where P_{opt} = optical power
  q = electron charge (1.6×10^{-19} C)
  η = quantum efficiency
  $h\nu$ = photon energy (equation 11.3)
  B = system bandwidth

and the second from dark current in the detector, i_{ND}^2,

$$i_{\text{ND}}^2 = 2qi_{\text{D}}B \tag{11.8}$$

where i_{D} = photodetector dark current,

$$\text{Signal-to-noise ratio } \frac{S}{N} = \frac{\text{signal power}}{\text{shot noise power + amplifier noise power}} \tag{11.9}$$

Signal power is a function of the mean-square value of the detector photocurrent, i_{s}^2,

$$i_{\text{s}}^2 = \frac{1}{2}\left(2P_{\text{opt}}\frac{\eta q}{h\nu}\right)^2 \tag{11.10}$$

Noise power, the two terms in the denominator in equation 11.9, was discussed above. The thermal noise power in this case is the preamplifier noise power. The signal-to-noise ratio equation 11.9 can now be written (Ref. 12):

$$\frac{S}{N} = \frac{2\left[P_{opt}(\eta q/h\nu)\right]^2}{\left[2qi_D + 4qP_{opt}(\eta q/h\nu) + 4kT_{eff}/R_L\right]B} \quad (11.11)$$

where R_L = load resistance and B has been factored out in the denominator. A bit error rate (BER) of 1×10^{-9}, a standard for fiber optic systems, requires a signal-to-noise ratio of 21.5 dB. The signal-to-noise ratio of the optical power incident on the detector is the square root of this figure, or 10.75 dB. For an APD the signal-to-noise ratio is given by (Ref. 12)

$$\frac{S}{N} = \frac{2\left[P_{opt}(\eta q/h\nu)\right]^2 M^2}{\left[\left[2qi_D + 4qP_{opt}(\eta q/h\nu)\right]M^2F(M) + 4kT_{eff}/R_L\right]B} \quad (11.12)$$

where M = avalanche gain of photocurrent and $F(M)$ = excess noise introduced by avalanche gain.

An APD has what is called optimum gain M_{opt}. The gain M of an APD cannot be increased indefinitely; there is a point where, as M is increased, the signal-to-noise ratio begins to degrade, and we have passed the point of M_{opt}. Ref. 13 describes a group of 53 APDs produced by Bell Telephone Laboratories with a calculated theoretical gain of 140 (equivalent to a power gain of about 21.5 dB), whereas the practical M_{opt} for the group, as measured, turned out to be around 80 (19 dB). Other sources specify no more than 15 dB of practical gain M_{opt} for an APD, and another, only 7 dB (Ref. 19).

In the noise analysis presented above we have seen that signal-to-noise ratios are a function of bandwidth B, which, in turn, is a function of bit rate (see Section 11.8). Table 11.1 summarizes detector thresholds in dBm for the standard BER of 1×10^{-9} for bit rates in common use for pulse code modulation (PCM) transmission in the telephone industry (Ref. 15). Noise equivalent power (NEP) is commonly used as the figure of merit of a photodiode. NEP is defined as the rms value of optical power required to

Table 11.1 Approximate receiver threshold levels[a] (dBm)

Receiver Type	Bit Rate (Mbps)							
	45	90	130	417	565	850	1700	4000
PIN receiver	−47.5	−45	−40	−34	−32.5	−30	−23	
APD receiver	−54	−62	−47	−41	−39.5	−37	−33	−28.5

Source: Ref. 19.

[a]BER = 1×10^{-9}.

produce a unity signal-to-noise ratio (i.e., $S/N = 1$) at the output of a light-detecting device. NEPs vary for specific diode detectors between 1×10^{-13} and 1×10^{-14} W/Hz$^{1/2}$.

Of the two types of photodiodes discussed here, the PIN is cheaper and requires less complex circuitry than its APD counterpart. The PIN diode has peak responsivity from about 800 to 900 nm for silicon devices. These responsivities range from about 300 to 600 μA/mW. The overall response time for the PIN is good for about 90% of the transient, but sluggish for the remaining 10%, which is a "tail." The poorer response of the tail portion of a pulse may limit the net bit rate on digital systems.

The PIN detector does not display gain, whereas the APD does. The response time of the APD is far better than that of the PIN, but the APD displays certain temperature instabilities where responsivity can change significantly with temperature. Compensation for temperature is usually required in APD detectors and is often accomplished by a feedback control of bias voltage. It should be noted that bias voltages for APDs are much higher than for PIN diodes, and some APDs require bias voltages as high 200 V or more. Both the temperature problem and the high-voltage bias supply complicate repeater design.

11.7 MODULATION AND WAVEFORM

The most widely used type of modulation of a light carrier is a form of amplitude modulation (AM) called *intensity modulation*. Both types of light sources, the LED and the ILD, can be conveniently modulated in the intensity mode. The detectors (PIN and APD) discussed above each respond directly to intensity modulation, producing a photocurrent proportional to the incident light intensity. Today's optical fiber communication systems are more suitable to digital than to analog operation.

Considering the previous chapters, let us review some advantages and disadvantages of optical fiber communication systems, particularly for digital operation:

- We are dealing with much higher frequencies
- Coherence bandwidths are much greater
- Component response times are inherently faster than anything discussed previously
- The linearity of optical fiber systems is comparatively poor

As mentioned above, both the LED and the ILD light sources are semiconductor devices that, in most cases, are directly modulated. Biasing of the source and the adjustment of the quiescent operational point are most important considerations. In view of this, the following guidelines should be

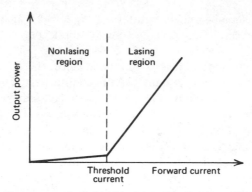

Figure 11.10   Power-current relationship for an injection laser diode (ILD) source.

followed:

1. The intensity of the driving source should vary directly with the bias current either in the lasing region (for the ILD, see Figure 11.10) or in the spontaneous emission region (for the LED).
2. In the case of continuous analog modulation the quiescent bias current must be established at a point such that the modulating signal causes an equal plus and minus swing about the quiescent value. It should also be in the most linear range of the intensity characteristic.
3. For digital modulation of the several variants of PAM the quiescent bias current is adjusted, given a specific source, as follows. In the case of an LED, it should be either near zero or at a quiescent point optimum for noise and/or response time. For the ILD case it should be adjusted at a point near or below threshold (Figure 11.10) providing that the transition noise is tolerable; slightly above threshold if transition noise must be reduced; or near zero if transition response time is adequate for the design modulation rate and if threshold transition noise is tolerable.
4. For continuous AM of an ILD, consideration must be given to the laser's life when operating in this mode.

An important digital system design consideration is the pulse format. For the discussion that follows, only binary digital systems will be covered (see Chapters 3 and 12). By definition, then, information such as on-line data, PCM, or delta modulation (DM) is transmitted serially as 1's and 0's. The manner (format) in which the 1's and 0's are presented to the modulator is important for a number of reasons.

First, amplifiers for fiber optic receivers are usually ac coupled. As a result, each light pulse that impinges on the detector produces a linear electrical output response with a low-amplitude negative tail of comparatively long duration. At high bit rates, tails from a sequence of pulses may tend to accumulate, giving rise to a condition known as baseline wander, and such tails cause intersymbol interference. If the number of "on" pulses and "off" pulses can be kept fairly balanced for periods that are short compared to the tail length, the effect of ac coupling is then merely to introduce a constant offset in the linear output of the receiver, which can be compensated for by adjusting the threshold of the regenerator. A line-signaling format can be selected that will provide such a balance. The selection is also important on synchronous systems for self-clocking at the receiver.

Figure 11.11 illustrates five commonly used binary formats. Each is briefly discussed below.

1. *NRZ (Nonreturn to Zero).* This signal format is discussed in Chapter 12, where by convention a 1 represents the active state and a 0 the passive state. A change of state only occurs when there is a 1-to-0 or 0-to-1 transition. A string of 1's is a continuous pulse or "on" condition, and a string of 0's is a continuous "off" condition. In NRZ information is extracted from transitions or lack of transitions in a synchronous format, and a single pulse completely occupies the designated bit interval.

2. *RZ (Return to Zero).* In this case there is a transition for every bit transmitted, whether a 1 or a 0, as shown in Figure 11.11, and, as a result, a pulse width is less than the bit interval to permit the return-to-zero condition.

3. *Bipolar NRZ.* This is similar to NRZ except that binary 1's alternate in polarity.

4. *Bipolar RZ.* The same as bipolar NRZ, but in this case there is a return-to-zero condition for each signal element, and again the pulse width is always less than the bit interval.

5. *Manchester Code.* This code format is commonly used in digital fiber optic systems. Here the binary information is carried in the transition which occurs at midpulse. By convention a logic 0 is defined as a positive-going transition and a logic 1 as a negative-going transition. The signal can be either unipolar or bipolar.

The choice of code format is important in fiber optic communication system design, and there are a number of trade-offs to be considered. For instance, RZ formats assist in reducing baseline wander. To extract timing on synchronous systems the Manchester code and the RZ bipolar codes are good candidates because of their self-clocking capability. However, it will be appreciated from Figure 11.11 that they will require at least twice the bandwidth as the NRZ unipolar code format. An advantage of the Manchester code is that it can be unipolar, which adapts well to direct intensity

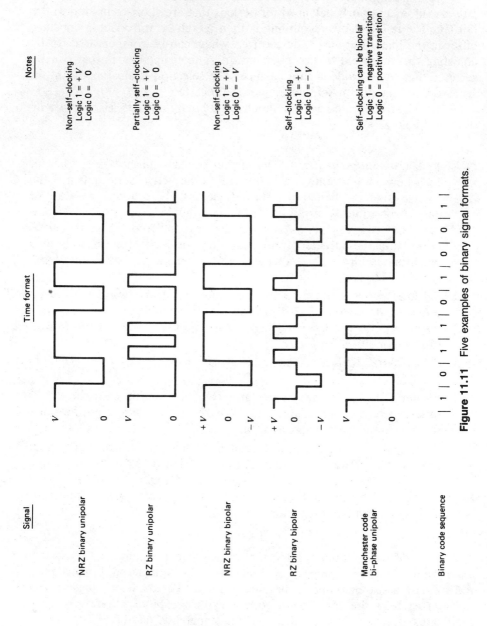

Figure 11.11 Five examples of binary signal formats.

modulation of LED or ILD sources and also provides at least one transition per unit interval (i.e., per bit) for self-clocking.

With the NRZ format we can attain the highest power per information bit if we wish to tolerate baseline wander. Achieving this power is especially desirable with LED sources. On the other hand, ILD sources can be driven to high power levels for short intervals, hence conserving the life of the laser diode, making the RZ format attractive. Longer life with the shorter duty cycle can be traded off for higher modulation rate, which will result in greater bandwidth requirements of the system. As we have seen, RZ systems require twice the bandwidth than NRZ systems for a given bit rate.

11.8 SYSTEM DESIGN

11.8.1 General Application

The design of a fiber optic communication system involves several steps. Certainly, the first consideration is to determine the feasibility of such a transmission system for a desired application. There are two aspects to this decision, economic and technical. Analog applications are for wide-band transmission of such information as video, particularly for CATV* trunks, studio-to-transmitter links, and multichannel FDM mastergroups. Throughout this chapter, however, we stress digital application for fiber optic transmission and in particular for

- On-premises data bus
- Higher level PCM or DM multiplex configurations for telephone trunks
- Radar data links
- Conventional data links well in excess of 9.6 kbps
- Digital video

The present trend is for the cost of fiber cable and components to go down, even with the impact of inflation. Fiber optic repeaters are more costly than their metallic PCM counterparts, but less repeaters are required for a given distance, and the powering of these repeaters is more involved, especially if the power is to be taken off the cable itself. In this case a metallic pair (or pairs) must be included in the cable sheath if the power is to be provided from trunk terminal points. Another approach is to supply power locally at the repeater site with a floating battery backup. Consider the following

---

*The cable television industry is beginning to implement fiber optic links on trunks connecting the head-end to distribution points. In some cases modulation is digital, in others FM with FM subcarriers.

features that make fiber optic transmission systems attractive to the telecommunication system design:

1. *Low transmission loss* as compared to wire pairs or coaxial cable for broadband transmission (i.e., in excess of several megahertz [or Mbps]), allowing a much greater distance between repeaters.
2. *Wide bandwidth* with present systems in the 1550-nm region using monomode transmission with bit rate distance products up to 60 Gbps/km. Future systems may see bit rate distance products up to 1000 Gbps/km.
3. *Small Bending Radius*. With proper cable design, a bending radius on the order of a few centimeters can have negligible effect on transmission.
4. *Nonradiative, Noninductive, Nonconductive, Low Crosstalk*. Complete isolation from nearby electrical systems, whether telecommunication or power, avoiding such problems as groundloops, radiative or inductive interference, and reducing lightning-induced interference.
5. *Lighter weight, smaller diameter* than its metallic equivalents.
6. *Growth Potential*. Low-bit-rate systems can easily be expanded in the future on fiber to much higher bit-rate systems using the same cable. This is being done today with coaxial cable by increasing the number of repeaters on older cable.

11.8.2   Design Procedure

The first step in designing a fiber optic communication system is to establish the basic input system parameters. Among these we would wish to know

- Signal to be transmitted
- Link length
- Growth requirements (i.e., additional circuits, increased bit rates)
- Tolerable signal impairment levels stated as signal-to-noise ratio or BER at the output of the terminal detector

Throughout the design procedure, when working with trade-offs, the designer must establish whether he is working in a power-limited domain or in a bandwidth- (bit-rate) limited* domain. For instance, with low bit rates such as the T1 carrier (1.544 Mbps), we would expect to be power limited in almost all circumstances. Once we get into higher bit rates, say above 45 Mbps using multimode fiber, the proposed system must be tested (by calculation) to determine if the system rise time is not bit-rate limited. This test will

---

*Perhaps more properly called *dispersion-limited*.

be described in detail. The designer then selects the most economic alternatives and trade-offs among the following:

- Fiber parameters: single mode or multimode, step index or graded index, number of fibers and cable makeup
- Source type: LED or ILD
- Wavelength (1330 or 1550 nm preferred)
- Detector type: PIN or APD
- Repeaters, if required, and their powering; use of erbium amplifiers
- Modulation type and waveform (code format)

If the system is power limited, splicing rather than fiber connectors should be considered because splicing loss is smaller than connector loss. In general we will find that for systems in excess of 2 km, the fiber cable becomes the cost driver, whereas with shorter links the components (i.e., source and detector) are the cost drivers. As an example, in nearly every case of on-premises data bus systems, low-cost components and cable can be selected such as LED sources, step-index rather high-attenuation fiber, and PIN detectors.

As in the design of any telecommunication system or subsystem, the bottom line is cost over the life of the system. The question then arises, What is the most economic fiber optic transmission subsystem that will meet requirements over the system life? For example, if the system is power limited, it would appear that fusion splices would be preferred because of their lower insertion loss when compared to connectors. We can argue for factory-installed connectors because the insertion loss difference is so small and we would have the savings of labor and splicing equipment.

We should establish which are the system cost drivers: the fiber and its laying or the electronic components (light sources and detectors) and their installation. Fiber cable cost has been dropping continuously. Present fiber cable consisting of from 2 to 200 fibers costs from $200 to $3000 per kilometer and an ILD/detector combination is about $6000 a pair. Of course, for short distances, the electronic components would consist of the more economic LED sources and PIN detectors. We may find the break point at about 30 km, where the cost of electronics and cost of fiber are equal. With long sections between repeaters using high-capacity fiber, in excess of 400 Mbps per strand, the cost of fiber and its laying may far exceed its accompanying electronics. We must also consider the powering of repeaters. Another consideration is availability. This might entail a one-for-one standby fiber and electronics or a one-for-n with appropriate switching. Route diversity may be another requirement to improve availability, and this can be a real cost multiplier. An example is power companies that depend so much on their telecommunication links to keep the power flowing that availability is a major issue. An alternative on very short links is to use plastic fiber as a cost-effective compromise. Selecting a reasonable link margin is another cost trade-off.

When systems are extended over several kilometers, much more care has to go into system design to optimize cost versus system performance. For instance, the available power into a fiber from an LED runs about 10–15 dB below the available power from an injection laser. Furthermore, for a high-bit-rate link (i.e., becoming bandwidth limited), the narrower spectral width of the laser is necessary to avoid material dispersion and improve system rise time. The more expensive APDs have 15–20 dB gain over the less expensive PIN diode detectors.

11.8.3   General Approach

11.8.3.1   Introduction   A fiber optic transmission link can be power limited or dispersion limited. In most practical cases with present technology, using monomode step-index fiber, links will be power limited. Dispersion limitations should be examined when transmitting over 500 Mbps on long monomode links. The type of dispersion will be chromatic dispersion, not modal dispersion. If we employ multimode fiber using laser diode sources, modal dispersion will predominate.

11.8.3.2   Power-Limited Operation   Figure 11.12 is a model of a fiber optics link. This may be a terminal-to-terminal repeaterless link or a terminal-to-first-repeater link. As shown in the figure, the link consists of a light source coupled to a pigtail fiber with a connector on the end, and fiber sections with connectors or splices and a light detector at the distant end, connecting to the fiber link via a connector/pigtail combination.

Conventionally, the output power of the optical source is specified at the end of the pigtail and the threshold of the detector at a BER of 1×10^{-9}, referenced at the detector pigtail.

Suppose the power output of an ILD is $+3$ dBm and the threshold of the detector is -41 dBm. The link power degradation from source to detector is

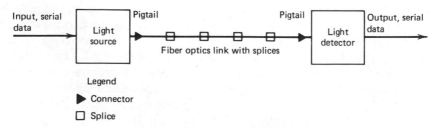

Figure 11.12   Fiber optics link model.

44 dB and we have to budget for this value. That is,

$$\text{Budget}_{dB} = \text{output}_{dBm} - \text{threshold}_{dBm}$$
$$= +3 \text{ dBm} - (-41 \text{ dBm})$$
$$= 44 \text{ dB} \tag{11.13}$$

The 44 dB is budgeted to the following:

- Connector insertion losses, at the source and at the detector
- Fiber loss
- Intervening connector or splice losses
- Extinction ratio penalty at light detector (in dB)
- Margin

If we measured the output power of an ILD with a photometer, the reading would be average power. Average power must be expressed in terms of the ILD duty cycle and we relate these in the two equations that follow:

$$P_{avg} = \frac{P_{max} + P_{min}}{M} \tag{11.14}$$

$$r_{ex} = \frac{P_{min}}{P_{max}} \tag{11.15}$$

where P_{avg} = average power
P_{max} = maximum or peak power
P_{min} = minimum power
r_{ex} = extinction ratio
M = 2 for an NRZ waveform and 4 for an RZ waveform

We must understand that an ILD is biased so that there is a P_{min} with a no-signal condition and P_{max} is the peak power during the signal condition.

From the value of r_{ex} we derive an extinction ratio penalty in decibels for the system based on using a PIN detector from (Ref. 4):

r_{ex}	Penalty (dB)
0.5	3
0.4	2.2
0.3	1.7
0.2	1.0
0.1	0.5
0.07	0.3
0.05	0.2
0.02	0.1

Example. How long a link can we have without repeaters if we have the 44 dB to budget? We assume that the fiber cable comes in 1-km reels, 1550-nm operation, and we use fusion splicing:

Connector losses at both ends	1 dB
Margin	9.5 dB
Extinction ratio penalty	0.5 dB
Subtotal	11.0 dB

We now have 33 dB remaining to allot to cable and splices. Let's assume 0.3 dB/km for monomode cable and 0.3 dB per splice, or a 0.6 dB/km total. The maximum length is 33/0.6, or approximately 55 km.

The following notes are added. The extinction ratio penalties for APDs are somewhat greater than for PIN detectors. Ref. 4 advises that, in any case, extinction ratios greater than 0.1 (power penalty, 0.5 dB) should be avoided except under special circumstances (Ref. 4).

11.8.4 Dispersion-Limited Domain and System Bandwidth

From the system rise time we can calculate the system bandwidth and the maximum data rate that the system can support. System rise time is a function of the root sum squares of the component rise times, namely,

- Source S
- Fiber due to multimode dispersion F_{mm}
- Fiber due to material dispersion F_{md}
- Detector d

For high-bit rate systems the rise time is measured in nanoseconds:

$$\text{System rise time (ns)} = 1.1\sqrt{S^2 + F_{mm}^2 + F_{md}^2 + d^2} \qquad (11.16)$$

$$\text{3-dB bandwidth (GHz)} = \frac{0.35}{\text{system rise time (ns)}} \qquad (11.17)$$

Material dispersion can be neglected when laser diode sources are used because of the laser's narrow spectral emission characteristic. For LED sources, if no other data is available, 5.5 ns/km can be used for material dispersion.

Fiber rise times are stated in nanoseconds per kilometer and therefore must be extended for the total cable length. For instance, with an 8-km system and a fiber multimode dispersion rise time of 7 ns/km, the value to be used in the system rise time calculation is 56 ns (i.e., 8 km × 7 ns/km). This linear extrapolation may be used for conservative system design. Others (Ref.

15) suggest an rms law, and thus bandwidth (BW) is calculated:

$$BW = \left[\left(\frac{1}{BW_1} \right)^2 + \left(\frac{1}{BW_2} \right)^2 + \cdots + \left(\frac{1}{BW_N} \right)^2 \right]^{-M} \qquad (11.18)$$

where M, for most cases, is between 0.5 and 0.6.

Suppose that we wished to verify whether we were in the dispersion-limited domain or just where we entered that domain for a system transmitting 45 Mbps of NRZ data. What would be the system rise time limits and the required cable bandwidth per kilometer properties required?

First calculate the pulse width at 45 Mbps. This is $1/45 \times 10^6$, or 22.22 ns. The maximum tolerable rise time for NRZ data is 70% of this figure, or 15.55 ns. For RZ data it is half this value, or 7.77 ns. These are values of dispersion on a system basis that we must not exceed.

The problem is now to find the maximum-length system acceptable with these figures in mind, or the point where we enter the dispersion-limited domain. Of course, to optimize system length with this high bit rate we would wish to use a laser diode source and an APD detector. With typical values used, calculate the system rise time with the rise time of the cable as the unknown X:

	Rise Time	Rise Time^2
Source (ILD)	1.5 ns	2.25 ns^2
Cable, graded index	X	X^2
Detector (APD)	2 ns	4 ns^2

$$15.55 \text{ ns} = 1.1\sqrt{X^2 + 2.25 + 4.0}$$
$$X^2 = 193.75$$
$$X = 13.9 \text{ ns}$$

Thus we can allow 13.9-ns rise time for the fiber cable portion of the system. This total quantity can be allotted to intermodal dispersion because we use an ILD source rather than an LED. Cable, such as the ITT T-211, displays 2-ns/km rise time. Using this cable to determine the maximum system length without repeaters (or between repeaters) working in the dispersion-limited domain, we divide 13.9 ns by 2 ns/km, and the length limit is 6.95 km using the conservative linear extrapolation method.

Allowing that a fiber's step function response resembles that of a low-pass filter, its electrical bandwidth can be estimated from its rise time, or

$$\text{Bandwidth } (-3 \text{ dB})(\text{MHz}) = \frac{350}{\text{rise time (ns)}}$$

In this case, with 13.9 ns for fiber rise time we have

$$\frac{350}{13.9} \text{ ns} = 25.2 \text{ MHz}$$

at the end of 6.95 km of fiber. When measured, we would probably find the bandwidth greater than this value and using the root mean square would give values in excess of the measured values.

11.9   SYNCHRONOUS OPTICAL NETWORK (SONET)

SONET is a digital interface standard for optical transmission, developed by ANSI (American National Standards Institute) T1X1 committee. It defines a basic signal of 51.480 Mbps and a byte interleaved multiplexing scheme. This results in a family of digital rates and formats defined at a rate of N times 51.840 Mbps where N is an integer. At present, N has a maximum value of 255. However, present SONET optical transmission systems support only certain values of N, which are 1, 3, 9, 12, 18, 24, 36, and 48. Table 11.2 lists these rates as optical carrier rates or OC rates (e.g., OC-1, OC-3, etc.). Ref. 17 states that values greater than 48 will be addressed in the future. 51.840 Mbps is the *synchronous transport signal* level 1, or STS-1, and is the basic modular signal. The optical counterpart of STS-1 is OC-1.

The basic frame structure of STS-1 is shown in Figure 11.13. It consists of 90 columns and 9 rows of 8-bit bytes for a total of 810 bytes (6480 bits). With a frame length of 125 ms (i.e., 8000 frames per second), the resulting STS-1 bit rate is 51.840 Mbps. The order of transmission is row by row, left to right. In each byte the most significant bit is sent first.

Table 11.2   Line rates for the allowable OC-N signals

OC Level	Line Rate (Mb/s)
OC-1	51.840
OC-3	155.520
OC-9	466.560
OC-12	622.080
OC-18	933.120
OC-24	1244.160
OC-36	1866.240
OC-48	2488.320

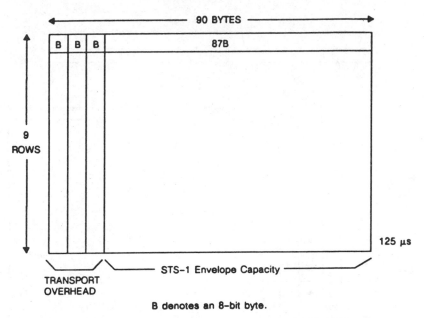

Figure 11.13   STS-1 frame. B denotes an 8-bit byte.

As shown in Figure 11.13, the first three columns are the transport overhead, which contains overhead bytes of section and line layers. Twenty-seven bytes have been assigned, with 9 bytes for section overhead and 18 bytes for line overhead. The remaining 87 columns constitute the STS-1 envelope capacity.

The STS-1 synchronous payload envelope (SPE) consists of 87 columns and 9 rows of bytes, for a total of 783 bytes. Column 1 contains 9 bytes and is designated as STS path overhead (POH). The remaining 774 bytes are available for payload. Figures 11.14–11.16 illustrate these concepts (Ref. 17).

The STS-1 SPE may begin anywhere in the STS envelope capacity. Typically, it begins in one frame and ends in the next, although it may be wholly contained in one frame. The STS-1 payload pointer contained in the transport overhead designates the location of the byte where the STS SPE begins.

The STS POH is associated with each payload and is used to communicate functions from the point where the service is mapped into the STS SPE to where it is delivered (see Figure 11.15).

The STS-N signal is formed by byte interleaving N STS-1 signals (see Figure 11.16). The transport overhead channels of the individual STS-1 signals are frame aligned before interleaving. The associated STS SPEs are not aligned because the STS-1 has a unique payload pointer to indicate the location of the SPE (Ref. 17).

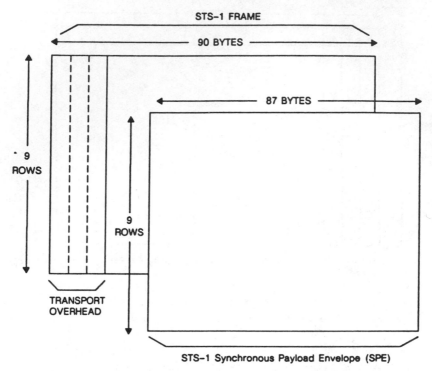

Figure 11.14   Synchronous payload envelope (SPE).

Each SONET STS-1 signal carries a payload pointer in its line overhead. The payload pointer is a key innovation of SONET and is used for multiplexing synchronization in a plesiochronous* environment and also to frame align STS-N signals.

Many digital schemes use fixed location mapping, such as DS1 for example. Fixed location mapping is the use of specific bit positions in a higher rate synchronous signal to carry lower rate synchronous signals. This method permits easy access to transported payloads, since no destuffing is required. Fixed location mapping requires 125-ms buffers to phase-align and slip the tributary signal. Slip means to repeat or delete frames of information to correct timing differences. Such buffers are undesirable because of the signal delay imposed and the error impact of slips (Ref. 18).

The payload pointer is a number carried in each STS-1 line overhead that indicates the starting byte location of the STS-1 SPE payload within the STS-1 frame. Consequently, the payload pointer is not locked to the STS-1

---

*plesiochronous: where all network nodes have high-precision, free-running clocks operating at the same nominal rate. The accuracy and stability of each clock is such that there is almost complete coincidence in timekeeping, and phase drift, in theory, is avoided, eliminating slips or keeping the phase drift in the network to an acceptable low level (Ref. 23).

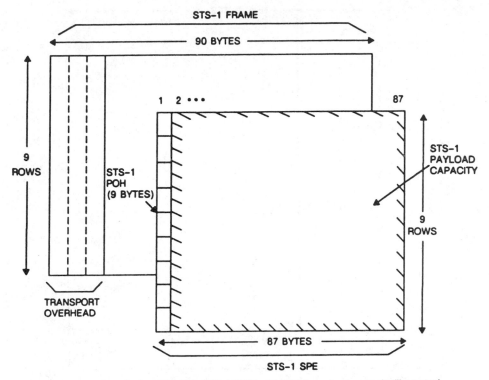

Figure 11.15   STS-1 SPE with STS-1 POH and STS-1 payload capacity illustrated.

frame structure as it is in fixed location mapping, but instead the payload pointer floats with respect to the STS-1 frame. The STS-1 section and line overhead byte positions determine the STS-1 frame structure such that an SPE payload can map across two 125-ms frames.

Any small timing variations of the STS-1 payload can be accommodated by either increasing or decreasing the pointer value. For example, if the STS-1 payload data rate is high with respect to the STS-1 frame rate, the payload pointer is decremented by one and the corresponding overhead byte (i.e., pointer action byte) is used to carry data for one frame. If the reverse is true, that is, if the payload data rate is slow with regard to the STS-1 frame rate, the data byte following the pointer action byte is nulled for one frame and the pointer is incremented by one. Accordingly, slips and the associated data loss are avoided.

Commonly, the STS-1 payload accommodates lower rate tributaries called *virtual tributaries* (VTs). There are four sizes of VTs: VT1.5 (1.728 Mbps), VT2 (2.304 Mbps), VT3 (3.456 Mbps), and VT6 (6.912 Mbps). Each VT has enough bandwidth to carry a DS1, CEPT 30 + 2, DS1C, and DS2 signal, respectively. Each VT occupies several 9-row columns within the SPE. For example, the VT1.5 uses 3 columns; the VT2, 4 columns; the VT3, 6 columns;

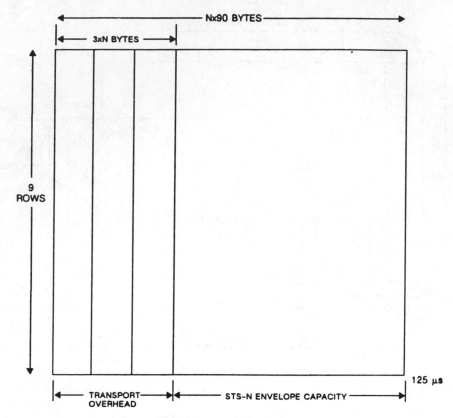

Figure 11.16 STS-*N* frame.

and the VT6, 12 columns. The VTs are delineated by pointers whose operation is similar to the STS-1 payload pointer operation (Ref. 18).

SONET is a four-layer system. (Note: OSI is a seven-layer system; see Chapter 12). The SONET layers are the

- Physical layer
- Section layer
- Line layer
- Path layer

The physical layer is responsible for the transport of bits as optical or electrical pulses across the physical medium. There is no overhead involved with the physical layer.

The section layer deals with the transport of STS-*N* frames across the physical medium. This layer uses the physical layer for transport. The functions of the section layer include framing, scrambling, section error

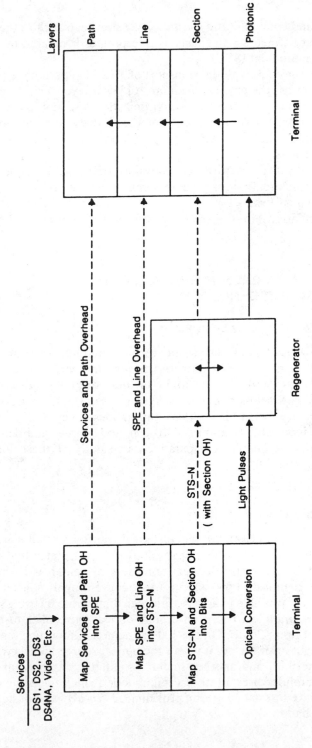

Figure 11.17 Optical interface layers.

monitoring, and communicating section layer overhead such as the orderwire. The overhead defined for this layer is interpreted or created by the section terminating equipment (STE).

The line layer deals with the transport of STS SPE path layer payload and its overhead across the physical medium. All lower layers exist to provide that transport. The line layer provides synchronization and multiplexing functions for the path layer. The overhead associated with these functions includes that for maintenance and protection purposes and is inserted into the line overhead channels.

The path layer deals with the transport of network services between SONET terminal multiplexing equipments. Examples of such services are DS1, DS3, and so forth.

The interrelationship of these layers is shown in Figure 11.17 (Ref. 17).

11.10   FUTURE TRENDS IN OPTICAL FIBER TRANSMISSION TECHNOLOGY

11.10.1   Extremely Low Loss Fiber

Silica- (glass-)based fiber has a theoretical minimum loss of about 0.15 dB/km at 1550 nm. New fiber materials now in the laboratory may have losses lower than 0.001 dB/km. Many of these are chloride-based, such as zinc chloride with a loss minimum at about 4 μm. Other materials holding great promise are based on fluorides and halides. They nearly all have loss minimums (windows) between 3 and 10 μm, and sources and detectors must be developed that operate comparatively efficiently at these longer wavelengths (Refs. 4 and 8).

11.10.2   Fiber Optic Amplifiers

British Telecom (BT) (Ref. 20) has sustained transmission of simulated data at rates of 20 Gbps over a fiber span of 100 km. BT states that rates of 50 Gbps are attainable in the longer term (several years) with this new approach. The demonstration system used conventional noncoherent light transmission. BT was able to achieve this high digital rate by using stand-alone in-line optical amplifiers based on erbium-doped active fiber to produce an optical gain of up to 20 dB. On the 100-km demonstration link, one amplifier was placed right after the point where the optical source launched the optical wave into the fiber, and another similar amplifier was placed at the 50-km point. The amplifier uses 10 m of erbium-doped silica fiber pumped with a continuous wave laser at 1480 μm, providing 12–20-dB gain for light waves in the 1550-nm band.

11.10.3   Coherent Fiber Optic Communication

Conventional direct-detection fiber optic communication systems have not been able to approach the fundamental limits of information capacity due to detector and preamplifier noise limitations. Better advantage of the capacity of optical systems can be taken through heterodyne or homodyne detection of the optical signal. Optical communication systems that use heterodyne or homodyne detection are commonly referred to as *coherent optical communication systems*. It should be noted that all receiving systems described in previous chapters used heterodyne techniques.

Heterodyning or homodyning techniques can provide significant improvement in receiver sensitivity, leading to substantial increases in repeater spacing over conventional optical systems. In addition, heterodyning would allow frequency division multiplexing (FDM) of several hundred or more optical carriers with very narrow separation, in contrast to the present wave division multiplexing (WDM) techniques used on noncoherent systems, which permit the multiplexing of only a few channels. Also, when a narrow spectral line width laser (e.g., < 0.5 nm) is used in an optical fiber system, it is possible to amplitude, phase, or frequency modulate the optical carrier (Refs. 4 and 21).

11.10.4   Mitigation of Chromatic Dispersion—Solitons

As we stated, a wavelength of about 1300 nm is called the *zero dispersion* wavelength. Unfortunately, even at this wavelength there is a spectral broadening of a light pulse as it travels down a pure silica light waveguide (fiber). The broadening is caused by silica's nonlinearity; its index of refraction is higher at a pulse's more intense peak than at its less intense tails. Thus the peak is retarded compared to its tails, which increases the frequency in the trailing half of the pulse and lowers the frequency in the leading half.

There is a transmission mode of a fiber that can be generated, called a soliton, which is essentially free of chromatic dispersion. Solitons cancel out the two properties that individually limit bit rate–distance product. These properties are the nonlinearity of the fiber and chromatic dispersion.

The soliton mode can be generated by slightly increasing the wavelength from its "zero dispersion" value. Now, where the higher frequencies travel faster than the lower ones, causing chromatic dispersion, the fiber's nonlinearity instead retards the leading frequencies and speeds up the trailing edge frequencies—the operation is analogous to an equalizer. The result is essentially dispersion-free transmission.

If we combine the use of the soliton transmission mode with the erbium-doped optical amplifiers described above, > 5-Gbps transmission can be supported for distances up to 10,000 km. Amplifier separation is on the order of 25–50 km (Refs. 4 and 24).

REVIEW EXERCISES

1. What is the basic commodity of transmission (i.e., just what are we selling?)?

2. If we compare optical fiber transmission to wire pair, coaxial cable, or radio transmission, describe the bandwidth available.

3. Name the basic elements we have to deal with in a fiber optics link. (Note: Some are active and some are passive.)

4. What is the basic, overriding advantage of optical fiber transmission over other media? Give at least four more advantages of fiber optics transmission.

5. Give the nominal wavelength band for the three minimum attenuation fiber optics transmission windows.

6. An optical fiber consists of a _____ and a _____ .

7. The notation we use for refractive index is n. If we let the core have a refractive index of n_1 and the cladding, n_2, what relationship of n_1 to n_2 is required for the optical fiber to act as a light waveguide? Can you give some idea of the typical difference between n_1 and n_2?

8. Define numerical aperture (NA) mathematically. What does NA mean in a practical sense?

9. What two very basic parameters can limit the length of a repeaterless fiber optic section (link)? (This question is aimed at understanding terminology.)

10. Dispersion is brought about by what two factors (dispersion types)? What is the result of excess dispersion on a derived digital bit stream?

11. Name the three types of optical fiber given in the text. Which type shows minimum dispersive properties?

12. What is about the best bandwidth product we can expect from multimode optical fiber when employing today's technology?

13. For monomode optical fiber operating at 1550 nm, what is the theoretical maximum bandwidth-distance product?

14. There is a dichotomy when employing silica-based optical fiber for digital transmission in that there is a dispersion null around 1330 nm and a minimum loss window at 1550 nm. What is one way to minimize dispersion at the more desirable 1550-nm window?

15. In general, we neglect dispersion with monomode fiber as a transmission impairment. However, on very high bit rate links, dispersion can corrupt performance. Discuss this type of dispersion and one method to mitigate it.

16. Give the approximate diameter in microns of a monomode optical fiber strand.

17. Compare LED and ILD light sources. Include at least six items in the comparison.

18. Compare hot-fusion splices with factory-installed connectors. The comparison should include at least three items.

19. Name and compare the two types of light detectors given in the text. The comparison should include at least four items.

20. Responsivity is comparable to _____ in radio receivers. What is (are) its unit(s) of measure?

21. What are the benefits of using a distributed feedback (DFB) laser?

22. What is the conventional bit error rate (BER) specified for an optical fiber communication link?

23. Compare RZ and NRZ waveforms in regard to transitions per second. Why would we resort to using RZ or Manchester coding on digital optical fiber links? When reviewing the remainder of the chapter, what might bode against using an RZ waveform?

24. As a first step in the design of an optical fiber link for communication, name at least four basic input parameters.

25. Identify at least six technical alternatives and trade-offs dealing with the most cost-effective design of a digital optical fiber link.

26. Where and why would we resort to route diversity in a fiber optic communication network?

27. An ILD has an output of $+1$ dBm and the far-end detector has a threshold (BER $= 1 \times 10^{-9}$) of -39 dBm. for this particular link we will have ___ dB to budget.

28. Name at least five items to which we must allocate a decibel value with the total decibels budgeted. (See question 27.)

29. What is the basic purpose of SONET?

30. What is the basic modular unit of SONET called and what is its bit rate?

31. Describe the basic SONET frame structure in regard to rows and columns. How many columns are devoted to overhead?

32. Define "envelope capacity" as used in SONET.

33. What is the time duration of an STS-1 frame?

34. What is the advantage of using "pointers"? In the discussion, use such terms as slips, bit stuffing, fixed location mapping, and small time variations.

35. Discuss virtual tributaries (VTs). How are they delineated?

36. Identify the four SONET layers and give the purpose of each.

37. Name at least three probable fiber optic communication advancements that will take place in the foreseeable future. Discuss the merits of each and the present drawbacks in development.

REFERENCES AND BIBLIOGRAPHY

1. S. E. Miller and A. G. Chynoweth, *Optical Fiber Telecommunication*, Academic Press, New York, 1979.
2. *Design Handbook for Optical Fiber Systems*, vols. I–III, Information Gatekeepers, Inv., Brookline, MA, 1979.
3. R. L. Galawa, *Optical Communications via Glass Fibers*, Institute of Telecommunication Sciences, Boulder, CO, 1978.
4. E. E. Basch, ed., *Optical-Fiber Transmission*, Howard W. Sams, Indianapolis, IN, 1987.
5. *ITT Optical Fiber Characteristics and Applications*, ITT Tech. Note R-5, Roanoke, VA, 1979.
6. *Optical Fiber Communications Link Design*, ITT Tech. Note R-1, ITT, Roanoke, VA, 1979.
7. Pierre Halley, *Fibre Optic Systems*, Wiley, New York, 1987.
8. Suzanne R. Nagel, "Optical Fiber—the Expanding Medium," *IEEE Commun. Mag.*, vol. 25, no. 4, Apr. 1987.
9. G. H. B. Yancy, "Fiber Optic Digital Telecommunication Systems," *Electronic Progress*, vol. 22, Raytheon Co., Lexington, MA, Spring 1980.
10. E. Randall and R. Lavelle, "Optimize Optical Modem Cost/Performance through Emitter, Detector and Fiber Selection," *Electron. Des.*, Apr. 12, 1980.
11. R. L. Galawa, Lecture notes on Optical Communication Via Glass Fiber Waveguides, IEEE Course Series, given at Raytheon Co., Equipment Division, Wayland, MA, Apr. 1979.
12. M. J. Howes and D. V. Morgan, eds., *Optical Fibre Communications*, Wiley, New York, 1980.
13. R. G. Smith, C. A. Brackett, and H. W. Reinbold, "Optical Detector Package," *Bell Syst. Tech. J.*, vol. 57, 1809–1822, July–Aug. 1978.
14. *DataPro*, vol. 2, *Fiber Optics*, McGraw-Hill, New York, July 1990.
15. E. E. Basch, H. A. Carnes, and R. F. Kearns, "Calculate Performance into Fiber-Optic Links," *Electron. Des.*, Aug. 16, 1980.
16. H. Melchior et al., "Planar Epitaxial Silicon Avalance Photodiode," *Bell Syst. Tech. J.*, vol. 57, 1791–1809, July–Aug. 1978.
17. *Synchronous Optical Network (SONET) Transport Systems: Common Generic Criteria*, Bellcore Tech. Reference TR-TSY-000253, issue 1, Sept. 1989.
18. Ralph Ballart and Yau-Cou Ching, "SONET: Now Its's the Standard Optical Network," *IEEE Commun. Mag.*, Mar. 1989.

19. Patrick R. Trischitta and Dixon T. S. Chinn, "Repeaterless Lightwave Systems," *IEEE Commun. Mag.*, Mar. 1989.

20. Peter Fletcher, "British Telecom Shows 20-Gbit Non-coherent Link," *Lightwave*, July 1990.

21. Richard A. Linke, "Optical Heterodyne Communication Systems," *IEEE Commun. Mag.*, Oct. 1989.

22. Siemens Aktiengessellschaft, *Optical Communications*, Wiley, New York, 1983.

23. Roger L. Freeman, *Telecommunication System Engineering*, 2nd ed., Wiley, New York, 1989.

24. Trudy E. Bell, "Light that Acts Like 'natural bits,'" *IEEE Spectrum*, Aug. 1990.

12

THE TRANSMISSION
OF DIGITAL DATA

12.1 INTRODUCTION

We must distinguish again between analog and digital transmission. An analog transmission system has an output at the far end which is a continuously variable quantity representative of the input. With analog transmission there is continuity (of waveform); with digital transmission there is discreteness.

The simplest form of digital transmission is binary, where an information element is assigned one of two possibilities. There are many binary situations in real life where only one of two possible values can exist; for example, a light may be either on or off, an engine is running or not, and a person is alive or dead.

An entire number system has been based on two values, which by convention have been assigned the symbols 1 and 0. This is the binary system, and its number base is 2. Our everyday number system has a base of 10 and is called the decimal system. Still another system has a base of 8 and is called the octal system.

The basic information element of the binary system is called the bit, which is an acronym for *b*inary dig*it*. The bit, as we know, may have the values 1 or 0.

A number of discrete bits can identify a larger piece of information, and we may call this larger piece a character. A code is defined by the IEEE as "a plan for representing each of a finite number of values or symbols as a particular arrangement or sequence of discrete conditions or events" (Ref. 1).

Binary coding of written information and its subsequent transmission have been with us for a long time. An example is teleprinter service (i.e., the transmission of a telegram).

The greater number of computers now in use operate in binary languages; thus binary transmission fits in well for computer-to-computer communication and the transmission of data.

This chapter introduces the reader to the transmission of binary information or data over telephone networks. It considers data on an end-to-end basis and the effects of the variability of transmission characteristics on the final data output. There is a review of the basics of the makeup of digital data signals and their application. Therefore the chapter endeavors to cover the entire field of digital data and its transmission over telephone network facilities. It includes a discussion of the nature of digital signals, coding, information theory, constraints of the telephone channel, modulation techniques, and the dc nature of data transmission.

No distinction is made between data and telegraph transmission. Both are binary, and often the codes used for one serve equally for the other. Likewise, the transmission problems of data apply equally to telegraph. Telegraph communication is in message format (but not always), transmitted at rates less than 110 words per minute (wpm), and is asynchronous. Data transmission is most often synchronous (but not necessarily so), and usually is an alphanumeric mix of information, much of the time destined for computers (electronic data processing [EDP]). Data often are transmitted at rates in excess of 110 wpm.

12.2 THE BIT AND BINARY CONVENTION

In a binary transmission system the smallest unit of information is the bit. As we know, either one or two conditions may exist, the 1 or the 0. We call one state a mark, the other a space. These conditions may be indicated electrically by a condition of current flow and no current flow. Unless some rules

Table 12.1 Equivalent binary designations

Binary 1	Binary 0
Mark or marking	Space or spacing
Perforation (paper tape)	No perforation
Negative voltage[a]	Positive voltage
Condition Z	Condition A
Tone on (amplitude modulation)	Tone off
Low frequency (frequency shift keying [FSK])	High frequency
Opposite to the reference phase	Reference phase (phase modulation)
No phase inversion	Inversion of the phase
(Differential two-phase modulation)	

Source: CCITT Rec. V.1 (Ref. 2). Courtesy of ITU–CCITT.

[a]CCITT Recs. V10 and V11 (Refs. 3 and 4, respectively) and EIA RS-232D (Ref. 5).

are established, an ambiguous situation would exist. Is the 1 condition a mark or a space? Does the no-current condition mean that a 0 is transmitted, or a 1? To avoid confusion and to establish a positive identity to binary conditions, CCITT Rec. V.1 recommends equivalent binary designations. These are shown in Table 12.1. If the table is adhered to universally, no confusion will exist as to which is a mark, which is a space, which is the active condition, which is the passive condition, which is 1, and which is 0. It defines the *sense* of transmission so that the mark and space, the 1 and 0, will not be inverted. Data transmission engineers often refer to such a table as a table of *mark-space convention*.

12.3   CODING

12.3.1   Introduction to Binary Coding Techniques

Written information must be coded before it can be transmitted over a digital system. The discussion of coding below covers only binary codes. But before launching into coding itself, the term *entropy* is introduced.

Operational telecommunication systems transmit information. We can say that information has the property of reducing the uncertainty of a situation. The measurement of uncertainty is called entropy. If entropy is large, then a large amount of information is required to clarify a situation; if entropy is small, then only a small amount of information is required for clarification. Noise in a communication channel is a principal cause of uncertainty. From this we now can introduce Shannon's noisy channel coding theorem, stated approximately (Ref. 6, pp. 41–42):

> If an information source has an entropy H and a noisy channel capacity C, then provided $H < C$, the output from the source can be transmitted over the channel and recovered with an arbitrarily small probability of error. If $H > C$, it is not possible to transmit and recover information with an arbitrarily small probability of error.

Entropy is a major consideration in the development of modern codes. Coding can be such as to reduce transmission errors (uncertainties) due to the transmission medium and even correct the errors at the far end. This is done by reducing the entropy per bit (adding redundancy). We shall discuss errors and their detection in greater detail in Section 12.4. Channel capacity is discussed in Section 12.10.

Now the question arises, how big a binary code? The answer involves yet another question, how much information is to be transmitted?

One binary digit (bit) carries little information; it has only two possibilities. If 2 binary digits are transmitted in sequence, there are four possibilities,

00   10
01   11

or four pieces of information. Suppose 3 bits are transmitted in sequence. Now there are eight possibilities:

000 100
001 101
010 110
011 111

Characters				Code Elements[a]						
Letters Case	Communi-cations	Weather	CCITT #2[b]	START	1	2	3	4	5	STOP
A	–	↑		▓	▓	▓				▓
B	?	⊕		▓	▓			▓	▓	▓
C	:	○		▓		▓	▓	▓		▓
D	$	⤢	WRU	▓	▓			▓		▓
E	3	3		▓	▓					▓
F	1	→	Unassigned	▓	▓		▓	▓		▓
G	&	↘	Unassigned	▓		▓		▓	▓	▓
H	STOP[c]	↧	Unassigned	▓			▓		▓	▓
I	8	8		▓		▓	▓			▓
J	'	⤢	Audible signal	▓	▓	▓		▓		▓
K	(	←		▓	▓	▓	▓	▓		▓
L	)	↖		▓		▓			▓	▓
M	.	.		▓			▓	▓	▓	▓
N	,	⊕		▓			▓	▓		▓
O	9	9		▓				▓	▓	▓
P	0	0		▓		▓	▓		▓	▓
Q	1	1		▓	▓	▓	▓		▓	▓
R	4	4		▓		▓		▓		▓
S	BELL	BELL	,	▓	▓		▓			▓
T	5	5		▓					▓	▓
U	7	7		▓	▓	▓	▓			▓
V	;	⊕	=	▓		▓	▓	▓	▓	▓
W	2	2		▓	▓	▓			▓	▓
X	/	/		▓	▓		▓	▓	▓	▓
Y	6	6		▓	▓		▓		▓	▓
Z	"	+	+	▓	▓				▓	▓
BLANK	–			▓						▓
SPACE				▓			▓			▓
CAR. RET.				▓				▓		▓
LINE FEED				▓		▓				▓
FIGURE				▓	▓	▓		▓	▓	▓
LETTERS				▓	▓	▓	▓	▓	▓	▓

[a] Blank, spacing element; crosshatched, marking element.

[b] This column shows only those characters which differ from the American "communications" version.

[c] Figures case H(COMM) may be stop or +.

Figure 12.1 Communication and weather codes, CCITT International Telegraph Alphabet no. 2. (From CCITT Rec. S.1 [Ref. 7]; courtesy of ITU–CCITT.)

We can now see that for a binary code the number of distinct information characters available is equal to 2 raised to a power equal to the number of elements or bits per character. For instance, the last example was based on a three-element code, giving eight possibilities or information characters.

Another more practical example is the Baudot teleprinter code. It has five bits or information elements per character. Hence the different or distinct graphics or characters available are $2^5 = 32$. The American Standard Code for Information Interchange (ASCII) has seven information elements per character, or $2^7 = 128$; so it has 128 distinct combinations of marks and spaces that are available for assignment as characters or graphic symbols.

The number of distinct characters for a specific code may be extended by establishing a code sequence (a special character assignment) to shift the system or machine to uppercase (as is done with a conventional typewriter). Uppercase is a new character grouping. A second distinct code sequence is then assigned to revert to lowercase. As an example, the CCITT International Telegraph Alphabet (ITA) no. 2 code (Figure 12.1) is a five-unit code with 58 letters, numbers, graphics, and operator sequences. The additional characters (additional above $2^5 = 32$) come from the use of uppercase. Operator sequences appear on a keyboard as *space* (spacing bar), *figures* (uppercase), *letters* (lowercase), *carriage return*, *line feed* (spacing vertically), and so on.

When we refer to a 5-unit, 6-unit, 7-unit, or 12-unit code, we refer to the number of information units or elements that make up a single character or symbol, that is, we refer to those elements assigned to each character that carry information and that make it distinct from all other characters or symbols of the code.

12.3.2 Some Specific Binary Codes for Information Interchange

In addition to the ITA no. 2 code, some of the more commonly used codes are the field data code, the IBM data transceiver code (Figure 12.2), the American Standard Code for Information Interchange (ASCII) (Figure 12.3), the International Alphabet no. 5 (Figure 12.4), the extended binary-coded decimal interchange code (EBCDIC) (Figure 12.5), the Hollerith code (Figure 12.6), and the binary-coded decimal (BCD) code.

The field data code was adopted by the U.S. Army in 1969 and is now being replaced by ASCII. The field data code is an 8-bit code consisting of 7 information bits and a control bit. With the 8 bits, 256 (2^8) bit patterns or permutations are available. Of these, 128 are available for assignment to characters. These 128 bit patterns are subdivided into two groups of 64 each. The groups are distinguished from one another by the value or state of a particular bit known as the control bit. Owing to this arrangement of bits and bit patterns, the field data code has parity check capability.

Bit Number	Code Assignments							
X →	0	0	0	0	1	1	1	1
O →	0	0	1	1	0	0	1	1
N →	0	1	0	1	0	1	0	1
R 7 4 2 1 ↓↓↓↓↓								
0 0 0 0 0								
0 0 0 0 1								TPH/TGR
0 0 0 1 0								@
0 0 0 1 1				(NA)			(NA)	Space
0 0 1 0 0								#
0 0 1 0 1				(NA)			(NA)	9
0 0 1 1 0				(NA)			(NA)	8
0 0 1 1 1		G	P		X			
0 1 0 0 0								(NA)
0 1 0 0 1				(NA)			(NA)	6
0 1 0 1 0				(NA)			(NA)	5
0 1 0 1 1		D	M		U			
0 1 1 0 0				(NA)			(NA)	3
0 1 1 0 1		B	K		S			
0 1 1 1 0		A	J		/			
0 1 1 1 1	0							
1 0 0 0 0								Restart
1 0 0 0 1				SOC/EOC		$\overset{+}{0}$	EOT	
1 0 0 1 0				•		%		
1 0 0 1 1		&	—		Ø			
1 0 1 0 0				$		,	.	
1 0 1 0 1		I	R		Z			
1 0 1 1 0		H	Q		Y			
1 0 1 1 1	7							
1 1 0 0 0				(NA)			(NA)	(NA)
1 1 0 0 1		F	O		W			
1 1 0 1 0		E	N		V			
1 1 0 1 1	4							
1 1 1 0 0		C	L		T			
1 1 1 0 1	2							
1 1 1 1 0	1							
1 1 1 1 1								

Figure 12.2 IBM data transceiver code. TPH/TGR = telephone/telegraph. SOC/EOC = start or end of card; EOT = end of transmission; (NA) = valid but not assigned; $\overset{+}{0}$ = plus zero; $\bar{0}$ = minus zero. Transmission order; bit X → bit 1.

Parity checks are one way to determine if a character contains an error after transmission. We speak of even parity and odd parity. On a system using an odd parity check, the total count of 1's or marks has to be an odd number per character (or block) (e.g., it carries 1, 3, 5, or 7 marks or 1's). Some systems, such as the field data code, use even parity (i.e., the total number of marks must be an even number, such as, 2, 4, 6, or 8). The code

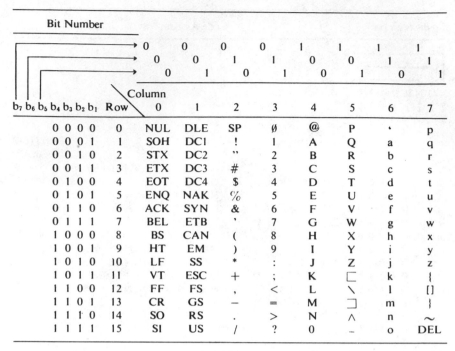

Figure 12.3 American Standard Code for Information Interchange.

Notes:

1. Columns 2, 3, 4, and 5 indicate the printable characters in the U.S. Defense Communication System automatic digital network (DCS Autodin).
2. Columns 6 and 7 fold over into columns 4 and 5, respectively, except DEL.

GENERAL DEFINITIONS

Communication control. A function character intended to control or facilitate transmission of information over communication networks.

Format effector. A functional character which controls the layout or positioning of information in printing or display devices.

Information separator. A character used to separate and qualify information in a logical sense. There is a group of four such characters, which are to be used in a hierarchical order. In order rank, highest to lowest, they appear as follows: FS-file separator; GS-group separator; RS-record separator; US-unit separator.

SPECIFIC CHARACTERS

NUL. The all-zeros character.

SOH (*start of heading*). A communication control character used at the beginning of a sequence is referred to as the *heading*. An STX character has the effect of terminating a heading.

(Continued)

STX (*start of text*). A communication control character which precedes a sequence of characters that is to be treated as an entity and thus transmitted through to the ultimate destination. Such a sequence is referred to as *text*. STX may be used to terminate a sequence of characters started by SOH.

ETX (*end of text*). A communication control character used to terminate a sequence of characters started with STX and transmitted as an entity.

EOT (*end of transmission*). A communication control character used to indicate the conclusion of a transmission, which may have contained one or more texts and any associated headings.

ENQ (*enquiry*). A communication control character used in data communication systems as a request for a response from a remote station.

ACK (*acknowledge*). A communication control character transmitted by a receiver as an affirmative response to a sender.

BEL. A character used when there is a need to call for human attention. It may trigger an alarm or other attention devices.

BS (*backspace*). A format effector which controls the movement of the printing position one printing space backward on the same printing line.

HT (*horizontal tabulation*). A format effector which controls the movement of the printing position to the next in a series of predetermined positions along the printing line.

LF (*line feed*). A format effector which controls the movement of the printing position to the next printing line.

VT (*vertical tabulation*). A format effector which controls the movement of the printing position to the next in a series of predetermined printing lines.

FF (*form feed*). A format effector which controls the movement of the printing position to the first predetermined printing line on the next form or page.

CR (*carriage return*). A format effector which controls the movement of the printing position to the first printing position on the same printing line.

SO (*shift out*). A control character indicating that the code combinations which follow shall be interpreted as outside of the character set of the standard code table until a shift in (SI) character(s) is (are) reached.

SI (*shift in*). A control character indicating that the code combinations which follow shall be interpreted according to the standard code table.

DLE (*data link escape*). A communication control character which will change the meaning of a limited number of contiguously following characters. It is used exclusively to provide supplementary controls in data communication networks. DLE is usually terminated by a shift in character(s).

DC1, DC2, DC3, DC4 (*device controls*). Characters for the control of ancillary devices associated with data processing or telecommunication systems, especially for switching devices on or off.

NAK (*negative acknowledgment*). A communication control character transmitted by a receiver as a negative response to the sender.

Figure 12.3   Notes (*Continued*)

Figure 12.3 Notes *(Continued)*

SYN *(synchronous idle)*. A communication control character used by a synchronous transmission system in the absence of any other character to provide a signal from which synchronism may be achieved or retained.

ETB *(end of transmission block)*. A communication control character used to indicate the end of a block of data for communication purposes.

CAN *(cancel)*. A control character used to indicate that the data with which it is sent is in error or is to be disregarded.

EM *(end of medium)*. A control character associated with the sent data which may be used to identify the physical end of the medium, or the end of the used or wanted portion of information recorded on a medium.

SS *(start of special sequence)*. A control character used to indicate the start of a variable-length sequence of characters which have special significance or which are to receive special handling. SS is usually terminated by a shift in (SI) character(s).

ESC *(escape)*. A control character intended to provide code extension (supplementary characters) in general information interchange. The escape character itself is a prefix affecting the interpretation of a limited number of contiguously following characters. ESC is usually terminated by a shift in (SI) character(s).

DEL *(delete)*. This character is used primarily to erase or obliterate erroneous or unwanted characters in perforated tape.

SP *(space)*. Normally a nonprinting graphic character used to separate words. It is also a format effector which controls the movement of the printing position, one printing position forward.

Diamond. A noncoded graphic which is printed by a printing device to denote the sensing of an error when such an indication is required.

Heart. A noncoded graphic which may be printed by a printing device in lieu of the symbols for the control characters shown in columns 0 and 1.

was used by the U.S. Army for communication in a common language between computing, input–output, and terminal equipments of the field data equipment family.

To explain parity and parity checks a little more clearly, let us look at some examples. Consider a 7-level* code with an extra parity bit. By system convention, even parity has been established. Suppose that a character is transmitted as 1111111. There are seven marks, so to maintain even parity we would need an even number of marks. Thus an eighth bit is added and must be a mark (1). Look at another bit pattern, 1011111. Here there are six marks, even; then the eighth (parity) bit must be a space. Still another example would be 0001000. To get even parity, a mark must be added on

---

*Levels and bits are often used interchangeably.

Bit Number							

				0	0	0	0	1	1	1	1
				0	0	1	1	0	0	1	1
				0	1	0	1	0	1	0	1
b7 b6 b5 b4 b3 b2 b1	Row \ Column	0	1	2	3	4	5	6	7
0 0 0 0	0	NUL	(TC₇)DLE	SP	0	((@)[3]	P	'[4]	p
0 0 0 1	1	(TC₁)SOH	DC₁	!	1	A	Q	a	q
0 0 1 0	2	(TC₂)STX	DC₂	"[6]	2	B	R	b	r
0 0 1 1	3	(TC₃)ETX	DC₃	£[3][7]	3	C	S	c	s
0 1 0 0	4	(TC₄)EOT	DC₄	$[3][7]	4	D	T	d	t
0 1 0 1	5	(TC₅)ENQ	(TC₈)NAK	%	5	E	U	e	u
0 1 1 0	6	(TC₆)ACK	(TC₉)SYN	&	6	F	V	f	v
0 1 1 1	7	BEL	(TC₁₀)ETB	'[6]	7	G	W	g	w
1 0 0 0	8	FE₀(BS)	CAN	(	8	H	X	h	x
1 0 0 1	9	FE₁(HT)	EM	)	9	I	Y	i	y
1 0 1 0	10	FE₂(LF)[1]	SUB	*	:[6]	J	Z	j	z
1 0 1 1	11	FE₃(VT)	ESC	+	;[6]	K	([)[3]	k	[3]
1 1 0 0	12	FE₄(FF)	IS₄(FS)	,	<	L	[6]	l	[3]
1 1 0 1	13	FE₅(CR)[1]	IS₃(GS)	—	=	M	(])[3]	m.	[3]
1 1 1 0	14	SO	IS₂(RS)	.	>	.N	∧[4][6]	n	-[4][6]
1 1 1 1	15	SI	IS₁(US)	/	?	O	—	o	DEL

Figure 12.4 International Alphabet no. 5. (From CCITT Recs. X.4 and T.50 [Refs. 9 and 10]; courtesy of ITU–CCITT.)

Notes:

1. The controls CR and LF are intended for printer equipment which requires separate combinations to return the carriage and to feed a line.
   For equipment which uses a single control for a combined carriage return and line feed operation, the function FE₂ will have the meaning of *new line* (NL). These substitutions must be in agreement between the sender and the recipient of the data.
   The use of this function NL is not allowed for international transmission on general switched telecommunication networks (telegraph and telephone networks).

2. For international information interchange, $ and £ symbols do not designate the currency of a given country. The use of these symbols combined with other graphic symbols to designate national currencies may be the subject to other recommendations.

3. Reserved for national use. These positions are intended primarily for alphabetic extensions. If they are not required for that purpose, they may be used for symbols, and a recommended choice is shown in parentheses in some cases.

4. Positions 5/14, 6/0, and 7/14 of the 7-bit set table normally are provided for the diacritical signs *circumflex*, *grave accent*, and *overline*. However, these positions may be used for other graphical symbols when it is necessary to have 8, 9, or 10 positions for national use.

(Continued)

Figure 12.4 Notes *(Continued)*

5. For international information interchange, position 7/14 is used for the graphical symbol − (overline), the graphical representation of which may vary
according to national use to represent ∼ (tilde) or another diacritical sign
provided that there is no risk of confusion with another graphical symbol
included in the table.

6. The graphics in positions 2/2, 2/7, 5/14 have respectively the significance of
quotation mark, *apostrophe*, and *upwards arrow*; however, these characters take
on the significance of the diacritical signs *diaeresis*, *acute accent*, and *circumflex
accent* when they precede or follow the backspace character.

7. For international information interchange, position 2/3 of the 7-bit code table
has the significance of the symbol £, and position 2/4 has the significance of the
symbol $.

8. If 10 and 11 as single characters are needed (for example, for Sterling currency
subdivision), they should take the place of *colon* (:) and *semicolon* (;), respectively. These substitutions require agreement between the sender and the
recipient of the data. On the general telecommunication networks, the characters *colon* and *semicolon* are the only ones authorized for international transmission.

transmission, and the character transmitted would be 00010001, maintaining
even parity. Suppose that, owing to some sort of signal interference, one
signal element was changed on reception. No matter which element was
changed, the receiver would indicate an error because we would no longer
have even parity. If two elements were changed, though, the error could be
masked. This would happen in the case of even or odd parity if two marks
were substituted for two spaces or vice versa at any element location in the
character.

The IBM data transceiver code (see Figure 12.2) is used for the transfer of
digital data information recorded on perforated cards. It is an 8-bit code
providing a total of 256 mark–space combinations. Only those combinations
or patterns having a "fixed count" of four 1's (marks) and four 0's (spaces)
are made available for assignment as characters. Therefore only 70 bit
patterns satisfy the fixed count condition, and the remaining 186 combinations are invalid. The parity (fixed count) or error checking advantage is
obvious. Of the 70 valid characters, 54 are assigned to alphanumerics and a
limited number of punctuation signs and other symbols. In addition, the code
includes special bit patterns assigned to control functions peculiar to the
transmission of cards such as "start card" and "end card."

The ASCII (see Figure 12.3) is the latest effort on the part of the U.S.
industry and common carrier systems, backed by the American National
Standards Institute, to produce a universal common language code. ASCII is
a seven-unit code with all 128 combinations available for assignment. Here
again, the 128 bit patterns are divided into two groups of 64. One of the
groups is assigned to a subset of graphic printing characters. The second

B	4	0	0	0	0	0	0	0	0	1	1	1	1	1	1	1	1
I	3	0	0	0	0	1	1	1	1	0	0	0	0	1	1	1	1
T	2	0	0	1	1	0	0	1	1	0	0	1	1	0	0	1	1
S	1	0	1	0	1	0	1	0	1	0	1	0	1	0	1	0	1
8 7 6 5																	
0 0 0 0		NUL				PF	HT	LC	DEL								
0 0 0 1						RES	NL	BS	IL								
0 0 1 0						BYP	LF	EOB	PRE		SM						
0 0 1 1						PN	RS	UC	EOT								
0 1 0 0		SP										¢	.	<	(	+	|
0 1 0 1		&										!	$	*	)	;	¬
0 1 1 0		-	/									^	,	%	—	>	?
0 1 1 1												‾:	#	@	'	=	"
1 0 0 0			a	b	c	d	e	f	g	h	i						
1 0 0 1			j	k	l	m	n	o	p	q	r						
1 0 1 0				s	t	u	v	w	x	y	z						
1 0 1 1																	
1 1 0 0			A	B	C	D	E	F	G	H	I						
1 1 0 1			J	K	L	M	N	O	P	Q	R						
1 1 1 0				S	T	U	V	W	X	Y	Z						
1 1 1 1		0	1	2	3	4	5	6	7	8	9						¤

PF – Punch Off	RES – Restore	BYP – Bypass
HT – Horiz. Tab	NL – New Line	LF – Line Feed
LC – Lower Case	BS – Backspace	EOB – End of Block
DEL – Delete	IL – Idle	PRE – Prefix
SP – Space	PN – Punch On	RS – Reader Stop
UC – Upper Case	EOT – End of Transmission	SM – Start Message

Figure 12.5 Extended binary-coded decimal interchange code (EBCDIC).

subset of 64 is assigned to control characters. An eighth bit is added to each character for parity check. ASCII is widely used in North America and has received considerable acceptance in Europe and Hispanic America.

CCITT Rec. T.50 offers a 7-level code as an international standard for information interchange. It is not intended as a substitute for CCITT ITA no. 2 code. CCITT no. 5, or the new alphabet no. 5, as the 7-level code is more commonly referred to is basically intended for data transmission.

Although International Alphabet no. 5 (CCITT no. 5) is considered a 7-level code, CCITT Rec. V.4 advises that an eighth bit may be added for parity. Under certain circumstances odd parity is recommended; on others, even parity.

Figure 12.4 shows International Alphabet no. 5. b_1 is the first signal element in serial transmission and b_7 is the last element of a character. Like the ASCII, International Alphabet no. 5 does not normally need to shift out (i.e., uppercase, lowercase as in CCITT ITA no. 2). However, like ASCII, it is provided with an escape, 1101100. Eight footnotes explaining peculiar char-

ZONES →	12				12	12		12
		11				11	11	11
DIGIT PUNCH ROWS ↓			0		0		0	0
	&	-	0	SP	(	|	}	11/10
1	A	J	/	1	a	j	~	13/9
2	B	K	S	2	b	k	s	13/10
3	C	L	T	3	c	l	t	13/11
4	D	M	U	4	d	m	u	13/12
5	E	N	V	5	e	n	v	13/13
6	F	O	W	6	f	o	w	13/14
7	G	P	X	7	g	p	x	13/15
8	H	Q	Y	8	h	q	y	14/0
9	I	R	Z	9	i	r	z	14/1
8-2	[	]	\	:	12/4	12/11	13/2	14/2
8-3	.	$	,	#	12/5	12/12	13/3	14/3
8-4	<	*	%	@	12/6	12/13	13/4	14/4
8-5	(	)	_	'	12/7	12/14	13/5	14/5
8-6	+	;	>	=	12/8	12/15	13/6	14/6
8-7	!	^	?	"	12/9	13/0	13/7	14/7
8-1	10/8	11/1	11/9	~	12/3	12/10	13/1	13/8
9-1	SOH	DC1	8/1	9/1	10/0	10/9	9/15	11/11
9-2	STX	DC2	8/2	SYN	10/1	10/10	11/2	11/12
9-3	ETX	DC3	8/3	9/3	10/2	10/11	11/3	11/13
9-4	9/12	9/13	8/4	9/4	10/3	10/12	11/4	11/14
9-5	HT	8/5	LF	9/5	10/4	10/13	11/5	11/15
9-6	8/6	BS	ETB	9/6	1G/5	10/14	11/6	12/0
9-7	DEL	8/7	ESC	EOT	10/6	10/15	11/7	12/1
9-8	9/7	CAN	8/8	9/8	10/7	11/0	11/8	12/2
9-8-1	8/13	EM	8/9	9/9	NUL	DLE	8/0	9/0
9-8-2	8/14	9/2	8/10	9/10	14/8	14/14	15/4	15/10
9-8-3	VT	8/15	8/11	9/11	14/9	14/15	15/5	15/11
9-8-4	FF	FS	8/12	DC4	14/10	15/0	15/6	15/12
9-8-5	CR	GS	ENQ	NAK	14/11	15/1	15/7	15/13
9-8-6	SO	RS	ACK	9/14	14/12	15/2	15/8	15/14
9-8-7	SI	US	BEL	SUB	14/13	15/3	15/9	15/15

NOTE:

A card code position that has not been assigned a corresponding ASCII Code character is designated with the corresponding, though not yet assigned, column/row of the ASCII Codes.

Figure 12.6 Hollerith punched card code. (From MIL-STD-188-100.)

acter usage are provided. Few differences exist between the ASCII and the CCITT no. 5 codes.

The reader's attention is called to what are known as computable codes, such as ASCII and CCITT no. 5 codes. Computable codes have the letters of the alphabet plus all other characters and graphics assigned values in continuous binary sequence. Thus these codes are in the native binary language of today's common digital computers. The CCITT ITA no. 2 is not, and when used with a computer, it often requires special processing.

The extended binary-coded decimal interchange code (EBCDIC) is similar to the ASCII, but it is a true 8-bit code. The eighth bit is used as an added bit to "extend" the code, providing 256 distinct code combinations for assignment. Figure 12.5 illustrates the EBCDIC code.

The Hollerith code was specifically designed for use with perforated (punched) cards. It attained wide acceptance in the business machine and computer fields. Hollerith is a 12-unit character code in that a character is represented on a card by one or more holes perforated in one column having 12 potential hole positions. It is most commonly used with the standard 80-column punched cards.

The theoretical capacity of a 12-unit binary code is very great, and by our definition, using all hole patterns available, it is 2^{12} or 4096 bit combinations. In the modern version of the Hollerith code only 64 of these, none using more than three holes, are assigned to graphic characters.

Because of its unwieldiness, the Hollerith code is seldom used directly for transmission. Most often it is converted to one of the more conventional transmission codes such as the ASCII, International Alphabet no. 5, BCD interchange, or EBCDIC. Figure 12.6 shows the Hollerith code as extended to cover ASCII equivalents. It should be noted that punched-card operation is phasing out in favor of other computer access methods such as the PC-based workstation.

12.3.3 Hexadecimal Representation and the BCD Code

The hexadecimal system is a numeric representation in the number base 16. The number base uses 0 through 9 as in the decimal base, and the letters A through F to represent the decimal numbers 10 through 15. The hexadecimal numbers can be translated to the binary base as follows:

Hexadecimal	Binary	Hexadecimal	Binary
0	0000	8	1000
1	0001	9	1001
2	0010	A	1010
3	0011	B	1011
4	0100	C	1100
5	0101	D	1101
6	0110	E	1110
7	0111	F	1111

Two examples of the hexadecimal notation are

Number Base 10	Number Base 16
21	15
64	40

The BCD is a compromise code assigning 4-bit binary numbers to the digits between 0 and 9. The BCD equivalents to decimal digits appear as follows:

Decimal Digit	BCD Digit	Decimal Digit	BCD Digit
0	1010	5	0101
1	0001	6	0110
2	0010	7	0111
3	0011	8	1000
4	0100	9	1001

To cite some examples, consider the number 16; it is broken down into 1 and 6. Therefore its BCD equivalent is 0001 0110. If it were written in straight binary notation, it would appear as 10000. The number 25 in BCD combines the digits 2 and 5 above as 0010 0101.

12.4 ERROR DETECTION AND ERROR CORRECTION

12.4.1 Introduction

In the transmission of data the most important goal in design is to minimize the error rate. Error rate may be defined as the ratio of the number of bits incorrectly received to the total number of bits transmitted. On many data circuits the design objective is an error rate no poorer than one error in 1×10^5 and on telegraph circuits, one error in 1×10^4.

One method to minimize the error rate is to provide a "perfect" transmission channel, one that will introduce no errors in the transmitted information by the receiver; unfortunately, the engineer designing a data transmission system can never achieve that perfect channel. Besides improvement of the channel transmission parameters themselves, the error rate can be reduced by forms of systematic redundancy. In old-time Morse code on a bad circuit words often were sent twice; this is redundancy in its simplest form. Of course, it took twice as long to send a message. This is not very economical if useful words per minute received is compared to channel occupancy.

This brings up the point of channel efficiency. Redundancy can be increased such that the error rate could approach zero. Meanwhile the information transfer or *throughput* across the channel also approaches zero. Hence unsystematic redundancy is wasteful and merely lowers the rate of useful communication. Maximum efficiency or throughput could be obtained in a digital transmission system if all redundancy and other code elements, such as start and stop elements, were removed from the code, and, in addition, if advantage were taken of the statistical phenomenon of our written language by making high-usage letters, such as E, T, and A, short in code length, and low-usage letters, such as Q and X, longer.

12.4.2 Throughput

The *throughput* of a data channel is the expression of how much data are put through. In other words, throughput is an expression of channel efficiency. The term gives a measure of useful data put through the data communication link. These data are directly useful to the computer or data terminal equipment (DTE).

Therefore on a specific circuit, throughput varies with the raw data rate; is related to the error rate and the type of error encountered (whether burst or random); and varies with the type of error detection and correction system used, the message handling time, and the block length, from which we must subtract the "nonuseful" bits such as overhead bits. Among overhead bits we have parity bits, flags, cyclic redundancy checks, and so forth.

12.4.3 The Nature of Errors

In data/telegraph transmission an error is a bit that is incorrectly received. For instance, a 1 is transmitted in a particular time slot and the element received in that slot is interpreted as a 0. Bit errors occur either as single random errors or as bursts of error. In fact, we can say that every transmission channel will experience some random errors, but on a number of channels burst errors may predominate. For instance, lightning or other forms of impulse noise often cause bursts of errors, where many contiguous bits show a very high number of bits in error. The IEEE defines error burst as "a group of bits in which two successive bits are always separated by less than a given number of correct bits" (Ref. 1).

12.4.4 Error Detection and Correction Defined

The data transmission engineer differentiates between error detection and error correction. Error detection identifies that a symbol; character, block*, packet*, or frame* has been received in error. As discussed above, parity is primarily used for error detection. Parity bits, of course, add redundancy and thus decrease channel efficiency or throughput.

Error correction corrects the detected error. Basically, there are two types of error correction techniques: forward-acting (FEC) and two-way error correction (automatic repeat request [ARQ]). The latter technique uses a return channel (backward channel). When an error is detected, the receiver signals this fact to the transmitter over the backward channel, and the block of information containing the error is transmitted again. FEC utilizes a type of coding that permits a limited number of errors to be corrected at the receiving end by means of special coding and software (or hardware) implemented at both ends of a circuit.

---

*A block, packet, or frame is a group of digits or data characters transmitted as a unit over which a coding procedure is usually applied for synchronization and error control purposes.

Error Detection There are various arrangements or techniques available for the detection of errors. All error detection methods involve some form of redundancy, those additional bits or sequences that can inform the system of the presence of error or errors. Parity, discussed above, was character parity, and its weaknesses were presented. Commonly, the data transmission engineer refers to such parity as *vertical redundancy checking* (VRC). The term *vertical* comes from the way characters are arranged on paper tape (i.e., hole positions).

Another form of error detection utilizes longitudinal redundancy checking (LRC), which is used in block transmission where a data message consists of one or more blocks. Remember that a block is a specific group of digits or data characters sent as a "package" (not to be confused with *packet*). In such circumstances an LRC character, often called a block check character (BCC), is appended at the end of each block. The BCC modulo-2 sums the 1's and 0's in the columns of the block (vertically). The receiving end also modulo-2 sums the 1's 0's in the block, depending on the parity convention for the system. If that sum does not correspond to the BCC, an error (or errors) exists in the block. The LRC ameliorates much of the problem of undetected errors that could slip through with VRC if used alone. The LRC method is not foolproof, however, as it uses the same thinking as VRC. Suppose that errors occur such that two 1's are replaced by two 0's in the second and third bit positions of characters 1 and 3 in a certain block. In this case the BCC would read correctly at the receive end and the VRC would pass over the errors as well. A system using both LRC and VRC is obviously more immune to undetected errors than either system implemented alone.

A more powerful method of error detection involves the use of the cyclic redundancy check (CRC). It is used with messages that are data blocks, frames, or packets. Such a data message can be simplistically represented as

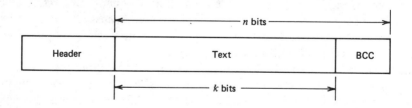

We let n equal the number of bits in the total message after the "header," and k is the number of bits in the message text over which we wish to check for errors. $n - k$ is the number of bits in the BCC. For most WANs (wide area networks) it is 16 bits long (2 bytes) and for most LANs (local area networks) it is 32 bits long (4 bytes).

The bit sequence (value) of the BCC is derived from two polynomials: the generating polynomial $P(X)$ and the message polynomial $G(X)$. To develop the 16-bit BCC, there are three standardized generating polynomials:

CRC-16 (ANSI) $X^{16} + X^{15} + X^2 + 1$
CRC (CCITT) $X^{16} + X^{12} + X^5 + 1$
CRC-12 $X^{12} + X^{11} + X^2 + X + 1$ (12-bit BCC)

In binary form we treat a polynomial as follows. A 1 is placed in each position that has a term; absence of a term is indicated by a 0. Of course, if the last term is a 1, that is X^0 power. For example, if a polynomial is given as $X^4 + X + 1$, its binary representation is 10011. The two consecutive zeros tell us that the second and third terms are not present. Let us restate the problem with several more examples.

Mathematically, a message block can be treated as a function such as

$$a_n X^n + a_{n-1} X^{n-1} + a_{n-2} X^{n-2} + \cdots + a_1 X + a_0$$

where coefficients a are set to represent a binary number. Consider the binary number 11011, which is represented by the polynomial

$$\begin{array}{ccccc} 1 & 1 & 0 & 1 & 1 \\ a_4 & a_3 & a_2 & a_1 & a_0 \end{array}$$

and then becomes

$$X^4 + X^3 + X + 1$$

or consider another example,

$$\begin{array}{ccccc} 0 & 1 & 1 & 0 & 1 \\ a_4 & a_3 & a_2 & a_1 & a_0 \end{array}$$

which then becomes

$$X^3 + X^2 + 1$$

Given now a message polynomial $G(X)$ and a generating polynomial $P(X)$, we wish to construct a code message polynomial $F(X)$ that is evenly divided by $P(X)$. This can be accomplished as follows:

1. Multiply the message $G(X)$ by X^{n-k} where $n - k$ is the number of bits in the BCC.
2. Divide the resulting product $X^{n-k}[G(X)]$ by the generating polynomial $P(X)$.
3. Disregard the quotient and add the remainder $C(X)$ to the product to yield the code message polynomial $F(X)$, which is represented as $X^{n-k}[G(X)] + C(X)$.

The division is binary division without carries and borrows. In this case, the remainder is always 1 bit less than the divisor. The remainder is the BCC and the divisor is the generating polynomial; hence the bit length of the BCC is always one less than the number of bits in the generating polynomial.

The code message polynomial as represented by the BCC is transmitted to the distant-end receiver. The receiving station divides it by the same generating polynomial. The division will produce no remainder (i.e., all zeros) if there is no error; if there is an error in the n bits of message text, there will be a remainder.

In Ref. 12 it is stated that CRC-12 provides error detection of bursts of up to 12 bits in length. Additionally, 99.955% of error bursts up to 16 bits in length can be detected. CRC-16 provides detection of bursts up to 16 bits in length, and 99.955% of error bursts greater than 16 bits can be detected.

Forward-Acting Error Correction (FEC)   FEC uses certain binary codes that are designed to be self-correcting for errors introduced by the intervening transmission media. In this form of error correction the receiving station has the ability to reconstitute messages containing errors.

FEC uses *channel encoding*, whereas the encoding covered previously in this chapter is generically called *source encoding*. The channel encoding used in FEC can be broken down into two broad categories, block codes and convolutional codes. These are discussed at considerable length in Section 6.7.

Key to this type of coding is the modulo-2 adder. Modulo-2 addition is denoted by the symbol $\oplus$. It is binary addition without the "carry," or $1 + 1 = 0$, and we do *not* carry the 1. Summing 10011 and 11001 in modulo-2, we get 01010.

A measure of the error detection and correction capability of a code is given by the Hamming distance. The distance is the minimum number of digits in which two encoded words differ. For example, to detect E digits in error, a code of a minimum Hamming distance of $(E + 1)$ is required. To correct E errors, a code must display a minimum Hamming distance of $(2E + 1)$. A code with a minimum Hamming distance of 4 can correct a single error *and* detect two digits in error.

Error Correction With Feedback Channel   Two-way or feedback error correction is used very widely today on data and some telegraph circuits. Such a form of error correction is called ARQ, which derives from the old Morse and telegraph signal, "automatic repeat request."

In most modern data systems block transmission is used, and the block is a convenient length of characters sent as an entity. There are two aspects in that "convenience" of length. One relates to the material that is being sent. For instance, the standard "IBM" card has 80 columns. With 8 bits per column, a block of 8×80, or 640 bits, would be desirable as data text so that we could transmit an IBM card in each block. In fact, one such operating system, Autodin, bases block length on that criterion, with blocks 672 bits long. The remaining bits, those in excess of 640, are overhead and check bits. On packet networks blocks are typically of uniform length.

The second aspect of block length is the trade-off for optimum length between length and error rate, or the number of block repeats that may be expected on a particular circuit. Longer blocks tend to amortize overhead bits better but are inefficient regarding throughput when an error rate is high. Under these conditions long blocks tend to tie up a circuit with longer retransmission periods.

ARQ, as we know, is based on the block transmission concept. There are three types of ARQ in use today:

- Stop-and-wait
- Continuous, sometimes called selective ARQ
- Go-back-n

Stop-and-wait ARQ is the most straightforward to implement and the least costly from an equipment standpoint. With stop-and-wait ARQ, a block (or frame or packet) is transmitted to the distant end. At completion of transmission of the block, transmission ceases. The receiving end of the link receives the block, stores the entire block, and then runs the block through CRC processing. If the CRC remainder is zero, an acknowledgment signal is sent to the far end, and the far-end transmitter sends the next data block. If an error is found (i.e., there is a CRC remainder at the receiver), the receiving station requests retransmission. The transmitter knows that it is the previous block sent that is wanted and retransmits that block. The drawback of stop-and-wait ARQ is the delay of waiting while the receiver does the processing and then transmits back the appropriate signal, which may be an acknowledgment (ACK) or a negative acknowledgment (NACK). It is particularly wasteful of expensive circuit time if propagation delays are long, such as on satellite circuits. It is, however, ideal for half-duplex operation.

Selective or continuous ARQ eliminates the quiescent nonproductive time of waiting and sending the appropriate signal after each block. With this type of ARQ, the transmit end sends a continuous string of blocks, each with an identifying number in the header of each block (or frame or packet). When the receive end detects a block in error, it requests that the distant-end transmitter repeat that block. We assume here, of course, full-duplex, four-wire operation. The request that the receiver sends to the transmitter includes the identifying number of the block in error. The transmit end pulls that block out of its buffer storage and retransmits it as another block in the continuous string with its appropriate identifying number. The receive end is responsible for rerunning a CRC check and then placing that block in its proper sequential order. Obviously, selective or continuous ARQ is more efficient regarding circuit usage. On the other hand, it requires more buffer memory size and a slight increase in overhead to accommodate message sequence numbering. It also requires full-duplex operation, although the return circuit may be only a slow 75-bps channel.

Go-back-*n* ARQ also permits continuous block transmission. When an errored block is discovered, the receiving end informs the far-end transmitter, which then transmits the errored block and all subsequent blocks, even though they have been transmitted before. This approach alleviates the problem at the receive end of inserting the errored block in its proper slot.

12.5 THE dc NATURE OF DATA TRANSMISSION

12.5.1 Loops

Binary data is transmitted on a dc loop. More correctly, the binary data end instrument delivers to the line and receives from the line one or several dc loops. In its most basic form a dc loop consists of a switch, a dc voltage source, and a termination. A pair of wires interconnects the switch and termination. The voltage source in data and telegraph work is called the battery, although the device is usually electronic, deriving the dc voltage from an ac power line source. The battery is placed in the line such that it provides voltage(s) consistent with the type of transmission desired. A simplified dc loop is shown in Figure 12.7.

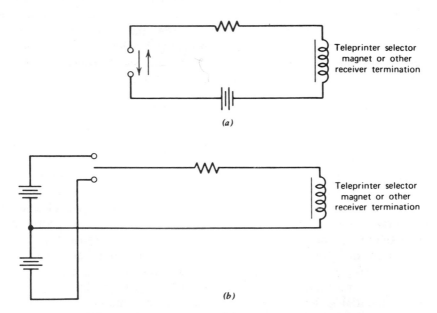

Teleprinter selector magnet or other receiver termination

(a)

Teleprinter selector magnet or other receiver termination

(b)

Figure 12.7 Simplified diagram of a dc loop. (*a*) With neutral keying. (*b*) With polar keying.

12.5.2 Neutral and Polar dc Transmission Systems

Nearly all dc data and telegraph systems functioning today are operated in either a neutral or a polar mode. However, the neutral mode has pretty much been phased out. The words *neutral* and *polar* describe the manner in which the battery is applied to the dc loop. On a neutral loop, following the mark–space convention in Table 12.1, the battery is applied during marking (1) conditions and is switched off during spacing (0). Current therefore flows in the loop when a mark is sent and the loop is closed. Spacing is indicated on the loop by a condition of no current. Thus we have the two conditions for binary transmission, an open loop (no current flowing) and a closed loop (current flows). Keep in mind that we could reverse this, change the convention (Table 12.1), and, say, assign spacing to a condition of current flowing or closed loop and marking to a condition of no current or open loop. This is sometimes done in practice and is called changing the sense. Either way, a neutral loop is a dc loop circuit where one binary condition is represented by the presence of voltage, flow of current, and the other by the absence of voltage/current. Figure 12.7*a* shows a simplified neutral loop.

Polar transmission approaches the problem a little differently. Two batteries are provided. One is called negative battery and the other positive. During a condition of marking, a positive battery is applied to the loop, following the convention of Table 12.1, and a negative battery is applied to the loop during the spacing condition. In a polar loop, current is always flowing. For a mark or binary 1 it flows in one direction and for a space or binary 0 it flows in the opposite direction. Figure 12.7*b* shows a simplified polar loop.

12.5.3 Some Common Digital Data Waveforms or Line Signals

In this section we discuss several basic concepts of "electrical" coding to develop line signals or specific line waveforms. Figure 12.8 graphically illustrates several line coding techniques.

Figure 12.8A shows what is still called by many today "neutral transmission." This was the principal method of transmitting telegraph signals until about 1960. First, this waveform is a non-return-to-zero (NRZ) format in its simplest form. "Non-return-to-zero" simply means that if a string of 1's (marks) is transmitted, the signal remains in the mark state with no transitions. Likewise, if a string of 0's is transmitted, there is no transition and the signal remains in the 0 state until a 1 is transmitted. As we can now see, with NRZ transmission we can transmit information without transitions.

Figures 12.8B and D show the typical "return-to-zero" (RZ) waveform, where, when a continuous string of marks (or spaces) is transmitted, the signal level (amplitude) returns to the zero-voltage condition at each element or bit. Obviously, RZ transmission is much richer in transitions than NRZ.

In Section 12.5.2 we discussed neutral and polar dc transmission systems. Figure 12.8A shows a typical neutral waveform where the two state condi-

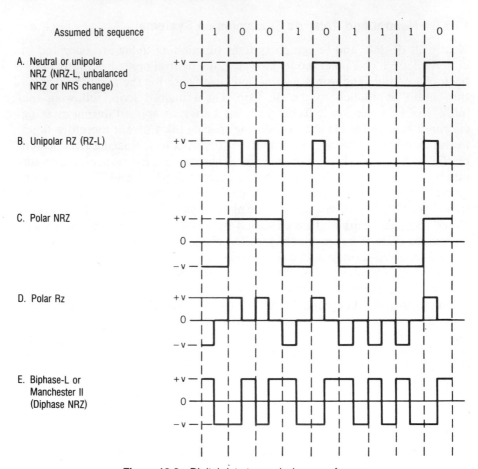

Figure 12.8 Digital data transmission waveforms.

tions are 0 V for the mark or 1 condition and some positive voltage for the space or 0 condition. On the other hand, in polar transmission, as shown in Figures 12.8C and D, a positive voltage represents a space, and a negative voltage, a mark. With NRZ transmission, the pulse width is the same as the duration of a unit interval or bit. Not so with RZ transmission, where the pulse width is less than the duration of a unit interval. This is because we have to allow time for the pulse to return to the zero condition.

Biphase-L or Manchester coding (Figure 12.8E) is a code format that is being used ever more widely on digital systems such as wire pair, coaxial cable, and fiber optics. Here the binary information is carried in the transition. By convention a logic 0 is defined as a positive-going transition and a logic 1 as a negative-going pulse. Note that Manchester coding has a signal transition in the middle of each unit interval (or bit), Manchester coding is a form of phase coding.

The reader should be cognizant of and be able to differentiate between two sets or ways of classifying binary digital waveforms. The first set is *neutral* and *polar*. The second set is *NRZ* and *RZ*. Manchester coding is still another way to represent binary digital data where the transition takes place in the middle of the unit interval. In Chapter 3 one other class of waveform was introduced: alternate mark inversion (AMI).

12.6 BINARY TRANSMISSION AND THE CONCEPT OF TIME

12.6.1 Introduction

Time and timing are most important factors in digital transmission. For this discussion consider a binary end instrument sending out in series a continuous run of marks and spaces. Those readers who have some familiarity with Morse code will recall that the spaces between dots and dashes told the operator where letters and words ended. With the sending device or transmitter delivering a continuous series of characters to the line, each consisting of 5, 6, 7, 8, or 9 elements (bits) to the character, let the receiving device start its print cycle when the transmitter starts sending. If the receiver is perfectly in step with the transmitter, ordinarily one could expect good printed copy and few, if any, errors at the receiving end.

It is obvious that when signals are generated by one machine and received by another, the speed of the receiving machine must be the same or very close to that of the transmitting machine. When the receiver is a motor-driven device, timing stability and accuracy are dependent on the accuracy and stability of the speed of rotation of the motors used.

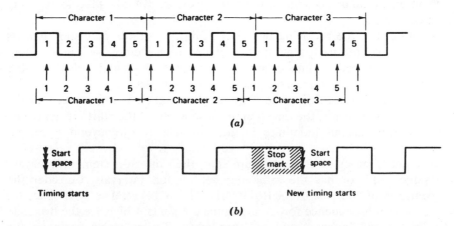

Figure 12.9 (*a*) 5-unit synchronous bit stream with timing error. (*b*) 5-unit start and stop stream of bits with a 1.5-unit stop element.

Most simple data/telegraph receivers sample at the presumed center of the signal element. It follows, accordingly, that whenever a receiving device accumulates a timing error of more than 50% of the period of 1 bit, it will print in error.

The need for some sort of synchronization is shown in Figure 12.9*a*. A 5-unit code is employed, and three characters transmitted sequentially are shown. Sampling points are shown in the figure as vertical arrows. Receiver timing begins when the first pulse is received. If there is a 5% timing difference between transmitter and receiver, the first sampling at the receiver will be 5% away from the center of the transmitted pulse. At the end of the 10th pulse or signal element the receiver may sample in error. The 11th signal element will indeed be sampled in error, and all subsequent elements are errors. If the timing error between transmitting machine and receiving machine is 2%, the cumulative error in timing would cause the receiving device to print all characters in error after the 25th bit.

12.6.2 Asynchronous and Synchronous Transmission

In the earlier days of printing telegraphy, start–stop transmission or asynchronous operation was developed to overcome the problem of synchronism. Here timing starts at the beginning of a character and stops at the end. Two signal elements are added to each character to signal the receiving device that a character has begun and ended.

As an example consider a five-element code, such as CCITT ITA no. 2 (see Figure 12.1). In the front of a character an element is added called a start space and at the end of each character a stop mark is inserted. Send the letter A in Figure 12.1. The receiving device starts its timing sequence on receiving element no. 1, a space or 0, then a 11000 is received; the A is selected, then the stop mark is received, and the timing sequence stops. On such an operation an accumulation of timing errors can take place only inside each character.

Suppose that the receiving device is again 5% slower or faster than its transmitting counterpart; now the fifth information element will be no more than 30% displaced in time from the transmitted pulse and well inside the 50% or halfway point for correct sampling to take place.

In start–stop transmission the *information* signal elements are each of the same duration, which is the duration or pulse width of the start element. The stop element has an indefinite length or pulse width beyond a certain minimum.

If a steady series of characters are sent, then the stop element is always the same width or has the same number of unit intervals. Consider the transmission of two A's, 0110001011000111111 → 11111. The start space (0) starts the timing sequence for six additional elements, which are the five code elements in the letter A and the stop mark. Timing starts again on the mark-to-space transition between the stop mark of the first A and the start space of the second. Sampling is carried out at pulse center for most asynchronous systems. One will note that at the end of the second A a continuous series of marks is sent. Hence the signal is a continuation of the stop element or just a continuous mark. It is the mark-to-space transition of the start element that tells the receiving device to start timing a character.

Minimum lengths of stop elements vary. The example discussed above shows a stop element of 1-unit interval duration (1 bit). Others are of 1.5- and 2-unit interval duration. The proper semantics of data/telegraph transmission would describe the code of the previous paragraph as a 5-unit start–stop code with a 1-unit stop element.

A primary objective in the design of telegraph and data systems is to minimize errors received or to minimize the error rate. There are three prime causes of errors. These are noise, intersymbol interference, and improper timing relationships. With start–stop systems a character begins with a mark-to-space transition at the beginning of the start space. 1.5-unit intervals later the timing causes the receiving device to sample the first information element, which simply is a mark or space decision. The receiver continues to sample at 1-bit intervals until the stop mark is received. In start–stop systems the last information bit is most susceptible to cumulative timing errors. Figure 12.9b is an example of a 5-unit start–stop bit stream with a 1.5-unit stop element.

Another problem in start–stop systems is that of mutilation of the start element. Once this happens the receiver starts a timing sequence on the next mark-to-space transition it sees and thence continues to print in error until, by chance, it cycles back properly on a proper start element.

Synchronous data/telegraph systems do not have start and stop elements but consist of a continuous stream of information elements or bits as shown in Figure 12.9a. The cumulative timing problems eliminated in asynchronous (start–stop) systems are present in synchronous systems. Codes used on synchronous systems are often 7-unit codes with an extra unit added for parity, such as the ASCII or CCITT no. 5 codes. Timing errors tend to be eliminated by virtue of knowing the exact rate at which the bits of information are transmitted.

If a timing error of 1% were to exist between transmitter and receiver, no more than 50 bits could be transmitted, then the synchronous receiving device would be 50% apart in timing from the transmitter and then all bits received would be in error. Even if timing accuracy were improved to 0.05%, the correct timing relationship between transmitter and receiver would exist for only the first 2000 bits transmitted. It follows that no timing error at all can be permitted to accumulate, since anything but absolute accuracy in timing would cause eventual malfunctioning. In practice the receiver is provided with an accurate clock which is corrected by small adjustments as explained below.

12.6.3 Timing

All currently used data transmission systems are synchronized in some manner. Start–stop synchronization has been discussed. Fully synchronous transmission systems all have timing generators or clocks to maintain stability. The transmitting device and its companion receiver at the far end of the data circuit must be a timing system. In normal practice the transmitter is the master clock of the system. The receiver also has a clock which in every case is corrected by one means or another to its transmitter equivalent at the far end.

Another important timing factor that must also be considered is the time it takes a signal to travel from the transmitter to the receiver. This is called propagation time. With velocities of propagation as low as 20,000 mi/s, consider a circuit 200 mi in length. The propagation time would then be 200/20,000 s or 10 ms, which is the time duration of 1 bit at a data rate of 100 bps. Thus the receiver in this case must delay its clock by 10 ms to be in step with its incoming signal.

Temperature and other variations in the medium may affect this delay. One can also expect variations in the transmitter master clock as well as other time distortions due to the medium.

There are basically three methods of overcoming these problems. One is to provide a separate synchronizing circuit to slave the receiver to the transmitter's master clock. This wastes bandwidth by expending a voice channel or subcarrier just for timing. A second method, which was used fairly widely up to 25 years ago, was to add a special synchronizing pulse for groupings of information pulses, usually for each character. This technique was similar to start–stop synchronization and lost its appeal largely owing to the wasted information capacity for synchronizing. The most prevalent system in use today is one that uses transition timing. With this type of timing the receiving device is automatically adjusted to the signaling rate of the transmitter, and adjustment is made at the receiver by sampling the transitions of the incoming pulses. This offers many advantages, most important of which is that it automatically compensates for variations in propagation time. With this type of synchronization the receiver determines the average repetition rate and phase of the incoming signal transition and adjusts its own clock accordingly.

In digital transmission the concept of a transition is very important. The transition is what really carries the information. In binary systems the space-to-mark and mark-to-space transitions (or lack of transitions) placed in a time reference contain the information. Decision circuits regenerate and retime in sophisticated systems and care only *if* a transition has taken place. Timing cares *when* it takes place. Timing circuits must have memory in case a long series of marks or spaces is received. These will be periods of no transition, but they carry meaningful information. Likewise, the memory must maintain timing for reasonable periods in case of circuit outage. Keep in mind that synchronism pertains to both frequency and phase and that the usual error in high-stability systems is a phase error (i.e., the leading edges of the received pulses are slightly advanced or retarded from the equivalent clock pulses of the receiving device).

High-stability systems once synchronized need only a small amount of correction in timing (phase). Modem internal timing systems may be as stable a 1×10^{-8} or better at both the transmitter and the receiver. Before a significant error condition can build up owing to a time rate difference at 2400 bps, the accumulated time difference between transmitter and receiver must exceed approximately 2×10^{-4} s. This figure neglects phase. Once the transmitter and receiver are synchronized and the circuit is shut down, then the clock on each end must drift apart by at least 2×10^{-4} s before significant errors take place. Again this means that the leading edge of the receiver clock equivalent timing pulse is 2×10^{-4} in advance or retarded from the leading edge of the received pulse from the distant end. Often an idling signal is sent on synchronous data circuits during conditions of no traffic to maintain timing. Other high-stability systems need to resynchronize only once a day.

Bear in mind that we are considering dedicated circuits only, not switched synchronous data. The problems of synchronization of switched data immediately come to light. Two such problems are that

- No two master clocks are in perfect phase synchronization
- The propagation time on any two paths may not be the same

Consequently, such circuits will need a time interval for synchronization at each switching event before traffic can be passed.

To sum up, synchronous data systems use high-stability clocks, and the clock at the receiving device is undergoing constant but minuscule corrections to maintain an in-step condition with the received pulse train from the distant transmitter by looking at mark–space and space–mark transitions.

12.6.4 Distortion

It has been shown that the key factor in data transmission is timing. The signal must be either a mark or a space, but that alone is not sufficient. The marks and spaces (or 1's and 0's) must be in a meaningful sequence based on a time reference.

In the broadest sense distortion may be defined as any deviation of a signal in any parameter, such as time, amplitude, or wave shape, from that of the ideal signal. For data and telegraph binary transmission, distortion is defined as a displacement in time of a signal transition from the time that the receiver expects to be correct. In other words, the receiving device must

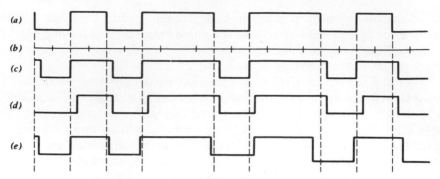

Figure 12.10   Three typical distorted data signals.

make a decision whether a received signal element is a mark or a space. It makes the decision during the sampling interval which is usually at the center of where the received pulse or bit should be. Therefore it is necessary for the transitions to occur between sampling times and preferably halfway between them. Any displacement of the transition instants is called distortion. The degree of distortion a data signal suffers as it traverses the transmission medium is a major contributor in· determining the error rate that can be realized.

Telegraph and data distortion is broken down into two basic types, systematic and fortuitous. Systematic distortion is repetitious and is broken down into bias distortion, cyclic distortion, and end distortion (more common in start–stop systems). Fortuitous distortion is random in nature and may be defined as a distortion type in which the displacement of the transition from the time interval in which it should occur is not the same for every element. Distortion caused by noise spikes in the medium or other transients may be included in this category. Characteristic distortion is caused by transients in the modulation process which appear in the demodulated signal.

Figure 12.10 shows some examples of distortion. Figure 12.10*a* is an example of a binary signal without distortion, and Figure 12.10*b* shows the sampling instants, which occur ideally in the center of the pulse to be sampled. From this we can see that the displacement tolerance is nearly 50%. This means that the point of sample could be displaced by up to 50% of a pulse width and still record the mark or space condition present without error. However, the sampling interval does require a finite amount of time so that in actual practice the displacement permissible is somewhat less than 50%. Figure 12.10*c* and *d* shows bias distortion. An example of spacing bias is shown in Figure 12.10*c*, where all the spacing impulses are lengthened at the expense of the marking impulses. Figure 12.10*d* shows the reverse of this; the marking impulses are elongated at the expense of the spaces. This latter is called marking bias. Figure 12.10*e* shows fortuitous distortion, which is a

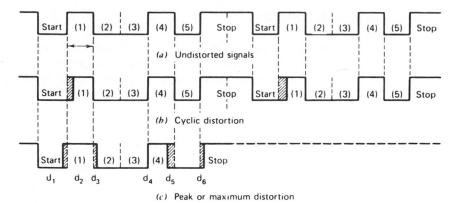

Figure 12.11 Distorted telegraph signals illustrating cyclic and peak distortion. The peak distortion in the character appears at transition d_5.

random type of distortion. In this case the displacement of the signal element is not the same as the time interval in which it should occur for every element.

Figure 12.11 shows distortion that is more typical of start and stop transmission. Figure 12.11a is an undistorted start and stop signal. Figure 12.11b shows cyclic or repetitive distortion typical of mechanical transmitters. In this type of distortion the marking elements may increase in length for a period of time, and then the spacing elements will increase in length. Figure 12.11c shows peak distortion. Identifying the type of distortion present on a signal often gives a clue to the source or cause of distortion. Distortion measurement equipment measures the displacement of the mark-to-space transition from the ideal of a digital signal. If a transition occurs too near to the sampling point, the signal element is liable to be in error. Standards dealing with distortion are published by the Electronic Industries Association (Refs. 13 and 14). Also see CCITT Rec. V.53 (Blue Books).

12.6.5 Bits, Baud, and Words per Minute

There is much confusion among transmission engineers in handling some of the semantics and simple arithmetic used in data and telegraph transmission.

The bit has been defined previously. Now the term *words per minute* will be introduced. A word in our telegraph and data language consists of six characters, usually five letters, numbers, or graphics and a space. All bit sequences transmitted must be counted, such as carriage return and line feed.

Let us look at some arithmetic:

1. A channel is transmitting at 75 bps using a 5-unit start and stop code with a 1.5-unit stop element. Thus for each character there are 7.5 unit intervals (7.5 bits). Accordingly, the channel is transmitting at 100 wpm:

$$\frac{75 \times 60}{6 \times 7.5} = 100 \text{ wpm}$$

2. A system transmits in CCITT no. 5 code at 1500 wpm with parity. How many bits per second are being transmitted?

$$1500 \times 8 \times \tfrac{6}{60} = 1200 \text{ bps}$$

The baud is the unit of modulation rate. In binary transmission systems baud and bits per second are synonymous. Therefore a modem in a binary system transmitting to the line 110 bps has a modulation rate of 110 baud. In multilevel or M-ary systems the number of bauds is indicative of the number of transitions per second. The baud is more meaningful to the transmission engineers concerned with the line side of a modem. This concept will be discussed more at length further on.

12.7 DATA INTERFACE

12.7.1 Introduction

To interface two or more data nodes satisfactorily is a fairly complex matter. One can argue that the transmission engineer should only be concerned with the electrical (physical) interface to assure compatibility of the electrical signals on a data link. Indeed, in this section, we will emphasize the electrical compatibility aspects. However, it is very shortsighted not to consider and appreciate the complexity and many facets of the complete interface. Many of these "facets" will impact transmission engineering directly or indirectly. To interface means to make compatible so interworking can be effected.

Some interface issues have already been discussed. Table 12.1 is just one example. Consider these other examples for a simple data link connecting two nodes or terminals. There will be no compatibility if one end is using the ASCII code and the other is using EBCDIC. Likewise, there will be incompatibility if the CRC used on one end is CRC-CCITT and on the other, CRC-ANSI, nor will there be compatibility if one end operates in the synchronous mode and the other expects to receive in the start–stop mode.

In this section we will first present a generalized picture of data communication compatibility and introduce "protocols." Then we will discuss how protocols are implemented on a layer basis using the OSI (open system interconnection) model. The remainder of the section deals with the electrical (physical) interface.

12.7.2   Initial Considerations for Data Network Access and Control Procedures

The term *communications protocol* defines a set of procedures by which communication is accomplished within standard constraints. As shown in the following outline of protocol topics, protocol deals with control functions.

1. Framing: frame makeup (format); block, message, or packet makeup.
2. Error control (note that this is an interface characteristic as well).
3. Sequence control: the numbering of messages (or blocks) to eliminate duplication, maintain proper sequence in the case of packet networks, and maintain a proper record of identification of messages, especially in ARQ systems or for message servicing.
4. Transparency of the communication links, link control equipment, multiplexers, concentrators, modems, and so on. Transparency allows the use of any bit pattern the user wishes to transmit, even though these patterns resemble control characters or prohibited bit sequences such as long series of 1's or 0's.
5. Line control: determination, in the case of a half-duplex or multipoint line, of which station is going to transmit and which station is going to receive.
6. Idle patterns to maintain network synchronization.
7. Time-out control: which procedures to follow if message (or block or packet) flow ceases entirely.
8. Start-up control: getting a network into operation initially or after some period of remaining idle for one reason or another.
9. Sign-off control: under normal conditions, the process of ending communication or transaction before starting the next transaction or message exchange.

At this juncture we distinguish data link control from user device control. The data link is defined as the configuration of equipment enabling end terminals in two different stations to communicate directly. The data link includes the paired data terminal equipment (DTE), modems, or other signal converters and the interconnecting facilities. The user device may be a central processing unit (CPU), workstation, or other data peripheral. Pictorially, we can illustrate the difference between data link control and user device control by the following diagram:

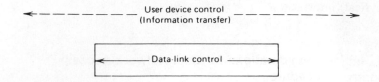

The current literature distinguishes *circuit connection* from *link connection*. Circuit connection is simply the establishment of an electrical path between two points (or multipoints) that want to communicate. We know from our previous discussions that the connection may be metallic (i.e., wire or cable), fiber optic, and/or radio in either the frequency or time domain (e.g., time slot in a frame). The mere establishment of an electrical connection does not mean that data communication can take place. Link establishment is a group of procedures that prepares the source to send data and the destination to receive that data.

In conventional telephony there are three distinct phases to a telephone call: (1) call setup; (2) information transfer, where the subscribers at each end of the connection carry on their conversation; and (3) call termination. A data link must essentially go through the same three procedures. The user control is analogous to the information transfer portion of the telephone call. Thus we introduce protocols.

12.7.3   Protocols

12.7.3.1 Basic Protocol Functions   Stallings (Ref. 15) lists some basic protocol functions. Typical among these functions are

- Segmentation and reassembly
- Encapsulation
- Connection control
- Ordered delivery
- Flow control
- Error control

In many respects these are really a restatement of the protocol topics listed in the previous section. A short description of each function is given below.

Segmentation and Reassembly. Segmentation refers to breaking up the data into blocks with some bounded size. Depending on the semantics or system, these blocks may be called frames or packets. Reassembly is the counterpart of segmentation, that is, putting the blocks or packets back into their original order. Another name used for a data block is *protocol data unit* (PDU).

Encapsulation. Encapsulation is the adding of control information on either side of the data *text* of a block. Typical control information is the *header*, which contains address information, sequence numbers, and error control.

Connection Control. There are three stages of connection control:

1. Connection establishment
2. Data transfer
3. Connection termination

Some of the more sophisticated protocols also provide connection interrupt and recovery capabilities to cope with errors and other sorts of interruptions.

Ordered Delivery. PDUs are assigned sequence numbers to ensure an ordered delivery of the data at the destination. In a large network, especially if it operates in the packet mode, PDUs (packets) can arrive at the destination out of order. With a unique PDU numbering plan using a simple numbering sequence, it is a rather simple task for a long data file to be reassembled at the destination in its original order.

Flow Control. Flow control refers to the management of data flow from source to destination such that buffers do not overflow but maintain full capacity of all facility components involved in the data transfer. Flow control must operate at several peer layers of a protocol.

Error Control. Error control is a technique that permits recovery of lost or errored PDUs. There are three possible functions involved in error control:

1. Acknowledgment of each PDU or string of PDUs
2. Sequence numbering of PDUs (e.g., missing numbers)
3. Error detection

Acknowledgment may be carried out by returning to the source the source sequence number of a PDU. This ensures delivery of all PDUs to the destination. Error detection initiates retransmission of errored PDUs.

12.7.3.2 Open System Interconnection (OSI)

Rationale. Interfacing data systems can be a complex matter. This is especially true when dealing with CPUs and workstations from different vendors as well as variations in software and operating systems. To accommodate the

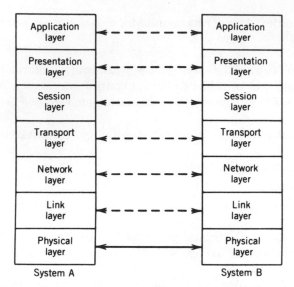

Figure 12.12 The open system interconnection (OSI) model.

multitude of processing-related equipment and software, the International Standards Organization (ISO) developed the open systems interconnection (OSI) seven-layer model. This means that there are seven layers of interface starting at the communication input–output ports of a data device. These seven layers are shown in Figure 12.12.

The purpose of the model is to facilitate communication among data entities. It takes at least two to communicate. Therefore we consider the model in twos, one entity at the left of the figure and one on the right. We use the term *peers*. Peers are corresponding entities on each side of Figure 12.12. A peer on one side (system A) communicates with its peer on the other side (system B) by means of a common protocol. For example, the transport layer system A communicates with its peer transport layer at system B. It is important to note that there is no *direct* communication between peer layers except at the physical layer (layer 1). That is, above the physical layer, each protocol entity sends data *down* to the next lower layer to get the data *across* to its peer entity on the other side. Even the physical layer need not be directly connected as in packet communication. Peer layers must share a common protocol to interface.

There are seven OSI layers, as shown in Figure 12.12. Any layer may be referred to as an N layer. Within a given system there are one or more active entities in each layer. An example of an entity is a process in a multiprocessing system. It could simply be a subroutine. Each entity communicates with entities above and below it across an interface. The interface is at a service access point (SAP). An $(N - 1)$ entity provides services to an N entity by use of primitives. A primitive (Ref. 15) specifies the function to be performed and is used to pass data and control information.

CCITT Rec. X.210 (Ref. 16) describes four types of primitive used to define the interaction between adjacent layers of the OSI architecture. A brief description of each of these primitives is given below:

Request. A primitive issued by a service user to invoke some procedure and to pass parameters needed to specify the service fully.

Indication. A primitive issued by a service provider either to invoke some procedure or to indicate that a procedure has been invoked by a service user at the peer service access point.

Response. A primitive issued by a service user to complete at a particular SAP some procedure invoked by an *indication* at that SAP.

Confirm. A primitive issued by a service provider to complete at a particular SAP some procedure previously invoked by a request at that SAP. CCITT Rec. X.210 (Ref. 16) adds this note: Confirms and responses can be positive or negative depending on the circumstances.

The data that passes between entities is a bit grouping called a *data unit*. We discussed protocol data units (PDUs) earlier. Data units are passed downward from a peer entity to the next OSI layer, called the $(N - 1)$ layer. The lower layer calls the PDU a *service data unit* (SDU). The $(N - 1)$ layer adds control information, transforming the SDU into one or more PDUs. However, the identity of the SDU is preserved to the corresponding layer at the other end of the connection. This concept is shown in Figure 12.13.

When we discussed throughput in Section 12.4.2, it became apparent that throughput must be viewed from the eyes of the user. With OSI some form of encapsulation takes place at every layer above the physical layer. To a greater or lesser extent OSI is used on every and all data connectivities. The concept

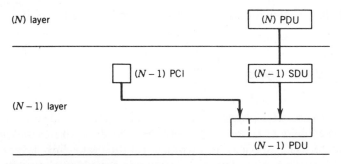

PCI = protocol control information
PDU = protocol data unit
SDU = service data unit

Figure 12.13   An illustration of mapping between data units in adjacent layers.

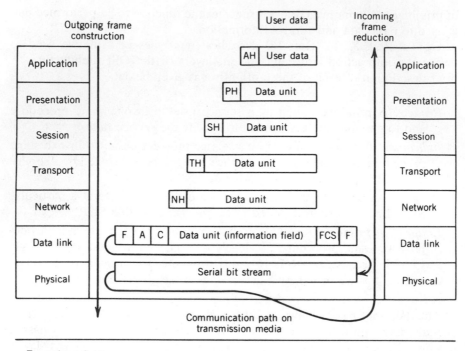

F = unique field
A = address
C = control
FCS = frame check sequence (BCC)
AH = application header
PH = presentation header

Figure 12.14 Buildup and breakdown of a data message following the OSI model. OSI encapsulates at every layer adding significant overhead.

of encapsulation, the adding of overhead, from layers 2 through 7 is shown in Figure 12.14.

Functions of OSI Layers

PHYSICAL LAYER. The physical layer is layer 1 and the lowest OSI layer. It provides the physical connectivity between two data end-users who wish to communicate. The services it provides to the data link layer are those required to connect, maintain, and disconnect the physical circuits that form the physical connectivity. The physical layer represents the traditional interface between data terminal equipment (DTE) and data communication equipment (DCE). (See Section 12.7.4.)

The physical layer has four important characteristics:

1. Mechanical
2. Electrical
3. Functional
4. Procedural

The mechanical aspects include the actual cabling and connectors necessary to connect the communication equipment to the media. Electrical characteristics cover voltage and impedance, balanced and unbalanced. Functional characteristics include connector pin assignments at thc interface and the precise meaning and interpretation of the various interface signals and data set controls. Procedures cover sequencing rules that govern the control functions necessary to provide higher layer services such as establishing a connectivity across a switched network.

Some applicable standards for the physical layer are

- EIA RS-232D, RS-449, RS-422, RS-423, and RS-530
- CCITT Recs. V.10, V.11, V.24, V.28, X.20, X.21, and X.21 bis
- ISO 2110, 2593, 4902, and 4903
- U.S. Fed. Stds. 1020A, 1030A, and 1031
- U.S. MIL-STD-188-114B

DATA LINK LAYER. The data link layer provides services for reliable interchange of data across a data link established by the physical layer. Link layer protocols manage the establishment, maintenance, and release of data link connections. These protocols control the flow of data and supervise error recovery. A most important function of this layer is recovery from abnormal conditions. The data link layer services the network layer or logical link control (LLC; in the case of LANs) and inserts a data unit into the INFO portion of the data frame or block. A generic data frame generated by the link layer is shown in Figure 12.15.

Some of the more common data link layer protocols are

- ISO HDLC, 3309, and 4375
- CCITT LAP-B and LAP-D
- IBM BSC and SDLC
- DEC DDCMP
- ANSI ADCCP (also U.S. government)

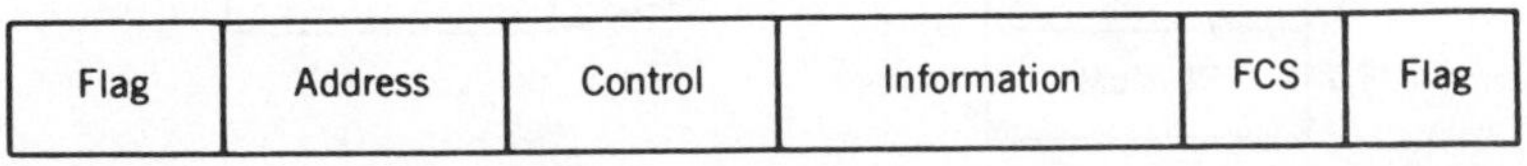

Flag	Address	Control	Information	FCS	Flag

Figure 12.15 Generalized data link layer frame. FCS = frame check sequence.

NETWORK LAYER. The network layer moves data through the network. At relay and switching nodes along the traffic route, layering concatenates. In other words, the higher layers (above layer 3) are not required and are utilized only at user end points.

The network layer carries out the functions of switching and routing, sequencing, logical channel control, flow control, and error recovery functions. We note the duplication of error recovery in the data link layer. However, in the network layer, error recovery is networkwide, whereas on the data link layer, error recovery is concerned only with the data link involved.

The network layer also provides and manages logical channel connections between points in a network such as virtual circuits across the public switched network (PSN). It will be appreciated that the network layer concerns itself with the network switching and routing function. On simpler data connectivities, where a large network is not involved, the network layer is not required and can be eliminated. Typical of such connectivities are point-to-point circuits, multipoint circuits, and LANS. A packet-switched network is a typical example where the network layer is required.

The best known standard for layer 3 is the CCITT Rec. X.25 layer 3 standard for packet operation. CCITT Rec. X.21 provides a standard for network layer functions for circuit-switched operation (Ref. 17).

LAYER 3.5: INTERNETWORK PROTOCOLS. Layer 3.5 is a sublayer of the OSI network layer and carries out the functions of interfacing two disparate networks. The sublayering is shown in Figure 12.16. This internet function is carried out by *gateways*.

There are two applicable protocols for internetworking: CCITT Rec. X.75 (Ref. 18) and IP (internet protocol) (Ref. 19). This latter protocol was

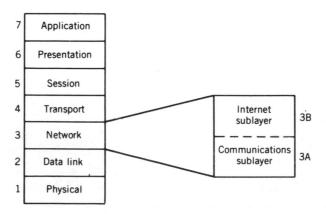

Figure 12.16   Sublayering of OSI layer 3 to achieve internetting.

initially developed by the U.S. Department of Defense Advanced Research Projects Agency (DARPA) and is now standardized by the Defense Department. A simpler standard has been developed by the ISO (ISO 8473).

The CCITT Rec. X.75 protocol is a subset of Rec. X.25 and assumes that all networks involved are based on the latter protocol. The IP provides datagram service; virtual-circuit service and fixed routing are typical of the Rec. X.75 protocol. The IP has broader application, and an IP gateway must know the access scheme of each network with which it interconnects.

TRANSPORT LAYER. The transport layer (layer 4) is the highest layer of the services associated with the provider of communication services. One can say that layers 1–4 are the responsibility of the communication system engineer. Layers 5–7 are the responsibility of the data end-user. However, we believe that the telecommunication system engineer should have a working knowledge of all seven layers.

The transport layer has the ultimate responsibility for providing a reliable end-to-end data delivery service for higher-layer users. It is defined as an end system function, located in the equipment using network service or services. In this way its operations are independent of the characteristics of all the networks that are involved. Services that a transport layer provide are

- *Connection Management*. This includes establishing and terminating connections between transport users. It identifies each connection and negotiates values of all needed parameters.
- *Data Transfer*. This involves the reliable delivery of transparent data between the users. All data is delivered in sequence, with no duplication or missing parts.
- *Flow Control*. This is provided on a connection basis to ensure that data is not delivered at a rate faster than the user's resources can accommodate.

The TCP (transport control protocol) was the first working version of a transport protocol and was created by DARPA for DARPANET. All the features in TCP have been adopted in the ISO version. TCP is often lumped with the internet protocol and referred to as TCP/IP.

The ISO transport protocol messages are called transport protocol data units (TPDUs). There are connection management TPDUs and data transfer TPDUs. The applicable ISO references are ISO 8073 OSI ("Transport Protocol Specification") and ISO 8072 OSI ("Transport Service Definition").

SESSION LAYER. The purpose of the session layer is to provide the means for cooperating presentation entities to organize and synchronize their dialogue and to manage the data exchange. The session protocol implements the services that are required for users of the session layer. It provides the following services for users:

1. The establishment of session connection with negotiation of connection parameters between users
2. The orderly release of connection when traffic exchanges are completed
3. Dialogue control to manage the exchange of session user data
4. A means to define activities between users in a way that is transparent to the session layer
5. Mechanisms to establish synchronization points in the dialogue and, in case of error, resumption from a specified point
6. Interruption of a dialogue and the resumption of it later at a specified point, possibly on a different session connection

Session protocol messages are called session protocol data units (SPDUs). The session protocol uses the transport layer services to carry out its function. A session connection is assigned to a transport connection. A transport connection can be reused for another session connection if desired. Transport connections have a maximum TPDU size. The SPDU cannot exceed this size. More than one SPDU can be placed on a TPDU for transmission to the remote session layer.

Reference standards for the session layer are ISO 8327 ("Session Protocol Definition" [CCITT Rec. X.225]) and ISO 8326 ("Session Services Definition" [CCITT Rec. X.215]).

PRESENTATION LAYER. The presentation layer services are concerned with data transformation, data formatting, and data syntax. These functions are required to adapt the information handling characteristics of one application process to those of another application process.

The presentation layer services allow an application to interpret properly the data being transferred. For example, there are often three syntactic versions of the information to be exchanged between end users A and B as follows:

• Syntax used by the originating application entity A
• Syntax used by the receiving application entity B
• Syntax used between presentation entities (this is called the transfer syntax) (Ref. 15)

Of course, it is possible that all three or any two of these may be identical. The presentation layer is responsible for translating the representation of information between the transfer syntax and each of the other two syntaxes as required.

The following standards apply to the presentation layer:

- ISO 8822   "Connection-Oriented Presentation Service Definition"
- ISO 8823   "Connection-Oriented Presentation Service Specification"
- ISO 8824   "Specification of Abstract Syntax Notation One"
- ISO 8824   "Specification of Basic Encoding Rules for Abstract Syntax Notation One"
- CCITT Rec. X.409   "Message Handling Systems: Presentation Transfer Syntax and Notation"

APPLICATION LAYER. The application layer is the highest layer of the OSI architecture. It provides services to the application processes. It is important to note that the applications do not reside in the application layer. Rather, the layer serves as a window through which the application gains access to the communication services provided by the model.

This highest OSI layer provides to a particular application all services related to communication in such a format that easily interfaces with the user application and is expressed in concrete quantitative terms. These include identifying cooperating peer partners, determining the availability of re- sources, establishing the authority to communicate, and authenticating the communication. The application layer also establishes requirements for data syntax and is responsible for overall management of the transaction.

Of course, the application itself may be executed by a machine, such as a CPU in the form of a program, or by a human operator at a workstation.

The following standards apply to the application layer:

ISO 8449/3   "Definition of Common Application Service Elements"

ISO 8650   "Specification of Protocols for Common Application Service Elements"

12.7.4   Interfacing the Physical Layer

Figure 12.17 illustrates the physical layer interfaces. The DTE (data terminal equipment) may be an I/O device such as a personal computer (PC) or workstation, a computer, mass storage device, or other peripheral equipment. The DCE (data communication equipment) is some type of device that

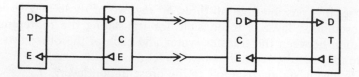

▷ Interface generator
▷ Interface load
≫ Telecommunication channel

Figure 12.17 Digital interface circuit illustrating DTE, DCE, generator, and load. DTE = data terminal equipment; DCE = data communication equipment.

conditions the data signal for transmission. It is a modem if it interfaces the conventional analog network.

We identify two interfaces in Figure 12.17:

- Between DTE and DCE
- Between the DCEs

The first interface has been well standardized; the second interface is less well standardized.

The U.S. Electronics Industries Association (EIA) has been a leader in establishing standards for the DTE–DCE interface. CCITT generally followed the EIA lead by issuing standards for this important area of data communication. Many of the CCITT standards are similar to (but not exactly identical to) the EIA standards. There are U.S. military and U.S. federal standards that must also be considered if we are working in those types of environments. It will be noted that even these standards are traceable to EIA counterparts, or are composites of EIA standards.

At present the leading DTE–DCE interface standard still is EIA RS-232D (Ref. 5). CCITT Rec. V.24 (Ref. 20) is its counterpart for that international agency. These two standards define an electromechanical interface, and, with several related standards, define signal levels, conditions, and polarity at each interface connection. The interface, in effect, is a 25-pin plug/socket. Each of the utilized 25 interface pins can be placed in one of four categories based on the dedicated function the particular pin performs:

1. Electrical ground
2. Data (interchange)
3. Control
4. Clock/timing

Table 12.2 gives RS-232D pin assignments, circuit numbers, and functions for each of the 25 pins with CCITT Rec. V.24 equivalence.

Table 12.2 EIA RS-232D pin assignments with CCITT Rec. V.24 equivalents

Pin No.	Circuit Mnemonic	Description[a]	Source	Type	Equivalent CCITT Rec. V.24
1	AA	Protective ground (chassis)	Ground	Ground	101
2	BA	Transmit data from terminal	Terminal	Data	103
3	BB	Receive data from terminal	Modem	Data	104
4	CA	Request to send (terminal on line)	Terminal	Control	105
5	CB	Clear to send (modem response to CA)	Modem	Control	106
6	CC	Data set ready (telephone function)	Modem	Control	107
7	AB	Signal ground (signal common return)	Ground	Ground	102
8	CF	Carrier detect (received line signal detect)	Modem	Control	109
9		(Positive test voltage/ data set testing)			
10		(Negative test voltage/ data set testing)			
11		Unassigned			
12	SCF/CI	Secondary carrier detect	Modem	Control	122
13	SCB	Secondary clear to send	Modem	Control	121
14	SBA	Secondary transmit data	Terminal	Data	118
15	DB	Transmit clock positive edge = signal element transition	Modem	Timing	114
16	SBB	Secondary receive data	Modem	Data	119
17	DD	Receive clock negative edge = signal element center	Modem	Timing	115
18	LL	Local loopback terminal test			141
19	SCA	Secondary request to send	Terminal	Control	120
20	CD	Data terminal ready (terminal on line)	Terminal	Control	108
21	CG	Signal quality detect (off for receive error)	Modem	Control	110
22	CE	Ring indicator	Modem	Control	125
23	CH/CI	Data signal rate selector	Terminal/ modem	Control	111/112
24	DA	Transmit clock negative edge = signal element center	Terminal	Timing	113
25	TM	Test mode modem test			142

[a]Terminal = DTE; modem = DCE.

Table 12.3   EIA RS-449 interchange circuits

Circuit Mnemonic	Circuit Name	Circuit Direction	Circuit Type
SG	Signal ground		Common
SC	Send common	To DCE	
RC	Receive common	From DCE	
IS	Terminal in service	To DCE	Control
IC	Incoming call	From DCE	
TR	Terminal ready	To DCE	
DM	Data mode	From DCE	
SD	Send data	To DCE	Primary channel
RD	Receive data	From DCE	data
TT	Terminal timing	To DCE	Primary channel
ST	Send timing	From DCE	timing
RT	Receive timing	From DCE	
RS	Request to send	To DCE	Primary channel
CS	Clear to send	From DCE	control
RR	Receiver ready	From DCE	
SQ	Signal quality	From DCE	
NS	New signal	To DCE	
SF	Select frequency	To DCE	
SR	Signal rate selector	To DCE	
SI	Signal rate indicator	From DCE	
SSD	Secondary send data	To DCE	Secondary channel
SRD	Secondary receive data	From DCE	data
SRS	Secondary request to send	To DCE	Secondary
SCS	Secondary clear to send	From DCE	channel
SRR	Secondary receiver ready	From DCE	Control
LL	Local loopback	To DCE	Control
RL	Remote loopback	To DCE	
TM	Test mode	From DCE	
SS	Select standby	To DCE	Control
SB	Standby indicator	From DCE	

Source: Ref. 21. Courtesy of EIA.

The following are the pertinent CCITT recommendations:

V.10 "Electrical Characteristics for Unbalanced Double-Current Interchange Circuits for General Use with Integrated Circuit Equipment in the Field of Data Communications" (Ref. 3)

V.11 "Electrical Characteristics for Balanced Double-Current Interchange Circuits for General Use with Integrated Circuit Equipment in the Field of Data Communications" (Ref. 4)

V.24 "List of Definitions for Interchange Circuits between Data-Terminal and Data Circuit-Terminating Equipment" (Ref. 20)

Table 12.4 EIA RS-449 equivalencies

EIA RS-449		RS-232D		CCITT Rec. V.24	
SG	Signal ground	AB	Signal ground	102	Signal ground
SC	Send common			102a	DTE common
RC	Receive common			102b	DCE common
IS	Terminal in service				
IC	Incoming call	CE	Ring indicator	125	Calling indicator
TR	Terminal ready	CD	Data terminal ready	108/2	Data terminal ready
DM	Data mode	CC	Data set ready	107	Data set ready
SD	Send data	BA	Transmitted data	103	Transmitted data
RD	Receive data	BB	Received data	104	Received data
TT	Terminal timing	DA	Transmitter signal element timing (DTE source)	113	Transmitter signal element timing (DTE source)
ST	Send timing	DB	Transmitter signal element timing (DTE source)	114	Transmitter signal element timing (DTE source)
RT	Receive timing	DD	Receiver signal element timing	115	Receiver signal element timing (DTE source)
RS	Request to send	CA	Request to send	105	Request to send
CS	Clear to send	CB	Clear to send	106	Ready for sending
RR	Receiver ready	CF	Received line signal detector	109	Data channel received line signal detector
SQ	Signal quality	CG	Signal quality detector	110	Data signal quality detector
NS	New signal				
SF	Select frequency			126	Select transmit frequency
SR	Signaling rate selector	CH	Data signal rate selector (DTE source)	111	Data signaling rate selector (DTE source)
SI	Signaling rate indicator	CI	Data signal rate selector (DCE source)	112	Data signaling rate selector (DCE source)
SSD	Secondary send data	SBA	Secondary transmitted data	118	Transmitted backward channel data
SRD	Secondary receive data	SBB	Secondary received data	119	Received backward channel data
SRS	Secondary request to send	SCA	Secondary request to send	120	Transmit backward channel line signal
SCS	Secondary clear to send	SCB	Secondary clear to send	121	Backward channel ready
SRR	Secondary receiver ready	SCF	Secondary received line signal detector	122	Backward channel received line signal detector
LL	Local loopback			141	Local loopback
RL	Remote loopback			140	Remote loopback
TM	Test mode			142	Test indicator
SS	Select standby			116	Select standby
SB	Standby indicator			117	Standby indicator

V.28 "Electrical Characteristics for Unbalanced Double-Current Interchange Circuits" (Ref. 22)

V.31 "Electrical Characteristics for Single-Current Interchange Circuits Controlled by Contact Closure" (Ref. 23)

V.230 "General Data Communications Interface Layer 1 Specification" (Ref. 24)

Note: Where CCITT refers to "double-current" we use polar keying or polar transmission; for "single-current" we use neutral keying or neutral transmission.

The relevant U.S. military standard is MIL-STD-188-114B, "Electrical Characteristics of Digital Interface Circuits" (Ref. 25).

EIA RS-232D will probably remain in force for some time, as well as its CCITT counterpart. However, during the mid-1990s EIA RS-232D may be phased out slowly and replaced by EIA RS-449, supplemented by EIA RS-422 and RS-423. With a few additional provisions for interoperability, equipment conforming to EIA RS-449 can interoperate with equipment designed to EIA RS-232D. Essentially, EIA RS-449 specifies the functional and mechanical characteristics of the interface between DTE and DCE, whereas EIA RS-422 and 423 specify the electrical characteristics of digital interface circuits, the former dealing with balanced voltage interface, and the latter, unbalanced (Refs. 14 and 27).

EIA RS-232D defines a single 25-pin plug socket and EIA RS-449 a 37-pin interface. Table 12.3 lists interchange circuits and their mnemonics relating to EIA RS-449 and Table 12.4 is an equivalency table for EIA RS-449/RS-232D and CCITT Rec. V.24.

Figures 12.18 and 12.19 illustrate the digital interface circuits for EIA RS-422 and RS-423, respectively. EIA RS-422, as mentioned, specifies a balanced interface and EIA RS-423 an unbalanced interface. As we would expect, the unbalanced circuit uses a common return (ground), which is shown in Figure 12.19. In either figure the load may be considered to be one or more receivers. Also see how the generator and load are configured in Figure 12.17.

Some basic guidelines for the application of balanced and unbalanced electrical connections DTE–DCE are presented below. EIA states in RS-422 that

> While the balanced interface is intended for use at the higher modulation rates, it may, in preference to the unbalanced interface circuit, generally be required where any of the following conditions prevail:
>
> • The interconnecting cable (i.e., DTE–DCE) is too long for effective unbalanced operation
>
> • The interconnecting cable is exposed to extraneous noise sources that may cause an unwanted voltage in excess of plus or minus 1 V measured

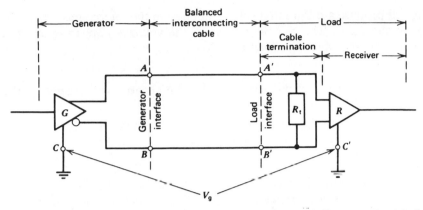

Figure 12.18 Balanced digital interface circuit, EIA RS-422. A, B = generator interface; A', B' = load interface; C = generator circuit ground; C' = load circuit ground; R = optional cable termination resistance; V_g = ground potential difference. (From Ref. 14; courtesy of EIA.)

between the signal conductor and the circuit common at the load end of the cable with a 50-Ω resistor substituted for the generator

- It is necessary to minimize interference with other signals
- Inversion of signals may be required, e.g., PLUS MARK to MINUS MARK, may be obtained by inverting the cable pair

RS-422 states, in essence (Figure 12.18), that a generator, as defined in the standard, results in a low-impedance (100 Ω or less) balanced voltage source

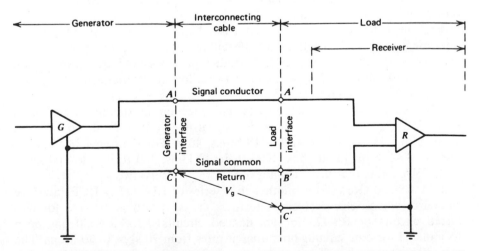

Figure 12.19 Unbalanced digital interface circuit, RS-423. A, C = generator interface; A', B' = load interface; C' = load circuit ground; C = generator circuit ground; V_g = ground potential difference. (From Ref. 27; courtesy of EIA.)

that will produce a differential voltage applied to the interconnecting cable in the range of 2-V. The signaling sense of the voltages appearing across the interconnection cable is defined as follows:

1. The *A* terminal of the generator shall be negative with respect to the *B* terminal for a binary 1 (MARK or OFF state)
2. The *A* terminal of the generator shall be positive with respect to the *B* terminal for a binary 0 (SPACE or ON state)

Note: Compare this signaling convention to that in Table 12.1.

EIA RS-423 states, in essence (Figure 12.19), that a generator circuit, as defined in the standard, results in a low-impedance (50 Ω or less) unbalanced voltage source that will produce a voltage range of 4–6 V. The signaling sense of the voltage appearing across the interconnecting cable is defined as follows:

1. The *A* terminal of the generator shall be negative with respect to the *C* terminal for a binary 1 (MARK or OFF state)
2. The *A* terminal of the generator shall be positive with respect to the *C* terminal for a binary 0 (SPACE or ON state)

The test load termination for a balanced circuit is two 50-Ω resistors in series between the two signal leads (*A* and *B* in Figure 12.18). The test lead termination for the unbalanced circuit is 450 Ω between the generator output terminal and "common."

For the balanced or unbalanced case, load characteristics during real operating conditions result in a differential receiver having a high input impedance (greater than 4 kΩ), a small input threshold transition region between −0.2 and +0.2 V, and an allowance for an internal bias voltage not to exceed 3 V in magnitude.

EIA RS-422 and RS-423 provide guidance on maximum cable length DTE–DCE. For the case of balanced signal lines (RS-422): up to 90 kbps, 4000 ft; to 1 Mbps, 380 ft; and to 10 Mbps, 40 ft. For the unbalanced signal lines (RS-423): up to 900 bps, 4000 ft; to 10 kbps, 380 ft; and to 100 kbps, 40 ft. EIA states that these lengths are on the conservative side.

EIA RS-530 (Ref. 28) is another in a series of EIA DTE–DCE interface standards whose electrical characteristics are specified in RS-422 for balanced circuits or RS-423 for unbalanced circuits. EIA RS-530 has been formulated for data circuits operating at rates from 20 kbps to 20 Mbps. The RS-530 connector is a 25-pin connector similar to that of RS-232D; however, it is D-shaped. Circuit names for the first eight pins in both standards are the

Table 12.5   Connector pin comparison: EIA RS-530 and RS-449

EIA RS-530			EIA RS-449		
Circuit Name	Circuit Mnemonic	Contact	Contact	Circuit Mnemonic	Circuit Name
Shield		1	1		Shield
Transmitted data	BA (A)	2	4	SD (A)	Send data
	BA (B)	14	22	SD (B)	
Received data	BB (A)	3	6	RD (A)	Receive data
	BB (B)	16	24	RD (B)	
Request to send	CA (A)	4	7	RS (A)	Request to send
	CA (B)	19	25	RS (B)	
Clear to send	CB (A)	5	9	CS (A)	Clear to send
	CB (B)	13	27	CS (B)	
DCE ready	CC (A)	6	11	DM (A)	Data mode
	CC (B)	22	29	DM (B)	
DTE ready	CD (A)	20	12	TR (A)	Terminal ready
	CD (B)	23	30	TR (B)	
Signal ground	AB	7	19	SG	Signal ground
Received line	CF (A)	8	13	RR (A)	Receiver ready
signal detector	CF (B)	10	31	RR (B)	
Transmit signal	DB (A)	15	5	ST (A)	Send timing
element timing	DB (B)	12	23	ST (B)	
(DCE source)					
Receiver signal	DD (A)	17	8	RT (A)	Receive timing
element timing	DD (B)	9	26	RT (B)	
(DCE source)					
Local loopback	LL	18	10	LL	Local loopback
Remote loopback	RL	21	14	RL	Remote loopback
Transmit signal	DA (A)	24	17	TT (A)	Terminal timing
element timing					
(DTE source)	DA (B)	11	35	TT (B)	
Test mode	TM	25	18	TM	Test mode

Source: EIA RS-530 (Ref. 28) and EIA RS-449 (Ref. 21).

same, but differ for pins 9–11, which are not used with RS-232D. Table 12.5 compares pin connector assignments and their functions for EIA RS-530 and RS-449. Eventually, it is expected that RS-530 will replace RS-449.

The following provides a cross-reference of interface standards (Ref. 58):

EIA RS-423A	U.S. Fed. Std. 1030A	CCITT V.10 (X.26)
EIA RS-422A	U.S. Fed. Std. 1020A	CCITT V.11 (X.27)
EIA RS-449	U.S. Fed. Std. 1031	CCITT V.24/V.10/V.11, ISO 4902
EIA RS-232D		CCITT V.24/V.28, ISO 2110
EIA RS-530		CCITT V.230

12.8 DATA INPUT–OUTPUT DEVICES

The following is meant to give the reader a broadbrush familiarity with data subscriber equipment, which is more often referred to as input–output devices. Such equipment converts user information (data or messages) into electrical signals, or vice versa. Many refer to the input–output devices as the human interface. Electrically, a data subscriber terminal consists of the end instrument, the DTE, and a device called a modem or communication set. The DTE has been discussed in Section 12.7. Modems are described in Section 12.13. The data source is the input device and the data destination is the output device.

Data input–output devices handle paper tape, punched cards, magnetic tape, drums, disks, visual displays, and printed page copy. Input devices* may be broken down into the following categories:

- Keyboard sending units
- Card readers (being phased out)
- Paper tape readers
- Magnetic tape, disk, drum readers, and hard drives
- Optical character readers (OCRs)

Output devices* are as follows:

- Printers (paper, hard copy)
- Card punches (being phased out)
- Paper tape punches
- Magnetic tape recorders, magnetic cores, disks, and hard drives
- Visual display units (VDUs) (cathode ray tubes and plasma)

Further, these devices may be used as on-line or off-line devices. Off-line devices are not connected directly to the communication system, but serve as auxiliary equipment.

Off-line devices such as a keyboard/display or workstation are used to store data messages on storage media such as mag tape disk, hard drive, or CD for later transmission.

---

*Compatible with the data rates under discussion.

The following table provides equivalence between telegraph and data terminology of input–output devices:

Data	Telegraph
Keyboard	Keyboard
Tape reader	Transmitter–distributor*
Printer	Teleprinter
Tape punch	Perforator, reperforator
Visual displays	

It is expected that the compact disk (CD) will shortly become an efficient input–output medium.

12.9   DIGITAL TRANSMISSION ON AN ANALOG CHANNEL

12.9.1   Introduction

There are two fundamental approaches to the practical problem of data transmission. The first is often to design and construct a complete new network expressly for the purpose of data transmission. The second approach is to adapt the many existing telephone facilities for data transmission. The paragraphs that follow deal with the second approach.

Transmission facilities designed to handle voice traffic have characteristics that make it difficult to transmit dc binary digits or bit streams. To permit the transmission of data over analog voice facilities (i.e., the telephone network), it is necessary to convert the dc data into a signal within the VF range. The equipment that performs the necessary conversion to the signal is generally termed a *data modem*. Modem is an acronym for modulator–demodulator.

12.9.2   Modulation–Demodulation Schemes

A modem modulates and demodulates. The types of modulation used by presentday modems may be one or combinations of the following:

- Amplitude modulation (AM)
- Frequency modulation (FM)
- Phase modulation (PM)

Amplitude Modulation   With this modulation technique, binary states are represented by the presence, or absence, of an audio tone or carrier. More often, it is referred to as on–off telegraphy. For data rates up to 1200 bps,

*The distributor performs the parallel-to-serial equivalent conversion of data transmission.

one such system uses a carrier frequency centered at 1600 Hz. For binary transmission AM has significant disadvantages which include (1) susceptibility to sudden gain change, and (2) inefficiency in modulation and spectrum utilization, particularly at higher modulation rates (see CCITT Rec. R.70). Binary modulation using AM is called ASK, or amplitude shift keying.

An improvement in the conventional technique results from the removal of one of the information-carrying sidebands. Since the essential information is present in each of the sidebands, there is no loss of content in the process. The carrier frequency must be preserved to recover the dc component of the information envelope. Therefore digital systems of this type use vestigial sideband (VSB) modulation in which one sideband, a portion of the carrier and a "vestige" of the other sideband, is retained. This is accomplished by producing a DSB signal and filtering out the unwanted sideband components. As a result, the signal takes only about three-fourths of the bandwidth required for a DSB system. Typical VSB data modems are operable up to 2400 bps in a telephone channel. Data rates up to 4800 bps are achieved using multilevel (*M*-ary) techniques. The carrier frequency is usually located between 2200 and 2700 Hz.

Frequency Modulation   A large number of data transmission systems utilize a digital form of frequency modulation, commonly called frequency shift keying (FSK). The two binary states are represented by two different frequencies and are detected by using two frequency tuned sections, one tuned to each of the two bit frequencies. The demodulated signals are then integrated over the duration of a bit, and upon the result a binary decision is based.

Digital transmission using frequency shift keying (FSK) has the following advantages. (1) The implementation is not much more complex than an AM system. (2) Since the received signals can be amplified and limited at the receiver, a simple limiting amplifier can be used, whereas the AM system requires sophisticated AGC in order to operate over a wide level range. Another advantage is that FSK can show a 3- or 4-dB improvement over AM in most types of noise environment, especially at distortion threshold (i.e., at the point where the distortion is such that good printing is about to cease). As the frequency shift becomes greater, the advantage over AM improves in a noisy environment.

Another advantage of FSK is its immunity from the effects of nonselective level variations even when extremely rapid. Thus it is used almost exclusively on worldwide HF radio transmission, where rapid fades are a common occurrence. In the United States it has nearly universal application for the transmission of data at the lower data rates (i.e., 1200 bps and below).

Phase Modulation   For systems using higher data rates, a form of phase modulation called phase shift keying (PSK) becomes more attractive. Various forms are used, such as two phase, relative phase, quadrature phase, and

8-phase. A two-phase system uses one phase of the carrier frequency for one binary state and the other phase for the other binary state. The two phases are 180° apart and are detected by a synchronous detector using a reference signal at the receiver that is of known phase with respect to the incoming signal. This known signal is at the same frequency as the incoming signal carrier and is arranged to be in phase with one of the binary signals. This is called coherent detection.

In the relative (differential) phase system, a binary 1 is represented by sending a signal burst of the same phase as that of the previous signal burst sent. A binary 0 is represented by a signal burst of a phase opposite to that of the previous signal transmitted. The signals are demodulated at the receiver by integrating and storing each signal burst of one bit period for comparison in phase with the next signal burst.

In the quadrature-phase system, two binary channels (2 bits) are phase multiplexed onto one tone by placing them in phase quadrature as shown in the sketch below. An extension of this technique places two binary channels on each of several tones spaced across the voice channel of a typical telephone circuit.

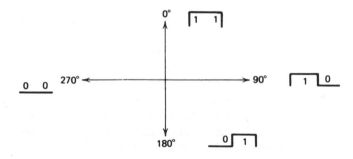

Some of the advantages of phase modulation are that

- All available power is utilized for intelligence conveyance
- The demodulation scheme has good noise rejection capability
- The system yields a smaller noise bandwidth

A disadvantage of such a system is the complexity of equipment required.

12.9.3 Critical Parameters

The effect of the various telephone circuit parameters on the capability of a circuit to transmit data is a most important consideration. The following discussion is to familiarize the reader with the problems most likely to be encountered in the transmission of data over analog circuits (e.g., the telephone network) and to make some generalizations in some cases, which can be used to help in planning the implementation of data systems.

Delay Distortion   Delay distortion "constitutes the most limiting impairment to data transmission, particularly over telephone voice channels" (Ref. 29). When specifying delay distortion, the terms *envelope delay distortion* and *group delay* are often used. The IEEE standard dictionary (Ref. 1) states that "envelope delay is often defined the same as group delay, that is, the rate of change, with angular frequency, of the phase shift between two points in a network" (see Section 1.9.4).

The problem is that in a band-limited analog system, such as the typical telephone voice channel, not all frequency components of the input signal will propagate to the receiving end in exactly the same elapsed time, especially on loaded cable circuits and frequency division multiplex (FDM) carrier systems. In carrier systems it is the cumulative effect of the many filters used in the FDM equipment. On long-haul circuits the magnitude of delay distortion is generally dependent on the number of carrier modulation stages that the circuit must traverse rather than the length of the circuit. Figure 12.20 shows a typical frequency–delay response curve in milliseconds of a voice channel due to FDM equipment only. For the voice channel (or any symmetrical passband for that matter), delay increases toward band edge and is minimum around the center portion of the passband.

In essence, therefore, we are dealing with the phase linearity of a circuit. If the phase–frequency relationship over the passband is not linear, distortion will occur in the transmitted signal. This distortion is best measured by a parameter called *envelope delay distortion* (EDD). Mathematically, envelope delay is the derivative of the phase shift with respect to frequency. The maximum difference in the derivative over any frequency interval is called EDD. Hence EDD is always the difference between the envelope delay at one frequency and that at another frequency of interest in a passband. The EDD unit of measurement is milliseconds or microseconds.

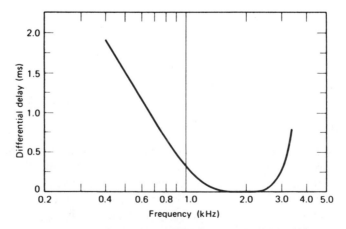

Figure 12.20   Typical differential delay across a voice channel. FDM equipment back-to-back.

When transmitting data, the shorter the pulse (or symbol) width (in the case of binary systems this would be the width of 1 bit), the more critical the EDD constraints become. As a rule of thumb, delay distortion in the passband should be below the period of 1 bit (or symbol).

Amplitude Response (Attenuation Distortion)   Another parameter that seriously affects the transmission of data and that can place very definite limits on the modulation rate is that of amplitude response. Ideally, all frequencies across the passband of the channel of interest should suffer the same attenuation. Place a −10-dBm signal at any frequency between 300 and 3400 Hz, and the output at the receiving end of the channel may be −23 dBm, for example, at any and all frequencies in the band; we would then describe a fully flat channel. Such a channel has the same loss or gain at any frequency within the band. This type of channel is ideal but would be unachievable in a real, working system. In Rec. G.132, CCITT recommends no more than 9 dB of amplitude distortion relative to 800 Hz between 400 and 3000 Hz. This figure of 9 dB describes the maximum variation that may be expected from the reference level at 800 Hz. This variation of amplitude response often is called attenuation distortion. A conditioned channel, such as a Bell System C-4 channel, will maintain a response of −2 to +3 dB from 500 to 3000 Hz and −2 to +6 dB from 300 to 3200 Hz. Channel conditioning and equalization are discussed in Section 12.12.

Considering tandem operation, the deterioration of amplitude response is arithmetically summed when sections are added. This is particularly true at band edge in view of channel unit transformers and filters which account for the upper and lower cutoff characteristics.

Amplitude response is also discussed in Section 1.9.3. Figure 12.21 illustrates a typical example of amplitude response across FDM carrier

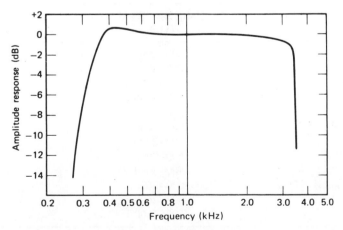

Figure 12.21   Typical amplitude versus frequency response across a voice channel. Channel modulator, demodulator back-to-back, FDM equipment.

Table 12.6 AT&T requirements for two-point or multipoint channel attenuation distortion

	Frequency Band (Hz)	Attenuation[a] (dB)
Basic requirements	500–2500	−2 to +8
	300–3000	−3 to +12
C1 conditioning	1000–2400	−1 to +3
	300–2700	−2 to +6
	2700–3000	−3 to +12
C2 conditioning	500–2800	−1 to +3
	300–3000	−2 to +6
C4 conditioning	500–3000	−2 to +3
	300–3200	−2 to +6
C5 conditioning	500–2800	−0.5 to +1.5
	300–3000	−3 to +3

Source: Ref. 30.
[a]Relative to 1000 Hz.

equipment (see Chapter 3) connected back-to-back at the voice channel input–output.

Tables 12.6 and 12.7 give the AT&T requirements for leased lines covering attenuation distortion and EDD.

Noise Another important consideration in the transmission of data is that of noise. All extraneous elements appearing at the voice channel output that

Table 12.7 AT&T requirements for two-point or multipoint channel envelope delay distortion

	Frequency Band (Hz)	EDD (μs)[a]
C1 conditioning	800–2600	1750
	1000–2400	1000
C2 conditioning	1000–2600	500
	600–2600	1500
	500–2800	3000
C4 conditioning	1000–2600	300
	800–2800	500
	600–3000	1500
	500–3000	3000
C5 conditioning	1000–2600	100
	600–2600	300
	500–2800	600

Source: Ref. 30.
[a]Maximum in-band envelope delay difference.

were not due to the input signal are considered to be noise. For convenience noise is broken down into four categories:

- Thermal
- Crosstalk
- Intermodulation
- Impulse

Thermal noise, often called resistance noise, white noise, or Johnson noise, is of a Gaussian nature or fully random. Any system or circuit operating at a temperature above absolute zero will inherently display thermal noise. It is caused by the random motions of discrete electrons in the conduction path.

Crosstalk is a form of noise caused by unwanted coupling from one signal path into another. It may be due to direct inductive or capacitive coupling between conductors or between radio antennas (see Section 1.9.6).

IM noise is another form of unwanted coupling, usually caused by signals mixing in nonlinear elements of a system. Carrier and radio systems are highly susceptible to IM noise, particularly when overloaded (see Section 1.9.6).

Impulse noise is a primary source of errors in the transmission of data over telephone networks. It is sporadic and may occur in bursts or discrete impulses called *hits*. Some types of impulse noise are natural, such as that from lightning. However, man-made impulse noise, such as from automobile ignition systems or power lines, is ever-increasing. Impulse noise may be of high level in analog telephone switching centers due to dialing, supervision, and switching impulses which may be induced or otherwise coupled into the data transmission channel.

For our discussion of data transmission, only two forms of noise will be considered: random or Gaussian noise and impulse noise. Random noise measured with a typical transmission measuring set appears to have a relatively constant value. However, the instantaneous value of the noise fluctuates over a wide range of amplitude levels. If the instantaneous noise voltage is of the same magnitude as the received signal, the receiving detection equipment may yield an improper interpretation of the received signal and an error or errors will occur. For a proper analytical approach to the data transmission problem, it is necessary to assume a type of noise that has an amplitude distribution that follows some predictable pattern. Thermal noise or random noise has a Gaussian distribution and is considered representative of the noise encountered on the analog telephone channel (i.e., the voice channel). From the probability distribution curve of Gaussian noise shown in Figure 12.22, we can make some accurate predictions. It may be noted from this curve that the probability of occurrence of noise peaks that have amplitudes 12.5 dB above the rms level is 1 in 10^5. Accordingly, if we wish to ensure an error rate of 10^{-5} in a particular system using binary polar

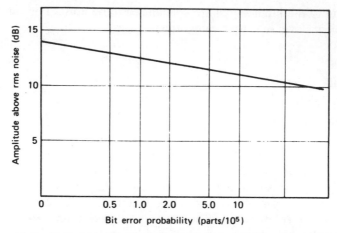

Figure 12.22   Probability of bit error in Gaussian noise, binary polar transmission.

modulation, the rms noise should be at least 12.5 dB below the signal level
(Ref. 29, p. 114; Ref. 31, p. 6). This simple analysis is valid for the type of
modulation used, assuming that no other factors are degrading the operation
of the system and that a cosine-shaped receiving filter is used. If we were to
interject EDD, for example, into the system, we could translate the degradation into an equivalent signal-to-noise ratio improvement necessary to restore
the desired error rate. For instance, if the delay distortion were the equivalent of one pulse width, the signal-to-noise ratio improvement required for
the same error rate would be about 5 dB, or the required signal-to-noise ratio
would now be 17.5 dB.

For reasons that will be discussed later, let us assume that the signal level
is -10 dBm at the zero transmission level point of the system. Then the rms
noise measured at the same point would be -27.5 dBm to retain the error
rate of 1 in 10^5.

In order for the above figure to have any significance, it must be related to
the actual noise found in a channel. CCITT recommends no more than
50,000 pW of noise psophometrically weighted on an international connection made up of six circuits in a chain. However, CCITT states (Recs. G.142A
and 142D) that for data transmission at as high a modulation rate as possible
without significant error rate, a reasonable circuit objective for maximum
random noise would be -40 dBm0p for leased circuits (impulse noise not
included) and -36 dBm0p for switched circuits without compandors. This
figure obviously appears quite favorable when compared to the -27 dBm0
(-29.5 dBm0p) required in the example above. However, other factors that
will be developed later will consume much of the noise margin that appears
available.

Whereas random noise has an rms value when we measure level, impulse
noise is another matter entirely. It is measured as the number of "hits" or

"spikes" per interval of time over a certain threshold. In other words, it is a measurement of the recurrence rate of noise peaks over a specified level. The word *rate* should not mislead the reader. The recurrence is not uniform per unit time, as the word *rate* may indicate, but we can consider a sampling and convert it to an average.

AT & T (Ref. 30) states the following:

> The impulse noise objective is specified in terms of the rate of occurrence of the impulse voltages above a specified magnitude. The objective is expressed as the threshold in dBrnc0 at which no more than 15 impulses in 15 minutes are measured by an impulse counter with a maximum counting rate of 7 counts per second. The overall objective of 71 dBrnc0 implies a 6-dB signal-to-impulse noise threshold in the presence of a -13-dBm0 signal.

CCITT states (Rec. Q.45) that

> in any four-wire international exchange the busy hour impulsive noise counts should not exceed 5 counts in 5 minutes at a threshold level of -35 dBm0.

Remember that random noise has a Gaussian distribution and will produce peaks at 12.5 dB over the rms value (unweighted) 0.001% of the time on a data bit stream for an equivalent error rate of 1×10^{-5}. It should be noted that some references use 12 dB, some 12.5 dB, and others 13 dB. The 12.5 dB above the rms random noise floor should establish the impulse noise threshold for measurement purposes. We should assume in a well-designed data transmission system traversing the telephone network that the signal-to-noise ratio of the data signal will be well in excess of 12.5 dB. Thus impulse noise may well be the major contributor to degrade the error rate.

Care must be taken when measuring impulse noise. A transient such as an impulse noise spike in a band-limited system (which our telephone network most certainly is) tends to cause "ringing." Here the initial impulse noise spike causes what we might call a main bang or principal spike followed by damped subsidiary spikes. If we are not careful, these subsidiary spikes, that ringing effect, may also be counted as individual hits in our impulse noise count total. To avoid this false counting, impulse noise meters have a built-in dead time after each count. It is a kind of damping. The Bell System, for example, specifies a 150-ms dead time after each count. This limits the counting capability of the meter to no more than 6 or 7 counts per minute (cpm).

In this damping or dead time period, missed (real) impulse noise hits may seem to be a problem. For instance, the Bell System suggests that the average improved (increased) sensitivity to measure "all" hits is only 0.9 dB, with a standard deviation of 0.76 dB (Ref. 32).

The period of measurement is also important. How long should the impulse noise measurement set remain connected to a line under test to give an accurate count? It appears empirically that 30 min is sufficient. However, a good estimate of error can be made and corrected for if that period of time is reduced to 5 min. This is done by reducing the threshold (on paper) of the measuring set. From Ref. 32, the standard deviation for a 5-min period is about 2.2 dB. Therefore 95% of all 5-min measurements will be within ±3.6 dB of a 30-min measurement period.

To clarify this, remember that impulse noise distributions are log normal and impulse noise level distributions are normal. With this in mind we can relate count distributions, which can be measured readily, to level distributions. The mean of the level distribution is the threshold value of which the impulse noise level meter was set to record the count distribution (in dBm, dBmp, dBrnC, or whatever unit). The set has a count associated with that threshold which is simply the median of the observed count distribution. The sigma σ_1 standard deviation of the impulse noise level distribution is estimated by the expression

$$\sigma_1 = m\sigma_D \tag{12.1}$$

where m = inverse slope of the peak amplitude distribution in decibels per decade of counts, and averages 7.0 dB, and σ_D = standard deviation of the log normal count distribution, which is the square root of the $\log_{10}$ of the ratio of the average number of counts to the median count, or

$$\sigma_D = \sqrt{\log_{10}\frac{\text{average count}}{\text{median count}}} \tag{12.2}$$

where the median is not equal to zero. For instance, if we measured 10 cpm at a given threshold, the 1-cpm threshold would be 7 dB above the 10-cpm threshold.

When an unduly high error rate has been traced to impulse noise, there are some methods for improving conditions. Noisy areas may be bypassed, repeaters may be added near the noise source to improve the signal-to-impulse noise ratio, or in special cases pulse smearing techniques may be used. This latter approach uses two delay distortion networks that complement each other such that the net delay distortion is zero. By installing the networks at opposite ends of the circuit, impulse noise passes through only one network and is hence smeared because of the delay distortion. The signal is unaffected because it passes through both networks.

Levels and Level Variations   The design signal levels of telephone networks traversing FDM carrier systems are determined by average talker levels, average channel occupancy, permissible overloads during busy hours, and so on. Applying constant amplitude digital data tone(s) over such an equipment at 0 dBm0 on each channel would result in severe overload and IM noise within the system.

Loading does not affect hard wire systems except by increasing crosstalk. However, once the data signal enters carrier multiplex (voice) equipment, levels must be considered carefully, and the resulting levels most probably have more impact on the final signal-to-noise ratio at the far end than anything else. CCITT (Recs. G.151C, H51, H23, and V2) recommends a level of -10 dBm0 in some cases, and -13 dBm0 when the proportion of nonspeech circuits on an international carrier circuit exceeds 10 or 20%. For multichannel telegraphy the composite level is -8.7 dBm0, or for 24 channels each channel would be adjusted for -22.5 dBm0. Even this loading may be too heavy if a high proportion of the voice channels are loaded with data. Depending on the design of the carrier equipment, cutbacks to -13 dBm0 or less may be advisable.

In a properly designed transmission system the standard deviation of the variation in level should not exceed 1.0 dB per circuit. However, data communication equipment should be able to withstand level variations in excess of 4 dB.

On digital systems, the proportion of data, facsimile, and speech channels is unimportant.

Frequency Translation Errors Total end-to-end frequency translation errors on a voice channel being used for data or telegraph transmission must be limited to 2 Hz (CCITT Rec. G.135). This is an end-to-end requirement. Frequency translations occur mostly owing to carrier equipment modulation and demodulation steps. FDM carrier equipment widely uses single-sideband suppressed carrier (SSBSC) techniques. Nearly every case of error can be traced to errors in frequency translation (we refer here to deriving the group, supergroup, mastergroup, and its reverse process; see also Chapter 3) and carrier reinsertion frequency offset, the frequency error being exactly equal to the error in translation and offset or the sum of several such errors. Frequency locked (e.g., synchronized) or high-stability master carrier generators (1×10^{-7} or 1×10^{-8}, depending on the system), with all derived frequency sources slaved to the master source, usually are employed to maintain the required stability.

Although 2 Hz seems to be a very rigid specification, when added to the possible back-to-back error of the modems themselves, the error becomes more appreciable. Much of the trouble arises with modems that employ sharply tuned filters. This is true of telegraph equipment in particular. But for the more general case, high-speed data modems can be designed to withstand greater carrier shifts than those that will be encountered over good telephone circuits.

Phase Jitter The unwanted change in phase or frequency of a transmitted signal due to modulation by another signal during transmission is defined as *phase jitter*. If a simple sinusoid is frequency or phase modulated during transmission, the received signal will have sidebands. The amplitude of these

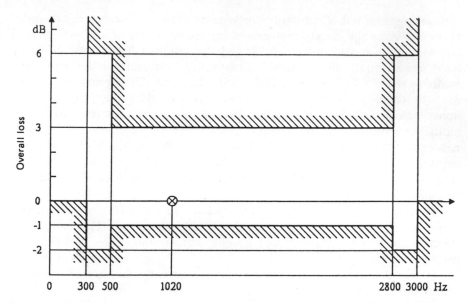

Figure 12.23 Limits of overall loss of the circuit relative to that at 1020 Hz. (From CCITT Rec. M.1020, Figure 1 / M.1020, Page 56, Fascicle IV-2, IXth Plenary Assembly, Melbourne 1988, Ref. 34.)

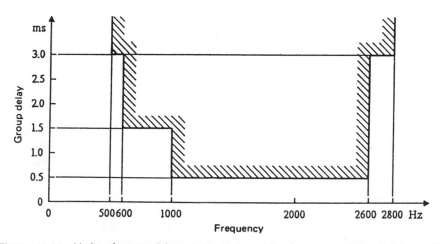

Figure 12.24 Limits of group delay relative to the minimum measured group delay in the 500 – 2800-Hz band. (From CCITT Rec. M.1020 Figure 2 / M.1020, Page 56, Fascicle IV-2, IXth Plenary Assembly, Melbourne 1988, Ref. 34.)

sidebands compared to the received signal is a measure of the phase jitter imparted to it during transmission.

Phase jitter is measured in degrees of variation peak to peak for each hertz of transmitted signal. Phase jitter shows up as unwanted variations in zero crossings of a received signal. It is the zero crossings that data modems use to distinguish marks and spaces. Consequently, the higher the data rate, the more jitter can affect the error rate on the receive bit stream.

The greatest cause of phase jitter in the telephone network is FDM carrier equipment, where it shows up as undesired incidental phase modulation. Modern FDM equipment derives all translation frequencies from one master frequency source by multiplying and dividing its output. To maintain stability, phase-lock techniques are used. Thus the low jitter content of the master oscillator may be multiplied many times. It follows, then, that we can expect more phase jitter in the voice channels occupying the higher baseband frequencies.

Jitter most commonly appears on long-haul systems at rates related to the power line frequency (e.g., 60 Hz and its harmonics and submultiples, or 50 Hz and its harmonics of submultiples) or derived from 20-Hz ringing frequency. Modulation components that we define as jitter usually occur close to the carrier, from about 0 to about ± 300 Hz maximum.

12.9.4 Special-Quality International Leased Circuits

12.9.4.1 Introduction CCITT offers two recommendations dealing with line conditioning for international leased circuits, Recs. M.1020 and M.1025 (Refs. 34 and 35). The two recommendations are similar except for amplitude and phase distortion. The latter is given as group delay. The following subsections give the highlights of the two recommendations. For Rec. M.1025, only amplitude and phase distortion parameters are given.

12.9.4.2 With Special Bandwidth Conditioning The following requirements are based on CCITT Rec. M.1020. Figure 12.23 shows the limits for overall loss for a voice channel relative to that at 1020 Hz. The figure, of course, also represents the limits of amplitude distortion, which can vary from -1 to $+3$ dB from 500 to 2500 Hz.

Figure 12.24 gives the limits of group delay (distortion) relative to the minimum measured group delay in the band from 500 to 2800 Hz.

The following summarizes other requirements of the recommendation:

- Amplitude hits greater than ± 2 dB should not exceed 10 in any 15-min measuring period
- Impulse noise: The number of impulse noise peaks exceeding -21 dBm0 should not exceed 18 in 15 min
- Phase jitter should not exceed 10° peak-to-peak under normal circumstances and 15° on particularly complex circuits

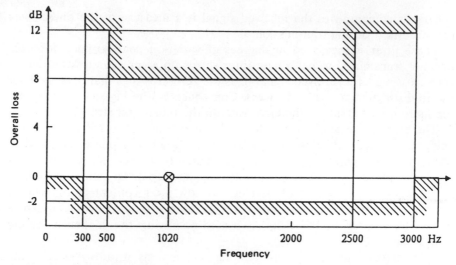

Figure 12.25 Limits for overall loss of the circuit relative to that at 1020 Hz. (From CCITT Rec. M.1025, Figure 1 / M.1025, Page 61, Fascicle IV-2, IXth Plenary Assembly, Melbourne 1988, Ref. 35.)

- Frequency error shall not exceed ±5 Hz
- Total distortion (including quantizing distortion): The signal-to-total-distortion ratio should be better than 28 dB using a sine wave signal at −10 dBm0

12.9.4.3 *With Basic Bandwidth Conditioning*   This subsection is based on CCITT Rec. M.1025. Figure 12.25 shows the limits of overall loss for a

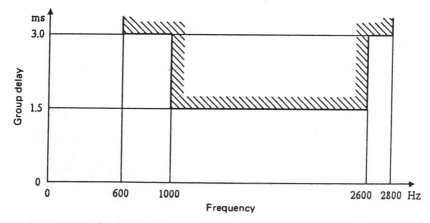

Figure 12.26 Limits for group delay relative to the minimum measured group delay in the 600–2800-Hz band. (From CCITT Rec. M.1025, Figure 2 / M.1025, Page 61, Fascicle IV-2, IXth Plenary Assembly, Melbourne 1988, Ref. 35.)

voice channel relative to 1020 Hz. As in the previous subsection, we can also interpret the figure for amplitude distortion. Figure 12.26 shows the limits of group delay relative to the minimum measured group delay in the band 600–2800 Hz.

12.10 CHANNEL CAPACITY

A leased or switched voice channel represents a financial investment. The goal of the system engineer is to derive as much benefit as possible from the money invested. For the case of digital transmission this is done by maximizing the information transfer across the system. This subsection discusses how much information in bits can be transmitted, relating information to bandwidth, signal-to-noise ratio, and error rate. An empirical discussion of these matters is carried out in Section 12.11.

First, looking at very basic information theory, Shannon stated in his bandwidth paper (Ref. 33) that if the input information rate to a band-limited channel is less than C (bps), a code exists for which the error rate approaches zero as the message length becomes infinite. Conversely, if the input rate exceeds C, the error rate cannot be reduced below some finite positive number.

The usual voice channel approximates to a Gaussian band-limited channel (GBLC) with additive Gaussian noise. For such a channel consider a signal wave of a mean power of S Watts applied at the input of an ideal low-pass filter having a bandwidth of W Hz and containing an internal source of mean Gaussian noise with a mean power of N Watts uniformly distributed over the passband. The capacity in bits per second is given by

$$C = W \log_2\left(1 + \frac{S}{N}\right) \qquad (12.3)$$

Applying Shannon's "capacity" formula (GBLC) to some everyday voice channel criteria, $W = 3000$ Hz and $S/N = 1023$, then

$$C = 30,000 \text{ bps}$$

(Remember that bits per second and baud are interchangeable in binary systems.)

Neither S/N or W is an unreasonable value. Seldom, however, can we achieve a modulation rate greater than 3000 baud. The big question in advanced design is how to increase the data rate and keep the error rate reasonable.

One important item that Shannon's formula did not take into consideration is intersymbol interference. A major problem of a pulse in a band-limited channel is that the pulse tends not to die out immediately, and a subsequent

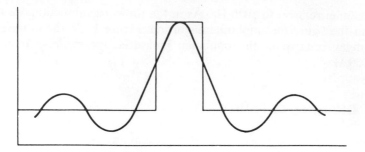

Figure 12.27   Pulse response through a Gaussian band-limited channel (GBLC).

pulse is interfered with by "tails" from the preceding pulse. This is shown in Figure 12.27.

Nyquist provided another approach to the data rate problem, this time using intersymbol interference (the tails in Figure 12.27) as a limit (Ref. 36). This resulted in the definition of the so-called Nyquist rate, which is $2W$ elements per second. W is the bandwidth in hertz of a band-limited channel as shown in Figure 12.27. In binary transmission we are limited to $2W$ bps. If we let $W = 3000$ Hz, the maximum data rate attainable is 6000 bps. Some refer to this as "the Nyquist 2-bit rule."

The key here is that we have restricted ourselves to binary transmission and we are limited to $2W$ bps no matter how much we increase the signal-to-noise ratio. The Shannon GBLC equation indicates that we should be able to increase the information rate indefinitely by increasing the signal-to-noise ratio. The way to attain a higher value of C is to replace the binary transmission system with a multilevel system, often termed an M-ary transmission system with $M > 2$. An M-ary channel can pass $2W \log_2 M$ bps with an acceptable error rate. This is done at the expense of the signal-to-noise ratio. As M increases (as the number of levels increases), so must the signal-to-noise ratio increase to maintain a fixed error rate.

12.11   VOICE CHANNEL DATA MODEMS VERSUS CRITICAL DESIGN PARAMETERS

The critical parameters that affect data transmission have been discussed. They are amplitude–frequency response (sometimes called amplitude distortion), envelope delay distortion (EDD), and noise. Now we relate these parameters to the design of data modems to establish some general limits or "boundaries" for equipment of this type. The discussion that follows purposely avoids HF radio considerations.

As stated earlier in the coverage of EDD, it is desirable to keep the transmitted pulse (bit) length equal to or greater than the residual differential EDD. Since about 1.0 ms is assumed to be a reasonable residual delay after equalization (conditioning), the pulse length should then be no less than approximately 1 ms. This corresponds to a modulation rate of 1000 pulses per second (binary). In the interest of standardization (CCITT Rec. V.22), this figure is modified to 1200 bps.

The next consideration is the usable bandwidth required for the transmission of 1200 bps. This figure is about 1800 Hz, using modulation methods such as PSK, FSK, or DSB-AM, and somewhat less for VSB-AM. Since delay distortion of a typical voice channel is at its minimum between 1700 and 1900 Hz, the required band, when centered about these points, extends from 800 to 2600 Hz or from 1000 to 2800 Hz. From the previous discussion we can see from Figure 12.20 and Table 12.7 that the EDD requirement is met easily over the range of 800–2800 Hz.

Bandwidth limits modulation rate (as will be discussed below). However, the modulation rate in bauds and the data rate in bits per second may not necessarily be the same. This is a very important concept.

Suppose a modulator looked at the incoming serial bit stream 2 bits at a time rather than the conventional 1 bit at a time. Now let four amplitudes of a pulse be used to define each of the four possible combinations of 2 consecutive bits, such that

$$A_1 = 00$$
$$A_2 = 01$$
$$A_3 = 11$$
$$A_4 = 10$$

where A_1, A_2, A_3, and A_4 represent the four pulse amplitudes. This form of treating 2 bits at a time is called dibit coding (see Section 12.9.2).

Similarly, we could let eight pulse levels cover all the possible combinations of 3 consecutive bits so that with a modulation rate of 1200 baud it is possible to transmit information at a rate of 3600 bps. Rather than vary amplitude to 4 or 8 levels, phase can be varied. A four-phase system (PSK) could be coded as follows:

$$F_1 = 0° = 00$$
$$F_2 = 90° = 01$$
$$F_3 = 180° = 11$$
$$F_4 = 270° = 10$$

Again, with a four-phase system using dibit coding, a tone with a modulation rate of 1200-baud PSK can be transmitting 2400 bps. An eight-phase PSK system at 1200 baud could produce 3600 bps of information transfer. Obviously, this process cannot be extended indefinitely. The limitation comes from channel noise. Each time the number of levels or phases is increased, it is necessary to increase the signal-to-noise ratio to maintain a given error rate. Consider the case of a signal voltage S and a noise voltage N. The maximum number of increments of signal (amplitude) that can be discerned is S/N (since N is the smallest discernible increment). Add to this the no-signal case, and the number of discernible levels now becomes $S/N + 1$ or $S + N/N$, where S and N are expressed as power. The number of levels becomes the square root of this expression. This formula shows that in going from 2 to 4 levels or from 4 to 8 levels, a roughly 6-dB noise penalty is incurred each time we double the number of levels.

A similar analysis is carried out for the multiphase case, the penalty in going from four phases to eight phases is 3 dB, for example. See Chapter 4.

Sufficient background has been developed to appraise the data modem for the voice channel. Now consider a data modem for a data rate of 2400 bps. By using quaternary phase-shift keying (QPSK), as described above, 2400 bps is transmitted with a modulation rate of 1200 baud. Assume that the modem uses differential phase detection wherein the detector decisions are based on the change in phase between the last transition and the preceding one.

Assume the bandwidth to be present for the data modem under consideration (for most telephone networks the minimum bandwidth discussed for the sample case is indeed present—1800 Hz). It is now possible to determine if the noise requirements can be satisfied. Figure 12.22 shows that a 12.5-dB signal-to-noise ratio (Gaussian noise) is required to maintain an error rate of 1×10^{-5} for a binary polar (AM) system. As is well established, PSK systems have about a 3-dB improvement. In this case only a 9.6-dB signal-to-noise ratio would be needed, all other factors held constant (no other contributing factors).

Assume the input from the line to be -10 dBm0 in order to satisfy loading conditions. To maintain the proper signal-to-noise ratio, the channel noise must be down to -19.6 dBm0.

To improve the modulation rate without the expense of increased bandwidth, QPSK (four-phase) is used. Allow a 3-dB general noise degradation factor, bringing the required noise level down to -22.6 dBm0.

Consider now the effects of EDD. It has been found that for a four-phase differential system, this degradation will amount to 6 dB if the permissible delay distortion is one pulse length. This impairment brings the noise requirement down to -28.6 dBm0 of average noise power in the voice channel. Allow 1 dB for frequency translation error or other factors, and the noise requirement is now down to -29.6 dBm0.

If the transmit level were -13 dBm0 instead of -10 dBm0, the numbers for noise must be adjusted another 3 dB such that it is now down to -32.6 dBm0. Therefore it can be seen that to achieve a certain error rate for a given modulation rate, several modulation schemes should be considered. It is safe to say that in the majority of these schemes the noise requirement will fall somewhere between -25 and -40 dBm0. This is safely inside the

CCITT figure of -43 dBm (see subsection entitled "Noise" under Section 12.9.3). More discussion on this matter may be found in Ref. 29.

12.12 EQUALIZATION

Of the critical circuit parameters mentioned in Section 12.9.3, two that have severe deleterious effects on data transmission can be reduced to tolerable limits by circuit conditioning, or equalization. These two are amplitude–frequency response (distortion) and EDD.

There are several methods of performing equalization. The most common is to use one or several networks in tandem. Such networks tend to flatten response. In the case of amplitude, they add attenuation increasingly toward channel center and less toward its edges. The overall effect is one of making the amplitude response flatter. The delay equalizer operates fairly similarly. Delay increases toward channel edges parabolically from the center. Delay is added in the center, much like an inverted parabola, with less and less delay added as the band edge is approached. Thus the delay response is flattened at some small cost to absolute delay, which, in most data systems, has no effect. However, care must be taken with the effect of a delay equalizer on an amplitude equalizer and, conversely, the amplitude equalizer on the delay equalizer. Their design and adjustment must be such that the flattening of the channel for one parameter does not entirely distort the channel for the other.

Another type of equalizer is the transversal type of filter. It is useful where it is necessary to select among or to adjust several attenuation (amplitude) and phase characteristics. The basis of the filter is a tapped delay line to which the input is presented. The output is taken from a summing network which adds or sums the outputs of the taps. Such a filter is adjusted to the desired response (equalization of both phase and amplitude) by adjusting the tap contributions.

If the characteristics of a line are known, another method of equalization is predistortion of the output signal of the data set. Some devices use a shift register and a summing network. If the equalization needs to be varied, then a feedback circuit from the receiver to the transmitter would be required to control the shift register. Such a type of active predistortion is valid for binary transmission only.

A major drawback of all the equalizers discussed (with the exception of the last with a feedback circuit) is that they are useful only on dedicated or leased circuits where the circuit characteristics are known and remain fixed. Obviously, a switched circuit would require a variable automatic equalizer, or conditioning would be required on every circuit in the switched system that would be transmitting data.

Circuits are usually equalized on the receiving end. This is called poste-qualization. Equalizers must be balanced and must present the proper impedance to the line. Administrations* may choose to condition (equalize) trunks and attempt to eliminate the need to equalize station lines; the economy of considerably fewer equalizers is obvious. In addition, each circuit that would possibly carry high-speed data in the system would have to be equalized, and the equalization must be good enough that any possible combination will meet the overall requirements. If equalization requirements become greater (i.e., parameters more stringent), then consideration may have to be given to the restriction of the maximum number of circuits (trunks) in tandem.

Equalization to meet amplitude–frequency response requirements is less exacting on the overall system than envelope delay. Equalization for envelope delay and its associated measurements are time-consuming and expensive. Envelope delay in general is arithmetically cumulative. If there is a requirement of overall EDD of 1 ms for a circuit between 1000 and 2600 Hz, then in three links in tandem, each link must be better than 333 μs between the same frequency limits. For four links in tandem, each link would have to be 250 μs or better. In practice, accumulation of delay distortion is not entirely arithmetical, resulting in a loosening of requirements by about 10%. Delay distortion tends to be inversely proportional to the velocity of propagation. Loaded cables display greater delay distortion than nonloaded cables. Likewise, with sharp filters a greater delay is experienced for frequencies approaching band edge than for filters with a more gradual roll-off.

In carrier multiplex systems, channel banks contribute more to the overall EDD than any other part of the system. Because channels 1 and 12 of the standard CCITT modulation plan, those nearest the group band edge, suffer additional delay distortion owing to the effects of group and, in some cases, supergroup filters, the system engineer should allocate channels for data transmission near group and supergroup centers. On long-haul critical data systems the data channels should be allocated to through-groups and through-supergroups, minimizing as much as possible the steps of demodulation back to voice frequencies (channel demodulation).

Automatic equalization for both amplitude and delay is widely used, especially for switched data systems. Such devices are self-adaptive and require a short adaptation period after switching, on the order of 1–2 s or less. This can be carried out during synchronization. Not only is the modem clock being "averaged" for the new circuit on transmission of a synchronous idle signal, but the self-adaptive equalizer adjusts for optimum equalization as well. The major drawback of adaptive equalizers is their expense.

Many texts use the term "conditioning" as a synonym for equalization. We like to differentiate the two terms. Conditioning is carried out by the common carrier or administration on a dedicated line to reduce its variablities such as EDD and attenuation distortion. Equalization is carried out at the user data set as described previously.

---

*Telephone companies.

12.13 PRACTICAL MODEM APPLICATIONS

12.13.1 Voice Frequency Carrier Telegraph (VFCT)

Narrow shifted FSK transmission of digital data goes under several common names. These are VFTG and VFCT, which stand for voice frequency telegraph and voice frequency carrier telegraph, respectively.

In practice VFCT techniques handle data rates up to 1200 bps by a simple application of FSK modulation. The voice channel is divided into segments or frequency bounded zones or bands. Each segment represents a data or telegraph channel, each with a frequency-shifted subcarrier.

For proper end-to-end system interface it is convenient to use standardized modulation plans, particularly on international circuits. In order for the far-end demodulator to operate with the near-end modulator, it must be tuned to the same center frequency and accept the same shift. Center frequency is the frequency in the center of the passband of the modulator–demodulator. The shift is the number of hertz that the center frequency is shifted up and down in frequency for the mark and space condition. From Table 12.1, by convention, the mark condition is the center frequency shifted downward, and the space upward. For modulation rates below 80 baud, bandpasses have either 170-* or 120-Hz bandwidths with frequency shifts of ± 42.5 or ± 30 Hz, respectively. CCITT recommends (Rec. R.70 bis [Ref. 37]) the 120-Hz channels for operating at 50 baud and below. However, some administrations operate these channels at higher modulation rates. Figure 12.28 shows graphically the partial modulation plan for 120-, 170-, and 240-Hz spacing. The 240-Hz channel is recommended by the CCITT for 100-baud operation with a ± 60-Hz frequency shift.

The number of tone telegraph or data channels that can be accommodated on a voice channel depends for one thing on the usable voice channel bandwidth. For HF radio with a voice channel limit on the order of 3 kHz, 16 channels may be accommodated using 170-Hz spacing (170 Hz between center frequencies). 24 VFCT channels may be accommodated between 390 and 3210 Hz with 120-Hz spacing, or 12 channels with 240-Hz spacing. This can easily meet standard telephone FDM carrier channels of 300–3400 Hz.

Some administrations use a combination of voice and telegraph/data simultaneously on a telephone channel. This technique is commonly referred to a "voice plus" or S + D (speech plus derived). There are two approaches to this technique. The first is recommended by CCITT and is used widely by INTELSAT orderwires. It places five telegraph channels (channels 20–24) above a restricted voice band with a roofing filter near 2500 Hz. Speech occupies a band between 300 and 2500 Hz. Above 2500 Hz appear up to five 50-baud telegraph channels.

---

*CCITT Rec. R.39 (Ref. 38).

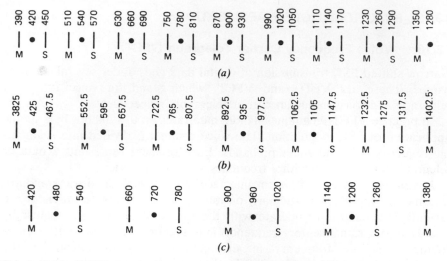

Figure 12.28   CCITT channel frequencies — VFCT. (*a*) 120-Hz spacing, ±30-Hz shift. (*b*) 170-Hz spacing, ±42.5-Hz shift. (*c*) 240-Hz spacing, ±60-Hz shift. Partial modulation plan shown. (From CCITT Rec. R.70 bis, Table 2, [Ref. 37]; courtesy of ITU – CCITT.)

The second approach removes a slot from the center of the voice channel into which up to two telegraph channels may be inserted. The slot is a 500-Hz band centered on 1275 Hz.

However, some administrations use a slot of telegraphy of frequencies 1680 and 1860 Hz by either AM or FM (FSK) (see CCITT Rec. R.43 [Ref. 39]).

The use of S + D should be avoided on trunks in large networks because it causes degradation to speech and also precludes the use of the channel for higher speed data. In addition the telegraph channels should be removed before going into two-wire telephone service (i.e., at the hybrid or term set); otherwise, service drops to half-duplex on telegraph.

12.13.2   Medium-Data-Rate Modems

In normal practice FSK is used for the transmission of data rates up to 1200 bps. The 120-Hz channel is nominally modified as in Figure 12.28*c* such that one 240-Hz channel replaces two 120-Hz channels. Administrations use the 240-Hz channel for modulation rates up to 150 bps. The same process can continue using 480-Hz channels for 300-bps FSK, and 960-Hz channels for 600 bps.

In the following paragraphs a synopsis is given for CCITT V recommendations dealing with modems that operate with the nominal analog 4-kHz voice channel.

CCITT Rec. V.21, "300 bits per second Duplex Modem Standardized for Use in the General Switched Telephone Networks" (Ref. 40), recommends

- Frequency shift ± 100 Hz
- Center frequency of channel 1, 1080 Hz
- Center frequency of channel 2, 1750 Hz
- In each case, space (0) is the higher frequency

It also provides for a disabling tone on echo suppressors, a very important consideration on long circuits.

CCITT Rec. V.22 (Ref. 41) defines a 1200-bps duplex modem for use on the general switched telephone network and on point-to-point two-wire leased-type circuits. In the full-duplex mode, channel separation is by frequency division. The modem uses QPSK modulation and the carrier frequencies are 1200 and 2400 Hz. There is a guard tone at 1800 Hz, which may be disabled at the discretion of users. The modem employs a fixed compromise equalizer and can operate at 1200 or 600 bps in the start–stop or synchronous modes. A scrambler is also included.

CCITT Rec. V.22 bis (Ref. 42) defines a modem for 2400-bps duplex operation using the frequency division technique standardized for use on the general switched telephone network (GSTN) and on point-to-point 2-wire leased circuits. It uses quadrature amplitude modulation. The carrier frequencies are the same as used in Rec. V.22 described above. It can operate either in the start–stop or synchronous modes at 2400 or 1200 bps. The signal constellation for this modem is shown in Figure 12.29. The scrambling algorithm uses the generating polynomial $1 + X^{-14} + X^{-17}$.

CCITT Rec. V.23 (Ref. 43) recommends a 600/1200-baud modem standardized for use in the general switched telephone network for application to synchronous or asynchronous systems. Provision is made for an optional backward channel for error control.

For the forward channel the following modulation rates and characteristic frequencies are presented:

	F_0 (Hz)	F_Z (Hz)	F_A (Hz)
Mode 1, up to 600 baud	1500	1300	1700
Mode 2, up to 1200 baud	1700	1300	2100

The backward channel for error control is capable of modulation rates up to 75 baud. Its mark and space frequencies are

$$
\begin{array}{cc}
F_Z & F_A \\
390 \text{ Hz} & 450 \text{ Hz}
\end{array}
$$

Refer to Table 12.1 for the mark–space convention (F_Z = mark or binary 1, F_A = space or binary 0).

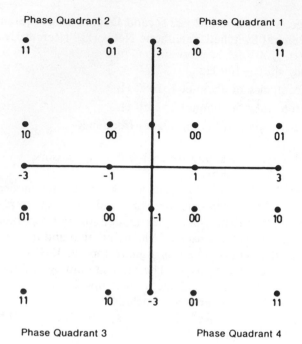

Figure 12.29   Signal constellation used with the CCITT Rec. V.22 bis modem. (From CCITT Rec. V.22 bis, Figure 2 / V.22, Page 84, Fascicle VIII.1, IXth Plenary Assembly, Melbourne 1988.)

CCITT Rec. V.26, "2400 bits per second Modem Standardized for Use on 4-Wire Leased Telephone-Type Circuits" (Ref. 44), involves

1. Carrier frequency 1800 Hz ± 1 Hz, four-phase modulation, synchronous mode of operation
2. Dibit coding as follows:

	Phase Change (deg)	
Dibit	Alternative A	Alternative B
00	0	+45
01	+90	+135
11	+180	+225
10	+270	+315

3. Data signaling rate 2400 bps $\pm 0.01\%$
4. Modulation rate 1200 baud $\pm 0.01\%$
5. Maximum frequency error at receiver, ± 7 Hz (allowing ± 1 Hz for modulator [transmitter] error)
6. Backward channel 75 bps (see Rec. V.23 [Ref. 43])

CCITT Rec. V.26 bis, "2400/1200 bits per second Modem for Use in the General Switched Telephone Network" (Ref. 45), is generally the same as Rec. V.26. At 1200 bps line operation, differential PSK is recommended where

$$0 = +90°$$

$$1 = +270°$$

CCITT Rec. V.27 is entitled "4800 bits per second Modem with Manual Equalizer for Use on Leased Telephone-Type Circuits" (Ref. 46). The principal characteristics of this modem are

1. Full-duplex or half-duplex operation
2. Differential eight-phase modulation, synchronous mode of operation
3. Possibility of backward (supervisory) channel with modulation rates to 75 baud in each direction of transmission
4. Inclusion of a manually adjustable equalizer
5. Carrier frequency 1800 Hz ± 1 Hz
6. Tribit coding as follows:

Tribit Values	Phase Change (deg)
001	0
000	45
010	90
011	135
111	180
110	225
100	270
101	315

7. Frequency tolerance same as Recs. V.26 and V.26 bis
8. Line signal characteristics: a 50% raised cosine energy spectrum is equally divided between transmitter and receiver
9. A self-synchronizing scrambler/descrambler having a generating polynomial $1 + X^{-6} + X^{-7}$ with additional guards against repeating patterns of 1, 2, 3, 4, 6, 9, and 12 bits shall be included in the modem

The basic characteristics of the modem in CCITT Rec. V.27 bis, "4800/2400 bits per second Modem with Automatic Equalizer Standardized for Use on Leased Telephone-Type Circuits" (Ref. 47), are the same as for that of Rec. V.27 except for

1. A fallback rate of 2400 bps with V.26 Alternative A characteristics.
2. An automatic equalizer with two types of turn-on sequences and a turn-off sequence. The first type of turn-on sequence is short in duration for comparatively good circuits, meeting CCITT Rec. M.1020 (Ref. 34), and a second is a longer sequence for relatively poor circuits, below the Rec. M.1020 standard. The sequences, which are transmitted on-line prior to traffic, provide for equalizer conditioning and serve to set descrambler synchronization into proper operation.

CCITT Rec. V.27 ter, "4800/2400 bits per second Modem Standardized for Use in the General Switched Telephone Network" (Ref. 48), is similar to Rec. V.27 bis except for the turn-on sequence which, in this case, includes the capability to protect against talker echo.

In CCITT Rec. V.29, "9600 bits per second Modem Standardized for Use on Point-to-Point 4-wire Telephone-Type Circuits" (Ref. 49), the main characteristics of the modem are

1. A fallback to rates of 7200 and 4800 bps
2. Full-duplex and half-duplex operation
3. Combined amplitude and phase modulation with synchronous mode operation
4. Automatic equalizer
5. Optional inclusion of a multiplexer for combining data rates of 7200, 4800, and 2400 bps
6. Line signal 1700 Hz ± 1 Hz
7. Signal space coding. At 9600 bps the scrambled data stream to be transmitted is divided into groups of four consecutive data bits (quadbits). The first bit Q_1 in time of each quadbit is used to determine the signal element amplitude to be transmitted. The second Q_2, third Q_3, and fourth Q_4 bits are encoded as a phase change relative to the phase of the immediately preceding element (Table 12.8). The phase encoding is identical to that of Rec. V.27.

The relative amplitude of the transmitted signal element is determined by the first bit Q_1 of the quadbit and the absolute phase of the signal element. (Table 12.9). The absolute phase is initially established by the synchronizing signal. The four possible signaling elements Q_1, Q_2, Q_3, and Q_4 represent 16 phase–amplitude possibilities ($2^4 = 16$), which are shown in Figure 12.30.

CCITT Rec. V.32 (Ref. 50) is a standard for a family of two-wire duplex modems operating at data signaling rates of up to 9600 bps for use on the general switched telephone network (GSTN) and on leased-type telephone

Table 12.8   Phase encoding

Q_2	Q_3	Q_4	Phase Change (deg)
0	0	1	0
0	0	0	45
0	1	0	90
0	1	1	135
1	1	1	180
1	1	0	225
1	0	0	270
1	0	1	315

Table 12.9   Amplitude – phase relationships

Absolute Phase (deg)	Q_1	Relative Signal Element Amplitude
0, 90, 180, 270	0	3
	1	5
45, 135, 225, 315	0	$\sqrt{2}$
	1	$3\sqrt{2}$

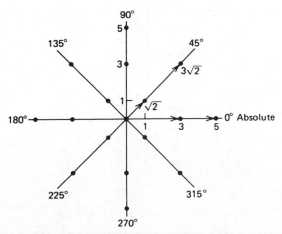

Figure 12.30 Signal space diagram, 9600-bps operation. (From CCITT Rec. V.29 [Ref. 49]; courtesy of ITU – CCITT.)

Table 12.10   Differential quadrant coding for 4800-bps operation and for nonredundant coding at 9600-bps operation

Inputs		Previous Outputs		Phase Quadrant Change (deg)	Outputs		Signal State for 4800 bps
$Q1_n$	$Q2_n$	$Y1_{n-1}$	$Y2_{n-1}$		$Y1_n$	$Y2_n$	
0	0	0	0	+90	0	1	B
0	0	0	1		1	1	C
0	0	1	0		0	0	A
0	0	1	1		1	0	D
0	1	0	0	0	0	0	A
0	1	0	1		0	1	B
0	1	1	0		1	0	D
0	1	1	1		1	1	C
1	0	0	0	+180	1	1	C
1	0	0	1		1	0	D
1	0	1	0		0	1	B
1	0	1	1		0	0	A
1	1	0	0	+270	1	0	D
1	1	0	1		0	0	A
1	1	1	0		1	1	C
1	1	1	1		0	1	B

Source: CCITT Rec. V.32, Table 1/V.32, Page 236, Fascicle VIII.1, IXth Plenary Assembly, Melbourne 1988.

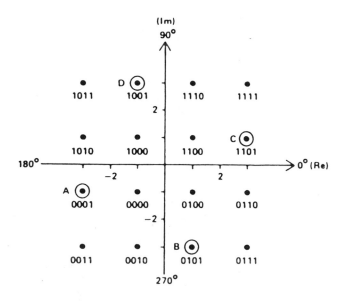

The binary numbers denote $Y1_n\, Y2_n\, Q3_n\, Q4_n$

Figure 12.31   16-point signal constellation with nonredundant coding for 9600-bps operation and subset A, B, C, and D states used for 4800-bps operation and for training. (From CCITT Rec. V. 32 Figure 1 / V.32, Page 236, Fascicle VIII.1, IXth Plenary Assembly, Melbourne 1988, Ref. 50.)

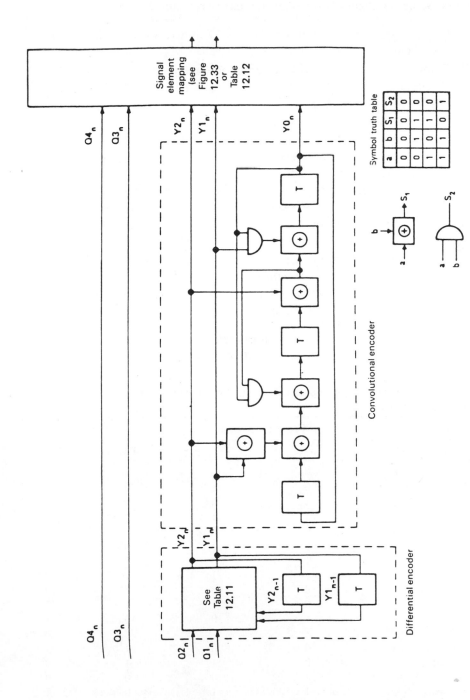

Figure 12.32 Trellis coding at 9600 bps. (From CCITT Rec. V.32, Figure 2 / V.32, Page 237, Fascicle VIII.1, IXth Plenary Assembly, Melbourne 1988, Ref. 50.)

circuits. Some highlights of the V.32 standard are listed below:

1. Duplex mode of operation on the GSTN and two-wire, point-to-point leased circuits.
2. Channel separation by echo cancellation techniques
3. Quadrature amplitude modulation for each channel with synchronous line transmission at 2400 baud.
4. Any combination of the following data signaling rates may be implemented: 9600-, 4800-, and 2400-bps synchronous
5. At 9600 bps, two alternative modulation schemes are provided. One uses 16-carrier states and the other uses trellis coding with 32-carrier states. However, modems providing the 9600-bps data signaling rate shall be capable of interworking using the 16-state alternative.
6. There is an exchange of rate sequences during start-up to establish the data rate, coding, and any other special facilities.
7. There is an optional provision for an asynchronous (start–stop) mode of operation in accordance with CCITT Rec. V.14.
8. Carrier frequency: 1800 Hz ± 1 Hz. The receiver can operate with frequency offsets up to ± 7 Hz.

The recommendation offers two alternatives for signal element coding at the 9600-bps data rate: (1) nonredundant and (2) trellis coding.

Table 12.11 Differential encoding for use with trellis coded alternative at 9600-bps operation

Inputs		Previous Outputs		Outputs	
$Q1_n$	$Q2_n$	$Y1_{n-1}$	$Y2_{n-1}$	$Y1_n$	$Y2_n$
0	0	0	0	0	0
0	0	0	1	0	1
0	0	1	0	1	0
0	0	1	1	1	1
0	1	0	0	0	1
0	1	0	1	0	0
0	1	1	0	1	1
0	1	1	1	1	0
1	0	0	0	1	0
1	0	0	1	1	1
1	0	1	0	0	1
1	0	1	1	0	0
1	1	0	0	1	1
1	1	0	1	1	0
1	1	1	0	0	0
1	1	1	1	0	1

Source: CCITT Rec. V.32, Table 2/V.32, Page 237, Fascicle VIII.1, IXth Plenary Assembly, Melbourne 1988, Ref. 50.

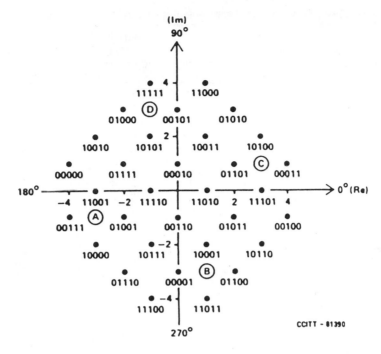

The binary numbers denote Y0$_n$ Y1$_n$ Y2$_n$ Q3$_n$ Q4$_n$

Figure 12.33 32-point signal constellation with trellis coding for 9600-bps operation and states A, B, C, and D used at 4800-bps operation and for training. (From CCITT Rec. V.32, Figure 3 / V.22, Page 239, Fascicle VIII.1, IXth Plenary Assembly, Melbourne 1988, Ref. 50.)

With nonredundant coding, the scrambled data stream to be transmitted is divided into groups of 4 consecutive bits. The first 2 bits in time $Q1_n$ and $Q2_n$ in each group, where the subscript n designates the sequence number of the group, are differentially encoded into $Y1_n$ and $Y2_n$ in accordance with Table 12.10. Bits $Y1_n$, $Y2_n$, $Q3_n$, and $Q4_n$ are then mapped into coordinates of the signal state to be transmitted according to the signal space diagram shown in Figure 12.31 and as listed in Table 12.12.

When using the second alternative with trellis coding, the scrambled data stream to be transmitted is divided into two groups of 4 consecutive data bits. As shown in Figure 12.32, the first 2 bits in time $Q1_n$ and $Q2_n$ in each group, where the subscript n designates the sequence number of the group, are first differentially encoded into $Y1_n$ and $Y2_n$ in accordance with Table 12.11. The two differentially encoded bits $Y1_n$ and $Y2_n$ are used as input to a systematic convolutional coder which generates redundant bit $Y0_n$. The redundant bit and the 4 information-carrying bits $Y1_n$, $Y2_n$, $Q3_n$, and $Q4_n$, are then mapped into the coordinates of the signal element to be transmitted according to the signal space diagram shown in Figure 12.33 and as listed in Table 12.12.

Table 12.12   Two alternative signal state mappings for 9600-bps operation

(Y0)	Y1	Y2	Q3	Q4	Nonredundant Coding Re	Nonredundant Coding Im	Trellis Coding Re	Trellis Coding Im
0	0	0	0	0	−1	−1	−4	1
	0	0	0	1	−3	−1	0	−3
	0	10	1	0	−1	−3	0	1
	0	0	1	1	−3	−3	4	1
	0	1	0	0	1	−1	4	−1
	0	1	0	1	1	−3	0	3
	0	1	1	0	3	−1	0	−1
	0	1	1	1	3	−3	−4	−1
	1	0	0	0	−1	1	−2	3
	1	0	0	1	−1	3	−2	−1
	1	0	1	0	−3	1	2	3
	1	0	1	1	−3	3	2	−1
	1	1	0	0	1	1	2	−3
	1	1	0	1	3	1	2	1
	1	1	1	0	1	3	−2	−3
	1	1	1	1	3	3	−2	1
1	0	0	0	0			−3	−2
	0	0	0	1			1	−2
	0	0	1	0			−3	2
	0	0	1	1			1	2
	0	1	0	0			3	2
	0	1	0	1			−1	2
	0	1	1	0	3			−2
	0	1	1	1	−1			−2
	1	0	0	0			1	4
	1	0	0	1			−3	0
	1	0	1	0			1	0
	1	0	1	1			1	−4
	1	1	0	0			−1	−4
	1	1	0	1			3	0
	1	1	1	0			−1	0
	1	1	1	1			−1	4

Source: CCITT Rec. V.32, Table 3/V.32, page 238, Fascicle VIII-1, IXth Plenary Assembly, Melbourne 1988.

[a]See Tables 12.10 and 12.11 and Figure 12.32.

For 4800-bps operation the data stream to be transmitted is divided into groups of 2 consecutive data bits. These bits, denoted $Q1_n$ and $Q2_n$, where $Q1_n$ is the first in time and the subscript n designates the sequence number of the group, are differentially encoded into $Y1_n$ and $Y2_n$, according to Table 12.10. Figure 12.31 shows the subset A, B, C, and D of signal states used for 4800-bps transmission.

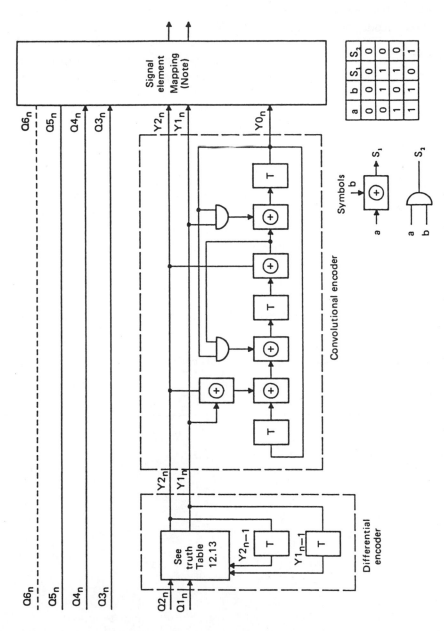

Figure 12.34 Trellis coder for 14,400- and 12,000-bps operation. (From CCITT Rec. V.33, Figure 1 /V.33, Page 253, Fascicle VIII.1, IXth Plenary Assembly, Melbourne 1988, Ref. 51.) Note: See Figure 12.35 for 14,400 bps rate and Figure 12.36 for 12,000 bps rate.

There are two scrambling generating polynomials:

- Call mode modem $= 1 + X^{-18} + X^{-23}$
- Answer mode modem $= 1 + X^{-5} + X^{-23}$

CCITT Rec. V.33 (Ref. 51) defines a 14,400-bps modem standardized for use on point-to-point four-wire leased telephone-type circuits. The modem is intended to be used primarily on special-quality leased circuits which are typically defined in CCITT Recs. M.1020 and M.1025 (see Section 12.9.4). The modem's principal characteristics are

- A fallback rate of 12,000 bps
- A capability of operating in a duplex mode with continuous carrier
- Combined amplitude and phase modulation with synchronous mode of operation
- Inclusion of an eight-state trellis-coded modulation
- Optional inclusion of a multiplexer for combining data rates of 12,000, 9600, 7200, 4800, and 2400 bps
- A carrier frequency of 1800 ± 1 Hz

At the 14,400-bps data rate, the scrambled data stream to be transmitted is

Table 12.13 Differential encoding for use with trellis coding

Inputs		Previous Outputs		Outputs	
$Q1_n$	$Q2_n$	$Y1_{n-1}$	$Y2_{n-2}$	$Y1_n$	$Y2_n$
0	0	0	0	0	0
0	0	0	1	0	1
0	0	1	0	1	0
0	0	1	1	1	1
0	1	0	0	0	1
0	1	0	1	0	0
0	1	1	0	1	1
0	1	1	1	1	0
1	0	0	0	1	0
1	0	0	1	1	1
1	0	1	0	0	1
1	0	1	1	0	0
1	1	0	0	1	1
1	1	0	1	1	0
1	1	1	0	0	0
1	1	1	1	0	1

Source: CCITT Rec. V.33, Table 1A/V.33, Page 253, Fascicle VIII.1, IXth Plenary Assembly, Melbourne 1988, Ref. 51.

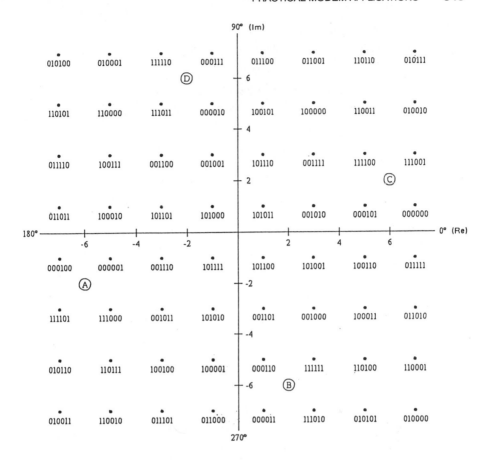

Binary numbers refer to $Q5_n$, $Q4_n$, $Q3_n$, $Y2_n$, $Y1_n$, $Y0_n$,
A, B, C, D refer to synchronizing signal elements

Figure 12.35 Signal constellation and mapping for trellis-coded modulation at 14,400 bps (From CCITT Rec. V.33, Figure 2 / V.33, Page 255, Fascicle VIII.1, IXth Plenary Assembly, Melbourne 1988, Ref. 51.)

divided into groups of 6 consecutive data bits. As illustrated in Figure 12.34, the first 2 bits in time $Q1_n$ and $Q2_n$ in each group are first differentially coded into $Y1$ and $Y2$ in accordance with Table 12.13. The two differentially encoded bits $Y1_n$ and $Y2_n$ are used as input to a systematic convolutional encoder which generates a redundant bit $Y0_n$. This redundant bit and the 6 information-carrying bits $Y1_n$, $Y2_n$, $Q3_n$, $Q4_n$, $Q5_n$, and $Q6_n$ are then mapped into the coordinates of the signal element to be transmitted in accordance with the signal space diagram shown in Figure 12.35. Figure 12.36 shows the signal constellation and mapping for trellis-coded modulation at 12,000 bps. The self-synchronizing scrambler generating polynomial is $1 + X^{-18} + X^{-23}$.

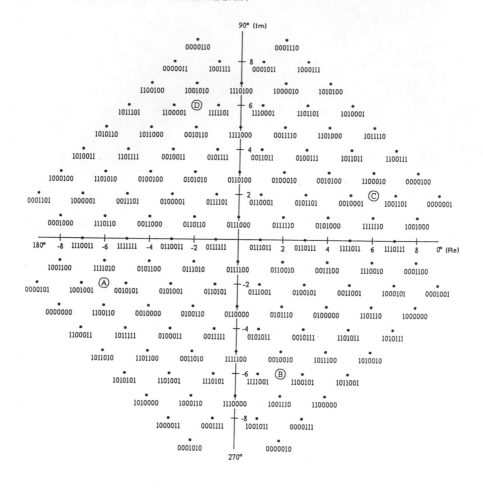

Binary numbers refer to $Q6_n$, $Q5_n$, $Q4_n$, $Q3_n$, $Y2_n$, $Y1_n$, $Y0_n$,
A, B, C, D refer to synchronizing signal elements

Figure 12.36 Signal space diagram and mapping for trellis-coded modulation at 12,000 bps. (From CCITT Rec. V.33, Figure 3 / V.33, Page 256, Fascicle VIII.1, IXth Plenary Assembly, Melbourne 1988.)

12.14 NOTES ON TRELLIS-CODED MODULATION (TCM)

In Section 6.7 we introduced forward error correction (FEC) coding and "coding gain." With FEC we can improve error rate performance, or, given a fixed bit error rate (BER), we can reduce power (level) because of coding gain. Depending on the coding/decoding design parameters, for a fixed BER, a lower E_b/N_0 will achieve the same performance if we used the same modulation scheme without coding. For example, with QPSK without coding, to achieve a BER of 1×10^{-5}, we would require an E_b/N_0 of 9.6 dB. Using

FEC, we may only need an E_b/N_0 of 5 dB for the same BER. The result is a coding gain of 9.6 − 5, or 4.6 dB. We pay for this by an increase in symbol rate (ergo bandwidth) due to the redundancy added for FEC.

When working with the voice channel we have severe bandwidth limitations, nominally 3 kHz. When implementing FEC and maintaining a high data rate, our natural tendency is to use bit packing schemes turning from binary modulation (two states) to M-ary modulation (M-states). Such bit packing schemes were described in Section 4.6.3.4. Typical types of M-ary modulation are QPSK, 8-ary PSK, 4-, 8-, 16-, and so on, quadrature amplitude modulation (QAM). QAM schemes, such as used in CCITT Rec. V.29 described above, have two-dimensional signal constellations; one dimension is phase and the other amplitude. The signal constellation shown in Figure 12.31 has 16 points. As the number of points increase, the receive modem becomes more prone to errors induced by Gaussian noise hits and intersymbol interference. This problem can be mitigated by the design of the M-ary signal set.

In our discussions in Chapter 4, coding and modulation were carried out separately. With trellis-coded modulation (TCM), there is a unified concept. Here coding gain is achieved without bandwidth expansion by expanding the signal set.

In conventional, multilevel (amplitude and/or phase) modulation systems, during each modulation interval, the modulator maps m binary symbols (bits) into one of $M = 2^m$ possible transmit signals, and the demodulator recovers the m bits by making an independent M-ary nearest-neighbor decision on each signal received. Previous codes were designed for maximum Hamming free distance. Hamming distance refers to the number of symbols in which two code symbols or blocks differ regardless of how these symbols differ (Ref. 52).

TCM broke with this tradition and maximized free Euclidian distance (distance between maximum likely neighbors in the signal constellation) rather than Hamming distance. The redundancy necessary for TCM would have to come from expanding the signal set to avoid bandwidth expansion. Ref. 52 states that, in principle, TCM can achieve coding gains of about 7–8 dB over conventional uncoded multilevel modulation schemes. Most of the achievable coding gain would be obtained by expanding the signal sets only by a factor or two. This means using signal set sizes of 2^{m+1} for transmission of m bits per modulation interval.

All TCM systems (Ref. 52) achieve significant distance gains with as few as 4, 8, and 16 states. Roughly speaking, it is possible to gain 3 dB with 4 states, 4 dB with 8 states, nearly 5 dB with 16 states, and up to 6 dB with 128 or more states. Doubling the number of states does not always yield a code with larger free distance. Generally, limited distance growth and increasing numbers of nearest neighbors, and neighbors with next-largest distances, are the two mechanisms that prevent realizing coding gains from exceeding the ultimate limit set by channel capacity. This limit can be characterized by the signal-to-noise ratio at which the channel capacity of a modulation system with $2m + 1$-ary signal set equals m bps/Hz (Ref. 52; also see Ref. 53). Practical TCM systems can achieve coding gains of 3–6 dB at spectral efficiencies equal to or larger than 2 bits/Hz.

12.15   DATA TRANSMISSION ON THE DIGITAL NETWORK

12.15.1   The Problem

To transmit telegraph or computer data on the digital network, two problems arise. One is quite simple to fix—the waveform. Computers often transmit nonreturn-to-zero (NRZ) serial data; pulse code modulation (PCM) uses alternate mark inversion (AMI) (Chapter 3). The second problem is data rate. Telegraph circuits and computer data sources have data rates related to 75×2^n such as $75, 150, 300, 600, 1200, \ldots, 9600$ bps. These rates have no relationship whatsoever to common digital network transmission rates of 64 kbps (North America 64/56 kbps) and the standard military rates of 16/32 kbps.

12.15.2   Some Solutions

One of the most common and least elegant approaches is to use analog modems such as those found in the CCITT V series recommendations discussed above. In essence here, we are impressing, say, 1200 bps on a 64-kbps DS0 or CEPT 30 + 20 channel. It seems a waste of bandwidth.

Another, more elegant approach, is AT & T's DDS (Digital Data System), described in Section 3.3.6. DDS delivers 56 kbps to a customer premise or, alternatively, 9600, 4800, or 2400 bps. These latter three rates derive from an aggregate of 48 kbps; the remaining 8 kbps $(56 - 48 = 8)$ is used for overhead. As shown in Section 3.3.6, these build up to or derive from the DS0 channel.

CCITT Rec. X.22 (Ref. 54) defines a multiplexing scheme of 75×2^n subrates to derive an aggregate of 48,000 bps.

Integrated services digital networks (ISDNs) dictate that data I/O and computer devices have a baseband output of 64 kbps (56 kbps in North America) or a subrate of 16 kbps. ISDN (and the non-ISDN digital network) can also turn to the rate adaption techniques suggested in CCITT Rec. V.110 (Ref. 55). The terminal adapter (TA) in Figure 12.38 is used for bit rate adaption functions in a two-step basis. First the user signaling rate is converted to an appropriate intermediate rate expressed by $2^k \times 8$ kbps, where $k = 0$, 1, or 2. The next stage performs the second conversion from the intermediate rates to 64 kbps.

12.16   INTEGRATED SERVICES DIGITAL NETWORK (ISDN) TRANSMISSION CONSIDERATIONS

12.16.1   Introduction

Integrated services digital networks have been developed to ease integration of all telecommunication services, except full motion video, on one basic digital channel, namely 64 kbps. Whereas 4 kHz is the basic building block of the analog network, 64 kbps is the basic building block of ISDN. The ISDN basic building block is designed to serve, among other services (Ref. 15):

- Digital voice
- High-speed data, both circuit and packet switched
- Telex/teletext
- Telemetry
- Facsimile
- Slow-scan video and compressed video

The goal of ISDN is to provide an integrated facility to incorporate each of the services listed on a common 64-kbps channel and/or a combination of 64- and 16-kbps channels. ISDN assumes that an all-digital network is in place up to and including a subscriber's serving exchange. It also assumes that the network has implemented CCITT Signaling System no. 7.

12.16.2   ISDN User Channels

Here we are looking from the end user into the network. We consider two user classes: residential and commercial. The following are standard transmission structures for user access links:

B-channel: 64 kbps
D-channel: 16 kbps
C-channel: 8 or 16 kbps
A-channel: 4-kHz analog VF channel

The B-channel is the basic user channel and serves any one of the following traffic types:

- PCM-based digital voice channel
- Computer digital data, either circuit or packet switched
- A mix of multiplexed lower data rate traffic, such as vocoded (digital) low data rate voice and lower data rate computer data

However, in this category, the traffic must have the same destination.

The D-channel is a 16-kbps channel. It serves not only as the user signaling channel but also as a lower speed data connectivity to the network. The A-channel serves as a transitional expedient to provide nominal 4-kHz analog connectivity to the network. The C-channel is associated with the A-channel to form a hybrid access arrangement.

There is an E-channel which is a 64-kbps channel that is primarily used to carry signaling information for circuit switching on ISDN. At the user–network interface, it is used in the primary rate multiplexed structures as an alternative arrangement for multiple access configurations. The *primary rate* is discussed below.

The H-channels have the following bit rates:

H0:   384 kbps
H1 which consists of the following two subsets:
  H11: 1536 kbps
  H12: 1920 kbps

The H-channel does *not* carry signaling information. Its purpose is to provide service for higher user data rates, such as digitized program channels, compressed video for teleconferencing, fast facsimile, and packet-switched data bit streams.

12.1.3 Basic and Primary User Interfaces

The *basic* interface structure is composed of two B-channels and a D-channel and is commonly referred to as 2B + D. The D-channel at this interface is 16 kbps. The *B*-channels may be used independently (i.e., two different simultaneous connections).

Appendix I to CCITT Rec. I.412 (Ref. 56) states that the basic access may also B + D or D.

The *primary* rate B-channel interface structures are composed of *N* B-channels and one D-channel, where the D channel in this case is 64 kbps. There are two primary B-channel data rates:

• 1.544 Mbps = 23B + D
• 2.048 Mbps = 30B + D

For the user–network access arrangement containing multiple interfaces, it is possible for the D-channel in one structure to carry signaling for B-channels in another primary rate structure without an activated D-channel. When a D-channel is not activated, the designated time slot may or may not be used to provide an additional B-channel, depending on the situation, such as 24B with 1.544 Mbps.

An alternative primary rate interface B-channel structure is composed of B-channels and an E-channel:

- 1.544 Mbps = 23B + E
- 2.048 Mbps = 30B + E

There are a number of H-channel interface structures covered in CCITT Rec. I.412 (Ref. 56). For the H0 structure, a D-channel may or may not be present, and if present, is always 64 kbps. At the 1.544-Mbps rate interface, the H0 channel structures are 4H0 or 3H0 + D. The H11 and H12 structures use D-channels from other structures to carry signaling, if required.

12.16.4 User Access and Interface

Figure 12.37 shows generic ISDN user connectivity to the network. Here we mean an interface with a link that connects a user to his/her digital serving exchange. We can select either basic or primary service (i.e., 2B + D, 23B + D, or 30B + D) to connect to the ISDN network.

The objectives of any digital interface design, and specifically of ISDN access and interface, are

1. Electrical and mechanical specification
2. Channel structure and access capabilities
3. User–network protocols
4. Maintenance and operation
5. Performance
6. Services

Figure 12.38 shows the ISDN reference model. It delineates interface points for the user. In the figure, NT1, or network termination 1, provides the physical layer interface; it is essentially equivalent to OSI layer 1. The

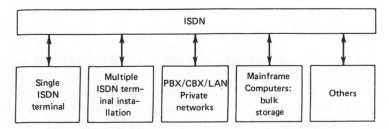

Figure 12.37   Generic ISDN user connectivity to the network.

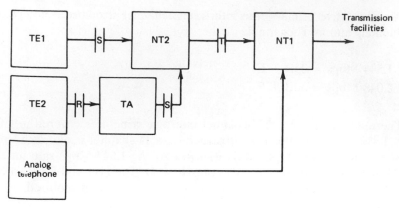

Figure 12.38 ISDN reference model.

functions of the physical layer include

- Transmission facility termination
- Layer-1 maintenance functions and performance monitoring
- Timing
- Power transfer
- Layer-1 multiplexing
- Interface termination, including multidrop termination employing layer-1 contention resolution

Network termination 2 (NT2) can be broadly associated with OSI layers 1, 2, and 3. Among the examples of equipment that provide NT2 functions are user terminal controllers, local area networks (LANs), and private automatic branch exchanges (PABXs). Among the NT2 functions are

- Layer-1, -2, and -3 protocols processing
- Multiplexing (layers 2 and 3)
- Switching
- Concentration
- Interface termination and other layer-1 functions
- Maintenance functions

A distinction must be drawn here between North American and European practice. As we are aware, in Europe the telecommunication administrations are, in general, national monopolies that are government controlled. In the United States and Canada they are private enterprises, often very competitive. Thus in Europe the NT1 and NT2 functions are combined and called NT12, and the equipment involved is the property of the telecommunication administration. Of course, in North America the functions are separated, and one belongs to the user and the other to the telephone company.

TE1 in Figure 12.38 is the terminal equipment and has an interface that must comply with the ISDN user–network interface specifications. Terminal equipment (TE) covers functions broadly belonging to OSI layer 1 and higher OSI layers. Among this equipment are digital telephones, computer workstations (data terminal equipment, DTE), and other devices in the user-end equipment category.

TE2 refers to equipment that does *not* meet ISDN terminal–network interface specifications and that requires interface modifications to adapt the equipment to ISDN. A terminal equipment adapter (TA) provides the necessary conversion functions to permit TE2-type terminal equipment to interface with ISDN.

Reference points S, T, and R are used to identify the interface available at those points. S and T are identical electrically, mechanically, and from the point of view of protocol. Point R relates to the TA interface or, in essence, is the interface of the nonstandard (i.e., non-ISDN) device.

12.16.5 Layer-1 Interface: Basic Rate

The S and T interface or layer-1 physical interface requires a balanced metallic transmission medium in each direction of transmission supporting 192 kbps. This is the NT interface shown in Figure 12.38. 192 kbps is made up of 2B + D, which equals 144 kbps; the remaining 48 kbps are overhead bits.

Layer 1 provides the following services to layer-2 ISDN operation:

- The transmission capability by means of appropriately encoded bit streams for both B- and D-channels and also any timing and synchronization functions that may be required
- The signaling capability and the necessary procedures to enable customer terminals and/or network terminating equipment to be deactivated when required, and reactivated when required
- The signaling capability and necessary procedures to allow terminals to gain access to the common resource of the D-channel in an orderly fashion while meeting performance requirements of the D-channel signaling system
- The signaling capability and procedures and necessary functions at layer 1 to enable maintenance functions to be performed
- An indication to the higher layers of the status of layer 1

The frame structure of 2B + D operation is given in Figure 12.39. Note in the figure that in both directions of transmission the bits are grouped into frames of 48 bits each. The frame structure is identical for all configurations,

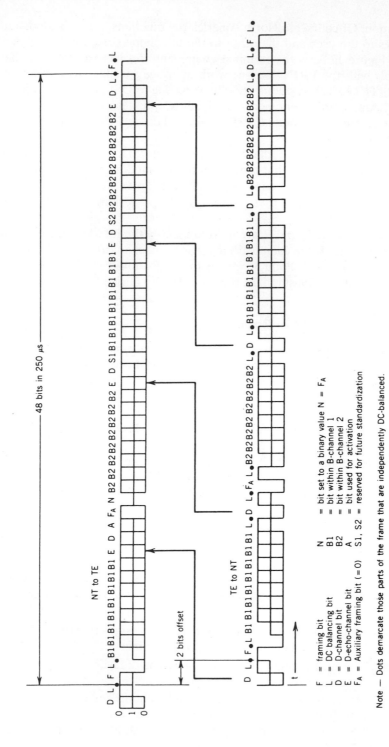

Figure 12.39 Frame structure at reference points S and T. (From CCITT Rec. I.430, Figure 3 / I.430, Page 177, Fascicle III.8, IXth Plenary Assembly, Melbourne 1988.)

F = framing bit
L = DC balancing bit
D = D-channel bit
E = D-echo-channel bit
F$_A$ = Auxiliary framing bit (=0)

N = bit set to a binary value N = F$_A$
B1 = bit within B-channel 1
B2 = bit within B-channel 2
A = bit used for activation
S1, S2 = reserved for future standardization

Note — Dots demarcate those parts of the frame that are independently DC-balanced.

Table 12.14   Notes on Bit Positions and Groups in Figure 12.39

Bit Position	Group
Terminal to Network: Each frame consists of the following group of bits; each individual group is dc balanced by its last bit (L-bit).	
1 and 2	Framing signal with balance bit
3–11	B1-channel with balance bit (first octet)
12 and 13	D-channel bit with balance bit
14 and 15	Auxiliary framing with balance bit
16–24	B2-channel with balance bit (first octet)
25 and 26	D-channel bit with balance bit
27–35	B1-channel with balance bit (second octet)
36 and 37	D-channel bit with balance bit
38–46	B2-channel with balance bit (second octet)
47 and 48	D-channel with balance bit
Network to terminal: Frames transmitted by the network (NT) contain an echo channel (E-bits) used to retransmit the D-bits received from the terminals. The D-echo channel is used for D-channel access control. The last bit of the frame (L-bit) is used for balancing each complete frame. The bits are grouped as follows:	
1 and 2	Framing signal with balance bit
3–10	B1-channel (first octet)
11	E-, D-echo-channel bit
12	D-channel bit
13	Bit A used for activation
14	F_A auxiliary framing bit
15	N bit[a]
16–23	B2-channel (first octet)
24	E-, D-echo-channel bit
25	D-channel bit
26	S1, reserved for future standardization[b]
27–34	B1-channel (second octet)
35	E-, D-echo-channel bit
36	D-channel bit
37	S2, reserved for future standardization[b]
38–45	B2-channel (second octet)
46	E-, D-echo-channel bit
47	D-channel bit
48	Frame balance bit

Source: From CCITT Rec. I.430, Tables 2 and 3/I.430, Page 177, Fascicle III.8, IXth Plenary Assembly, Melbourne, 1988.

[a]As defined in Section 6.3 of Ref. 57.
[b]S1 and S2 are set to binary 0.

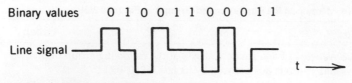

Figure 12.40 ISDN pseudoternary line code.

whether point-to-point or point-to-multipoint. However, the frame structures are different for each direction of transmission. Explanatory notes to Figure 12.39 are given in Table 12.14.

The line code for both directions of transmission is pseudoternary coding with 100% pulse width, as shown in Figure 12.40. Coding is performed such that a binary 1 is represented by a no-line signal, whereas a binary 0 is represented by a positive or negative pulse. The first binary signal following the framing balance bit is the same polarity as the framing balance bit. Subsequent binary 0's alternate in polarity. A balance bit is a 0 if the number of 0's following the previous balance bit is odd. A balance bit is binary 1 if the number of binary 0's following the previous balance bit is even.

The NT derives its timing from the network clock. The TE synchronizes its bit, octet, and frame timing from the received bit stream from the NT and uses the derived timing to synchronize its transmitted signal.

12.17 SERIAL-TO-PARALLEL CONVERSION FOR TRANSMISSION ON IMPAIRED MEDIA

Often the transmission medium, in most cases the voice channel, cannot support a high data rate even with conditioning and equalizers. The impairments may be due to poor amplitude–frequency response, envelope delay distortion, (EDD), or excessive impulse noise.

One step that may be taken in these circumstances is to convert the high-speed serial bit stream at the dc level (e.g., demodulated) to a number of lower speed parallel bit streams. One technique widely used on HF radio systems is to divide a 2400-bit serial stream into 16 parallel streams, each carrying 150 bps. If each of the slower streams is dibit coded (2 bits at a time, discussed in Section 12.11) and applied to a QPSK tone modulator, the modulation rate on each subchannel is reduced in this case to 75 baud. The equivalent period for a dibit interval is $\frac{1}{75}$ s, or 13 ms.

There are two obvious advantages of this technique. First, each subchannel has a comparatively small bandwidth and therefore looks at a small and tolerable segment of the total delay across the channel. The impairment of EDD is less on slower speed channels. Second, there is less of a chance of a noise burst or hit of impulse noise to smear the subchannel signal beyond recognition. If the duration of the noise burst is less than half the pulse width, the data pulse can be regenerated, and the pulse will not be in error. The longer we can make the pulse width, the less chance there is of disturbance from impulse noise. In this case the interval or pulse width has had an equivalent lengthening by a factor of 32.

12.18   PARALLEL-TO-SERIAL CONVERSION FOR IMPROVED ECONOMY OF CIRCUIT USAGE

Long high-quality (conditioned) toll telephone circuits are costly to lease or are a costly investment. The user is often faced with a large number of slow-speed circuits (50–300 bps) that originate in one general geographic location, with a general destination to another common geographic location. If we assume that these are 75-bps circuits (100 wpm), which are commonly encountered in practice, only 18–24 can be transmitted on a high-grade telephone channel by conventional voice frequency carrier telegraph (VFCT) techniques (see Section 12.13.1).

Circuit economy can be affected using a data/telegraph time division multiplexer (TDM). A typical application of this type is illustrated in Figure 12.41. It shows one direction of transmission only. Here incoming slow-speed VFCT channels are converted to equivalent dc bit streams. Up to 32 of these bit streams serve as input to a TDM in the application that is illustrated in the figure. The output of the TDM is a 2400-bit synchronous serial bit stream. This output is fed to a conventional 2400-bit modem. At the far end the 2400-bps serial stream is demodulated to dc and fed to the equivalent demultiplexer. The demultiplexer breaks the serial stream back down to the original 75-bps circuits. Figure 12.41 illustrates the concept. It does not show clocking or other interconnect circuitry. By use of a TDM a savings of up to 2:1 can be effected. Whereas by means of a conventional VFCT, only about 18 75-bps circuits can be transmitted on a good telephone channel, by means of the multiplexer up to 32 such circuits can now be transmitted on the same channel.

TDMs are also available with line data rates of 4800 and 9600 bps, accepting inputs for multiplexing from 75 bps up to 2400 bps for the 4800-bps line rate and up to 4800 bps for the 9600-bps line rate.

The statistical TDM is an interesting variation of the conventional TDM and is particularly useful where the input port data traffic is "bursty" (i.e., where data messages are relatively short and where there are periods where there is no traffic being transmitted). The conventional TDM creates a permanently dedicated time slot or subchannel for each input port in the sharing group. A statistical TDM, by contrast, dynamically allocates the subchannels or time slots on a statistical basis to increase efficiency by providing time slots only for ports actively transmitting data. In effect these devices not only multiplex, but concentrate as well. Of course, it will be appreciated that in the case of a true statistical TDM the number of 75-bps

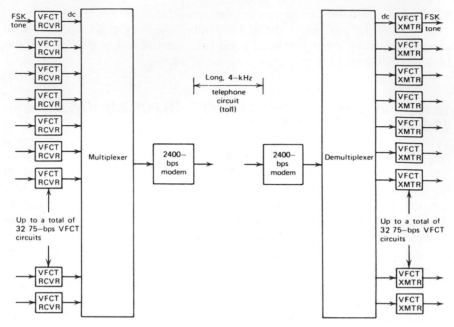

Figure 12.41 Typical application of parallel-to-serial conversion. VFCT = voice frequency carrier telegraph; RCVR = receiver (converter); XMTR = transmitter (keyer); FSK = frequency shift keying.

input ports shown in Figure 12.41 can exceed 32, whereas for the conventional TDM, 32 is the limit.

The amount by which the statistical TDM can exceed the limiting number of input ports (whatever the data rate) depends largely on its storage capabilities as well as on the nature and intensity of the data traffic appearing at the input ports.

Data multiplexers frame traffic when placing the signal on-line, and we must then expect overhead bits in this case. Many multiplexers provide error control and other housekeeping duties in accordance with particular protocols.

12.19   DATA TRANSMISSION SYSTEM — FUNCTIONAL BLOCK DIAGRAM

Figure 12.42 is a simplified functional block diagram showing a typical data transmission system end to end. The diagram is meant to be representative and conceptual. The system shown uses an 8-level code operating in a synchronous mode. Sections 12.3.2 and 12.6.2 cover 8-level codes and synchronous transmission, respectively, as well as clocking requirements. The

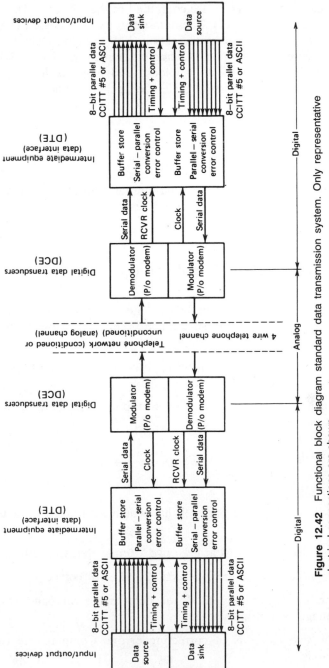

Figure 12.42 Functional block diagram standard data transmission system. Only representative electrical connections are shown.

857

receive clock in this case is corrected by averaging transitions on the incoming bit stream. Master clocks are contained in the modems. Timing problems are covered in Section 12.6.3. The data interface is discussed in Section 12.7, as are input–output devices in Section 12.8. Analog (telephone) channel criteria are covered in Section 12.9 and subsequently. Figure 12.42 shows the functions of the various building blocks and how the higher speed, more sophisticated data transmission system differs from the more conventional telegraph system.

REVIEW EXERCISES

1. Distinguish between analog and digital transmission and extend the discussion to binary digital transmission.

2. Distinguish between a binary 1 and binary 0: voltage, mark/space, paper tape, and frequency shift keying.

3. What important function in the telephone network is carried out by a 1-bit code?

4. A certain binary source code has 6 bits. How many distinct characters/functions can such a code represent? How many for an 8-bit code?

5. Name at least three nonprinting functions that are represented by a binary sequence.

6. In a particular case we are using even parity with ASCII. Give the value of bit number 8 in each case:
   0 1 1 0 1 1 0
   1 1 0 1 0 1 1
   0 0 1 0 0 0 0
   0 0 0 1 0 1 0
   Suppose we use odd parity. What would be the value of bit 8 in that case?

7. Define throughput. List at least five "items" that will reduce throughput.

8. Distinguish between error detection and error correction.

9. Define two generic methods of detecting errors and compare their advantages and disadvantages.

10. Define two methods of correcting errors. Compare the two and cite trade-offs.

11. When using cyclic redundancy check (CRC), what does the block check count/character (BCC) contain?

12. Define and compare the three types of automatic repeat request (ARQ) described in the text. Give advantages and disadvantages of each.

13. Draw a neutral waveform for the sequence 1 0 1 0 1 1 0 1. Directly underneath draw a polar waveform, using the same binary values.

14. Draw the same sequence for a return-to-zero (RZ) waveform.

15. Distinguish between start–stop and synchronous transmission.

16. The stop element in start–stop transmission has three standard durations in use today. Give the value of "unit intervals" (bits) of each. Define the standard start element.

17. What are the three basic causes of errors in a data transmission system?

18. Describe how transition timing works.

19. Analyze clock stability versus bit rate versus error rate for the following:

Bit Rate	Bit Error Rate (BER)	Clock Stability
75 bps	1×10^{-5}	?
?	1×10^{-6}	$1 \times 10^{-11}/mo$
2400 bps	?	$1 \times 10^{-8}/mo$
10 Mbps	?	$1 \times 10^{-9}/mo$
100 Mbps	1×10^{-10}	?

20. What are the two basic types of distortion we must deal with in data/telegraph transmission?

21. Where, in a pulse, do we assume sampling takes place?

22. Explain the difference between bits and bauds. When are the two synonymous?

23. Name at least seven topics that deal with data interface.

24. What are the three distinct phases of a telephone call? Relate these to a data connection.

25. What do segmentation and reassembly involve? What is encapsulation?

26. List the seven open system interconnection (OSI) layers.

27. What is a service access point (SAP)?

28. Give the four most important characteristics of the physical layer.

29. Draw a generalized data link layer frame. Relate the "flag" field to the start and stop elements of asynchronous transmission.

30. With EIA RS-232D and CCITT Recs. V.10/V.11, in a certain bit period we find +4-V dc on the data signal lead. Is it a mark or space, a 1 or a 0?

31. CCITT uses the term *double current*. What does the term signify?

32. What is the primary difference between EIA RS-422 and RS-423 (CCITT V.10 and V.11)?

33. How do we transmit a nonreturn-to-zero (NRZ) digital data waveform on a standard analog channel?

34. Name some advantages of frequency shift keying (FSK) over amplitude shift keying (ASK); over phase shift keying (PSK).

35. Describe quaternary PSK (QPSK) in relation to binary PSK (BPSK). With QPSK, how many bits are transmitted per transition?

36. There are three basic ways of modulating a signal. What are they?

37. Give at least three advantages of phase modulation when transmitting a digital waveform compared to amplitude modulation (AM) and frequency modulation (FM).

38. What is the difference between coherent PSK and differential PSK when viewed from the receive side?

39. If the envelope delay distortion (EDD) between 800 and 2600 Hz is 1000 μs, all other impairments disregarded, what is the maximum modulation rate the channel will support?

40. List the four types of noise described in the text. Define each in one sentence.

41. How is impulse noise specified?

42. Why is the CCITT specification on frequency translation error so stringent?

43. Shannon (Ref. 33) stated that the capacity of a channel in bits per second was a function of what two parameters?

44. A modem transmits 2400 bps using QPSK modulation. What is the modulation rate in baud?

45. What is the cause of intersymbol interference? Could there be still another cause?

46. An equalizer on a voice channel modem tends to nullify the effects of what two basic impairments on a voice channel? How does it accomplish this?

47. Describe how voice frequency carrier telegraph (VFCT) operates.

48. Why would a modem using phase shift keying (PSK) use 1800 Hz as the tone frequency rather than some other tone frequency, such as 1000 Hz?

49. How can we have full-duplex data operation on a two-wire facility?

50. What is the modulation rate in baud of a CCITT V.29 modem operating at 9600 bps?

51. Define trellis-coded modulation. What are its merits?

52. What is bit packing?

53. Digital data conventionally is transmitted with bit rates related to 75×2^n, where n is an integer including zero. What is (are) the problem(s) of transmitting such data rates on the public switched digital network?

54. Supply at least two approaches on how one might transmit computer data on the digital network when such data operates at a line bit rate related to 75×2^n?

55. How does integrated services digital network (ISDN) address the problem expressed in question 53?

56. What are the two underlying service rates (bit rates) offered by ISDN? How are they expressed?

57. Give at least two applications of the D-channel.

58. Identify at least five telecommunication user services ISDN will support.

59. What is the bit rate of a B-channel?

60. What type of digital network signaling is so vital to the operation of ISDN?

61. In the ISDN reference model, what would the interface NT12 signify? Discuss NT1, NT2, and why we would have NT12.

62. How is noncompatible ISDN user equipment interface handled? Name the appropriate element in the ISDN user model.

63. Describe the ISDN baseband line code.

64. What is the gross bit rate of 2B + D at the user interface? Allocate the bits to their functions.

65. How does an ISDN network termination (NT) derive its timing?

REFERENCES AND BIBLIOGRAPHY

1. IEEE Std. 100-1977, *Dictionary of Electrical and Electronic Terms*, 2nd ed., IEEE, New York, 1977.

2. "Equivalence between Binary Notation Symbols and the Significant Conditions of a Two-Condition Code," CCITT Rec. V.1, IXth Plenary Assembly, Melbourne, 1988, vol. VIII.1.

3. "Electrical Characteristics for Unbalanced Double-Current Interchange Circuits for General Use with Integrated Circuit Equipment in the Field of Data Communications," CCITT Rec. V.10, IXth Plenary Assembly, Melbourne, 1988, vol. VIII.1.

4. "Electrical Characteristics for Balanced Double-Current Interchange Circuits for General Use with Integrated Circuit Equipment in the Field of Data Communications," CCITT Rec. V.11, IXth Plenary Assembly, Melbourne, 1988, vol. VIII.1.

5. "Interface between Data Terminal Equipment and Data Circuit Terminating Equipment Employing Serial Binary Data Interchange," EIA RS-232D, EIA, Washington, DC, Jan. 1987.

6. *Reference Data for Radio Engineers*, 6th ed., Howard W. Sams, Indianapolis, IN, 1977.

7. "International Telegraph Alphabet No. 2," CCITT Rec. S.1, IXth Plenary Assembly, Melbourne, 1988, vol. VII.1.

8. "Procedures for the Use of the Communication Control Characters of American Standard Code for Information Interchange in Specified Data Communication Links," ANSI X.3.28 1976, ANSI, New York, 1976.

9. "General Structure of Signals of International Alphabet No. 5 for Character Oriented Data Transmission over Data Networks," CCITT Rec. X.4, IXth Plenary Assembly, Melbourne, 1988, vol. VIII.1.

10. "International Alphabet No. 5," CCITT Rec. T.50. IXth Plenary Assembly, Melbourne, 1988, vol. VII.3.

11. Roger L. Freeman, *Telecommunication System Engineering*, 2nd ed., Wiley, New York, 1989.

12. J. E. McNamara, *Technical Aspects of Data Communications*, 2nd ed., Digital Equipment, Corp., Maynard, MA, 1982.

13. "Standard for Specifying Signal Quality for Transmitting and Receiving Data Processing Terminal Equipments Using Serial Data Transmission at the Interface of Nonsynchronous Data Communications Equipment," EIA RS-363, EIA, Washington, DC, May 1969.

14. "Electrical Characteristics of Balanced Voltage Digital Interface Circuits," EIA RS-422, EIA, Washington, DC, Apr. 1973.

15. W. Stallings, *Handbook of Computer Communication Standards*, vol. I, Macmillan, New York, 1987.

16. "Open Systems Interconnection (OSI) Service Definition Conventions," CCITT Rec. X.210, IXth Plenary Assembly, Melbourne, 1988, vol. VIII.3.

17. "Data Communication Networks Services and Facilities: Interfaces," CCITT Recs. X.1–X.32, IXth Plenary Assembly, Melbourne, 1988, vol. VIII.3.

18. "Data Communication Networks, Transmission, Signaling and Switching, Network Aspects, Maintenance and Administration Arrangements," CCITT Recs. X.40–X.181, IXth Plenary Assembly, Melbourne, 1988, vol. VIII.4.

19. James W. Conard, *Standards and Protocols for Communication Networks*, Carnegie Press, Madison, NJ, 1982.

20. "List of Definitions of Interchange Circuits between Data Terminal Equipment (DTE) and Data Circuit-Terminating Equipment (DCE)," CCITT Rec. V.24, IXth Plenary Assembly, Melbourne, 1988, vol. VIII.1.

21. "General Purpose 37-position and 9-position Interface for Data Terminal Equipment and Data Circuit-Terminating Equipment Employing Serial Binary Data Interchange," EIA RS-449, EIA, Washington, DC, 1977.

22. "Electrical Characteristics for Unbalanced Double-Current Interchange Circuits," CCITT Rec. V.28, IXth Plenary Assembly, Melbourne, 1988, vol. VIII.1.

23. "Electrical Characteristics for Single-Current Interchange Circuits Controlled by Contact Closure," CCITT Rcc. V.31, IXth Plenary Assembly, Melbourne, 1988, vol. VIII.1.

24. "General Data Communications Interface Layer 1 Specification," CCITT Rec. V.230, IXth Plenary Assembly, Melbourne, 1988, vol. VIII.1.

25. "Electrical Characteristics of Digital Interface Circuits," MIL-STD-188-114B, U.S. Department of Defense, Washington, DC, 1989.

26. D. R. Doll, *Data Communications*, Wiley, New York, 1978.

27. "Electrical Characteristics of Unbalanced Voltage Digital Interface Circuits," EIA RS-423, EIA, Washington, DC, 1975.

28. "High Speed 25-position Interface for Data Terminal Equipment and Data Circuit-Terminating Equipment," EIA RS-530, EIA, Washington, DC, Mar. 1987.

29. W. R. Bennett and J. R. Davey, *Data Transmission*, McGraw-Hill, New York, 1965.

30. *Telecommunication Transmission Engineering*, vol. 2, 2nd ed., AT & T, New York, 1977.

31. *Data Transmission*, *Parameters and Capabilities*, ITT Federal Labs, Nutley, NJ, 1961.

32. Bell System Tech. Reference, "Transmission Parameters Affecting Voiceband Data Transmission—Description of Parameters," Pub. 41008, AT&T, New York, 1974.

33. C. E. Shannon, "A Mathematical Theory of Communication," *Bell Syst. Tech. J.*, vol. 27, 379–423, July 1948; 623–656, Oct. 1948.

34. "Characteristics of Special Quality International Leased Circuits with Special Bandwidth Conditioning," CCITT Rec. M.1020, IXth Plenary Assembly, Melbourne, 1988, vol. IV-2.

35. "Characteristics of Special Quality International Leased Circuits with Basic Bandwidth Conditioning," CCITT Rec. M.1025, IXth Plenary Assembly, Melbourne, 1988, vol. IV-2.

36. H. Nyquist, "Certain Topics in Telegraph Transmission Theory," *Trans. AIEE*, vol. 47, 617–644, Apr. 1928.

37. "Numbering of International VFT Channels," Table 2/R.70 bis, CCITT Rec. R.70 bis, IXth Plenary Assembly, Melbourne, 1988, vol. VII.

38. "Voice Frequency Telegraphy on Radio Circuits," CCITT Rec. R.39, IXth Plenary Assembly, Melbourne, 1988, vol. VII.

39. "Simultaneous Communication by Telephone and Telegraph on a Telephone-Type Circuit," CCITT Rec. R.43, IXth Plenary Assembly, Melbourne, 1988, vol. VII.

40. "300 bits per second Duplex Modem Standardized for Use in the General Switched Telephone Network," CCITT Rec. V.21, IXth Plenary Assembly, Melbourne, 1988, vol. VIII.1.

41. "1200 bits per second Duplex Modem Standardized for Use in the General Switched Telephone Network or 2-Wire Leased Telephone-Type Circuits," CCITT Rec. V.22, IXth Plenary Assembly, Melbourne, 1988 vol. VIII.1.

42. "2400 bits per second Duplex Modem Using Frequency Division Technique Standardized for Use on the General Switched Telephone Network and on Point-to-Point 2-wire Leased Telephone-Type Circuits," CCITT Rec. V.22 bis, IXth Plenary Assembly, Melbourne, 1988, vol. VIII.1.

43. "600/1200 Baud Modem Standardized for Use on the General Switched Telephone Network," CCITT Rec. V.23, IXth Plenary Assembly, Melbourne, 1988, vol. VIII.1.

44. "2400 bits per second Modem Standardized for Use on 4-Wire Leased Telephone-Type Circuits," CCITT Rec. V.26, IXth Plenary Assembly, Melbourne, 1988, vol. VIII.1

45. "2400/1200 bits per second Modem Standardized for Use in the General Switched Telephone Network," CCITT Rec. V.26 bis, IXth Plenary Assembly, Melbourne, 1988, vol. VIII.1.

46. "4800 bits per second Modem with Manual Equalizer Standardized for Use on Leased Telephone-Type Circuits," CCITT Rec. V.27, IXth Plenary Assembly, Melbourne, 1988, vol. VIII.1.

47. "4800/2400 bits per second Modem with Automatic Equalizer Standardized for Use on Leased Telephone-Type Circuits," CCITT Rec. V.27 bis, IXth Plenary Assembly, Melbourne, 1988, vol. VIII.1.

48. "4800/2400 bits per second Modem Standardized for Use in the General Switched Telephone Network," CCITT Rec. V.27 ter, IXth Plenary Assembly, Melbourne, 1988, vol. VIII.1.

49. "9600 bits per second Modem Standardized for Use on Point-to-Point 4-wire Telephone-Type Circuits," CCITT Rec. V.29, IXth Plenary Assembly, Melbourne, 1988, vol. VIII.1.

50. "A Family of 2-wire, Duplex Modems Operating at Data Signaling Rates up to 9600 bits/s for Use on the General Switched Telephone Network or Leased Telephone-Type Circuits," CCITT Rec. V.32, IXth Plenary Assembly, Melbourne, 1988, vol. VIII.1.

51. "14,400 bits per second Modem Standardized for Use on Point-to-Point 4-wire Leased Telephone-Type Circuits," CCITT Rec. V.33, IXth Plenary Assembly, Melbourne, 1988, vol. VIII.1.

52. Gottfried Ungerboeck, "Trellis-Coded Modulation with Redundant Sets—Parts I and II," *IEEE Commun. Mag.*, vol. 27, no. 2, Feb. 1987.

53. Andrew J. Viterbi et al., "A Pragmatic Approach to Trellis-Coded Modulation," *IEEE Mag.*, vol. 27, no. 7, July 1989.

54. "Multiplex DTE/DCE Interface for User Classes 3-6," CCITT Rec. X.22, IXth Plenary Assembly, Melbourne, 1988, vol. VIII.2.

55. "Support of Data Terminal Equipments (DTEs) with V-Series Type Interfaces by an Integrated Services Digital Network (ISDN)," CCITT Rec. V.110, IXth Plenary Assembly, Melbourne, 1988, vol. VIII.1.

56. "ISDN User–Network Interface Structure and Access Capabilities," CCITT Rec. I.412, IXth Plenary Assembly, Melbourne, 1988, vol. III.8.

57. "Basic User–Network Interface: Layer 1 Specification," CCITT Rec. I.430, IXth Plenary Assembly, Melbourne, 1988, vol. III.8.

58. H. C. Folts, "A Powerful Standard Replaces the Old Interface Standby," *Data Commun.*, May 1980.

13

VIDEO TRANSMISSION

13.1 GENERAL

This chapter provides the basic essentials for designing point-to-point video transmission systems. To understand the video problem the transmission system engineer must first have an appreciation of video and how the standard TV video signal is developed. The discussion that follows provides an explanation of the "what and why" of video. There follows a review of critical video transmission parameters, black and white and color transmission standards, video program channel transmission, and the transmission of video over specific media. There is a brief discussion of basic tests of video point-to-point facilities. Finally, digital TV is treated, including freeze-frame, compression techniques and video conferencing. TV broadcast problems are covered only where they specifically interact with point-to-point transmission.

13.2 AN APPRECIATION OF VIDEO TRANSMISSION

A video transmission system must deal with four factors when transmitting images of moving objects:

- A perception of the distribution of luminance or simply the distribution of light and shade
- A perception of depth or a three-dimensioned perspective
- A perception of motion relating to the first two factors above
- A perception of color (hues and tints)

Monochrome TV deals with the first three factors. Color TV includes all four factors.

A video transmission system must convert these three (or four) factors into electrical equivalents. The first three factors are integrated to an equivalent electric current or voltage whose amplitude is varied with time. Essentially, at any one moment it must integrate luminance from a scene in the three dimensions (i.e., width, height, and depth) as a function of time. And time itself is still another variable, for the scene is changing in time.

The process of integration of visual intelligence is carried out by *scanning*. The horizontal detail of a scene is transmitted continuously and the vertical detail discontinuously. The vertical dimension is assigned discrete values that become the fundamental limiting factor in a video transmission system.

The scanning process consists of taking a horizontal strip across the image on which discrete square elements called pels or pixels (picture elements) are scanned from left to right. When the right-hand end is reached, another, lower, horizontal strip is explored, and so on until the whole image has been scanned. Luminance values are translated on each scanning interval into

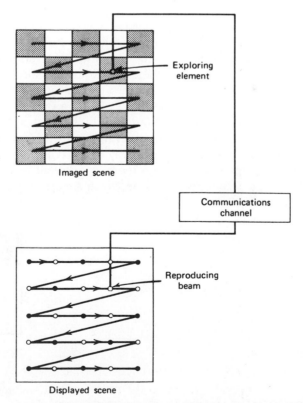

Figure 13.1   Scanning process from TV camera to receiver display.

voltage and current variations and are transmitted over the system. The concept of scanning by this means is shown in Figure 13.1.

The National Television Systems Committee (U.S.) (NTSC) practice divides an image into 525 horizontal scanning lines. It is the number of scanning lines that determines vertical detail or resolution of a picture.

When discussing picture resolution, the aspect ratio is the width-to-height ratio of the video image. The aspect ratio used almost universally is 4:3. In other words, a TV image 12 in. wide would necessarily be 9 in. high. Thus an image divided into 525 (491) vertical elements would then have 700 (652) horizontal elements to maintain an aspect ratio of 4:3. The numbers in parentheses represent the practical maximum active lines and elements. Therefore the total number of elements approaches something on the order of 250,000. We reach this number because, in practice, vertical detail reproduced is 64–87% of the active scanning lines. A good halftone engraving may have as many as 14,400 elements per square inch, compared to approximately 3000 elements per square inch for a 9 by 12 in. TV image.

Motion is another variable factor that must be transmitted. The sensation of continuous motion in standard TV video practice is transmitted to the viewer by a successive display of still pictures at a regular rate similar to the method used in motion pictures. The regular rate of display is called the *frame rate*. A frame rate of 25 frames per second will give the viewer a sense of motion, but on the other hand he/she will be disturbed by luminance flicker (bloom and decay), or the sensation that still pictures are "flicking" on screen one after the other. To avoid any sort of luminance flicker sensation, the image is divided into two closely interwoven (interleaving) parts, and each part is presented in succession at a rate of 60 frames per second, even though *complete* pictures are still built up at a 30 frame-per-second rate. It should be noted that interleaving improves resolution as well as apparent persistence of the cathode ray tube (CRT) by tending to reinforce the scanning spots. It has been found convenient to equate flicker frequency to power line frequency. Hence in North American practice, where

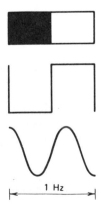

Figure 13.2 Development of a sinusoid wave from the scan of adjacent squares.

1 Hz

power line frequency is 60 Hz, the flicker is 60 frames per second. In Europe it is 50 frames per second to correspond to the 50-Hz line frequency used there.

Following North American practice, some other important parameters derive from the previous paragraphs:

1. A field period is 1/60 s. This is the time that is required to scan a full picture on every horizontal line.
2. The second scan covers the lines not scanned on the first period, offset one-half horizontal line.
3. Thus 1/30 s is required to scan all lines on a complete picture.
4. The transit time of exploring and reproducing scanning elements or spots along each scanning line is 1/15,750 s (525 lines in 1/30 s) = 63.5 μs.
5. Consider that about 16% of the 63.5 μs is consumed in flyback and synchronization. Accordingly, only about 53.3 μs are left per line of picture to transmit information.

What will be the bandwidth necessary to transmit images so described? Consider the worst case, where each scanning line is made up of alternate black and white squares, each the size of the scanning element. There would be 652 such elements. Scan the picture, and a square wave will result with a positive-going square for white and a negative for black. If we let a pair of adjacent square waves be equivalent to a sinusoid (see Figure 13.2), then the baseband required to transmit the image will have an upper cutoff of about 6.36 MHz, permitting no degradation in the intervening transmission system. The lower limit will be a dc or zero frequency.

13.3 THE COMPOSITE SIGNAL

The word *composite* is confusing in the TV industry. On one hand, composite may mean the combination of the full video signal plus the audio subcarrier; the meaning here is narrower. Composite in this case deals with the transmission of video information as well as the necessary synchronizing information.

Consider Figure 13.3. An image made up of two black squares is scanned. The total time for the line is 63.5 μs, of which 53.3 μs are available for the transmission of actual video information and 10.2 μs are required for synchronization and flyback.

During the retrace time or flyback it is essential that no video information be transmitted. To accomplish this, a blanking pulse is superimposed on the video at the camera. The blanking pulse carries the signal voltage into the reference black region. Beyond this region in amplitude is the "blacker than black" region, which is allocated to the synchronizing pulses. The blanking level (pulse) is shown in Figure 13.3.

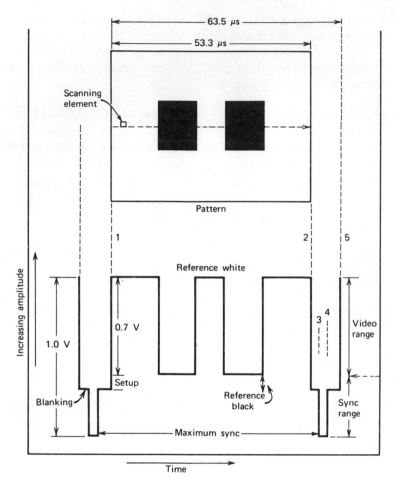

Figure 13.3   Breakdown in time of a scan line.

The maximum signal excursion of a composite video signal is 1.0 V. This 1.0 V is a video/TV reference and is always taken as a peak-to-peak measurement. The 1.0 V may be reached at maximum synchronizing voltage and is measured between synchronizing "tips."

Of the 1.0 V peak, 0.25 V is allotted for the synchronizing pulses and 0.05 V for the setup, leaving 0.7 V to transmit video information. Therefore the video signal varies from 0.7 V for the white-through-gray tonal region to 0 V for black. The best way to describe the actual video portion of a composite signal is to call it a succession of rapid nonrepeated transients.

The synchronizing portion of a composite signal is exact and well defined. A TV/video receiver has two separate scanning generators to control the position of the reproducing spot. These generators are called the horizontal and vertical scanning generators. The horizontal one moves the spot in the X or horizontal direction, and the vertical in the Y direction. Both generators

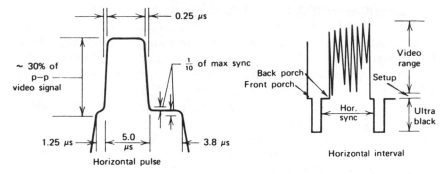

Figure 13.4 Sync pulses and porches.

control the position of the spot on the receiver and must in turn be controlled from the camera (transmitter) synchronizing generator to keep the receiver in step (synchronization).

The horizontal scanning generator in the video receiver is synchronized with the camera synchronizing generator at the end of each scanning line by means of horizontal synchronizing pulses. These are the synchronizing pulses shown in Figure 13.3, and they have the same polarity as the blanking pulses.

When discussing synchronization and blanking, we often refer to certain time intervals. These are discussed as follows:

- The time at the horizontal blanking pulse, 2–5 in Figure 13.3, is called the *horizontal synchronizing interval*.
- The interval 2–3 in Figure 13.3 is called the *front porch*.
- The interval 4–5 is the *back porch*.

The intervals are important because they provide isolation for overshoots of video at the end of scanning lines. Figure 13.4 illustrates the horizontal synchronizing pulses and corresponding porches.

The vertical scanning generator in the video/TV receiver is synchronized with the camera (transmitter) synchronizing generator at the end of each field by means of vertical synchronizing pulses. The time interval between successive fields is called the vertical interval. The vertical synchronizing pulse is built up during this interval. The scanning generators are fed by differentiation circuits. Differentiation for the horizontal scan has a relatively short time constant (RC) and that for the vertical a comparatively long time constant. Thus the long-duration vertical synchronization may be separated from the comparatively short-duration horizontal synchronization. This method of separation of synchronization, known as *waveform separation*, is standard in North America.

In the composite video signal (North American standards) the horizontal synchronization has a repetition rate of 15,750 frames per second, and the vertical synchronization has a repetition rate of 60 frames per second.

13.4 CRITICAL VIDEO TRANSMISSION IMPAIRMENTS

The nominal video baseband is divided into two segments: the high-frequency (HF) segment, above 15,750 Hz, and the low-frequency (LF) segment, below 15,750 Hz. Impairments in the HF segment are operative (in general) along the horizontal axis of the received video image. LF impairments are generally operative along the vertical image axis. Table 13.1 lists critical impairments and their corresponding causes.

The preceding sections present a short explanation of the mechanics of video transmission for a video camera directly connected to a receiving device for display. The primary concern of the chapter, however, is to describe and discuss the problems of point-to-point video transmission. Often the entity responsible for the point-to-point transport of TV programs is not the same entity that originated the image transmitted, except in the case of that link directly connecting the studio to a local transmitter (STL links). A great deal of the "why" has now been covered. The remaining parts of the chapter discuss the video transmission problem (the "how") on a point-to-point basis. The medium employed for this purpose may be radiolink, satellite link, coaxial cable, fiber optics cable, or specially conditioned wire pairs.

Table 13.1 Critical video transmission impairments

Impairment	Cause
HF Segment	
Undistorted echos	Cyclic gain and phase deviation throughout the passband
Distorted echos	Nonlinear gain and phase deviations, especially in the higher end of the band
HF cutoff effects, ringing	1. Limited bandpass distorts, and shows picture transitions, causing overshoot and undershoot 2. This type of distortion (ringing) may also show up on test pattern from lack of even energy distribution and reduced resolution
Porch distortion (poor reproduction of porches)	Poor attenuation and phase distortion
Porch displacement	Zero wander, dc-restored devices such as clamper circuits
Smearing (blurring of the vertical edges of objects)	1. Coarse variations in attenuation and phase 2. Quadrature distortion
LF Segment	
DC suppression, zero wander, distorted image	Lack of clamping

Table 13.1 *(Continued)*

Impairment	Cause

LF Segment

LF roll-off (gradual shading from top to bottom)	Poor clamping, deterioration of coupling networks
Streaking	Phase and attenuation distortion, usually from transmission medium

Nonlinear Distortion

Nonlinearity in the extreme negative region, resulting in horizontal striations or streaking	Compression of synchronizing pulses

Impairments Due to Noise

Noise, in this case, may be considered an undesirable visual sensation. Noise is considered to consist of three types: single frequency, random, and impulse.

Unwanted pattern in received picture	Single-frequency noise
Pattern in alternate fields	Single-frequency noise as an integral multiple of field frequency (50 or 60 Hz)
Horizontal or vertical bars	Single-frequency noise
It should be noted that the LF region is very sensitive to single-frequency noise	
Picture graininess, "snow"	As random noise increases, graininess of picture increases

Note: System design objective of 47-dB signal-to-weighted noise ratio or 44-dB signal-to-flat noise ratio, based on 4.2-MHz video bandwidth, meets Television Allocation Study Organization (TASO) rating of excellent picture. Peak noise should be 37 dB below peak video. Signal-to-noise ratios are taken at peak synchronizing tips to rms noise.

Noise "hits," momentary loss of synchronization, momentary rolling, momentary masking of picture	Impulse noise

Note: Bell System limit at receiver input -20-dB reference to 1-V peak-to-peak signal level point, 1 hit/min. Large amplitudes of impulse noise often masked in black.

Weak, extraneous image superimposed on main image	Strong crosstalk
Nonsynchronization of two images causes violent horizontal motion	
Effect is most noticeable in line and field synchronization intervals	

Note: Limiting loss in crosstalk coupling path should be 58 dB or greater for equal signal levels, design objective 61 dB. Crosstalk for video may be defined as the coupling between two TV channels

13.5 CRITICAL VIDEO PARAMETERS

13.5.1 General

Raw video baseband transmission requires excellent frequency response, in particular from dc to 15 kHz and extending to 4.2 MHz for North American systems and to 5 MHz for European systems. Equalization is extremely important. Few point-to-point circuits are transmitted at baseband because transformers are used for line coupling which deteriorate low-frequency response and make phase equalization very difficult.

To avoid low-frequency deterioration, cable circuits transmitting video have resorted to the use of carrier techniques and frequency inversion using vestigial sideband (VSB) modulation. However, if raw video baseband is transmitted, care must be taken in preserving its dc component.

13.5.2 Transmission Standard—Level

Standard power levels have developed from what is roughly considered to be the input level to an ordinary TV receiver for a noise-free image. This is 1 mV across 75 Ω. With this as a reference, TV levels are given in dBmV. For RF and carrier systems carrying video, the measurement refers to rms voltage. For raw video it is 0.707 of instantaneous peak voltage, usually taken on synchronizing tips.

The signal-to-noise ratio is normally expressed for video transmission as

$$\frac{S}{N} = \frac{\text{peak signal (dBmV)}}{\text{rms noise (dBmV)}} \tag{13-1}$$

TASO picture ratings (4-MHz bandwidth) are related to the signal-to-noise ratio (RF) as follows:

1. Excellent (no perceptible snow)   45 dB
2. Fine (snow just perceptible)   35 dB
3. Passable (snow definitively perceptible
   but not objectionable)   29 dB
4. Marginal (snow somewhat objectionable)   25 dB

13.5.3 Other Parameters

For black and white video systems there are four critical transmission parameters:

1. Amplitude–frequency response
2. EDD (group delay)

3. Transient response
4. Noise (thermal, intermodulation distortion (IM), crosstalk, and impulse)

Color transmission requires consideration of two additional parameters:

5. Differential gain
6. Differential phase

A description of amplitude–frequency response may be found in Section

Table 13.2   End-to-end performance summary

Characteristic	Standard
Amplitude distortion (frequency response)	See Figure 13.5
Chrominance-to-luminance gain inequality	±7 IRE units
Chrominance-to-luminance delay inequality	±60 ns
Field-time waveform distortion	3 IRE units
Line-time waveform distortion	2 IRE units
Short-time waveform distortion (bounce)	8 IRE units peak, 3-s settling time
Insertion gain variation	±0.5 dB hourly ±0.3 dB over 1 s
Luminance nonlinearity	10%
Differential gain	10%
Differential phase	3°
Chrominance-to-luminance intermodulation	4%
Chrominance nonlinear gain	5%
Chrominance nonlinear phase	5°
Dynamic gain of picture signal	6%
Dynamic gain of synchronizing signal	7%
Transient synchronizing signal nonlinearity	5%
Signal-to-noise ratio (10 kHz–5.0 MHz)	54 dB
Signal-to-noise ratio (0–10 kHz)	43 dB
Signal-to-periodic noise ratio (300 Hz–4.2 MHz)	57 dB
Continuity of video service·	99.99% (objective)
AUDIO SIGNAL CHANNEL	
Harmonic distortion	1%
Signal-to-noise ratio	56 dB
Insertion gain	±0.5 dB
Continuity of audio service	99.99% objective
Audio–video time differential	25-ms lead, 40-ms lag

Source: EIA RS-250B (Ref. 12). Courtesy of EIA.

Note: An IRE unit is 1/100 of the luminance range (blanking to reference white). The zero IRE unit shall be at the blanking level.

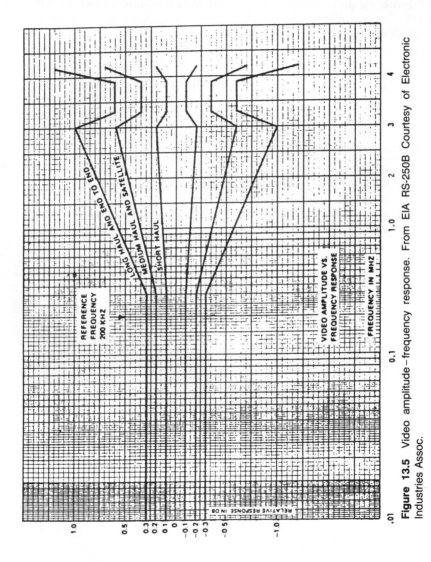

Figure 13.5 Video amplitude–frequency response. From EIA RS-250B Courtesy of Electronic Industries Assoc.

1.9.3. Because video transmission involves such wide bandwidths compared to the voice channel and because of the very nature of video itself, both phase and amplitude requirements are much more stringent.

Transient response is the ability of a system to "follow" sudden, impulsive changes in signal waveform. It usually can be said that if the amplitude–frequency and phase characteristics are kept within design limits, the transient response will be sufficiently good.

Noise is described in Section 1.9.6. Differential gain is the variant in the gain of the transmission system as the video signal level varies (i.e., as it traverses the extremes from black to white). Differential phase is any variation in phase of the color subcarrier as a result of a changing luminance level. Ideally, variations in the luminance level should produce no changes in either the amplitude or the phase of the color subcarrier. Table 13.2 summarizes end-to-end critical transmission parameters for video.

13.6 VIDEO TRANSMISSION STANDARDS (CRITERIA FOR BROADCASTERS)

The following outlines video transmission standards from the point of view of broadcasters (i.e., as emitted from TV broadcast transmitters). Figure 13.6 illustrates the components of the emitted wave (North American practice).

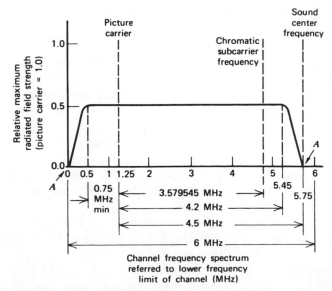

Figure 13.6 RF amplitude characteristics of TV picture transmission. Field strength at points A shall not exceed 20 dB below picture carrier. Drawing not to scale.

13.6.1 Basic Standards

Tables 13.3A and B give a capsule summary of some national standards as taken from CCIR Rep. 624-3 (Ref. 29). Table 13.3C serves as a supplement to these tables.

13.6.2 National Standards

UNITED STATES STANDARD (REF. 2, SEC. 30-13)

Channel width	
(transmission)	6 MHz
Video	4.2 MHz
Aural	±25 kHz
(see Figure 13.6)	
Picture carrier location	1.25 MHz above lower boundary of channel
Modulation	AM composite picture and synchronizing signal on visual carrier together with FM audio signal on audio carrier
Scanning lines	525 per frame, interlaced 2:1
Scanning sequence	Horizontally from left to right, vertically from top to bottom
Horizontal scanning frequency	15,750 Hz for monochrome, or 2/455 × chrominance subcarrier, = 15,734.264 ± 0.044 Hz for NTSC color transmission
Vertical scanning frequency	60 Hz for monochrome, or 2/525 × horizontal scanning frequency for color = 59.94 Hz
Blanking level	Transmitted at 75 ± 25% of peak carrier level
Reference black level	Black level is separated from blanking level by 7.5 ± 2.5% of video range from blanking level to reference white level
Reference white level	Luminance signal of reference white is 12.5 ± 2.5% of peak carrier
Peak-to-peak variation	Total permissible peak-to-peak variation in one frame due to all causes is less than 5%
Polarity of transmission	Negative; a decrease in initial light intensity causes an increase in radiated power
Transmitter brightness response	For monochrome TV, RF output varies in an inverse logarithmic relation to brightness of scene

Aural transmitter power	Maximum radiated power is 20% (minimum 10%) of peak visual transmitter power

BASIC EUROPEAN STANDARD

Channel width (transmission)	7 or 8 MHz
Video	5, 5.5, and 6 MHz
Aural	FM, ± 50 kHz
Picture carrier location	1.25 MHz above lower boundary of channel

Note: VSB transmission is used, similar to North American practice

Modulation	AM composite picture and synchronizing signal on visual carrier together with FM audio signal on audio carrier
Scanning lines	625 per frame, interlaced 2:1
Scanning sequence	Horizontally from left to right, vertically from top to bottom
Horizontal scanning frequency	15,625 Hz $\pm$ 0.1%
Vertical scanning frequency	50 Hz
Blanking level	Transmitted at 75 $\pm$ 2.5% of peak carrier level
Reference black level	Black level is separated from blanking by 3–6.5% of peak carrier
Peak white level as a percentage of peak carrier	10–12.5%
Polarity of transmission	Negative; a decrease in initial light intensity causes an increase in radiated power
Aural transmitter power	Maximum radiated power is 20% of peak visual power

See Tables 13.3A and B for variances.

13.6.3 Color Transmission

Three color transmission standards exist:

NTSC	National Television System Committee (North America, Japan)
SECAM	Sequential color and memory (Europe)
PAL	Phase alternation line (Europe)

Table 13.3A   Basic characteristics of video and synchronizing signals

Item	Characteristics	System (see Table 13.3C for country code)								
		M	N[1]	B, G	H	I	D, K	K1	L	Rec. 472[2]
1	Number of lines per picture (frame)	525	625	625	625	625	625	625	625	625
2	Field frequency, nominal value (fields/s)[3]	60 (59.94)	50	50	50	50	50	50	50	50
3	Line frequency f_H and tolerance when operated nonsynchronously (Hz)[3,4]	15,750 (15,734.264 ±0.0003%)	15,625 ±0.15% (±0.00014%)	15,625[5] ±0.02% (±0.0001%)	15,625 ±0.02% (±0.0001%)	15,625 ±0.00002% [6]	15,625[5] ±0.02% (±0.0001%)	15,625 ±0.03% (±0.0001%)	15,625 ±0.02% (±0.0001%)	15,625 ±0.02% (±0.0001%)
3(a)	Maximum variation rate of line frequency valid for monochrome transmission (%/s)[7,8]	0.15[9]		0.05	0.05	0.05	0.05	0.05	0.05	
4[10]	Blanking level (reference level)	0	0	0	0	0	0	0	0	
	Peak white level	100	100	100	100	100	100	100	100	
	Synchronizing level	−40	−40 (−43)	−43	−43	−43	−43	−43	−43	
	Difference between black and blanking level	7.5 ± 2.5	7.5 ± 2.5 (0)	0	0	0	0–7	0 (color) 0–7 (mono.)	9 (color) 0–7 (mono.)	0$^{+5}_{-0}$
	Peak level including chrominance signal	120		133[11]		133	115[12]	115[12]	124[12]	
5	Assumed gamma of display device for which precorrection of monochrome signal is made	2.2	2.2 (2.8)		2.8[13]				[14]	
6	Nominal video bandwidth (MHz)	4.2	4.2	5	5	5.5	6	6	6	5.0 or 5.5 or 6.0
7	Line synchronization	See Table 1-1 of table source								
8	Field synchronization	See Table 1-2 of table source								

Notes:

(1) The values in parentheses apply to the combination N/PAL used in Argentina (PAL = phase alternation line).

(2) Figures are given for comparison.

(3) Figures in parentheses are valid for color transmission.

(4) In order to take full advantage of precision offset when the interfering carrier falls in the sideband of the upper video range (greater than 2 MHz) of the wanted signal, a line-frequency stability of at least 2×10^{-7} is necessary.

(5) The exact value of the tolerance for line frequency when the reference of synchronism is being changed requires further study.

(6) When the reference of synchronism is being changed, this may be relaxed to $15{,}625 \pm 0.02\%$.

(7) These values are not valid when the reference of synchronism is being changed.

(8) Further study is required to define maximum variation rate of line frequency valid for color transmission. See in this regard CCIR, 1978–82. In the UK this is 0.1 Hz/s (CCIR 1982–86b).

(9) The values used in Japan are ± 0.1.

(10) It is also customary to define certain signal levels in 625-line systems, as follows:

Synchronizing level = 0
Blanking level = 30
Peak white level = 100

For this scale, the peak level including chrominance signal for system D, K/SECAM (SECAM = sequential color and memory) = 110.7. (See [CCIR, 1982–86a].)

(11) Value applies to PAL signals.

(12) Values apply to SECAM signals. For program exchange the value is 115.

(13) Assumed value for overall gamma approximately 1.2. The gamma of the picture tube is defined as the slope of the curve giving the logarithm of the luminance reproduced as a function of the logarithm of the video signal voltage when the brightness control of the receiver is set so as to make this curve as straight as possible in a luminance range corresponding to a contrast of at least 1/40.

(14) In Rec. 472, a gamma value for the picture signal is given as approximately 0.4.

Source: CCIR Rep. 624-3 (Ref. 29, Table I, pp. 2–3). Vol. X, XVth Plenary assembly, Dubrovnik 1986, Ref. 29.

881

Table 13.3B Characteristics of the radiated signals (monochrome and color)

Item	Characteristics	M	N [1]	B, G	H	I	D, K	K1	L
1	Nominal radio-frequency channel bandwidth (MHz)	6	6	B: 7 G: 8	8	8	8	8	8
2	Sound carrier relative to vision carrier (MHz)	+4.5 [2]	+4.5	+5.5 ±0.001 [3,4,5]	+5.5	+5.9996 ±0.0005	+6.5 ±0.001	+6.5	+6.5
3	Nearest edge of channel relative to vision carrier (MHz)	−1.25	−1.25	−1.25	−1.25	−1.25	−1.25	−1.25	−1.25
4	Nominal width of main sideband (MHz)	4.2	4.2	5	5	5.5	6	6	6
6	Minimum attenuation of vestigial sideband (dB at MHz) [6]	20 (−1.25) 42 (−3.58)	20 (−1.25) 42 (−3.5)	20 (−1.25) 20 (−3.0) 30 (−4.43) [7]	20 (−1.75) 20 (−3.0)	20 (−3.0) 30 (−4.43)	20 (−1.25) 30 (−4.33 ±0.1) [8,9]	20 (−2.7) 30 (−4.3) Ref.: 0 (+0.8)	15 (−2.7) 30 (−4.3) Ref.: 0 (+0.8)
7	Type and polarity of vision modulations	C3F neg	C3F neg	C3F neg	C3F neg	C3F neg	C3F neg	C3F neg	C3F pos
8	Synchronizing level	100	100	100	100	100	100	100	< 6
	Blanking level	72.5–77.5	72.5–77.5 (75 ± 2.5)	75 ± 2.5 [10]	72.5–77.5	76 ± 2	75 ± 2.5	75 ± 2.5	30 ± 2
	Difference between black level and blanking level	2.88 to 6.75	2.88 to 6.75	0–2 (nominal) [10,12]	0–7	0 (nominal)	0–4.5 [11]	0–4.5	0–4.5
	Peak white level	10–15	10–15 (10–12.5)	10–12.5 [10,12]	10–12.5	20 ± 2	10–12.5 [13,14]	10–12.5	100 (≈ 110) [15]
9	Type of sound modulation	F3E	F3E	F3E	F3E	F3E	F3E	F3E	A3E
10	Frequency deviation (kHz)	±25	±25	±50	±50	±50	±50	±50	

Levels in the radiated signal (% of peak carrier)

Frequency spacing (see Figure 10 of table source)

11	Preemphasis for modulation (μs)	75	75	50	50	50	50	50	
12	Ratio of effective radiated powers of vision and sound [16]	10/1 to 5/1 [17]	10/1 to 5/1	20/1 to 10/1 [3,18,19]	5/1 to 10/1	5/1 to 10/1 [20]	10/1 to 5/1 [21]	10/1	10/1
13	Precorrection for receiver group-delay characteristics at medium video frequencies (ns) (see also Figure 3 of table source)	0	(1 MHz 0 ±100 / 1 MHz 0 ±100 / 1 MHz 0 ± 60)	[22]			[23a]		
14	Precorrection for receiver group-delay characteristics at color sub-carrier frequency (ns) (see also Figure 3 of table source)	−170 (nominal)	(−170 +60 / −170 −40)	−170 (nominal) [22]			[23b]		

Source: CCIR Rep. 624-3 (Table III, pp. 17–19, Assembly, Dubrovosik 1986, Ref. 29.)

Notes:

[1] The values in parentheses apply to the combination N/PAL used in Argentina (PAL = phase alternation line [Europe]).

[2] In Japan, the values +4.5 ± 0.001 are used.

[3] In the Federal Republic of Germany a system of two sound carriers is used, the frequency of the second carrier being 242.1875 kHz above the frequency of the first sound carrier. The ratio between vision/sound effective radiated power (ERP) for this second carrier is 100/1. For further information on this system see Rep. 795. For stereophonic sound transmissions a similar system is used in Australia with vision/sound power ratios being 20/1 and 100/1 for the first and second sound carriers, respectively.

[4] New Zealand uses a sound carrier displaced 5.4996 MHz from the vision carrier.

[5] The sound carrier for single carrier sound transmissions in Australia may be displaced 5.5 ± 0.005 MHz from the vision carrier.

[6] In some cases, low-power transmitters are operated without vestigial sideband filter.

[7] For B/SECAM and G/SECAM (SECAM = sequential color and memory [Europe]): 30 dB at −4.33 MHz, within the limits of ±0.1 MHz.

[8] In some countries, members of the OIRT, additional specifications are in use:
(a) Not less than 40 dB at −4.286 MHz ± 0.5 MHz
(b) 0 dB from −0.75 to +6.0 MHz
(c) Not less than 20 dB at ±6.375 MHz and higher
Reference: 0 dB at +1.5 MHz

[9] In the People's Republic of China, the attenuation value at the point (−4.33 ± 0.1) has not yet been determined.

Table 13.3B *(Continued)*

(10) Australia uses the nominal modulation levels specified for system I.

(11) In the People's Republic of China, the values 0–5 have been adopted.

(12) Italy is considering the possibility of controlling the peak white-level after weighting the video frequency signal by a low-pass filter, so as to take account only of those spectrum components of the signal that are likely to produce intercarrier noise in certain receivers when the nominal level is exceeded. Studies should be continued with a view to optimizing the response of the weighting filter to be used.

(13) The USSR has adopted the value $15 \pm 2\%$.

(14) A new parameter "white level with subcarrier" should be specified at a later date. For that parameter, the USSR has adopted the value of $7 \pm 2\%$.

(15) The peak white level refers to a transmission without color subcarrier. The figure in parentheses corresponds to the peak value of the transmitted signal, taking into account the color subcarrier of the respective color television system.

(16) The values to be considered are: A) The rms value of the carrier at the peak of the modulation envelope for the vision signal. For system L, only the luminance signal is to be considered (see note 15 above); B) The rms value of the unmodulated carrier for amplitude-modulated and frequency-modulated sound transmissions

(17) In Japan, a ratio of 1/0.15 to 1/0.35 is used. In the United States, the sound carrier ERP is not to exceed 22% of the peak authorized vision ERP.

(18) It may be that the Austrian Administration will continue to use a 5/1 power ratio in certain cases, when necessary.

(19) Recent studies in India (CCIR, 1982–86f) confirm the suitability of a 20/1 ratio of effective radiated powers of vision and sound. This ratio still enables the introduction of a second sound carrier.

(20) The ratio 10/1 is used in the Republic of South Africa.

(21) In the People's Republic of China, the value 10/1 has been adopted.

(22) In the Federal Republic of Germany and the Netherlands the correction for receiver group-delay characteristics is made according to curve B in Figure 3a). Tolerances are shown in the table under Figure 3a). From CCIR, 1966–69, it is learned that Spain uses curve A. The OIRT countries using the B/SECAM and G/SECAM systems use a nominal precorrection of 90 ns at medium video frequencies. In Sweden, the precorrection is 0 ± 40 ns up to 3.6 MHz. For 4.43 MHz, the correction is -170 ± 20 ns and for 5 MHz it is -350 ± 80 ns. In New Zealand the precorrection increases linearly from 0 ± 20 ns at 0 MHz to 60 ± 50 ns at 2.25 MHz, follows curve A of Figure 3a from 2.25 MHz to 4.43 MHz, and then decreases linearly to -300 ± 75 ns at 5 MHz. In Australia, the nominal precorrection follows curve A up to 2.5 MHz, then decreases to 0 ns at 3.5 MHz, -170 ns at 4.43 MHz, and -280 ns at 5 MHz. Based on studies on receivers in India, the receiver group-delay preequalization proposed to be adopted in India at 1 MHz, 2 MHz, 3 MHz, 4.43 MHz, and 4.8 MHz is $+125$ ns, $+150$ ns, $+142$ ns, -75 ns, and -200 ns, respectively.

(23a) Not yet determined. The Czechoslovak Socialist Republic proposes $+90$ ns (nominal value).

(23b) Not yet determined. The Czechoslovak Socialist Republic proposes $+25$ ns (nominal value).

Table 13.3C   Supplement to Tables 13.3A and B

Country	System Used in Bands[a]	
	VHF	UHF
Germany	B/PAL	G/PAL
Argentina	N/PAL	N/PAL
Australia	B/PAL	B/PAL
Austria	B/PAL	G/PAL
Belgium	B/PAL	G/PAL
Brazil	M/PAL	M/PAL
Canada	M/NTSC	M/NTSC
Chile	M/NTSC	M/NTSC
China (PRC)	D/PAL	D/PAL
Colombia	M	M
Korea, Costa Rica	M/NTSC	M/NTSC
Denmark	B/PAL	G/PAL
Egypt	B/SECAM	G/SECAM
Spain	B/PAL	G/PAL
United States	M/NTSC	M/NTSC
Finland	B/PAL	G/PAL
France	L/SECAM	L/SECAM
Greece	B/SECAM	G/SECAM
Hungary	D/SECAM	K/SECAM
India, Indonesia	B/PAL	
Ireland	I/PAL	I/PAL
Iceland	B/PAL	G
Israel and Italy	B/PAL	G/PAL
Japan	M/NTSC	M/NTSC
Malaysia	B/PAL	G/PAL
Morocco	B/SECAM	G/SECAM
Mexico	M/NTSC	M/NTSC
Norway, New Zealand, Pakistan	B/PAL	G/PAL
Netherlands	B/PAL	G/PAL
Peru	M/NTSC	M/NTSC
Poland	B/PAL	G/PAL
United Kingdom	(Does not use VHF)	I/PAL
South Africa	I/PAL	I/PAL
Sweden, Switzerland	B/PAL	G/PAL
Czechoslovakia	D/SECAM	K/SECAM
Turkey	B/PAL	G/PAL
Uruguay	N/PAL	
Venezuela	M	
Yugoslavia	B/PAL	G/PAL

[a]   PAL = Phase line alternation (Europe).
    NTSC = National Television System Committee (North America and Japan).
    SECAM = Sequential color and memory (Europe).

The systems are similar in that they separate the luminance and chrominance information and transmit the chrominance information in the form of two color difference signals which modulate a color subcarrier transmitted within the video band of the luminance signal. The systems vary in the processing of chrominance information.

In the NTSC system, the color difference signals I and Q amplitude-modulate subcarriers that are displaced in phase by $\pi/2$, giving a suppressed carrier output. A burst of the subcarrier frequency is transmitted during the horizontal back porch to synchronize the color demodulator.

In the PAL system, the phase of the subcarrier is changed from line to line, which requires the transmission of a switching signal as well as a color burst.

In the SECAM system, the color subcarrier is frequency modulated alternately by the color difference signals. This is accomplished by an electronic line-to-line switch. The switching information is transmitted as a line-switching signal.

13.6.4 Standardized Transmission Parameters* (Point-to-Point TV)

Interconnection at video frequencies:

Impedance	75 Ω unbalanced or 124 Ω balanced (resistive)
Return loss	No less than 30 dB
Nominal signal amplitude	1 V peak to peak (monochrome)
Nominal signal amplitude	1.25 V peak to peak, maximum (composite color)
Polarity	Black-to-white transitions, positive going

Interconnection at intermediate frequency (IF):

Impedance	75 Ω unbalanced
Input level	0.3 V rms
Output level	0.5 V rms
IF up to 1 GHz	35 MHz
IF above 1 GHz	70 MHz

Signal-to-weighted noise: 53 dB

13.7 METHODS OF PROGRAM CHANNEL TRANSMISSION FOR VIDEO

Composite transmission normally is used on broadcast and community antenna television (CATV [cable]) distribution. Video and audio carriers are "combined" before being fed to the radiating antenna for broadcast. These audio subcarriers are described in Section 13.6.

*Based on CCIR Rec. 567-2.

For point-to-point transmission on coaxial cable, radiolink, and earth station systems, the audio program channel is generally transmitted separately from its companion video providing the following advantages:

- Individual channel level control
- Greater control over crosstalk
- Increased guard band between video and audio
- Saves separation at broadcast transmitter
- Leaves TV studio as separate channel
- Permits individual program channel prcemphasis

13.8   VIDEO TRANSMISSION OVER COAXIAL CABLE

13.8.1   The L3 System of North America

The L3 carrier system was designed for use on 12-MHz coaxial cable. In most applications when L3 is used for video transmission, it shares the same cable with up to 600 frequency division multiplex (FDM) telephone channels. The FDM voice channel segment occupies the band from 564 to 3084 kHz. The video signal occupies the region from 3639 to 8500 kHz. The modulation of the video signal, translating it to the indicated segment of the spectrum, is carried out by vestigial sideband (VSB) methods. The virtual carrier as transmitted on-line is at 4139 kHz and the VSB occupies the space of 3639–4139 kHz. VSB is used to avoid some of the problems encountered with the normally used envelope detection regarding video, such as the production of a spurious envelope wherein video signals that exceed a certain value are inverted. In the L3 system, as in most video transmission systems of this type, homodyne detection is used. Here the demodulator is driven by a locally generated carrier, which is synchronous in phase angle and frequency with the carrier component of the transmitted wave. Homodyne detection also makes possible the necessary suppression of the quadrature distortion associated with VSB transmission. Figure 13.7 illustrates the L3 frequency allocation and modulation processes to develop the line frequency.

To ameliorate somewhat the effects of second harmonic distortion, a preemphasis network is used in the transmitting terminal to accentuate the amplitude of the HF components of the signal before transmission. At the receiving terminal a deemphasis network introduces a complementary frequency characteristic to make the overall transmission characteristic constant with frequency.

Delay and amplitude equalizers are incorporated in the transmit section with an objective of maintaining amplitude in the band of interest to vary no more than ± 0.02 dB. Phase shift objectives are on the order of $\pm 0.1°$. Mop-up equalizers also are used on receiver terminals. Repeater spacing for the L3 system is 4 mi (6.4 km).

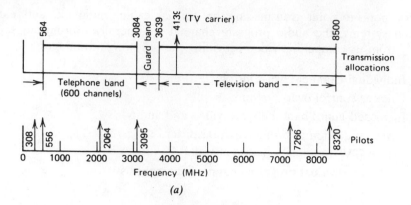

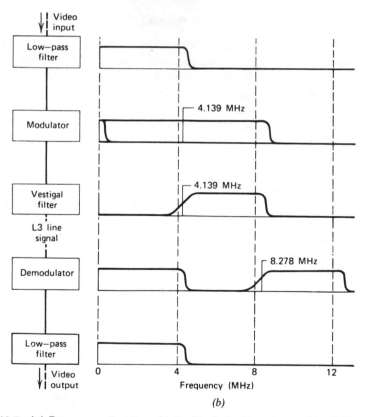

Figure 13.7 (*a*) Frequency allocations for the North American L3 combined TV–telephone transmission system. (*b*) L3 TV terminal modulation processes. (From Ref. 15, copyright © 1953 by American Telephone and Telegraph Company.)

13.8.2 A 12-MHz European System

12-MHz coaxial cable systems, if used for video transmission, almost always transmit combined FDM telephone channels with the video. One such system modulates a 5.5-MHz video sideband with a 6.799-MHz carrier. A baseband signal is produced in the band of 6.3–12.3 MHz, using the upper sideband of the modulation process and a vestigial portion of the lower sideband. 1200 FDM telephone channels are transmitted in the lower portion of the band.

The required flatness in envelope delay and amplitude response is maintained in the video transmission band (i.e., 6.3–12.3 MHz) by equalizers built into the coupling and separating filter units for contributions from those units (i.e., each unit is provided with equalizers to flatten response of its own filters). The envelope delay distortion (EDD) caused by the coaxial line itself and its associated equipment is equalized at the receiving end.

The television baseband for a modulator-demodulator back-to-back is maintained as follows:

Attenuation distortion	0.2 dB
EDD	25 ns
Differential gain	0.05 dB .
Differential phase	0.2°
Signal-to-hum ratio	60 dB
Continuous random noise, unweighted	66 dB (down)

CCITT Rec. G.332 covers this type of transmission under "mixed systems."

13.9 TRANSMISSION OF VIDEO OVER RADIOLINKS

13.9.1 General

Telephone administrations increasingly are expanding their offerings to include other services. One such service is to provide point-to-point broadcast-quality video relay on a lease basis over radiolinks. As covered earlier in this section, video transmission requires special consideration.

The following paragraphs summarize the special considerations a planner must take into account for video transmission over radiolinks.

Raw video baseband modulates the radiolink transmitter. The aural channel is transmitted on a subcarrier well above the video portion. The overall subcarriers are themselves frequency modulated. Recommended subcarrier frequencies may be found in CCIR Rec. 402-2 and Rep. 289-4.

13.9.2 Bandwidth of the Baseband and Baseband Response

One of the most important specifications in any radiolink system transmitting video is frequency response. A system with cascaded hops should have essentially a flat bandpass in each hop. For example, if a single hop is 3 dB down at 6 MHz in the resulting baseband, a system of five such hops would be 15 dB down. A good single hop should be ±0.25 dB or less out to 8 MHz. The most critical area in the baseband for video frequency response is in the low-frequency area of 15 kHz and below. Cascaded radiolink systems used in transmitting video must consider response down to 10 Hz.

Modern radiolink equipment used to transport video operates in the 2-GHz band and above. 525-line video requires a baseband in excess of 4.2 MHz plus available baseband above the video for the aural channel. Desirable characteristics for 525-line video then would be a baseband at least 6 MHz wide. 8 MHz would be required for 625-line TV, assuming that the aural channel would follow the channelization recommended by CCIR Rec. 402-2.

13.9.3 Preemphasis

Preemphasis–deemphasis characteristics are described in CCIR Rec. 405-1 (also see Figure 4.15).

13.9.4 Differential Gain

Differential gain is the difference in gain of the radio relay system as measured by a low amplitude, high frequency (chrominance) signal at any two levels of a low frequency (luminance) signal on which it is superimposed. It is expressed in percent of maximum gain. Differential gain shall not exceed the amounts indicated below at any value of APL (average picture level) between 10 and 90%:

- Short haul   2%
- Medium haul   5%
- Satellite   4%
- Long haul   8%
- End-to-end   10%

Based on EIA RS-250B (Ref.12). Also see CCIR Rec. 567-2.

13.9.5 Differential Phase

Differential phase is the difference in phase shift through the radio relay system exhibited by a low amplitude, high frequency (chrominance) signal at any two levels of a low frequency (luminance) signal on which it is superimposed. Differential phase is expressed as the maximum phase change between any two levels. Differential phase, expressed in degrees of the high-frequency sine wave, shall not exceed the amounts indicated below at any value of APL (average picture level) between 10 and 90%:

- Short haul   0.5°
- Medium haul 1.3°
- Satellite   1.5°
- Long haul   2.5°
- End-to-end   3.0°

Based on EIA RS-250B (Ref. 12).

13.9.6   Signal-to-Noise Ratio (10 kHz – 5.0 MHz)

The video signal-to-noise ratio is the ratio of the total luminance signal level (100 IRE units) to the weighted rms noise level. The noise referred to is predominantly thermal noise in the 10 kHz–5.0 MHz range. Synchronizing signals are not included in the measurement. The EIA states that there is a difference of less than 1 dB between 525-line systems and 625-line systems.

As stated in the EIA RS-250B standard, the signal-to-noise ratio shall not be less than

- Short haul   67 dB
- Medium haul 60 dB
- Satellite   56 dB
- Long haul   54 dB
- End-to-end   54 dB

and, for the low-frequency range (0–10 kHz), the signal-to-noise ratio shall not be less than

- Short haul   53 dB
- Medium haul 48 dB
- Satellite   50 dB
- Long haul   44 dB
- End-to-end   43 dB

13.9.7   Radiolink Continuity Pilot

For video transmission the continuity pilot is always above the baseband. CCIR recommends an 8.5-MHz pilot. (Refer to CCIR Rec. 401-2 and Table 4.8.)

13.10   TV TRANSMISSION BY SATELLITE RELAY

Table 13.4 provides general guidance on the basic performance requirements for the transmission of broadcast-type TV signals via satellite relay based on CCIR recommendations.

Table 13.4 Satellite relay TV performance

Parameters	Space Segment	Terrestrial Link[a]	End-to-End Values
Nominal impedance	75 Ω		
Return loss	30 dB		
Nonuseful dc component	0.5 V		
Nominal signal amplitude	1 V		
Insertion gain	0 ± 0.25 dB	0 ± 0.3 dB	0 ± 0.5 dB
Insertion gain variation (1 s)	± 0.1 dB	± 0.2 dB	± 0.3 dB
Insertion gain variation (1 h)	± 0.25 dB	± 0.3 dB	± 0.5 dB
Signal-to-continuous random noise	53 dB[b]	58 dB	51 dB
Signal-to-periodic noise (0–1 kHz)	50 dB	45 dB	39 dB
Signal-to-periodic noise (1 kHz–6 MHz)	55 dB	60 dB	53 dB
Signal-to-impulsive noise	25 dB	25 dB ([2])	25 dB[c]
Crosstalk between channels (undistorted)	58 dB	64 dB	56 dB
Crosstalk between channels (differentiated)	50 dB	56 dB	48 dB
Luminance nonlinear distortion	10%	2%	12%
Chrominance nonlinear distortion (amplitude)	3.5%	2%	5%
Chrominance nonlinear distortion (phase)	4°	2°	6°
Differential gain (x or y)	10%	5%	13%
Differential phase (x or y)	3°	2°	6°
Chrominance–luminance intermodulation	± 4.5%	± 2%	± 5%
Steady-state sync pulse nonlinear distortion	+ 5–10%	± 5%	+ 10–15%
Transient sync pulse nonlinear distortion	20%		
Field-time waveform distortion	6%	2%	10%
Line-time waveform distortion	3%	2%	4%
Short-time waveform distortion (pulse/bar)	100 ± 12%	100 ± 6%[c]	100 ± 18%[c]
Short-time waveform distortion (pulse lobes)	3%[d]	1.5%[d]	4.5%[d]
Chrominance–luminance gain inequality	± 10%	± 6%	± 13%
Chrominance–luminance delay inequality	± 50 ns	± 60 ns	± 90 ns
Gain–frequency characteristic (0.15–6 MHz)	± 0.5 dB	± 0.5 dB[e]	± 1.0 dB[e]

Table 13.4 *(Continued)*

Parameter	Space Segment	Terrestrial Kink[a]	End-to-End Values
Delay–frequency characteristic (0.15–6 MHz)	± 50 ns	± 50 ns[e]	± 105 ns[e]

Source: CCIR Rep. 965 (Ref. 22, p. 45). Courtesy of ITU–CCIR.

[a]Connecting earth station to national technical control center.

[b]In cases where the receive earth station is collocated with the broadcaster's premises, a relaxation of up to 3 dB in video signal-to-weighted-noise ratio may be permissible. In this context, the term *colocated* is intended to represent the situation where the noise contribution of the local connection is negligible.

[c]Law of addition not specified in CCIR. Rec. 567.

[d]The pulse lobes are contained within a mask of the type shown in Fig. 29a in CCIR Rec. 567. The figure in the space segment column is the amplitude of the mask in Fig. 29a for times ≤ -800 ns and ≥ 800 ns. The figures in the terrestrial link and end-to-end values columns are corresponding amplitudes of scaled versions of the mask.

[e]Highest frequency: 5 MHz.

13.11   TRANSMISSION OF VIDEO OVER CONDITIONED PAIRS

13.11.1   General

Broadcast-quality video transmission over conditioned pairs has application to interconnect broadcast facilities and the long-distance transmission system. The broadcaster may lease these facilities from a telephone administration to interconnect a master control point and outlying studios or remote program pickup points. Normally, only the video is transmitted on the conditioned pair. The audio is usually transmitted separately on its own program facilities. The cable is designed for installation in ducts and the repeaters in duct-type cabinets or racks.

13.11.2   Cable Description

A conditioned-pair video transmission system is composed of a shielded wire pair, terminal equipment, and repeaters. We will now describe a system that has had wide use in North America. Such a description of a typical system will point up the advantages and limitations of video transmission over conditioned pairs.

The line facilities consist of 16-gauge polyethylene-insulated pairs generally referred to as PSV.* At 75°F the loss at 4.5 MHz is 3.52 dB/1000 ft and 18.6 dB/mi. The normal slope variation with temperature is approximately 0.1%/°F. The loss at zero frequency is taken as 0 dB.

Because of the effective shielding of the PSV pairs, there is no limitation as to the direction of transmission or the number of circuits obtainable within any given size of cable. Noise considerations require that the PSV pairs be

*Pair shield video.

separated from the remainder of the cable conductors at building entrances. In this case the shielded video pair is brought to the video equipment under a separate sheath.

The characteristic impedance of the video cable is nearly purely resistive, 124 Ω, at frequencies above 500 kHz. The resistive component increases to 1000 Ω at 60 Hz. The reactive component is about the value of the resistive component at 60 Hz, and drops to nearly zero at the higher frequencies.

13.11.3 Terminal Equipment and Repeaters

A transmitting terminal is provided to match the video output of the line, secure the proper level, and predistort the signal as a first step in conditioning the PSV cable. Equalization is basically one of amplitude. The impedance is from 75 Ω unbalanced to 124 Ω balanced.

Amplitude equalization for this type of transmission facility is such that levels are described as fractions. The level unit is dBV (decibels relative to 1 V). For instance, a voltage level may be expressed as $-10/+5$ dBV. Such a "fraction" describes the attenuation–frequency characteristic or slope. The numerator refers to a level of zero frequency, and the denominator to the reference HF, in this case 4.5 MHz. Voltages here are peak to peak. Zero frequency may be taken to mean a very low frequency, 30 Hz.

This method of designating level is a useful tool when the amplitude–frequency response and its equalization are of primary concern. At the transmitting end of a PSV link we would expect a zero slope, and the transmitter output would be 1.0 V peak to peak (0 dBV) or 0/0 dBV. To equalize the line, the transmitter must predistort the signal. After equalization the level may be described, assuming 15-dB equalization, as $-10/+5$ dBV. After traversing 33 dB of cable, the level may then be described as $-10/-28$ dBV. Repeaters provide both amplification and equalization. It would appear that most systems would require custom design. To simplify engineering and standardize components, PSV systems often are built in blocks. The system is configured with fixed-length repeater sections, plug-in equalizers, and standardized receivers. Repeater spacing is on the order of 4.5 mi (7.2 km).

The receiver provides both gain and equalization. It also includes a clamper to correct for low-frequency distortion. Equalization is usually variable when the block approach is used for residual gain deviations.

13.12 BASIC TESTS FOR VIDEO QUALITY

13.12.1 Window Signal

The window signal when viewed on a picture monitor is a large square or rectangular white area with a black background. The signal is actually a sine-squared pulse. As such it has two normal levels, reference black and reference white. The signal usually is adjusted so that the white area covers one-fourth to one-half the total picture width and one-fourth to one-half the total picture height. This is done in order to locate the maximum energy content of the signal in the lower portion of the frequency band.

A number of useful checks derive from the use of a window signal and a picture monitor. These include the following:

1. *Continuity or Level Check.* With a window signal of known white level, the peak-to-peak voltage of the signal may be read on a calibrated oscilloscope using a standard roll-off characteristic (i.e., EIA, etc.).

2. *Sync Compression or Expansion Measurements.* Comparison of locally received window signals with that transmitted from the distant end with respect to white level and horizontal synchronization on calibrated oscilloscopes using standard roll-off permits evaluation of linearity characteristics.

3. *Test and Adjustment.* Can be made at clamper amplifiers and low-frequency equalizers to minimize streaking by observing the test signal on scopes using the standard roll-off characteristics at both the vertical and the horizontal rates.

4. *Indication of Ringing.* With a window signal the presence of ringing may be detected by using properly calibrated wide-band oscilloscopes and adjusting the horizontal scales to convenient size. Both amplitude and frequency of ringing may be measured by this method.

13.12.2 Sine-Squared Test Signal

The sine-squared test signal is a pulse type of test signal that permits an evaluation of amplitude–frequency response, transient response, envelope delay, and phase. An indication of the HF amplitude characteristic can be determined by the pulse width and height, and the phase characteristic by the relative symmetry about the pulse axis. However, this test signal finds its principal application in checking transient response and phase delay. The sine-squared signal is far more practical than a square wave test signal to detect overshoot and ringing. The pulse used for checking video systems should have a repetition rate equal to the line frequency, and a duration, at half amplitude, equal to one-half the period of the nominal upper cutoff frequency of the system.

13.12.3 Multiburst

This test signal is used for a quick check of gain at a few determined frequencies. A common form of multiburst consists of a burst of peak white (called white flag) that is followed by bursts of six sine wave frequencies from 0.5 to 4.0 MHz (for NTSC systems) plus a horizontal synchronizing pulse. All

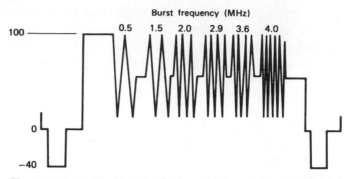

Figure 13.8 Multiburst signal (horizontal frequencies normally used).

these signals are transmitted during one-line intervals. The peak white or white flag serves as a reference. For system checks a multiburst signal is applied to the transmit end of a system.

At the receiving point the signal is checked on an oscilloscope. Measurements of peak-to-peak amplitudes of individual bursts are indicative of gain. A multiburst image on an oscilloscope gives a quick check of amplitude–frequency response and changes in setup. Figure 13.8 illustrates a typical NTSC-type multiburst signal. The y-axis of the figure is in IRE units.*

13.12.4 Stair Steps

For the measurement of differential phase and gain, a 10-step stair-step signal is often used. Common practice (in the United States) is to superimpose 3.6 MHz on the 10 steps that extend progressively from black to white level. The largest amplitude sine wave block is adjusted on the oscilloscope to 100 standard (i.e., IRE) divisions and is made a reference block. Then the same 3.6-MHz sine waves from the other steps are measured in relation to the reference black. Any difference in amplitudes of the other blocks represents differential gain. By the use of a color analyzer in conjunction with the above, differential phase may also be measured.

The stair-step signal may be used as a linearity check without the sine wave signal added. The relative height between steps is in direct relation to signal compression or nonlinearity.

13.12.5 Vertical Interval Test Signals (VITS)

The VITS makes use of the vertical retrace interval for the transmission of test signals. The FCC specifies the interval for use in the United States as the

---

*IRE standard scale is linear scale for measuring, in arbitrary IRE units, the relative amplitude of the various components of a television signal. The scale varies from −40 (sync peaks, max. carrier) to +120 (zero carrier). Reference white is +100 and blanking is 0.

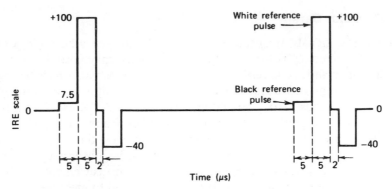

Figure 13.9 Typical vertical-interval reference signal.

last 12 μs of lines 17 through 20 of the vertical blanking interval of each field. For whichever interval boundaries specified, test signals transmitted in the interval may include reference modulation levels, signals designed to check performance of the overall transmission system or its individual components, and cue and control signals related to the operation of TV broadcast stations. These signals are used by broadcasters because, by necessity, they are inserted at the point of origin. Standard test signals are used as described above or with some slight variation, such as multiburst, window, and stair step. Some broadcasters use vertical interval reference signals. Figure 13.9 shows one currently in use in the United States.

13.12.6 Test Patterns

Standard test patterns, especially those inserted at a point of program origination, provide a simple means of determining transmission quality. The distant viewer, knowing the exact characteristics of the transmitted image, can readily detect distortion(s). Standard test patterns such as the EIA test pattern used widely in the United States, with a properly adjusted picture monitor, can verify

- Horizontal linearity
- Vertical linearity
- Contrast
- Aspect ratio
- Interlace
- Streaking
- Ringing
- Horizontal and vertical resolution

13.12.7 Color Bars

Color bar test signals are used by broadcasters for the adjustment of their equipment, including color monitors. Color bars may also be sent over transmission facilities for test purposes. The color bar also may be used to test color transmission using a black and white monitor by examining gray densities of various bars, depending on individual colors. A wide-band A-scope horizontal presentation can show whether or not the white reference of the luminance signal and the color information have the proper amplitude relationships. The color bar signal may further be observed on a vector display oscilloscope (chromascope), which allows measurement of absolute amplitude and phase angle values. It also can be used to measure differential phase and gain.

13.13 DIGITAL TELEVISION

13.13.1 Introduction

Our concern in this chapter is television transmission. The world's telecommunication trunk network will be all digital within the time frame of the 1990s. Television, on the other hand, remains nearly all analog as described in the previous sections of this chapter.

The trend to digitize television is to use 8-bit pulse code modulation (PCM) (Chapter 3) with linear quantization. If we sample a 4.2-MHz signal at the Nyquist rate and assign 8 bits per sample, we derive a 67.2-Mbps data rate. 70 Mbps rates are discussed in CCIR Rep. 646-3 (Ref. 28). If we are to transmit TV on the digital network, it would be highly desirable to have the TV bit rate match a rate in the present digital hierarchies (see Section 3.3.5).

In this section we will briefly discuss some digitizing schemes for TV video and some introductory methods of bit rate reduction. This will be followed by an overview of three practical systems now on the market for the transmission of TV over the public switched telecommunication network. The first of these is called *freeze-frame* video, where we can transmit video images over data circuits at 2400, 4800, or 9600 bps; in the treatment of the second practical system, we discuss a scheme to transmit video at DS1 (CEPT 30 + 2) subrates for video teleconferencing; the third practical system is a method of transmitting broadcast-quality television at the DS3 nominal 45-Mbps rate.

13.13.2 Two Digital Coding Schemes Suggested by CCIR

CCIR Rec. 601 (Ref. 31) describes two distinct digital coding concepts for color television: component and composite coding.

For this discussion there are four components that make up a color video signal. These are: R for red, G for green, B for blue, and Y for luminance.

The output signals of a TV camera are converted by a linear matrix into luminance (Y) and two color-difference signals, R–Y and B–Y.

With the component method these signals are individually digitized by an A/D converter. The resulting digital bit streams are then combined with overhead and timing by means of a multiplexer for transmission over a single medium such as a wire pair or coaxial cable.

Composite coding, as the term implies, directly codes the entire video baseband. The derived bit stream has a notably lower bit rate than that for component coding.

CCIR Rep. 646-3 (Rcf. 28) compares the two coding techniques. The advantages of separate-component coding are that

- The input to the circuit is provided in separate component form by the signal sources (in the studio)
- The component coding is adopted generally for studios, and the inherent advantages of component signals for studios must be preserved over the transmission link in order to allow downstream processing at a receiving studio
- The country receiving the signals via an international circuit uses a color system different from that used in the source country
- The transmission path is entirely digital, which fits in with the trend toward all digital systems, that is expected to continue

The advantages of transmitting in the composite form are

- The input to the circuit is provided in the composite form by the signal sources (at the studio)
- The color system used by the receiving country, in the case of an international circuit, is the same as that used by the source country
- The transmission path consists of mixed analog and digital sections

13.13.3 Bit Rate Reduction

CCIR discusses digital television systems that are PCM-based with bit rates from 120 to 240 Mbps. When bandwidth is at a premium, the transmission over long distances may require bit-rate reduction methods in order to reduce transmission costs. The choice of technique depends on relative performance of the technique and equipment costs. CCIR Rep. 646-3 (Ref. 28) states that three basic bit-rate reduction methods may be employed:

- Removal of horizontal and/or vertical blanking intervals
- Reduction of sampling frequency
- Reduction of the number of bits per sample

We will only cover the last method.

13.13.3.1 *Reduction of the Number of Bits per Sample* There are three methods that may be employed for bit rate reduction of digital television by reducing the bits per sample. These may be used singly or in combination:

- Predictive coding, sometimes called differential PCM
- Entropy coding
- Transform coding

Differential PCM, according to CCIR (Ref. 28), has so far emerged as the most popular method. The prediction process required can be classified into two groups. The first one is called *intraframe* or *intrafield* and is based only on the reduction of spatial redundancy. The second group is called *interframe* or *interfield*, and is based on the reduction of temporal redundancy as well as spatial redundancy.

13.13.3.2 *Specific Bit Rate Reduction Techniques* Ref. 21 provides the following information on some of the more advanced digital video compression techniques:

> *Intraframe coding*: Intraframe coding techniques provide compression by removing redundant information within each video frame. These techniques rely on the fact that images typically contain a great deal of similar information; for example, a one-color background wall may occupy a large part of each frame. By taking advantage of this redundancy, the amount of data necessary to accurately reproduce each frame may be reduced.
>
> *Interframe coding*: Interframe coding is a technique that adds the dimension of time to compression by taking advantage of the similarity between adjacent frames. Only those portions of the picture that have changed since the previous picture frame are communicated. Interframe coding systems do not transmit detailed information if it has not changed from one frame to the next. The result is a significant increase in transmission efficiency.
>
> *Intraframe and interframe coding used in combination*: Intraframe and interframe coding used together provide a powerful compression technique. This is achieved by applying intraframe coding techniques to the image changes that occur from frame to frame. That is, by subtracting image elements between adjacent frames, a new image remains that contains only the differences between the frames. Intraframe coding, which removes similar information within a frame, is applied to this image to provide further reduction in redundancy.
>
> *Motion compensation coding*: To improve image quality at low transmission rates, a specific type of interframe coding motion compensation is commonly used. Motion compensation applies the fact that most changes between frames of a video sequence occur because objects move. By focusing on the motion that

has occurred between frames, motion compensation coding significantly reduces the amount of data that must be transmitted.

Motion compensation coding compares each frame with the preceding frame to determine the motion that has occurred between the two frames. It compensates for this motion by describing the magnitude and direction of an object's movement (e.g., a head moving right). Rather than completely regenerating any object that moves, motion compensation coding simply commands that the existing object be moved to a new location.

Once the motion compensation techniques estimate and compensate for the motion that takes place from frame-to-frame, the differences between frames are smaller. As a result, there is less image information to be transmitted. Intraframe coding techniques are applied to this remaining image information.

Motion Compensation Transform Coding and Hierarchical Vector Quantization Coding. Motion Compensated Transform (MCT) is proprietary technology of PictureTel Corp. This technology has significantly improved upon traditional technology described above. MCT combines the interframe coding technique, motion compensation, with the intraframe technique, transform coding, to greatly reduce the amount of information that must be transmitted to achieve a given level of video quality. PictureTel's Hierarchical Vector Quantization (HVQ) Coding further advanced compression technology. For example, HVQ transmitted at 112 kbps is comparable to the video quality of MCT transmitted at 224 kbps. Like MCT, HVQ compression technology employs motion compensation coding. However, it replaces cosine transform coding with hierarchical vector quantization.

In transform coding, images are divided into blocks prior to image compression. Under certain conditions, transmission blocks can result in image distortion or breaking of the image into a block pattern, particularly in background portions of the picture.

HVQ greatly improves upon transform coding significantly enhancing image quality by eliminating the artifacts and other distortions inherent in transform coding. With HVQ, the image is coded not in terms of discrete blocks, but on the basis of different levels of image resolution, arranged in a hierarchy from low to high. For each frame, HVQ transmits the image in as much detail as possible. In areas of the image containing motion, any image alterations resulting from compression are much less easily perceived than the more visible effects of transform coding (e.g., blocking) because, with HVQ, they are only the result of reductions in resolution.

Experiments in Broadcast-Quality Compression Techniques. CCIR Rep. 646-3 (Ref. 28) describes some experimental digital TV systems. One, using intrafield predictive codecs, operating directly on composite NTSC TV signals, achieved 32 and 44 Mbps. An alternative coding method for NTSC composite signals employing the Hadamard transform technique shows that, although there is no significant difference in coding performance between differential pulse code modulation (DPCM) and transform coding for most of the natural television images at the same bit rate, DPCM gave better picture quality for pictures containing many saturated color components.

Another NTSC system described has a 32-Mbps transmission rate with composite intrafield PCM coding, which uses two-dimensional prediction and dual word length coding (entropy coding). A 28-Mbps intrafield DPCM coding system for electronic news-gathering is also reported.

Another codec reported by CCIR (Ref. 28) utilizes interfield and intrafield adaptive prediction. This codec transmits a color signal, sound, and forward error correction codes, at a 30-Mbps PCM line rate.

13.13.4   The Basics of a Freeze-Frame Video Transmission System

Freeze-frame is a method of transmitting video images at standard voice grade telephone channel data rates such as 2400, 4800, and 9600 bps or it can be transmitted in analog form using FM.

Colorado Video provided the following description of how freeze-frame transmission works (Ref. 17):

> The video from a camera goes to two locations. The sync signal on the video signal is used to create all the timing and memory addressing signals necessary to operate a digital memory at real-time video rates. The video or picture portion of the signal is fed to a high speed A/D converter. The parallel digital data stream is stored in a digital memory. This is the "freeze-frame" portion of the system. It is now a still picture.
>
> The high speed addressing and timing circuit reads the data out of the memory and the data is fed to a D/A converter that converts the data back to a standard analog television signal that can be displayed on a monitor.
>
> A slow speed timing and addressing circuit module is responsible for the actual transmission of the picture. This circuit can be timed by a sub-multiple of the high speed timing and addressing circuit or by the actual transmission circuit itself. The slow speed timing and addressing circuit is responsible for reading out the picture in memory at a speed compatible with the speed of the transmission path to be used.
>
> The low data rate coming out of memory can be treated in several ways before it is presented to the transmission path. The parallel data can be converted to an analog signal and be modulated in a form suitable for the transmission path. Typically, an FM signal is used for analog telephone VF channels. The parallel data can also be converted to a serial data stream for presentation to an analog voice channel using a standard data modem or to a digital subrate channel.
>
> The data can be processed through an optional compression circuit to speed up the transmission time on the path used.
>
> At the receiving end, the slow speed serial data is converted back to parallel data using the reverse of the processes used at the transmitting end of the circuit. This data is written into a digital memory using the same slow speed timing and addressing circuit used at the transmitter. It reads out the memory

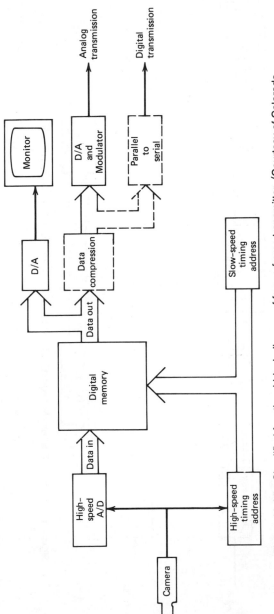

Figure 13.10 Simplified functional block diagram of freeze-frame transmitter. (Courtesy of Colorado Video, Inc., Ref. 17.)

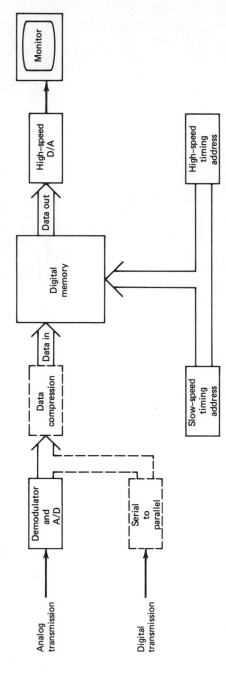

Figure 13.11 Simplified block diagram of freeze-frame receiver. (Courtesy of Colorado Video, Inc. Ref. 17.)

to a high speed D/A converter that converts the memory's digital data to an analog television signal suitable for viewing on a television monitor.

There are many variables to be considered in a freeze-frame video transmission system. The resolution of the images to be transmitted is controlled mainly by the memory size. How many pixels (picture elements) is the picture divided into? The more pixels, the higher the resolution and the longer the transmission time for a given data rate. Whether ... color or just monochrome [is transmitted] will impact transmission time. In general, color transmission requires more transmission time than black-and-white transmission.

The available bandwidth of the transmission medium or bit rate it can support will affect transmission time. For a given pixel matrix size (i.e., 256 × 256), a 64 kbps channel can transmit many more images per unit time than a 4800 bps channel.

Compression algorithms can also reduce the transmission times. However, some algorithms may cause various forms of degradation of image quality. Usually the time reduction and the picture artifacts will depend upon picture content.

Figure 13.10 is a functional block diagram of a typical freeze-frame transmitter and Figure 13.11 is a similar diagram for a freeze-frame receiver.

Performance of Typical Freeze-Frame Equipment (Colorado Video, Inc. Digital Transceiver 286 [Ref. 33]

Images can be transmitted using the "286" system with a refresh rate of between 0.75 and 220 s. We'll call "refresh rate" the time to send an image or TV frame. Refresh rate is a function of the transmission bit rate and the desired image resolution.

For the 286 system, resolution depends on memory size. Gray scale may be represented by 6 or 8 bits per pixel. Tables 13.5A, B, and C give comparative memory requirements for the 286 system versus resolution versus bit rate versus refresh rate (time to send a frame). Suppose a frame is 256 × 480. This means that there are 256 lines and 480 elements (pixels) per line. The

Table 13.5A   NTSC color scan time in seconds

Clock Rate (bits/s)	Memory Configuration			
	Single Field 512 × 240		Full Frame 512 × 480	
	6-bit	8-bit	6-bit	8-bit
4800	166	221	330	439
9600	83	111	165	220
56 K	14	19	28	38
200 K	4.0	5.3	7.9	10.5
500 K	1.6	2.1	3.2	4.2

Source: Ref. 33 Courtesy of Colorado Video, Inc.

Table 13.5B 525-Line black & white scan time in seconds

Clock Rate (bits/s)	Memory Configuration			
	Single Field 256 × 240		Full Frame 256 × 480	
	6-bit	8-bit	6-bit	8-bit
4800	78	104	155	206
9600	39	52	77	102
56 K	6.7	8.9	13	18
500 K	0.75	1.0	1.5	2.0

Source: Ref. 33. Courtesy of Colorado Video, Inc.

Table 13.5C 625-Line black and white scan time in seconds

Clock Rate (bits/s)	Memory Configuration			
	Single Field 256 × 256		Full Frame 256 × 512	
	6-bit	8-bit	6-bit	8-bit
4800	83	111	165	220
9600	41	55	82	110
48 K	8.3	11.0	16.5	22
64 K	6.1	8.3	12.2	16.6
500 K	0.8	1.1	1.6	2.1

Source: Ref. 33. Courtesy of Colorado Video, Inc.

total pixels per frame are 256 × 480 or 122,880 pixels. If each pixel is represented by 8 bits, we will have 983,040 bits per frame (image). Suppose we transmit this image at 4800 bps; how long will it take to transmit one image or frame? 983,040/4800 = 205 s (206 s in the table). If we transmit at 9600 bps, naturally it will take half the time; and at a 56-kbps rate it will take 983,040/56,000 = 18 s.

To put things in perspective, the theoretical resolution of a 525-line NTSC video system is 525 × 700 pixels.

13.13.5 Video Conferencing Using DS1/CEPT 30 + 2 Digital Subrates

A number of manufacturers produce video conferencing systems which transmit full-motion video at subrates of DS1/CEPT 30 + 2. These subrates, of course, are multiples of 56 or 64 kbps. PictureTel of Peabody, MA, provided the following data regarding their equipment (Ref. 20):

- 56–2048 kbps
- Line resolution: video graphics, 512 × 480; full-motion video, 256 × 240 or 352 × 288—both provided with 6.5-kHz audio
- NTSC or PAL compatible
- HVQ and SG3 compression technology

13.13.6 Television Transmission at the Digital DS3 Rate

13.13.6.1 Introduction In Section 13.13.3 we discussed the desirability of transmitting digital television at a bit rate that equated a line rate of one of the two digital hierarchies (Chapter 3) and, at the same time, achieve compression. The compression, of course, is desirable so that less bandwidth is required for transmission. In this section we describe one system on the market in the United States that will transmit full composite color TV at the nominal 45 Mbps DS3 rate (44.736-Mbps). The unit described below is the ABL Engineering VT45A (Ref. 19). The VT45A has the capability of simultaneous transmission of a full-motion broadcast-quality color video signal and two DS1 1.544-Mbps carriers. One or both channels can be configured as an audio system, each with either a two-channel 20-kHz linear system or one four-channel 15-kHz audio system utilizing 14:11 companding techniques.

13.13.6.2 System Description Figure 13.12 is a functional block diagram of the VT45A. Following the signal flow through the block diagram, the applied NTSC video signal is fed to an analog-to-digital (A/D) converter via a low-pass filter. The filter is used to eliminate any out-of-band frequencies that will create aliased products to the A/D converter. The signal is sampled at a rate of about 18 MHz and quantized to a 10-bit linear binary representation.

The linear PCM samples are applied to a three-stage video encoder, which encodes the video into a data stream of approximately 38 Mbps. The first stage of the encoder is composed of a high-order digital filter and a resampling circuit. The second encoding stage is a combination of a differential PCM encoder and selection of statistical quantizers. The DPCM encoder utilizes a three-dimensional complex (14-element) predictor. The third encoder state is a statistical encoder, which consists of a variable-word-length (VWL) encoder and buffer circuit.

The encoded video information is fed via a forward error correction (FEC) unit to a 45-Mbps multiplexer. The FEC utilizes a $(63, 59)$ Reed–Solomon code interleaved six ways, which corrects errors with an error rate as poor as 1×10^{-4}. The 45-Mbps multiplexer time division multiplexes the video information with two 1.544-Mbps DS1 bit streams. The unit uses the stan-

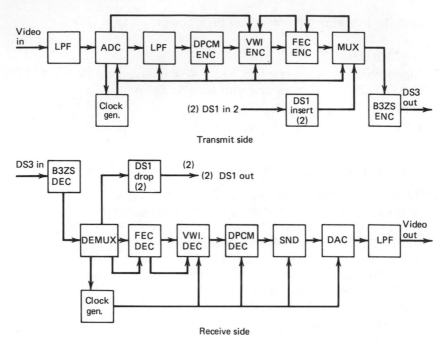

Figure 13.12 Functional block diagram of a VT45A. (From Ref. 18, courtesy of ABL Engineering, Inc.).

dard DS3 multiplex format but does not use the "C" bits, which, therefore, can be used for signaling or other services at the discretion of the user.

The two DS1 1.544-Mbps digital bit stream ports can be equipped with either standard DS1 interfaces or audio systems as described above.

The formatted DS3 45-Mbps bit stream is then encoded to a B3ZS code for the standard interface at the DS3 level. The receiver section performs the inverse process of the encoder to restore the analog video signal and the DS1/audio information.

13.13.6.3   *Framing and Bit Assignments*   The VT45A codec operates on a multiple of the standard DS3 frame. CCITT Rec. G.702 specifies for the DS3 frame eight sets each with 85 bits. The DS3 frame totals 680 bits. The DS3 *multiframe* consists of seven DS3 frames, which totals 4760 bits. The VT45A uses a *superframe* consisting of 10 *multiframes,* for a total of 47,600 bits. This superframe is used to synchronize the FEC circuitry in the receiver with that of the transmitter. The data rates and bit assignments are shown in Table 13.6. The FEC is so designed to handle a bit error rate (BER) as poor as 1×10^{-4} and improves the error performance to 1×10^{-9}. The FEC can handle error bursts as long as 67 bits. (From Refs. 18 and 19)

Table 13.6   Bit rates in the VT45A Codec

Function/Type	Bits Per Superframe	Approximate Bit Rate (Mbps)
DS3 M bits	70	0.066
c_{ji} bits	210	0.197
F bits	280	0.263
Video data	43,092	40.499
DS1-A	1820	1.710
DS1-B	1820	1.710
PLL sync	70	0.066
Aux serial	70	0.066
FEC sync	20	0.019
0 always	20	0.019
Available	80	0.075
E bits	48	0.045
Total	47,600	47,736

13.14   HIGH DEFINITION TELEVISION (HDTV)

We are told that high definition television (HDTV) is the wave of the future. The veracity of this statement certainly can be questioned.

Ref. 5 states:

> It is generally accepted that HDTV should have twice the vertical and twice the horizontal resolution of the conventional TV systems with an aspect ratio of 5:3 (e.g., 16:9) rather than 4:3 used with present home TV systems. This would give a spatial resolution close to that of 35 mm film and an aspect ratio recognizing the subjective improvement achieved by the wider screens now used in the cinema. The field frequencies used for TV would of course give a much better temporal resolution than achieved with the 24 frames per second used for film, particularly if sequential scanning were used for HDTV production.

Ref. 5 continues:

> ...The HDTV production standard would have twice the number of lines per field compared with conventional standard or even effectively four times the number, if sequentially scanned (i.e., non-interlaced) pictures were sent at a picture frequency equivalent to the present field frequency. In order to double the spatial resolution along the line, there would have to be an additional luminance sample between every one of the 720 samples [Ref. 31] in the orthogonal pattern of CCIR Rec. 601 and additional samples because of the wider aspect ratio. For an aspect ratio of 16:9 an additional luminance sample would need to be added to the 1440 required to double the resolution of a 4:3 aspect ratio picture.

We are speaking now of sampling frequencies above 70 MHz and data rates possibly above 2 GHz.

One HDTV system successfully demonstrated in Japan uses 1125 scan lines, more than double present U.S./Japan broadcast standards. These initial Japanese systems required 32 MHz of bandwidth and plan to use this technology on forthcoming DBS (direct broadcast by satellite) systems.

We again see a lack of standardization—the Japanese will have a different system than the European, and the United States at this time is far from establishing a standard. The FCC may go one way and cable TV interests may go another.

Consumer demand (or lack thereof) may well drive whether HDTV becomes ubiquitous. We see its use basically limited to wide-screen applications for theater, commercial establishments, and the home, although there will also be a military use because of the high resolution, and likewise a scientific use, especially medical.

We must consider the extremely high density infrastructure of conventional TV based on 6-MHz RF assignments in most parts of the world. The consumer will decide whether the enhanced quality is worth the expense. The U.S. manufacturer Zenith proposes a simulcasting system that permits broadcasters to use the 6-MHz channel besides their current NTSC channel to transmit a totally separate HDTV-only signal. By digitizing and compressing the signal, 30 MHz of video and sound can be transmitted in 6 MHz of bandwidth. Another proposal uses adjacent 6-MHz channels for the increased bandwidth required for HDTV. 12 MHz may accommodate the 1125- or 1050-line systems proposed in Japan and in the United States (ACTV-2).

Cost estimates for commercially available HDTV receivers run from $1500 to $3000 in 1990 money. It is the author's opinion that the incremental quality will not justify the price in the domestic TV market. If we are to rely on the specialized markets, receiver costs may be much greater.

REVIEW EXERCISES

1. What are the four factors a color video transmission system must deal with when transmitting images of moving objects?

2. Describe scanning: (1) horizontally and (2) vertically.

3. Define a pel or pixel (beyond the meaning of the acronym).

4. What do we mean by an aspect ratio of 4:3? Suppose the width of an image is 12 in.; what is its height?

5. NTSC divides an image into how many horizontal lines? How many horizontal lines are there in European practice, by and large?

6. Given an aspect ratio of 4:3, an image with 525 vertical lines (elements) would have how many horizontal elements?

7. How do we achieve the sensation of motion in TV? Relate this to frame rate and flicker.

8. Describe interleaving (regarding TV video) and explain why it is used.

9. Define a field period regarding North American (NTSC) practice.

10. In North American practice the time to scan a line is 63.5 μs. This time interval consists of two segments. What are they?

11. What is the standard maximum voltage excursion of a video signal? Just what are we measuring here? (What TV signal component?)

12. How is frame rate related to power line frequency or "flicker frequency?"

13. Give two definitions of composite signal.

14. At a TV receiver, what signal-to-noise ratio S/N (dB) is required for an excellent picture?

15. How is S/N normally measured and in what units? (We mean here the signal and the noise.)

16. What are the two major additional impairments that must be considered when transmitting a color signal?

17. What type of modulation is used to transmit video; an audio subcarrier; and a chrominance subcarrier?

18. Give four generalized differences in video transmission between European practice and North American/Japanese practice.

19. How is a program channel commonly handled when TV is transported by line-of-sight (LOS) microwave, satellite, and other media when specifically to be used by broadcasters (not necessarily cable TV)?

20. Why is a video signal inverted when transmitted on coaxial cable and not inverted when transmitted on LOS microwave or satellite?

21. Why is S/N end-to-end specified as 54 dB when only 45 dB is required at a user TV receiver for an excellent picture?

22. Differentiate composite coding from component coding. Name one major advantage of component coding.

23. What type of coding is nearly universally used to digitize video for transmission?

24. Give the three basic bit-rate reduction methods for digitized TV.

25. What are the desirable digital line rates for TV? Why those rates and not some other rate(s)?

26. What are the two basic types of redundancy we will encounter in a TV video system?

27. Distinguish intraframe coding from interframe coding. What type of redundancy does each address?

28. 0 dBmV is what voltage value? When we give a value in dBmV, we must also state another parameter. What is this important parameter?

29. How does DPCM (alone) achieve bit rate reduction?

30. How does motion compensation coding reduce bit rate?

31. What is the general range of bit rates achieved for broadcast-quality TV when effective bit-rate reduction techniques are employed?

32. Describe how freeze-frame video transmission operates.

33. Define *refresh rate*. Refresh rate is basically a function of three parameters. What are they?

34. We have a 512 × 480 pixel frame and each element (pixel or pel) is assigned an 8-bit PCM word. The data rate available on the circuit is 9600 bps. What is the refresh rate in this case?

35. What are some typical bit rates used for video conferencing? What advantage is there to using higher bit rates?

36. Name three practical applications of HDTV.

37. Identify at least three drawbacks to the wide implementation of HDTV —let's say an HDTV receiver in every home in the next 15 or 20 years.

REFERENCES AND BIBLIOGRAPHY

1. *Reference Data for Engineers*: *Radio, Electronic, Computer and Communication*, Howard W. Sams, Indianapolis, IN, 1985.
2. *Reference Data for Radio Engineers*, 6th ed., Howard W. Sams, Indianapolis, IN, 1977.
3. "Fundamentals of Television Transmission," *Bell System Practices*, Section AB 96. 100, American Telephone and Telegraph Co., New York, Mar. 1954.
4. "Television Systems Descriptive Information—General Television Signal Analysis," *Bell System Practices*, Section 318-015-100, no. 3, American Telephone and Telegraph Co., New York, Jan. 1963.
5. C. P. Sandbank, ed., *Digital Television*, John Wiley, New York, 1990.
6. "Television Systems A2A Video Transmission System Description," *Bell System Practices*, Section 318-200-100, no. 5, American Telephone and Telegraph Co., New York, 1952.
7. *Transmission Systems for Communications*, 5th ed., Bell Telephone Laboratories, AT&T, New York, 1982.
8. *Lenkurt Demodulator*, Lenkurt Electric Corp., San Carlos, CA, Feb. 1962, Oct. 1963, Jan. 1965, Mar. 1966, and Feb. 1971.
9. "Transmission Performance of Television Circuits Designed for Use in International Connections," CCIR Rec. 567-2, XVIth Plenary Assembly, Dubrovnik, 1986, vol. XII.
10. J. Herbstreit and H. Pouliquen, "International Standards for Colour Television," paper, ITU, Geneva, 1967.

11. K. Simons, *Technical Handbook for CATV Systems*, 3rd ed., General Instrument —Jerrold Electronics Corp., Hatboro, PA, 1980.

12. "Electrical Performance Standards for Television Relay Facilities," EIA RS-250B, EIA, Washington, DC, Sept. 1976.

13. D. Kirk, Jr., "Video Microwave Specifications for System Design," reprint from *Broadcast Engineering*, Jerrold Electronics Corp., Philadelphia, PA, 1966.

14. K. Blair, ed., *Television Engineering Handbook*, McGraw-Hill, New York, 1986.

15. J. W. Ricke and R. S. Graham, "The L3 Coaxial System Television Terminals," *Bell Syst. Tech. J.*, July 1953.

16. W. von Guttenberg and E. Kugler, "Modulation of TV Signals for Combined Telephone and Television Transmission over Cables," *NTZ-CJJ*, no. 2, 1965.

17. Larry McLelland, *Basics of Freeze Frame Video Transmission*, Colorado Video, Inc., Boulder, CO, Sept. 1990.

18. *The VT45A, A Technical Description*, ABL Engineering, Mentor, Ohio, Dec. 1989.

19. *Specifications VT45A*, ABL Engineering, Mentor, Ohio, Feb. 1990.

20. *Specification C-3000 Video CODEC System*, PictureTel Corp., Peabody, MA, June 1990.

21. *Technical Backgrounder: Hierarchical Vector Quantization*, PictureTel Corp., Peabody, MA, June 1990.

22. "Transmission Performance of Television Circuits over Systems in the Fixed-Satellite Service," CCIR Rep. 965, XVIth Plenary Assembly, Dubrovnik, 1986, vol. XII.

23. "Single Value of Signal-to-Noise Ratio for All Television Systems," CCIR Rec. 568, XVIth Plenary Assembly, Dubrovnik, 1986, vol. XII.

24. "Characteristics, Methods of Measurement and Design Objectives for International Television Circuits," CCIR Rep. 816-2, XVIth Plenary Assembly, Dubrovnik, 1986, vol. XII.

25. "Transmission of High Definition Television Signals," CCIR Rep. 1092, XVIth Plenary Assembly, Dubrovnik, 1986, vol. XII.

26. "Hypothetical Reference Chain for Television Transmissions over Very Long Distances," CCIR Rep. 603, XVIth Plenary Assembly, Dubrovnik, 1986, vol. XII.

27. "Digital Transmission—General Principles," CCIR Rec. 604-1, XVIth Plenary Assembly, Dubrovnik, 1986, vol. XII.

28. "Digital or Mixed Analog-and-Digital Transmission of Television Signals," CCIR Rep. 646-3, XVIth Plenary Assembly, Dubrovnik, 1986, vol. XII.

29. "Characteristics of Television Systems," CCIR Rep. 624-3, XVIth Plenary Assembly, Dubrovnik, 1986, vol. XI, part 1.

30. "The Present State of High-Definition Television," CCIR Rep. 624-3, XVIth Plenary Assembly, Dubrovnik, 1986, vol. XI, part 1.

31. "Encoding Parameters of Digital Television for Studios," CCIR Rec. 601-1, XVIth Plenary Assembly, Dubrovnik, 1986, vol. XI, part 1.

32. "Digital Coding of Color Television Signals," CCIR Rep. 629-3, XVIth Plenary Assembly, Dubrovnik, 1986, vol. XI, part 1.

33. *1990 Colorado Video Communications Catalog*, Colorado Video, Inc., Boulder, CO, 1990.

14

FACSIMILE
COMMUNICATION

14.1 APPLICATION

Facsimile is a method of electrical communication of graphic information. It is used to transmit pictorial or printed matter from one location to another with reasonably faithful copy permanently recorded at the receiving end. Facsimile (fax) has been designed primarily to operate over comparatively narrow-band media, lending itself well for transmission over the telephone network and on HF radio.

From the 1920s until sometime after World War II, facsimile was used almost exclusively for the transmission of weather maps and news media picture transmission. Today it has much more extensive usage. It is finding ever wider application in such areas as

- Bank verification of signatures
- The transmission of fingerprints and "mug shots" in the area of law enforcement
- In the commercial world for the delivery of waybills and invoices
- For the production of newspapers and magazines to dispatch news copy from satellite offices or bureaus to main newsrooms and to eliminate duplication of typesetting efforts
- In industry for the transmission of engineering drawings, parts lists, and so forth
- For "electronic mail," especially as conventional mail service is becoming less efficient and more costly
- As a principal means of point-to-point record traffic communication

14.2   ADVANTAGES AND DISADVANTAGES

On first analysis it would appear that conventional data/telegraph methods of transmitting graphic information are faster than facsimile. For instance, a standard printed page may take less than 6 sec to transmit over the telephone network using digital data techniques (assuming a throughput of 400 characters per second). The same page may take up to 30 sec to transmit by facsimile.* However, there is an operational difference that is often overlooked. To the 6 sec required to transmit the data or telex page must be added the operator time to keystroke each character, whereas for facsimile the operator just inserts the printed page in the facsimile machine. In addition, operationally facsimile is less prone to error. With the case of digital telegraph or data, the operator can (and often does) cause errors by unknowingly stroking the wrong key. In the case of facsimile no such errors can enter because there is no operator transcription from the original copy. Optical character readers, of course, can eliminate this source of error. Nevertheless, once we enter the domain of computer-to-computer or smart-data-terminal-to-computer operations, such communication should be left to the technique of data transmission discussed in Chapter 12.

14.3   BASIC FACSIMILE OPERATION

14.3.1   General

A facsimile system consists of some method of converting graphic copy on paper to an electrical equivalent signal suitable for transmission on a telephone pair (or other narrow-band media), the connection of the pair/telephone circuit and transmission to the desired distant-end user, and the recording/printing of the copy on paper by that user. Three basic elements or processes are involved:

- Scanning
- Switching/transmission
- Recording

Basically, we are dealing with analog technology. However, lately there is a marked trend toward the use of digital techniques. Figure 14.1 is a simplified block diagram of a facsimile system.

   The scanner is a photoelectric transducer converting the reflected light, as a document is scanned, to an electrical signal that varies in intensity in accordance with the light intensity. The modem converts or conditions the

*Group 3 facsimile systems can transmit a standard page in less than 30 s.

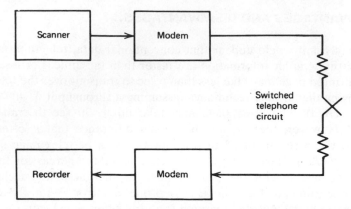

Figure 14.1 Simplified functional diagram of a facsimile system.

electrical signal from the scanner, making it compatible for transmission over the telephone network or other media. On the receive side in Figure 14.1 the modem converts or conditions the line signal entering from the telephone network, making it compatible for the facsimile recorder. The recorder prints the received image by various methods that are discussed below. Often the modem is incorporated with the scanner or recorder, as the case may be.

14.3.2 Scanning

The scanner is a photoelectric transducer that produces an electrical analog signal representing the graphic copy to be transmitted. With conventional facsimile, scanning is carried out by one of three basic methods, which are essentially mechanical in variation:

- A spot of light scans a fixed graphic copy
- The copy moves across a fixed spot of light
- Both the copy and the scan spot may move, usually at right angles to each other and at different speeds

There are two common approaches of "lighting" the copy. Both employ the technique of bouncing (reflecting) light off the graphic material to be transmitted. One approach to scanning projects a tiny spot of light onto the surface of the printed copy. The reflection of the spot is then picked up directly by a photoelectric cell or other transducer.

The other approach is the flood projection technique whereby the printed copy is illuminated with diffuse light in the area of scan. The reflected light is then optically projected through a very small aperture onto the cathode of a photoelectric transducer.

In modern facsimile systems using electromechanical methods, scanning is either done with the copy attached to a cylinder that rotates, or placed on a flatbed. In the case of cylinder scanning, the cylinder is rotated such as to effect a continuous helical scan of the entire copy in a period of time depending on system constraints, possibly up to 6 min. Flatbed scanners use a feed mechanism. The copy to be transmitted is fed into a slot where a feed mechanism takes over, slowly advancing the copy through the machine. Another type of flatbed scanner uses a flying spot scan where the copy remains at rest and the spot does all the movement.

The electronics of facsimile scanners is based on the photoelectric cell, the photomultiplier tube, or the photodiode. They differ primarily in the relative signal strength of their outputs in relation to scan spot intensity.

The charge-coupled-device (CCD) electronic scanning, which previously had been exclusively in the realm of cathode ray tubes (CRTs), now offers promise in the area of facsimile scanning, making it particularly attractive for direct digital transmission of a facsimile signal.

The electrical output of a conventional scanner may also be digital. In fact, the trend is more and more toward digital facsimile systems. In its simplest form of digital transmission, where the signal represents either black or white, a decision circuit may provide a series of marks or 1's for black copy and spaces or 0's for white copy. If, in addition, shades of gray are to be transmitted as well, a level quantizing scheme with subsequent coding is required. This, of course, in most respects is similar to the quantizing (based on fixed signal thresholds) and coding described in Chapter 3.

14.3.3 Recording

Recording in a facsimile system is the reproduction (e.g., printing) of visual copy of graphic material from an electric signal. Recorders remain electromechanical, just as their scanner counterparts are. And like scanners, recorders are available in cylinder configurations and flatbed. Certain desktop versions are on the market that are semicylindrical, allowing continuous paper feed. There are four basic electromechanical recording methods in use today:

- Electrolytic
- Electrothermal
- Electropercussive
- Electrostatic

Two types of facsimile recording do not fit into the above categories. One prominent German manufacturer uses an offset process. The other process involves the use of modulation of a fine spray of ink directed to the surface of plain paper. Yet another process, which is well established, is not electromechanical and is used in the facsimile reproduction of newspapers.

Electrolytic Recording   Electrolytic recording is one of the most popular methods of facsimile recording, yet one of the oldest. It requires a special type of recording paper that is actually an electrolyte-saturated material. When an electric current passes through it, the material tends to discolor. The amount of discoloration or darkness is a function of the current passing through the paper.

To record an image, the electrolytic paper is passed between two electrodes. One is a fixed electrode, a backplate or platen on the machine; the other is a moving stylus. Horizontal lines of varying darkness appear on the paper as the stylus sweeps across the sheet. As each recorded horizontal line is displaced one line width per sweep of the stylus, a "printed" pattern begins to take shape. This pattern is a facsimile of the original pattern transmitted from the distant-end scanner.

The more practical helix–blade technique of electrolytic recording is now favored over the stylus–backplate method. The concept is basically the same, except that the two electrodes in this case consist of a special drum containing a helix at the rear and a stationary blade in the front. The drum–helix makes one complete revolution per scanning line. The rotating drum moving the helix carries out the same function as the moving stylus of the more conventional electrolytic recorders.

Electrothermal (Electroresistive) Recording   Electrothermal facsimile recording is misnamed "thermal" because the recording process gives the appearance of being a "heat" process or burning. More properly, it should be called electroresistive. It is similar to the electrolytic process in that the recording paper is interposed between two electrodes (i.e., stylus and backplate). The recorded pattern is made by an electric arc passing through the paper from one electrode to the other. A special recording paper is required, the type varying from equipment to equipment. The paper has a white coating that is decomposed by the current passing through it, the amount of decomposition being a function of the current passing between the electrodes at any moment in time. A major characteristic of electrothermal recording is the high contrast that can be achieved. A true electrothermal process is evolving using specially treated paper and a resistive heat element that responds to rapid temperature changes as a function of signal current.

Electropercussive Recording   Electropercussive recording is a technique similar to that of recording audio on a record. In this case, an amplified facsimile signal is fed to an electromagnetic transducer that actuates a stylus in response to the electrical signal variations of the facsimile signal. When a sheet of carbon paper is interposed between the stylus and a sheet of plain white paper, a carbon copy impression is made on the plain paper by the vibrations of the stylus in accordance with the signal variations. The intensity of the darkness of the copy varies in proportion to the variation in strength of the picture signal. An advantage of this type of recording is that no special recording paper is required. Other names for electropercussive recording are *impact* or *impression* recording. It is also known by the term *pigment transfer*.

Electrostatic Recording   There are essentially two types of electrostatic recording used in facsimile. One kind is based on printing the facsimile image from a CRT by means of xerography techniques requiring only plain paper. The other kind is a direct copy method requiring a specially coated paper. One type of electrostatic recorder is called a *pin* printer. The advantage of direct electrostatic recording is its exceptional capability of reproducing gray tonal qualities.

14.4 FUNDAMENTAL SYSTEM INTERFACE

14.4.1 General

If we connect a scanner and recorder back to back, will they interoperate correctly? Three requirements must be met to assure this interoperation:

- Phasing compatibility
- Synchronization
- Index of cooperation

14.4.2 Phasing and Synchronization

Proper phasing and synchronization are vital factors in conventional facsimile transmission. Both are time-domain functions. Phasing and synchronization of the far-end facsimile receiver with the near-end transmitting scanner permit the reassembly of picture elements in the same spatial order as when the picture was scanned by the transmitter:

- Phasing assures that the receiving recorder stylus coincides with the transmitter in time and position on the copy at the start of transmission
- Synchronization keeps the two this way throughout the transmission of a single graphic copy

In most conventional analog systems phasing is carried out by what may be termed a stop-and-start technique. The receiving end in this case is not really stopped, only retarded until a start-of-stroke recording coincides with the start of scanning stroke at the transmitting end. To indicate start-of-stroke, a phasing signal is generated at the scanner and transmitted on-line. The signal is a pulse of full amplitude equivalent to full black or full white, depending on signal polarity. The pulse duration is from 5 to 7% of line length. The duration of the phasing sequence following CCITT Rec. T.2 is 15 s for the standard 6-min transmission of an (ISO) A4 page size and 6 s for the 3-min transmission of an A4 page.

Synchronization assures that the facsimile recorder remains in step with the transmitting scanner during the transmission period. With conventional analog facsimile systems, synchronization may be carried out by one of the following three methods:

- Both machines are tied to a common ac power source frequency
- Each machine operates with a stabilized frequency power source or with built-in frequency standards
- The system operates with the transmission of synchronizing signals during picture transmission

The first two methods of synchronization are applicable where facsimile scanners and recorders use synchronous drive motors. Of course, the first method can only be used where the scanners and recorders derive ac power from the same ac grid. In this case synchronization is simple and automatic after initial phasing.

If the two ends of a facsimile system are not on a common ac grid, such as on numerous international circuits, military systems, air-to-ground, shore-to-ship, and other mobile and portable operations, then we must resort to one of the two latter methods. With the second method synchronous drive motors operate from stabilized frequency sources or from frequency standards. In this case frequency stability should be 1×10^{-5} or better. CCITT Rec. T.1 (Ref. 8) states:

> The speed of transmitters must be maintained as nearly as possible to the nominal speed and in any case within ± 10 parts in 10^6 of nominal speed. The speed of receivers must be adjustable and the range of adjustment should be at least ± 30 parts in 10^6. After regulation, the speeds of the transmitting and receiving sets should not differ by more than 10 parts in 10^6.

Out-of-synchronization conditions cause skewing of the received picture.

Slaving the facsimile recorder to the distant-end scanner essentially avoids the stability problems of the first two methods. There are two ways synchronization can be accomplished. The first uses a synchronization tone, conveniently a multiple of 50 or 60 Hz which can be divided down to drive the recorder synchronous drive motor. The tone is transmitted above or below the picture signal in the VF passband. Effective filtering is then required to separate the picture from the synchronizing signal.

The other method is pulse synchronization, which is continuous during picture transmission. It uses short between-the-line synchronizing pulses which provide a check on the speed of the recorder drive motor. The first method we can denote as operating in the frequency domain and the second in the time domain.

Referring back to CCITT, Rec. T.1 calls for a 1020-Hz synchronizing tone, and Rec. T.3 (Ref. 19) states:

> During transmission of document information the transmitter should transmit full amplitude carrier during lost time. The phase of the carrier may be reversed at the end of this signal. Both transmitter and receiver should align the end of lost time to this phase reversal with an accuracy of $\pm 2\%$ in this case.

The concept of lost time (flatbed facsimile) or dead sector (cylinder facsimile) is analogous to flyback as used in video transmission (see Section 13.2).

The synchronization of digital facsimile systems is similar to the synchronization of data systems discussed in Chapter 12. There are asynchronous and synchronous systems. Also, a similarity to video transmission exists in that the scan and recording strokes have to start and end in synchronization. There is also the requirement of picture element synchronization. The element, which is a code word, identifies the quantized level of the picture signal.

Line advance in digital facsimile systems is usually carried out on a stepping basis. This is triggered by a start of line code word. Of course, it is similar to carriage return (CR) and line feed (LF) used in data/telegraph printer operation (Chapter 12). Digital facsimile operation varies greatly from manufacturer to manufacturer, depending greatly on the amount of processing involved to reduce redundancy. We find that the normal printed page is 95% white. As a consequence, to reduce transmission time, digital facsimile systems have been developed that are called *white space skipping* systems. White space skipping can be carried out by fairly rudimentary processing.

The distinction between asynchronous and synchronous digital facsimile systems lies in that for asynchronous systems synchronism is only maintained on a line-for-line basis, the line advance acting much like the stop-to-start transition of asynchronous data systems described in Chapter 12. Synchronous digital facsimile systems, on the other hand, maintain synchronization throughout the entire transmission period, similar to the synchronism on a synchronous data link.

14.4.3 Index of Cooperation

On a facsimile circuit where the scanner and recorder have compatible phasing arrangements and synchronization can be maintained, assume that both machines have the same scan rate as measured in lines per minute (LPM). Now they must also have the same index of cooperation for correct interoperability. The index of cooperation is defined by the IEEE (Ref. 3) as follows:

- For rotating systems it is the drum diameter times the lines per unit length.
- The international definition is the product of the scanning or recording line (length) by the scanning or recording lines per unit length divided by π. (That is the product of the line length across the sheet by the lines per inch [LPI] [vertical]).

An older IEEE definition, which is still much in use, is the product of the total line length by the number of lines per unit length. To convert this older standard to the CCITT standard, multiply the IEEE index of cooperation by 0.318. For instance, CCITT Rec. T.1 recommends an index of cooperation of 352, alternatively 264. These convert to IEEE standards of 1105 and 829, respectively. CCITT is now using a term called *factor of cooperation*, which is indeed equivalent to the older IEEE definition of index of cooperation where the conversion factor of 0.318 is valid. The World Meteorological Organization (WMO) specifies indices of cooperation of 576 and 288. These are equivalent to IEEE indices of 1809 and 904, respectively. The U.S. Electronics Industries Association (EIA) recommends an index of 829 (IEEE), equivalent to a CCITT index of 264.

What happens when the indices of cooperation are not the same? This is shown in Figure 14.2. Figures 14.2*a* and *c* show examples of transmission between scanner and recorder with indices of 352 and 264. Figure 14.2*b* is an example where both machines are operating with an index of 264 and the received picture is not distorted. When there is a lengthening distortion in the received copy, such as in Figure 14.2*a*, the scanner index is greater than the recorder index. When this situation is reversed, the distortion appears as

a b c

Figure 14.2 Distortion due to different indices of cooperation. (From Ref. 5; courtesy of IEEE.)

a fattening of the received picture, as in Figure 14.2*c*. When the indices of cooperation are the same at each end, it does not necessarily follow that the picture size at the recorder is the same as that of the scanned picture. The size will be the same if the scanner and recorder have the same line lengths.

14.5 FACSIMILE TRANSMISSION

14.5.1 General

The signal output of the conventional scanner operating in the analog mode contains frequency components from subaudio to several thousand hertz. The amplitude variations as a scan spot moves across a scan line consist of rapid transitions in accordance with the white, black, and tonal hues of gray of the line. The transmission problem that the telecommunication engineer faces is to condition or convert this signal to make it compatible with the band-limited media it will traverse (e.g., the telephone network).

14.5.2 Modulation Techniques

Because of the critical subaudio content of a facsimile signal extending down to about 20 Hz, the signal, as is, is not compatible for transmission over the telephone network. To overcome the frequency response problem, simple carrier techniques have been adopted. (See Chapter 3.) A common approach is to use vestigial-sideband (VSB) modulation with an 1800-Hz carrier. The principal information content is carried in the lower sideband with about 1400 Hz of information bandwidth. The vestige of the upper sideband extends out to about 2300 Hz. If single sideband (SSB) were used where the information bandwidth is greater than half of the carrier frequency, it would give rise to a form of distortion known as the *Kendall effect*. This manifests itself as a fuzziness at the edges of the recorded picture. Thus some Kendall effect must be expected, but the recording quality is still adequate for most applications.

FM is generally more desirable than AM (VSB as described above) for transmission over the switched telephone network and on radio systems, especially HF. There are two reasons for this. First, FM is nearly impervious to noise because most noise is AM in nature. Second, FM is much less sensitive to level variations, which are fairly common on a telephone network, particularly on switched connections. For HF radio using the skywave phenomenon, the receive level is constantly varying.

A common standard for FM transmission of facsimile is to assign 1500 Hz for white and 2300 Hz for black, with a mean frequency of 1900 Hz (CCITT Rec. T.1). Regarding frequency stability, CCITT recommends that "the stability of transmission be such that the frequency corresponding to a given tone does not vary by more than 8 Hz in a period of 1 s and by more than 16

Hz in a period of 15 m." Further, CCITT Rec. T.1 states that the receiver–recorder be capable of operating correctly when the drift of black and white frequencies does not exceed their nominal values by more than ± 32 Hz. It is assumed that the frequency deviation varies linearly with the photocell (or equivalent) voltage or, in the case of conversion from AM to FM, with the amplitude of the amplitude-modulated carrier.

CCITT Rec. T.2 has provisionally assigned 1700 Hz as center frequency (mean frequency), and white then corresponds to 1300 Hz and black to 2100 Hz. Rec. T.2 has been specifically established for document transmission, transmitting an A4 page in approximately 6 min.

CCITT Rec. T.3 was prepared for A4-page-size document transmission in 3 min. It recommends a 2100-Hz (± 10 Hz) carrier frequency using VSB modulation. A white signal is represented by maximum carrier and a black signal by minimum or no carrier. For proper operation the minimum carrier state must be at least 26 dB below the maximum carrier state. Receiver frequency drift should be no more than ± 16 Hz. Receivers should be capable of functioning correctly with signal levels as low as -40 dBm (white signal).

14.6 CRITICAL TRANSMISSION PARAMETERS

14.6.1 General

The effects of various telephone circuit parameters on the capability of such circuits to transmit facsimile signals are important engineering considerations in system design. Principally, the facsimile transmitter output looks into a standard 4-kHz telephone channel.* The circuit characteristics are as follows:

- Effective bandwidth
- Amplitude–frequency response
- Envelope delay distortion (EDD)
- Noise
- Echo
- Level stability
- Phase jitter

Each characteristic has been discussed at length previously in this text. However, a brief review is carried out below. Reference paragraphs in other chapters are shown in parentheses.

---

*Exceptions are noted. For instance, the U.S. common carriers and certain specialized common carriers have special digital offerings, as do some telecommunication administrations in foreign countries, where the digital facsimile transmitter looks directly into a digital circuit of fixed data rate and the discussion below is essentially not applicable.

14.6.2 Effective Bandwidth

The CCITT voice channel has effective bandwidth limits of 300 and 3400 Hz. An AM facsimile signal requires an information bandwidth of about 1400 Hz. Therefore a double-sideband AM signal would occupy twice the frequency, or 2800 Hz of the total 3100 Hz available. As we shall see below, occupying this much of the available bandwidth is impractical due to other constraining factors. Hence we have had to resort to other types of modulation occupying less bandwidth, such as VSB and FM. (See Section 1.9.1.)

14.6.3 Amplitude – Frequency Response

CCITT Rec. T.11 (Ref. 24) refers the facsimile user to CCITT Rec. G.151. It further states that it is desirable to have an amplitude–frequency distortion in the desired band between end-user equipments of less than 8.7 dB.

A basic telephone channel in North America between 500 and 2500 Hz referenced to 1004 Hz can expect the worst amplitude–frequency response to be from -2 to $+8$ dB, or 10 dB of attenuation distortion (Ref. 12). (See Sections 1.9.3 and 12.9.3.)

14.6.4 Envelope Delay Distortion

This is a most important parameter in facsimile transmission. The reader is advised not to equate EDD with or directly derive it from the more familiar delay distortion. The effect on a facsimile received picture is the same and the cause is the same. Parameters expressed as EDD are preferred simply because it is easier to measure. However, the references used below use delay or delay distortion. Ref. 7 states:

> Experience has shown that for effective facsimile transmission within the voice band, delay must be held uniform (± 300 μs) for all frequencies within the full frequency range of the transmitted signal.

CCITT Rec. T.12 gives recommended limits of delay distortion as a function of drum rotation speed. This is shown in Figure 14.3.

Excessive delay distortion or EDD between an analog facsimile scanner and a far-end recorder(s) gives rise to received picture distortion, which manifests itself in smears or ghosts not necessarily spread uniformly over the received copy. On digital facsimile circuits it results in intersymbol interference degrading the error rate. If delay distortion becomes very severe, then digital facsimile operation becomes impossible. Automatic delay equalizers are recommended on circuits transmitting 4800 bps and above. (See Sections 1.9.4, 12.9.3 and 12.12.)

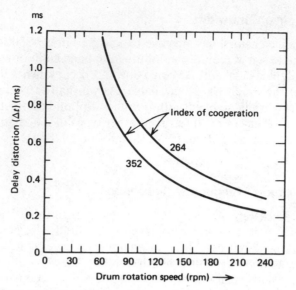

Figure 14.3 Permissible delay distortion in the transmitted frequency band as a function of the phototelegraph transmission speed. *Note:* The scanning spot is assumed to have the same dimensions in both directions (square or circular). (From CCITT Rec. T.12 [Ref. 25]; courtesy of ITU – CCITT.)

14.6.5 Noise

On analog facsimile transmission systems the signal-to-noise ratio at the receiver input should be 30 dB or better. On digital facsimile circuits the receiver input signal-to-noise ratio (probably expressed in E_b/N_0) should be such as to meet the system BER requirements.

Impulse noise is momentary, and when severe, it can obliterate certain essential details of a received picture, especially when there are multiple "hits" of some duration. Single-frequency interference, such as from power-line harmonics, is continuous, showing up on received copy as a herringbone pattern. Crosstalk also gives a herringbone effect. Whereas the single-frequency interference shows symmetry and continuity in its interference pattern, crosstalk shows up as noncontinuous and nonsymmetrical. (See Sections 1.9.6 and 12.9.3.)

14.6.6 Echo

On facsimile systems with AM transmission, echo manifests itself with ghosts. On FM systems it shows up as ripple distortion in the received picture. (See Section 2.6.4.)

14.6.7 Level Variation

A drop in the received level (AM systems) of 1 dB can affect facsimile recording, whereas at least a 3-dB level variation is required to affect voice communication. A level reduction will deteriorate the signal-to-noise ratio and, as a result, reduces tonal response. Level variations affect AM (VSB) facsimile systems much more than FM systems. (See Sections 1.9.5 and 12.9.3.)

14.6.8 Phase Jitter

Excessive phase jitter shows up on a received picture as a herringbone pattern similar to that of single-frequency interference. On digital circuits, excessive phase jitter can have much greater deleterious effects. (Refer to Section 12.9.3.)

14.7 END-TO-END QUALITY

The quality of a received facsimile document, map, or picture is a subjective matter, as is the "quality" of a received voice signal. It is difficult to quantify facsimile quality in hard and fast numerical parameters. We do judge picture quality by its resolution, legibility, contrast, and gray-scale reproduction. Each criterion is discussed below.

14.7.1 Resolution

As in any image communication system, such as radar, TV, and facsimile, resolution is defined as the degree to which adjacent elements of an image are distinguishable as being separate. There is horizontal resolution and vertical resolution. We mean here our ability to define picture elements first horizontally and then vertically.

The resolution of any picture is basically determined by the picture elements per unit area, called *pixels* or *pels* (see Section 13.2). Elements on a scanned facsimile page are a function of lines per inch, or LPI (or mm), working vertically down the page, and spot size. Spot size can be related to LPI. Let us assume an LPI of 100. Then the optimum spot size would be the inverse, or $\frac{1}{100}$ (0.25 mm). Thus a square inch would have 100×100, or 10,000 picture elements (pixels or pels). A 10×10-in page in this case would have $100 \times 100 \times 100$, or 1×10^6 picture elements. More commonly, a practical facsimile system with 96 LPI has a $\frac{1}{200}$-in spot size, and for an $8\frac{1}{2} \times 11$-in page the resulting total number of elements (pixels) is about 500,000. The actual number of elements for analog facsimile systems is reduced by about 30% due to the Kell factor or, in this case, 0.70 × 500,000, or 350,000 pixels.

The Kell factor corrects the ideal picture resolution to the practical. The reason is that scan lines are of finite width. Consequently, there can be no variation in light intensity values across a scan line. Stated another way, a mark reproduced by a facsimile recorder will always occupy at least the full width of a recording stroke, regardless of the dimensions of the actual mark on the scanned original. On recorded copy there is a certain spreading of line width from the scanned original because effectively the number of elements resolvable per inch is less than the LPI. The amount of reduction is the Kell factor.

The size and shape of the scan aperture are also factors determining resolution in a facsimile system. Just as in photography where the smallest aperture possible for a given set of light conditions produces the "sharpest" picture (i.e., has the best resolution), so in facsimile scanners, the smaller the opening, the sharper the picture. And as in photography, the limiting factor on aperture size is the intensity of the light that will pass through the opening sufficient for system operation.

A round aperture will serve for most facsimile applications but does not give optimum resolution. However, to achieve nearly equal resolution in both the horizontal and the vertical dimensions on a scanned page without unduly degrading fidelity, the scan aperture should have a height somewhat greater than its width.

14.7.2 Gray Scale and Contrast

For the transmission of engineering line drawings and document (printed) copy, we would want good sharp contrast, but the tonal qualities of gray are relatively insignificant. Of course, a lot has to do with the contrast of the copy to be scanned. Where that contrast is weak, the output of the scanner can be increased. But at some point, background noise will begin to increase, deteriorating contrast. The amount of black or the blackness of a print is described and measured by density. Whiteness is described by reflectance, and the standard of measurement is the percentage of reflectance. Table 14.1 is taken from Stafford's series of articles on facsimile (Ref. 6). It provides a good guide to reflectivity. With optimum contrast, whitest white has 100% reflectance and black nearly 0%, often assigned 1%. When measuring in density units, blackest black is assigned 2.00, and the same white as above, 0.00. Tonal gray is then some number between 0.0 and 2.0. If it had 50% reflectance, it then would have 0.30 density units.

Contrast is a measure of the faithful reproduction of scanned prints in facsimile communication; resolution is a measure of how well we can distinguish closely spaced objects or identify small items on a print. Up to some point there is a one-to-one trade-off available between contrast and resolu-

Table 14.1 Guide to reflectance

Type of Paper	Reflectivity (%)
White	85
Slight cream tint	83
Deep cream or very light buff	82
Slight sepia cream tint	82
Very light sepia tint	81
Fairly saturated yellow	79
Light yellowish-green tint	74
Reddish buff verging on salmon	71
Light blue–green tint	70

Source: Ref. 6 (Courtesy of G. F. Stafford, Alden Electronic & Impulse Recording Co., Inc.)

tion. However, some subjective tests indicate that contrast may affect legibility to a greater extent than resolution.

14.7.3 Legibility

Legibility has a lot to do with the type of copy to be transmitted. Legibility criteria, for instance, would be more severe for photos and drawings than for printed documents. Under some marginal situations we may not be able to identify certain letters in a word if they were isolated, but given the entire word or phrase there would be no trouble reading it.

For proper legibility of a character (i.e., a letter, number, or other common graphic symbol) at least 10 scan lines are required. Therefore, with a 96-LPI facsimile system, minimum letter height on copy to be transmitted must be $\frac{10}{96}$-in. or about one-tenth of an inch. For word legibility only about half that number of scan lines is required. A $\frac{1}{8}$-in. high character requires about 80 LPI, and a word $\frac{1}{8}$ in. high, 40 LPI. On the other hand, photo transmission, depending on the original picture quality, requires from 200 to 700 LPI.

14.7.4 Test Charts

Facsimile test charts provide the system engineer and maintenance technician basic information on the quality of the system regarding contrast, legibility, and resolution. It is an aid to troubleshooting where specific extraneous patterns and types of chart distortion on the receive side can immediately red flag a problem such as noise and its type, Kendall effect, poor synchronization or phasing, or incompatible index of cooperation.

There are three standard charts issued by two agencies: the IEEE and CCITT. The IEEE chart is identified by a picture of a woman, the chart

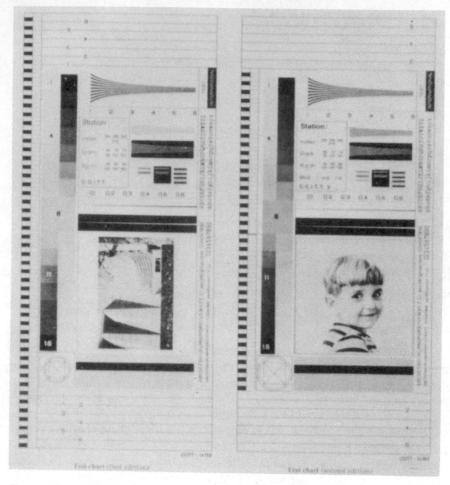

Figure 14.4 CCITT facsimile test charts. (From CCITT Rec. T.20 [Ref. 23]; courtesy of ITU – CCITT.)

edition 1 of the CCITT with a picture of the UNESCO building (Paris), and the edition 2 with a picture of an Argentinian boy. Any of the three charts will provide the user with the same basic information. The CCITT charts are shown in Figure 14.4.

14.8 FACSIMILE TRANSMISSION STANDARDS

14.8.1 General

Telecommunication standards are issued to ensure system interoperability. If a facsimile scanner and distant-end recorder are made by the same manufacturer and are designed to interoperate, and the intervening transmission medium is provided as specified by the manufacturer, then system compatibility can be expected.

Under many circumstances, particularly where a large community of diverse users are serviced by a system, we probably cannot expect such ideal pairing arrangements. Typical examples of large diverse communities to be served are wire services (news), weather/meterological organizations, military forces, and marine and large corporate systems.

Another factor that has stimulated one form or another of standardization is a common transmission medium. The international telephone network is the best example, with constraints on bandwidth, level, noise, amplitude, and delay response. HF radio is another example.

The following sections review some of the basic standards. Several are what might be termed domestic U.S. standards, but they are used so widely that they have taken on worldwide significance. The standard document size varies: for the U.S. civilian market it is $8\frac{1}{2} \times 11$ in., for the U.S. military, 8×10 in.; the international paper size is DIN A4 ($8\frac{1}{4} \times 11\frac{11}{16}$ in.). Military machines accept widths up to $8\frac{1}{2}$ in. The older standard document transmission speed for a page is 6 min; CCITT has standards for a 3-min (A4) page, a 1-minute page, and a 25-second page.

14.8.2 Electronic Industries Association (EIA)

EIA has issued several standards dealing with facsimile operation. EIA RS-328 (Ref. 11), "Message Facsimile Equipment for Operation on Switched Voice Facilities Using Data Communication Terminal Equipment," is summarized below:

Spectral sensitivity	RMA S-4 photo surface
Scan line length (total)	18.85 in. (478 mm)
Scanning direction	Normal (corresponding to left-hand helix)
Dead sector	0.56–0.94 in. (14.2–23.9 mm) (a sector at the end of the scanning line which is 3–5% of the scanning line length; coincides with time position of phasing signal)
Scan speeds	60, 90, or 120 strokes per minute, selectable
Line advance	$\frac{1}{96}$ in. (0.26 mm)
Scanning spot size	0.0104 in. (0.26 mm) $\times$ 0.0104 in. (0.26 mm)
Index of cooperation	576 (CCITT definition)
Signal contrast	20 dB (± 2 dB)
Synchronization	Built-in frequency standard with stability of 3 parts in 10^6
Standard frequency	300 Hz or multiple thereof

Start signal	Alternating black and white levels modulated at a rate of 300/s for a period of 5 s
Phasing signal	30-s transmission of alternating black and white.
Stop signal (on completion of scanning)	Alternating black and white interrupted at a rate of 450/s for 5 s, then black transmission for a period of 10 s
Control functions	Start command, phase signal, and stop command
Modulation characteristics	
AM	Maximum amplitude black with carrier frequency 1800 or 2400 Hz (when VSB used, carrier 2400 Hz and upper sideband completely attenuated)
FM	1500 Hz black, 2300 Hz white

The standard refers to the ANSI IEEE Std. 167-1966 (Reaff. 1971 [Ref. 5]), "Test Procedure for Facsimile." Other related EIA standards are RS-357 (Ref. 15), "Interface between Facsimile Terminal Equipment and Voice Frequency Data Communications Terminal Equipment" (June 1968), and RS-373 (Ref. 16), "Unattended Operation of Facsimile Equipment (as Defined in EIA Standard RS-328)" (June 1970).*

14.8.3   U.S. Military Standards

The following excerpts have been taken from MIL-STD-188-100 with Notices 1, 2, and 3. These standards deal with analog devices and systems. Section 14.11 gives a brief overview of portions of MIL-STD-188-161A for digital facsimile.

Meteorological Equipment

Transmitter

Original copy size	
Drum scanner	$18\frac{5}{8}$ in. (473 mm) × 12 in. (305 mm)
Flatbed scanner	$18\frac{5}{8}$ in. (473 mm) by any length (continuous scanners)
Continuous recorder	400-ft roll wound on 1-in. core (121.9 m on 25.4-mm core), $18\frac{5}{8}$ in. (473 mm) wide
Recorded line length	18.85 in. (478 mm)

*These standards may be ordered from the Electronic Industries Association, 2001 Eye St. NW, Washington, DC 20006.

The remainder of the specification is essentially the same as that for the transmitter.

Meteorological Receiver, Large Format

Recorded copy size	
Drum recorder	$18\frac{5}{8}$ in. (473 mm) × 12 in. (305 mm) or integral multiples of 12 in. (305 mm)
Scanning rate	180 LPM
Index of cooperation	805–889, eventually to be fixed at 829 to agree with CCITT/CCIR
Total line length	8.5–9.2 in.
Available line length	8.0 in. minimum, eventually to be fixed at 8.5 in.
Phasing time	15 s
Signal sense	Maximum on black
Synchronization	1. 60 Hz ($\pm$0.5 Hz), transmitted synchronizing signal
	2. Frequency standard accurate to 1 part in 10^5
LPI	96
Paper speed	$1\frac{7}{8}$ in. min
Phasing pulse	25 $\pm$ 3 ms. Phasing signal is a black signal with a white pulse once per scanning line, the black-to-white transition occurring at the right-hand edge of an $8\frac{1}{2}$-in. wide sheet

Meteorological Receiver, Small Format

General	Continuous recording; able to resolve at least 200 LPI (25.4 mm)
Recorded copy size	$8\frac{1}{2}$ in. (216 mm) wide; length dependent on duration of transmission; 400-ft roll on 1 in. core (121.9 m on 25.4-mm core)
Recording line length	8.64 in. (220 mm)
Recording direction	Normal (corresponding to left-hand helix)
Recording speed	60, 90, or 120 strokes/min, selectable
Index of cooperation	576 and 264 (CCITT definition) (index of 288 will work with specified index of 264 with negligible distortion)
Line advance	1/209.5 in. (0.12 mm) for index of cooperation of 576 or $\frac{1}{96}$ in. (0.26 mm) for index of cooperation of 264, selectable

Recording spot size $1/209.5 \times 1/209.5$ in. (0.12×0.12 mm)
for index of cooperation of 576 or
$\frac{1}{96} \times \frac{1}{96}$ in. (0.26×0.26 mm) for index of
cooperation of 264, with marking stylus or
element being changed for the index employed

Dead sector   The signal transmitted during interval that the
transmitter is scanning dead sector may be
blanked if desired

Input power level   For high signal contrast between -9 and -36
dBm

14.8.4   U.S. National Weather Service

For service over the telephone network voice channel:

Scan speed	120 LPM
Resolution	48 or 96 LPI
Index of cooperation	576 for 96 LPI and 288 for 48 LPI (CCITT definition)
Transmission/modulation	AM, 2400-Hz carrier
Start signal	Carrier modulated by 300-Hz tone
Phasing signal	Black signal interrupted by 12.5-ms white pulse twice per second prior to recording
Stop signal	Carrier modulated by 450-Hz tone

14.8.5   CCITT Facsimile Recommendations

The CCITT T recommendations discuss three distinct groups of facsimile equipment:

- *Group 1*. Equipment that uses double-sideband (DSB) AM modulation without resorting to special measures to compress bandwidth. Equipment in this group is designed for document transmission of (ISO) A4 size paper at nominally 4 lines per millimeter in about 6 min via a telephone-type circuit. Covered by Rec. T.2.
- *Group 2*. Equipment that uses bandwidth compression techniques to achieve a transmission time for an A4 size document at nominally 4 lines per millimeter via a telephone circuit. Bandwidth compression in this context includes encoding and/or vestigial-sideband (VSB) AM but does not include processing to reduce redundancy. Covered by Rec. T.3.
- *Group 3*. Equipment that incorporates a means of reducing redundancy prior to modulation to achieve a transmission time for an A4 size document of 1 min via the telephone network. Bandwidth compression of the line signal may be used. Under study by CCITT Study Group XIV.

Rec. T.1, "Standardization of Phototelegraph Apparatus"

Index of cooperation	352, alternatively 264
Drum scanning	
Drum diameter	66, 70, and 88 mm
Drum factor	2.4 (sending and receiving)
Flatbed scanning, line length	207, 220, and 276 mm, of which 15 mm is not used for effective transmission
Phasing and synchronization	See Section 14.4.2
Modulation	
AM	Carrier frequency 1300 Hz; for systems operating over carrier systems, 1900 Hz is recommended; high amplitude black, low white with at least 30 dB between nominal white and black levels
FM	Mean frequency 1900 Hz; white frequency 1500 Hz; black frequency 2300 Hz

Table 14.2 gives the corresponding values of the index of cooperation M, the factor of cooperation C, the drum diameter, D, the total length L of scanning line, the scanning pitch P, and the scanning density F for equipment in most common use.

Table 14.3 gives the normal and approved alternative combinations of drum rotation speeds or scanning line frequencies and indices of cooperation.

Table 14.2   Common standards[a]

M	C	D (mm)	L (mm)	P (mm)	F (lines/mm)
264	829	66	207	1/4	4
264	829	70	220	1/3.77	3.77
264	829	88	276	1/3	3
350	1099	70	220	1/5	5
352	1105	66	207	3/16	16/3
352	1105	88	276	1/4	4

Source: CCITT T recommendations.

[a]The maximum dimensions of the pictures to be transmitted result from the parameters given in the table.

Table 14.3 Standard drum rotation speeds and scanning line frequencies versus indices of cooperation

	Drum Rotation Speed (rpm) or Scanning Line Frequency	Index of Cooperation	
		Metallic Circuits	Combined Metallic and Radio Circuits
Normal conditions	60	352	352
	90		264
Alternatives for use when	90	264 and 352	
the phototelegraph	120	264 and 352	
apparatus and metallic	150	264	
circuits are suitable			

Notes:

1. In the case of transmitters operating on metallic circuits, the index 264 is not intended to be used with an 88-mm drum. In the case of transmitters operating on combined metallic and radio circuits, the index 264 associated with a drum diameter of 88 mm is intended to be used only exceptionally.

2. The provisions given in the table are not intended to require the imposition of such standards on users who use their own equipment for the transmission of pictures over leased circuits. However, the characteristics of the apparatus used should be compatible with the characteristics of the circuits used.

Rec. T.2, "Standardization of Group 1 Facsimile Apparatus for Document Transmission"

Index of cooperation	264; when a lower vertical resolution is acceptable, 176
Document size	Minimum ISO A4 (210 × 297 mm)
Total scan line length	215 mm (active sector plus dead sector)
Total scan lines per document	1144 for an index of cooperation of 264, and 762 for an index of 176
Scanning density	3.85 lines/mm
Scanning frequency	180 LPM, alternatively 240 LPM with mutual agreement between both ends
Scanning frequency stability	± 1 part in 10^5
Phasing and synchronization	See Section 14.4.2
Modulation	
AM	High-level signal is black and carrier frequency should range between 1300 and 1900 Hz, depending upon circuit characteristics
FM	Center frequency 1700 Hz, shift ± 400 Hz with the higher frequency corresponding to black
Power at the receiver input	Receiver functions correctly with input levels from 0 to -40 dBm (for AM these correspond to black levels)

Rec. T.3, "Standardization of Group 2 Facsimile Apparatus for Document Transmission"

Equipment dimensions	
Factor of cooperation	829 ± 1%
Total line scanning	215 mm length
Usable scanning line	200 mm length
Document size	Minimum ISO A4 (210 × 297 mm)
Index of cooperation	264
Scanning density	3.85 lines/mm
Number of scanning lines in a document 297 mm long	1145
Scanning frequency	360 LPM, alternatively 300 LPM with mutual agreement between both ends
Phasing and synchronization	See Section 14.4.2
Modulation, AM	VSB-phase modulation with a carrier at 2100 Hz ± 10 Hz. White signal represented by maximum amplitude, and black should be at least 26 dB below white level; the phase of the carrier may be reversed after each transition through black
Receiver input level	White signal between 0 and −40 dBm

(*Note*: CCITT Rec. T.4 is reviewed in Section 14.9.)

Rec. T.30, "Procedures for Document Facsimile in the General Switched Telephone Network"

This recommendation describes the procedures and signals to be used for facsimile equipments operated over the switched network. Two separate signaling systems are described, a simple system using single-frequency tones and a second binary-coded system that offers a wide range of signals for more complex operational procedures. Automatic operation is implied with the latter, manual and semiautomatic operation with the former.

The binary-coded signaling system is based on a high-level data link control (HDLC) format developed for data transmission procedures. The basic HDLC structure consists of a number of frames, each of which is subdivided into a number of fields. It provides for frame labeling, error checking, and confirmation of correctly received information, and the frames can be easily extended if this should be required in the future.

14.8.6   CCIR Recommendations

In the recommendations below CCIR generally treats facsimile transmission over HF radio circuits and stipulates the use of direct FM or subcarrier FM for these applications.

Rec. 343-1, "Facsimile Transmission of Meteorological Charts over Radio Circuits"   This is the same as CCITT Rec. T.15.

Rec. 344-2, "Standardization of Phototelegraph Systems for Use on Combined Radio and Metallic Circuits

Modulation	See CCITT Rec. T.15
Index of cooperation	352, alternatively 264
Drum speed	60 rpm, alternatively 90/45 rpm

Note: In due course the alternative methods will become obsolete.

14.8.7   World Meteorological Organization (WMO)

The WMO has set forth certain standards for the collection and dissemination of weather data by facsimile (Ref. 10). A partial review of these standards is given below:

Index of cooperation	576 with minimum picture elements of 0.4 mm, and 288 for minimum picture elements of 0.7 mm (CCITT definition)
Drum speed	60, 90, 120, or 240 strokes/min (rpm)
Drum diameter	152 mm
Scanning density	4 lines/mm for index of cooperation of 576, and 2 lines/mm for index of 288
Length of drum	55 cm minimum
Stability (scanning speed)	Within 5 parts in 10^6 of its normal value
Modulation	
AM	Maximum carrier amplitude corresponds to black; carrier at 1800 Hz for drum speeds of 60, 90, and 120 rpm; for drum speed of 240 rpm the carrier is at 2600 Hz and VSB must be used.
FM	Center frequency 1900 Hz; black frequency 1500 Hz; white frequency 2300 Hz
Contrast ratio	Between 12 and 25 dB

14.9 DIGITAL TRANSMISSION OF FACSIMILE

14.9.1 Introduction

CCITT Rec. T.4 (Ref. 20) is a widely accepted standard for the transmission of digital facsimile. It was primarily designed for the transmission of the European (ISO) A4 page but works equally as well with the North American standard $8\frac{1}{2} \times 11$-in. page. Rec. T.4 describes what is called "Group 3 Apparatus," which will transmit a page in 1 min or less. Using standard resolution of 3.85 lines/mm (vertical), an A4 page can be transmitted in about 25 s. This is dependent, of course, on the data rate used, whether 2400, 4800, or 9600 bps. It is also based on "minimum" line transmission time.

14.9.2 Equipment Dimensions — DIN A4

In the vertical direction the standard resolution is 3.85 lines/mm $\pm 1\%$. The optional higher resolution is 7.7 lines/mm $\pm 1\%$.

The standard scan line length is 215 mm and is made up of 1728 pixels (black and white elements). There are two optional scan line lengths, 255 and 303 mm, which are made up of 2048 and 2432 pixels, respectively.

14.9.3 Transmission Time Per Total Coded Scan Line

The standard minimum time duration to scan a standard line is 20 ms. A line consists of Data bits plus any required Fill bits and End-of-line (EOL) bits. The Data bits give the information on the line's pixel content; Fill bits complete unfinished lines out to 215 mm, the standard line length, and ensure minimum line transmission time is met. The EOL bit sequence carries out the line feed and carriage return function used in teleprinter service or the "soft-return" command on computers.

There is an optional two-dimensional coding scheme where the total scan line is defined as the sum of the Data bits plus the Fill bits plus the EOL and a tag bit.

Rec. T.4 has optional 10- and 5-ms minimum transmission times of the total scan line. It notes that with the 10-ms option, the minimum transmission time of the total coded scan line is the same both for standard resolution and for the optional higher resolution.

14.9.4 Coding Scheme

In CCITT Rec. T.4 two coding schemes are given: one-dimensional and an optional two dimensional. We will only briefly describe the one-dimensional scheme here. Both schemes use what is called *run length coding*.

Table 14.4 Terminating codes

White Run Length	Code Word	Black Run Length	Code Word
0	00110101	0	0000110111
1	000111	1	010
2	0111	2	11
3	1000	3	10
4	1011	4	011
5	1100	5	0011
6	1110	6	0010
7	1111	7	00011
8	10011	8	000101
9	10100	9	000100
10	00111	10	0000100
11	01000	11	0000101
12	001000	12	0000111
13	000011	13	00000100
14	110100	14	00000111
15	110101	15	000011000
16	101010	16	0000010111
17	101011	17	0000011000
18	0100111	18	0000001000
19	0001100	19	00001100111
20	0001000	20	00001101000
21	0010111	21	00001101100
22	0000011	22	00000110111
23	0000100	23	00000101000
24	0101000	24	00000010111
25	0101011	25	0000001100
26	0010011	26	000011001010
27	0100100	27	000011001011
28	0011000	28	000011001100
29	00000010	29	000011001101
30	00000011	30	000001101000
31	00011010	31	000001101001
32	00011011	32	000001101010
33	00010010	33	000001101011
34	00010011	34	000011010010
35	00010100	35	000011010011
36	00010101	36	000011010100
37	00010110	27	000011010101
38	00010111	38	000011010110
39	00101000	39	000011010111
40	00101001	40	000001101100
41	00101010	41	000001101101
42	00101011	42	000011011010

Table 14.4 *(Continued)*

White run Length	Code Word	Black Run Length	Code Word
43	00101100	43	000011011011
44	00101101	44	000001010100
45	00000100	45	000001010101
46	00000101	46	000001010110
47	00001010	47	000001010111
48	00001011	48	000001100100
49	01010010	49	000001100101
50	01010011	50	000001010010
51	01010100	51	000001010011
52	01010101	52	000000100100
53	00100100	53	000000110111
54	00100101	54	000000111000
55	01011000	55	000000100111
56	01011001	56	000000101000
57	01011010	57	000001011000
58	01011011	58	000001011001
59	01001010	59	000000101011
60	01001011	60	000000101100
61	00110010	61	000001011010
62	00110011	62	000001100110
63	00110100	63	000001100111

Source: CCITT Rec. T.4. Table 1/T.4, Page 24, Fascicle VII.3, IXth Plenary Assembly, Melbourne 1988, Ref. 20.

A line of data is composed of a series of variable-length code words. Each code word represents a run length of either all white or all black. White runs and black runs alternate. Again, 1728 picture elements (pixels) represent one horizontal scan line 215 mm long.

To ensure black–white synchronization at the far-end receiver, all Data lines begin with a white-run-length code word. If the actual line begins with a black run, a white run length of zero is sent. Black or white run lengths up to the maximum length of one scan line (1728 pixels) are defined by the code words shown in Tables 14.4 and 14.5. There are two types of code words used: terminating code words and make-up code words. Each run length is represented by either one terminating code word or one make-up code word followed by a terminating code word.

Run lengths in the range of 0–63 pels (pixels) are encoded with their appropriate terminating code word. Note that there is a different list of code words for black and white run lengths (see Table 14.4).

Table 14.5 Make-up codes

White Run Lengths	Code Word	Black Run Lengths	Code Word
64	11011	64	0000001111
128	10010	128	000011001000
192	010111	192	000011001001
256	0110111	256	000001011011
320	00110110	320	000000110011
384	00110111	384	000000110100
448	01100100	448	000000110101
512	01100101	512	0000001101100
576	01101000	576	0000001101101
640	01100111	640	0000001001010
704	011001100	704	0000001001011
768	011001101	768	0000001001100
832	011010010	832	0000001001101
896	011010011	896	0000001110010
960	011010100	960	0000001110011
1024	011010101	1024	0000001110100
1088	011010110	1088	0000001110101
1152	011010111	1152	0000001110110
1216	011011000	1216	0000001110111
1280	011011001	1280	0000001010010
1344	011011010	1344	0000001010011
1408	011011011	1408	0000001010100
1472	010011000	1472	0000001010101
1536	010011001	1536	0000001011010
1600	010011010	1600	0000001011011
1664	011000	1664	0000001100100
1728	010011011	1728	0000001100101
EOL	000000000001	EOL	000000000001

Note: It is recognized that machines exist that accommodate larger paper widths maintaining the standard horizontal resolution. This option has been provided for by the addition of the make-up code set defined as follows:

Run Length (Black and White)	Make-Up Codes
1792	00000001000
1856	00000001100
1920	00000001101
1984	000000010010
2048	000000010011
2112	000000010100
2176	000000010101
2240	000000010110
2304	000000010111
2368	000000011100
2432	000000011101
2496	000000011110
2560	000000011111

Source: CCITT Rec. T.4, Table 2/T.4, page 25, Fascicle VII.3, IXth Plenary Assembly, Melbourne, 1988.

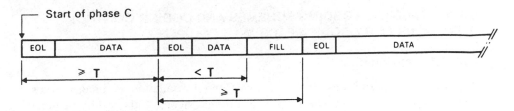

T   Minimum transmission time of a total coded scan line

Figure 14.5 Several scan lines of data starting at the beginning of a transmitted page. (From CCITT Rec. T.4, Figure 1/T.4, page 26; Fascicle VII.3, IXth Plenary Assembly, Melbourne 1988 Ref. 20.)

Run lengths in the range of 64–1728 pixels are encoded first by the make-up code word representing the run length that is equal to or shorter than required. This then is followed by the terminating code word representing the difference between the required run length represented by the make-up code word (see Table 14.5).

The end-of-line (EOL) sequence is a unique bit sequence that can never be found within a valid line of Data. If it is imitated due to an error burst, resynchronization may be required. This same bit sequence is used prior to the first Data line of a page. The bit sequence format is 000000000001.

When there is a pause in the message flow, Fill is transmitted. Fill may be inserted between a line of data and an EOL, but never within a line of Data. Fill must be added to ensure that the transmission time of Data, Fill, and EOL is not less than the minimum transmission time of the total coded scan line established in the premessage control procedure. The format of Fill is a variable length string of 0's.

When scanning reaches the end of a document, an RTC (return-to-control) signal is sent. An RTC signal consists of six consecutive EOLs. Following the RTC signal, the transmitter will send the postmessage commands in the frame format and the data signaling rate of the control signals, which are defined in CCITT Rec. T.30 (Ref. 18).

Figures 14.5 and 14.6 illustrate the relationship of the signals described above. Figure 14.5 shows several scan lines of data starting at the beginning of a transmitted page, and Figure 14.6 the last coded scan line of a page.

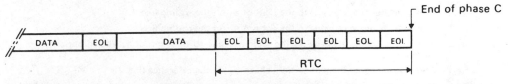

Figure 14.6 The last coded scan line of a page. (From CCITT Rec. T.4, Figure 2/T.4, page 26, Fascicle VII.3, IXth Plenary Assembly, Melbourne 1988, Ref. 20.)

14.10 FACSIMILE CODING SCHEMES AND CODING CONTROL FUNCTIONS FOR GROUP 4 FACSIMILE EQUIPMENT

14.10.1 Introduction

Group 4 document transmission involves major reduction of image redundancy by means of a two-dimensional coding scheme. This results in much faster transmission. However, at present, CCITT Rec. T.6 (Ref. 21) only covers black and white transmission; there is no provision for gray scales. The coding schemes utilized assume forward error correction (FEC) coding to maintain a bit error rate (BER) of 1×10^{-5} or better (Ref. 14). CCITT Rec. T.6 appears essentially unchanged in U.S. Standard FIPS 150 (Ref. 17) and in Section 4.2 of CCITT Rec. T.4 (Ref. 20).

14.10.2 Basic Coding Scheme

14.10.2.1 Principle of the Coding Scheme The coding scheme uses a two-dimensional line-by-line coding method in which the position of each changing picture element on the current coding line is coded with respect to the position of a corresponding reference element situated on either the coding line or the reference line that is immediately above the coding line. After the coding line has been coded, it becomes the reference line for the next coding line. The reference line for the first coding line on a page is an imaginary white line.

A changing picture element is defined as an element whose *color* (i.e., black or white) is different from that of the previous element along the scan line. An example of this concept is shown in Figure 14.7.

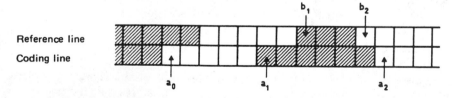

a_0 : The reference or starting changing element on the coding line. At the start of the line a_0 is set on an imaginary white changing element situated just before the first element on the line. During the coding of the coding line, the position of a_0 is defined by the previous coding mode.

a_1 : The next changing element to the right of a_0 on the coding line.

a_2 : The next changing element to the right of a_1 on the coding line.

b_1 : The first changing element on the reference line to the right of a_0 and of opposite colour to a_0.

b_2 : The next changing element to the right of b_1 on the reference line.

Figure 14.7 Changing picture elements. (From CCITT Rec. T.6, page 49, Figure 1/T.6, Fascicle VII.3, IXth Plenary Assembly, Melbourne 1988, Ref. 21.)

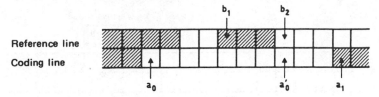

Figure 14.8   Pass mode. (From CCITT Rec. T.6, Figure 2/T6, page 50, Fascicle VII.3, IXth Plenary Assembly, Melbourne 1988, Ref. 21.)

14.10.2.2 Coding Modes   There are three coding modes. The mode selected depends on the coding procedure, as described in Section 14.10.3, which is used to code the position of each changing element along the coding line. Examples of the three coding modes are given in Figures 14.8, 14.9, and 14.10.

The *pass mode* is shown in Figure 14.8. This mode is identified when the position of b_2 lies to the left of a_1. However, the state where b_2 occurs just above a_1, as shown in Figure 14.9, is not considered a pass mode.

When the *vertical mode* is identified, the position of a_1 is coded relative to the position of b_1. The relative distance a_1b_1 can take on one of seven values: $V(0)$, $V_R(1)$, $V_R(2)$, $V_R(3)$, $V_L(1)$, $V_L(2)$, and $V_L(3)$, each of which is represented by a separate code word. The subscripts R and L indicate that a_1 is to the right or left, respectively, of b_1; the number in brackets indicates the value of the distance a_1b_1 (see Figure 14.10).

When the *horizontal mode* is identified, both run lengths a_0a_1 and a_1a_2 are coded using code words $H + M(a_0a_1) + M(a_1a_2)$. H is the flag code word 001 taken from the two-dimensional code table, Table 14.6. $M(a_0a_1)$ and $M(a_1a_2)$ are code words that represent the length and "color" of the runs a_0a_1 and a_1a_2, respectively, and are taken from the appropriate white or black run-length code tables (Table 14.4).

14.10.3 Coding Procedure

The coding procedure (shown in the coding flow diagram in Figure 14.11), identifies the coding mode that is to be used to code each changing element along the coding line. When one of the three coding modes has been identified according to step 1 or step 2 as shown in the figure, an appropriate code word is selected from the code table given in Table 14.6. The steps of

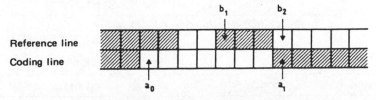

Figure 14.9   An example not corresponding to a pass mode. (From CCITT Rec. T.6, page 50, Figure 3/T6, Fascicle VII.3, IXth Plenary Assembly, Melbourne 1988, Ref. 21.)

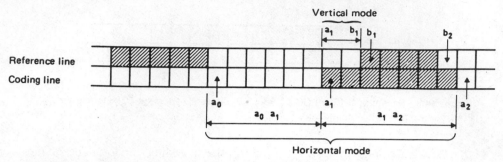

Figure 14.10 Vertical mode and horizontal mode. (From CCITT Rec. T.6, Figure 4/T6, page 50, Fascicle VII.3, IXth Plenary Assembly, Melbourne 1988, Ref. 21.)

the coding procedure are described below:

Step 1:

1. If a pass mode is identified, this is coded using the word 0001 (Table 14.6). After this processing, picture element a_0' just under b_2 is regarded as the new starting picture element a_0 for the next coding (see Figure 14.8).
2. If a pass mode is not detected, then proceed to step 2.

Step 2:

1. Determine the absolute value of the relative distance $a_1 b_1$.
2. If $[a_1 b_1] \leq 3$, as shown in Table 14.6, $a_1 b_1$ is coded by the vertical mode, after which position a_1 is regarded as the new starting picture element a_0 for the next coding.

Table 14.6   Code table

Mode	Elements to Be Coded			Notation	Code Word
Pass	b_1, b_2			P	0001
Horizontal	$a_0 a_1, a_1 a_2$			H	$001 + M(a_0 a_1) + M(a_1 a_2)^a$
Vertical	a_1 just under b_1	$a_1 b_1 = 0$	$V(0)$	1	
	a_1 to the right of b_1	$a_1 b_1 = 1$	$V_R(1)$	011	
		$a_1 b_1 = 2$	$V_R(2)$	000011	
		$a_1 b_1 = 3$	$V_R(3)$	0000011	
	a_1 to the left of b_1	$a_1 b_1 = 1$	$V_L(1)$	010	
		$a_1 b_1 = 2$	$V_L(2)$	000010	
		$a_1 b_1 = 3$	$V_L(3)$	0000010	
Extension					$0000001 xxx$

Source: CCITT Rec. T.6, Table 1/T.6, page 51, Fascicle VII.3, IXth Plenary Assembly, Melbourne, 1988. Ref. 21.

[a]Code M( ) of the horizontal mode represents the code words in Table 14.5.

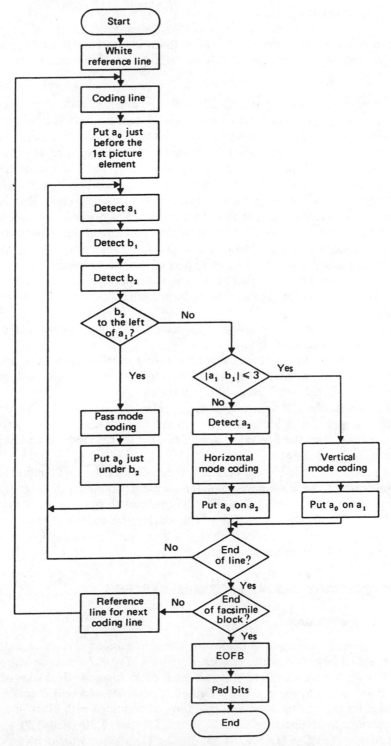

Figure 14.11 Coding flow diagram. (From CCITT Rec. T.6, Figure 6/T.6, page 52, Fascicle VII.3, IXth Plenary Assembly, Melbourne 1988, Ref. 21.)

3. If $[a_1 b_1] > 3$, as shown in Table 14.6, following horizontal mode code 001, $a_0 a_1$ and $a_1 a_2$ are respectively coded by one-dimensional run-length coding.

Run lengths in the range of 0–63 pixels are encoded with their appropriate terminating code word from Table 14.4. Note that there is a different list of code words for black and white run lengths. Run lengths in the range of 64–2623 pixels are encoded first by the make-up code word representing the run length that is nearest, not longer, to that required. This is then followed by a terminating code word representing the difference between the required run length and the run length represented by the make-up code. Run lengths in the range of lengths longer than or equal to 2624 pixels are coded first by the make-up code of 2560. If the remaining part of the run (after the first make-up code of 2560) is 2560 pixels or greater, additional make-up code(s) of 2560 are issued until the remaining part of the run becomes less than 2560 pixels. Then the remaining part of the run is encoded by terminating code or by make-up code plus terminating code according to the range mentioned above.

After this processing, position a_2 is regarded as the new starting picture element a_0 for the next coding.

To process the first picture element, the first starting element a_0 on each coding line is imaginarily set at a position just before the first picture element, and is regarded as a white picture element.

The first run length on a line $a_0 a_1$ is replaced by $a_0 a_1 - 1$. Therefore, if the first actual run is black and is deemed to be coded by the horizontal mode coding, then the first code word $M(a_0 a_1)$ corresponds to an imaginary white run of zero length.

To process the last picture element, the coding of the line continues until the position of the imaginary changing element situated just after the last actual element has been coded. This may be coded as a_1 or a_2. Also, if b_1 and/or b_2 are not detected at any time during the coding of the line, they are positioned on the imaginary changing element situated just after the last actual picture element on the reference line.

14.11 MILITARY DIGITAL FACSIMILE SYSTEMS

14.11.1 Definitions

Type I facsimile equipment provides for the transmission and reception of an image with black and white information only. *Type II facsimile equipment* provides for the transmission and reception of an image with shades of gray, as well as black and white. *CCITT Group 3 facsimile equipment* (Section 14.9) provides for the transmission and reception of an image with black and white information, as defined by CCITT Recs. T.4 and T.30 (Refs. 20 and 18, respectively) (FED-STD-1062 and FED-STD-1063), which incorporates means for reducing redundant information in the document signal prior to the modulation process.

Table 14.7 Code words and signaling sequences for Type I and Type II facsimile equipment

Name	Make Up
Beginning of intermediate line pair (BILP)	0000000000000011
Beginning of line pair (BOLP)	0000000000000010
End of line (EOL)	000000000001
End of message (EOM)	16 consecutive S_1 code words
Not end of message ($\overline{\text{EOM}}$)	16 consecutive inverted S_1 code words
Return to control (RTC)	EOL EOL EOL EOL EOL EOL
Start of message (SOM)	$S_1 S_0 X$ clock periods $S_0 S_1$ (where X is the number of clock periods between the pairs of code words)
S_0	111100010011010
S_1	111101011001000
Fill	Variable-length string of 0's
Stuffing	Variable-length string of 1's
Preamble	Variable-length string of all 1's or all 0's

Source: MIL-STD-188-161A, with Notice 1 (Ref. 14).

Basic mode is where the transmitter for Type I and Type II facsimile does not pause after calling the receiving unit to wait for an acknowledgment before transmitting an image in the simplex and broadcast mode of operation. *Handshake mode* is where the transmitter for Type I, Type II, and CCITT Group 3 facsimiles pauses after calling the receiving unit to wait for an acknowledgment before transmitting an image in the duplex mode of operation.

Synchronization code words and signaling sequences used in Type I and Type II facsimile are defined in Table 14.7.

Table 14.8 Functional interchange circuits

Circuit	Direction[a]
Request to send	From DTE to DCE
Clear to send	From DCE to DTE
Receive input control	From DTE to DCE
Send data	From DTE to DCE
Receive data	From DCE to DTE
Send timing	From DCE to DTE
Receive timing	From DCE to DTE
Send common	Return
Receive common	Return
Signal ground	Ground

[a]DTE = data terminal equipment; DCE = data communication equipment.

14.11.2   Technical Requirements for Type I and Type II Facsimile Equipment

Transmission rates are bit-by-bit synchronous at data rates of 2400, 4800, and 9600 bps, and 16 and 32 kbps, with timing provided by an external clock. Digital interfaces shall comply with the applicable requirements of MIL-STD-188-114 (Chapter 12). The functional interchange circuits are shown in Table 14.8.

Image Parameters:

Scan line length: 215 mm, left justified

Resolution: three switch-selectable standards for horizontal and vertical resolutions: These are

- 3.85 lines/mm (vertical) by 1728 black and white pixels along the horizontal scan line (*Note*: This is a nominal medium resolution of 100×200 lines per inch)
- 3.85 lines/mm (vertical) by 864 black and white pixels along the horizontal scan line (*Note*: This is the nominal low resolution of 100×100 lines per inch)
- 7.7 lines/mm by 1728 pixels along the horizontal scan line (*Note*: This is a nominal high resolution of 200×200 lines per inch)

Tolerance of image parameters: $\pm 1\%$

Scanning direction: from left to right and from top to bottom

Scanning line transmission time: the minimum scanned line transmission time is 20 ms

Contrast levels: black and white

Document dimensions: input documents up to a maximum of 215 mm wide and 1000 mm long. Documents up to 230 mm wide may be accepted into scanner, but only 215 mm of the document will be scanned

14.11.3   Image Coding Schemes

The facsimile equipment shall be capable of operating in three modes: uncompressed, compressed, and compressed with forward error correction (FEC).

In the uncompressed mode, facsimile data is transmitted pixel by pixel, with logic 1 representing black. Each line of the output data consists of a synchronization code followed by the number of pixels as specified in "Resolution" above.

In the compressed mode, facsimile data is transmitted after compression by the redundancy algorithm (see Section 14.9, CCITT Rec. T.4). A line of data is composed of a series of variable-length code words. Each code word represents a run length of either all white or all black. White runs and black runs alternate. All data lines begin with a white-run-length code word to ensure that the receiver maintains color synchronization. A white run length of zero is sent if the actual scan line begins with a black run. Black or white run lengths may be up to a maximum of one scanning line (1728 pixels) and are defined in Tables 14.4 and 14.5 in Section 14.9.4. For descriptions of run lengths dealing with EOL (end of line) and Fill, see Section 14.9.

In the compressed mode with FEC, facsimile data is further processed by a channel coder and bit interleaving buffer to provide this FEC. The channel coder uses a Bose–Chaudhuri–Hocquenghen (BCH) code which, in this case, has the capability of correcting 2 errored bits per block. Section 14.11 is excerpted from MIL-STD-188-161A.

REVIEW EXERCISES

1. Give at least five applications of facsimile as it is used today.

2. Why is facsimile tending to replace Telex?

3. What are the three basic steps in facsimile communication?

4. Describe the function of a scanner. Is the signal developed from a scanner analog or digital? If we were to assume only black and white facsimile transmission (no gray scale), discuss analog versus digital output of a scanner.

5. Name four basic facsimile recording techniques.

6. Differentiate phasing and synchronization and state the purpose of each.

7. There are two definitions of index of cooperation. Give each and show how we may convert one to the other.

8. What do out-of-synchronization conditions cause in the receive copy?

9. What happens when the indices of cooperation are not the same between the transmitter and receiver?

10. Describe the Kendall effect.

11. Give at least six critical parameters affecting the transmission of conventional analog facsimile. Considering these parameters and reviewing Chapter 12, identify other parameters that would affect digital transmission performance.

12. What is a pixel (pel)? (Don't simply define the acronym.)

13. What is the Kell factor?

14. If we have an LPI (lines per inch) of 100, what is the approximate spot size?

15. About how many pixels are there in a practical $8\frac{1}{2} \times 11$-in. page transmitted by facsimile generally meeting CCITT T recommendations? (In other words, not high resolution facsimile.)

16. Define contrast and resolution.

17. In general, the trend is toward digital facsimile—in fact, nearly all document transmission today is digital. What is the great advantage of digital, especially considering the very nature of a printed page?

18. Older analog facsimile systems used what kind of modulation almost exclusively?

19. For Group 3 "apparatus," define Data, Fill, and EOL bits.

20. What does run-length encoding do for us?

21. Distinguish between make-up codes and terminating codes.

22. What is a return-to-control (RTC) signal and what does it do?

23. Group 4 facsimile equipment uses what kind of coding to reduce redundancy? What is the *present* limitation of Group 4 equipment?

24. Differentiate between Type I and Type II facsimile equipment.

25. Distinguish between basic mode and handshake mode when dealing with CCITT Recs. T4 and T6.

REFERENCES AND BIBLIOGRAPHY

1. *Reference Data for Radio Engineers*, 6th ed., Howard W. Sams, Indianapolis, IN, 1977.

2. MIL-STD-188-100, including Notice 3, U.S. Department of Defense, Washington, DC, Nov. 1976.

3. IEEE Std. 100-1977, *Dictionary of Electrical and Electronic Terms*, 3rd ed., IEEE, New York, 1984.

4. "Definition of Terms on Facsimile," IEEE Std. 168-1956, equivalent to ANSI C16.37-1972, (reaffirmed 1971 by IEEE), IEEE, New York, 1977.

5. "Test Procedure for Facsimile," ANSI IEEE Std. 167-1966, equivalent to ANSI C16.37-1971, IEEE, New York, reaff. 1971.

6. G. Stafford, "The Facsimile Series," reprint from *Communications*. Reference provided by G. Stafford.

7. D. M. Costigan, *Electronic Delivery of Documents and Graphics*, Van Nostrand Reinhold, New York, 1978.

8. "Standardization of Phototelegraph Apparatus," CCITT Rec. T.1, IXth Plenary Assembly, Melbourne, 1988, Fascicle VII.3.

9. "Standardization of Group 1 Facsimile Apparatus for Document Transmission," CCITT Rec. T.2, IXth Plenary Assembly, Melbourne, 1988, Fascicle VII.3.

10. *Manual on Global Telecommunications System*, vol. I, part III, sect. 7, World Meteorological Organization (WMO) Pub. 386, Geneva, 1972.

11. "Message Facsimile Equipment for Operation on Switched Voice Facilities Using Data Communication Terminal Equipment," EIA RS-328, EIA, Washington, DC, Oct. 1966.

12. *Data Communications Using Voiceband Private Line Channels*, Bell System Tech. Reference Pub. 41004, American Telephone and Telegraph Co., New York, Oct. 1973.

13. Roger L. Freeman, *Telecommunication System Engineering*, 2nd ed., Wiley, New York, 1989.

14. "Interoperability and Performance Standards for Digital Facsimile Equipment," MIL-STD-188-161A with Notice 1, U.S. Department of Defense, Washington, DC, Mar. 1989.

15. "Interface between Facsimile Terminal Equipment and Voice Frequency Data Communications Terminal Equipment," EIA RS-357, EIA, Washington, DC, June 1968.

16. "Unattended Operation of Facsimile Equipment (as Defined in EIA Standard RS-328)," EIA RS-373, EIA, Washington, DC, June 1970.

17. "Facsimile Coding Schemes and Coding Control Functions for Group 4 Facsimile Apparatus," FIPS Pub. 150, U.S. Department of Commerce, NIST, Washington, DC, Nov. 1988 (EIA-538).

18. "Procedures for Document Transmission in the General Switched Telephone Network," CCITT Rec. T.30., IXth Plenary Assembly, Melbourne, 1988, Fascicle VII.3.

19. "Standardization of Group 2 Facsimile Apparatus for Document Transmission," CCITT Rec. T.3, IXth Plenary Assembly, Melbourne, 1988, Fascicle VII.3.

20. "Standardization of Group 3 Facsimile Apparatus for Document Transmission," CCITT Rec. T.4, IXth Plenary Assembly, Melbourne, 1988, Fascicle VII.3.

21. Facsimile Coding Schemes and Coding Control Functions for Group 4 Facsimile Apparatus," CCITT Rec. T.6, IXth Plenary Assembly, Melbourne, 1988, Fascicle VII.3. (See also Ref. 17.)

22. "Phototelegraph Transmission over Combined Radio and Metallic Circuits," CCITT Rec. T.15, IXth Plenary Assembly, Melbourne, 1988, Fascicle VII.3.

23. "Standardized Test Chart for Facsimile Transmissions," CCITT Rec. T.20, IXth Plenary Assembly, Melbourne, 1988, Fascicle VII.3.

24. "Phototelegraph Transmission on Telephone-Type Circuits," CCITT Rec. T.11, IXth Plenary Assembly, Melbourne, 1988, Fascicle VII.3.

25. "Range of Phototelegraph Transmissions on a Telephone-Type Circuit," CCITT Rec. T.12, IXth Plenary Assembly, Melbourne, 1988, Fascicle VII.3.

APPENDIX A

ASYNCHRONOUS TRANSFER MODE (ATM)

A.1 INTRODUCTION

The present digital network is based on a rigid structure of channels (e.g., 64 kbps) and hierarchy. All channels and groups at a given level of digital hierarchy are of the same size. Let us call this the *synchronous transfer mode* (STM). Our present network is evolving into one with a dynamically changing mix of services at a variety of *fixed* channel rates. A new demand is being placed on the network to support what is broadly called *bursty services*. Our present system based on STM cannot transfer these services efficiently, and such channels would be dimensioned to accommodate peak transfer rates. ATM will provide the digital network much greater flexibility in handling the large variety of services that a broadband network may offer, such as on the broadband integrated services digital network (B-ISDN). ATM will have one of its greatest impacts in switching.

A.2 CCITT PERSPECTIVE

CCITT Rec. I.121 (Ref. 1) calls ATM a target transfer mode solution for implementing B-ISDN. It will influence the standardization of digital hierarchies and multiplexing structures, switching, and interface of broadband digital signals.

Rec. I.121 speaks of ATM as a specific packet-oriented transfer mode using the *asynchronous time division multiplexing* technique: the multiplexed information flow is organized in fixed-sized blocks, called cells. A cell consists of a user information field and a header, as shown in Figure A.1. The

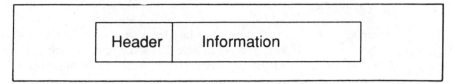

Figure A.1 ATM cell format.

primary role of the header is to identify cells belonging to the same virtual channel on an asynchronous time division multiplex. Cells are assigned on demand, depending on the source activity and available resources. Call sequence integrity on a virtual channel is preserved by the ATM layer.

ATM is a connection-oriented technique. Header values are assigned to each section of a connection when required, and released when no longer required. The connection identified by the headers remains unchanged during the lifetime of a call. As with ISDN, signaling and user information are carried on separate virtual channels. ATM offers a flexible transfer capability common to all services, including connectionless service.

Figure A.2 shows the ATM layer model as applied to B-ISDN. The physical layer has been covered at length in Chapter 12. Here we are concerned with the two upper layers in Figure A.2:

- ATM layer, which is common to all services and provides cell-transfer capabilities
- The adaptation layer that is service dependent

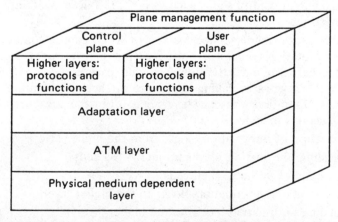

Figure A.2 B-ISDN protocol model for ATM. (From CCITT Rec. I.121, Page 41, Figure 5/I.121, Fascicle III.7 IXth Plenary Assembly, Melbourne 1988.)

A.2.1 The Adaptation Layer

The adaptation layer supports higher layer functions, a control module, and connections between ATM and non-ATM interfaces. Here user information is mapped into ATM cells. At the originating end, information units such as LAPD (link access protocol, D-channel) or other frames are segmented, or information units such as PCM voice channels are collected and inserted into ATM cells. At the destination end, information units are reassembled (e.g., data frames) or read-out (e.g., pulse code modulation [PCM] voice channels) from ATM cells. There may be adaptation-layer-specific information, such as sequence numbers and length of data field that is peer-to-peer adaptation-field-oriented, both of which are contained in the cell's information field.

In the case of ISDN (or B-ISDN), the adaptation field could be terminated in a network termination (NT), terminal adapter (TA), terminal equipment (TE) and so forth (See Section 12.16.)

The following is a listing of adaptation layer functions from CCITT Rec. I.121:

- For common-bit-stream-oriented (CBO)) services:
- CBO adaptation functions. CBO-oriented services are those that involve an uninterrupted flow of digital information, for example, 64-kbps PCM voice. The CBO adaptation functions support these services over an ATM network. Within the adaptation layer, the following functions may be performed:

  1. Cell assembly and disassembly
  2. Compensation for the variable delay of the ATM network
  3. Handling of lost cell conditions
  4. Clock recovery. Some alternatives are to synchronize the output bit stream to the network clock or to the source bit stream
  5. Mapping of control signal such as in CCITT Rec. V.35 into the ATM cell stream

- Existing packet mode services (such as LAPD) can be supported by the CBO adaptation functions. This does not take advantage of the idle periods between data transmission. The packet mode layer provides bandwidth savings by taking advantage of the bursty nature of packet services. Operations that may be carried out by the packet mode adaptation functions include

  1. Detection of information blocks from the higher layer
  2. Dividing information blocks up into ATM cells
  3. Handling of partially filled cells
  4. Reassembling information blocks to the higher layer
  5. Sending information blocks to the higher layer
  6. Rate adaptation
  7. Action on loss of cells

A.3 ASYNCHRONOUS TRANSFER MODE CHARACTERISTICS

A cell, as we are aware, consists of a header and an information field. The information field is transported transparently by the ATM layer. There is no processing performed on the information field at the ATM layer. For instance, there is no error control processing. The header and the information field each consist of a fixed number of octets at a given reference point. The information field length is the same for all connections at all reference points where the ATM technique is applied.

A.3.1 The Header

The size of the header is minimized. It contains just enough information to transfer its associated information field through the ATM network. CCITT Rec. I.121 states that application-oriented or service-oriented information does not appear in the header. There are three mandatory functions of a header:

- Virtual channel identification (VCI)
- Error detection on the header
- Unassigned cell indication

CCITT is investigating additional header functions. The following candidates have been identified:

- Error correction on the header
- Quality of service identification, such as delay or loss priority
- Cell loss detection
- Access control at the user–network interface (UNI)
- Cell sequence numbering
- Terminal identifier
- Virtual path identification
- Line equipment identification

Header Size The size of the header should be in the range of 3–8 octets. The appropriate size is to be determined by CCITT. As an objective, CCITT recommends that the header size be the same at all reference points.

Information Field Size CCITT (Ref. 1) recommends that the size of the information field should be in the range of 32–120 octets. The final optimum size of the information field will largely be governed by

- End-to-end quality of service covering acceptable end-to-end delay and loss of information
- Transmission efficiency: the information field size-of-header size ratio should allow all existing and envisaged services to be efficiently supported on the transmission media

A.4   BROADBAND CHANNEL RATES

ATM is primarily directed toward B-ISDN. Thus, of interest, are channel rates that can be supported by B-ISDN. In addition to the B-, N0- and H1-channel, B-ISDN will support broadband channels H2 and H4 having the following bit rates:

1. H21 broadband channel: 32,768 kbps
2. • In the approximate range of 43–45 Mbps
   • An integer of 64 kbps
   • Not greater than the payload of existing third level asynchronous transmission systems of the 1.5-Mbps based hierarchy
     Consistent with these requirements, one objective is to maximize the bit rate of the H22 broadband channel.
3. H4 broadband channel:
   • In the range of 132–138.24 Mbps
   • An integer multiple of 64 kbps

A.5   VIRTUAL PATH CONCEPT AND BUNDLING

Ref. 2 states that the implementation of the *virtual path* concept will simplify network architecture and node processing. Introduction of the virtual path concept into an ATM-based network allows management of virtual circuits by grouping them into bundles. Consequently, virtual circuits can be transported, processed, and managed in bundles, reducing node costs and simplifying architecture.

Ref. 3 defines a virtual path as a labeled path (bundles of multiplexed circuits) which is defined between virtual path terminators. Here, a virtual path is defined in a more limited sense as a labeled path that is terminated by virtual path terminators, which can identify each connection from or to the network elements (terminals) included in the segment network to which the virtual path terminators belong. The segment networks, for example, can be local switching networks, local area networks (LANs), or private networks.

Virtual path terminators can be switching systems, LAN gateways, or private network gateways. The capacity of a virtual path can be deterministic or statistical according to requirements. Hence a virtual path provides a logical direct link between virtual path terminators.

Virtual paths are implemented by employing virtual path terminators and labeled cross-connect and/or add/drop multiplexers. Virtual paths are established or released by setting the path connect tables of the labeled cross-connect and/or add/drop multiplexers on the path, and the transmission capacity is reserved or cleared along the path at the same time.

Virtual paths are established or released dynamically, based on long-term service provisioning, short-term demand, or immediate requirements due to failure at one or more points in the network. Call setup is performed at an access node. During the call setup process, the access node identifies the appropriate path or establishes a path if necessary. A call is routed by a *virtual path identifier* (VPI) appended to the header of each cell, which has a specific *virtual channel identifier* (VCI) corresponding to the call. As a call traverses the network, only the VPI is necessary to route the call at transit nodes. The VCI is used at the access node to route the call further (Ref. 3).

ATM is a glimpse of the future. It will be a major improvement over our present network structure, providing a flexible digital format. It will accommodate voice, bursty data with rapid turnaround, long data files, all types of video, and facsimile in an optimal cost-effective manner.

REFERENCES AND BIBLIOGRAPHY

1. "Broadband Aspects of ISDN," CCITT Rec. I.121, IXth Plenary Assembly, Melbourne, 1988, vol. III, Fascicle III.7.
2. Steven E. Minzer, "Broadband ISDN and Asynchronous Transfer Mode (ATM)," *IEEE Commun. Mag.*, vol. 27, no. 9, Sept. 1989.
3. Ken Ichi-Sato, Satoru Ohta, and Ikuo Tokizawa, "Broad-band ATM Network Architecture Based on Virtual Paths," *IEEE Trans. Commun.* vol. 38, no. 8, Aug. 1990.

GLOSSARY

ac Alternating current

ACK Acknowledgment

A/D or **ADC** Analog-to-digital (conversion or converter)

ADP Automatic data processing

ADPCM Adaptive differential pulse code modulation

AF Audio frequency, as distinguished from RF, radio frequency

AFC Automatic frequency control

AGC Automatic gain control

A-**law** A pulse code modulation companding law used with CEPT PCM systems (see also μ-law)

ALBO Automatic line build-out

ALC Automatic load control; automatic level control

ALE Automatic link establishment (HF)

AM Amplitude modulation

AMD Automatic message display

AMI Alternate mark inversion

AMP A connector manufacturer

ANSI American National Standards Institute

APC Automatic phase control

APD (1) Avalanche photodiode; (2) amplitude probability distribution

APK Amplitude phase shift keying

APL Average picture level

ARQ Automatic repeat request. A feature in data links that allows for requesting the retransmission of data blocks, frames or packets in which errors have been detected. Acronym derives from pre–WWII automatic teleprinter service

ASCII American Standard Code for Information Interchange

ASK Amplitude shift keying

ATB All trunks busy

ATM Asynchronous transfer mode

Autodin Automatic digital network (U.S. Department of Defense)

Autovon Automatic voice network. A switched telephone network for U.S. Department of Defense users

AWG American wire gauge

AWGN Additive white Gaussian noise

AZ Azimuth. Refers to orienting an antenna system in the horizontal plane, usually referenced to true north

balun Balanced-to-unbalanced. An RF transformer that converts balanced-to-unbalanced transmission (and vice-versa)

baud The unit of digital modulation rate measured in transitions per second. Only in binary transmission are baud and bit synonymous

BCC Block check count or block check character (see also FCS)

BCD Binary-coded decimal

BCH Bose–Chaudhuri–Hocquenghen. A class of error-correcting block codes with well-defined decoding algorithms

BCI Bit count integrity

BER Bit error rate

BERT Bit error rate test (or test set)

BFSK Binary frequency shift keying

BH Busy hour. A term used in telephone traffic engineering

B_{IF} Intermediate-frequency (IF) bandwidth

BILP Beginning of intermediate line pair (facsimile)

BINR Baseband intrinsic noise ratio

BISDN or **B-ISDN** Broadband integrated services digital network

bit Binary digit

BIT Built-In-Test

BITE Built-in test equipment

BNZS Bipolar with N zero substitution (PCM term)

BOLP Beginning of line pair (facsimile)

BPF Bandpass filter

BPO British Post Office, now British Telecomm (BT). The largest British telephone company

bps Bits per second

BPSK Binary phase shift keying

BR Bit rate

B_{RF} Radio frequency (RF) bandwidth

BSS Broadcast satellite service

BT British Telecom

BTM Bell Telephone Manufacturing Company. An ALCATEL subsidiary in Belgium

BW Bandwidth

BWR Bandwidth ratio =

$$10 \log \frac{\text{occupied baseband bandwidth}}{\text{voice-frequency channel bandwidth}}$$

It is used to determine signal-to-noise ratio (S/N) from noise power ratio (NPR)

B3ZS Bipolar with 3 zero substitution (used in PCM)

B6ZS Bipolar with 6 zero substitution (used in PCM)

CARS Community antenna radio service

CATV Community antenna television, another name for cable television

CBO Common-bit-stream oriented

CBX Computer-based exchange (a private automatic branch exchange [PABX] that is computer controlled)

CCD Charge coupled device

CCIR International Consultive Committee for Radio. It is a subsidiary organization to the ITU (International Telecommunication Union), Geneva

CCITT International Consultive Committee for Telephone and Telegraph. It is a subsidiary organization to the ITU (International Telecommunication Union), Geneva

ccs Cent call-second (a traffic intensity of 100 call-seconds)

CD Compact disk

CDMA Code division multiple access

CEPT Conférence Européenne des Postes et Télécommunication, the regional European telecommunication conference committee

CF Conversion factor

CFSK Coherent frequency shift keying

CLR Circuit loudness rating

cm Centimeter

CMI Coded mark inversion

C-msg C-message (weighting). Telephone weighting used in North America

cm region As designated by the ITU, the band of frequencies from 3 to 30 GHz

C/N Carrier-to-noise ratio

C/N_0 Carrier-to-noise spectral density ratio

CNT   Canadian National Telecommunications. Previously called Canadian National Telephone Company. Based in Toronto

codec   Coder–decoder, an acronym used in pulse code modulation

compandor   Acronym for compressor–expandor

CPE   Customer premise equipment

CPES   Customer premises earth station

CP-FSK   Continuous-phase FSK

CPSK   Coherent-phase shift keying

CPM   Counts per minute

CPU   Central processing unit

CR   Carriage return

CR/BTR   Carrier recovery/bit timing recovery

CRC   (1) Cyclic redundancy check; (2) communications relay center (HF)

CRE   Corrected reference equivalent

CREG   Concentrated range extension with gain

CRPL   Central Radio Propagation Laboratory, Boulder, CO (USA)

CRT   Cathode ray tube

CSC   Common signaling channel

C/T Carrier-to-thermal noise power ratio, usually expressed in dBW per kelvin

CU   Crosstalk unit

CVSD   Continuous variable-slope delta modulation

CW   Continuous wave or carrier wave. Sometimes used to mean "Morse" operation

CXR   Carrier (equipment)

D1A   An early North American pulse code modulation system in which a 7-bit PCM code word was used and the 8th bit was a signaling bit

DA   Dependent amplifier

D/A or **DAC**   Digital-to-analog (conversion or converter)

DAMA   Demand assignment multiple access

DARPA   (U.S. Dept. of Defense) Defense Advanced Research Projects Agency

DASS   Demand assignment signaling and switching (unit)

dB   Decibel

dBc   Decibels referenced to the carrier level

dBi   Decibels referenced to an isotropic antenna

dBm   Decibels referenced to 1 mW

DBM   Data block mode

dBm0   An absolute power unit referenced to the 0 TLP (test level point)

dBm0p   dBm0 psophometrically weighted

dBmp   A noise measurement unit based on the dBm using psophometric weighting

dBmV   An absolute voltage level measurement unit based on the dB and referenced to 1 mV across 75 Ω

DBPSK   Differential binary phase shift keying

dBr   Decibels reference. The number of decibels of level above or below a specified reference point in a network. A minus sign indicates the level to be below or less than the reference, and a plus sign, above or more than the reference level

dBrn   dB reference noise (now obsolete)

dBrnC   A North American noise measurement unit using C-msg weighting

dBrnC0   dBrnC referenced to the 0 TLP (test level point)

DBS   Direct broadcast by satellite

dBV   Decibels referenced to 1 V (common impedances)

dBW   Decibels referenced to 1 W

dBx   The crosstalk coupling in dB above reference coupling, which is a crosstalk coupling loss of 90 dB

dc   Direct current

DCA   Defense Communications Agency, an agency of the U.S. Department of Defense

DCE   Data communication equipment

DCS   Defense Communications System (U.S.)

DDS   Digital Data System (AT & T)—data transmission over a pulse code modulation network

DEC   Digital Equipment Corporation

DE-PSK   Differentially encoded phase shift keying

DFB   Distributed feedback (laser)

dibit coding   Coding such that each transition of the modulated data signal represents 2 bits

DL   Data link

DLC   Digital loop carrier

DM   Delta modulation

DNHR   Dynamic nonhierarchical routing

dNp   Decineper, or one-tenth of a neper

DPCM   Differential pulse code modulation

DQPSK   Differential quaternary phase shift keying

DS0   64-kbps basic pulse code modulation channel rate (North American parlance)

DS1, DS2, DS3, etc.   North American pulse code modulation line rates

DSB   Double sideband (amplitude modulation with or without the carrier suppressed)

DSCS   Defense Satellite Communication System

DSI   Digital speech interpolation

DTE   Data terminal equipment

DTM   Data text mode

DTU   Direct to user

duobinary   A three-level modulation scheme

EBCDIC   Extended binary-coded decimal interchange code

EC   (1) Earth curvature; (2) earth coverage

EDD   Envelope delay distortion

EDP   Electronic data processing

EHF   Extremely high frequency (30–300 GHz)

EIA   Electronics Industries Association (U.S.)

EIRP   Effective isotropically radiated power

EL   Elevation. Refers to the orientation of an antenna in the vertical plane, especially an earth station antenna (see AZ)

EMI   Electromagnetic interference

E_b/N_0 Signal energy per bit per hertz of thermal noise

EOFB   End of facsimile block

EOL   End of line

EOM   End of message

erfc   The complementary error function

ERL   Echo return loss

erlang   Dimensionless unit of traffic intensity. One erlang is the intensity in a traffic path continuously occupied or in one or more paths carrying an aggregate traffic of 1 call-hour per hour, 1 call-minute, and so forth. When based on the hour, 36 ccs = 1 erlang

ERP   Effective radiated power

ESF   Extended superframe

F1A   An obsolete noise weighting used in the United States. Noise units used with this weighting were dBa

FAS   Frame alignment signal

FCC   Federal Communications Commission, the U.S. federal telecommunications and radio regulatory authority

FCS   Frame check sequence (see also BCC)

FDM Frequency division multiplex(er)

FDMA Frequency division multiple access

FEC Forward error correction

FET Field-effect transistor

FH Frequency hop. A method of spread spectrum transmission

FLTSAT Fleet satellite (communication) (U.S. Navy)

FM Frequency modulation

FOT "Fréquence optimum de travail," a high-frequency propagation term from the French. We more often call it the OWF, for optimum working frequency

Fox "Fox" broadcast—one-way shore–ship message transmission system to U.S. Navy vessels

FSK Frequency shift keying

FSL Free-space loss

FSS Fixed satellite service

ft Foot

GaAs Gallium arsenide

GBLC Gaussian band-limited channel

Gbps Gigabits per second

GHz Gigahertz, Hz $\times 10^9$

GPS Geographical positioning system (satellite based)

GSTN General switched telephone network

G/T The figure of merit of an earth station receiving system. It is expressed in dB/K. $G/T = G_{dB} - 10 \log T_{sys}$, where G is the gain of the antenna at a specified reference point and T_{sys} is the receiving system effective noise temperature in kelvin

HDB3 High-density binary 3. Similar to B3ZS

HDLC High-level data link control. An ISO datalink layer protocol

HDTV High-definition television

HE Mode of propagation in a waveguide

HF High frequency. ITU definition: "The band of frequencies between 3 and 30 MHz"

highway A common path over which signals from a plurality of channels pass with separation achieved in time division

HPA High-power amplifier

HRC Hypothetical reference circuit (radio—CCIR)

HRX Hypothetical reference connection (digital—CCITT)

HU High usage (route)

Hydro Broadcast to ships at sea with latest notices to mariners (derived from U.S. Navy Hydrographic Office)

HVQ Hierarchical vector quantization

Hz Hertz, the unit of measurement of frequency

I and Q In-phase and quadrature (channels)

IBM International Business Machines (Company)

IBS INTELSAT Business Service

ICL Inserted connection loss

ICW Interrupted continuous wave

IDR Intermediate data rate (INTELSAT)

IEEE Institute of Electrical and Electronic Engineers, a U.S. engineering society

IESS INTELSAT earth station standards

IF Intermediate frequency

IFRB International Frequency Registration Board, a unit of the ITU

ILD Injection laser diode

IM Intermodulation distortion

INMARSAT International Marine Satellite (Organization)

inside plant See outside plant

INTELSAT International Telecommunication Satellite, a series of satellites under an international consortium with the same name

I/O Input–output (device). A generic group of equipment that serves as input and/or output for a telecommunication system such as a keyboard or a display

IONCAP Ionospheric communication analysis and prediction program

IP Internet protocol

IR Infrared

IRE Institute of Radio Engineers, a predecessor of the IEEE

IRL Isotropic receive level

ISB Independent sideband

ISDN Integrated services digital network(s)

ISI Intersymbol interference

ISL Intersatellite link

ISO International Standards Organization

ITA International Telegraph Alphabet

ITT International Telephone and Telegraph (Co.)

ITU International Telecommunication Union. This organization issues the radio regulations. CCITT and CCIR are subsidiary organizations of the ITU

IXC Interexchange carrier (North American parlance)

junction See trunk

kbps Kilobits per second

K-factor A constant used in path profiling for the calculation of ray beam bending caused by the atmosphere in radiolink (line-of-sight [LOS] microwave) or tropospheric scatter radio system path

kHz Kilohertz, Hz $\times$ 10^3

km Kilometer

kW Kilowatt

LAN Local area network

LAPB Link Access Protocol-B

LAPD Link Access Protocol-D channel

LATA Local access and transport area

L-carrier The FDM carrier series developed by the U.S. Bell System for long-haul transmission over broadband media. There are L1, L2, L3, L4, and L5 systems

LAP Link access protocol such as LAP, LAP-B, and LAP-D

LBO Line build-out. A method of extending the length of a line electrically, usually by means of capacitors

LEC Local exchange carrier (North American term)

LED Light-emitting diode

LF (1) Low frequency; (2) line feed

LHCP Left-hand circularly polarized or left-hand circular polarization

Lincompex Link compression–expansion (HF)

LLC Logical link control (higher layer protocol used on local area networks [LANs])

LNA Low-noise amplifier

LO Local oscillator

LOS Line-of-sight, generally in reference to line-of-sight microwave or radiolink

LP (1) Log periodic (antenna); (2) Laws and Parsons (raindrop-size distribution)

LPA Linear-power amplifier

LPI (1) Lines per inch; (2) low probability of intercept

LPM Lines per minute

LPSS Line protection switching system

LQA Link quality assessment (HF)

LR Loudness rating

LRC Longitudinal redundancy check(ing)

LRD Long route design (subscriber loop)

LSA mode Limited space charge accumulation mode

LSB Lower sideband

LST Local sidereal time

LUF Lowest usable frequency (HF)

MAN Metropolitan area network

***M*-ary** Multilevel digital modulation. A 2-level modulation system is binary

MBA Multiple-beam antenna (satellite)

MBC Meteor burst communication

Mbps Megabits per second

MCT Motion Compensated Transform (proprietary technology of Picture-Tel Corp.)

MERCAST Merchant marine broadcast

MFSK M-ary frequency shift keying

MHz Megahertz, Hz $\times$ 10^6

mi Mile(s)

MIL, Mil Military, U.S. military, often used in the nomenclature of U.S. military standards

MILSTAR An advanced U.S. military strategic and tactical satellite system

MLRD Modified long route design

mm Millimeter(s)

mm region As designated by the ITU, the frequency band from 30 to 300 GHz

modem Acronym for modulator–demodulator

MP Multipath value

MPSK M-ary phase shift keying

ms Millisecond(s)

MSE Mobile subscriber equipment—U.S. Army tactical cellular radio communication system for the field forces

MSL Mean sea level

MTBF Mean time between failures

MTTR Mean time to repair

MUF Maximum usable frequency (HF)

mV Millivolt(s)

N_0 (1) Sea level refractivity; (2) spectral noise density

n_1 and n_2 Indices of refraction for two media

NA Numerical aperture

NACK or **NAK** Negative acknowledgment

nanometer (nm) 10^{-9} m

NARS North Atlantic Radio System. A broadband radio system, principally tropospheric scatter, connecting the UK to the United States via the

Faeroes, Iceland, Greenland, Baffin Island, and Canada. The system is now being dismantled

NATO North Atlantic Treaty Organization

NBS National Bureau of Standards (U.S.). Now called NIST, or National Institute of Standards and Technology

NEP Noise equivalent power

NF Noise figure

NL Nonloaded (in reference to voice frequency telephone cables)

NLR Noise load ratio (or noise load factor)

nm nanometer(s)

NOSFER Laboratory standard for reference equivalent

NPR Noise power ratio

NRZ Nonreturn to zero (a digital baseband waveform)

ns nanosecond(s)

NT Network termination

NTSC National Television Systems Committee (U.S.)

NVI Near vertical incidence (HF)

NVIS Near vertical incidence skywave (HF)

O & M Operation and maintenance

OC (OC-1, OC-2, etc.) Optical carrier (data rates for SONET)

OCL Overall connection loss (via net loss [VNL] concept)

OCR Optical character reader

off-hook A telephone handset that has been taken off-hook closes the loop and "busies" the line

OK-QPSK Offset-keyed quaternary phase shift keying

OLR Overall loudness rating

on-hook A telephone handset that is on-hook is not in use. Placing the handset on-hook opens the loop, making the line "idle"

OOK On–off keying

ORE Overall reference equivalent. It is the sum of the TRE (transmit reference equivalent) and the RRE (receive reference equivalent) of a telephone connection

OSI Open system interconnection (data interface)

outside plant A telephone operating company term or derived from telephone operating company usage. It has various meanings. The total telephone plant may be considered the sum of "inside plant" and "outside plant" facilities. One may consider "outside plant" as all telephone facilities that are out of doors. In this text outside plant is that part of the telephone plant that takes the signal from the local switch to the subscriber, including the subscriber equipment as well as the local trunk (junction) plant. Another connotation of outside plant is any telephone plant activity that involves civil (construction) engineering

OW   Orderwire

O-wave   Ordinary wave (ray)—HF propagation

OWF   Optimum working frequency (HF) (same as FOT)

PABX   Private automatic branch exchange

PAL   Phase alternation line. One of two European color television systems; the other is SECAM

PAM   Pulse amplitude modulation

PBX   Private branch exchange

PC   Personal computer

PCM   Pulse code modulation

PDU   Protocol data unit

pel, pixel   Picture element (image transmission)

PEP   Peak envelope power

PF   Power factor

PIN   A type of fiber optic detector. Terminology derives from the semiconductor construction of a device where an intrinsic (I) material is used between the p–n junction of a diode

PLL   Phase-lock(ed) loop

PM   Phase modulation

PN   Pseudo-noise. A spread spectrum transmission technique

POH   Path overhead (SONET)

PPM   Pulse position modulation

ppm   Pulses per minute

pps   Pulses per second

PRBS   Pseudo-random-binary sequence

PRF   Pulse repetition frequency

PROPHET   A high-frequency propagation prediction program prepared for the U.S. Coast Guard. It can operate on a personal computer

PSK   Phase shift keying

PSTN   Public switched telephone network

PSV   Pair shield video

pW   Picowatt(s), $W \times 10^{-12}$; 1 pW = -90 dBm

pWp   Picowatts psophometrically weighted

QAM   Quadrature amplitude modulation

QPR   Quadrature partial response (keying)

QPSK   Quaternary phase shift keying, 4-level PSK

QSY   An international Q signal indicating that a change in frequency is required or is being carried out

QVI   Quasi-vertical incidence—a mode of high-frequency propagation (see NVI and NVIS)

RA   Random access

RBER   Error ratio in absence of fading (CCIR definition)

RC   Resistance–capacitance product. Determines the time constant of a circuit

Rcv, rcvr, recvr   Receive or receiver

RD   Resistance design

RE   Reference equivalent

Rec.   Recommendation, especially of CCITT and CCIR

Rep.   Report, especially from CCIR

RF   Radio frequency

RFI   Radio-frequency interference

RH   Relative humidity

RHCP   Right-hand circularly polarized or right-hand circular polarization

RLR   Receive loudness rating

rms   Root mean square

rpm   Revolutions per minute

RRD   Revised resistance design

RRE   Receive reference equivalent. The reference equivalent (in dB) of the receiving portion of a telephone connection

RSL   Receive signal level

RTC   Return to control (facsimile)

RZ   Return to zero. Refers to a type of digital modulation baseband waveform

SAP   Service access point (data communication)

SAR   Search and rescue

SBS   Satellite Business Systems. A U.S. domestic satellite communication system. Company no longer in business under that name

SCPC   Single channel per carrier

S + D   Speech plus derived

S/D   Signal-to-distortion ratio

SDLC   Synchronous data link control (IBM data link layer protocol)

SDU   Service data unit

SECAM   Sequential color with memory. One of two European color TV transmission systems; the other is PAL

SG   Supergroup

SHF   Superhigh frequency. That band of frequencies from 3000 to 30,000 MHz

SIAM   Society of Industrial and Applied Mathematics

SIC   Station identification code

SID   Sudden ionospheric disturbance (HF propagation)

SINAD   Signal + noise + distortion-to-noise + distortion ratio

SLR   Send loudness rating

S/N, **SNR**   Signal-to-noise ratio

SOM   Start of message (facsimile, teleprinter and data)

SONET   Synchronous optical network

SPADE   Single channel per carrier multiple access demand assignment equipment

SPC   Stored program control (computer-controlled switch)

SPDU   Session protocol data unit

SPE   Synchronous payload envelope (SONET)

SQPSK   Staggered QPSK

SSA   Solid-state amplifier

SSB   Single sideband

SSBSC   Single-sideband suppressed carrier

SSN #7   CCITT signaling system no. 7

SSPA   Solid-state power amplifier

SS/TDMA   Switched-satellite TDMA

STC   Standard Telephone and Cable. A large British telecommunication firm

STE   Section terminating equipment (SONET)

STL   (1) studio-to-transmitter link (TV); (2) Standard Telephone Laboratories (see STC)

STM   Synchronous transfer mode

STS   Synchronous transport signal

sync   Synchronization, synchronizing

T1, T2, etc.   North American series of PCM line arrangements

TA   Terminal (equipment) adapter

T_{ant} Antenna noise temperature

TASI   Time-assigned speech interpolation

TASO   Television Allocation Study Organization (U.S.)

TCI   Technology for Communications, Inc.

TCM   Trellis-coded modulation

TCP   Transport control protocol

TDM   Time division multiplex(er)

TDMA   Time division multiple access

T_e Effective noise temperature

TE   (1) terminal equipment; (2) transverse electric (mode—i.e., TE_{11} mode [In circular waveguide the TE_{11} mode is the dominant wave])

TELESAT   A Canadian domestic satellite system

THz   Terahertz, Hz $\times 10^{12}$

Time availability   The percentage of time a link (typically, a radiolink) meets certain minimum performance requirements

TLP   Test level point, such as the 0 TLP, or zero test level point

TO   Technical order. A U.S. Air Force technical manual

TOA   Take-off angle (refers to antennas)

TOD   Time of day

TPDU   Transport protocol data unit

T_r Receiver noise temperature (see T_{ant})

T/R   Transmit–receive (switch). Permits a common antenna to be used for both transmit and receive functions in the half-duplex mode

TRE   Transmit reference equivalent. The reference equivalent of the transmit portion of a telephone connection

TRI-TAC   Tri-service tactical communication system. An advanced three-service military tactical communication system developed for the U.S. forces

Trunk   In the local area, those circuits connecting one exchange to another. This is called a junction in the UK. In the long distance area (inter-LATA), any circuit connecting one long-distance exchange to another

T_{sys} System noise temperature (usually refers to receive systems)

TT/N Test tone-to-noise ratio

TT & C   Telemetry, tracking, and command (control) (deals with earth satellites)

TV   Television

TVRO   Television receive only (satellite systems)

TWT   Traveling wave tube

TWTA   Traveling wave tube amplifier

UG   Unigauge design

UHF   Ultrahigh frequency. The band of frequencies encompassing 300–3000 MHz

U_{ndp} Outage probability (unavailability)

UNI   User–network interface

USB   Upper sideband

USITA   U.S. Independent Telephone Association (now called USTA)

UTC   Universal coordinated time

UVR   Useful volume range

UW   Unique word

VCI   Virtual channel identification/identifier

VDU   Video display unit

VF   Voice frequency. The nominal VF channel is 0–4000 Hz; the CCITT VF channel is 300–3400 Hz

VFCT   Voice frequency carrier telegraph (synonymous with next term)

VFTG   Voice frequency telegraph

VHF   Very high frequency. The band of frequencies encompassing 30–300 MHz

VITS   Vertical interval test signal(s) (TV)

VNL   Via net loss. The basis of the variable transmission loss plan in North America

VNLF   Via net loss factor. Term used in computing VNL; the unit is dB/mi

VOGAD   Voice-operated gain adjust device (HF)

VPI   Virtual path identifier

VRC   Vertical redundancy check(ing)

VSAT   Very small aperture terminal (satellite communication)

VSB   Vestigial sideband (amplitude modulation)

VSWR   Voltage standing wave ratio

VT   Virtual tributary (SONET)

VU   Volume unit. A unit for the measure of voice (speech) level

VWL   Variable word length (encoder)

WAN   Wide area network

WARC   World Administrative Radio Conference (ITU)

WBHF   Wide-band HF

WDM   Wave division multiplex(er)

WMO   World Meteorological Organization

wpm   Words per minute

xmt, XMT, xmtr   Transmit or transmitter

x-wave (ray)   Extraordinary wave (ray)—HF propagation

Z_0 Characteristic impedance

0 TLP   Zero test level point. A single reference point in a circuit or system at which we can expect to find the level to be 0 dBm (test level). From the 0 TLP, other points in the circuit or network can be referenced using the unit dBr or dBm0, such that dBm = dBm0 + dBr

ε Relative dielectric constant

λ Notation for wavelength

μ 10^{-6}

μ-law   The companding law used on North American pulse code modulation systems (see also A-law)

μm   micron, micrometer, 1×10^{-6} m

Σ sum, summation

σ sigma, standard deviation

Ω Ohms

η Antenna efficiency (in percentage or decimal)

INDEX